M-ary orthogonal signals, ML coherent receiver:

$$P_e = 1 - \frac{1}{\sqrt{2\pi}} \int_{-\infty}^{\infty} \left[\Phi\left(v + \sqrt{2\,\mathcal{E}/N_0}\,\right) \right]^{M-1} e$$

Binary orthogonal signals, suboptimum noncoherent receiver:

$P_{e,i} = \frac{1}{2} \exp\{-\alpha_i^2/4\sigma^2\}$, where $\alpha_i = (m_i,\, w_i)$, $\|w_0\| = \|w_1\|$, and $\sigma^2 = N_0\|w_i\|^2/2$

U_0, V_0, U_1, V_1 jointly Gaussian, independent, $u_i = E\{U_i\}$, $v_i = E\{V_i\}$, $u_1 = v_1 = 0$, $\sigma_i^2 = \text{Var}\{U_i\} = \text{Var}\{V_i\}$:

$$P\left(U_1^2 + V_1^2 > U_0^2 + V_0^2\right) = \frac{\sigma_1^2}{\sigma_0^2 + \sigma_1^2}\, \exp\left\{ \frac{-(u_0^2 + v_0^2)}{2(\sigma_0^2 + \sigma_1^2)} \right\}$$

Binary orthogonal signals, ML noncoherent receiver: $\quad P_e = \frac{1}{2}\exp\{-\mathcal{E}/2N_0\}$

DBPSK, differentially coherent receiver: $\quad P_e = \frac{1}{2}\exp\{-\mathcal{E}/N_0\}$

M-ary orthogonal signals, ML noncoherent receiver:

$$P_e = \frac{1}{M} \sum_{n=2}^{M} \binom{M}{n}(-1)^n \exp\left\{ -\left(1 - n^{-1}\right)\mathcal{E}/N_0 \right\}.$$

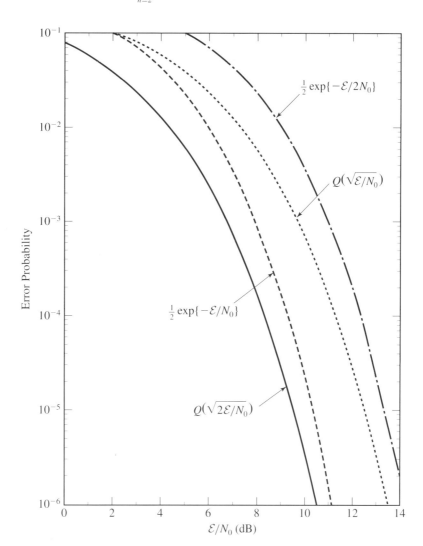

Frequently Used Equations

Gaussian distribution function:

$$\Phi(x) = \int_{-\infty}^{x} \frac{1}{\sqrt{2\pi}} \exp\left(-y^2/2\right) dy$$

$$Q(x) = \Phi(-x) = 1 - \Phi(x) = \int_{x}^{\infty} \frac{1}{\sqrt{2\pi}} \exp\left(-y^2/2\right) dy$$

Binary baseband communications:

$$Z_i(T_0) = \hat{s}_i(T_0) + \hat{X}_i(T_0), \quad i = 0, 1; \quad \hat{s}_0(T_0) > \hat{s}_1(T_0)$$

$$\mu_i(T_0) = E\{Z_i(T_0)\} = \hat{s}_i(T_0), \quad \sigma^2 = \mathrm{Var}\{Z_i(T_0)\} = R_{\hat{X}}(0), \quad i = 0, 1$$

$$P_{e,0} = P[Z_0(T_0) \le \gamma] = \Phi([\gamma - \mu_0(T_0)]/\sigma) = Q([\mu_0(T_0) - \gamma]/\sigma)$$

$$P_{e,1} = P[Z_1(T_0) > \gamma] = \Phi([\mu_1(T_0) - \gamma]/\sigma) = Q([\gamma - \mu_1(T_0)]/\sigma)$$

$$\mathrm{SNR} = [\mu_0(T_0) - \mu_1(T_0)]/2\sigma, \quad \gamma_m = [\mu_0(T_0) + \mu_1(T_0)]/2, \quad P_{e,m}^* = Q(\mathrm{SNR})$$

$$\bar{\gamma} = \gamma_m + \{\sigma^2 \ln(\pi_1/\pi_0)/[\mu_0(T_0) - \mu_1(T_0)]\}$$

$$\overline{P_e^*} = Q[\mathrm{SNR} - (2\,\mathrm{SNR})^{-1} \ln(\pi_1/\pi_0)]\,\pi_0 + Q[\mathrm{SNR} + (2\,\mathrm{SNR})^{-1} \ln(\pi_1/\pi_0)]\,\pi_1$$

Matched filter for AWGN:

$$h(t) = c[s_0(T_0 - t) - s_1(T_0 - t)], \quad c > 0$$

$$\bar{\mathcal{E}} = (\mathcal{E}_0 + \mathcal{E}_1)/2, \quad \rho = \int_{-\infty}^{\infty} s_0(t)\,s_1(t)\,dt, \quad r = \rho/\bar{\mathcal{E}}$$

$$d^2 = \int_{-\infty}^{\infty} [s_0(t) - s_1(t)]^2\,dt, \quad \mathrm{SNR} = \sqrt{\frac{\bar{\mathcal{E}}(1-r)}{N_0}} = \frac{d}{\sqrt{2N_0}}$$

BPSK, phase error θ: $\quad P_e(\theta) = Q\left(\sqrt{2\mathcal{E}/N_0}\,\cos(\theta)\right)$

Regular M-ASK, ML coherent receiver:

$$\bar{P}_e = \frac{2(M-1)}{M}\,Q\left(\sqrt{\frac{6\,\bar{\mathcal{E}}_b \log_2 M}{(M^2-1)N_0}}\right)$$

$$P_{e,0} = P_{e,M-1} = Q\left(d/\sqrt{2N_0}\right); \quad P_{e,i} = 2Q\left(d/\sqrt{2N_0}\right), \quad 1 \le i \le M-2$$

Regular M-QASK, ML coherent receiver:

$$P_{e,n} = 4Q\left(d/\sqrt{2N_0}\right)\left[1 - Q\left(d/\sqrt{2N_0}\right)\right], \quad \text{interior point}$$

$$P_{e,n} = Q\left(d/\sqrt{2N_0}\right)\left[2 - Q\left(d/\sqrt{2N_0}\right)\right], \quad \text{corner point}$$

$$P_{e,n} = Q\left(d/\sqrt{2N_0}\right)\left[3 - 2Q\left(d/\sqrt{2N_0}\right)\right], \quad \text{other exterior point}$$

Regular M-QASK, M an even power of 2, ML coherent receiver:

$$Q\left(d/\sqrt{2N_0}\right) = Q\left(\sqrt{\frac{3\,\bar{\mathcal{E}}_b \log_2 M}{(M-1)N_0}}\right)$$

Introduction
to Digital Communications

Michael B. Pursley

Holcombe Professor of Electrical and Computer Engineering
Clemson University

Pearson Education International

Vice President and Editorial Director, ECS: *Marcia J. Horton*
Vice President and Director of Production and Manufacturing, ESM: *David W. Riccardi*
Executive Managing Editor: *Vince O'Brien*
Managing Editor: *David A. George*
Production Editors: *Barbara A. Till; Scott Disanno*
Director of Creative Services: *Paul Belfanti*
Art Director: *Jayne Conte*
Cover Designer: *Bruce Kenselaar*
Art Editor: *Greg Dulles*
Manufacturing Manager: *Trudy Pisciotti*
Manufacturing Buyer: *Lisa McDowell*
Senior Marketing Manager: *Holly Stark*

© 2005 by Pearson Education, Inc.
Pearson Prentice Hall
Pearson Education, Inc.
Upper Saddle River, NJ 07458

Printed in the United States of America

10 9 8 7 6 5 4 3 2 1

ISBN   0-13-123392-0

Pearson Education Ltd.
Pearson Education Australia Pty. Ltd.
Pearson Education Singapore, Pte. Ltd.
Pearson Education North Asia Ltd.
Pearson Education Canada, Inc.
Pearson Educación de Mexico, S.A. de C.V.
Pearson Education—Japan
Pearson Education—Malaysia, Pte. Ltd.
Pearson Education, Inc., *Upper Saddle River, New Jersey*

To my wife, Lou Ann, our daughter, Jessica,
my mother, Evelyn Pursley,
and the memory of my father, Bader Pursley

Contents

Preface

This book provides an introduction to the basic concepts in digital communications for students with little or no previous exposure to either digital or analog communications. The intent is to help the student develop a firm understanding of digital communication system engineering in order that he or she will be able to conduct system-level design and analysis for digital communication systems of the future. As a result the basic principles of digital communications theory and techniques are emphasized rather than specific technologies for implementation.

No one book can encompass all aspects of digital communications. The focus in this book is on modulation and demodulation. Other important issues in digital communications, such as error-correction coding and synchronization, are discussed only briefly. Such topics are appropriate for more advanced courses that traditionally follow the first course on digital communications.

The level of the presentation is appropriate for advanced undergraduates and beginning graduate students in electrical and computer engineering. A good background in linear systems, including the use of convolution and Fourier transforms in linear systems analysis, is required as a prerequisite. The student is expected to have a good understanding of probability and random variables from a previous course. A brief review of probability and random variables is included in Chapter 1, but this material is intended primarily to serve as a convenient reference for some of the basic properties of random variables and to introduce the notation for subsequent chapters. An adequate understanding of the concepts requires approximately 25 to 30 hours of instruction based on a text such as *A First Course in Probability* by Sheldon Ross or *Introduction to Probability and Its Applications* by Richard Scheaffer.

Chapters 2–4 are devoted to second-order random processes, emphasizing correlation functions, spectral densities, and their role in the analysis of random processes in linear systems. Understanding of this material is a requirement for subsequent chapters on digital communications. At some universities, a course that includes basic material on random processes is a prerequisite to the first course in communication systems. Consequently, the book is written in a way that Chapter 2, Chapter 3, and parts of Chapter 4 may be used for review or skipped entirely for courses in digital communications that have such a prerequisite. The latter sections of Chapter 4 are less commonly included in a course on probability and random processes, so these sections should be covered at the beginning of the course or as the need for this material arises in Chapters 6 and 7.

The basic principles of digital communications are presented in Chapters 5–7, which deal with baseband communications, coherent radio-frequency communications, and noncoherent radio-frequency communications. It is expected that these chapters will provide the core material for any introductory course on digital communications. More advanced and more specialized topics are covered in the remaining two chapters, and the inclusion in the course of material from these chapters is at the discretion of the instructor. Chapters 5–7 also prepare the student for subsequent courses that deal with advanced topics in digital communications.

The approach to teaching digital communication theory followed in the book is to begin with baseband communications, because it is free of the complications caused by the sinusoidal carriers that are required in radio-frequency communications. The basic principles of matched filtering, optimum correlation receivers, and statistical decision theory are introduced in the simpler setting of baseband communications in Chapter 5. In this chapter, we impose a specific structure on the communication receiver, and the optimum elements for this receiver are derived. This permits postponement of the proof of the optimality of the receiver structure until Chapter 6, where we can use the Fourier series to derive the optimum receiver structure for binary phase-shift keying, the most popular binary modulation technique for coherent communications. Restriction of the derivation to a sinusoidal signal set avoids the need for general orthogonal expansions, yet it gives the student the essential concepts needed to understand more general derivations. An intuitive approach, which avoids the need for orthogonal expansions, is provided in Appendix D. An important feature of Chapter 5 is the thorough explanation of methods for the analysis of suboptimum filters in communication receivers. An introduction to detection theory is provided, and discussions of minimax, Bayes, and maximum-likelihood decision rules are included.

The problem of extracting a phase reference and the degradation that results from an imperfect phase reference are discussed in Chapter 6. Coherent communication receivers are examined, and performance analyses are provided for binary and quaternary phase-shift keying, minimum-shift keying, quadrature amplitude modulation, and nonbinary orthogonal signal sets. The spectral efficiencies of various modulation techniques are also presented in Chapter 6.

Chapter 7 is devoted to noncoherent communications, and again we exploit the student's familiarity with the Fourier series to derive the optimum noncoherent receiver for binary frequency-shift keying. Analyses are given for optimum and suboptimum receivers. Noncoherent demodulation of differentially encoded binary phase-shift-key modulation is described, and the performance of nonbinary orthogonal signaling with noncoherent reception is derived.

The primary topics covered in Chapter 8 are intersymbol interference and its effect on the performance of a digital communication system. An introduction to equalization for channels with known transmission characteristics is also provided. Spread-spectrum communications is the topic of Chapter 9, and the basic properties of Hamming and Reed–Solomon codes are provided in Appendix A and Appendix B, respectively. The complex representation of communication signals is introduced in Appendix C, the sampling method for deriving the optimum receiver is presented in Appendix D, and an alternative receiver structure for coded signals is derived in Appendix E.

The book is designed to be suitable for self-study by engineers and beginning graduate students. The derivations and discussions are sufficiently detailed to walk the reader through the applications of the concepts and techniques that are presented. Several examples and exercises with solutions are provided to test the reader's understanding along the way. Each chapter has a set of problems that further test the reader's understanding and extend some of the topics presented in the text.

I wish to thank each of the instructors who taught from the manuscript for the book and supplied suggestions and corrections. Special thanks are due Professors Dilip Sarwate and Bruce Hajek of the University of Illinois, Professor James Lehnert of Purdue University, and Professors John Komo and Daniel Noneaker of Clemson University. Each was kind enough to teach from one or more versions of the manuscript and provide extensive feedback that improved the book. I also wish to express my appreciation to the students who suffered through numerous revisions of the manuscript and furnished lists of corrections. Finally, I would like to thank Thomas Royster for his capable assistance in reviewing several sets of page proofs.

MICHAEL B. PURSLEY

Introduction

In this book, the term *communication* system refers to an electronic system whose purpose is to move information from one location, known as the *source*, to another location, known as the *destination*. The information can be in the form of a *digital message*, in which case the message consists of a finite number of variables, referred to as *symbols*, each of which takes values in a common finite set known as the *message alphabet*. It can also be in the form of an *analog message*, in which case there may be infinitely many variables in the message. A digital message is always a discrete-amplitude, discrete-time sequence, while a typical analog message is a continuous-time waveform.

It is clear from recent technological developments that digital communication systems are the communication systems of the future. In many current applications, there are no alternatives, because the communication system is required to handle a message that is inherently digital. An example of this arises in the transfer of a computer file. For other applications, the original message may not be digital (e.g., a speech signal recorded in analog format), but there are advantages to converting the message to a digital format and employing a digital communication system. The digital format is more amenable to storage at the destination, encoding for privacy, and, most importantly, encoding for protection against noise that inevitably arises in the communication channel.

The requirement for a communication link between two computers is a good example of an application for digital communications, and the transfer of a file from one computer to the other is a good scenario for illustrating the function of a digital communication system. In this example, the digital information is to be sent from one computer, the source, to another, the destination, by way of a communication channel that is made available to connect the two computers. The available channel may be an optical fiber, a telephone line, a wireless link, or a combination of such media. Suppose the information is in the form of *binary data* (i.e., a sequence of variables, each of which takes on one of *two* possible values). The function of the communication system is to reproduce at the destination the binary data provided by the source. At the source computer, the binary data must be converted into a signal that is suitable for transmission over the available communication channel. As with all communication channels, the presence of noise in the available channel means that the output of the channel is not a perfect reproduction of its input. At the destination, the channel output, which is a noise-corrupted version of the transmitted signal, must be converted to binary data for use in the destination computer.

Errors may be made in the conversion process as a result of noise and distortion. The goal in designing and building a digital communication system is to make the frequency of errors as low as possible for the cost, power, and complexity constraints imposed on the system designer.

The Role of a Digital Communication System

We now return to the general situation and provide a slightly more detailed description of the role of the digital communication system. Recall that a digital message is a sequence of variables, each of which takes values in a common finite set. The binary data sequence mentioned previously for communications between two computers is a digital message for which the common set has two values only. A *digital communication system* accepts a digital message as its input at the source, and it produces a digital message as its output at the destination. The communication subsystem at the source is referred to as the *transmitter*, and the communication subsystem at the destination is called the *receiver*. A communication channel connects the two subsystems of the digital communication system. The features of the channel (e.g., distortion, noise, or interference) are such that its output differs from its input in a way that is usually random and difficult to predict. Whenever a particular symbol in the message delivered to the destination differs from the corresponding symbol in the original digital message at the source, we say that an *error* has occurred in the reception of the symbol and the received symbol itself is *erroneous*. The goal of a digital communication system is to reproduce the message at the destination as reliably as possible, and the measure of reliability for a particular reproduced message is usually related to the total number of errors or the relative frequency of errors in the message. This leads to the consideration of the *probability of error* for the symbols in the reproduced message as the performance measure for a digital communication system.

It is impossible for any book of reasonable size to include meaningful expositions of all of the elements of digital communications. The focus in this book is on modulation and demodulation. The modulation and demodulation processes have a common goal, so they must be compatible. They are generally designed as a single subsystem even though they are almost always physically separated, often by a great distance. The modulator and demodulator together are often referred to as the *modem* (*mo*dulator–*dem*odulator).

Modulation and Demodulation in Digital Communication Systems

Modulation is the process whereby a digital message is converted into a continuous-time signal that is suitable for transmission over the communication channel. *Demodulation* is the process in which a received continuous-time waveform, typically consisting of a signal component and a noise component, is converted into a digital data sequence. One would like to have the digital data sequence at the output of the demodulator be identical to the digital message at the input to the modulator; this represents error-free

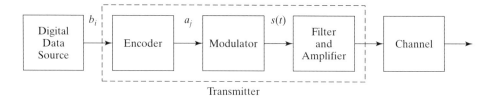

Figure 1: The transmitter for a digital communication system.

communication. For actual communication channels, error-free communication is not possible, and the best we can strive for is a high probability that a given data symbol at the demodulator output agrees with the corresponding data symbol at the modulator input.

As illustrated in Figure 1, the modulation process takes place in the transmitter, where other functions such as encoding and filtering are also performed. The type of signal into which the data sequence is converted depends on the communication channel that is available. For example, if the communication channel is a telephone line, then the modulated signal may be a baseband signal that is constrained in frequency to less than a few kilohertz (kHz). Communication systems that utilize baseband modulation are known as *baseband transmission systems*. On the other hand, if the communication channel is a mobile radio channel, the modulated signal may be a bandpass signal that has a center frequency of several hundred megahertz (MHz) and a bandwidth that may be as small as a few kilohertz or as large as several megahertz. Such signals are known as radio-frequency (RF) signals, and the corresponding communication systems are referred to as RF communication systems. Baseband digital data transmission is the subject of Chapter 5, and RF digital communication systems are covered primarily in Chapters 6 and 7.

The encoder and modulator of Figure 1 do not necessarily operate on individual symbols. In the simplest systems that use no error-correction coding, the encoder maps each individual source output to an individual symbol, and the modulator maps each individual symbol to a waveform. For more sophisticated systems, it may be that a block $\mathbf{b} = (b_1, b_2, \ldots, b_k)$ of source outputs is mapped into a block $\mathbf{a} = (a_1, a_2, \ldots, a_n)$ of symbols from the message alphabet. (The lengths of these two blocks may be different.) The process is repeated enough times to map the entire source output sequence into a digital message. The modulator then maps the message sequence into a sequence of waveforms to produce the signal $s(t)$. Because the effects of the filter and amplifier in the transmitter are often negligible, and because any nonnegligible effects can be included in the model for the channel, it is common to refer to the modulator output $s(t)$ as the *transmitted signal* for the digital communication system.

As an example, consider again the transmission of binary data. The basic element of binary data is the binary digit, referred to as a *bit*. Digital messages can be formed from binary data in several different ways. At one extreme, each bit is treated as a separate symbol, and each symbol is mapped into a separate waveform. This is the approach that is taken in Chapter 5. On the other hand, there are advantages to grouping together subsets of bits to form the symbols. In the latter situation, a digital message consists of a sequence of symbols, and each symbol represents a group of bits. This approach is employed in parts of Chapters 6 and 7.

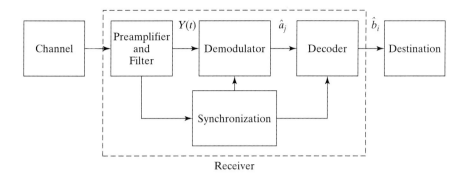

Figure 2: The receiver for a digital communication system.

We pause here for a brief word about the notation for filtered signals and estimates of parameters. If $v(t)$ is a deterministic signal, $\hat{v}(t)$ denotes a linearly filtered version of that signal. The same is true of a random process: If $X(t)$ is a suitable random process, then $\hat{X}(t)$ denotes the output of a linear filter when $X(t)$ is its input. If θ is a deterministic constant, $\hat{\theta}$ denotes an estimate of this constant, and if Θ is a random variable, $\hat{\Theta}$ denotes an estimate of Θ. In a communication system, the constant or random variable may represent a parameter of the transmitted signal, and this parameter may be unknown to the receiver. A typical example is the phase angle of a sinusoidal signal. An estimate of this phase angle may be required for use in the demodulation of a RF communication signal that consists of a message signal modulated onto a sinusoidal carrier.

The demodulation process takes place in the receiver, as illustrated in Figure 2. Other operations in the receiver include the synchronization subsystem that extracts timing information and information about certain parameters of the received signal. For a linear channel with additive noise, the input to the demodulator is

$$Y(t) = \hat{s}(t) + X(t),$$

which is referred to as the received signal. The term $\hat{s}(t)$ represents a filtered version of the transmitted signal, and $X(t)$ is a random process that represents the noise, including thermal noise that arises in the receiver itself. In most digital communication systems, the primary sources of thermal noise are in the front end of the receiver (e.g., the preamplifier and filter of Figure 2). The random process that is used to model the noise is often referred to as the *noise process*. With the aid of the synchronization subsystem, the goal of the demodulator is to produce a reliable estimate $\hat{a}_1, \hat{a}_2, \hat{a}_3, \ldots$ of the message stream $a_1, a_2, a_3, \ldots$. The filtering operations in the receiver alter the characteristics of the noise, and these characteristics influence the performance of the demodulator. Mathematical characterizations of random processes are discussed in Chapter 2, and the effects of linear filtering on random processes are described in Chapters 3 and 4.

We can combine Figures 1 and 2 into a single block diagram that contains the essential elements for studying the modulator and demodulator. The diagram of Figure 3 is sufficient for most of the material in this book. In this simplified block diagram, the

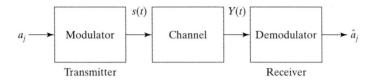

Figure 3: Model for a digital communication system.

input is the digital message and the output is another digital message; the encoding, decoding, and synchronization subsystems are viewed as external to the system under consideration. Any filtering and amplification is absorbed into the channel, and the channel is viewed as the source of the noise as well. In effect, we have combined the transmitter's filters and amplifiers, the physical channel, and the receiver's filters and amplifiers into a single block that we now call the *channel*. The most common channel model is based on the assumption that the effects of channel filtering on the signal can be ignored and the noise process $X(t)$ is a Gaussian random process. This model is employed in Chapters 5–7, but we depart from it in Chapter 8, where some effects of signal distortion due to channel filtering are described and evaluated.

In the simplified model of Figure 3, all that is left in the transmitter is the modulator; as a result, the terms *transmitter* and *modulator* are often used interchangeably in the book. Similarly, in many situations, the terms *receiver* and *demodulator* are used interchangeably. Our goal in Chapters 5–7 is to analyze the performance of modulation and demodulation schemes, study the effects of various parameter variations on the performance, and find demodulation schemes that give the best possible performance for a given modulation technique. Such a demodulation scheme is often referred to as the *optimum receiver* for the given modulator.

Chapter 1

Probability and Random Variables: Review and Notation

1.1 Purpose of the Chapter

The purpose of this chapter is to review the essential facts from probability and random variables that are necessary for the material on random processes in Chapters 2–4 and for subsequent use in the analysis of digital communication systems in Chapters 5–9. Because it is assumed that the reader has taken a previous course in probability and random variables, this chapter is considerably more terse than those that follow, and no claim is made regarding the completeness of this review. There are several excellent texts on probability and random variables, such as those listed at the end of this chapter, that provide more information on the topics included in the review. These texts are also good sources for examples and exercises that illustrate the material.

1.2 Probability Spaces

A *probability space* is a triple $(\Omega, \mathcal{F}, P)$ consisting of a set Ω, a collection $\mathcal{F}$ of subsets of Ω, and a function P defined on $\mathcal{F}$. The set Ω is the *sample space*, $\mathcal{F}$ is the *event class*, and P is the *probability measure*. The elements of the collection $\mathcal{F}$ are the events, which are the subsets of Ω to which we wish to assign probabilities. In order to be an event class, a collection $\mathcal{F}$ of subsets of Ω must satisfy certain conditions so that we can assign probabilities in a consistent way to the events of interest. To say that the set A is an event is equivalent to saying that A is an element of $\mathcal{F}$, which is denoted by $A \in \mathcal{F}$. The basic properties of an event class $\mathcal{F}$ are (1) $\Omega \in \mathcal{F}$; (2) if $A \in \mathcal{F}$, then so is its complement A^c; and (3) if each of the sets in a countable collection $A_1, A_2, A_3, \ldots$ is in $\mathcal{F}$, so is the union of all of the sets in this collection. A *countable* collection or set is one whose elements can be counted; that is, the elements can be put in a one-to-one correspondence with a subset of the positive integers. Such a collection or set is often referred to as a *discrete* collection or set. Any finite set is countable, but many infinite

1

sets are also countable, such as the set of all integers (positive and negative), the set of all fractions of the form $1/n$ where n is a positive integer, and the set of all rational numbers. An example of a set that is *not* countable is the set of all real numbers.

To summarize, the event class $\mathcal{F}$ includes the sample space Ω (Ω is an event), is closed under complementation (the complement of an event is an event), and is closed under countable unions (the union of a countable collection of events is an event). It can be shown that an event class is closed under countable intersections as well; that is, countable intersections of events are also events.

The probability measure P assigns probabilities to the events in a consistent way. If A is an event, then $P(A)$, the probability of the event A, is a nonnegative number that does not exceed unity (i.e., $0 \leq P(A) \leq 1$ for each $A \in \mathcal{F}$). The probability of the entire sample space is unity (i.e., $P(\Omega) = 1$). If $A_1, A_2, A_3, \ldots$ is a countable collection of disjoint events, the probability of the union of all of these events is the sum of the probabilities: $P(A_1 \cup A_2 \cup A_3 \cup \ldots) = P(A_1) + P(A_2) + P(A_3) + \cdots$. Recall that events A and B are *disjoint* if their intersection $A \cap B$ is empty. Disjoint events are also referred to as being *mutually exclusive*.

1.3   Random Variables

For the time being, it is sufficient to consider real random variables and vectors whose components are such random variables. In an engineering problem, a random variable may model the number of phone calls received during a particular time interval, the voltage at a certain point in a noisy electronic system, the number of bits in error in a word stored in a computer memory, or the number of failures of a certain type of electronic component. In a communication receiver, random variables arise in many ways. For example, a random variable is obtained if the output of the receiver's filter is sampled at a single time instant. The randomness in this output is due in part to the thermal noise in the receiver. If the filter output is sampled several times, the resulting samples can be grouped together to form a vector whose components are random variables. For some applications, it is convenient to consider complex random variables. A complex random variable consists of a real part and an imaginary part, each of which is a real random variable. In this book, unless it is stated otherwise, a random variable is assumed to be a real random variable.

The formal definition of a real random variable X on a probability space $(\Omega, \mathcal{F}, P)$ is obtained by viewing X as a real-valued function defined on the sample space Ω. For each point ω in the sample space, $X(\omega)$ is a real number. Not just any such function will do. For mathematical and physical reasons, we must require that X satisfy the condition

$$\{\omega \in \Omega : X(\omega) \leq u\} \in \mathcal{F}$$

for each choice of the real number u. In words, X is a random variable if and only if for each real number u, the set S is an event, where S consists of all points ω for which $X(\omega)$ does not exceed u. The set S is illustrated in Figure 1-1. Notice that the point ω shown in Figure 1-1 is in the set S, while the point ω' is not.

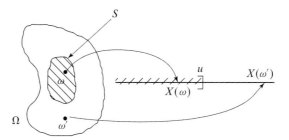

Figure 1-1: The event $S = \{\omega \in \Omega : X(\omega) \leq u\}$.

If X is to be of any value in engineering applications, we must be able to inquire about the probability that X does not exceed some number. For example, we may be interested in the probability that the number of failures does not exceed the design limit for a fault-tolerant system, the probability that the voltage does not exceed a certain threshold in an electronic system, or the probability that the number of incoming phone calls does not exceed the number of operators at the switchboard. We cannot even pose questions involving such probabilities unless $S = \{\omega \in \Omega : X(\omega) \leq u\}$ is an event for each u. This is precisely the requirement for X to be a random variable.

The probability that we seek in the preceding paragraph is $P(\{\omega \in \Omega : X(\omega) \leq u\})$. The notation $\{X \leq u\}$ is shorthand notation for $\{\omega \in \Omega : X(\omega) \leq u\}$, and $P(X \leq u)$ is shorthand notation for the probability $P(\{\omega \in \Omega : X(\omega) \leq u\})$. Such probabilities can be defined and their values sought if X is a random variable.

If $X_1, X_2, \ldots, X_n$ are random variables on the same probability space $(\Omega, \mathcal{F}, P)$, then $\{X_k \leq u_k\}$ is an event for each k in the range $1 \leq k \leq n$ and each choice of the real numbers $u_1, u_2, \ldots, u_n$. It follows that

$$\bigcap_{k=1}^{n} \{X_k \leq u_k\} \in \mathcal{F},$$

and so probabilities of the form

$$P(X_1 \leq u_1, X_2 \leq u_2, \ldots, X_n \leq u_n) = P\left(\bigcap_{k=1}^{n} \{X_k \leq u_k\}\right)$$

are defined for each positive integer n and each choice of $u_1, u_2, \ldots, u_n$. It is often useful to consider the n random variables $X_1, X_2, \ldots, X_n$ to be an n-dimensional random vector $\mathbf{X} = (X_1, X_2, \ldots, X_n)$. Each component of $\mathbf{X}$ is a real random variable.

In engineering, we are concerned with the application of probabilistic methods to the solution of practical problems. Consequently, we usually need not concern ourselves with the detailed mathematical structure of the probability space or the random variable. In particular, we can usually assume that the random phenomenon encountered in practice can be modeled in terms of valid random variables for *some* choice of the probability space $(\Omega, \mathcal{F}, P)$, and so we study the detailed structure of $X(\omega)$ only if it is convenient to do so and only when it is helpful in solving the problem at hand. We are usually able to derive all of the necessary information from the distribution or density functions, the subjects of the next section.

1.3.1   Distribution and Density Functions for Random Variables

The *distribution function*, also referred to as the cumulative distribution function, for a random variable X is denoted by F_X and defined by

$$F_X(u) = P(X \le u) = P(\{\omega \in \Omega : X(\omega) \le u\})$$

for each real number u. For each positive integer n, the distribution function for the random vector $\mathbf{X} = (X_1, X_2, \ldots, X_n)$ is the n-dimensional distribution function denoted by $F_{\mathbf{X},n}$ and defined by

$$F_{\mathbf{X},n}(u_1, u_2, \ldots, u_n) = P(X_1 \le u_1, X_2 \le u_2, \ldots, X_n \le u_n),$$

which is equivalent to

$$F_{\mathbf{X},n}(u_1, u_2, \ldots, u_n) = P\left(\bigcap_{k=1}^{n}\{X_k \le u_k\}\right) \tag{1.1}$$

for each choice of the real numbers u_k ($1 \le k \le n$). The fact that each X_k is a random variable guarantees that the intersection in (1.1) is an event.

The rationale for the notation $F_{\mathbf{X},n}$ is as follows: The first subscript identifies the random vector in question, and the second subscript gives the length of the random vector, which is also the dimension of the distribution. This notation is particularly useful when considering samples of a random process, as in Chapters 2 and 3. In such applications, the number of samples may vary throughout the discussion.

The concept of the distribution function for a random variable is a special case ($n = 1$) of the concept of a distribution function for a random vector. Rather than using $F_{\mathbf{X},1}$, we employ the simpler notation F_X for the distribution function for a one-dimensional random vector (i.e., a random vector $\mathbf{X}$ whose only component is the random variable X). If $\mathbf{X}$ is given by $\mathbf{X} = (X_1, X_2)$, an alternative notation for $F_{\mathbf{X},2}$ is F_{X_1,X_2}, and the latter is usually referred to as the joint distribution function for the random variables X_1 and X_2.

If the distribution function $F_{\mathbf{X},n}$ is continuous and has an associated density function $f_{\mathbf{X},n}$, then we say that the random vector $\mathbf{X} = (X_1, X_2, \ldots, X_n)$ is a continuous random vector with density $f_{\mathbf{X},n}$. This means that

$$F_{\mathbf{X},n}(u_1, \ldots, u_n) = \int_{-\infty}^{u_n} \cdots \int_{-\infty}^{u_1} f_{\mathbf{X},n}(y_1, \ldots, y_n)\, dy_1 \ldots dy_n \tag{1.2a}$$

for each n and each choice of $\mathbf{u} = (u_1, \ldots, u_n)$. It follows that we can let

$$f_{\mathbf{X},n}(y_1, \ldots, y_n) = \partial^n F_{\mathbf{X},n}(u_1, \ldots, u_n)/\partial u_1 \ldots \partial u_n \Big|_{\mathbf{u}=\mathbf{y}} \tag{1.2b}$$

for all $\mathbf{y} = (y_1, \ldots, y_n)$ for which the derivative exists. As with distribution functions, we usually drop the subscript "1" in the notation for the (one-dimensional) density function of a random variable. Thus, if X is a continuous random variable having distribution function F_X, its density f_X satisfies

$$F_X(u) = \int_{-\infty}^{u} f_X(y)\, dy \qquad (1.3a)$$

for each choice of u. It follows that we can let

$$f_X(y) = dF_X(u)/du \Big|_{u=y} \qquad (1.3b)$$

for all values of y for which the derivative exists.

Next we consider a certain class of random variables that do not have continuous distribution functions. Let $\mathbb{Z}$ be the set of all integers (positive, negative, and zero), and let $\mathbb{I}$ be any subset of $\mathbb{Z}$. Consider random variables that take on values in a discrete ordered set $S = \{s_k : k \in \mathbb{I}\}$. In this book, by *ordered* we mean that $s_i < s_j$ whenever $i < j$. We can assume that $\{X(\omega) : \omega \in \Omega\} = S$, but it usually suffices to require only the slightly weaker condition that $P(X \in S) = 1$. Two important examples of discrete sets that fit our definition of *ordered* are $S = \mathbb{Z}$ and $S = \{0, 1, 2, \dots\}$, the set of nonnegative integers. A random variable of this type is called a *discrete random variable*. For a discrete random variable, we can always define a *discrete density function* (also known as a probability mass function) from its distribution function. The discrete density function for such a random variable X is given by

$$f_X(s_i) = F_X(s_i) - F_X(s_{i-1}), \qquad (1.4a)$$

which is just $P(X = s_i)$, the probability that X is exactly s_i. The expression on the right-hand side is the difference between the value of the distribution function at s_i and the value at the next smaller element in the set S. On the other hand, given the discrete density function f_X, the corresponding distribution function can be written as

$$F_X(s_i) = \sum_{k \le i} f_X(s_k). \qquad (1.4b)$$

If $S = \mathbb{Z}$, the sum in (1.4b) can be written as

$$F_X(s_i) = \sum_{k=-\infty}^{i} f_X(s_k);$$

if $S = \{0, 1, 2, \dots\}$, then (1.4b) is equivalent to

$$F_X(s_i) = \sum_{k=0}^{i} f_X(s_k).$$

A typical example of a discrete random variable is the number of successes in a sequence of independent trials of an experiment that can have only two outcomes. The simplest illustration is a sequence of n tosses of a biased coin that comes up heads with probability p and tails with probability $1 - p$. The probability of the occurrence of a *particular sequence* of k heads and $(n - k)$ tails is $p^k(1 - p)^{n-k}$. If X denotes the random variable representing the number of heads out of n independent tosses of the coin, then

$$f_X(k) = \binom{n}{k} p^k(1 - p)^{n-k}$$

for $0 \leq k \leq n$. This is just a result of multiplying the number of sequences that have exactly k heads times the probability of each such sequence. A random variable X with this discrete density function is said to have a *binomial distribution* with parameters n and p.

If a two-dimensional random vector $\mathbf{X} = (X_1, X_2)$ is such that X_1 and X_2 take values in the same discrete ordered set, the discrete density is

$$f_{\mathbf{X},2}(s_i, s_j) = F_{\mathbf{X},2}(s_i, s_j) - F_{\mathbf{X},2}(s_{i-1}, s_j) - F_{\mathbf{X},2}(s_i, s_{j-1}) + F_{\mathbf{X},2}(s_{i-1}, s_{j-1}). \quad (1.5a)$$

On the other hand, given the two-dimensional discrete density function $f_{\mathbf{X},2}$, the corresponding distribution function can be obtained from

$$F_{\mathbf{X},2}(s_i, s_j) = \sum_{k \leq i} \sum_{m \leq j} f_{\mathbf{X},2}(s_k, s_m). \quad (1.5b)$$

Similar equations can be obtained for general n-dimensional discrete density functions.

It is clear that a random variable cannot have both a density function (as in (1.3)) and a discrete density function (as in (1.4)). Thus, when referring to the *density of a random variable*, we mean the form of density that is appropriate for the particular random variable in question. Obviously, for a discrete random variable, only a discrete density function is appropriate; for a continuous random variable, the only density that can be considered is a density of the form given in (1.3).

1.3.2   Means, Variances, and Moments

Suppose the continuous random variable X has the density function f_X. The *mean* or *expected value* of X is given by

$$E\{X\} = \int_{-\infty}^{\infty} u\, f_X(u)\, du, \quad (1.6)$$

provided that the integral exists. By the phrase *the integral exists*, we mean the integrand is absolutely integrable. Thus, for the mean to exist, it is required that

$$\int_{-\infty}^{\infty} |u|\, f_X(u)\, du$$

exist and be finite. The *second moment* of X exists and is defined by

$$m = E\{X^2\} = \int_{-\infty}^{\infty} u^2\, f_X(u)\, du \quad (1.7)$$

if the integral exists. It is known that the existence of the second moment implies the existence of the mean. (See Exercise 2-6 in Section 2.4.1.) If the second moment exists and $\mu = E\{X\}$, then the *variance* of X is defined by

$$\sigma^2 = \text{Var}\{X\} = E\{(X - \mu)^2\} = \int_{-\infty}^{\infty} (u - \mu)^2 \, f_X(u) \, du. \tag{1.8}$$

If X is a discrete random variable taking values in the set $\mathcal{S} = \{s_k : k \in \mathbb{I}\}$, then the mean of X is given by

$$E\{X\} = \sum_{k \in \mathbb{I}} s_k \, f_X(s_k), \tag{1.9}$$

provided that

$$\sum_{k \in \mathbb{I}} |s_k| \, f_X(s_k) < \infty.$$

Expressions for the second moment and variance of a discrete random variable can be obtained by replacing integrals by sums in the corresponding expressions for a continuous random variable.

For any random variable X that has a second moment, the second moment, mean, and variance are related by

$$\sigma^2 = E\{X^2\} - [E\{X\}]^2 = m - \mu^2. \tag{1.10}$$

This follows from (1.8) if we expand $(X - \mu)^2$ and use the fact that the expected value of a sum is the sum of the expected values. In this case, we obtain

$$\sigma^2 = E\{X^2 - 2\mu X + \mu^2\} = E\{X^2\} - 2\mu E\{X\} + \mu^2,$$

which gives (1.10). It is easy to see that (1.10) is equivalent to

$$\sigma^2 = E\{X(X - \mu)\}. \tag{1.11}$$

The *standard deviation* of X is defined as $\sigma = \sqrt{\text{Var}\{X\}}$.

If X is a random variable and g is a well-behaved function, then $Y = g(X)$ is also a random variable. If X is a continuous random variable, the expected value of $g(X)$ can be evaluated from the expression

$$E\{g(X)\} = \int_{-\infty}^{\infty} g(u) \, f_X(u) \, du, \tag{1.12}$$

provided that the integral exists. If X is a discrete random variable, $E\{g(X)\}$ is given by a similar expression in which the integral is replaced by a sum. The mean, variance, and moments of the continuous random variable X can all be expressed as special cases of (1.12). The mean of X is just (1.12) with $g(u) = u$ for each u; the second moment of X corresponds to the function $g(u) = u^2$ for each u; and the kth moment for any positive integer k is $E\{X^k\}$, which corresponds to $g(u) = u^k$ for each u. The variance of X corresponds to $g(u) = (u - \mu)^2$ for each u.

1.3.3 Some Inequalities

We begin with two simple inequalities that involve only the moments of a random variable. From Section 1.3.2, we know that if X has a finite second moment, its variance can be written as

$$\text{Var}\{X\} = E\{X^2\} - [E\{X\}]^2. \tag{1.13}$$

Notice that (1.13) implies that

$$\text{Var}\{X\} \le E\{X^2\}, \tag{1.14}$$

and equality can hold in (1.14) if and only if the mean is zero. Since $\text{Var}\{X\} \ge 0$, (1.13) also gives the inequality

$$[E\{X\}]^2 \le E\{X^2\}, \tag{1.15}$$

and equality can hold in (1.15) if and only if $\text{Var}\{X\} = 0$.

Markov's inequality provides a bound on the distribution function of a nonnegative random variable in terms of the mean of the random variable. Suppose the random variable X is nonnegative, which means that $P(X < 0) = 0$. Let u be any positive number, and define the function g by

$$g(v) = \begin{cases} u, & v \ge u, \\ 0, & v < u. \end{cases}$$

Notice that $g(v) \le v$ for any $v \ge 0$. It follows that $E\{g(X)\} \le E\{X\}$. Now define the random variable Y by $Y = g(X)$, so that

$$E\{Y\} \le E\{X\}. \tag{1.16}$$

Because Y takes on only the two values 0 and u, (1.9) implies that the mean of Y is given by

$$E\{Y\} = 0\,P(Y = 0) + u\,P(Y = u) = u\,P(Y = u).$$

Since $u > 0$, we have

$$P(Y = u) = \frac{E\{Y\}}{u}.$$

Therefore, (1.16) implies that

$$P(Y = u) \le \frac{E\{X\}}{u}. \tag{1.17}$$

Because

$$P(Y = u) = P(g(X) = u) = P(X \ge u),$$

(1.17) is equivalent to

$$P(X \geq u) \leq \frac{E\{X\}}{u}, \tag{1.18}$$

which holds for for any positive number u. The inequality in (1.18) is known as *Markov's inequality*. It is valid for any nonnegative random variable. The bound may be too loose to be of much value in some situations. For example, if the mean of X is larger than u, the right-hand side of (1.18) exceeds one, so the bound tells us nothing new since the probability of any event cannot exceed one. Nevertheless, Markov's inequality is very useful in some situations, and it is particularly easy to evaluate since it requires computation of only the mean of the random variable. The promised bound on the distribution function is

$$F_X(u) \geq 1 - \frac{E\{X\}}{u},$$

which is simple to derive from (1.18).

Markov's inequality can be used to derive other important inequalities, the best known of which is the *Chebyshev inequality*, which states that if Z is a random variable with mean μ, then

$$P(|Z - \mu| \geq v) \leq \frac{\text{Var}\{Z\}}{v^2} \tag{1.19}$$

for any positive number v. A straightforward application of Markov's inequality with $X = (Z - \mu)^2$ gives

$$P((Z - \mu)^2 \geq u) \leq \frac{E\{(Z - \mu)^2\}}{u} = \frac{\text{Var}\{Z\}}{u}$$

for any $u > 0$. We can obtain (1.19) by letting $u = v^2$ and using the fact that $v > 0$ implies that

$$P((Z - \mu)^2 \geq v^2) = P(|Z - \mu| \geq v).$$

One consequence of Chebyshev's inequality is that a random variable whose variance is zero must be trivial in the sense that it is equal to its mean value with probability one. To see this, notice that for $\text{Var}\{Z\} = 0$, the Chebyshev inequality implies that $P(|Z - \mu| \geq v) = 0$ for any $v > 0$. It follows that $P(|Z - \mu| = 0) = 1$, and hence $P(Z = \mu) = 1$. Because of (1.14), it is also true that $E\{Z^2\} = 0$ implies that Z is trivial. In fact, according to (1.14) and (1.15), $E\{Z^2\} = 0$ implies both $\text{Var}\{Z\} = 0$ and $\mu = 0$, so that in this case $P(Z = 0) = 1$. Thus, if the second moment of a random variable is zero, the random variable is zero with probability one.

1.4   Probabilities and Moments Involving Multiple Random Variables

Let X and Y be random variables on the same probability space $(\Omega, \mathcal{F}, P)$. X and Y are *statistically independent* if

$$P(X \in A, Y \in B) = P(X \in A)P(Y \in B)$$

for all suitable sets A and B. For example, it is sufficient if the sets A and B are intervals, but they can also be unions of intervals. A necessary and sufficient condition for the independence of X and Y is

$$F_{X,Y}(u, v) = F_X(u) F_Y(v), \quad -\infty < u < \infty, -\infty < v < \infty.$$

If the random variables X and Y are jointly continuous and have joint density function $f_{X,Y}$, they are statistically independent if and only if their joint density function factors into the product of the marginal densities f_X and f_Y; that is,

$$f_{X,Y}(u, v) = f_X(u) f_Y(v), \quad -\infty < u < \infty, -\infty < v < \infty.$$

The same condition is also necessary and sufficient for jointly discrete random variables to be independent if $f_{X,Y}$ is their joint discrete density function and f_X and f_Y are their marginal discrete density functions. Conditions for mutual independence of a collection of random variables are straightforward extensions of the conditions for pairs of random variables. For example, the random variables $X_1, X_2, \ldots, X_n$ are mutually independent if and only if their n-dimensional distribution function $F_{\mathbf{X},n}(u_1, u_2, \ldots, u_n)$ factors as the product of the marginal distribution functions for the individual random variables.

Suppose the independent random variables X and Y are continuous and have density functions f_X and f_Y, respectively. If the random variable Z is given by $Z = X + Y$, then Z is also a continuous random variable and its density function satisfies

$$f_Z(v) = \int_{-\infty}^{\infty} f_X(v - u) f_Y(u) \, du = \int_{-\infty}^{\infty} f_Y(v - u) f_X(u) \, du. \tag{1.20}$$

Thus, the density for Z is the convolution of the densities for X and Y, which is denoted by $f_Z = f_X * f_Y$. As illustrated in (1.20), $f_X * f_Y = f_Y * f_X$. If the independent random variables are discrete, then the integrals in (1.20) are replaced by sums.

1.4.1   Correlation and Covariance

If X and Y are random variables and g is a well-behaved function, then $Z = g(X, Y)$ is also a random variable. If X and Y are jointly continuous, the mean of Z is given by

$$E\{g(X, Y)\} = \int_{-\infty}^{\infty} \int_{-\infty}^{\infty} g(u, v) \, f_{X,Y}(u, v) \, du \, dv, \tag{1.21}$$

provided that the integral exists. If X and Y are discrete random variables, then the integrals in (1.21) are replaced by sums.

In the remainder of this section, it is assumed that the second moment exists for each random variable that is considered. The covariance and correlation for the pair X and Y can be defined as special cases of $E\{g(X, Y)\}$. If $g(u, v) = uv$ for all real numbers u and v, then $E\{g(X, Y)\}$ is the *correlation* $E\{X Y\}$. If $g(u, v) = (u - \mu_1)(v - \mu_2)$, where $\mu_1 = E\{X\}$ and $\mu_2 = E\{Y\}$, then $E\{g(X, Y)\}$ is the *covariance*

$$\text{Cov}\{X, Y\} = E\{(X - \mu_1)(Y - \mu_2)\}. \tag{1.22}$$

By expanding the product in the right-hand side of (1.22) and taking the expectation term by term, we find that the covariance and correlation are related by

$$\text{Cov}\{X, Y\} = E\{X\,Y\} - E\{X\}\,E\{Y\}. \tag{1.23}$$

If $g(u, v) = (u - \mu_1)(v - \mu_2)/\sigma_1\,\sigma_2$, where $\sigma_1 = \sqrt{\text{Var}\{X\}}$ and $\sigma_2 = \sqrt{\text{Var}\{Y\}}$, then $E\{g(X, Y)\}$ is the *correlation coefficient*, which we denote by ρ. It follows that

$$\rho = \frac{\text{Cov}\{X, Y\}}{\sigma_1\,\sigma_2}. \tag{1.24}$$

The *Schwarz inequality* states that for any random variables U and V,

$$|E\{UV\}| \le \sqrt{E\{U^2\}\,E\{V^2\}}. \tag{1.25}$$

By letting $U = (X - \mu_1)/\sigma_1$ and $V = (Y - \mu_2)/\sigma_2$, we see that (1.25) implies that $-1 \le \rho \le 1$.

Exercise 1-1

Show that for any random variables X and Y that have finite second moments,

$$|E\{X\,Y\}| \le E\{|X\,Y|\} \le \sqrt{E\{X^2\}\,E\{Y^2\}}. \tag{1.26}$$

Solution.   First, observe that $|X(\omega)|\,|Y(\omega)| = |X(\omega)Y(\omega)|$ for each $\omega \in \Omega$, and therefore $E\{|X|\,|Y|\} = E\{|X\,Y|\}$. Next, let $U = |X|$ and $V = |Y|$ in (1.25), and replace $E\{|X|\,|Y|\}$ with $E\{|X\,Y|\}$ to conclude that

$$E\{|X\,Y|\} \le \sqrt{E\{X^2\}\,E\{Y^2\}}. \tag{1.27}$$

Then observe that $-|X(\omega)Y(\omega)| \le X(\omega)Y(\omega) \le |X(\omega)Y(\omega)|$ for each $\omega \in \Omega$, which implies that

$$|E\{X\,Y\}| \le E\{|X\,Y|\}. \tag{1.28}$$

The desired result follows from (1.27) and (1.28).   ■

If $E\{X\,Y\} = E\{X\}\,E\{Y\}$, then the random variables X and Y are said to be *uncorrelated*. If X and Y are independent, it follows that they are uncorrelated. However, uncorrelated random variables are not necessarily independent. If X and Y are independent and g and h are any suitably well-behaved functions, then the random variables $U = g(X)$ and $V = h(Y)$ are also independent. In particular, independence of X and Y implies that

$$E\{g(X)\,h(Y)\} = E\{g(X)\}\,E\{h(Y)\}. \tag{1.29}$$

However, (1.29) is not true in general if X and Y are only uncorrelated.

If $E\{X\,Y\} = 0$, then the random variables X and Y are said to be *orthogonal*. Note that if X and Y are uncorrelated and at least one of them has a zero mean, they are also orthogonal. In general, uncorrelated random variables need not be orthogonal, and orthogonal random variables need not be uncorrelated. However, if X and Y are uncorrelated and have means μ_1 and μ_2, respectively, then the random variables $W = X - \mu_1$ and $Z = Y - \mu_2$ are orthogonal and uncorrelated.

Notice that (1.23) implies that the random variables X and Y are uncorrelated if and only if $\text{Cov}\{X,Y\} = 0$, which is equivalent to $\rho = 0$. Thus, if X and Y are jointly Gaussian and uncorrelated, their joint density function (Section 1.5) simplifies greatly, and it factors as the product of the two marginal densities, each of which is Gaussian. This proves that jointly Gaussian random variables that are uncorrelated are also independent.

It is often necessary to evaluate the expected value of powers of sums of random variables. To begin with a simple example, consider $E\{(X + Y)^2\}$. First, expand the square to obtain

$$E\{(X + Y)^2\} = E\{X^2 + 2XY + Y^2\},$$

and then use the fact that the expected value of a sum is equal to the sum of the expected values to get

$$E\{(X + Y)^2\} = E\{X^2\} + 2\,E\{XY\} + E\{Y^2\},$$

which is the sum of the second moments plus twice the correlation of X and Y. A similar argument shows that

$$\text{Var}\{X + Y\} = \sigma_1^2 + 2\,\text{Cov}\{X, Y\} + \sigma_2^2,$$

which is equivalent to

$$\text{Var}\{X + Y\} = \sigma_1^2 + 2\sigma_1\,\sigma_2\,\rho + \sigma_2^2.$$

It follows that $E\{(X + Y)^2\} = E\{X^2\} + E\{Y^2\}$ if and only if X and Y are orthogonal, and $\text{Var}\{X + Y\} = \text{Var}\{X\} + \text{Var}\{Y\}$ if and only if X and Y are uncorrelated.

1.4.2   Sums and Sequences of Random Variables

Suppose that $X_1, X_2, \ldots, X_N$ are random variables on the same probability space $(\Omega, \mathcal{F}, P)$, and let the random variable Z be defined by

$$Z = \sum_{n=1}^{N} X_n.$$

If we wish to find $E\{Z^2\}$, we can use the same procedure as for the expected value of the square of the sum of two random variables. First, observe that

$$Z^2 = \left[\sum_{n=1}^{N} X_n\right]^2 = \sum_{n=1}^{N} X_n \sum_{m=1}^{N} X_m = \sum_{n=1}^{N}\sum_{m=1}^{N} X_n\,X_m.$$

Using the fact that the expected value of a sum of random variables is the sum of the expected values of the individual random variables, we find that

$$E\{Z^2\} = \sum_{n=1}^{N} \sum_{m=1}^{N} E\{X_n X_m\}.$$

Because $E\{X_n X_m\} = E\{X_m X_n\}$, the second moment of Z can be determined from the correlations $E\{X_n X_m\}$, $1 \le n \le m \le N$. If $n \ne m$ implies that X_n and X_m are orthogonal, then the expression for $E\{Z^2\}$ simplifies to

$$E\{Z^2\} = \sum_{n=1}^{N} E\{X_n^2\}.$$

In particular, the second moment of the sum of zero-mean, uncorrelated random variables is equal to the sum of the second moments of the individual random variables. Similarly, the variance of the sum of uncorrelated random variables is equal to the sum of the variances of the individual random variables.

Let v be an arbitrary positive number and let (Y_n) denote a sequence $Y_n, n = 1, 2, \ldots$, of random variables, all defined on the same probability space. Suppose that Y is a random variable on this probability space and

$$\lim_{n \to \infty} E\{(Y - Y_n)^2\} = 0. \tag{1.30}$$

If (1.30) holds, we say that Y is the *mean-square limit* of the sequence (Y_n) and the form of convergence is referred to as *mean-square convergence*. Mean-square convergence plays a very important role in the study of linear filtering of random processes in Chapter 3. (See especially Section 3.3.)

An application of Markov's inequality (1.18) with $u = v^2$ gives

$$P\left[(Y - Y_n)^2 \ge v^2\right] \le \frac{E\{(Y - Y_n)^2\}}{v^2}. \tag{1.31}$$

It follows from (1.31) that if Y is the mean-square limit of the sequence (Y_n), then

$$\lim_{n \to \infty} P(|Y - Y_n| \ge v) = 0. \tag{1.32}$$

If (1.32) holds for every $v > 0$, we say that the sequence of random variables (Y_n) *converges in probability* to the random variable Y. Thus, mean-square convergence implies convergence in probability.

In one important situation, the convergence is to a deterministic constant rather than a random variable. Let (X_i) be a sequence of independent random variables, each of which has mean μ and variance σ^2. Define the sequence (Y_n) by letting

$$Y_n = \frac{X_1 + X_2 + \cdots + X_n}{n} \tag{1.33}$$

for each positive integer n. The right-hand side of (1.33) is the average or sample mean of the random variables $X_1, X_2, \ldots, X_n$. Observe that $E\{Y_n\} = \mu$ and $\mathrm{Var}\{Y_n\} = \sigma^2/n$ for each n. It follows from Chebyshev's inequality that

$$P(|Y_n - \mu| \geq v) \leq \frac{\sigma^2}{nv^2}.$$

Substituting for Y_n in terms of the sample mean in (1.33) and taking the limit, we see that

$$\lim_{n \to \infty} P\left(\left| \left(\frac{1}{n} \sum_{i=1}^{n} X_i \right) - \mu \right| \geq v \right) = 0 \tag{1.34}$$

for each $v > 0$. Thus, the sequence of the sample means of the X_i's converges in probability to μ, the common mean value for the X_i's. This result is known as the *weak law of large numbers*. If the random variables in the sequence (X_i) also have the same distribution function for each i and we define Z_n by

$$Z_n = \frac{1}{\sigma \sqrt{n}} \sum_{i=1}^{n} (X_i - \mu),$$

then

$$\lim_{n \to \infty} P(Z_n \leq z) = \Phi(z) \tag{1.35}$$

for any real number z. The function Φ in (1.35) is the standard Gaussian distribution function, which is defined in the next section. The result in (1.35) is known as the *central limit theorem*. It states that the distribution function for an appropriately scaled sum of a large number of independent, identically distributed, zero-mean random variables is approximately the standard Gaussian distribution function, regardless of the distribution of the random variables in the sum.

1.5   Gaussian Random Variables

A *Gaussian random variable* X is a continuous random variable with density function of the form

$$f_X(u) = \frac{\exp\{-(u - \mu)^2/2\sigma^2\}}{\sqrt{2\pi\sigma^2}}, \tag{1.36}$$

where μ is the mean of the random variable X and σ^2 is the variance of X. The corresponding distribution function cannot be written in closed form, but it can be expressed in terms of the tabulated integral

$$\Phi(x) = \int_{-\infty}^{x} \frac{\exp(-y^2/2)}{\sqrt{2\pi}}\, dy, \quad -\infty < x < \infty. \tag{1.37}$$

The function Φ is referred to as the *standard Gaussian distribution function*. It is the distribution function for a zero-mean, unit-variance, Gaussian random variable. If X is a Gaussian random variable with mean μ and variance σ^2, then

$$F_X(u) = \Phi[(u - \mu)/\sigma], \quad -\infty < u < \infty.$$

Error probabilities for digital communication systems are best expressed in terms of the complementary distribution function, which is defined by

$$Q(x) = \int_x^\infty \frac{\exp(-y^2/2)}{\sqrt{2\pi}}\, dy, \quad -\infty < x < \infty. \tag{1.38}$$

It should be clear from (1.37) and (1.38) that $Q(x) + \Phi(x) = 1$. Also, replacing x with $-x$ in (1.37), followed by the change of variable $u = -y$, shows that

$$\Phi(-x) = \int_x^\infty \frac{\exp(-u^2/2)}{\sqrt{2\pi}}\, du = Q(x).$$

It follows that the relationship between the two functions is $Q(x) = 1 - \Phi(x) = \Phi(-x)$ for all x in the range $-\infty < x < \infty$.

Although there is no exact closed-form expression for the function Q, several good bounds and approximations can be computed easily. In [1.2], it is shown that the family of functions

$$G_{a,b}(x) = \left\{ (1-a)x + a\sqrt{x^2 + b} \right\}^{-1} \exp(-x^2/2)/\sqrt{2\pi} \tag{1.39}$$

is very useful in obtaining bounds and approximations for $Q(x)$ that are appropriate for performance analyses of digital communication systems. In particular, if $a = 1/\pi$ and $b = 2\pi$, a lower bound on $Q(x)$ is obtained. By substituting for a and b in (1.39), we can easily show that this lower bound is given by

$$Q_1(x) = \sqrt{\pi/2} \left\{ (\pi - 1)x + \sqrt{x^2 + 2\pi} \right\}^{-1} \exp(-x^2/2). \tag{1.40}$$

An upper bound, which we denote by $Q_2(x)$ is obtained if the values of a and b in (1.39) are $a = 0.344$ and $b = 5.334$. It is also pointed out in [1.2] that a good approximation, which we denote by $Q_3(x)$, results from (1.39) by letting $a = 0.339$ and $b = 5.510$. Thus, we have $Q_1(x) \leq Q(x) \leq Q_2(x)$ and $Q(x) \approx Q_3(x)$ for values of x that are of interest in the evaluation of the performance of digital communication systems. See [1.2] for further discussion and additional references on this family of bounds and approximations.

A family of series approximations for the function Q is given in [1.1]. For one choice of parameters, this approximation can be written as

$$\tilde{Q}(x) = \tfrac{1}{2} - \left(\tfrac{2}{\pi}\right) H(x), \tag{1.41a}$$

where

$$H(x) = \sum_{\substack{n=1 \\ n \text{ odd}}}^{33} n^{-1} \exp\{-(n\pi)^2/392\} \sin(n\pi x/14). \tag{1.41b}$$

There are many situations in engineering in which two or more random variables are Gaussian, not only individually, but also collectively. This concept is made precise by the consideration of the *joint* distribution for the random variables. One way to define jointly Gaussian random variables is to give their joint density function: The random variables X and Y are said to be *jointly Gaussian* if their joint density function is of the form

$$f_{X,Y}(u, v) = \frac{\exp\left\{\frac{-1}{2(1-\rho^2)}\left[\left(\frac{u-\mu_1}{\sigma_1}\right)^2 - 2\rho\left(\frac{u-\mu_1}{\sigma_1}\right)\left(\frac{v-\mu_2}{\sigma_2}\right) + \left(\frac{v-\mu_2}{\sigma_2}\right)^2\right]\right\}}{2\pi\sigma_1\sigma_2\sqrt{(1-\rho^2)}},$$

where μ_1 and σ_1 are the mean and standard deviation for X, μ_2 and σ_2 are the mean and standard deviation for Y, and

$$\rho = \frac{E\{(X-\mu_1)(Y-\mu_2)\}}{\sigma_1\sigma_2} = \frac{\text{Cov}\{X, Y\}}{\sigma_1\sigma_2}$$

is the correlation coefficient. An important feature of jointly Gaussian random variables is that their joint density function is completely determined by the five parameters $\mu_1, \mu_2, \sigma_1, \sigma_2$, and ρ. Also, it is easy to show that jointly Gaussian random variables are independent if and only if their correlation coefficient is zero ($\rho = 0$).

For generalization to more than two random variables, it is more convenient to define a pair of random variables X and Y to be jointly Gaussian if each linear combination of X and Y is a Gaussian random variable; that is, X and Y are jointly Gaussian if $\alpha X + \beta Y$ is a Gaussian random variable for each choice of the real numbers α and β. It can be shown that this definition is equivalent to the condition that X and Y have the joint density function $f_{X,Y}$ given in the previous paragraph. (See pp. 46–49 of [1.9] or pp. 156 and 210 of [1.10].)

Consider a set of n random variables $X_1, X_2, \ldots, X_n$. We may treat these collectively by defining the random vector $\mathbf{X} = (X_1, X_2, \ldots, X_n)$, and then the density function for the random vector $\mathbf{X}$ is just the joint density function for the random variables $X_1, X_2, \ldots, X_n$. We say that a set of n random variables $X_1, X_2, \ldots, X_n$ is jointly Gaussian if every linear combination of these random variables is a Gaussian random variable. It follows from this definition that the n random variables are jointly Gaussian if and only if they have a joint density function that is of the Gaussian form. Specifically, if Λ is the $n \times n$ matrix with

$$\Lambda_{i,j} = E\{(X_i - \mu_i)(X_j - \mu_j)\}$$

as the element in the ith row and jth column, and if μ_i is the mean of X_i ($1 \le i \le n$), then the joint density function for $\mathbf{X}$ is given by

$$f_{\mathbf{X},n}(u) = (2\pi)^{-n/2}|\det(\Lambda)|^{-1/2}\exp\left\{-\tfrac{1}{2}(\mathbf{u} - \boldsymbol{\mu})\Lambda^{-1}(\mathbf{u} - \boldsymbol{\mu})^T\right\},$$

where $\det(\Lambda)$ is the determinant of the matrix Λ, Λ^{-1} is the inverse of the matrix Λ, $\mathbf{u} = (u_1, u_2, \ldots, u_n)$, $\boldsymbol{\mu} = (\mu_1, \mu_2, \ldots, \mu_n)$, and $(\mathbf{u} - \boldsymbol{\mu})^T$ is the transpose of the vector $(\mathbf{u} - \boldsymbol{\mu})$. The matrix Λ is known as the covariance matrix for the random vector $\mathbf{X}$, and the vector $\boldsymbol{\mu}$ is the mean vector for $\mathbf{X}$. For $n = 2$, this n-dimensional density function reduces to the two-dimensional density

$$f_{X,Y}(u, v) = \frac{\exp\left\{\frac{-1}{2(1-\rho^2)}\left[\left(\frac{u-\mu_1}{\sigma_1}\right)^2 - 2\rho\left(\frac{u-\mu_1}{\sigma_1}\right)\left(\frac{v-\mu_2}{\sigma_2}\right) + \left(\frac{v-\mu_2}{\sigma_2}\right)^2\right]\right\}}{2\pi\sigma_1\sigma_2\sqrt{(1-\rho^2)}},$$

if we set $X_1 = X$, $X_2 = Y$, $\Lambda_{1,1} = \sigma_1^2$, $\Lambda_{2,2} = \sigma_2^2$, and $\Lambda_{2,1} = \Lambda_{1,2} = \rho\sigma_1\sigma_2$.

Several important problems that arise in digital communications deal with a pair of independent Gaussian random variables. If X and Y are jointly Gaussian and independent, they have the joint density function just given, with $\rho = 0$. But if $\rho = 0$, then this density is

$$f_{X,Y}(u, v) = \frac{\exp\{-(u - \mu_1)^2/2\sigma_1^2\}\exp\{-(v - \mu_2)^2/2\sigma_2^2\}}{2\pi\sigma_1\sigma_2}. \qquad (1.42)$$

Of course, X and Y are individually Gaussian, so they have densities

$$f_X(u) = \exp\{-(u - \mu_1)^2/2\sigma_1^2\}/\sqrt{2\pi}\,\sigma_1 \qquad (1.43)$$

and

$$f_X(v) = \exp\{-(v - \mu_2)^2/2\sigma_2^2\}/\sqrt{2\pi}\,\sigma_2. \qquad (1.44)$$

Notice that (1.42)–(1.44) show that $\rho = 0$ implies $f_{X,Y}(u, v) = f_X(u)\, f_Y(v)$ for each value of u and v, which establishes that uncorrelatedness implies independence for a pair of jointly Gaussian random variables.

The problem that we wish to address is the determination of the probability that the random pair (X, Y) falls in a rectangular region of the form

$$S = \{(u, v) : x_1 \le u \le x_2, y_1 \le v \le y_2\}$$

for an arbitrary choice of x_1, x_2, y_1, and y_2, subject only to the constraint that $x_1 < x_2$ and $y_1 < y_2$. This set is illustrated in Figure 1-2.

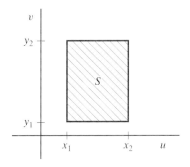

Figure 1-2: The region S.

The easiest way to evaluate $P((X, Y) \in S)$ is to observe that

$$P((X, Y) \in S) = P(x_1 \leq X \leq x_2, y_1 \leq Y \leq y_2)$$

and, because X and Y are independent continuous random variables,

$$P(x_1 \leq X \leq x_2, y_1 \leq Y \leq y_2) = P(x_1 \leq X \leq x_2)P(y_1 \leq Y \leq y_2)$$
$$= [F_X(x_2) - F_X(x_1)][F_Y(y_2) - F_Y(y_1)].$$

Since X and Y are Gaussian, it follows that

$$F_X(x) = \Phi[(x - \mu_1)/\sigma_1]$$

for each x and

$$F_Y(y) = \Phi[(y - \mu_2)/\sigma_2]$$

for each y. Consequently,

$$P((X, Y) \in S) =$$
$$\{\Phi[(x_2 - \mu_1)/\sigma_1] - \Phi[(x_1 - \mu_1)/\sigma_1]\}\{\Phi[(y_2 - \mu_2)/\sigma_2] - \Phi[(y_1 - \mu_2)/\sigma_2]\}. \quad (1.45)$$

In (1.45), it is permissible to let x_2 or y_2 be $+\infty$ and x_1 or y_1 be $-\infty$. For example, if $y_2 = +\infty$ and $x_2 = +\infty$, then

$$\Phi[(x_2 - \mu_1)/\sigma_1] = 1$$

and

$$\Phi[(y_2 - \mu_2)/\sigma_2] = 1,$$

so that

$$P[(X, Y) \in S] = \{1 - \Phi[(x_1 - \mu_1)/\sigma_1]\}\{1 - \Phi[(y_1 - \mu_2)/\sigma_2]\}$$
$$= Q[(x_1 - \mu_1)/\sigma_1] \, Q[(y_1 - \mu_2)/\sigma_2]. \quad (1.46)$$

References and Suggestions for Further Reading

[1.1] N. C. Beaulieu, "A simple series for personal computer computation of the error function $Q(\cdot)$," *IEEE Transactions on Communications*, vol. 37, September 1989, pp. 989–991.

[1.2] P. O. Börjesson and C.-E. W. Sundberg, "Simple approximations of the error function $Q(t)$ for communications applications," *IEEE Transactions on Communications*, vol. COM-27, March 1979, pp. 639–643.

[1.3] G. R. Cooper and C. D. McGillem, *Probabilistic Methods of Signal and System Analysis* (2nd ed.). New York: Holt, Rinehart, and Winston, 1986.

[1.4] C. W. Helstrom, *Probability and Stochastic Processes for Engineers* (2nd ed.). New York: Macmillan, 1991.

[1.5] P. G. Hoel, S. C. Port, and C. J. Stone, *Introduction to Probability Theory*. Boston: Houghton Mifflin, 1971.

[1.6] J. J. Komo, *Random Signal Analysis in Engineering Systems*. New York: Academic Press, 1987.

[1.7] A. Leon-Garcia, *Probability and Random Processes for Electrical Engineering* (2nd ed.). Reading, MA: Addison-Wesley, 1994.

[1.8] S. Ross, *A First Course in Probability* (6th ed.). New York: Macmillan, 2002.

[1.9] E. Wong and B. Hajek, *Stochastic Processes in Engineering Systems*. New York: Springer-Verlag, 1985.

[1.10] J. M. Wozencraft and I. M. Jacobs, *Principles of Communication Engineering*. New York: Wiley, 1965.

Problems

1.1 A sequence of binary digits is transmitted in a certain communication system. Any given digit is received in error with probability p and received correctly with probability $1 - p$. Errors occur independently from digit to digit. Out of a sequence of n transmitted digits, what is the probability that no more than j of these are received in error?

1.2 A binary error-correcting code has code words consisting of n binary digits (e.g., the (7,4) Hamming code described in Appendix A, for which $n = 7$). Suppose this code is applied to the communication system described in Problem 1.1.

   (a) The code can correct any single error, but it cannot correct any pattern of two or more errors. What is the probability that the code can correct the error pattern that occurs when n digits are transmitted? Express your answer in terms of n and p.

   (b) The code can detect any pattern of two or fewer errors. What is the probability that the code can detect the error pattern that occurs when n digits are transmitted?

1.3 The Reed–Solomon code described in Appendix B can correct multiple errors per word. For example, words consisting of eight symbols are transmitted if the (8,4) Reed–Solomon code is employed, and bounded-distance decoding corrects any pair of errors that occur in a word. Suppose each symbol is represented by three binary digits, and each eight-symbol word is sent to the receiver as a corresponding sequence of 24 binary digits. Suppose the channel described in Problem 1.1 is used for the transmission of the binary digits. For each of the following, express your answer in terms of the parameter p:

   (a) What is the probability of error for each received symbol?

   (b) What is the probability that a received word is not decoded correctly?

1.4 Suppose that the random variables X and Y satisfy $E\{X^2\} = E\{Y^2\}$. Define the random variables U and V by $U = X + Y$ and $V = X - Y$.

(a) Show that U and V are orthogonal.

(b) If, in addition, X and Y have the same mean, are they uncorrelated? Prove your answer.

1.5 Define the functions g and h by

$$g(u) = \begin{cases} 1, & 0 \le u < 1/2, \\ 0, & 1/2 \le u \le 1, \end{cases}$$

and $h(u) = 1 - g(u)$ for $0 \le u \le 1$. Suppose the random variable U is uniformly distributed on the unit interval $[0, 1]$, and the random variables X and Y are defined by $X = g(U)$ and $Y = h(U)$, respectively.

(a) Find the mean and variance of X.

(b) Find the mean and variance of Y.

(c) Are X and Y statistically independent? Prove your answer.

(d) Are X and Y orthogonal? Prove your answer.

(e) Are X and Y uncorrelated? Prove your answer.

1.6 The random variables $X_1, X_2, \ldots, X_{10}$ each have a finite second moment, and they satisfy

$$E\{X_i X_j\} = \beta(9 - |i - j|), \quad 1 \le i \le 10, 1 \le j \le 10.$$

In addition, $P(X_i = 0) < 1$ for each i.

(a) Give an expression for $E\{[X_3 + X_9]^2\}$ in terms of the parameter β.

(b) What restrictions are there on the possible values for β?

1.7 The random variable U is uniformly distributed on the interval $[0, \pi]$, and the random variables V and W are defined by $V = \sin(U)$ and $W = \cos(U)$. The following can be answered without using any trigonometric identities or evaluating any integrals. You may find it beneficial to rely on sketches of certain trigonometric functions to guide your solutions.

(a) Find the mean of W.

(b) Are V and W uncorrelated? Prove your answer.

(c) Are V and W statistically independent? Prove your answer.

1.8 Two random variables, X_1 and X_2, are produced in a digital communication receiver. An error occurs if $X_1 > X_2$; otherwise, the receiver output is correct. The random variables X_1 and X_2 are independent, Gaussian random variables, with means μ_1 and μ_2 and standard deviations σ_1 and σ_2, respectively. Find the probability of error for this receiver. First express your answer in terms of the function Φ and the parameters $\mu_1, \mu_2, \sigma_1,$ and σ_2. Convert your answer so that it is in terms of the function Q. (*Hint:* perhaps the best approach is to let $X = X_1 - X_2$ so that X is a Gaussian random variable and $P(X > 0)$ is the probability of error for the communication receiver.)

1.9 Suppose the random variables X_1, X_2, and X_3 are jointly Gaussian with means $\mu_i = E\{X_i\}$ and covariances $\Lambda_{i,j} = \text{Cov}\{X_i, X_j\}$ for $1 \leq i \leq 3$ and $1 \leq j \leq 3$. Let $Y = X_1 + X_2 + X_3$. Express the distribution function for Y in terms of the standard Gaussian distribution function Φ and the parameters μ_i and $\Lambda_{i,j}$.

1.10 The decision statistic in a certain communication receiver is a Gaussian random variable Z, which has mean μ and standard deviation σ. The decision made by the receiver depends on the value of Z^2. Find the probability that $Z^2 < 2$. Express your answer in terms of the function Φ and the parameters μ and σ. Note that Z^2 is not Gaussian!

1.11 Suppose that X_1 and X_2 are Gaussian random variables with means μ_1 and μ_2, respectively. Assume that $\mu_1 \neq \mu_2$. The variance for each of the two random variables is σ^2. Find the value of x for which $f_{X_1}(x) = f_{X_2}(x)$.

1.12 The random variables X_1 and X_2 are defined as in Problem 1.11. Show that for each value of y, $f_{X_1}(y)$ is larger than $f_{X_2}(y)$ if and only if $|y - \mu_1|$ is smaller than $|y - \mu_2|$. This observation is employed in communication receivers that use maximum-likelihood statistical decisions.

1.13 The random variables X_1 and X_2 are Gaussian and independent with means μ_1 and μ_2, respectively. Each has variance σ^2. What is the probability that the point (X_1, X_2) is in the square with vertices $(1,1)$, $(1,2)$, $(2,1)$, and $(2,2)$? For what values of μ_1 and μ_2 is this probability maximized?

1.14 Let the parameter α represent a measure of the signal-to-noise ratio in a certain binary digital communication system. In this system, the probability of a given binary digit being in error is $Q(\sqrt{2\alpha})$. Define α_{dB} to be $10 \log_{10}(\alpha)$.
(a) Evaluate $Q_1(\sqrt{2\alpha})$, $Q_2(\sqrt{2\alpha})$, and $Q_3(\sqrt{2\alpha})$ for $\alpha_{\text{dB}} = 10.5$.
(b) Write a computer program to evaluate $Q_1(\sqrt{2\alpha})$, $Q_2(\sqrt{2\alpha})$, and $Q_3(\sqrt{2\alpha})$ for values of α_{dB} ranging from 7.0 to 12.0 in steps of 0.1. Give your answers by filling in the following table:

α_{dB}	$Q_1(\sqrt{2\alpha})$	$Q_2(\sqrt{2\alpha})$	$Q_3(\sqrt{2\alpha})$
7.0			
7.1			
7.2			
$\vdots$			
12.0			

1.15 It is often necessary to determine the value of α_{dB} for which $Q(\sqrt{2\alpha}) = 10^{-n}$ for some integer n. Use the bounds and approximations to estimate to the nearest tenth of a dB the values of α_{dB}, for which $Q(\sqrt{2\alpha}) = 10^{-n}$ for $n = 1, 2, \ldots, 8$.

Chapter 2

Introduction to Random Processes

2.1 Origins of Random Processes in Electronic Systems

Random processes are present in all electronic systems. Such a process may arise as a result of random motion of electrons in a resistive component of the system, or it may be due to some kind of human-made disturbance. Often, we model a signal generated by an electronic system as a random process, even though it may be deterministic in the sense that *if* we had an accurate model of every detail of the signal generation process and *if* we could solve all of the mathematical problems that arise in the analysis of that model, we could determine the values of the signal exactly at any given time. In practice, we rarely have such a model and we could not solve the analytical and computational problems required for a complete deterministic description of the signal. Moreover, for many applications, a deterministic characterization of the signal is unnecessary and would be far too complicated to be of any practical utility.

One example of a random process is the message process that is generated at the transmitter and conveys information to the receiver in a communication system such as a telephone system or a satellite broadcast link. The message process could be an analog video signal, a digitized speech signal, telemetry from a satellite or deep-space probe, or a sequence of binary digits from a digital computer. The message process may be modulated onto a radio-frequency (RF) signal (known as the carrier) that is suitable for transmission through the available communication medium. If so, the resulting RF signal represents a random process, not only because of the randomness in the message, but also because there may be a random phase associated with the unmodulated carrier itself.

In thinking of communication signals as represented by random processes, we may wish to take the point of view of the intended receiver, in which case the modeling of a signal as a random process reflects the fact that, before reception of the signal, the receiver does not know precisely the message that is being transmitted. If it did, there would be no need to transmit the signal at all! Sometimes it is simply the value of a parameter, such as the amplitude or phase of a sinusoidal signal, that is unknown to the intended receiver; if so, the randomness can be expressed via one or more random variables. In such a situation, modeling the signal as a random process may be more for convenience than necessity. Often, however, the unknown portion of the signal is more complex than this, and the randomness cannot be expressed in terms of a finite number of random variables. In such cases, modeling the signal as a random process is essential.

The various kinds of noise processes that arise in electrical and electronic devices are examples of highly complicated random processes that cannot be described in terms of finite numbers of random variables. Such noise processes may be due to automobile ignition, lightning, electromagnetic radiation from electronic equipment, radar signals from aircraft, or thermal noise from various components in a communication receiver. Strictly speaking, all electronic systems generate noise within the system and radiate noise to nearby systems. In order to be of any value, electronic equipment must be capable of satisfactory operation in the presence of such noise; therefore, it is necessary to understand noise processes and their effects in order to be able to design efficient electronic systems.

In this chapter, basic mathematical models for random processes are introduced, and many properties of random processes that arise in engineering problems are presented. We begin with some examples.

2.2   Examples of Random Processes

Before giving a precise definition of a random process, we present some of the basic intuitive notions that are needed in the study of random processes. This is accomplished by means of four examples that are typical of the random processes encountered in electrical and systems engineering. As a substitute for a precise definition of a *continuous-time random process*, consider such a process to be a collection of waveforms defined on a common interval (e.g., on $[0, \infty)$, (a, b), or $(-\infty, \infty)$). Randomness is introduced into this conceptual model by considering a hypothetical experiment in which one waveform is drawn at random from the collection of waveforms. Similarly, a *discrete-time random process* can be thought of as a collection of sequences, and the experiment is to draw one sequence at random from this collection.

Example 2-1   An Oscillator with a Random Phase
Consider an experiment in which an oscillator is switched on at some time T_0, and its output is observed during the time interval $0 \leq t \leq T$ (where $T_0 < 0 < T$). Assume that $T = f_0^{-1}$, where f_0 is the frequency of the oscillator, and let α be the amplitude of the sinusoidal signal at the output of the oscillator. The time T_0 is selected at random, but we assume that $|T_0|$ is sufficiently large that any transients have decayed.

Because T_0 is random (and typically, the oscillator's initial phase at time T_0 is also random), the sinusoidal signal at the output has a random phase Θ at time $t = 0$. That is, the output is

$$X(t) = \alpha \sin(2\pi f_0 t + \Theta)$$

for $0 \leq t \leq T$, as shown in Figure 2-1. The phase Θ is a random variable, and the nature of the random process $X(t)$ depends on the distribution of Θ, which in turn depends on

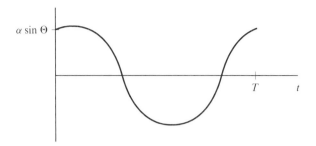

Figure 2-1: A sample function for $X(t) = \alpha \sin(2\pi f_0 t + \Theta)$.

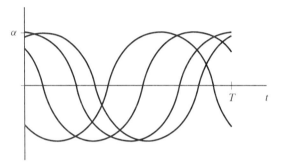

(a) Typical sample functions if Θ is uniform on $[0, 2\pi]$

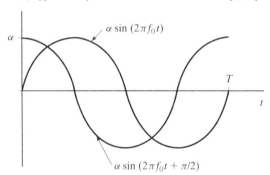

(b) Possible sample functions if Θ takes values 0 and $\pi/2$ only

Figure 2-2: Typical sample functions for $X(t) = \alpha \sin(2\pi f_0 t + \Theta)$.

the distribution of T_0. We can consider the sinusoidal waveform $\alpha \sin(2\pi f_0 t + \Theta)$ for a particular value of the random variable Θ; such a waveform is called a *sample function* for the random process $X(t)$.

We can think of an experiment in which Θ is selected at random according to some distribution and the resulting sample function is sketched. If the experiment is repeated several times, we might observe waveforms such as those shown in Figure 2-2(a). These waveforms are typical outcomes if, for example, the random variable Θ is uniformly distributed on the interval $[0, 2\pi]$. On the other hand, if Θ takes on the values 0 and $\pi/2$ only, the two waveforms illustrated in Figure 2-2(b) are the *only* ones possible.

Notice that for each value of t, $X(t)$ is a function of the random variable Θ; hence, $X(t)$ is itself a random variable. We are interested in statistical characterizations of the random variable $X(t)$, including the distribution function for $X(t)$, and these characterizations depend on the distribution function for Θ.   ∎

Example 2-2   Thermal Noise

Suppose we conduct an experiment in which a resistor of R ohms is placed in a controlled environment with the temperature held constant at T kelvins (K). As shown in Figure 2-3(a), the terminals of the resistor are connected to the input of an ideal bandpass filter (BPF) with center frequency f_0 Hz and bandwidth B Hz. (See Figure 2-3(b).)

The output $Y(t)$ of the bandpass filter obtained in this experiment will be a random-looking waveform that does not appear to exhibit much statistical regularity, as illustrated in Figure 2-4. If, however, we observe the average power $W(t)$, we will see that it always settles down to a value close to $w_0 = 4kTRB$, also illustrated in the figure, where $k \approx 1.38 \times 10^{-23}$ joule/K is Boltzmann's constant. With R in ohms and B in hertz, $w_0 = 4kTRB$ is in volts squared. The reason $W(t)$ is referred to as "power" is that it represents the power dissipated if $Y(t)$ is the voltage across a *one-ohm* resistor. The resulting power is then referred to as the *power on a one-ohm basis*. A more precise statement is that $\sqrt{W(t)}$ is the rms value, averaged from time 0 to time t, of the voltage $Y(t)$ at the filter output.

The phenomenon just described was observed experimentally by Johnson and derived analytically by Nyquist in 1928. The model is valid for frequencies f_0 up to at least 10^{11} Hz. For an idea of the magnitude of the numbers involved, consider the following typical values of the parameters: $T = 293$ K (approximately room temperature),

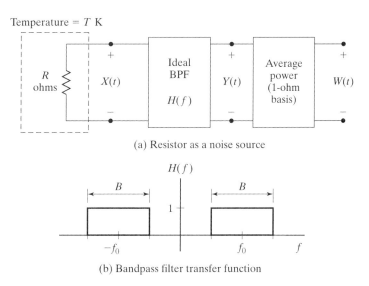

(a) Resistor as a noise source

(b) Bandpass filter transfer function

Figure 2-3: Generation of thermal noise.

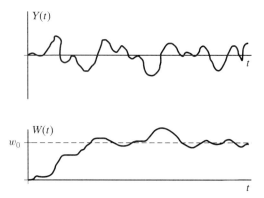

Figure 2-4: Sample functions for the processes $Y(t)$ and $W(t)$.

$B = 10\,\text{kHz}$, and $R = 10,000$ ohms. The result is that $w_0 \approx 1.6 \times 10^{-12}$ volts squared. This corresponds to an rms voltage of about $1.3\ \mu\text{V}$.

   The experiment can be repeated several times, in which case a variety of different random-looking waveforms will be obtained at the output of the filter. The set of all such waveforms that could result from this experiment is the set of sample functions of the process $Y(t)$. In spite of the differences between different sample functions, the average power $W(t)$ will stay close to $w_0 = 4\,kTRB$ for sufficiently large t.

   The foregoing discussion illustrates just one example in which the process of interest may appear not to exhibit any statistical regularity, but upon closer examination it will be seen to have well-defined second-order properties. In this example, the second-order property considered is the time-averaged power. ∎

Example 2-3   Counting Process

Suppose that at time $t = 0$ we begin counting the number of phone calls coming into a switchboard. For $t > 0$, let $X(t)$ be the total number of phone calls received at or before time t. The counting process is initialized by setting $X(0) = 0$. If this experiment is repeated several times (e.g., once each day for several days) and the results are recorded in the form of graphs of $X(t)$ vs. t, the outcomes will be several different increasing step functions (a typical version of which is shown in Figure 2-5).

Figure 2-5: A typical sample function for the counting process.

If we postulate that no more than one call can be received at a time, all of the steps (jump discontinuities) in the recorded graphs are of size one. The randomness is in the location of the steps, which are the arrival times of the phone calls. For a fixed time interval $[0, T]$, both the total number of steps and the location of the steps are random, and both the number and location of the steps can be determined from a knowledge of $X(t)$ for all t in the range $0 \leq t \leq T$.   ∎

Example 2-4   A Discrete-Time Filtering Problem

Suppose that $X_0, X_1, X_2, \ldots$ is a sequence of independent random variables whose distribution is defined by $P(X_n = 0) = P(X_n = 1) = 1/2$ for each n. Define a discrete-time random process $X(t)$, $t = 0, 1, 2, \ldots$, by setting $X(t) = X_t$. Let $X(t)$ be the input to a discrete-time filter for which the output at time t is the sum of the input at time t and the input at time $t - 1$. Hence, the output is $Y(t) = X(t) + X(t - 1)$. The output process is then a discrete-time random process $Y(t)$ consisting of a sequence $Y_1, Y_2, Y_3, \ldots$ of random variables for which

$$P(Y_k = 0) = P(Y_k = 2) = \frac{P(Y_k = 1)}{2} = \frac{1}{4}$$

for each k. However, the sequence $Y_1, Y_2, Y_3, \ldots$ is *not* a sequence of independent random variables. For instance,

$$P(Y_1 = 2, Y_2 = 0) = 0 \neq P(Y_1 = 2)\, P(Y_2 = 0) = \frac{1}{16}.$$ ∎

2.3   Definitions and Basic Concepts

Thus far, we have been using the term "random process" in an intuitive way, without a formal mathematical description. To proceed further, it is necessary to give a precise definition. A *random process* on a probability space $(\Omega, \mathcal{F}, P)$ is defined as an indexed collection $\{X_t : t \in \mathbb{T}\}$ of random variables on $(\Omega, \mathcal{F}, P)$. Random processes are also often called *stochastic processes* in the literature.

The parameter t typically denotes time, and the index sets of interest are usually intervals of the real line $\mathbb{R}$ (e.g., $\mathbb{T} = [0, \infty)$, $\mathbb{T} = (a, b)$, or $\mathbb{T} = (-\infty, \infty)$), subsets of set $\mathbb{Z}$ of all integers (e.g., $\mathbb{T} = \mathbb{Z}$, $\mathbb{T} = \{0, 1, 2, \ldots\}$, or $\mathbb{T} = \{1, 2, \ldots, n\}$), or sets that can be indexed by a subset of $\mathbb{Z}$ (e.g., $\mathbb{T} = \{1/n, 2/n, 3/n, \ldots\}$ for some positive integer n). For convenience, in what follows, let $\mathbb{N}$ denote a set of consecutive integers. For example, $\mathbb{N}$ could be $\mathbb{Z}$, $\{0, 1, 2, \ldots\}$, $\{1, 2, \ldots, n\}$ for some integer n, or $\{-9, \ldots, -1, 0, 1, \ldots, 9\}$. These sets are all countable (i.e., they can be indexed by a subset of the integers), and they are ordered in the sense defined in Section 1.3.1. We refer to ordered countable sets as *discrete* sets in this book. An example of a set that is *not* countable is the set of all real numbers between 0 and 1.

Various types of random processes are encountered in engineering applications, and it is helpful to divide them into four categories, determined by the type of index set $\mathbb{T}$ and by the set of values that can be taken on by the random variables that make up the random processes. If $\mathbb{T}$ is an interval of the real line, $\{X_t : t \in \mathbb{T}\}$ is a *continuous-time* random process. If $\mathbb{T}$ is of the form $\{t_k : k \in \mathbb{N}\}$ and $t_k < t_i$ for all k and i in the set $\mathbb{N}$ such that $k < i$, then $\mathbb{T}$ is a *discrete-time* random process; that is, the index set for a discrete-time random process is a discrete set. In general, the set S of values taken on by a random process is called the *state space* of the process. Unless stated otherwise, the random processes in this book are assumed to be real valued; that is, $S \subset \mathbb{R}$. A real-valued random process is simply an indexed collection of real-valued random variables. If the state space of a random process is a discrete set, we say that the *state space is discrete* and the process is a *discrete-amplitude* random process. If $S \subset \mathbb{R}$, but S is not discrete, such as when S is an interval of $\mathbb{R}$, the process is said to be a *continuous-amplitude* random process.

Notice that the processes in Examples 2-1 and 2-2 are continuous-amplitude, continuous-time processes. The process in Example 2-3 is a discrete-amplitude, continuous-time process, and the process in Example 2-4 is a discrete-amplitude, discrete-time process. For an example of the only remaining type—a continuous-amplitude, discrete-time process—consider a random process formed from a sequence of independent Gaussian random variables.

The statement "$\{X_t : t \in \mathbb{T}\}$ is a collection of random variables on $(\Omega, \mathcal{F}, P)$" means that for each $t \in \mathbb{T}$, $\{X_t \leq u\} \in \mathcal{F}$. This latter condition is just shorthand notation for the requirement that $\{\omega \in \Omega : X_t(\omega) \leq u\}$ be an event for each real number u. (See Figure 2-6.)

Because X_t is a random variable for each t,

$$\{X_t \leq u\} = \{\omega \in \Omega : X_t(\omega) \leq u\} \in \mathcal{F}$$

for each choice of t and u. It follows that

$$\bigcap_{k=1}^{n} \{X_{t_k} \leq u_k\} \in \mathcal{F},$$

so that probabilities of the form

$$P\left(X_{t_1} \leq u_1, X_{t_2} \leq u_2, \ldots, X_{t_n} \leq u_n\right) = P\left[\bigcap_{k=1}^{n} \{X_{t_k} \leq u_k\}\right]$$

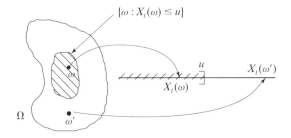

Figure 2-6: The event $\{\omega \in \Omega : X_t(\omega) \leq u\}$.

are defined for each positive integer n; each set of points $t_1, t_2, \ldots, t_n$; and each choice of $u_1, u_2, \ldots, u_n$. Such probabilities are fundamental building blocks for the analysis of random processes in communication systems and other applications. We consider them further in Section 2.3.1.

From an engineering point of view, it is sometimes helpful to think of a continuous-time random process as a collection of waveforms and a discrete-time random process as a collection of sequences, as discussed in Section 2.2. Mathematically, this point of view is derived from the consideration of $X_t(\omega)$ as a function of t for fixed ω. We then refer to $X_t(\omega)$ as a *sample function* of the process. The notion of repeating an experiment several times, which is mentioned in the examples of Section 2.2, can be viewed as drawing several points at random from Ω according to the probability assignment P. If the outcomes of these drawings are $\omega_1, \omega_2, \ldots, \omega_k$, then the experimentally observed waveforms are $X_t(\omega_1), X_t(\omega_2), \ldots, X_t(\omega_k)$, each considered as a function of t. Hence, we can either think of drawing waveforms from a large collection of waveforms or drawing points from Ω and letting $X_t(\omega)$ generate the waveforms. Thus, on the kth drawing, the point ω_k is selected from Ω, and the sample function $X_t(\omega_k)$ is generated. This point of view is illustrated in Figure 2-7 for $k = 3$.

Thus, we may interpret a random process as a collection of waveforms that are indexed by the parameter $\omega \in \Omega$. This amounts to a reversal of the roles of "variable" and "parameter" from the point of view adopted in the original mathematical definition. In the definition of a random process given at the beginning of this section, the "parameter" is t and the "variable" is ω. In order to view a random process as a collection of waveforms, the "variable" becomes t and the "parameter" is ω, in which case it is more natural to use the notation $X(t, \omega)$ than $X_t(\omega)$.

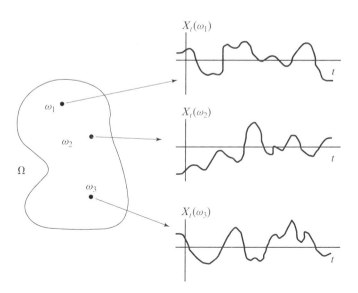

Figure 2-7: Three sample functions for the random process X_t.

Most engineering textbooks use $X(t)$, $t \in \mathbb{T}$, or simply $X(t)$, to denote a random process; that is, the variable ω in the notation $X(t, \omega)$ is suppressed. We adopt that convention in most of what follows, except for situations in which it is important to display the functional dependence on ω for clarity. In such cases, the notation will be $X(t, \omega)$ or $X_t(\omega)$, whichever is more appropriate. In using the engineering notation $X(t)$, the reader must keep in mind that for each t, $X(t)$ is a *random variable* on the probability space $(\Omega, \mathcal{F}, P)$ and its value at the point $\omega \in \Omega$ is $X(t, \omega)$. For a fixed value of t, $X(t)$ is actually a function $X_t : \Omega \to \mathbb{R}$, a mapping from Ω into the real line, and this function satisfies $\{\omega \in \Omega : X(t, \omega) \leq u\} \in \mathcal{F}$ for each choice of t and u.

Example 2-5   A Ramp Signal with a Random Slope

Let the sample space Ω for the random process consist of all real numbers (i.e., $\Omega = \mathbb{R}$, the real line), and let the index set $\mathbb{T}$ be the set of nonnegative real numbers (i.e., $\mathbb{T} = [0, \infty)$). Suppose a random process is defined by $X_t(\omega) = X(t, \omega) = \omega t$ for each $\omega \in \mathbb{R}$ and each $t \geq 0$. For a fixed value of ω, $X_t(\omega)$ is a straight line of slope ω that starts at the origin. Some sample functions for this random process are shown in Figure 2-8.

As is true of any random process, the random process $X_t(\omega) = \omega t$ has the property that, for any fixed value of t, X_t is a random variable. We can think of fixing $t = t'$, drawing a point at random from Ω, and examining the value of $X_{t'}(\omega) = \omega t'$. This procedure is illustrated in Figure 2-9. Notice that what we observe will depend on the value of t' that was chosen. For example, "most" sample functions give a value of $\omega t'$ that is small in magnitude if t' is small, but "most" sample functions give a value of $\omega t'$ that is large in magnitude if the value of t' is large. A precise interpretation of "most" depends on the probability measure P on the sample space; probability distributions associated with random processes are discussed in the next subsection. ∎

The two exercises that follow are optional. The concepts illustrated by these exercises are useful, but they are not required for the material presented elsewhere in the text.

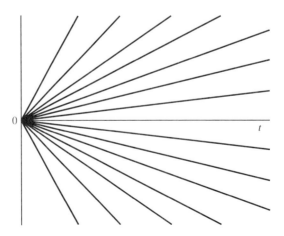

Figure 2-8: Sample functions for the random process $X(t, \omega) = \omega t$.

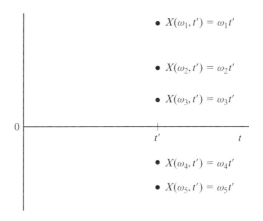

Figure 2-9: Sample values for the random process $X(t, \omega) = \omega t$ at time t'.

Exercise 2-1

Suppose that $\Omega = (0, \infty)$ and we define

$$X(t, \omega) = \omega t$$

for all $t > 0$ and all $\omega \in \Omega$. This exercise is similar to Example 2-5, except that only positive slopes are allowed and we now have the restriction that $t > 0$. Suppose that $\mathcal{F}$ contains all intervals $(a, b]$ for which $0 \le a < b$, and suppose the probability measure P is such that $P((a, b]) = e^{-a} - e^{-b}$. Show that $X(t, \omega)$ is a random process, and find $P(X(t) \le u)$ for each $t > 0$ and each value of u.

Solution.   To show that $X(t, \omega)$ is a random process, we let t be arbitrary, but fixed, and show that $\Omega_u \in \mathcal{F}$ for each u, where $\Omega_u = \{\omega \in \Omega : X(t, \omega) \le u\}$. We then replace $X(t, \omega)$ by ωt in the expression for the set Ω_u and notice that, for $u > 0$,

$$\Omega_u = \{\omega : 0 < \omega t \le u\} = \{\omega : 0 < \omega \le t^{-1}u\} = (0, t^{-1}u],$$

and for $u \le 0$,

$$\Omega_u = \{\omega : 0 < \omega t \le u\} = \varnothing.$$

Now, $P(X(t) \le u) = P(\Omega_u) = 1 - \exp(-t^{-1}u)$ for $u > 0$, and $P(X(t) \le u) = P(\varnothing) = 0$ for $u \le 0$.   ∎

Exercise 2-2

Suppose the amplitude of the oscillator output in Example 2-1 is $\alpha = 1$ and the random variable Θ is defined by $\Theta(\omega) = \omega$ for each ω in $\Omega = [0, 2\pi]$. Show that $X(t, \omega) = \sin[2\pi f_0 t + \Theta(\omega)]$ is a random process if $\mathcal{F}$ is an event class that includes all intervals of $[0, 2\pi]$.

Solution.   Since, for $0 \leq t \leq T$,

$$X(t, \omega) = \sin(2\pi f_0 t + \omega), \quad \text{for } 0 \leq \omega \leq 2\pi,$$

it follows that the set $\Omega_u = \{\omega \in \Omega : X(t, \omega) \leq u\}$ is empty if $u < -1$ and that Ω_u is just the entire sample space Ω if $u \geq 1$. On the other hand, if $-1 \leq u < 1$, then the set

$$\Omega_u = \{\omega \in \Omega : \sin(2\pi f_0 t + \omega) \leq u\}$$

is a union of not more than two intervals, as illustrated in Figure 2-10 for two specific values of t: $t = 0$ and $t = t_1 = (4f_0)^{-1}$. If $\mathcal{F}$ contains all intervals of $\Omega = [0, 2\pi]$, then it must also contain all countable unions of such intervals. Hence, $\Omega_u \in \mathcal{F}$ for each u.   ∎

In engineering, we are concerned with the application of probabilistic methods to the solution of practical problems involving random processes. Consequently, we usually need not concern ourselves with the detailed mathematical structure of the random process. In particular, we can assume that each random phenomenon encountered in practice is a valid random process for some choice of the probability space $(\Omega, \mathcal{F}, P)$. Hence, we study the detailed structure of $X(t, \omega)$ only if it is convenient to do so and only when it is helpful in solving the problem at hand. For instance, an examination of the structure of the random process at the level of Exercises 2-1 and 2-2 is not necessary for the solution of the vast majority of engineering problems. Instead, we are usually able to derive all of the information we require from the finite-dimensional distribution functions, and these are the subject of the next subsection.

2.3.1   Distribution and Density Functions for Random Processes

The *one-dimensional distribution function* for a random process $X(t), t \in T$, is denoted by $F_{X,1}$ and defined by

$$F_{X,1}(u; t) = P[X(t) \leq u] = P[\{\omega : X(t, \omega) \leq u\}]$$

for each real number u and each $t \in \mathbb{T}$. This distribution function is also known as the univariate distribution function for the random process. For each positive integer n, the *n-dimensional distribution function* is denoted by $F_{X,n}$ and defined by

$$F_{X,n}(u_1, \ldots, u_n; t_1, \ldots, t_n) = P[X(t_1) \leq u_1, \ldots, X(t_n) \leq u_n],$$

which is equivalent to

$$F_{X,n}(u_1, \ldots, u_n; t_1, \ldots, t_n) = P\left[\bigcap_{k=1}^{n} \{X(t_k) \leq u_k\}\right]$$

$$= P\left[\bigcap_{k=1}^{n} \{\omega : X(t_k, \omega) \leq u_k\}\right] \qquad (2.1)$$

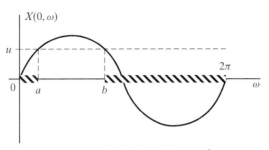

(a) $t=0, 0<u<1, \Omega_u=[0,a] \cup [b,2\pi], a=\sin^{-1}(u), b=\pi-a$

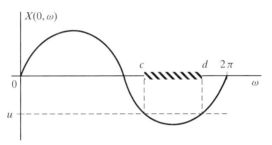

(b) $t=0, -1 \leq u \leq 0, \Omega_u=[c,d], d=2\pi+\sin^{-1}(u), c=3\pi-d$

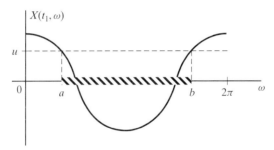

(c) $t=t_1=(4f_0)^{-1}, 0<u<1, \Omega_u=[a,b], a=(\pi/2)-\sin^{-1}(u), b=2\pi-a$

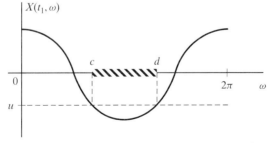

(d) $t=t_1=(4f_0)^{-1}, -1 \leq u \leq 0, \Omega_u=[c,d], c=(\pi/2)-\sin^{-1}(u), d=2\pi-c$

Figure 2-10: $\Omega_u = \{\omega \in [0, 2\pi] : \sin(2\pi f_0 t + \omega) \leq u\}$.

for each $u_k \in \mathbb{R}$ and $t_k \in \mathbb{T}$ $(1 \le k \le n)$. For our purposes, a random process is specified completely by giving its n-dimensional distribution function for each positive integer n.

Just as with random variables, if each of the distribution functions $F_{X,n}$ has an associated density function $f_{X,n}$, then we say that each random vector $[X(t_1), \dots, X(t_n)]$ is a *continuous random vector* with density $f_{X,n}$. This means that

$$F_{X,n}(x_1, \dots, x_n; t_1, \dots, t_n) = \int_{-\infty}^{x_n} \cdots \int_{-\infty}^{x_1} f_{X,n}(y_1, \dots, y_n; t_1, \dots, t_n) \, dy_1 \dots dy_n$$

$$(2.2)$$

for each n and each choice of $\mathbf{x} = (x_1, \dots, x_n)$ and $\mathbf{t} = (t_1, \dots, t_n)$. It follows that we can let

$$f_{X,n}(y_1, \dots, y_n; t_1, \dots, t_n) = \partial^n F_{X,n}(x_1, \dots, x_n; t_1, \dots, t_n)/\partial x_1 \dots \partial x_n \big|_{\mathbf{x}=\mathbf{y}}$$

for each $\mathbf{y} = (y_1, \dots, y_n)$ for which the derivative exists.

Next, we consider a certain class of discrete-amplitude random processes: the class of random processes for which the state space $\mathcal{S} = \{s_k : k \in \mathbb{N}\}$ is a discrete ordered set according to the definition in Section 1.3.1; that is, $s_i < s_j$ whenever $i < j$. Two important examples are $\mathcal{S} = \mathbb{Z}$ and $\mathcal{S} = \{0, 1, 2, \dots\}$. For a random process in this class, we can always define a *discrete density function* (also known as a probability mass function) from the distribution function for the process. For example, the one-dimensional discrete density function for such a process $X(t)$ is given by

$$f_{X,1}(s_i; t) = F_{X,1}(s_i; t) - F_{X,1}(s_{i-1}; t),$$

and the two-dimensional discrete density is

$$\begin{aligned} f_{X,2}(s_i, s_j; t_1, t_2) &= F_{X,2}(s_i, s_j; t_1, t_2) - F_{X,2}(s_{i-1}, s_j; t_1, t_2) \\ &\quad - F_{X,2}(s_i, s_{j-1}; t_1, t_2) + F_{X,2}(s_{i-1}, s_{j-1}; t_1, t_2). \end{aligned}$$

On the other hand, for the n-dimensional discrete density function $f_{X,n}$, the corresponding distribution function is given by

$$F_{X,n}(s_{i_1}, \dots, s_{i_n}; t_1, \dots, t_n) = \sum_{j_1 \le i_1} \cdots \sum_{j_n \le i_n} f_{X,n}(s_{j_1}, \dots, s_{j_n}; t_1, \dots, t_n). \qquad (2.3)$$

Of course, a random process cannot have both a density function (as in (2.2)) and a discrete density function (as in (2.3)). Thus, in referring to the *density of a random process*, we mean the form of density that is appropriate for the particular random process in question. Obviously, for a discrete-amplitude process, only a discrete density function is appropriate, and the only density that can be considered for a continuous-amplitude process is a density of the form given in (2.2).

The solution to the next exercise illustrates how to determine the distribution and density functions for a discrete-amplitude random process.

Exercise 2-3

Find the one- and two-dimensional distribution and density functions for the discrete-amplitude, discrete-time process $Y(t)$ of Example 2-4.

Solution.   The one-dimensional functions follow easily from the observation that, for $t \in \{1, 2, 3 \dots \}$,

$$P[Y(t) = 0] = P[Y(t) = 2] = \frac{P[Y(t) = 1]}{2} = \frac{1}{4},$$

as noted in Example 2-4. Clearly, the discrete density function is given by

$$f_{Y,1}(y; t) = \begin{cases} \frac{1}{4}, & y = 0 \text{ or } y = 2, \\ \frac{1}{2}, & y = 1, \\ 0, & \text{otherwise,} \end{cases}$$

for $t \in \{1, 2, 3 \dots \}$; thus, the distribution function is

$$F_{Y,1}(y; t) = \begin{cases} 0, & y < 0, \\ \frac{1}{4}, & 0 \le y < 1, \\ \frac{3}{4}, & 1 \le y < 2, \\ 1, & y \ge 2. \end{cases}$$

For two dimensions, we give only the density function, because the two-dimensional distribution function is somewhat cumbersome to describe for this example. The two-dimensional distribution function can, of course, be obtained from the two-dimensional density function via (2.3).

First, we assume that t_1 and t_2 are positive integers for which $|t_1 - t_2| \ge 2$. This condition guarantees that

$$Y(t_1) = X(t_1) + X(t_1 - 1)$$

and

$$Y(t_2) = X(t_2) + X(t_2 - 1)$$

are independent. The condition $|t_1 - t_2| \ge 2$ implies that the values of $t_1, t_1 - 1, t_2$, and $t_2 - 1$ are all different, so the independence of $Y(t_1)$ and $Y(t_2)$ follows from the fact that the random process $X(t)$ of Example 2-4 has the property that $X(t)$ and $X(\tau)$ are independent if $t \ne \tau$. Thus, for $|t_1 - t_2| \ge 2$,

$$\begin{aligned} f_{Y,2}(y_1, y_2; t_1, t_2) &= P[Y(t_1) = y_1, Y(t_2) = y_2] = P[Y(t_1) = y_1] P[Y(t_2) = y_2] \\ &= f_{Y,1}(y_1; t_1) f_{Y,1}(y_2; t_2). \end{aligned}$$

For $|t_1 - t_2| \le 1$, $Y(t_1)$ and $Y(t_2)$ are *dependent*. First notice that if $t_1 = t_2$ then

$$f_{Y,2}(y_1, y_2; t_1, t_2) = P[Y(t_1) = y_1, Y(t_1) = y_2] = \begin{cases} 0, & y_1 \ne y_2, \\ f_{Y,1}(y_1; t_1), & y_1 = y_2. \end{cases}$$

If $|t_1 - t_2| = 1$, then either $t_1 - 1 = t_2$ or $t_1 = t_2 - 1$ (but not both, of course), so we can write

$$Y(t_1) = Z_1 + Z_2$$

and

$$Y(t_2) = Z_2 + Z_3,$$

where Z_1, Z_2, and Z_3 are independent and satisfy $P(Z_i = 0) = P(Z_i = 1) = 1/2$. If $t_1 - 1 = t_2$, then $Z_1 = X(t_1)$, $Z_2 = X(t_2) = X(t_1 - 1)$, and $Z_3 = X(t_2 - 1)$. If $t_1 = t_2 - 1$, then $Z_2 = X(t_1) = X(t_2 - 1)$, $Z_1 = X(t_1 - 1)$, and $Z_3 = X(t_2)$. In either case, we see that $P[Y(t_1) = y_1, Y(t_2) = y_2]$ is equal to $1/8$ for all pairs (y_1, y_2) in the set

$$S_1 = \{(0, 0), (2, 2), (0, 1), (1, 0), (2, 1), (1, 2)\},$$

is equal to $1/4$ for $y_1 = y_2 = 1$, and is equal to zero otherwise. Thus, for $|t_1 - t_2| = 1$,

$$f_{Y,2}(y_1, y_2; t_1, t_2) = \begin{cases} \frac{1}{8}, & (y_1, y_2) \in S_1, \\ \frac{1}{4}, & y_1 = y_2 = 1, \\ 0, & \text{otherwise.} \end{cases}$$

In particular, note that since it is impossible that both $Y(t) = 0$ and $Y(t+1) = 2$ or both $Y(t) = 2$ and $Y(t+1) = 0$, it must be that, for $|t_1 - t_2| = 1$,

$$f_{Y,2}(0, 2; t_1, t_2) = 0 = f_{Y,2}(2, 0; t_1, t_2),$$

and hence,

$$f_{Y,2}(y_1, y_2; t_1, t_2) \neq f_{Y,1}(y_1; t_1)\, f_{Y,1}(y_2; t_2) = \tfrac{1}{16}$$

whenever $y_1 = 0$ and $y_2 = 2$ (or $y_1 = 2$ and $y_2 = 0$). Clearly, $Y(t_1)$ and $Y(t_2)$ are statistically dependent, in contrast to the previous results for $|t_1 - t_2| \geq 2$. ■

2.3.2   Gaussian Random Processes

Next, we consider a very important class of continuous-amplitude random processes known as Gaussian random processes. A random process $X(t)$, $t \in \mathbb{T}$, is a *Gaussian random process* if, for any positive integer n, any choice of real coefficients a_k $(1 \leq k \leq n)$, and any choice of sampling times $t_k \in \mathbb{T}$ $(1 \leq k \leq n)$, the random variable $a_1 X(t_1) + a_2 X(t_2) + \cdots + a_n X(t_n)$ is a Gaussian random variable. As a consequence, for any $t_k \in \mathbb{T}$ $(1 \leq k \leq n)$, the vector

$$\mathbf{X}(t) = [X(t_1), \ldots, X(t_n)]$$

has the n-dimensional Gaussian density. In particular, the one-dimensional density function for a Gaussian random process is the Gaussian density function. The solution to the next exercise illustrates some methods for working with Gaussian random processes.

Exercise 2-4

Consider the continuous-amplitude, continuous-time random process $X(t)$, $t \in \mathbb{R}$, defined by

$$X(t) = Y_1 + t\, Y_2,$$

where Y_1 and Y_2 are independent Gaussian random variables, each having zero mean and variance σ^2. Find the one- and two-dimensional density functions for this random process.

Solution.   For each fixed t, $X(t)$ is a linear combination of the independent Gaussian random variables Y_1 and Y_2, so $X(t)$ is a Gaussian random variable. The mean of the random variable $X(t)$ is zero and the variance of $X(t)$ is

$$\mathrm{Var}\{Y_1\} + \mathrm{Var}\{t\, Y_2\} = \sigma^2 + t^2 \sigma^2 = \sigma^2(1 + t^2).$$

Hence,

$$f_{X,1}(x; t) = \left[2\pi\sigma^2 (1 + t^2) \right]^{-1/2} \exp\left\{ -x^2 / \left[2\sigma^2 \left(1 + t^2 \right) \right] \right\}$$

is the one-dimensional density function for the process $X(t)$.

For fixed t_1 and t_2, $X(t_1)$ and $X(t_2)$ are jointly Gaussian with zero means and with variances given by

$$\mathrm{Var}\{X(t_k)\} = \sigma^2 \left(1 + t_k^2 \right) \tag{2.4}$$

for $k = 1, 2$. Because the means are zero, the covariance is

$$\begin{aligned}
\mathrm{Cov}\{X(t_1), X(t_2)\} &= E\{X(t_1)\, X(t_2)\} \\
&= E\{(Y_1 + t_1\, Y_2)(Y_1 + t_2\, Y_2)\} \\
&= \sigma^2 + t_1 t_2 \sigma^2 = \sigma^2(1 + t_1 t_2).
\end{aligned}$$

The correlation coefficient for $X(t_1)$ and $X(t_2)$ is defined by

$$\rho(t_2, t_1) = \frac{\mathrm{Cov}\{X(t_1), X(t_2)\}}{\sqrt{\mathrm{Var}\{X(t_1)\}\mathrm{Var}\{X(t_2)\}}}.$$

For the random process given in this example, the correlation coefficient reduces to

$$\rho(t_2, t_1) = \frac{1 + t_1 t_2}{\left[\left(1 + t_1^2 \right) \left(1 + t_2^2 \right) \right]^{1/2}}. \tag{2.5}$$

Therefore, the two-dimensional density function for the process $X(t)$ is given by

$$f_{X,2}(x_1, x_2; t_1, t_2) = \frac{\exp\left\{ -\left[\left(1 + t_2^2 \right) x_1^2 - 2 \left(1 + t_1 t_2 \right) x_1 x_2 + \left(1 + t_1^2 \right) x_2^2 \right] / \left[2\sigma^2 \left(t_1 - t_2 \right)^2 \right] \right\}}{2\pi\sigma^2 |t_1 - t_2|},$$

provided that $t_1 \neq t_2$. If $t_1 = t_2$, then the joint distribution of $X(t_1)$ and $X(t_2)$ is degenerate, since $\rho(t_1, t_2) = 1$. ∎

For the random process $X(t)$ of Exercise 2-4, we can show directly that the random variable

$$Z = \sum_{k=1}^{n} a_k \, X(t_k) = a_1 \, X(t_1) + \cdots + a_n \, X(t_n)$$

is a Gaussian random variable (without appealing to the definition of a Gaussian random process). First observe that Z can be expressed as $c_1 \, Y_1 + c_2 \, Y_2$, where

$$c_1 = \sum_{k=1}^{n} a_k \quad \text{and} \quad c_2 = \sum_{k=1}^{n} a_k \, t_k.$$

That is, Z is a linear combination of two jointly Gaussian random variables Y_1 and Y_2, and therefore, Z is Gaussian.

Some additional discussion of Gaussian random processes is given in Section 2.8. Many of the properties of Gaussian random processes that are used in this book are derived from the fact that each of the n-dimensional distribution functions ($n = 1, 2, 3, \ldots$) is completely specified by the appropriate means and covariances. (The variance is a special case of the covariance; for example, $\text{Var}\{X(t)\} = \text{Cov}\{X(t), X(t)\}$.) The reader should observe that the foregoing solution to Exercise 2-4 demonstrates this fact for the two-dimensional distribution of the Gaussian random process in question.

2.3.3   Conditional Densities for Random Processes

Once we have defined n-dimensional density functions, we can obtain conditional density functions in a straightforward manner, since conditional density functions are just ratios of joint density functions. We shall engage in a slight abuse of notation at this point by using the same notation $f_{X,n}$ for both the conditional and joint densities. The appearance of a vertical bar (e.g., as in $f_{X,n}(\cdot|\cdot)$) will, as usual, inform the reader that the function should be interpreted as a *conditional* density.

If, for each k ($1 \le k \le n$), the function $f_{X,k}$ is the k-dimensional density function for a continuous-amplitude random process $X(t)$, then the conditional density function for $X(t_1), \ldots, X(t_m)$ given $X(t_{m+1}), \ldots, X(t_n)$ is

$$f_{X,n}(x_1, \ldots, x_m; t_1, \ldots, t_m | x_{m+1}, \ldots, x_n; t_{m+1}, \ldots, t_n)$$
$$= \frac{f_{X,n}(x_1, \ldots, x_n; t_1, \ldots, t_n)}{f_{X,n-m}(x_{m+1}, \ldots, x_n; t_{m+1}, \ldots, t_n)} \qquad (2.6)$$

for $n \ge 2$ and $1 \le m < n$.

Exercise 2-5

In Exercise 2-4, find the conditional density function for $X(t_1)$ given $X(t_2)$ if $t_1 \ne t_2$.

Solution.   By simply substituting into the expression

$$f_{X,2}(x_1; t_1 | x_2; t_2) = \frac{f_{X,2}(x_1, x_2; t_1; t_2)}{f_{X,1}(x_2; t_2)}$$

for the densities $f_{X,2}$ and $f_{X,1}$ determined in Exercise 2-4, we find that

$$f_{X,2}(x_1; t_1 | x_2; t_2) = \frac{\exp\left\{-\left[x_1\left(1+t_2^2\right)-(1+t_1\,t_2)\,x_2\right]^2 \big/ 2\sigma^2\,(t_1-t_2)^2\left(1+t_2^2\right)\right\}}{\left[2\pi\sigma^2\,(t_1-t_2)^2 \big/ \left(1+t_2^2\right)\right]^{1/2}}.$$

(2.7)

Alternatively, we can simply use the fact that $X(t_1)$ and $X(t_2)$ are jointly Gaussian, so that $f_{X,n}(\cdot|\cdot)$ has the form of a Gaussian density. The conditional mean is

$$\alpha = E\{X(t_1)|X(t_2) = x_2\} = \left[\frac{\mathrm{Var}\{X(t_1)\}}{\mathrm{Var}\{X(t_2)\}}\right]^{1/2} \rho(t_1,t_2)\,x_2$$

(since $E\{X(t_1)\} = E\{X(t_2)\} = 0$) and the conditional variance is

$$\beta^2 = \mathrm{Var}\{X(t_1)|X(t_2)\} = \mathrm{Var}\{X(t_1)\}\left(1 - [\rho(t_1,t_2)]^2\right).$$

Substituting into these two expressions from (2.4) and (2.5), we have

$$\alpha = \left[\frac{1+t_1\,t_2}{1+t_2^2}\right]x_2$$

and

$$\beta^2 = \frac{\sigma^2(t_1-t_2)^2}{1+t_2^2}.$$

Since the conditional density is then of the form

$$\frac{\exp\{-[x_1-\alpha]^2/2\beta^2\}}{\sqrt{2\pi\beta^2}},$$

we arrive at (2.7) after some rearrangement of terms. ∎

2.4   Mean, Autocorrelation, and Autocovariance

It is not always practical (or even possible) to determine all of the n-dimensional distribution functions for a random process. Furthermore, it is usually unnecessary to have that much information about the process in order to obtain a satisfactory solution to an engineering problem. In most of the problems encountered in engineering, a great deal can be learned from the two-dimensional distribution functions. The most important "parameters" for the two-dimensional distributions are the mean, autocorrelation, and autocovariance functions.

2.4.1   Definitions and Basic Properties

We define the mean, autocorrelation, and autocovariance functions for random processes $X(t)$ that satisfy $E\{[X(t)]^2\} < \infty$ for each t. Such processes are called *second-order* processes. All processes in this section are assumed to be second-order, real-valued random processes, unless stated otherwise. If $X(t)$ represents a physical signal, such as the voltage or current in an electrical system, then $E\{[X(t)]^2\}$ represents the expected value of the instantaneous power (on a one-ohm basis) in the process $X(t)$. When the random process $X(t)$ arises in such a manner—and this is the situation of primary interest to us in electrical engineering—it is reasonable to assume that $E\{[X(t)]^2\}$ is finite. In the next exercise, we establish that $E\{|X(t)|\}$ is also finite for a second-order process $X(t)$. In this exercise and for the remainder of the book, we often omit the braces from the expressions involving expectations whenever the meaning is clear. Thus, $E\{[X(t)]^2\}$ is written more simply as $E[X(t)]^2$. Similarly, $E\{X(t)\}$ and $E\{|X(t)|\}$ are often denoted by $EX(t)$ and $E|X(t)|$, respectively. In addition, the index set $\mathbb{T}$ is frequently not displayed explicitly; in such cases, the statement "for each t" should be interpreted as "for each t in $\mathbb{T}$."

Exercise 2-6

Show that a second-order random process $X(t)$ always satisfies $E|X(t)| < \infty$ for each t.

Solution.   First observe that for any real number u, $|u| \leq 1 + u^2$. The easiest way to show this is to consider the two cases $|u| < 1$ and $|u| \geq 1$ separately. If $|u| < 1$, then $|u|$ is certainly no larger than $1 + u^2$; if $|u| \geq 1$, then $|u| \leq u^2$. An application of this inequality gives

$$|X(t)| \leq 1 + [X(t)]^2. \tag{2.8}$$

Therefore, for each t,

$$E|X(t)| \leq 1 + E[X(t)]^2 < \infty.$$

We should point out that what is really meant by (2.8) is that

$$|X(t, \omega)| \leq 1 + [X(t, \omega)]^2$$

for each t and each ω in Ω. Following the standard convention in engineering, we have omitted the ω-dependence in (2.8), since the meaning is clear. ∎

The *mean function* (or simply the *mean*) of a second-order random process $X(t)$, $t \in \mathbb{T}$, is defined by

$$\mu_X(t) = E\{X(t)\}$$

for each $t \in \mathbb{T}$. The *autocorrelation function* is defined by

$$R_X(t_1, t_2) = E\{X(t_1)\, X(t_2)\}$$

for t_1 and t_2 in $\mathbb{T}$, and the *autocovariance function* is defined by

$$C_X(t_1, t_2) = E\{[X(t_1) - \mu_X(t_1)][X(t_2) - \mu_X(t_2)]\}.$$

As will be shown later, $R_X(t_1, t_2)$ and $C_X(t_1, t_2)$ are finite for a second-order process.

In a sense, the mean $\mu_X(t)$ is the average value of the random process at time t, where "average" refers to the probabilistic average or expectation. The autocorrelation $R_X(t_1, t_2)$ and autocovariance $C_X(t_1, t_2)$ are quantitative measures of a certain type of statistical "coupling" between $X(t_1)$ and $X(t_2)$. For instance, if $X(t_1)$ and $X(t_2)$ are *statistically independent*, there is no such "coupling" and, in fact, $C_X(t_1, t_2) = 0$, which is equivalent to saying that the two random variables $X(t_1)$ and $X(t_2)$ are uncorrelated.

The autocorrelation and autocovariance functions play a very important role in the analysis of random processes in linear systems. One of the important topics in this subject is the study of noise in linear systems, and one of the most fundamental parameters of noise is the power in the noise. This is needed, for instance, in order to determine the signal-to-noise ratios at the input and output of a linear system. A filter is often employed in an electronic system to decrease the noise power (i.e., increase the signal-to-noise ratio). In such applications, it is important to be able to obtain analytical expressions for the noise power at the output of the filter as a function of the filter characteristics. The results developed in Chapters 2–4 enable the reader to obtain such expressions, which can then be used to perform engineering tradeoff studies in the design of the filter.

If the random process $X(t)$ represents noise, then, as we have discussed, the power in the noise is $E[X(t)]^2$. From the definition of the autocorrelation function, we see that this power is $R_X(t, t)$. Suppose $X(t)$ is the noise at the input to a linear system, and ignore the signal for the present discussion. In this case, the output $Y(t)$ is also a random process (i.e., noise in gives noise out). In order to determine the output noise power $E[Y(t)]^2$, it turns out that we need to know more than the input noise power. In fact, except for trivial linear systems, it is necessary to know the autocorrelation function $R_X(t_1, t_2)$ for the input random process for several values of its arguments t_1 and t_2, even if all we want to know about the output noise process is its power.

Generally, the mean, autocorrelation, and autocovariance functions are relatively easy to determine, compared with other parameters of a random process. This is because these functions can all be determined from the two-dimensional distribution function $F_{X,2}$. In fact, the mean can be determined even if all that we know is the one-dimensional distribution function. Notice that

$$C_X(t_1, t_2) = R_X(t_1, t_2) - \mu_X(t_1)\,\mu_X(t_2), \tag{2.9}$$

so it suffices to consider the mean together with only *one* of the two functions R_X or C_X. Notice also that $C_X(t_1, t_2) = 0$ if and only if $R_X(t_1, t_2) = \mu_X(t_1)\,\mu_X(t_2)$.

Next, we establish that the mean, autocorrelation, and autocovariance functions are finite for a process with finite power. First, we use the fact that

$$-|X(t)| \le X(t) \le |X(t)|,$$

which implies that both $+E\{X(t)\}$ and $-E\{X(t)\}$ are not greater than $E|X(t)|$, so that

$$|\mu_X(t)| = |EX(t)| \le E|X(t)|. \tag{2.10}$$

This fact and Exercise 2-6 establish that $\mu_X(t)$ is finite for any second-order process. Also, we will see in Exercise 2-7 that

$$|R_X(t_1, t_2)| \le \frac{E[X(t_1)]^2 + E[X(t_2)]^2}{2}, \tag{2.11}$$

so $R_X(t_1, t_2)$ is also finite for any second-order process. Finally, observe that (2.9) implies that $C_X(t_1, t_2)$ is finite whenever the mean and autocorrelation functions are finite, so the autocovariance function is finite for any second-order process.

Exercise 2-7
Show that

$$|R_X(t_1, t_2)| \le \frac{R_X(t_1, t_1) + R_X(t_2, t_2)}{2}. \tag{2.12}$$

Notice that (2.11) and (2.12) are the same inequality written in different notation.

Solution.   Since the expected value of the square of a random variable is nonnegative, the following inequalities are valid for all t_1 and t_2:

$$0 \le E[X(t_1) + X(t_2)]^2 = R_X(t_1, t_1) + 2R_X(t_1, t_2) + R_X(t_2, t_2); \tag{2.13}$$
$$0 \le E[X(t_1) - X(t_2)]^2 = R_X(t_1, t_1) - 2R_X(t_1, t_2) + R_X(t_2, t_2). \tag{2.14}$$

But these two inequalities imply that both $+R_X(t_1, t_2)$ and $-R_X(t_1, t_2)$, and hence $|R_X(t_1, t_2)|$, are not greater than the right-hand side of (2.12). ∎

At the beginning of this subsection, it is pointed out that if $X(t_1)$ and $X(t_2)$ are *independent*, the autocovariance function satisfies $C_X(t_1, t_2) = 0$. If, on the other hand, $X(t_1) = X(t_2)$, which is, intuitively, as dependent as two random variables can be, then

$$C_X(t_1, t_2) = E[X(t_1) - \mu_X(t_1)]^2 = E[X(t_2) - \mu_X(t_2)]^2$$
$$= \text{Var}\{X(t_1)\} = \text{Var}\{X(t_2)\}.$$

In order to explore this relationship further, fix t_1 and t_2 and let

$$\sigma_i = \sqrt{\text{Var}\{X(t_i)\}}$$

for $i = 1$ and $i = 2$. Now, define the *normalized autocovariance function* K_X by

$$K_X(t_1, t_2) = \frac{C_X(t_1, t_2)}{\sigma_1 \sigma_2}.$$

Later in this section it is shown that $|C_X(t_1, t_2)| \le \sigma_1 \sigma_2$, so the normalized auto-covariance function is always between -1 and $+1$. Notice that our example $X(t_1) = X(t_2)$ gives the *largest* possible positive normalized autocovariance, and if $X(t_1)$ and

$X(t_2)$ are independent, we get $K_X(t_1, t_2) = 0$, the smallest possible magnitude for the normalized autocovariance. Notice also that the latter equation is equivalent to $C_X(t_1, t_2) = 0$, which, from (2.9), is in turn equivalent to $R_X(t_1, t_2) = \mu_X(t_1) \mu_X(t_2)$. As a side issue, note that the value -1, the smallest possible value for the normalized autocovariance, is achieved if $X(t_1) = -X(t_2)$. An example of a random process $X(t)$ with the property that $X(t_1) = X(t_2)$ for certain values of t_1 and t_2 is the sinusoid with a random phase angle, as described in Example 2-1: simply let $t_1 = 0$ and $t_2 = T$. Similarly, letting $t_1 = 0$ and $t_2 = T/2$ in this example gives $X(t_1) = -X(t_2)$. The full autocorrelation function for the sinusoidal signal with a random phase angle is derived in Exercise 2-8.

Often, we are dealing with random processes that have rather simple descriptions, such as the sinusoid with a random phase. In such cases, it is usually a straightforward exercise to determine the mean, autocorrelation, and autocovariance functions. In addition to Example 2-1, this is also the situation for Examples 2-4 and 2-5. The mean and autocorrelation functions for a sinusoidal signal with a random phase angle are found in the next exercise.

Exercise 2-8

Find the mean and autocorrelation function for the random process

$$X(t) = \alpha \sin(2\pi f_0 t + \Theta),$$

which was introduced in Example 2-1. The amplitude α is a deterministic constant, but the phase Θ is a random variable.

Solution.   Let $\omega_0 = 2\pi f_0$, and observe that

$$X(t) = \alpha[\sin \Theta \cos \omega_0 t + \cos \Theta \sin \omega_0 t].$$

Thus, the mean of $X(t)$ is given by

$$\mu_X(t) = \alpha[E\{\sin \Theta\} \cos \omega_0 t + E\{\cos \Theta\} \sin \omega_0 t].$$

The autocorrelation function is

$$\begin{aligned}
R_X(t_1, t_2) &= \alpha^2 E\{\sin(\omega_0 t_1 + \Theta) \sin(\omega_0 t_2 + \Theta)\} \\
&= \frac{\alpha^2}{2}[E\{\cos[\omega_0(t_1 - t_2)]\} - \cos[\omega_0(t_1 + t_2) + 2\Theta]\}] .
\end{aligned}$$

It follows that

$$R_X(t_1, t_2) = \left(\frac{\alpha^2}{2}\right)[\cos[\omega_0(t_1 - t_2)] + E\{\sin 2\Theta\} \sin[\omega_0(t_1 + t_2)] \\ - E\{\cos 2\Theta\} \cos[\omega_0(t_1 + t_2)]] .$$

This is the general result. Now suppose that

$$E\{\sin \Theta\} = E\{\cos \Theta\} = E\{\sin 2\Theta\} = E\{\cos 2\Theta\} = 0,$$

which is a much weaker restriction on the distribution of Θ than is being uniformly distributed on $[0, 2\pi]$. It follows from the preceding condition that $\mu_X(t) = 0$ for each t and

$$R_X(t_1, t_2) = \left(\frac{\alpha^2}{2}\right)\cos[\omega_0(t_1 - t_2)]$$

for each t_1 and t_2. In particular, if $t_1 = 0$ and $t_2 = T$, it follows that

$$R_X(0, T) = \left(\frac{\alpha^2}{2}\right)\cos(\omega_0 T).$$

Using the fact that $T = f_0^{-1}$, we see that $\omega_0 T = 2\pi$, so $R_X(0, T) = \alpha^2/2$. Notice that $R_X(t, t) = \alpha^2/2$ for *any* time t, which can be deduced from the expression

$$R_X(t_1, t_2) = \left(\frac{\alpha^2}{2}\right)\cos[\omega_0(t_1 - t_2)]$$

by setting $t_1 = t_2 = t$. The fact that $R_X(t, t) = \alpha^2/2$ should not be surprising, because $R_X(t, t)$ is $E\{[X(t)]^2\}$, the expected value of the instantaneous power in the random process at time t. The average power in a sinusoidal signal of amplitude α is $\alpha^2/2$.

An interesting feature of this random process is that the power at time t need not be $\alpha^2/2$ for all choices of t if the distribution of the random phase angle Θ does not satisfy

$$E\{\sin \Theta\} = E\{\cos \Theta\} = E\{\sin 2\Theta\} = E\{\cos 2\Theta\} = 0.$$

For example, if $P(\Theta = 0) = 1$, then

$$E\{[X(0)]^2\} = E\{[X(T/2)]^2\} = E\{[X(T)]^2\} = 0,$$

which results from the expression for $X(t)$ if $\Theta = 0$ (or it can be seen from Figure 2-2). Notice that if Θ is uniformly distributed on $[0, 2\pi]$, then because the random process $X(t)$ has zero mean, $R_X(t_1, t_2) = C_X(t_1, t_2)$, and the normalized autocovariance is given by

$$K_X(t_1, t_2) = \frac{2R_X(t_1, t_2)}{\alpha^2} = \cos[\omega_0(t_1 - t_2)].$$

As is true in general, the maximum magnitude of the normalized autocovariance is unity. For this particular process, the maximum is achieved for several choices of t_1 and t_2, such as if $t_1 = t_2$ for any t_2 in the range $0 \leq t_2 \leq T$ or if t_1 and t_2 satisfy $|t_2 - t_1| = T/2$. Another choice is $t_1 = 0$ and $t_2 = T$ (i.e., $K_X(0, T) = +1$).   ∎

The next exercise gives a simple illustration of a very important general procedure. The setting is as follows: The random process $X(t)$ is the input to a linear system, and the random process $Y(t)$ is the output. Given some information about the input process, we wish to determine certain information about the output process. Examples of the type of information we are given or wish to determine are the mean, second moment, variance, autocorrelation function, and autocovariance function.

Exercise 2-9

Find the mean, autocorrelation, and autocovariance functions for the discrete-amplitude, discrete-time process $Y(t)$ defined in Example 2-4. Recall that $Y(t) = X(t) + X(t-1)$; $X(t) = X_t$ for $t = 0, 1, 2, \ldots$; and $X_0, X_1, X_2, \ldots$ is a sequence of independent random variables for which $P(X_t = 0) = P(X_t = 1) = 1/2$ for each nonnegative integer t.

Solution.   First notice that $E\{X_t\} = 1/2$ and $E\{[X_t]^2\} = 1/2$ for each nonnegative integer t. Since $Y(t) = X(t) + X(t-1)$ for $t = 1, 2, 3, \ldots$, the mean function for $Y(t)$ is

$$\mu_Y(t) = E\{Y(t)\} = E\{X(t) + X(t-1)\} = \mu_X(t) + \mu_X(t-1).$$

But $\mu_X(t) = E\{X_t\} = 1/2$ for all t, so $\mu_Y(t) = 1$ for all t. The autocorrelation function for $Y(t)$ is given by

$$
\begin{aligned}
R_Y(t_1, t_2) &= E\{[X(t_1) + X(t_1 - 1)][X(t_2) + X(t_2 - 1)]\} \\
&= R_X(t_1, t_2) + R_X(t_1, t_2 - 1) + R_X(t_1 - 1, t_2) + R_X(t_1 - 1, t_2 - 1).
\end{aligned}
$$

Because $X(t)$ and $X(s)$ are independent whenever $t \neq s$, $R_X(s, t) = \mu_X(t)\mu_X(s) = 1/4$ for $t \neq s$. Furthermore, $R_X(s, t) = E[X(t)]^2 = 1/2$ for $s = t$. Consequently, the expression for $R_Y(t_1, t_2)$ reduces to

$$
R_Y(t_1, t_2) = \begin{cases} \frac{3}{2}, & |t_1 - t_2| = 0, \\ \frac{5}{4}, & |t_1 - t_2| = 1, \\ 1, & |t_1 - t_2| \geq 2. \end{cases}
$$

Notice that $R_Y(t_1, t_2)$ can also be evaluated directly from the two-dimensional density function obtained in Exercise 2-3. We simply use the fact that

$$
\begin{aligned}
R_y(t_1, t_2) &= E[Y(t_1)\, Y(t_2)] \\
&= \sum_{i=0}^{2} \sum_{j=0}^{2} ij f_{Y,2}(i, j; t_1\, t_2).
\end{aligned}
$$

By applying (2.9), we find that the autocovariance function for $Y(t)$ is given by

$$
C_Y(t_1, t_2) = \begin{cases} \frac{1}{2}, & |t_1 - t_2| = 0, \\ \frac{1}{4}, & |t_1 - t_2| = 1, \\ 0, & |t_1 - t_2| \geq 2. \end{cases}
$$

Alternatively, we can evaluate

$$
C_y(t_1, t_2) = \sum_{i=0}^{2} \sum_{j=0}^{2} (i-1)(j-1)\, f_{Y,2}(i, j; t_1\, t_2).
$$

Notice that the autocorrelation and autocovariance functions for the process $Y(t)$ do not depend on t_1 and t_2 individually, but only on their difference. Such processes are very important in the study of random processes in linear systems and are discussed further in Section 2.5 and in Chapters 3 and 4. Notice also that the autocorrelation and autocovariance functions are symmetric functions of t_1 and t_2 in the sense that t_1 and t_2 can be interchanged without changing the value of the function. We will see that this latter property is true for any real-valued random process.   ∎

Recall that, for a given second-order random process $X(t), t \in \mathbb{T}$, the autocorrelation function for the process is a function of two variables:

$$R_X(t_1, t_2) = E\{X(t_1)\, X(t_2)\}.$$

However, not every function of two variables can be an autocorrelation function for some random process. The fact that an autocorrelation function is a very special function has important implications in the analysis of second-order random processes. First we present some of the important properties of an autocorrelation function for a given second-order random process, and then we discuss the necessary and sufficient conditions for a given function of two variables to be a valid autocorrelation function for some second-order random process.

Given any second-order random process $X(t), t \in \mathbb{T}$, the autocorrelation function for $X(t)$ satisfies

$$R_X(t, t) \geq 0, \tag{2.15}$$

$$R_X(t_1, t_2) = R_X(t_2, t_1), \tag{2.16}$$

and

$$|R_X(t_1, t_2)| \leq \sqrt{R_X(t_1, t_1)\, R_X(t_2, t_2)}, \tag{2.17}$$

for each choice of t, t_1, and t_2. If (2.16) is satisfied for each t_1 and t_2, the function $R_X(\cdot, \cdot)$ is said to be *symmetric*.

In addition to satisfying (2.15)–(2.17), $R_X(\cdot, \cdot)$ is *nonnegative definite*; that is, for *any* positive integer n, *any* points $t_1, \ldots, t_n$ in $\mathbb{T}$, and *any* complex numbers $\alpha_1, \ldots, \alpha_n$,

$$\sum_{i=1}^{n} \sum_{k=1}^{n} \alpha_i\, \alpha_k^*\, R_X(t_i, t_k) \geq 0, \tag{2.18}$$

where α_k^* denotes the complex conjugate of α_k. By (2.18) we mean, of course, that the indicated sum is *real* and *nonnegative*. An alternative term for nonnegative definite is *positive semidefinite*, and an equivalent definition to (2.18) is that, for any n and any $t_1, \ldots, t_n$ in $\mathbb{T}$, the $n \times n$ matrix that has the element $R_X(t_i, t_k)$ in the ith row and kth column is a nonnegative definite matrix.

Actually, it turns out that nonnegative definiteness implies (2.15)–(2.17), as is shown in Section 2.4.2. Moreover, it can be shown that, given a function $R(\cdot, \cdot)$ that is nonnegative definite, there always exists a second-order random process that has $R(\cdot, \cdot)$ as its autocorrelation function. Thus, nonnegative definiteness is a necessary and sufficient condition for a function $R(\cdot, \cdot)$ to be an autocorrelation function for some second-order random process.

Autocovariance functions must also satisfy all the properties of autocorrelation functions. The reason is that, given a second-order process $X(t)$, the autocovariance function for $X(t)$ is just the autocorrelation function for the process $Y(t) = X(t) - \mu_X(t)$. Hence, any autocovariance function is an autocorrelation function. Furthermore, it can be shown that any symmetric, nonnegative definite function $C(\cdot, \cdot)$ is an autocovariance function for some second-order process. Consequently, the collection of all autocovariance functions is identical to the collection of all autocorrelation functions.

Exercise 2-10

Show that the following functions are not valid autocorrelation functions:

(a) $R(t_1, t_2) = |t_1 - t_2| \exp\{-3|t_1 - t_2|\}$, $t_1 \in \mathbb{R}$, $t_2 \in \mathbb{R}$;

(b) $R(t_1, t_2) = \begin{cases} 1, & |t_1 - t_2| \leq T, \\ 2 - (|t_1 - t_2|/T), & T < |t_1 - t_2| \leq 2T, \\ 0, & |t_1 - t_2| > 2T. \end{cases}$

Solution.   The function in part **(a)** cannot be an autocorrelation function, because it does not satisfy (2.17). Notice that $R(t_1, t_1) = R(t_2, t_2) = 0$, but $R(t_1, t_2) > 0$, except for the special case $t_1 = t_2$.

The function in part **(b)** is not nonnegative definite. (Notice, however, that it does satisfy (2.15)–(2.17).) For instance, suppose $n = 3$, $t_1 = 0$, $t_2 = T$, $t_3 = 2T$, $\alpha_1 = \alpha_3 = +1$, and $\alpha_2 = -1$. For these values of the parameters, (2.18) is not satisfied. Alternatively, for the same choice of t_1, t_2, and t_3, we can simply observe that the matrix

$$\begin{bmatrix} R(0,0) & R(0,T) & R(0,2T) \\ R(T,0) & R(T,T) & R(T,2T) \\ R(2T,0) & R(2T,T) & R(2T,2T) \end{bmatrix} = \begin{bmatrix} 1 & 1 & 0 \\ 1 & 1 & 1 \\ 0 & 1 & 1 \end{bmatrix}$$

is not nonnegative definite. In fact, it has a negative determinant (-1). ■

The mean, autocorrelation, and autocovariance functions play a particularly important role in Gaussian processes. Recall that if $X(t)$ is Gaussian, the vector $\mathbf{X}(t) = [X(t_1), \ldots, X(t_n)]$ has the n-dimensional Gaussian density for any integer n and any $t_1, \ldots, t_n$. Thus, for a Gaussian random process $X(t)$ we need to know only $E\{X(t)\}$ and $\text{Cov}\{X(t), X(s)\}$ for all t and s in order to determine all of the n-dimensional distributions of $X(t)$. Therefore, the *mean* and *autocovariance* functions (or *mean* and *autocorrelation* functions) are sufficient to completely specify a Gaussian process. (See Section 2.8 for some further remarks on this point.)

2.4.2   Derivations of the Properties of Autocorrelation Functions

In this section, we show that (2.15)–(2.18) are satisfied by any autocorrelation function and, therefore, by any autocovariance function as well. This material is included for the reader who is interested in the detailed derivations of the key properties of autocorrelation and autocovariance functions. The material is not required for subsequent portions of the book, so it may be best to skip the section entirely during the first reading of this chapter.

The first step is to observe that (2.16) follows immediately from the definition of $R_X(t_1, t_2)$:

$$R_X(t_1, t_2) = E\{X(t_1)\, X(t_2)\} = E\{X(t_2)\, X(t_1)\} = R_X(t_2, t_1).$$

Next, notice that, for arbitrary choices of n; $t_1, \ldots, t_n$ in $\mathbb{T}$; and complex numbers $\alpha_1, \ldots, \alpha_n$, we have

$$0 \le E\left\{ \left| \sum_{i=1}^{n} \alpha_i\, X(t_i) \right|^2 \right\} = \sum_{i=1}^{n} \sum_{k=1}^{n} \alpha_i\, \alpha_k^*\, R_X(t_i, t_k),$$

which is just (2.18). Finally, observe that (2.15) is just a special case of (2.18): Let $n = 1$, $\alpha_1 = 1$, and $t_1 = t$.

The next step is to show that (2.18) implies (2.17). Recall that the latter is the inequality

$$|R_X(t_1, t_2)| \le \sqrt{R_X(t_1, t_1)\, R_X(t_2, t_2)}\,.$$

To derive this inequality from (2.18), it is helpful to begin by generalizing the approach used in the solution of Exercise 2-7. If $n = 2$ and α_1 and α_2 are real, (2.18) reduces to

$$\alpha_1^2 R_X(t_1, t_1) + \alpha_2^2 R_X(t_2, t_2) \ge -2\alpha_1\, \alpha_2 R_X(t_1, t_2). \tag{2.19}$$

Notice that the left-hand side of (2.19) does not depend on the sign of α_1 or of α_2. Hence, if λ_1 and λ_2 are arbitrary *nonnegative* numbers, then

$$\lambda_1^2 R_X(t_1, t_1) + \lambda_2^2 R_X(t_2, t_2) \ge 2\lambda_1\, \lambda_2\, |R_X(t_1, t_2)|. \tag{2.20}$$

Inequality (2.20) can be derived from (2.19) by first letting $\alpha_1 = \lambda_1$ and $\alpha_2 = \lambda_2$ and then letting $\alpha_1 = \lambda_1$ and $\alpha_2 = -\lambda_2$. This step is a generalization of the approach used to obtain (2.13) and (2.14). Looking back at the derivation of those equations, we see that substituting $\alpha_1 = 1$ and $\alpha_2 = 1$ in (2.19) gives (2.13), while substituting $\alpha_1 = 1$ and $\alpha_2 = -1$ gives (2.14).

To proceed further, it is necessary to consider separately the cases in which $R_X(t_i, t_i) = 0$ for either $i = 1$ or $i = 2$ (or both). In a sense, such cases are degenerate, because they correspond to a random process for which the expected value of the instantaneous power is zero at time t_i. Nevertheless, we consider them for completeness. Both $R_X(t_1, t_1)$ and $R_X(t_2, t_2)$ are nonzero if and only if

$$R_X(t_1, t_1)\, R_X(t_2, t_2) > 0,$$

and this is the case that will be handled first. Recall that λ_1 and λ_2 can be any nonnegative numbers in (2.20). Assuming that $R_X(t_1, t_1)\, R_X(t_2, t_2) > 0$, we can substitute

$$\lambda_1^2 = \frac{R_X(t_2, t_2)}{2\sqrt{R_X(t_1, t_1)\, R_X(t_2, t_2)}}$$

and

$$\lambda_2^2 = \frac{R_X(t_1, t_1)}{2\sqrt{R_X(t_1, t_1)\, R_X(t_2, t_2)}}$$

into (2.20) to obtain (2.17). On the other hand, if at least one of these correlations is zero, then $R_X(t_1, t_1)\, R_X(t_2, t_2) = 0$, so the procedure will not work. However, if at least one of the two correlations is zero, we claim that the validity of (2.20) for all $\lambda_1 \geq 0$ and $\lambda_2 \geq 0$ implies that $R_X(t_1, t_2) = 0$. If this claim is true, then (2.17) must hold, since the left-hand side is zero and the right-hand side is nonnegative.

It is particularly easy to establish that the claim is true if $R_X(t_i, t_i) = 0$ for *both* $i = 1$ and $i = 2$; we can just set $\lambda_1 = \lambda_2 = 1$ in (2.20) to deduce $|R_X(t_1, t_2)| \leq 0$. If $R_X(t_i, t_i) = 0$ for *only one* value of i, then we have to work a little harder. First, suppose that $R_X(t_1, t_1) = 0$ and $R_X(t_2, t_2) \neq 0$ (i.e., $i = 1$). Letting $\lambda_2 = 1$, we see that (2.20) reduces to

$$R_X(t_2, t_2) \geq 2\lambda_1 |R_X(t_1, t_2)|.$$

Since this inequality must hold no matter how large λ_1 is, it must be that $R_X(t_1, t_2) = 0$; otherwise, as $\lambda_1 \to \infty$, the inequality will eventually be violated. For instance, if $R_X(t_1, t_2) \neq 0$ and we let

$$\lambda_1 = \frac{R_X(t_2, t_2)}{|R_X(t_1, t_2)|},$$

then the inequality becomes the assertion that $1 \geq 2$, which, of course, is false, so it must be that $R_X(t_1, t_2) = 0$, which establishes the claim for the case in which $R_X(t_1, t_1) = 0$ and $R_X(t_2, t_2) \neq 0$. The same argument with an interchange of appropriate 1s and 2s establishes the claim for the case in which $R_X(t_1, t_1) \neq 0$ and $R_X(t_2, t_2) = 0$.

We have shown that (2.15) and (2.17) follow from the nonnegative definiteness of the autocorrelation function. Although we obtained (2.16) directly from the definition of the autocorrelation function, it also follows from the nonnegative definiteness of the autocorrelation function. We simply let $n = 2$, $\alpha_1^2 = -1$ (recall that α_1 can be complex), and $\alpha_2 = 1$, so that the *imaginary* part of the sum in (2.18) is $R_X(t_1, t_2) - R_X(t_2, t_1)$, which must be zero, since the sum must be a *real* number.

Thus, nonnegative definiteness implies all of the properties that we have considered. This is not surprising in view of the fact that nonnegative definiteness is a *necessary* and *sufficient* condition for a function $R(\cdot, \cdot)$ to be a valid autocorrelation. We already showed necessity when we derived (2.18). It turns out, however, that, given any non-negative definite function $R(\cdot, \cdot)$ defined on the set of all (t_1, t_2) such that $t_1 \in \mathbb{T}$ and $t_2 \in \mathbb{T}$, there exists a random process $X(t)$, $t \in \mathbb{T}$, that has autocorrelation function $R(\cdot, \cdot)$; that is,

$$R_X(t_1, t_2) = R(t_1, t_2)$$

for each t_1 and t_2. The proof of this statement is not given here, but a related result is proved in Exercise 4-3. In fact, there always exists a *Gaussian* random process that has autocorrelation function $R(\cdot, \cdot)$.

Finally, we should point out that (2.17) can be obtained as an immediate consequence of the *Schwarz inequality*, which states that, for any random variables Y_1 and Y_2,

$$|E\{Y_1 \, Y_2\}| \leq \sqrt{E\{Y_1^2\} \, E\{Y_2^2\}}.$$

To derive (2.17) from the Schwarz inequality, simply let $Y_i = X(t_i)$ for both $i = 1$ and $i = 2$. One reason for deriving (2.17) from (2.20) rather than from the Schwarz inequality is that the derivation from (2.20) helps to bring out the importance of the nonnegative definiteness property: We have shown that the other properties can be deduced from the fact that a correlation function is nonnegative definite.

2.5   Stationary Random Processes

A random process $X(t)$, $t \in \mathbb{T}$, is said to be *stationary* (or *strictly stationary*) if, for each n and each choice of $t_1, \ldots, t_n$ in $\mathbb{T}$,

$$F_{X,n}(x_1, \ldots, x_n; t_1, \ldots, t_n) = F_{X,n}(x_1, \ldots, x_n; t_1 + t_0, \ldots, t_n + t_0) \qquad (2.21)$$

for all t_0 in $\mathbb{T}$. In other words, $X(t)$ is stationary if all of its finite-dimensional distribution functions remain unchanged under all possible shifts in the time origin. The n-dimensional distributions, therefore, depend only on *relative* time (or time *differences*) for stationary random processes. This is a very important property for engineering applications, since the selection of a time origin is often a very artificial exercise. For many problems, it is important that the answer not depend on this arbitrary choice of a time origin. For obvious reasons, stationary processes are often called *shift-invariant* processes.

In (2.21), we must require that $t_i + t_0$ be in the index set $\mathbb{T}$ for all i; otherwise, the right-hand side of (2.21) is undefined. From this point on, then, assume that the index set is closed under addition (i.e., if t and s are in $\mathbb{T}$, then so is $t + s$). The most important examples of index sets that are closed under addition are the real line $\mathbb{R}$, the set of nonnegative real numbers $[0, \infty)$, the set of all integers $\mathbb{Z}$, and the set of nonnegative integers $\{0, 1, 2, \ldots\}$. It is convenient to make the additional assumption that the number 0 is in $\mathbb{T}$, which is true for all four of the preceding examples.

Example 2-6   A Satellite Beacon
Suppose a satellite transmits a tracking beacon to a ground station. The signal received at the ground station is a random process, even if the thermal noise in the receiver is ignored. A model for the received signal that is appropriate for some applications is the random process

$$X(t) = A \cos(2\pi f_0 t + \Theta).$$

The amplitude A of the signal is a random variable because of the random or unpredictable attenuation of the beacon signal as it propagates through the atmosphere. The phase angle Θ is also a random variable that may represent the randomness in the initial phase of the satellite's oscillator (as in Example 2-1) or the randomness in a phase shift that is introduced as the signal propagates through the atmosphere. In general, the random process $X(t)$ is not stationary unless conditions are placed on the distributions of A and Θ. For instance, suppose A and Θ are independent, $P(\Theta = 0) = P(\Theta = \pi) = \frac{1}{2}$, and A is a continuous random variable (e.g., Gaussian). Then $P[X(0) = 0] = 0$, but $P\left[X\left(\frac{1}{4} f_0\right) = 0\right] = 1$. Hence, even the one-dimensional distribution function depends on the parameter t, which violates (2.21).

However, for certain distributions of the random variable Θ, the random process $X(t)$ is stationary. For instance, if Θ is uniformly distributed on the interval $[0, 2\pi]$ and A is independent of Θ, then $X(t)$ is stationary. In practice, it is often true that Θ and A are independent and Θ is uniformly distributed on $[0, 2\pi]$.   ∎

Example 2-7   A Nonstationary Process

Certain classes of engineering problems lend themselves to an obvious and natural choice for the time origin. For instance, if $X(t)$ is the number of telephone calls received at a switchboard, $t = 0$ represents the time at which we start counting calls. In this case, $X(t)$ is the number of calls received during the time interval $[0, t]$ (for $t \geq 0$). This is one example of a process that is inherently nonstationary according to its physical description. We would not expect $X(t_1)$ and $X(t_1 + \tau)$ to have the same distribution for $\tau > 0$, since we expect that, with high probability, there will be more calls received during $[0, t_1 + \tau]$ than during $[0, t_1]$. Hence, it is reasonable to conjecture that, for such a random process,

$$F_{X.1}(x_1; t_1) > F_{X.1}(x_1; t_1 + \tau), \quad \tau > 0.   ∎$$

For a second-order random process $X(t)$, a number of important consequences of stationarity are extremely useful in engineering. For the moment, think of t as being fixed, and consider the random variable $X(t)$. The mean of this random variable depends only on the distribution function $F_{X.1}(\cdot\,; t)$. For example, if $F_{X.1}$ has derivative $F'_{X.1} = f_{X.1}$, then

$$\mu_X(t) = E\{X(t)\} = \int_{-\infty}^{\infty} u\, f_{X.1}(u; t)\, du.$$

If $X(t)$ is a discrete random variable that takes values in $\mathbb{Z}$, then

$$\mu_X(t) = E\{X(t)\} = \sum_{i=-\infty}^{\infty} i \left[F_{X.1}(i; t) - F_{X.1}(i - 1; t)\right].$$

If the process $X(t)$ is stationary, (2.21) implies that $F_{X.1}(x; t)$ does not depend on t; thus, $\mu_X(t)$ does not depend on t. The conclusion is that *the mean of a stationary random process is a constant.* Whenever $\mu_X(t)$ does not depend on t, we denote its value by μ_X. This abuse of notation should cause no confusion, because it will be clear from the context whether the mean does or does not depend on the parameter t. If $\mu_X(t)$ depends on t, the process $X(t)$ is said to have a time-dependent or time-varying mean; otherwise, it is said to have a constant mean. For a process $X(t)$ with a time-varying mean, μ_X is a *function* with value $\mu_X(t)$ at time t. For a process $X(t)$ with a constant mean, μ_X is a real number.

If the process $X(t)$ is stationary, then (2.21) implies that $F_{X,2}(x_1, x_2; t + \tau, t)$ does not depend on t, although it may depend on τ. In other words, if $t_1 = t + \tau$ and $t_2 = t$, then $F_{X,2}(x_1, x_2; t_1, t_2)$ may depend on $t_1 - t_2$, but not on t_1 and t_2 individually. Since the autocorrelation and autocovariance functions depend only on the two-dimensional distribution functions, it follows that if $X(t)$ is stationary, then $R_X(t+\tau, t)$ and $C_X(t + \tau, t)$ do not depend on t.

The properties of the mean, autocorrelation, and autocovariance functions for a stationary, second-order random process are summarized as follows: If $X(t)$ is such a random process, then, for all t in $\mathbb{T}$,

$$\mu_X(t) = \mu_X(0), \tag{2.22}$$
$$R_X(t + \tau, t) = R_X(\tau, 0), \tag{2.23}$$

and

$$C_X(t + \tau, t) = C_X(\tau, 0), \tag{2.24}$$

for all τ in the index set $\mathbb{T}$. If (2.22) holds for all t, the mean is constant and is denoted by μ_X. Whenever $R_X(t + \tau, t)$ does not depend on t, but is instead a function of τ only, we write $R_X(\tau)$ in place of $R_X(t + \tau, t)$. Similarly, we denote $C_X(t + \tau, t)$ by $C_X(\tau)$ whenever (2.24) holds for all t and τ.

2.5.1   Wide-Sense and Covariance Stationary Processes

It may be that the functions μ_X, R_X, and C_X are shift invariant in the sense of equations (2.22)–(2.24), yet the process $X(t)$ is not stationary. Since those equations represent properties of a second-order process that are of interest in their own right, we define two forms of stationarity that are, in general, much weaker than (strict) stationarity as defined in (2.21). A second-order random process is said to be *wide-sense stationary* (WSS) if (2.22) and (2.23) hold for all t and τ; the process is said to be *covariance stationary* if (2.24) holds for all t and τ.

Letting $t_1 = t + \tau$ and $t_2 = t$ in (2.9), we find that

$$C_X(t + \tau, t) = R_X(t + \tau, t) - \mu_X(t + \tau)\mu_X(t).$$

From this relationship, we see that the right-hand side does not depend on t for a wide-sense stationary process, and, therefore, neither does the left-hand side. So $C_X(t + \tau, t)$ does not depend on t for a wide-sense stationary process. This establishes the following important fact:

A wide-sense stationary process is also covariance stationary.

However, there are important processes that are covariance stationary, but *not* wide-sense stationary. One example arises in the investigation of a signal plus noise. If $X(t)$ is wide-sense stationary and $v(t)$ is a deterministic signal, then the random process

$$Y(t) = v(t) + X(t)$$

is covariance stationary; however, it is wide-sense stationary only in the special case where $v(t)$ is a constant. The process $Y(t)$ might represent the voltage at some point in an electronic system, and this voltage consists of a signal component $v(t)$ plus a noise component $X(t)$. Only in the trivial situation in which the signal does not change with time is the resulting random process wide-sense stationary, and signals that do not change with time are of little interest in practice. This establishes another important fact:

A covariance stationary process need not be wide-sense stationary.

It is helpful to consider $Y(t) = v(t) + X(t)$ further to understand how such random processes can be covariance stationary, but not wide-sense stationary. Recall that the covariance function is defined as

$$C_Y(t, s) = E\{[Y(t) - \mu_Y(t)][Y(s) - \mu_Y(s)]\},$$

and notice that the first step in computing this function is to subtract out the mean of the process $Y(t)$, which is given by

$$\mu_Y(t) = v(t) + \mu_X(t).$$

Although $\mu_X(t)$ is constant ($X(t)$ is WSS), the signal component $v(t)$ is not. So, it is the *mean* that keeps this particular process $Y(t)$ from being wide-sense stationary.

In determining covariance stationarity, we ignore the mean by subtracting it out. We then focus on the autocorrelation function for the new process that results when the mean is subtracted from the original process. That is, we form the random process $W(t) = Y(t) - \mu_Y(t)$ and then evaluate the autocorrelation function

$$R_W(t, s) = E\{W(t) W(s)\}$$

for that process. If $Y(t) = v(t) + X(t)$, the resulting random process $W(t)$ is wide-sense stationary, as can be seen from the following: First, observe that $W(t)$ is a zero-mean process. Second, because $E\{Y(t)\} = v(t) + E\{X(t)\}$,

$$W(t) = Y(t) - \mu_Y(t) = Y(t) - [v(t) + \mu_X(t)] = X(t) - \mu_X(t),$$

so

$$R_W(t, s) = C_Y(t, s) = C_X(t, s) = R_X(t, s) - \mu_X(t)\mu_X(s).$$

Since $X(t)$ is wide-sense stationary, $\mu_X(t)$ does not depend on t, $\mu_X(s)$ does not depend on s, and $R_X(t, s)$ depends on t and s via the difference $t - s$ only. Thus, $R_W(t, s)$ also depends on t and s via the difference $t - s$ only. Because the mean of $W(t)$ is constant, $W(t)$ is wide-sense stationary, and because $R_W(t, s) = C_Y(t, s)$, it follows that $Y(t)$ is covariance stationary, which is what we promised to show.

If a random process is not wide-sense stationary, then either the mean or the autocorrelation function (or both) must depend on t. This requires the two-dimensional distribution function to depend on t, so a process that is not wide-sense stationary cannot be stationary. However, even if a random process $X(t)$ is wide-sense stationary, that does not guarantee, for example, that higher order moments $E\{[X(t)]^m\}$, $m > 2$, are constant with respect to t. Thus, wide-sense stationarity is a much weaker property than stationarity. We have already established that covariance stationarity does not even imply wide-sense stationarity, let alone stationarity. The relationships between wide-sense stationarity and stationarity for second-order random processes are summarized as follows:

A stationary process is also wide-sense stationary.

A wide-sense stationary process need not be stationary.

For a wide-sense stationary process $X(t)$, (2.15)–(2.17) become

$$R_X(0) \geq 0, \tag{2.25}$$
$$R_X(\tau) = R_X(-\tau), \tag{2.26}$$

and

$$|R_X(\tau)| \leq R_X(0), \tag{2.27}$$

for any τ. Notice that the inequality

$$|R_X(t, s)| \leq [R_X(t, t) + R_X(s, s)]/2,$$

which is just (2.12), and the inequality

$$|R_X(t, s)| \leq \sqrt{R_X(t, t)\, R_X(s, s)}\,,$$

which is (2.17), are equivalent if $X(t)$ is wide-sense stationary.

Example 2-8   Sinusoidal Signal with Random Amplitude and Phase
It was pointed out in Example 2-6 that the process $X(t) = A\cos(2\pi f_0 t + \Theta)$ is not stationary in general. In fact, the process may not even be wide-sense stationary, as we illustrate next. Suppose that A and Θ are independent and $E\{A^2\} < \infty$. Then

$$E\{X(t_1)\, X(t_2)\} = \tfrac{1}{2} E\{A^2\}\{\cos[2\pi f_0(t_1 - t_2)] \ + \ E\{\cos 2\Theta\}\cos[2\pi f_0(t_1 + t_2)]$$
$$- \ E\{\sin 2\Theta\}\sin[2\pi f_0(t_1 + t_2)]\},$$

which, in general, depends not only on $t_1 - t_2$, but also on $t_1 + t_2$. An important special case is when $E\{\cos 2\Theta\} = E\{\sin 2\Theta\} = 0$, in which case $R_X(t, t) = E\{A^2\}/2$ and

$$R_X(t + \tau, t) = \tfrac{1}{2} E\{A^2\}\cos(2\pi f_0 \tau),$$

which depends on τ only (not on t). If, in addition, $E\{\cos \Theta\} = E\{\sin \Theta\} = 0$, then $\mu_X(t) = 0$ for each t, so that $X(t)$ is wide-sense stationary.

Thus, the process $X(t) = A\cos(2\pi f_0 t + \Theta)$ is a second-order, wide-sense stationary random process if A and Θ are independent, $E\{A^2\} < \infty$, and

$$E\{\cos \Theta\} = E\{\sin \Theta\} = E\{\cos 2\Theta\} = E\{\sin 2\Theta\} = 0.$$

The latter condition holds if, for instance, Θ is uniformly distributed on $[0, 2\pi]$. ∎

Exercise 2-11
Show that Example 2-8 can be used to construct a random process that is wide-sense stationary, but not stationary.

Solution.   Let $A = 1$ in Example 2-8, so that $X(t) = \cos(2\pi f_0 t + \Theta)$. Let Θ take on each of the values 0, $\pi/2$, π, and $3\pi/2$ with probability $1/4$. It is easy to show that

$$E\{\cos \Theta\} = E\{\sin \Theta\} = E\{\cos 2\Theta\} = E\{\sin 2\Theta\} = 0,$$

so that, according to Example 2-8, the resulting random process is wide-sense stationary. There are several ways to show that $X(t)$ is not stationary, such as observing that $P[X(0) = 0] = 1/2$ and $P[X(0) = +1] = P[X(0) = -1] = 1/4$, but if $t_0 = (8 f_0)^{-1}$, the random variable $X(t_0)$ takes on each of the two values $+1/\sqrt{2}$ and $-1/\sqrt{2}$ with probability $1/2$. Clearly, $X(0)$ and $X(t_0)$ do not have the same distribution. ∎

Example 2-9   Simple Nonstationary Process
Let $X(t) = Y + t Z$, where Y and Z are random variables with finite second moments. The random process $X(t)$ is therefore a second-order random process. It is clear from the beginning that, unless Z is a trivial random variable (i.e., $P(Z = 0) = 1$), the process $X(t)$ is not stationary. To prove that it is not, it suffices to prove that $X(t)$ is not wide-sense stationary. Toward that end, let μ_Y and μ_Z be the mean values of Y and Z, respectively. Straightforward evaluation then shows that the autocorrelation function is

$$R_X(t + \tau, t) = E\{Y^2\} + (2t + \tau) E\{YZ\} + t(t + \tau) E\{Z^2\}$$

and the mean function is $\mu_X(t) = \mu_Y + \mu_Z t$. The process $X(t)$ is not, in general, wide-sense stationary, because both the mean and the autocorrelation depend on t, rather than having a constant mean and an autocorrelation that depends on the time difference τ alone. Even if $\mu_Z = 0$, so that $\mu_X(t) = \mu_Y$ (a constant) for all t, the process is still not necessarily wide-sense stationary. For instance, if $\mu_Z = 0$ and the random variables Y and Z are uncorrelated, then

$$R_X(t + \tau, t) = E\{Y^2\} + t(t + \tau) \operatorname{Var}\{Z\},$$

which depends on t. ∎

A Gaussian random process is completely specified by its mean and autocorrelation functions. The reason for this is that if $X(t)$ is a Gaussian random process, the joint distribution function for $X(t_1)$, $X(t_2)$, ..., $X(t_n)$ depends only on the means and covariances of these n random variables. The means are given by

$$\mu_X(t_i) = E\{X(t_i)\}, \quad 1 \le i \le n,$$

and the covariances are given by

$$C_X(t_i, t_j) = \operatorname{Cov}\{X(t_i), X(t_j)\}, \quad 1 \le i \le n, 1 \le j \le n.$$

Thus, the n-dimensional distributions for a Gaussian process are independent of the time origin if and only if the mean of the process is constant and its autocovariance function does not depend on the time origin. (See Section 2.8.) From the relationship

$$C_X(t, s) = R_X(t, s) - \mu_X(t)\,\mu_X(s),$$

it follows that a necessary and sufficient condition for the n-dimensional distributions for a Gaussian process to be independent of the time origin is that the mean and auto-correlation do not depend on the time origin.

We conclude that a Gaussian random process $X(t)$ satisfies

$$F_{X,n}(x_1, \ldots, x_n; t_1, \ldots, t_n) = F_{X,n}(x_1, \ldots, x_n; t_1 + t_0, \ldots, t_n + t_0)$$

for all $t_0, t_1, \ldots, t_n$ in $\mathbb{T}$ if and only if it satisfies $\mu_X(t) = \mu_X(0)$ for all t in $\mathbb{T}$ and, for each τ, $R_X(t + \tau, t) = R_X(\tau, 0)$ for all t in $\mathbb{T}$. This is equivalent to the statement that a Gaussian random process is stationary whenever it is wide-sense stationary.

The relationship between stationary and nonstationary second-order random pro-cesses and the special role of Gaussian processes are summarized as follows:

> *A stationary random process is always wide-sense stationary. A wide-sense stationary process need not be stationary. However, for a Gaussian random process, stationarity and wide-sense stationarity are equivalent.*

2.5.2   Examples of Autocorrelation Functions for Wide-Sense Stationary Processes

In this section, we give some common examples of autocorrelation functions that depend on time differences only. Some of these autocorrelation functions arise in subsequent sections and in homework problems as the autocorrelation functions for certain random processes. For now, we simply introduce them without connection to specific processes. The parameters α and β that appear in the autocorrelation functions are positive in each case. The reader should verify that each of the autocorrelation functions presented in Table 2.1 satisfies (2.25)–(2.27).

From these examples, it is possible to obtain the autocorrelation functions for some other random processes by using certain analytical results that give the autocorrelation

Table 2.1: Some Examples of Autocorrelation Functions

---

1. $R(\tau) = \beta \exp(-\alpha|\tau|)$ [illustrated in Figure 2-11(a)]

2. $R(\tau) = \begin{cases} \beta(T - |\tau|)/T, & -T \le \tau \le T, \\ 0, & \text{otherwise} \end{cases}$ [illustrated in Figure 2-11(b)]

3. $R(\tau) = \beta \exp(-\alpha|\tau|) \cos(\omega_0 \tau)$

4. $R(\tau) = 2\,W\{\sin(2\pi W \tau)/2\pi W \tau\}$

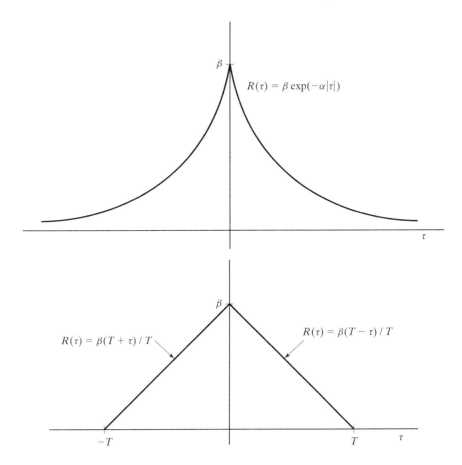

$$R(\tau) = \beta \exp(-\alpha|\tau|)$$

$$R(\tau) = \beta(T + \tau)/T$$

$$R(\tau) = \beta(T - \tau)/T$$

Figure 2-11: (a) The exponential autocorrelation function. (b) The triangular auto-correlation function.

function for a random process defined in terms of one or more other random processes. For example, suppose that $Y(t) = cX(t)$ for some constant c. The autocorrelation function for $Y(t)$ is given by

$$R_Y(t, s) = E\{Y(t)\,Y(s)\} = E\{[c\,X(t)][c\,X(s)]\} = c^2\,E\{X(t)\,X(s)\} = c^2\,R_X(t, s).$$

Clearly, if $X(t)$ is wide-sense stationary, then so is $Y(t)$, and the identity becomes

$$R_Y(\tau) = c^2\,R_X(\tau).$$

Other examples will be given after we introduce the notion of the crosscorrelation function for two random processes.

2.6   Crosscorrelation and Crosscovariance Functions

Let $X(t)$, $t \in \mathbb{T}$, and $Y(t)$, $t \in \mathbb{T}$, be two second-order random processes defined on the same probability space. The *crosscorrelation function* $R_{X,Y}$ for these two random processes is defined by

$$R_{X,Y}(t, s) = E\{X(t) Y(s)\}$$

for all t and s. The crosscorrelation function satisfies

$$R_{X,X}(t, s) = R_X(t, s), \tag{2.28}$$
$$R_{X,Y}(t, s) = R_{Y,X}(s, t), \tag{2.29}$$

and

$$|R_{X,Y}(t, s)| \leq \sqrt{R_X(t, t) \, R_Y(s, s)}, \tag{2.30}$$

for all t and s. The *crosscovariance function* $C_{X,Y}$ is defined by

$$C_{X,Y}(t, s) = E\{[X(t) - \mu_X(t)][Y(s) - \mu_Y(s)]\}$$

for all t and s. Notice that the crosscovariance function and crosscorrelation function are related by

$$C_{X,Y}(t, s) = R_{X,Y}(t, s) - \mu_X(t) \, \mu_Y(s).$$

The two second-order random processes $X(t)$, $t \in \mathbb{T}$, and $Y(t)$, $t \in \mathbb{T}$, are said to be *uncorrelated* if, for all t in $\mathbb{T}$ and all s in $\mathbb{T}$,

$$R_{X,Y}(t, s) = \mu_X(t) \, \mu_Y(s).$$

Observe from the relationship between the crosscovariance and crosscorrelation functions that $X(t)$ and $Y(t)$ are uncorrelated if and only if $C_{X,Y}(t, s) = 0$ for all t and s.

The processes $X(t)$, $t \in \mathbb{T}$, and $Y(t)$, $t \in \mathbb{T}$, are said to be *independent random processes* if, for each positive integer n and each choice of $t_1, t_2, \ldots, t_n, s_1, s_2, \ldots, s_n$ in $\mathbb{T}$, the two random vectors $[X(t_1), X(t_2), \ldots, X(t_n)]$ and $[Y(s_1), Y(s_2), \ldots, Y(s_n)]$ are (statistically) independent. Two random vectors $\mathbf{X}$ and $\mathbf{Y}$ are independent if and only if their joint distribution function factors; that is, $\mathbf{X}$ and $\mathbf{Y}$ are independent if and only if $F_{\mathbf{X},\mathbf{Y}}(\mathbf{x}, \mathbf{y}) = F_{\mathbf{X}}(\mathbf{x}) \, F_{\mathbf{Y}}(\mathbf{y})$ for each pair $\mathbf{x}$ and $\mathbf{y}$.

If $X(t)$ and $Y(t)$ are independent random processes, then, for each choice of t_0 and s_0 in $\mathbb{T}$, $X(t_0)$ and $Y(s_0)$ are independent random variables. Because independent random variables are uncorrelated, it follows that independent random processes are also uncorrelated. In general, however, $X(t)$, $t \in \mathbb{T}$, and $Y(t)$, $t \in \mathbb{T}$, can be uncorrelated random processes without being independent random processes, as is demonstrated by the next example.

Example 2-10   A Pair of Uncorrelated Random Processes

Let Z be a zero-mean Gaussian random variable, and define the random processes $X(t)$, $-\infty < t < \infty$, and $Y(t)$, $-\infty < t < \infty$, by

$$X(t) = Z t$$

for all t and

$$Y(t) = Z^2 t$$

for all t. First, observe that these random processes are not independent. For instance, $Y(1) = [X(1)]^2$, so that $X(1)$ and $Y(1)$ are not even independent. More generally, for any $t_0 \neq 0$, $Y(t) = [X(t_0)/t_0]^2 t$ for all t. Thus, if we observe the process $X(t)$ at any nonzero time, we can determine the value of the process $Y(t)$ for all t. This is a very strong form of statistical dependence!

Next, we show that these random processes are uncorrelated. Because Z is a zero-mean Gaussian random variable, its density function is symmetric; that is, $f_Z(z) = f_Z(-z)$ for all z. Thus, all odd-order moments of Z are zero. In particular, $E\{Z\} = 0$, so $\mu_X(t) = 0$ for all t, and $E\{Z^3\} = 0$, so

$$R_{X,Y}(t, s) = E\{X(t)\,Y(s)\} = E\{(Zt)(Z^2s)\} = t\,s\,E\{Z^3\} = 0$$

for all t and s. Recall that the two processes are uncorrelated if, for all t and s,

$$R_{X,Y}(t, s) = \mu_X(t)\,\mu_Y(s).$$

Clearly, this relationship is satisfied, because each side of the equation is equal to zero. Hence, $X(t)$ and $Y(t)$ are uncorrelated random processes that are not independent. ■

The random processes $X(t)$, $t \in \mathbb{T}$, and $Y(t)$, $t \in \mathbb{T}$, are said to be *jointly Gaussian* random processes if, for each positive integer n and each choice of $t_1, t_2, \ldots, t_n$, $s_1, s_2, \ldots, s_n$ in $\mathbb{T}$, the two random vectors $[X(t_1), X(t_2), \ldots, X(t_n)]$ and $[Y(s_1), Y(s_2), \ldots, Y(s_n)]$ are jointly Gaussian. That is, the single random vector

$$[X(t_1), X(t_2), \ldots, X(t_n), Y(s_1), Y(s_2), \ldots, Y(s_n)]$$

has a distribution governed by the $2n$-dimensional Gaussian density. Because uncorrelatedness and independence are equivalent for jointly Gaussian *random variables*, jointly Gaussian *random processes* that are uncorrelated are also independent. Notice that, although the random process $X(t)$ in Example 2-10 is Gaussian, the random process $Y(t)$ is not, so the two processes are certainly not jointly Gaussian. The relationship between uncorrelatedness and independence for two random processes is summarized as follows:

Independent random processes are uncorrelated. In general, uncorrelated random processes need not be independent. However, jointly Gaussian random processes that are uncorrelated are also independent.

The random processes $X(t)$ and $Y(t)$ are said to be *jointly wide-sense stationary* if $X(t)$ is wide-sense stationary, $Y(t)$ is wide-sense stationary, and $R_{X,Y}(t, s)$ depends on $t - s$ only. Clearly, joint wide-sense stationarity of a pair of random processes is, in general, a stronger condition than wide-sense stationarity of each of the processes. The condition that $R_{X,Y}(t, s)$ depends on $t - s$ only does not arise in determining the wide-sense stationarity of the individual random processes. This condition can be expressed as $R_{X,Y}(t, s) = R_{X,Y}(t - s, 0)$ for all t and s in $\mathbb{T}$. If the processes $X(t)$ and $Y(t)$ are jointly wide-sense stationary, we usually write $R_{X,Y}(t - s)$ in place of $R_{X,Y}(t, s)$ and $R_{X,Y}(\tau)$ in place of $R_{X,Y}(t + \tau, t)$.

Recall that if $X(t), t \in \mathbb{T}$, and $Y(t), t \in \mathbb{T}$, are uncorrelated random processes, then $C_{X,Y}(t, s) = 0$ for all t and s, so that $R_{X,Y}(t, s) = \mu_X(t)\,\mu_Y(s)$ for all t and s. If these random processes are each wide-sense stationary, then $\mu_X(t)$ and $\mu_Y(s)$ are the constants μ_X and μ_Y, respectively. So, if $X(t)$ and $Y(t)$ are uncorrelated random processes that are each wide-sense stationary, the product $\mu_X(t)\,\mu_Y(s)$ is just the constant $\mu_X\,\mu_Y$ for all t and s. Hence, $R_{X,Y}(t, s) = \mu_X\,\mu_Y$ does not depend on t or s at all. Thus, *uncorrelated* random processes that are *individually* wide-sense stationary are also *jointly* wide-sense stationary. In general, however, two wide-sense stationary processes are not necessarily jointly wide-sense stationary.

We can now return to the topic of the autocorrelation function for a random process defined in terms of two (or more) other random processes. For the remainder of this section, let $X(t), t \in \mathbb{T}$, and $Y(t), t \in \mathbb{T}$, be random processes defined on the same probability space. The autocorrelation function for the random process defined by $Z(t) = X(t) + Y(t)$ is given by

$$
\begin{aligned}
R_Z(t, s) &= E\{[X(t) + Y(t)][X(s) + Y(s)]\} \\
&= R_X(t, s) + R_{X,Y}(t, s) + R_{Y,X}(t, s) + R_Y(t, s).
\end{aligned}
\tag{2.31}
$$

If each random process has zero mean and the two random processes are uncorrelated, then (2.31) reduces to

$$
R_Z(t, s) = R_X(t, s) + R_Y(t, s).
\tag{2.32}
$$

Thus, for zero-mean, uncorrelated random processes, the autocorrelation function for the sum of two random processes is the sum of the individual autocorrelation functions. It should be clear that this relationship extends to the sum of any finite number of zero-mean, uncorrelated random processes. If, in addition, *each* of the two random processes is *wide-sense stationary* (note that this plus the fact that they are uncorrelated implies that they are jointly wide-sense stationary), then (2.32) becomes $R_Z(t - s) = R_X(t - s) + R_Y(t - s)$, or, simply,

$$
R_Z(\tau) = R_X(\tau) + R_Y(\tau).
\tag{2.33}
$$

It follows that the sum of any two autocorrelation functions is a valid autocorrelation function.

Next, we consider the product of two random processes: $Z(t) = X(t)\,Y(t)$ for $t \in \mathbb{T}$. The autocorrelation function for the random process $Z(t)$ is given by

$$
R_Z(u, s) = E\{X(u)\,Y(u)\,X(s)\,Y(s)\}.
$$

If $X(t)$, $t \in \mathbb{T}$, and $Y(t)$, $t \in \mathbb{T}$, are independent random processes, then $X(u)\,X(s)$ and $Y(u)\,Y(s)$ are independent for each choice of u and s in $\mathbb{T}$. It follows that

$$E\{X(u)\,Y(u)\,X(s)\,Y(s)\} = E\{X(u)\,X(s)\}E\{Y(u)\,Y(s)\} = R_X(u, s)\,R_Y(u, s)$$

for all u and s, and therefore,

$$R_Z(u, s) = R_X(u, s)\,R_Y(u, s) \tag{2.34}$$

for all u and s. That is, for independent random processes, the autocorrelation function for the product of two random processes is the product of the individual autocorrelation functions. This relationship also extends easily to the product of any finite number of independent random processes. The reader should verify that it is not sufficient that the random processes be uncorrelated. A stronger condition, such as independence, is required in order that $E\{X(u)\,Y(u)\,X(s)\,Y(s)\}$ can be written as the product of $E\{X(u)\,X(s)\}$ and $E\{Y(u)\,Y(s)\}$. Note that $E\{X(u)\,Y(u)\,X(s)\,Y(s)\}$ is a fourth-order moment, and uncorrelatedness tells us something about second-order moments only. If the random processes are independent and each is wide-sense stationary, (2.34) can be written as $R_Z(u - s) = R_X(u - s)\,R_Y(u - s)$, or, simply,

$$R_Z(\tau) = R_X(\tau)\,R_Y(\tau). \tag{2.35}$$

It follows that the product of any two autocorrelation functions is a valid autocorrelation function.

Example 2-11   Amplitude Modulation
Suppose that the wide-sense stationary random process $X(t)$ is amplitude modulated onto an RF carrier to produce the signal

$$Z(t) = \sqrt{2}\,X(t)\cos(2\pi f_c t + \Theta),$$

where Θ is uniformly distributed on $[0, 2\pi]$ and is independent of $X(t)$. If we let

$$Y(t) = \sqrt{2}\cos(2\pi f_c t + \Theta),$$

we see from Exercise 2-8 that

$$R_Y(\tau) = \cos(2\pi f_c \tau).$$

If Θ is independent of $X(t)$, any deterministic function of Θ is also independent of $X(t)$. Also, $X(t)$ and $Y(t)$ are each wide-sense stationary, so that (2.35) applies and establishes that the autocorrelation function for $Z(t)$ is

$$R_Z(\tau) = R_X(\tau)\cos(2\pi f_c \tau).$$

■

2.7   White Noise

In the previous sections, we have considered only random processes that satisfy $E[X(t)]^2 < \infty$ for all t. This is simply a requirement that the process have finite power, which we expect to be true of all processes that arise in practical problems. Such processes are what we have referred to as *second-order* random processes. All of the processes that we will be concerned with are second-order processes, with one very important exception: white noise.

White noise is a mathematical idealization of the thermal noise process discussed in Example 2-2. In fact, as described there, the random process that models thermal noise has infinite power, since it has power equal to $4kTRB$ in each frequency band of B Hz. If we integrate this power over all frequencies, we see that the random process has infinite power. However, two important points should be considered: First, the model discussed in the example is not valid for arbitrarily large frequencies; second, no physical system has infinite bandwidth, and we can observe the thermal noise process only at the output of some physical system. Even if the input process is modeled as white noise, the output of a physical system will not have a fixed amount of power in each part of the frequency band. In fact, it will decrease as the frequency increases in a way that gives finite power. Thus, it can be *assumed* (and it is mathematically convenient to do so) that thermal noise is as described in Example 2-2 for all values of f_0, even though this leads to an assumption of infinite power. In most situations, it makes no difference whether we restrict f_0 to be less than 10^{11} Hz or 10^{12} Hz or allow all finite values of f_0, because noise at frequencies greater than about 10^{11} Hz will have no effect on the output of most electronic systems.

The autocorrelation function of a stationary white-noise process is taken to be a delta function centered at the origin; that is, $E\{X(t+\tau)X(t)\} = c\,\delta(\tau)$ for some positive constant c. As we will see later, this actually follows from the earlier assumption that the noise power is the same in all frequency bands of a given bandwidth.

We conclude that white noise is *not* a second-order process, since it has an infinite second moment. Strictly speaking, $E\{[X(t)]^2\}$ is not even defined, according to the delta-function autocorrelation model. The reason for wanting to employ the white-noise model we have described is that it leads to a considerable simplification of the analysis of thermal noise in linear systems.

2.8   Stationary Gaussian Random Processes

We begin with the following fact about jointly Gaussian random variables:

The distribution of a collection of n jointly Gaussian random variables $X_1, X_2, \ldots, X_n$ depends only on the mean values $\mu_i = E\{X_i\}$ and the covariances $\text{Cov}\{X_i, X_j\}$ for the n random variables.

This fact follows from the discussion of Gaussian random variables in Section 1.5, where it is pointed out that if Λ is the $n \times n$ matrix that has

$$\Lambda_{i,j} = E\{(X_i - \mu_i)(X_j - \mu_j)\} = \text{Cov}\{X_i, X_j\} \tag{2.36}$$

as the element in the ith row and jth column, and if μ_i is the mean of X_i ($1 \le i \le n$), then the joint density function for $\mathbf{X} = (X_1, X_2, \ldots, X_n)$ is given by

$$f_{\mathbf{X},n}(\mathbf{u}) = (2\pi)^{-n/2} |\det(\Lambda)|^{-1/2} \exp\left\{-\tfrac{1}{2}(\mathbf{u} - \boldsymbol{\mu})\Lambda^{-1}(\mathbf{u} - \boldsymbol{\mu})^T\right\}, \qquad (2.37)$$

where $\det(\Lambda)$ is the determinant of the matrix Λ, Λ^{-1} is the inverse of Λ, $\mathbf{u} = (u_1, u_2, \ldots, u_n)$, $\boldsymbol{\mu} = (\mu_1, \mu_2, \ldots, \mu_n)$, and $(\mathbf{u} - \boldsymbol{\mu})^T$ is the transpose of the vector $(\mathbf{u} - \boldsymbol{\mu})$. The matrix Λ is the covariance matrix for the random vector $\mathbf{X}$, and the vector $\boldsymbol{\mu}$ is the mean vector for $\mathbf{X}$. We see from (2.37) that $\boldsymbol{\mu}$ and Λ completely characterize the density function; furthermore, $\boldsymbol{\mu}$ and Λ depend only on the means and covariances of the random variables $X_1, X_2, \ldots, X_n$.

To apply this fact to a Gaussian random process $X(t)$, let $X_i = X(t_i)$ for each i, and let $\mathbf{X} = (X_1, X_2, \ldots, X_n)$. Then

$$\mu_i = E\{X(t_i)\} = \mu_X(t_i) \qquad (2.38)$$

and

$$\Lambda_{i,j} = \mathrm{Cov}\{X(t_i), X(t_j)\} = C_X(t_i, t_j). \qquad (2.39)$$

If the Gaussian random process $X(t)$ is wide-sense stationary, the mean $\mu_X(t_i)$ does not depend on t_i, and the autocovariance $C_X(t_i, t_j)$ depends on the time differences $t_i - t_j$ only. Consequently, the n-dimensional density function for the wide-sense stationary Gaussian random process $X(t)$ depends only on the differences between the sample times. In fact, this density is given in terms of the density of (2.37) by

$$f_{X,n}(u_1, u_2, \ldots, u_n; t_1, t_2, \ldots, t_n) = f_{\mathbf{X}}(\mathbf{u}). \qquad (2.40)$$

Since all this is true for all choices of the integer n, we conclude that a wide-sense stationary Gaussian random process is also (strictly) stationary.

Problems

2.1 Let $g(t)$ denote the periodic "sawtooth" signal defined in the interval $[-T/2, T/2]$ by

$$g(t) = 1 - \frac{4t}{T} \quad \text{for } 0 \le t \le T/2$$

and

$$g(t) = 1 + \frac{4t}{T} \quad \text{for } -T/2 \le t \le 0.$$

For other intervals, $g(t)$ is defined in a way to make it periodic with period T; that is, for $(2n-1)T/2 \le t \le (2n+1)T/2$, $g(t) = g(t - nT)$. A random process is given by $X(t) = g(t - V)$, where V is a random variable.

(a) Sketch some typical sample functions for the random process $X(t)$ if V is uniformly distributed on the interval $[0, T]$.

(b) Sketch *all* possible sample functions if V is distributed according to $P(V - mT/4) = 1/4$ for $m = 0, 1, 2$, and 3.

2.2 Consider a checkout line at a local grocery store. Let $N(t)$ denote the number of customers (including the one being served) waiting in line at time t. Sketch three typical sample functions for this random process. What is the primary difference between these sample functions and that of Figure 2-5?

2.3 Consider a packet communication network with K communication terminals, each of which can simultaneously transmit and receive. The network has the capability of allowing any subset of these terminals to transmit packets at a given time (i.e., packet transmissions can overlap). Each packet requires T seconds of transmission time. Let $N(t)$ denote the number of packets being transmitted at time t. Sketch three typical sample functions for each of the following situations:

(a) Slotted transmission is used, and the clocks at all terminals are completely synchronized. That is, time is divided into T-second slots, each transmission must start at the beginning of a slot and stop at the end of the same slot, and each terminal "knows" precisely when the slots begin and end.

(b) The terminals are operated asynchronously and the transmissions are not slotted. In this case, a terminal begins its transmission anytime it has a packet ready to transmit.

2.4 What are the *two* major differences between the sample functions of Problem 2.3(**b**) and those of Problem 2.2? (*Hint*: Consider the maximum value and the duration of "service.")

2.5 Determine the classification of the random processes in each of the situations that follow. In each case, indicate whether the process is a continuous- or discrete-time process and whether it is a continuous- or discrete-amplitude process.

(a) A manufacturing process begins at time 0, and we are interested in the number of defects that have occurred up to time t for all positive values of t.

(b) A computation is carried out in a sequence of steps in a special-purpose digital computer, and the content of a particular shift register is converted to decimal form and recorded at the end of each clock cycle.

(c) A continuous-time signal is sampled and quantized every T_0 seconds, and we are interested in the quantization error at the sampling times.

(d) A continuous-time signal is sampled and quantized every T_0 seconds, and we are interested in the quantized value of the signal at each sampling time.

(e) Because of interference and thermal noise in an analog FM receiver, the demodulated audio signal differs from the transmitted audio signal. We are interested in the error signal (the difference between the transmitted and demodulated audio signals).

2.6 In this problem, we make a minor modification to Example 2-4. The output process is $Y(t) = X(t) + X(t-1) + X(t-2)$ for $t \geq 2$. Find the one-dimensional density function for the random process $Y(t), t \geq 2$. Also, find $P[Y(t) = 3, Y(t+1) = 2]$ for $t \geq 2$.

2.7 Suppose X is a random variable that is uniformly distributed on $[0, 1]$. The random process $Y(t), t > 0$, is defined by $Y(t) = \exp\{-Xt\}$. Find the one-dimensional distribution function $F_{Y,1}(u; t)$ for the random process $Y(t)$. Find the one-dimensional density function $f_{Y,1}(u; t)$.

2.8 The random process $X(t)$, $t \in \mathbb{R}$, is wide-sense stationary with mean μ_X and auto-correlation function $R_X(\tau)$. Suppose that for $|\tau| > \tau_0$, the random variables $X(t_0)$ and $X(t_0 + \tau)$ are uncorrelated for all t_0. What is the value of $R_X(\tau)$ for $\tau > \tau_0$? In particular, what is the value of $\lim_{\tau \to \infty} R_X(\tau)$?

2.9 Suppose that $X(t)$ is a zero-mean, wide-sense stationary, continuous-time Gaussian random process with autocorrelation function $R_X(\tau)$. Let the random process $Y(t)$ be given by $Y(t) = c_1 X(t) + c_2 X(t - T)$. Find the probability that $Y(t_0)$ is greater than some threshold γ. Express your answer in terms of the standard Gaussian distribution function Φ, the autocorrelation function R_X, and the parameters c_1, c_2, γ, and T.

2.10 The wide-sense stationary random process $X(t)$, $t \in \mathbb{R}$, is a Gaussian random process with zero mean and autocorrelation function $R_X(\tau)$. Find the one-dimensional distribution function for the random process $Y(t) = \Phi\left[X(t)/\sqrt{R_X(0)} \right]$ for $-\infty < t < \infty$. Find the one-dimensional density function for $Y(t)$. Is the process $Y(t)$ stationary?

2.11 Consider the function R given by $R(\tau) = 1$, $-T \leq \tau \leq T$, and $R(\tau) = 0$, otherwise. Does this function satisfy (2.25)–(2.27)? Is it a valid autocorrelation function? Prove your answer to each question.

2.12 Does the function $R(t) = \beta \exp(-\alpha \tau^2)$ satisfy (2.25)–(2.27) if β and α are positive real numbers? Prove your answer.

2.13 State whether $R(\tau) = \exp\{-(\tau - 1)^2\}$ does or does not satisfy (2.25)–(2.27), and prove your answer.

2.14 A random process $X(t)$ has a constant mean and an autocorrelation function

$$R_X(t, s) = \cos(\omega_0 t) \cos(\omega_0 s) + \sin(\omega_0 t) \sin(\omega_0 s).$$

Is the random process wide-sense stationary? Prove your answer.

2.15 Let $X(t) = A \cos(2\pi \Lambda t + \Theta)$, $-\infty < t < \infty$. This is a sinusoidal signal with random amplitude, frequency, and phase. Suppose that A, Λ, and Θ are mutually independent random variables. Suppose also that A is a nonnegative random variable with density $f_A(a) = 0.1 \exp(-a/10)$, $a > 0$, Λ is uniformly distributed on the interval $[-W, W]$, and Θ is uniformly distributed on $[0, 2\pi]$. Find the mean and autocorrelation functions for the random process $X(t)$. Is $X(t)$ wide-sense stationary?

2.16 Suppose $X(t)$ and $Y(t)$ are zero-mean, wide-sense stationary, continuous-time random processes. If $X(t)$ and $Y(t)$ are independent, find the autocorrelation function for $Z(t)$ in terms of the autocorrelation functions for $X(t)$ and $Y(t)$ in each of the cases that follow. In each case, determine whether the random process $Z(t)$ is wide-sense stationary.

(a) $Z(t) = c X(t) Y(t) + d$, where c and d are deterministic constants.

(b) $Z(t) = X(t) \cos(\omega_0 t) + Y(t) \sin(\omega_0 t)$.

2.17 Consider the random processes

$$X(t) = \alpha_1 \cos(2\pi f_0 t + \Theta)$$

and

$$Y(t) = \alpha_2 \sin(2\pi f_0 t + \Theta).$$

where α_1 and α_2 are deterministic constants and Θ is uniformly distributed on $[0, 2\pi]$.

(a) Is each of these random processes wide-sense stationary?

(b) Find the crosscorrelation function for the two random processes.

(c) Are these random processes jointly wide-sense stationary?

(d) Are the two random processes uncorrelated?

(e) Are the two random processes independent?

2.18 Suppose V is a zero-mean Gaussian random variable, and define the random processes $X(t) = V t$ and $Y(t) = V^2 t$ for $-\infty < t < \infty$.

(a) Find the crosscorrelation function for these two random processes.

(b) Are these random processes jointly wide-sense stationary?

2.19 A wide-sense stationary Gaussian random process $X(t)$, $-\infty < t < \infty$, has autocorrelation function given by $R_X(\tau) = \beta(T - |\tau|)/T$ for $|\tau| \leq T$ and $R_X(\tau) = 0$ otherwise, as illustrated in Figure 2-11(b). The mean of this random process is zero.

(a) Show that $\beta \geq 0$, and show that $\beta = 0$ only if the process is trivial in some sense. Define in what sense it is trivial if $\beta = 0$.

(b) Find the one-dimensional distribution function for this random process.

(c) Find the two-dimensional density function $f_{X,2}(u_1, u_2; t_1, t_2)$ for $t_1 = T$ and $t_2 = 2T$.

(d) Repeat (c) for $t_1 = T/2$ and $t_2 = T$.

(e) Find the probability that $X(0) + X(T/2) + X(T)$ exceeds 10.

In (c) and (d), your answers should be expressed in terms of the parameters β and T only (not t_1 and t_2). In (b)–(e), simplify your answers as much as possible.

2.20 A wide-sense stationary random process $X(t)$ has autocorrelation function

$$R_X(\tau) = c_1 \exp(-c_2|\tau|) \cos(\omega_0 \tau), \quad -\infty < \tau < \infty.$$

(a) The fact that this is a valid autocorrelation function imposes certain restrictions on c_1 and c_2. What is the range of possible values for c_1? What is the range of possible values for c_2?

(b) What is the expected value of the instantaneous power in $X(t)$?

(c) Consider the random process $Y(t)$ given by $Y(t) = 3 X(4t)$, $-\infty < t < \infty$. Is $Y(t)$ wide-sense stationary? Justify your answer.

(d) Find the autocorrelation function for $Y(t)$.

2.21 Consider the system shown in Figure 2-3 of Example 2-2. Suppose that, for the duration of the experiment, the temperature is not held constant, but is instead increased linearly from some initial temperature. The initial temperature at time $t = 0$ is b, so the temperature at time t is given by $T = ct + b$ for some positive constant c.

(a) What is the mean of the random process $Y(t)$ shown in the figure?

(b) Give an expression for the variance of $Y(t)$.

(c) Give an expression for the one-dimensional density function for the random process $Y(t)$.

(d) Does this random process have a constant mean? Is the process stationary?

Chapter 3

Linear Filtering of Random Processes

3.1 The Need for Filtering of Random Processes

In electronic systems, it is often necessary to filter random processes in order to improve certain features or remove undesirable characteristics. For example, a particular random process may have "spikes" that will produce unwanted transients in an electronic system. In this situation, filtering of the random process will make it smoother in some sense, such as by eliminating the spikes in the waveform or by decreasing their amplitudes.

If the random process consists of a signal and noise, it is usually necessary to filter the random process to reduce the effects of the noise while minimizing the distortion of the signal. In this case, the filtering can be employed to enhance the signal-to-noise ratio, or it can be used to produce a better quality output according to some other criteria. The need for filtering a random process that consists of a signal and noise arises in every radar, navigation, and communication system.

There are other applications in which the engineer may not have complete control of the filtering operation. An example occurs in the transmission of a signal through the atmosphere, over an optical fiber, or in a waveguide. The propagation medium acts as a filter in these situations, and it is necessary to analyze the effects of the filtering induced by the medium in order to design an efficient receiver. Analysis is also required in order to predict the performance of a system that must operate with such a dispersive medium.

In this chapter, we present the fundamental analytical methods needed to design linear filters and systems to accomplish the various objectives just described. These methods are employed by engineers to determine optimum filters for specific applications, to conduct tradeoff studies and design suboptimum filters, to analyze the performance of existing and proposed electronic systems, and determine the effects of various types of noise and other interference.

The most common performance criteria are based on second-order quantities, such as the mean-squared error, noise power, and signal-to-noise ratio. These are all related to the correlation functions or spectral densities of the noise processes involved. As a result, the relevant analysis of linear filters deals with the relationships between the input and output autocorrelation functions or spectral densities. Accordingly, the chapter is concerned primarily with the determination of the autocorrelation function of the output random process in terms of the autocorrelation function of the input random process and the response function of the linear system. Chapter 4 deals with the spectral densities of the input and output random processes.

3.2   Discrete- and Continuous-Time Linear Systems

We begin with a brief review of linear filtering of deterministic signals. Recall that if $x(t)$ is a deterministic continuous-time signal and $h(t)$ is the impulse response of a time-invariant, continuous-time linear system, then the output of the system when x is the input is given by

$$y(t) = \int_{-\infty}^{\infty} h(t - \tau) x(\tau) \, d\tau = \int_{-\infty}^{\infty} h(\tau) x(t - \tau) \, d\tau. \qquad (3.1)$$

That is, y is the convolution of x with h, denoted by $y = x * h$. According to (3.1),

$$x * h = h * x.$$

If $x(k)$ is a deterministic discrete-time signal and $h(k)$ is the pulse response of a time-invariant discrete-time linear system, then

$$y(k) = \sum_{n=-\infty}^{\infty} h(k - n) x(n) = \sum_{n=-\infty}^{\infty} h(n) x(k - n) \qquad (3.2)$$

is the output of the system when x is the input. Again, y is the convolution of x with h and $x * h = h * x$. In writing discrete-time signals, we always normalize the time scale so that $x(k)$ physically represents the signal at time $k\,t_0$. That is, time is measured as multiples of t_0 time units.

Example 3-1   A Simple Linear Discrete-Time Filter
Suppose that the pulse response of a discrete-time system is given by

$$h(k) = \begin{cases} 1, & k = 0 \text{ or } 1, \\ 0, & \text{otherwise.} \end{cases}$$

Then, for any input signal $x(k)$, (3.2) implies that

$$\begin{aligned} y(k) &= \sum_{n=0}^{1} h(n) x(k - n) \\ &= x(k) + x(k - 1). \end{aligned}$$

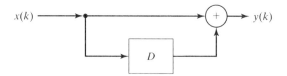

Figure 3-1: A simple discrete-time, linear, time-invariant filter.

Notice that this is the same linear system as in Example 2-4; however, the input is a random process in that example, whereas it is a deterministic signal here. This linear filter is illustrated in Figure 3-1.   ∎

   The input to a discrete-time linear system is often a sampled version $x_s(k)$ of a continuous-time signal $x(t)$. That is, $x_s(k) = x(k\,t_0)$ for each integer k, where t_0 is the sampling interval (i.e., the time between samples). Similarly, a continuous-time signal $x_c(t)$ may be constructed from a discrete-time signal $x(k)$. For example, $x_c(t) = x(k)$ for $k\,t_0 \leq t < (k+1)t_0$; that is, $x_c(t)$ is a sequence of rectangular pulses, each of duration t_0, and the amplitude of the kth pulse is $x(k)$. The continuous-time signal $x_c(t)$ can be written in terms of the discrete-time signal $x(k)$ as

$$x_c(t) = \sum_{k=-\infty}^{\infty} x(k)\, p_{t_0}(t - k\,t_0),$$

where $p_\lambda(t)$ denotes the rectangular pulse of unit amplitude and duration λ that begins at time $t = 0$: $p_\lambda(u) = 1$ for $0 \leq u < \lambda$ and $p_\lambda(u) = 0$ otherwise.
   With the preceding relationship between continuous-time signals and discrete-time signals, continuous-time signals can be inputs to discrete-time systems, and discrete-time signals can be inputs to continuous-time systems. Of course, the signals must be converted to the appropriate form first (e.g., the continuous-time signal is initially sampled, and then it is processed by a discrete-time system).
   In general, a linear system need not be time invariant. That is, the response at time t due to an input impulse occurring at time λ may depend on both t and λ, rather than on the difference $t - \lambda$ only, as is true for time-invariant linear systems. If the continuous-time signal x is the input to a linear system for which the response at time t to an impulse occurring at time λ is $h(t, \lambda)$, then

$$y(t) = \int_{-\infty}^{\infty} h(t, \tau)\, x(\tau)\, d\tau \tag{3.3}$$

is the output. As a check, let the input be an impulse occurring at time λ; the corresponding output is, of course, the impulse response of the system. Accordingly, we replace $x(\tau)$ with $\delta(\tau - \lambda)$ in the integrand of (3.3), and this gives $y(t) = h(t, \lambda)$ as the output of the filter, as it should be. If the linear system is time invariant, then $h(t, \tau)$ depends on the time difference $t - \tau$ only, and the impulse response $h(t, \tau)$ is denoted by $h(t - \tau)$. Hence, (3.3) reduces to (3.1) if the linear filter is time invariant.

Figure 3-2: Integrate-and-dump filter.

Example 3-2   The Integrate-and-Dump Filter

The integrate-and-dump filter is a continuous-time, time-varying, linear filter. It is an active filter that can be implemented as an integrator together with a pair of switches. One switch, which is internal to the circuit, closes at time $t = 0$ to eliminate any energy stored from previous intervals, and the other switch, which is at the output of the integrator, closes at time $t = T$ to sample the output. Mathematically, the integrate-and-dump filter is just a finite-time-duration integration process: The output is the integral of the input over the interval $[0, T]$, as illustrated in Figure 3-2.

The input–output relation for the integrate-and-dump filter is as follows: If x is the input signal, the output is given by $y(t) = 0$ for $t \neq T$ and

$$y(T) = \int_0^T x(u) \, du.$$

What is the impulse response for this filter? By definition, the impulse response $h(t, \tau)$ is the response at time t to an impulse at time τ, so if $x(u) = \delta(u - \tau)$ is the input, the corresponding output is $h(t, \tau)$. First, observe from the foregoing description that $h(t, \tau) = 0$ for $t \neq T$. For $t = T$, the output $h(T, \tau)$ is given by the preceding integral expression with $x(u)$ replaced by $\delta(u - \tau)$; that is,

$$h(T, \tau) = \int_0^T \delta(u - \tau) \, du.$$

But the value of this integral is zero unless the impulse occurs within the limits of integration. So $h(T, \tau) = 0$ for $\tau < 0$ and for $\tau > T$. For τ within the limits of integration (i.e., $0 < \tau < T$), the value of the integral is unity, so $h(T, \tau) = 1$ for $0 < \tau < T$. For $\tau = 0$ or $\tau = T$, the value depends on the convention being followed regarding the value of an integral of a delta function located at one of the endpoints of the interval of integration. Common conventions are to define the value to be zero at one or both endpoints, unity at one or both endpoints, or $1/2$ at each endpoint. For our purposes, it makes no difference how the value of the integral is defined, and it is sufficient to leave it undefined for now. Summarizing these results, we have shown that the impulse response of the integrate-and-dump filter is

$$h(t, \tau) = \begin{cases} 1, & t = T \text{ and } 0 < \tau < T, \\ 0, & t \neq T \text{ or } \tau < 0 \text{ or } \tau > T. \end{cases} \qquad \blacksquare$$

For a general *discrete-time* system, the output $h(k, i)$ at time k for a unit-amplitude input pulse at time i depends on k and i. Thus, the more general version of (3.2) is

$$y(k) = \sum_{n=-\infty}^{\infty} h(k, n) x(n). \tag{3.4}$$

Again as a check, suppose x is a unit pulse occurring at time i. That is, $x(i) = 1$ and $x(n) = 0$ for $n \neq i$. Then (3.4) implies that $y(k) = h(k, i)$ as required.

A linear system is *stable* if, for any input that is bounded, the output of the system is also bounded. This is actually just one form of stability, known as *bounded-input, bounded-output stability*, but it is the only form considered in this book. Notice that the filtering operations characterized by (3.1) and (3.3) represent stable continuous-time linear systems if, for any finite positive number M_x, $|x(\tau)| \leq M_x$ for all τ implies that there is a number $M_y < \infty$ such that $|y(t)| \leq M_y$ for all t. Similarly, (3.2) and (3.4) represent stable discrete-time linear systems if, for any $M_x < \infty$, $|x(n)| \leq M_x$ for all n implies that there is a number $M_y < \infty$ such that $|y(k)| \leq M_y$ for all k.

Notice that if $|x(\tau)| \leq M_x$ in (3.3), then

$$|y(t)| \leq \int_{-\infty}^{\infty} |h(t, \tau)| \, |x(\tau)| \, d\tau$$

$$\leq M_x \int_{-\infty}^{\infty} |h(t, \tau)| \, d\tau.$$

Thus, a sufficient condition for stability is the existence of a number $M_h < \infty$ such that

$$\int_{-\infty}^{\infty} |h(t, \tau)| \, d\tau \leq M_h, \quad \text{for all } t. \tag{3.5}$$

If (3.5) holds, we can simply let $M_y = M_x M_h$. On the other hand, if the system is stable, then it must be that

$$\int_{-\infty}^{\infty} |h(t, \tau)| \, d\tau < \infty, \quad \text{for all } t. \tag{3.6}$$

Otherwise, for a fixed t, we can simply let $x(\tau) = +1$ for values of τ for which $h(t, \tau) \geq 0$ and $x(\tau) = -1$ for those τ for which $h(t, \tau) < 0$. For this bounded input, the output at time t is infinite.

Notice that if a system is time invariant, then (3.5) and (3.6) are equivalent, and they reduce to

$$\int_{-\infty}^{\infty} |h(\lambda)| \, d\lambda < \infty. \tag{3.7}$$

That is, (3.7) is both necessary and sufficient for the stability of a time-invariant, continuous-time, linear system.

The discrete-time counterpart to (3.5) is the existence of a number $M_h < \infty$ such that

$$\sum_{n=-\infty}^{\infty} |h(k, n)| \leq M_h, \quad \text{for all } k, \tag{3.8}$$

which is a sufficient condition for the stability of a discrete-time linear system. A necessary condition is

$$\sum_{n=-\infty}^{\infty} |h(k, n)| < \infty, \quad \text{for all } k. \tag{3.9}$$

These two conditions reduce to

$$\sum_{i=-\infty}^{\infty} |h(i)| < \infty \tag{3.10}$$

if the discrete-time linear system is time invariant. Thus, (3.10), which is analogous to (3.7), is a necessary and sufficient condition for the stability of a time-invariant, discrete-time, linear system.

An important observation to be made concerning (3.1)–(3.4) is that the mathematical operation that appears on the right-hand side of each of these equations is a *limit of finite linear combinations* of values of the input signal. For instance, (3.4) is just the statement that, for each k,

$$y(k) = \lim_{N \to \infty} y_N(k), \tag{3.11}$$

where

$$y_N(k) = \sum_{n=-N}^{N} h(k, n) x(n). \tag{3.12}$$

Similarly, (3.3) actually means that, for each t,

$$y(t) = \lim_{N \to \infty} y_N(t), \tag{3.13}$$

where

$$y_N(t) = \int_{-N}^{N} h(t, \tau) x(\tau) \, d\tau, \tag{3.14}$$

which implies that

$$y_N(t) = \lim_{m \to \infty} m^{-1} \sum_{k=-Nm}^{Nm} h(t, km^{-1}) x(km^{-1}). \tag{3.15}$$

Notice that (3.15) is just the definition of the Riemann integral that appears in (3.14); it may look more familiar if m^{-1} is replaced by $\Delta\tau$. The limits in (3.11), (3.13), and (3.15) always exist for stable systems with bounded inputs.

The reason for discussing the mathematical meaning of (3.1)–(3.4) is to emphasize that these linear operations are defined as *limits of finite linear combinations* of values of the input signal. That is, both (3.3) and (3.4) involve limits of sequences of quantities of the form

$$\sum_{k=1}^{K} c_k \, x(t_k),$$

where the coefficient c_k is a real number and each t_k is an integer (for discrete time) or a real number (for continuous time). The coefficients c_k depend on the impulse response of the system, but not on the input signal; the parameters t_k represent the times at which the signal is sampled in order to form the finite linear combination.

3.3   Linear Operations on Random Processes

Suppose now that the input to the discrete-time linear system described by (3.4) is a discrete-time random process $X(k), k \in \mathbb{Z}$. The output random process $Y(k), k \in \mathbb{Z}$, can be described by an infinite series, which we denote by

$$Y(k) = \sum_{n=-\infty}^{\infty} h(k, n) \, X(n). \tag{3.16}$$

However, simply writing (3.16) does not define $Y(k)$, because the right-hand side is an infinite series of random variables, which is a limit of finite sums of random variables. Does such a limit exist? If so, in what sense?

We would like to define the infinite series in (3.16) in a manner analogous to the way we defined it in (3.11) and (3.12), except that the approximating finite sums

$$Y_N(k) = \sum_{n=-N}^{N} h(k, n) \, X(n) \tag{3.17}$$

are now *random variables*. It turns out that under certain conditions, the limit does exist in the sense that there exists a second-order random process $Y(k)$ for which

$$\lim_{N\to\infty} E\{[Y(k) - Y_N(k)]^2\} = 0. \tag{3.18}$$

According to (3.18), the *mean* of the *square* of the difference between $Y(k)$ and $Y_N(k)$ must converge to zero as $N \to \infty$. Hence, the infinite series in (3.16) is defined as a mean-square infinite series. If (3.18) holds, we say that, for each k, $Y_N(k)$ converges to $Y(k)$ in a *mean-square sense*.

For deterministic signals, we require only that

$$\lim_{N\to\infty} |y(k) - y_N(k)| = 0, \tag{3.19}$$

which follows from (3.11). Therefore, it is reasonable to ask why we employ (3.18) to define convergence, rather than

$$\lim_{N \to \infty} E\{|Y(k) - Y_N(k)|\} = 0. \qquad (3.20)$$

In fact, it turns out that (3.18) implies (3.20), but the converse is not true. More importantly, (3.20) is not sufficient to handle the type of problems that we treat in subsequent sections. We shall be concerned primarily with second-order properties (e.g., properties derived from correlation functions), and the study of second-order properties of the output of a linear system with a random-process input requires that the linear operation be defined in a *mean-square sense*.

The mean-square limit of the sequence $Y_N(k)$ need not exist for all input random processes and discrete-time linear filters. However, the conditions sufficient for the existence of a second-order random process $Y(k)$ satisfying (3.16) are met by most random processes and filters encountered in practical applications. These conditions are given in terms of the function $S(\cdot)$ defined by

$$S(k) = \sum_{i=-\infty}^{\infty} \sum_{n=-\infty}^{\infty} |h(k, n) \, h(k, i) \, R_X(n, i)|, \quad \text{for each } k. \qquad (3.21a)$$

It can be shown that the mean-square limit $Y(k)$ exists for each k if and only if the process $X(k)$ has an autocorrelation function that satisfies

$$S(k) < \infty, \quad \text{for each } k. \qquad (3.21b)$$

If the system is stable and if there exists a real number r for which

$$R_X(n, n) \leq r, \quad \text{for each } n, \qquad (3.22)$$

then (3.21) is satisfied.* To show that this is true, we first use the bound of (2.17), along with (3.22), to conclude that

$$|R_X(n, i)| \leq \sqrt{R_X(n, n) \, R_X(i, i)} \leq r. \qquad (3.23)$$

It follows from (3.21a) and (3.23) that

$$S(k) \leq r \left\{ \sum_{i=-\infty}^{\infty} |h(k, i)| \right\} \left\{ \sum_{n=-\infty}^{\infty} |h(k, n)| \right\},$$

which must be finite for all k because of (3.9). Hence, for an input random process $X(k)$ with bounded second moment, the limit required by (3.16) always exists in a mean-square sense if the discrete-time system is stable. Notice that (3.22) is always satisfied for $r = R_X(0)$ if $X(k)$ is a second-order, wide-sense stationary random process.

---

*The statement that r is a real number is meant to imply that r is finite. The numbers $+\infty$ and $-\infty$ are usually referred to as *extended* real numbers

If the input to a continuous-time system is a continuous-time random process $X(t)$, then the output is a continuous-time process described by the stochastic integral

$$Y(t) = \int_{-\infty}^{\infty} h(t, \tau) X(\tau) \, d\tau. \tag{3.24}$$

This integral is defined by taking limits of finite linear combinations, just as in (3.13)–(3.15), except that the limits are defined in a mean-square sense analogous to the way they are defined in (3.16)–(3.18). A sufficient condition for $Y(t)$ in (3.24) to be defined as a mean-square limit is that

$$\int_{-\infty}^{\infty} \int_{-\infty}^{\infty} |h(t, t_1) h(t, t_2) R_X(t_1, t_2)| \, dt_1 \, dt_2 < \infty, \quad \text{for each } t, \tag{3.25}$$

by which we mean that this Riemann integral must exist and be finite.

For random processes encountered in practice, the stability of the system and the condition that there is a real number r for which

$$R_X(t, t) \le r, \quad \text{for each } t, \tag{3.26}$$

are enough to ensure that the Riemann integral in (3.25) is defined. Given that the integral is defined, these two conditions also guarantee that the integral is finite. If $X(t)$ is a second-order wide-sense stationary process, then (3.26) is always satisfied for $r = R_X(0)$.

In general, by a *linear operation on a random process*, we mean either a finite linear combination of the random variables that make up the process or a mean-square limit of a sequence of such linear combinations. Most of the linear operations that are of interest to us are of the form of (3.16), (3.17), or (3.24). There are a few important operations not easily described by one of these three forms, however. One example is the differentiation of a continuous-time random process,

$$Y(t) = X'(t) = \frac{dX(t)}{dt},$$

which is defined as a mean-square limit of difference quotients. That is, $Y(t) = X'(t)$ satisfies

$$\lim_{n \to \infty} E \left[Y(t) - \frac{X(t + s_n) - X(t)}{s_n} \right]^2 = 0$$

for any sequence $\{s_n\}$ converging to zero. Clearly, if

$$Y_n(t) = \frac{X(t + s_n) - X(t)}{s_n},$$

then $Y(t)$ is the mean-square limit of $Y_n(t)$, and $Y_n(t)$ is a finite linear combination of the random variables $\{X(t) : -\infty < t < \infty\}$. Thus, the differentiation of $X(t)$ is a linear operation easily described as a mean-square limit of finite linear combinations of the random variables in the collection $\{X(t) : -\infty < t < \infty\}$, but it is not conveniently expressible as (3.16), (3.17), or (3.24).

3.4   Time-Domain Analysis of Second-Order Random Processes in Linear Systems

In most engineering investigations, such as the study of noise in electronic systems, the information about a random process that is most readily obtained and, for many problems, most useful is that which is provided by the mean and autocorrelation functions of the random process. Knowledge of these functions is sufficient to permit evaluation of the *noise power* at some point in a circuit or system or the *signal-to-noise ratio* in a radar, navigation, or communication receiver. These functions are also required in order to obtain certain types of *predictions* of a future value of a random process based on observations of past values or *estimates* of one random process based on observations of another random process.

As discussed at the beginning of this chapter, linear filtering of the random process is often required in order to decrease the power in the random process, improve the signal-to-noise ratio, or accomplish some other desirable objective. To design a linear filter for this purpose, it is necessary to be able to determine the characteristics of the output of the filter as a function of the characteristics of the input and the filter.

In this section, the focus is on the *time-domain* characteristics of the input and output random processes and the linear filter. Consequently, we will be working with the mean and autocorrelation functions for the random processes and the appropriate response functions for the linear filters (i.e., impulse response for continuous-time filters or pulse response for discrete-time filters). Specifically, we will show that if the random process $X(t)$ is the input to a linear filter with known response function and $Y(t)$ is the corresponding output, the mean and autocorrelation functions for the output process can be determined, even if only the mean and autocorrelation functions for $X(t)$ are known. In particular, the output mean function can be determined from the input mean function, and the output autocorrelation function can be determined from the input autocorrelation function. Moreover, for the important special case in which $X(t)$ is a Gaussian random process, we can determine a great deal more, as will be discussed in Section 3.5.

We begin by considering the linear filtering of a *continuous-time* random process $X(t)$, $-\infty < t < \infty$. Recall that, under suitable conditions, the output is a continuous-time random process

$$Y(t) = \int_{-\infty}^{\infty} h(t, \tau) \, X(\tau) \, d\tau.$$

Given *only* the mean and autocorrelation functions for the input process $X(t)$, can the mean and autocorrelation functions for the output process $Y(t)$ be determined? Fortunately, the answer is that they can. In this section, expressions are developed that give the mean function for the output process in terms of the mean function for the input process and the impulse response of the linear system. Similarly, expressions are developed that give the autocorrelation function for the output process in terms of the autocorrelation function for the input process and the impulse response of the linear system.

Analogous expressions will be given for discrete-time linear systems in which the input–output relationship is given by

$$Y(k) = \sum_{n=-\infty}^{\infty} h(k, n) X(n).$$

For a discrete-time linear system with a given pulse response, knowledge of the mean function for the input is sufficient to determine the mean function for the output. Furthermore, in order to find the autocorrelation function for the output process, it suffices to know only the input autocorrelation function.

Before we develop the general expressions, it may be helpful to look at a specific example. In the solution of the exercise that follows, expressions for the mean and autocorrelation functions for the output are developed that turn out to be special cases of the general results presented later in this section. The primary simplification in this exercise stems from the fact that only finite sums are needed, rather than the integrals or infinite sums that are required in the analysis of general continuous-time or discrete-time linear systems.

Exercise 3-1

Suppose a discrete-time, time-invariant, linear system has pulse response

$$h(n) = \begin{cases} 1, & n = 1, \\ 1/2, & n = 0 \text{ or } n = 2, \\ 0, & \text{otherwise.} \end{cases}$$

Let $Y(k)$ be the output of this system when the input is a second-order random process $X(k)$ for which the mean is $\mu_X(k) = E\{X(k)\}$ and the autocorrelation is $R_X(k, i) = E\{X(k) X(i)\}$. Find the mean and autocorrelation functions for the output.

Solution.   Since $h(n) = 0$ except for n in the range $0 \le n \le 2$, the output is given by

$$Y(k) = \sum_{n=0}^{2} h(n) X(k - n) = \left(\tfrac{1}{2}\right) X(k) + X(k - 1) + \left(\tfrac{1}{2}\right) X(k - 2).$$

Thus, the mean of the output is

$$\mu_Y(k) = E\{Y(k)\} = \sum_{n=0}^{2} h(n) E\{X(k - n)\}$$

$$= \sum_{n=0}^{2} h(n) \mu_X(k - n)$$

$$= \left(\tfrac{1}{2}\right) \mu_X(k) + \mu_X(k - 1) + \left(\tfrac{1}{2}\right) \mu_X(k - 2).$$

The autocorrelation function for the output process is

$$R_Y(k, i) = E\{Y(k)\,Y(i)\}$$

$$= E\left\{\sum_{n=0}^{2} h(n)\,X(k-n) \sum_{m=0}^{2} h(m)\,X(i-m)\right\}$$

$$= E\left\{\sum_{n=0}^{2}\sum_{m=0}^{2} h(n)\,h(m)\,X(k-n)\,X(i-m)\right\}$$

$$= \sum_{n=0}^{2}\sum_{m=0}^{2} h(n)\,h(m)\,E\{X(k-n)\,X(i-m)\}.$$

Replacing the expectation of the product by the autocorrelation function, we find that

$$R_Y(k, i) = \sum_{n=0}^{2}\sum_{m=0}^{2} h(n)\,h(m)\,R_X(k-n, i-m)$$

$$= \left(\tfrac{1}{4}\right)\left[R_X(k, i) + R_X(k, i-2) + R_X(k-2, i) + R_X(k-2, i-2)\right]$$

$$+ \left(\tfrac{1}{2}\right)\left[R_X(k, i-1) + R_X(k-1, i) + R_X(k-1, i-2)\right.$$

$$\left. + R_X(k-2, i-1)\right] + R_X(k-1, i-1).$$

If the input process is wide-sense stationary, the expressions for the mean and auto-correlation functions simplify to $\mu_Y(k) = \mu_Y = 2\mu_X$ and

$$R_Y(k, i) = R_Y(k-i) = \left(\tfrac{1}{4}\right)\left[2R_X(k-i) + R_X(k-i+2) + R_X(k-i-2)\right]$$

$$+ \left(\tfrac{1}{2}\right)\left[2R_X(k-i+1) + 2R_X(k-i-1)\right] + R_X(k-i)$$

$$= \left(\tfrac{3}{2}\right)R_X(k-i) + \left(\tfrac{1}{4}\right)\left[R_X(k-i+2) + R_X(k-i-2)\right]$$

$$+ R_X(k-i+1) + R_X(k-i-1). \tag{3.27}$$

Since $\mu_Y(k)$ does not depend on k and $R_Y(k, i)$ depends on the difference $k-i$ only, the output process is also wide-sense stationary.

As a special case of (3.27), obtained by setting $i = k$, we have

$$E[Y(k)]^2 = R_Y(0) = \left(\tfrac{3}{2}\right)R_X(0) + 2R_X(1) + \left(\tfrac{1}{2}\right)R_X(2), \tag{3.28}$$

where we have used the property $R_X(\tau) = R_X(-\tau)$. From (3.28), we see that the variance of the output process can be written as

$$\sigma_Y^2 = E[Y(k) - \mu_Y]^2 = \left(\tfrac{3}{2}\right)R_X(0) + 2R_X(1) + \left(\tfrac{1}{2}\right)R_X(2) - 4\mu_X^2$$

$$= \left(\tfrac{3}{2}\right)C_X(0) + 2\,C_X(1) + \left(\tfrac{1}{2}\right)C_X(2),$$

where we have used the fact that $R_X(\tau) - \mu_X^2 = C_X(\tau)$ in the second step. The result for the variance is equivalent to

$$\sigma_Y^2 = \left(\tfrac{3}{2}\right)\sigma_X^2 + 2\,C_X(1) + \left(\tfrac{1}{2}\right)C_X(2). \tag{3.29}$$

Notice that although the output mean depends only on the input mean and the output autocorrelation function depends only on the input autocorrelation function, *the output variance does not depend only on the input variance.* As shown in (3.29), certain values of the input autocovariance function must be known in order to compute the output variance. Similarly, (3.28) shows that the second moment of the output depends on certain values of the input autocorrelation function. This fact is one of the reasons that much of our attention has been focused on the analysis of the mean and autocorrelation function (or the mean and autocovariance function) for the input and output random processes. ∎

We now develop the general relationships between input and output mean and autocorrelation functions. The complete derivation is given for continuous-time random processes and continuous-time linear systems. The derivations are analogous for discrete-time linear systems—just replace integrals by infinite sums—so only the final results are given. The results are for general, linear, time-varying systems, from which the results for time-invariant systems are obtained as a special case. It is assumed throughout that (3.21) and (3.25) are satisfied, so that the input–output equations (i.e., (3.16) and (3.24)) are well defined in a mean-square sense.

3.4.1   Mean of the Output of a Linear System

If the input process $X(t)$ to a continuous-time linear system with impulse response $h(t, \tau)$ has mean $\mu_X(t)$, then the output process

$$Y(t) = \int_{-\infty}^{\infty} h(t, \tau)\, X(\tau)\, \mathrm{d}\tau$$

has mean

$$
\begin{aligned}
\mu_Y(t) = E\{Y(t)\} &= E\left\{ \int_{-\infty}^{\infty} h(t, \tau)\, X(\tau)\, \mathrm{d}\tau \right\} \\
&= \int_{-\infty}^{\infty} h(t, \tau)\, E\{X(\tau)\}\, \mathrm{d}\tau \\
&= \int_{-\infty}^{\infty} h(t, \tau)\, \mu_X(\tau)\, \mathrm{d}\tau.
\end{aligned}
\tag{3.30}
$$

A *sufficient* condition for the interchange of expectation and integration to be valid is that the linear system be stable and the input be a second-order random process. In fact, this interchange is valid for virtually all linear systems and random processes that are actually encountered in engineering problems.

Notice in particular that $\mu_Y(t)$ is the output of the system when $\mu_X(t)$ is the (deterministic) input, a result that is certainly not surprising, either from an intuitive point of view or from purely mathematical arguments. The expected value of a sum (or integral) of random variables (or random processes) is simply the sum (or integral) of the expected values of the random variables (or random processes). This fact is an illustration of a situation in which "the expectation operation can be moved inside the integral," a principle that is valid for the vast majority of engineering applications.

More generally, we say that "the order of the expectation operation and a *linear* operation can be interchanged without changing the result." If L denotes such a linear operation, this statement can be written in shorthand form as $E\{L(X(t))\} = L(E\{X(t)\})$. The notation used here is not very precise and is intended only as an aid to intuition. For example, it is more accurate to write $L(\{X(t) : t \in T\})$ or $L(\{X_t : t \in T\})$ rather than $L(X(t))$, because, in general, L operates on the entire process $\{X(t) : t \in T\}$. The linear operations of interest to us in this section are those described by (3.16), (3.17), or (3.24).

If the linear system is time invariant, (3.30) reduces to

$$\mu_Y(t) = \int_{-\infty}^{\infty} h(t - \tau)\,\mu_X(\tau)\,d\tau = \int_{-\infty}^{\infty} h(\tau)\,\mu_X(t - \tau)\,d\tau. \qquad (3.31)$$

If, in addition, the mean of the random process $X(t)$ is not a function of t (i.e., $E\{X(t)\} = \mu_X$, a constant), then (3.31) reduces further to

$$\mu_Y(t) = \mu_Y = \mu_X \int_{-\infty}^{\infty} h(\tau)\,d\tau. \qquad (3.32)$$

Thus, given an input process $X(t)$ with a finite, constant mean, a sufficient condition for the output of a time-invariant linear system with impulse response $h(t)$ to have a finite, constant mean is

$$\int_{-\infty}^{\infty} |h(t)|\,dt < \infty. \qquad (3.33)$$

This is just the condition for stability of the time-invariant linear system.

On the basis of the results of the preceding analysis, we can state some general properties of time-invariant linear systems with random-process inputs. These properties are valid for both discrete- and continuous-time systems and processes:

1. *The output mean can be determined from the input mean alone; higher order moments are not required.*

2. *If the input mean is constant, so is the output mean.*

3. *If the input mean is constant and finite and the system is stable, the output mean is constant and finite.*

These properties do *not* hold for general nonlinear systems.

If the mean of the input process $X(t)$ is not constant and the linear system is not time invariant, the condition for finiteness of the output mean $\mu_Y(t)$ is

$$\int_{-\infty}^{\infty} |h(t, \tau)|\, |\mu_X(\tau)|\, d\tau < \infty. \tag{3.34}$$

Condition (3.34) reduces to (3.33) if the system is time invariant, in which case $h(t, \tau) = h(t - \tau)$ for all t and τ, and if $\mu_X(t) = \mu_X$ is a finite constant. Condition (3.34) is also sufficient to ensure the validity of the interchange in the order of expectation and integration that is employed in (3.30).

Exercise 3-2
A time-invariant linear system has impulse response

$$h(t) = \begin{cases} 0, & t < 0, \\ 1, & 0 \le t < T_0, \\ 0, & t \ge T_0. \end{cases}$$

Suppose the input to this system is the random process defined by $X(t) = Y_1 + tY_2$ for $-\infty < t < \infty$, where Y_1 and Y_2 are random variables with finite means ν_1 and ν_2, respectively. Find the mean of the output process $Y(t)$.

Solution.   The mean of $Y(t)$ is given by

$$\mu_Y(t) = \int_0^{T_0} \mu_X(t - \tau)\, d\tau.$$

We use the fact that $\mu_X(t) = \nu_1 + \nu_2 t$ to deduce

$$\mu_Y(t) = \nu_1 T_0 + \int_0^{T_0} (t - \tau)\, \nu_2\, d\tau.$$

Evaluating the integral, we see that $\mu_Y(t) = [\nu_1 T_0 - (1/2)\, \nu_2\, T_0^2] + (\nu_2\, T_0)\, t$. Notice that the output mean is of the form $a + tb$ where $a = \nu_1 T_0 - (1/2)\, \nu_2\, T_0^2$ and $b = \nu_2 T_0$. The special case $\nu_2 = 0$ gives a constant input mean $\mu_X(t) = \nu_1$ and a constant output mean $\mu_Y(t) = \nu_1 T_0$. Notice also that both (3.33) and (3.34) are satisfied in this example (even if $\nu_2 \ne 0$). ∎

The key results for *continuous-time* linear systems and processes are (3.30), (3.31), and (3.32). For *discrete-time* linear systems, the analogous expressions are as follows: In general,

$$\mu_Y(k) = \sum_{i=-\infty}^{\infty} h(k, i)\, \mu_X(i). \tag{3.35}$$

For time-invariant systems, (3.35) reduces to

$$\mu_Y(k) = \sum_{i=-\infty}^{\infty} h(k - i)\, \mu_X(i) = \sum_{i=-\infty}^{\infty} h(i)\, \mu_X(k - i), \tag{3.36}$$

which in turn yields

$$\mu_Y(k) = \mu_Y = \mu_X \sum_{i=-\infty}^{\infty} h(i) \tag{3.37}$$

if $X(k)$ has constant mean μ_X. The condition for the finiteness of $\mu_Y(k)$ in (3.35) is

$$\sum_{i=-\infty}^{\infty} |h(k,i)|\,|\mu_X(i)| < \infty. \tag{3.38}$$

The mean μ_Y in (3.37) is finite if μ_X is finite and the system is stable.

Example 3-3   A Time-Varying Discrete-Time Linear Filter

Consider a discrete-time linear system with pulse response

$$h(k,i) = \begin{cases} k\, 2^{-(k-i)}, & k \geq i, \\ 0, & k < i. \end{cases}$$

The input to this system is a wide-sense stationary process $X(k)$ with mean $\mu_X > 0$. The mean of the output process $Y(k)$ is given by (3.35), with $\mu_X(i) = \mu_X$. That is,

$$\mu_Y(k) = \mu_X \sum_{i=-\infty}^{\infty} h(k,i) = \mu_X \sum_{i=-\infty}^{k} k\, 2^{-(k-i)},$$

so the output mean is given by

$$\mu_Y(k) = k\mu_X \sum_{i=-\infty}^{k} 2^{-(k-i)}.$$

Making the substitution $n = k - i$, we find that

$$\mu_Y(k) = k\mu_X \sum_{n=0}^{\infty} 2^{-n} = 2k\mu_X.$$

Clearly, the output mean $\mu_Y(k)$ is not constant, so the process $Y(k)$ is not wide-sense stationary. In fact, the output mean is not even bounded ($\mu_Y(k) \to \infty$ as $k \to \infty$). This is because the given pulse response does not satisfy (3.8), although it does satisfy (3.9). The system is not stable, as can be seen by considering the bounded input $x(i) = \mu_X$ for all i. The corresponding output is $y(k) = \mu_Y(k) = 2k\mu_X$, which increases without bound as $k \to \infty$ if $\mu_X > 0$. However, the output mean is finite for each k, which is always true if (3.38) is satisfied. (Note that (3.38) is indeed satisfied in this example.) ■

Exercise 3-3

Suppose a continuous-time linear system has impulse response

$$h(t, \tau) = \begin{cases} \exp\{-\alpha(t - \tau)\}, & t \geq \tau, \\ 0, & t < \tau. \end{cases}$$

If the input process $X(t)$ has constant mean μ_X, find the mean of the output process $Y(t)$. Assume that $\alpha > 0$.

Solution.   Since the system is time invariant and the input mean is constant, we employ (3.32) and conclude that

$$\mu_Y = \mu_X \int_0^\infty e^{-\alpha \tau} \, d\tau = \alpha^{-1} \mu_X.$$

■

Exercise 3-4

A discrete-time linear system is described as follows: For any input $x(k)$, the output is the signal

$$y(k) = \left(\tfrac{1}{2}\right) y(k - 1) + x(k). \tag{3.39}$$

Suppose the input to this system is a random process $X(k)$ with constant mean μ_X. What is the mean of the output process $Y(k)$?

Solution.   We give two methods:

(a) The pulse response of the system described by (3.39) is found by letting $x(i) = 1$ and $x(n) = 0$ for all $n \neq i$. For this input, $y(k) = h(k, i)$. From (3.39), we see that $y(k) = 0$ for $k < i$, $y(i) = x(i) = 1$, and $y(k) = (1/2) y(k - 1)$ for $k > i$. Thus, $y(i) = 1$, $y(i + 1) = 1/2$, $y(i + 2) = 1/4, \ldots, y(i + m) = 2^{-m}$. The pulse response is therefore

$$h(k, i) = \begin{cases} 2^{-(k-i)}, & k \geq i, \\ 0, & k < i. \end{cases}$$

Hence, the system is time invariant, so (3.37) can be applied and we conclude that the output has constant mean

$$\mu_Y = \mu_X \sum_{n=0}^\infty 2^{-n} = 2\mu_X.$$

(b) For the second method, we notice first that the linear system described by (3.39) is time invariant (since, if $w(k)$ is the output signal for input $v(k)$, then $w(k - m)$ is the output signal whenever $v(k - m)$ is the input). This implies that the output

$$Y(k) = \left(\tfrac{1}{2}\right) Y(k - 1) + X(k)$$

has constant mean μ_Y. Taking expectations, we find that $\mu_Y = (1/2) \mu_Y + \mu_X$, so $\mu_Y = 2\mu_X$.

■

3.4.2   Autocorrelation Function for the Output of a Linear System

If $Y(t)$ is the output of a continuous-time linear system with impulse response $h(t, \tau)$ and input $X(t)$, then the crosscorrelation function for the processes $Y(t)$ and $X(t)$ is given by

$$
\begin{aligned}
R_{Y,X}(t, s) &= E\{Y(t)\,X(s)\} \\
&= E\left\{\int_{-\infty}^{\infty} h(t, \tau)\,X(\tau)\,\mathrm{d}\tau\,X(s)\right\} \\
&= E\left\{\int_{-\infty}^{\infty} h(t, \tau)\,X(\tau)\,X(s)\,\mathrm{d}\tau\right\}.
\end{aligned}
$$

Interchanging the order of expectation and integration, we see that

$$
R_{Y,X}(t, s) = \int_{-\infty}^{\infty} h(t, \tau)\,R_X(\tau, s)\,\mathrm{d}\tau. \tag{3.40}
$$

The autocorrelation function for the output process $Y(t)$ is

$$
\begin{aligned}
R_Y(t, s) &= E\{Y(t)\,Y(s)\} \\
&= E\left\{Y(t)\int_{-\infty}^{\infty} h(s, \lambda)\,X(\lambda)\,\mathrm{d}\lambda\right\} \\
&= \int_{-\infty}^{\infty} h(s, \lambda)\,E\{Y(t)\,X(\lambda)\}\,\mathrm{d}\lambda \\
&= \int_{-\infty}^{\infty} h(s, \lambda)\,R_{Y,X}(t, \lambda)\,\mathrm{d}\lambda. \tag{3.41}
\end{aligned}
$$

Combining (3.40) and (3.41), we find that

$$
R_Y(t, s) = \int_{-\infty}^{\infty}\int_{-\infty}^{\infty} h(s, \lambda)\,h(t, \tau)\,R_X(\tau, \lambda)\,\mathrm{d}\tau\,\mathrm{d}\lambda. \tag{3.42}
$$

Note in particular that for the determination of the autocorrelation function for the output process, knowledge of the system impulse response and the autocorrelation function for the input process is sufficient. The same is true for autocovariance functions, since the foregoing derivation can easily be modified to establish the relationship

$$
C_Y(t, s) = \int_{-\infty}^{\infty}\int_{-\infty}^{\infty} h(s, \lambda)\,h(t, \tau)\,C_X(\tau, \lambda)\,\mathrm{d}\tau\,\mathrm{d}\lambda. \tag{3.43}
$$

Alternatively, (3.43) can be obtained from (3.42), (3.30), and (2.9).

If the input process is wide-sense stationary, (3.40) can be written as

$$
\begin{aligned}
R_{Y,X}(t, s) &= \int_{-\infty}^{\infty} h(t, \tau)\,R_X(\tau - s)\,\mathrm{d}\tau \\
&= \int_{-\infty}^{\infty} h(t, \tau)\,R_X(s - \tau)\,\mathrm{d}\tau \\
&= \int_{-\infty}^{\infty} h(t, s - \alpha)\,R_X(\alpha)\,\mathrm{d}\alpha,
\end{aligned}
$$

which is a convolution integral: For each t, the function $R_{Y,X}(t, \cdot)$ is the convolution of the function $R_X(\cdot)$ with the function $h(t, \cdot)$.

Another important special case results if the system is time invariant. In this case, (3.40) can be written as

$$R_{Y,X}(t, s) = \int_{-\infty}^{\infty} h(t - \tau) \, R_X(\tau, s) \, d\tau$$

$$= \int_{-\infty}^{\infty} h(\alpha) \, R_X(t - \alpha, s) \, d\alpha, \tag{3.44}$$

and (3.41) reduces to

$$R_Y(t, s) = \int_{-\infty}^{\infty} h(s - \lambda) \, R_{Y,X}(t, \lambda) \, d\lambda$$

$$= \int_{-\infty}^{\infty} h(\beta) \, R_{Y,X}(t, s - \beta) \, d\beta. \tag{3.45}$$

Thus, for time-invariant systems, (3.42) becomes

$$R_Y(t, s) = \int_{-\infty}^{\infty} \int_{-\infty}^{\infty} h(s - \lambda) \, h(t - \tau) \, R_X(\tau, \lambda) \, d\tau \, d\lambda$$

$$= \int_{-\infty}^{\infty} \int_{-\infty}^{\infty} h(\alpha) \, h(\beta) \, R_X(t - \alpha, s - \beta) \, d\alpha \, d\beta.$$

Finally, if $X(t)$ is wide-sense stationary and the system is time invariant, then

$$R_{Y,X}(t, s) = \int_{-\infty}^{\infty} h(t - \tau) \, R_X(\tau - s) \, d\tau$$

$$= \int_{-\infty}^{\infty} h(t - s - \alpha) \, R_X(\alpha) \, d\alpha. \tag{3.46}$$

Notice that $R_{Y,X}(t, s)$ is a function of t and s via the difference $t - s$ only; that is, $R_{Y,X}(t, s) = R_{Y,X}(t - s, 0) = R_{Y,X}(t - s)$ for all t and s. Also, we see that in this case (3.45) and (3.46) imply that

$$R_Y(t, s) = \int_{-\infty}^{\infty} h(\beta) \, R_{Y,X}(t, s - \beta) \, d\beta$$

$$= \int_{-\infty}^{\infty} \int_{-\infty}^{\infty} h(\alpha) \, h(\beta) \, R_X(t - \alpha, s - \beta) \, d\alpha \, d\beta$$

$$= \int_{-\infty}^{\infty} \int_{-\infty}^{\infty} h(\alpha) \, h(\beta) \, R_X(t - s + \beta - \alpha) \, d\alpha \, d\beta, \tag{3.47}$$

so that $R_Y(t, s) = R_Y(t - s, 0) = R_Y(t - s)$ for all t and s. Notice that (3.32), (3.46), and (3.47) establish the following important fact:

If the input to a time-invariant linear system is a wide-sense stationary random process, the output is also a wide-sense stationary random process; moreover, the input and output processes are jointly wide-sense stationary.

For this situation, we can write (3.46) as

$$R_{Y,X}(\tau) = \int_{-\infty}^{\infty} h(\tau - \alpha)\, R_X(\alpha)\, d\alpha$$

$$= \int_{-\infty}^{\infty} h(\alpha)\, R_X(\tau - \alpha)\, d\alpha \qquad (3.48)$$

and (3.47) as

$$R_Y(\tau) = \int_{-\infty}^{\infty} h(\beta) \left\{ \int_{-\infty}^{\infty} h(\alpha)\, R_X[(\tau + \beta) - \alpha]\, d\alpha \right\} d\beta. \qquad (3.49)$$

Notice that (3.48) is just the convolution of h with R_X; that is, $R_{Y,X} = h * R_X$.

The outer integral in (3.49) is not quite a convolution integral; it is given by

$$R_Y(\tau) = \int_{-\infty}^{\infty} h(\beta)\, R_{Y,X}(\tau + \beta)\, d\beta, \qquad (3.50)$$

where we have replaced τ by $\tau + \beta$ in (3.48) and substituted for the term in brackets in (3.49). However, if we let $\tilde{h}$ be the *time-reverse* impulse response (i.e., $\tilde{h}(t) = h(-t)$ for all t), then (3.50) becomes

$$R_Y(\tau) = \int_{-\infty}^{\infty} \tilde{h}(-\beta)\, R_{Y,X}(\tau + \beta)\, d\beta$$

$$= \int_{-\infty}^{\infty} \tilde{h}(\lambda)\, R_{Y,X}(\tau - \lambda)\, d\lambda.$$

That is,

$$R_Y = \tilde{h} * R_{Y,X} = \tilde{h} * (h * R_X).$$

Since the order in which the convolutions are performed does not alter the answer, we write

$$R_Y = \tilde{h} * (h * R_X) = (\tilde{h} * h) * R_X. \qquad (3.51)$$

So, in fact, we can write $R_Y = \tilde{h} * h * R_X$ with no fear of ambiguity. The preceding considerations suggest one approach to finding the output autocorrelation: First, convolve the impulse response with its time reverse, and then convolve the result with the input autocorrelation function. Notice that the definition of $f = \tilde{h} * h$ is the integral

$$f(u) = \int_{-\infty}^{\infty} \tilde{h}(u - v)\, h(v)\, dv,$$

but this is equivalent to

$$f(u) = \int_{-\infty}^{\infty} h(v - u)\, h(v)\, dv, \qquad (3.52)$$

because $\tilde{h}(t) = h(-t)$ for any t. Replacing the variables u and v by τ and t, respectively, and reversing the order of the terms in the integrand, we see that (3.52) becomes

$$f(\tau) = \int_{-\infty}^{\infty} h(t)\,h(t - \tau)\,dt, \qquad (3.53)$$

which is just the integral of the product of the impulse response with a delayed version of the impulse response. The expression in (3.53) is much easier to work with than the general convolution integral, because (3.53) requires only the integration of the product of a function and a time-shifted version of itself; there is no requirement to "flip" one of the functions about the origin, as is necessary in the general convolution procedure. This gives us the following important fact:

$f(\tau)$ is just the area under the product of $h(t)$ and $h(t - \tau)$.

The term "area" in this statement refers to integration with respect to the variable t (not τ).

Next, we show that f is an *even* function; that is, it is symmetrical about the origin: $f(u) = f(-u)$ for all real numbers u. The fact that f is a symmetrical function is useful, both for the evaluation of the function and for the subsequent convolution of f with the input autocorrelation function to find the output autocorrelation function. To show that f is symmetrical, we begin with

$$f(-u) = \int_{-\infty}^{\infty} \tilde{h}(-u - v)\,h(v)\,dv$$

$$= \int_{-\infty}^{\infty} h(u + v)\,h(v)\,dv.$$

The change of variable $\lambda = u + v$ gives

$$f(-u) = \int_{-\infty}^{\infty} h(\lambda)\,h(\lambda - u)\,d\lambda$$

$$= \int_{-\infty}^{\infty} h(\lambda - u)\,h(\lambda)\,d\lambda.$$

A comparison of the last integral and (3.52) shows that this is just $f(u)$. Hence, we have established the relationship

$$f(-u) = f(u).$$

One consequence of this relationship is that it suffices to evaluate the integral in (3.53) for $\tau \geq 0$ only. We then set $f(\tau) = f(-\tau)$ for $\tau < 0$.

If $h(t)$ happens to be time limited, the integral in (3.53) is equivalent to an integral with *finite* limits, as illustrated by the next example.

Example 3-4   A Time-Limited Impulse Response
Suppose T is a (finite) positive number and the impulse response h satisfies $h(t) = 0$ for all values of t that are not in the range $0 \leq t \leq T$. Such an impulse response is said

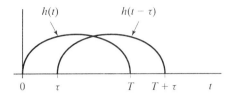

Figure 3-3: An illustration of $h(t)$ and $h(t - \tau)$ if h is time limited to $[0, T]$.

to be *time limited—in this case to the interval* $[0, T]$. For an impulse response that is time limited to $[0, T]$, $h(t - \tau) = 0$ for $t < \tau$ and $h(t - \tau) = 0$ for $t > T + \tau$. If $\tau > 0$, the product $h(t) h(t - \tau)$ is zero for all t outside the range $\tau \leq t \leq T$, as can be seen from Figure 3-3. So if h is time limited to $[0, T]$ and $0 < \tau \leq T$, then (3.53) reduces to

$$f(\tau) = \int_{\tau}^{T} h(t) h(t - \tau) \, dt.$$

Notice that, because the impulse response is time limited to $[0, T]$, $f(\tau) = 0$ for $\tau > T$.

The procedure for evaluating the function f is illustrated in the next three exercises. Analytical methods and graphical methods can be used in this evaluation. In Exercise 3-5, a purely analytical derivation is given for a specific example. For most impulse responses, however, it is best to use a combination of analytical and graphical methods in the evaluation of f. That is, we use sketches of the functions, such as those in Figure 3-3, to help set up the necessary integrals. For very simple functions, graphical methods alone suffice, as is illustrated later by Exercise 3-6.

Exercise 3-5
The rectangular pulse of duration $T > 0$ is defined by $p_T(t) = 1$ for $0 \leq t < T$ and $p_T(t) = 0$, otherwise. Using analytical methods, determine the function $f = \tilde{h} * h$ for the linear filter with impulse response $h(t) = p_T(t)$.

Solution.   The impulse response is time limited to $[0, T]$, so the expression developed in Example 3-4 applies. That is, for $h(t) = p_T(t)$ and $0 < \tau < T$,

$$f(\tau) = \int_{\tau}^{T} h(t) h(t - \tau) \, dt = \int_{\tau}^{T} p_T(t) p_T(t - \tau) \, dt.$$

As mentioned in that example, $f(\tau) = 0$ for $\tau > T$. The product $p_T(t) p_T(t - \tau)$ is zero for a given value of t and τ if *either* pulse function takes the value zero for that t and τ. The product is nonzero only if both pulse functions are nonzero, in which case the product is unity. Now, $p_T(t)$ is nonzero for $0 \leq t < T$, and $p_T(t - \tau)$ is nonzero for $\tau \leq t < \tau + T$, so the product is nonzero for $\tau \leq t < T$. That is, for $0 < \tau < T$, $p_T(t) p_T(t - \tau) = 1$ for all t in the range $\tau \leq t < T$. It follows that

$$f(\tau) = \int_{\tau}^{T} 1 \, dt = T - \tau, \quad \text{for } 0 < \tau < T.$$

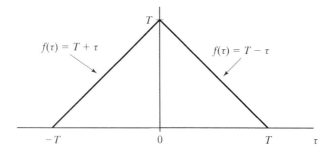

Figure 3-4: The function $f = \tilde{h} * h$ for $h(t) = p_T(t)$.

We have already established that $f(\tau) = 0$ for $\tau > T$ (Example 3-4). Next, we use the symmetry of the function f to conclude that $f(\tau) = 0$ for $\tau < -T$ and $f(\tau) = T + \tau$ for $-T \leq \tau < 0$. Thus,

$$f(\tau) = 0 \quad \text{for } |\tau| \geq T$$

and

$$f(\tau) = T - |\tau| \quad \text{for } |\tau| < T.$$

$f(\tau)$ is the triangular function illustrated in Figure 3-4.   ■

Exercise 3-6
Using graphical methods, determine the function $f = \tilde{h} * h$ for the linear filter with impulse response $h(t) = p_T(t)$.

Solution.   The solution presented here is unfair in one respect: We use the fact that f is an even function, which was proved analytically. However, the reader should be able to make a few sketches to provide a convincing "graphical proof" that f is an even function. Given that f is an even function, we can restrict attention to $\tau \geq 0$. First, we sketch $h(t)$, $h(t - \tau)$, and the product $h(t) h(t - \tau)$, as shown in Figure 3-5. Next, we determine the area under the product. As can be seen in the figure, this area is just the area of a rectangle of width $T - \tau$ and unity height, provided that $\tau < T$. Thus, we see that $f(\tau) = T - \tau$ for $0 \leq \tau < T$. Inspection of the graphs of $h(t)$ and $h(t - \tau)$ shows that the product $h(t) h(t - \tau)$ is identically zero if $\tau > T$. This takes care of all nonnegative values of τ. The values of $f(\tau)$ for negative values of τ are determined from the fact that f is an even function.   ■

Exercise 3-7
Evaluate the function $f = \tilde{h} * h$ for a time-invariant linear system with impulse response $h(t) = \beta \exp(-\alpha t)$ for $t \geq 0$ and $h(t) = 0$ for $t < 0$. (Assume that α and β are positive constants.)

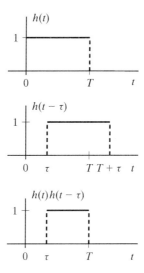

Figure 3-5: Illustration of the graphical method for convolving two rectangular pulses.

Solution.   First, observe that the impulse response can be written as

$$h(t) = \beta \exp(-\alpha t)\, u(t),$$

where $u(\cdot)$ is the unit step function (i.e., $u(\lambda) = 1$ for $\lambda \geq 0$ and $u(\lambda) = 0$ for $\lambda < 0$). Note that h is not time limited, so the expression developed in Example 3-4 does not apply. From the more general expression in (3.53), we see that the function $f = \tilde{h} * h$ is given by

$$
\begin{aligned}
f(\tau) &= \int_{-\infty}^{\infty} h(t)\, h(t - \tau)\, \mathrm{d}t \\
&= \beta^2 \int_{-\infty}^{\infty} \mathrm{e}^{-\alpha t}\, u(t)\, \mathrm{e}^{-\alpha(t-\tau)}\, u(t - \tau)\, \mathrm{d}t.
\end{aligned}
$$

Illustrations of the functions appearing in the integrands of these two integrals are given in Figure 3-6 for $\tau \geq 0$.

Because we can restrict our attention to $\tau \geq 0$, we see that the preceding integral reduces to

$$f(\tau) = \beta^2 \, \mathrm{e}^{\alpha\tau} \int_{\tau}^{\infty} \mathrm{e}^{-2\alpha t}\, \mathrm{d}t.$$

Evaluating this integral, we find that

$$f(\tau) = \left(\frac{\beta^2}{2\alpha} \right) \mathrm{e}^{-\alpha\tau}, \quad \tau \geq 0,$$

which implies that

$$f(\tau) = f(-\tau) = \left(\frac{\beta^2}{2\alpha} \right) \mathrm{e}^{+\alpha\tau}, \quad \tau < 0.$$

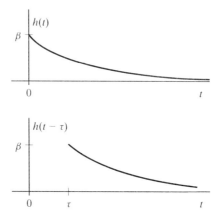

Figure 3-6: Original and delayed versions of the impulse response.

Thus, $f(\tau) = (\beta^2/2\alpha)\exp(-\alpha|\tau|)$ for $-\infty < \tau < \infty$.   ◼

Next, we examine in greater detail the evaluation of the output autocorrelation function for a time-invariant linear system that has a wide-sense stationary random process as its input. Equations (3.32) and (3.47) imply that the output random process is wide-sense stationary, so the output autocorrelation can be obtained from $R_Y = (\tilde{h} * h) * R_X$. In working with convolutions, it is important to recall that the convolution of two functions is a *function*, not a number. The convolution of $\tilde{h}$ and h is the function f defined by (3.53) for each value of its argument. To find R_Y, the function f is convolved with the function R_X. That is,

$$R_Y(\tau) = \int_{-\infty}^{\infty} f(\tau - u)\, R_X(u)\, du$$

for each real number τ. Because $f(\tau - u) = f(u - \tau)$, the preceding equation is equivalent to

$$R_Y(\tau) = \int_{-\infty}^{\infty} R_X(u)\, f(u - \tau)\, du. \tag{3.54}$$

Just like the integral in (3.53), the integral in (3.54) is simpler than the general convolution integral. All that is needed is to integrate the product of the autocorrelation function and a delayed version of the function f.

The simplest procedure for evaluating the output autocorrelation function for most filter impulse responses and input autocorrelation functions is to first determine the function f and then convolve f with the input autocorrelation function R_X to get the output autocorrelation function. That is, we first evaluate $f = \tilde{h} * h$ and then evaluate $R_Y = f * R_X$. The second part of this procedure can be carried out by applying (3.54), whose evaluation is simplified further by the fact that each of the functions f, R_X, and R_Y is even. Earlier in this section, we proved that f is even, and we proved in Chapter 2

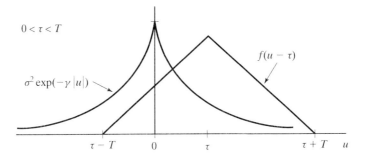

Figure 3-7: The functions $\sigma^2 \exp(-\gamma|u|)$ and $f(u - \tau)$ for $0 < \tau < T$.

that all autocorrelation functions for wide-sense stationary processes are even, so both R_X and R_Y are even. This implies that (3.54) need be evaluated for $\tau \geq 0$ only. The values of $R_Y(\tau)$ for $\tau < 0$ can be obtained from $R_Y(\tau) = R_Y(-\tau)$. The procedure for evaluating $R_Y = f * R_X$ via (3.54) is illustrated in the next exercise.

Exercise 3-8
Consider the linear system of Exercise 3-5 with impulse response $h(t) = p_T(t)$. The input is a zero-mean, wide-sense stationary random process $X(t)$ with autocorrelation function

$$R_X(\tau) = \sigma^2 \exp(-\gamma|\tau|), \quad -\infty < \tau < \infty,$$

where γ is a positive constant. Find the autocorrelation function for the output process $Y(t)$.

Solution.   In Exercise 3-5, the function f is shown to be the triangular function $f(\tau) = T - |\tau|$ for $|\tau| < T$ and $f(\tau) = 0$, otherwise. All that remains is to evaluate the convolution $R_Y = f * R_X$. From (3.54), we see that the corresponding convolution integral is equivalent to

$$R_Y(\tau) = \int_{-\infty}^{\infty} \sigma^2 \exp(-\gamma|u|)\, f(u - \tau)\, du.$$

The functions appearing in the integrand are illustrated in Figure 3-7 for $0 < \tau < T$. Notice that for any $\tau > 0$, the integrand is identically zero on the intervals $(-\infty, \tau - T)$ and $(\tau + T, \infty)$. Therefore, for $\tau > 0$, we have

$$R_Y(\tau) = \int_{\tau-T}^{\tau+T} \sigma^2 \exp(-\gamma|u|)\, f(u - \tau)\, du$$

$$= \int_{\tau-T}^{\tau+T} \sigma^2 \exp(-\gamma|u|)\, \{T - |u - \tau|\}\, du$$

$$= \int_{\tau-T}^{\tau+T} \sigma^2 T \exp(-\gamma|u|)\, du - \int_{\tau-T}^{\tau+T} \sigma^2 |u - \tau| \exp(-\gamma|u|)\, du.$$

Next, for $0 < \tau < T$, we write the first of these two integrals as the sum of two integrals and the second as the sum of three integrals. (This part will be different for $\tau \geq T$.)

The regions of integration for the new integrals are regions over which the functional forms of the integrands are constant. The term $|u - \tau|$ changes form (from a positive slope to a negative slope) at $u = \tau$, and the term $\exp(-\gamma|u|)$ changes form at $u = 0$. Therefore, for $0 < \tau < T$, the foregoing expression for $R_Y(\tau)$ is equivalent to

$$
R_Y(\tau) = \sigma^2 T \int_{\tau-T}^{0} \exp(+\gamma u)\, du + \sigma^2 T \int_{0}^{\tau+T} \exp(-\gamma u)\, du
$$
$$
- \sigma^2 \int_{\tau-T}^{0} (-u + \tau)\exp(+\gamma u)\, du - \sigma^2 \int_{0}^{\tau} (-u + \tau)\exp(-\gamma u)\, du
$$
$$
- \sigma^2 \int_{\tau}^{\tau+T} (u - \tau)\exp(-\gamma u)\, du.
$$

All that remains is to evaluate these integrals for τ in the range $0 < \tau < T$, a task that is left to the reader, who should also derive the expression for $R_Y(\tau)$ for $\tau \geq T$. ∎

Frequently, it is necessary to determine the power in the output process from the autocorrelation function of the input process and the impulse response of the linear system. Recall that the expected value of the instantaneous power in the process $Y(t)$ is equal to the second moment $E\{[Y(t)]^2\}$. If the input process $X(t)$ is wide-sense stationary and the linear filter is time invariant, then

$$
E\{[Y(t)]^2\} = R_Y(0).
$$

If all we need to compute is the second moment of the output process, rather than its complete autocorrelation function, the evaluation is considerably simpler. In particular, no convolutions are required once the function f has been obtained. To see this, consider the fact that

$$
R_Y(0) = \int_{-\infty}^{\infty} f(\lambda)\, R_X(\lambda)\, d\lambda, \tag{3.55a}
$$

which follows from (3.54) with $\tau = 0$. Note that if $X(t)$ is *white noise*, $R_X(\tau) = c\delta(\tau)$ for some positive constant c, and thus,

$$
R_Y(0) = c\, f(0) = c \int_{-\infty}^{\infty} [h(t)]^2\, dt.
$$

From (3.55a), we see that the second moment of the output random process is just the integral of the product of the functions f and R_X. Because these functions are both *even*, it follows that if f and R_X have no delta functions at the origin, then

$$
R_Y(0) = 2 \int_{0}^{\infty} f(\lambda)\, R_X(\lambda)\, d\lambda. \tag{3.55b}
$$

Applications of this result are illustrated in the next two exercises.

Exercise 3-9

Suppose the random process $X(t)$ has zero mean and autocorrelation function

$$R_X(\tau) = \begin{cases} (\tau_0 - |\tau|)/\tau_0, & 0 \le |\tau| \le \tau_0, \\ 0, & |\tau| > \tau_0. \end{cases}$$

Let $X(t)$ be the input to a time-invariant linear system with impulse response

$$h(t) = \begin{cases} 1, & 0 \le t < t_0, \\ 0, & \text{otherwise}, \end{cases}$$

where $\tau_0 > t_0$. Find the power in the output process $Y(t)$.

Solution.   The process is wide-sense stationary, so its power is $R_Y(0)$. To find $R_Y(0)$, we first determine the function f, which is related to the impulse response of the filter by $f = \tilde{h} * h$. Thus, to find the function f, we must convolve two rectangular pulses, each of which has duration t_0. But the convolution of any two rectangular pulses of duration t_0 is a triangular pulse of duration $2t_0$, so $\tilde{h} * h$ is such a triangular pulse. In particular, because we know that $f = \tilde{h} * h$ is a symmetrical function, the triangular pulse must be centered at the origin. In addition, the height of the triangle is equal to $f(0)$, which is just the area under the square of the impulse response. (To see this, let $\tau = 0$ in (3.53).) For the given impulse response, the said area is t_0. Putting all these facts together, we conclude that $f(t)$ is a triangular pulse of duration $2t_0$, the peak of the triangle must be at the origin, and $f(0) = t_0$. Thus,

$$f(t) = \begin{cases} t_0 - |t|, & -t_0 \le t \le t_0, \\ 0, & |t| > t_0. \end{cases}$$

Given the function f, we can find $R_Y(0)$ by substituting into (3.55b) for $f(\lambda)$ to obtain

$$R_Y(0) = 2 \int_0^{t_0} (t_0 - \lambda) R_X(\lambda) \, d\lambda.$$

Because $\tau_0 > t_0$, this equation is equivalent to

$$R_Y(0) = 2\,\tau_0^{-1} \int_0^{t_0} (t_0 - \lambda)(\tau_0 - \lambda) \, d\lambda = t_0^2 (3\,\tau_0)^{-1} [3\,\tau_0 - t_0].$$

If t_0 were greater than τ_0, the upper limit on the integral would be τ_0 (i.e., in general, the upper limit is the minimum of t_0 and τ_0). ∎

Exercise 3-10

Let $X(t)$ be a zero-mean wide-sense stationary process. Define a random process $Z(t)$, $t \ge 0$, by the mean-square integral

$$Z(t) = \int_0^t X(\tau) \, d\tau.$$

Find the variance of $Z(t)$ in terms of the autocorrelation function for the process $X(t)$.

Solution.   Consider an arbitrary $t_0 > 0$ and evaluate $\mathrm{Var}\{Z(t_0)\}$. First, we define an impulse response h for a hypothetical linear time-invariant system by

$$h(\tau) = \begin{cases} 1, & 0 \leq \tau \leq t_0, \\ 0, & \text{otherwise.} \end{cases}$$

Next, notice that, since $0 \leq t_0 - \tau \leq t_0$ if and only if $0 \leq \tau \leq t_0$, it follows that

$$h(\tau) = h(t_0 - \tau) = \begin{cases} 1, & 0 \leq t_0 - \tau \leq t_0, \\ 0, & \text{otherwise.} \end{cases}$$

The reason we are interested in the impulse response $h(t)$ is the fact that

$$Z(t_0) = \int_0^{t_0} X(\tau)\,d\tau = \int_{-\infty}^{\infty} h(\tau)\,X(\tau)\,d\tau = \int_{-\infty}^{\infty} h(t_0 - \tau)\,X(\tau)\,d\tau,$$

for $t_0 > 0$. That is, if $X(t)$ is the input to this filter, $Z(t_0)$ can be obtained by sampling the output of the filter at time t_0. If we define the random process $Y(t)$, $-\infty < t < \infty$, by

$$Y(t) = \int_{-\infty}^{\infty} h(t - \tau)\,X(\tau)\,d\tau,$$

then $Z(t_0) = Y(t_0)$, and $Y(t)$ is wide-sense stationary even though $Z(t)$ is not.

From the preceding results, we see that the mean of $Z(t_0)$ is given by

$$\mu_Z(t_0) = \mu_Y = \mu_X \int_{-\infty}^{\infty} h(t_0 - \tau)\,d\tau = 0,$$

since $\mu_X = 0$. Furthermore,

$$\mathrm{Var}\{Z(t_0)\} = \mathrm{Var}\{Y(t_0)\} = R_Y(t_0, t_0) = R_Y(0).$$

From (3.55) and Exercise 3-9, we see that for this particular impulse response h,

$$R_Y(0) = \int_{-t_0}^{t_0} (t_0 - |\lambda|)\,R_X(\lambda)\,d\lambda = 2 \int_0^{t_0} (t_0 - \lambda)\,R_X(\lambda)\,d\lambda,$$

where the first integral is valid in general, but the second integral requires that R_X not have a delta function at the origin. Because t_0 is arbitrary, we conclude from the foregoing results that, for each $t > 0$,

$$\mathrm{Var}\{Z(t)\} = \int_{-t}^{t} (t - |\lambda|)\,R_X(\lambda)\,d\lambda = 2 \int_0^{t} (t - \lambda)\,R_X(\lambda)\,d\lambda, \qquad (3.56)$$

with the same restriction that there must not be a delta function at the origin in the rightmost integral.

As an example, suppose $X(t)$ has the autocorrelation function given by

$$R_X(\tau) = \sigma^2 \exp(-\gamma|\tau|), \quad -\infty < \tau < \infty,$$

which is examined in Exercise 3-8. For this autocorrelation function,

$$\begin{aligned} \text{Var}\{Z(t)\} &= 2\int_0^t (t-\lambda)\sigma^2 e^{-\gamma\lambda}\, d\lambda \\ &= 2\sigma^2 \gamma^{-2}[\gamma t - 1 + \exp(-\gamma t)] \end{aligned}$$

for $t > 0$.

An alternative approach is to work directly with the definition of the process $Y(t)$ as follows: For $t > 0$ and $s > 0$,

$$\begin{aligned} R_Z(t,s) &= E\{Z(t)\,Z(s)\} \\ &= E\left\{\int_0^t X(\tau)\,d\tau \int_0^s X(\lambda)\,d\lambda\right\} = \int_0^t \int_0^s E\{X(\tau)\,X(\lambda)\}\,d\lambda\,d\tau \\ &= \int_0^t \int_0^s R_X(\tau-\lambda)\,d\lambda\,d\tau. \end{aligned}$$

As a special case, for $t > 0$,

$$\text{Var}\{Z(t)\} = R_Z(t,t) = \int_0^t \int_0^t R_X(\tau-\lambda)\,d\lambda\,d\tau,$$

which can be shown to reduce to (3.56) by the change of variable $u = \tau$, $v = \tau - \lambda$. (This change of variable is illustrated in detail in Section 4.4.)   ∎

Before proceeding, it is beneficial to summarize certain key results obtained so far in this subsection. These results provide the methods for evaluating the correlation functions that arise in the analysis of random processes in continuous-time linear systems. First, for *general* second-order continuous-time random processes and linear systems, the primary results are

$$R_{Y,X}(t,s) = \int_{-\infty}^{\infty} h(t,\tau)\,R_X(\tau,s)\,d\tau$$

for the crosscorrelation between the input and output random processes and

$$R_Y(t,s) = \int_{-\infty}^{\infty} h(s,\lambda)\,h(t,\tau)\,R_X(\tau,\lambda)\,d\tau\,d\lambda$$

for the autocorrelation function for the output random process. For the special case of continuous-time *wide-sense stationary* random processes and *time-invariant* linear systems, these results simplify to give

$$\begin{aligned} R_{Y,X}(\tau) &= \int_{-\infty}^{\infty} h(\tau-\alpha)\,R_X(\alpha)\,d\alpha \\ &= \int_{-\infty}^{\infty} h(\alpha)\,R_X(\tau-\alpha)\,d\alpha \end{aligned}$$

and

$$R_Y(\tau) = \int_{-\infty}^{\infty} h(\beta) \left\{ \int_{-\infty}^{\infty} h(\alpha) R_X[(\tau + \beta) - \alpha] \, d\alpha \right\} d\beta.$$

These last two results can be written in terms of convolutions: $R_{Y,X} = h * R_X$ and $R_Y = \tilde{h} * h * R_X$. In most cases, the simplest procedure for evaluating the autocorrelation function R_Y is to find $f = \tilde{h} * h$ and then use $R_Y = f * R_X$. Each of the functions involved in this last convolution is an even function. In order to find the function f, it is usually simpler to use the alternative expression

$$f(\tau) = \int_{-\infty}^{\infty} h(t) h(t - \tau) \, dt,$$

which expresses $f(\tau)$ as the integral of the product of the impulse response and a delayed version of itself. The amount of the delay is τ, the argument of the function. This integral is simpler than the general convolution integral.

Analogous equations can be obtained for linear *discrete-time* systems. Generally, these equations correspond to the results for continuous-time systems if the integrals are replaced by sums. For the results that follow, we assume that (3.21) is satisfied. First, for general discrete-time random processes and linear systems,

$$R_{Y,X}(k, i) = \sum_{n=-\infty}^{\infty} h(k, n) R_X(n, i) \qquad (3.57)$$

and

$$R_Y(k, i) = \sum_{m=-\infty}^{\infty} h(i, m) R_{Y,X}(k, m), \qquad (3.58)$$

or

$$R_Y(k, i) = \sum_{m=-\infty}^{\infty} \sum_{n=-\infty}^{\infty} h(i, m) h(k, n) R_X(n, m). \qquad (3.59)$$

If the input process is wide-sense stationary and the linear system is time invariant, then the output process is also wide-sense stationary and has autocorrelation function

$$R_Y(k) = \sum_{j=-\infty}^{\infty} h(j) R_{Y,X}(k + j), \qquad (3.60)$$

or

$$R_Y(k) = \sum_{n=-\infty}^{\infty} \tilde{h}(n) R_{Y,X}(k - n), \qquad (3.61)$$

where $\tilde{h}(n) = h(-n)$ for each n and

$$R_{Y,X}(i) = \sum_{m=-\infty}^{\infty} h(m) R_X(i - m). \qquad (3.62)$$

That is,

$$R_Y = \tilde{h} * h * R_X, \tag{3.63}$$

where the convolutions are now *discrete* and h is the pulse response of the discrete-time linear time-invariant system. Equations (3.60) and (3.62) can be combined to give a single equation for R_Y in terms of R_X. This equation,

$$R_Y(k) = \sum_{j=-\infty}^{\infty} \sum_{m=-\infty}^{\infty} h(j) h(m) R_X(k + j - m), \tag{3.64}$$

is just the discrete-time counterpart to (3.49). However, for the evaluation of R_Y, it is easier in most cases to evaluate $f = \tilde{h} * h$ from

$$
\begin{aligned}
f(i) &= \sum_{n=-\infty}^{\infty} \tilde{h}(i - n) h(n) \\
&= \sum_{n=-\infty}^{\infty} h(n - i) h(n)
\end{aligned}
\tag{3.65}
$$

and then evaluate $R_Y = f * R_X$ from

$$R_Y(k) = \sum_{n=-\infty}^{\infty} f(k - n) R_X(n). \tag{3.66}$$

We have already given one exercise (Exercise 3-1) that illustrates the evaluation of the output autocorrelation function for a discrete-time linear system. Other examples are provided by the exercises that follow.

Exercise 3-11
Consider the discrete-time linear system described in Exercise 3-4. Let the input process $X(k)$ have autocorrelation function $R_X(k) = \sigma^2$ for $k = 0$ and $R_X(k) = 0$ for $k \neq 0$. Find the output autocorrelation function and the input–output crosscorrelation function.

Solution.   It is shown in the solution to Exercise 3-4 that the system is time invariant with impulse response

$$
h(n) = \begin{cases} 2^{-n}, & n \geq 0, \\ 0, & n < 0. \end{cases}
$$

Therefore, from (3.62),

$$
R_{Y,X}(i) = \sum_{m=0}^{\infty} 2^{-m} R_X(i - m) = \begin{cases} 2^{-i} \sigma^2, & i \geq 0, \\ 0, & i < 0. \end{cases}
$$

We can then use (3.60) to see that

$$
R_Y(k) = \sum_{j=0}^{\infty} 2^{-j} R_{Y,X}(k + j).
$$

For $k \geq 0$,

$$R_Y(k) = \sum_{j=0}^{\infty} 2^{-j} \, 2^{-(k+j)} \, \sigma^2 = \sigma^2 \, 2^{-k} \sum_{j=0}^{\infty} 4^{-j}$$
$$= \left(\tfrac{4}{3}\right) \sigma^2 \, 2^{-k},$$

and for $k < 0$,

$$R_Y(k) = R_Y(-k) = \left(\tfrac{4}{3}\right) \sigma^2 \, 2^{k}.$$

Therefore, for each k,

$$R_Y(k) = \left(\tfrac{4}{3}\right) \sigma^2 \, 2^{-|k|}.$$

An alternative approach is to first find $f = \tilde{h} * h$, and then determine $R_Y = f * R_X$. Using (3.65), we see that

$$f(i) = \sum_{n=0}^{\infty} h(n - i) \, 2^{-n} = \left(\tfrac{4}{3}\right) 2^{-|i|}.$$

Then, (3.66) gives

$$R_Y(k) = \sum_{n=-\infty}^{\infty} f(k - n) \, R_X(n) = f(k) \, R_X(0) = f(k) \, \sigma^2$$
$$= \left(\tfrac{4}{3}\right) \sigma^2 \, 2^{-|k|}.$$

Exercise 3-12

Consider a discrete-time system for which the output consists of differences between successive values of the input. That is, if x is the input signal, the output is

$$y(k) = x(k) - x(k - 1). \tag{3.67}$$

Suppose the input is a discrete-time, wide-sense stationary random process with autocorrelation function

$$R_X(n) = \sigma^2 \, e^{-\beta|n|}, \tag{3.68}$$

where $\beta > 0$. Find the autocorrelation function of the output.

Solution.   Two approaches are presented:

(a) For input process $X(k)$, the output process is given by

$$Y(k) = X(k) - X(k-1)$$

for all k. Therefore, from $R_Y(k, i) = E\{Y(k) Y(i)\}$, it follows that

$$
\begin{aligned}
R_Y(k, i) &= E\{[X(k) - X(k-1)][X(i) - X(i-1)]\} \\
&= R_X(k-i) - R_X(k-i+1) - R_X(k-1-i) + R_X(k-i).
\end{aligned}
$$

Thus, $R_Y(k, i)$ depends on $k - i$ only. That is,

$$
\begin{aligned}
R_Y(k, i) &= R_Y(k-i, 0) = R_Y(k-i) \\
&= 2R_X(k-i) - R_X(k-i+1) - R_X(k-i-1) \\
&= \sigma^2 \left\{ 2e^{-\beta|k-i|} - e^{-\beta|k-i+1|} - e^{-\beta|k-i-1|} \right\}.
\end{aligned}
$$

This equation can be simplified somewhat by writing it as

$$R_Y(n) = \sigma^2 \left\{ 2e^{-\beta|n|} - e^{-\beta|n+1|} - e^{-\beta|n-1|} \right\}. \tag{3.69}$$

Next, observe that, for $n > 0$,

$$|n+1| = |n| + 1$$

and

$$|n-1| = |n| - 1,$$

and for $n < 0$,

$$|n+1| = |n| - 1 \quad \text{and} \quad |n-1| = |n| + 1.$$

In either case,

$$e^{-\beta|n+1|} + e^{-\beta|n-1|} = e^{-\beta|n|} \left\{ e^{-\beta} + e^{\beta} \right\}$$

(for $n \neq 0$). Thus,

$$
R_Y(n) =
\begin{cases}
2\sigma^2\{1 - e^{-\beta}\}, & n = 0, \\
\sigma^2\{2 - (e^{-\beta} + e^{\beta})\} \exp\{-\beta|n|\}, & n \neq 0.
\end{cases}
$$

(b) For the *second* approach, we first observe that the system described by (3.67) has pulse response

$$
h(n) =
\begin{cases}
1, & n = 0, \\
-1, & n = 1, \\
0, & \text{otherwise.}
\end{cases}
$$

We then use (3.65) to conclude that $f = \tilde{h} * h$ is given by

$$f(i) = \begin{cases} 2, & i = 0, \\ -1, & |i| = 1, \\ 0, & \text{otherwise.} \end{cases}$$

Hence, (3.66) implies that

$$\begin{aligned} R_Y(k) &= \sum_{n=k-1}^{k+1} f(k-n)\, R_X(n) \\ &= -R_X(k-1) + 2R_X(k) - R_X(k+1). \end{aligned}$$

Using (3.68) to substitute into the preceding expression gives (3.69), which can then be simplified, as in part (a). ∎

Exercise 3-13
Show that the inequality

$$\sum_{i=-\infty}^{\infty} |h(k,i)|\, |\mu_X(i)| < \infty$$

(which is (3.38)) is satisfied whenever (3.9) and (3.22) hold. Thus, if the discrete-time linear system is stable and the second moment of the random process $X(k)$ is bounded, then the mean of the random process $Y(k)$ is finite.

Solution.   We use the fact that $|x| \le x^2 + 1$ for any real number x (see the solution to Exercise 2-6) to obtain

$$|\mu_X(i)| \le [\mu_X(i)]^2 + 1 \le \text{Var}\{X(i)\} + [\mu_X(i)]^2 + 1.$$

Since $\text{Var}\{X(i)\} + [\mu_X(i)]^2 = R_X(i,i)$, the preceding inequality and (3.22) imply that

$$|\mu_X(i)| \le R_X(i,i) + 1 \le r + 1. \tag{3.70}$$

Combining (3.70) and (3.9) we have

$$\sum_{i=-\infty}^{\infty} |h(k,i)|\, |\mu_X(i)| \le (r+1) \sum_{i=-\infty}^{\infty} |h(k,i)| < \infty,$$

which establishes the desired inequality and proves that the mean of $Y(k)$ is finite.

It should be clear at this point that the same arguments can be applied to continuous-time processes and systems to show that (3.6) and (3.26) imply (3.34), which in turn implies that the mean of the output of the continuous-time system is finite. ∎

On the basis of the results of the analysis of this section, we can state some general properties of *time-invariant* linear systems with inputs that are second-order random processes. These properties hold for both discrete- and continuous-time systems and processes. We assume, of course, that the system pulse or impulse response is known and satisfies conditions required to guarantee the output is a second-order random process (e.g., (3.21) or (3.25)). The properties are as follows:

1. *The output autocorrelation function can be determined from knowledge of the input autocorrelation function only.*

2. *The crosscorrelation function for the input and output processes can be determined from knowledge of the input autocorrelation function only.*

3. *If the input process is wide-sense stationary, the input and output processes are jointly wide-sense stationary; hence, the output process is wide-sense stationary.*

4. *If the input process is covariance stationary, the output process is also covariance stationary.*

Properties **1** and **2** are also true for time-varying linear systems, but properties **3** and **4** depend critically on the time invariance of the system.

3.5   Gaussian Random Processes in Linear Systems

It is pointed out in Chapter 2 that the mean and autocorrelation functions completely specify the finite-dimensional distributions for a *Gaussian* random process. (This is *not* true for general random processes.) Thus, if we know the mean and autocorrelation functions for a Gaussian random process $X(t)$, we can, in principle, compute the probabilities of any events involving a finite number of samples of $X(t)$. If $X(t)$ is a Gaussian random process, it suffices to know the mean and autocorrelation function for $X(t)$ in order to compute such probabilities as the following:

1. *the probability that $X(t_0)$ is negative;*

2. *the probability that $X(t_1) + X(t_2)$ is less than 2;*

3. *the probability that $[X(t_1) + X(t_2)]^2$ is between 10 and 20;*

4. *the probability that $X(t_1) + X(t_2) + \cdots + X(t_{100})$ is greater than some threshold value γ.*

All of these examples involve a finite number of random variables only; hence, the finite-dimensional distributions for the random process $X(t)$ provide enough information. The first example requires knowledge of the one-dimensional distribution only, and the second and third require knowledge of the two-dimensional distribution only; however, the fourth example requires knowledge of the 100-dimensional distribution.

Fortunately, because $X(t)$ is a Gaussian random process, all of these distributions can be determined from the mean and autocorrelation function for $X(t)$.

In many cases, we do not have to actually write down the higher dimensional distributions in order to solve the problem. If $X(t)$ is a Gaussian random process, then any linear combination of samples of $X(t)$ is a Gaussian random variable. That is, given the sampling times $t_1, t_2, \ldots, t_n$ and the coefficients $\alpha_1, \alpha_2, \ldots, \alpha_n$, the random variable

$$Z = \sum_{i=1}^{n} \alpha_i X(t_i) \tag{3.71}$$

is a Gaussian random variable. So, for instance,

$$P(Z \leq u) = \Phi\left(\frac{u - \mu}{\sigma}\right),$$

where $\mu = E\{Z\}$ and $\sigma = \sqrt{\mathrm{Var}\{Z\}}$. Thus, it suffices to find the mean and variance of Z by using (3.71). Of course, if we wish to find $P(Z \geq u)$, rather than $P(Z \leq u)$, we can use the fact that

$$P(Z \geq u) = 1 - P(Z < u) = 1 - [P(Z \leq u) - P(Z = u)].$$

Then, because the Gaussian distribution function is continuous, $P(Z = u) = 0$ for each u. It follows that if Z is Gaussian, then

$$P(Z \geq u) = 1 - P(Z \leq u) = 1 - \Phi\left(\frac{u - \mu}{\sigma}\right).$$

Similarly,

$$P(v \leq Z < u) = P(Z \leq u) - P(Z < v),$$

and because Z is Gaussian,

$$P(v \leq Z \leq u) = \Phi\left(\frac{u - \mu}{\sigma}\right) - \Phi\left(\frac{v - \mu}{\sigma}\right).$$

The general procedure just illustrated permits the computation of probabilities such as the foregoing **1–4**. However, with number **3**, do not make the unforgivable mistake of assuming that $[X(t_1) + X(t_2)]^2$ is Gaussian. Although it is true that $Z = X(t_1) + X(t_2)$ is Gaussian, *the square of a Gaussian random variable is definitely not Gaussian.* However, we can use the fact that for $0 < v < u$,

$$P(v \leq Z^2 \leq u) = P(-\sqrt{u} \leq Z \leq -\sqrt{v}) + P(\sqrt{v} \leq Z \leq \sqrt{u}).$$

If Z is a continuous random variable, this is valid for $v = 0$ as well. Next, the fact that Z is Gaussian can be used to express $P(v \leq Z^2 \leq u)$ in terms of the function Φ.

Examples such as number **4** in the list do not arise often in practice, and even when they do, it is rarely, if ever, necessary to determine the 100-dimensional density or distribution function in order to carry out the calculation. Normally, one should determine the mean μ and variance σ^2 of the sum $W = X(t_1) + X(t_2) + \cdots + X(t_{100})$ and then use the fact that a linear combination of jointly Gaussian random variables is Gaussian. Hence, W is a Gaussian random variable with distribution function

$$F_W(u) = P(W \le u) = \Phi\left(\frac{u - \mu}{\sigma}\right),$$

so

$$P(W > \gamma) = 1 - P(W \le \gamma) = 1 - F_W(\gamma) = 1 - \Phi\left(\frac{\gamma - \mu}{\sigma}\right).$$

Now, μ, the mean of the sum, is just the sum of the means, but, in general, the variance of the sum is not just the sum of the variances. (Why?) Nevertheless, σ^2 can be determined from the autocorrelation function of the random process $X(t)$ with the use of standard results on sums of random variables (e.g., see the solution to Exercise 3-1).

The preceding examples involve functions of a finite number of samples of the random process $X(t)$. In many engineering investigations, such as the study of noise in electronic systems, it is necessary to be able to work with functions of an infinite number of samples of $X(t)$. In fact, the number of samples may be uncountably infinite; for example, we may have to deal with random variables and random processes that depend on $X(t)$ as t ranges over the set of all real numbers. This is the case, for example, if the random process $Y(t)$, $-\infty < t < \infty$, is the output of a linear system with impulse response h when the input is the random process $X(t)$, $-\infty < t < \infty$—that is, when

$$Y(t) = \int_{-\infty}^{\infty} h(t, \tau) X(\tau) \, d\tau.$$

In general, questions concerning the random process $Y(t)$, or even a finite number of samples of $Y(t)$, cannot be answered if our knowledge of the input random process $X(t)$ is limited to its mean and autocorrelation functions. If $X(t)$ is not Gaussian, then, in general, knowledge of $\mu_X(t)$ and $R_X(t, s)$ for all t and s is not sufficient to determine the distribution of the input process $X(t)$, let alone the distribution of the output process. As we have already demonstrated in Section 3.4, at least the mean and autocorrelation functions for the output process $Y(t)$ can be determined, even if our knowledge of the input $X(t)$ is limited to its mean and autocorrelation functions. Specifically, the output mean can be determined from the input mean, and the output autocorrelation function can be determined from the input autocorrelation function.

For the important special case in which $X(t)$ is a Gaussian random process, we can determine a great deal more about the output than its mean and autocorrelation functions. Suppose $X(t)$, the input to a linear system, is a *Gaussian* random process. Given *only* the mean and autocorrelation functions for $X(t)$, can the finite-dimensional distributions for the output process $Y(t)$ be determined? Fortunately, the answer is that they can. This is because the class of Gaussian random processes is closed under linear operations: *A linear operation on a Gaussian process produces another Gaussian process.* Thus, if the input to a linear filter is a Gaussian random process $X(t)$ with mean $\mu_X(t)$ and autocorrelation $R_X(t, s)$, then the output is a Gaussian random process $Y(t)$ with mean $\mu_Y(t)$ and autocorrelation $R_Y(t, s)$. This information completely specifies the finite-dimensional distributions for the output process. One consequence is that $Y(t)$ is stationary whenever it is wide-sense stationary.

To summarize, for a linear system with a known impulse response (in continuous time) or pulse response (in discrete time), knowledge of the input mean and auto-correlation functions is sufficient to determine the finite-dimensional distributions for the output process *if* the input is a *Gaussian* random process. In principle, we can therefore compute the probabilities of any events involving a finite number of samples of $Y(t)$ from the mean and autocorrelation functions for $X(t)$ and the mathematical characterization of the linear system. In most situations, the easiest way to compute such probabilities is to first determine the mean and autocorrelation functions for $Y(t)$.

Recall from Section 3.4 that for continuous-time processes and systems, the output mean and autocorrelation functions are given by

$$\mu_Y(t) = \int_{-\infty}^{\infty} h(t, \tau) \, \mu_X(\tau) \, d\tau$$

and

$$R_Y(t, s) = \int_{-\infty}^{\infty} \int_{-\infty}^{\infty} h(s, \lambda) \, h(t, \tau) \, R_X(\tau, \lambda) \, d\tau \, d\lambda,$$

respectively. For discrete-time processes and systems, these functions are given by

$$\mu_Y(k) = \sum_{i=-\infty}^{\infty} h(k, i) \, \mu_X(i)$$

and

$$R_Y(k, i) = \sum_{m=-\infty}^{\infty} \sum_{n=-\infty}^{\infty} h(i, m) \, h(k, n) \, R_X(n, m).$$

All of these equations have simpler forms, as presented in Section 3.4, if the input random process is wide-sense stationary or the linear filter is time invariant (or both).

The very important fact that the output of a linear system is a Gaussian random process whenever the input random process is Gaussian is stated in nearly every engineering text on random processes. However, the proof of this result is rarely given in such texts. In fact, the proof is quite straightforward, although it does make use of certain convergence properties of sequences of random variables that have not been discussed in this book. However, it is instructive to at least go through an outline of the proof. In this outline, we will consider only discrete-time random processes and systems.

Recall from Section 3.3 that the output random process $Y(k)$ is related to the input random process by

$$Y_N(k) = \sum_{n=-N}^{N} h(k, n) \, X(n)$$

and

$$\lim_{N \to \infty} E\{[Y(k) - Y_N(k)]^2\} = 0.$$

That is, $Y(k)$ is a limit (in the mean-square sense) of a finite linear combination of the random variables $X(n)$, $n = 0, \pm 1, \pm 2, \ldots$, provided that this limit exists. To ensure that the mean-square limit does exist, we assume that

$$\sum_{i=-\infty}^{\infty} \sum_{n=-\infty}^{\infty} |h(k,n) h(k,i) R_X(n,i)| < \infty, \quad \text{for each } k.$$

In particular, this implies that the input process must be a second-order process (because it guarantees that $R_X(n,n) < \infty$ for each n).

To prove that the discrete-time process $Y(k)$ is a Gaussian process, we must show that for an arbitrary positive integer m and arbitrary integers $k_1, k_2, \ldots, k_m$, the random variable

$$Z = \sum_{j=1}^{m} a_j Y(k_j) \tag{3.72}$$

has a Gaussian distribution function for each choice of the coefficients $a_1, a_2, \ldots, a_m$. Accordingly, we note that for each j, $Y(k_j)$ is a mean-square limit of a sequence of random variables $Y_N(k_j)$; that is, $E\{[Y(k_j) - Y_N(k_j)]^2\} \to 0$ as $N \to \infty$, where

$$Y_N(k_j) = \sum_{n=-N}^{N} h(k_j, n) X(n), \tag{3.73}$$

as in (3.17). Analogously to (3.72), we define

$$Z_N = \sum_{j=1}^{m} a_j Y_N(k_j) \tag{3.74}$$

for each positive integer N. Combining (3.73) and (3.74), we find that

$$Z_N = \sum_{n=-N}^{N} \sum_{j=1}^{m} a_j h(k_j, n) X(n).$$

Since $X(k)$ is a Gaussian random process, this last equation implies that Z_N is a Gaussian random variable for each N.

Next, we note that because each term of the finite sum in (3.74) converges in a mean-square sense to the corresponding term of the sum in (3.72) as $N \to \infty$, Z_N must converge in a mean-square sense to Z as $N \to \infty$. Let G_N be the distribution function for Z_N. Mean-square convergence of Z_N to Z implies that the sequence of distribution functions G_N must converge to the distribution function F_Z for the random variable Z; that is, for each z,

$$\lim_{N \to \infty} G_N(z) = F_Z(z). \tag{3.75}$$

But Z_N is a Gaussian random variable, so

$$G_N(z) = \Phi\left(\frac{z - \mu_N}{\sigma_N}\right), \tag{3.76}$$

where $\mu_N = E\{Z_N\}$, $\sigma_N = \sqrt{\text{Var}\{Z_N\}}$, and Φ is the standard Gaussian distribution function.

Now, recall that

$$\lim_{N \to \infty} Z_N = Z \text{ (mean square)}$$

means that

$$\lim_{N \to \infty} E\{[Z_N - Z]^2\} = 0.$$

It should be intuitively clear (and it can be proven rigorously) that this implies that

$$\lim_{N \to \infty} E\{Z_N\} = E\{Z\}$$

and

$$\lim_{N \to \infty} \text{Var}\{Z_N\} = \text{Var}\{Z\}.$$

That is, if $\mu = E\{Z\}$ and $\sigma^2 = \text{Var}\{Z\}$, then

$$\lim_{N \to \infty} \mu_N = \mu$$

and

$$\lim_{N \to \infty} \sigma_N = \sigma.$$

The latter two results imply that for each z,

$$\lim_{N \to \infty} \frac{(z - \mu_N)}{\sigma_N} = \frac{(z - \mu)}{\sigma}.$$

Because Φ is a continuous function, it follows that

$$\lim_{N \to \infty} \Phi\left(\frac{z - \mu_N}{\sigma_N}\right) = \Phi\left[\lim_{N \to \infty}\left(\frac{z - \mu_N}{\sigma_N}\right)\right] = \Phi\left(\frac{z - \mu}{\sigma}\right). \tag{3.77}$$

Finally, we combine (3.76) and (3.77) to obtain

$$\lim_{N \to \infty} G_N(z) = \lim_{N \to \infty} \Phi\left(\frac{z - \mu_N}{\sigma_N}\right) = \Phi\left(\frac{z - \mu}{\sigma}\right).$$

In view of (3.75), it must be that

$$F_Z(z) = \Phi\left(\frac{z - \mu}{\sigma}\right).$$

That is, Z is Gaussian with mean μ and variance σ^2. This concludes the outline of the proof.

The foregoing result is one of the most important properties of Gaussian random processes. Stated concisely, this key property is as follows:

If the input to a linear system is a Gaussian random process, the output is also a Gaussian random process.

This statement is valid for both continuous-time and discrete-time linear systems and random processes. All that is required is that the output process be well defined (i.e., (3.21) or (3.25) must be satisfied). In fact, under the same conditions, the following property holds:

If the input to a linear system is a Gaussian random process, the input and output processes are jointly Gaussian.

The latter property means that for any $t_1, t_2, \ldots, t_n$, and any $m < n$, the collection $X(t_1), \ldots, X(t_m), Y(t_{m+1}), \ldots, Y(t_n)$ of random variables has an n-dimensional Gaussian density function.

In particular, for any t and any s, the joint density function for the random variables $X(t)$ and $Y(s)$ is the bivariate Gaussian density with parameters

$$\mu_1 = E\{X(t)\}, \quad \mu_2 = E\{Y(s)\},$$

$$\sigma_1^2 = \text{Var}\{X(t)\}, \quad \sigma_2^2 = \text{Var}\{Y(s)\},$$

and

$$\rho = \frac{\text{Cov}\{X(t), Y(s)\}}{\sigma_1 \sigma_2} = \frac{C_{X,Y}(t, s)}{\sqrt{C_X(t, t) \, C_Y(s, s)}}.$$

All of these parameters can be evaluated by using the methods of Section 3.4. Of course, the marginal density of $Y(s)$ is the univariate Gaussian density with mean $\mu_Y(s)$ and variance $C_Y(s, s)$. Consequently, the univariate (i.e., one-dimensional) distribution function for $Y(s)$ is given by

$$F_{Y,1}(y; s) = \Phi\left(\frac{y - \mu_Y(s)}{\sqrt{C_Y(s, s)}}\right)$$

for all y. A typical application is given in the next exercise.

Exercise 3-14

In Exercise 3-9, suppose that the input process is Gaussian and that $\tau_0 = 3t_0$. Find the probability that the output at time t exceeds the threshold value $\gamma = 2$.

Solution. Since the input is Gaussian, the output is Gaussian. The mean of the input is zero, so $Y(t)$ has mean $\mu_Y(t) = 0$. According to the solution to Exercise 3-9,

$$\text{Var}\{Y(t)\} = (\sigma_Y)^2 = t_0^2 (3 \tau_0)^{-1} [3 \tau_0 - t_0].$$

Notice that neither the mean nor the variance of $Y(t)$ depends on t. In fact, the process $Y(t)$ is wide-sense stationary, since the input process is wide-sense stationary and the linear system is time invariant. Since $Y(t)$ is also Gaussian, it is (strictly) stationary. As a result, $P[Y(t) > \gamma]$ does not depend on t. In fact, we can conclude that

$$P[Y(t) > \gamma] = 1 - \Phi(\gamma/\sigma_Y).$$

For $\tau_0 = 3t_0$, $\sigma_Y = 2\sqrt{2}\,t_0/3$. For $\gamma = 2$,

$$P[Y(t) > \gamma] = P[Y(t) > 2] = 1 - \Phi[3/(\sqrt{2}\,t_0)].$$

For instance, if $t_0 = \sqrt{2}$, then $P[Y(t) > 2] = 1 - \Phi(1.5) \approx 1 - (0.9332) = 0.0668.$

∎

An important situation that arises in practice, but is not covered by the preceding results, is when the continuous-time input process is *not* a second-order random process. Suppose a continuous-time linear system has an impulse response that satisfies

$$\int_{-\infty}^{\infty} [h(t, \lambda)]^2 \, d\lambda < \infty, \quad \text{for each } t. \tag{3.78}$$

It is known that if the input to this system is thermal noise, the output is a *Gaussian* random process with zero mean and autocorrelation function

$$R_Y(t, s) = \frac{N_0}{2} \int_{-\infty}^{\infty} h(s, \lambda) \, h(t, \lambda) \, d\lambda, \tag{3.79}$$

where $N_0/2$ is called the *two-sided power spectral density* of the thermal noise process. In terms of the parameters of Example 2-2, $N_0 = 4kTR$, which is usually expressed in units of watts per hertz (i.e., joules), treating N_0 as a power density, where the power is measured on a one-ohm basis.

Notice that (3.78) is just the necessary and sufficient condition for $R_Y(t, s) < \infty$ for all t and s. This fact follows from (3.79) and (2.17). Notice also that (3.79) can be derived from (3.42) if we set

$$R_X(\tau, \lambda) = \frac{N_0}{2} \delta(\tau - \lambda).$$

To show this, we use the defining property of the Dirac delta function: For any "well-behaved" function g,

$$g(\lambda) = \int_{-\infty}^{\infty} g(\tau) \, \delta(\tau - \lambda) \, d\tau, \quad \text{for each } \lambda.$$

As pointed out in Chapter 2, a white-noise process (Gaussian or not) is not a second-order random process. Hence, we must be careful in working with white noise in linear systems. Just as with the delta function in deterministic signal analysis, the autocorrelation function for white noise causes no real difficulties if it appears in the integrand of a mathematical expression.

Because the second moment of white noise is infinite, we have no guarantee that the mean exists. Hence, we do not define the mean of white noise, but we acknowledge that white noise produces a zero-mean output when it is the input of a linear system that satisfies (3.78). Hence, in such a linear system, white noise *behaves* as though it were a *zero-mean* random process. Notice that the autocorrelation function for the white-noise process is a function of the time difference only. As a result, we usually write the autocorrelation function as

$$R_X(\tau) = \frac{N_0}{2} \delta(\tau), \quad -\infty < \tau < \infty.$$

Thus, we can think of white noise as a wide-sense stationary random process, even though it is not a second-order random process.

The foregoing discussion motivates the following definition:

> White Gaussian noise is a random process which has the property that when it is the input to a linear system that satisfies (3.78), the output is a zero-mean Gaussian process with autocorrelation function given by (3.79).

As a practical matter, it is more convenient to think of white Gaussian noise as a random process with the following characteristics:

> For engineering applications, a white Gaussian noise process $X(t)$ can be viewed as a stationary Gaussian random process with autocorrelation function given by

$$R_X(\tau) = \frac{N_0}{2} \delta(\tau).$$

To illustrate the kind of manipulations that are valid in dealing with white noise, we derive the crosscorrelation function for a continuous-time linear system that has a white-noise input. First, in (3.40), let

$$R_X(\tau, s) = \frac{N_0}{2} \delta(\tau - s)$$

to obtain

$$R_{Y,X}(t, s) = \frac{N_0}{2} \int_{-\infty}^{\infty} h(t, \tau) \delta(\tau - s) \, d\tau.$$

Next, we use the defining property of the delta function to conclude that

$$R_{Y,X}(t, s) = \frac{N_0}{2} h(t, s).$$

This result is the basis for an interesting approach to the experimental determination of the impulse response of a linear system (i.e., system identification), a topic that is discussed in Section 3.7.

The preceding results all simplify somewhat for time-invariant systems. In particular, (3.78) and (3.79) become

$$\int_{-\infty}^{\infty} [h(t)]^2 \, dt < \infty$$

and (for $u = \lambda - s$)

$$
\begin{aligned}
R_Y(t, s) &= \frac{N_0}{2} \int_{-\infty}^{\infty} h(-u)\, h(t - s - u)\, du \\
&= \frac{N_0}{2} \int_{-\infty}^{\infty} \tilde{h}(u)\, h(t - s - u)\, du,
\end{aligned}
$$

where $\tilde{h}(t) = h(-t)$. Because $Y(t)$ has zero mean, it follows that $Y(t)$ is wide-sense stationary and has autocorrelation function

$$
\begin{aligned}
R_Y(\tau) &= \frac{N_0}{2} \int_{-\infty}^{\infty} h(-u)\, h(\tau - u)\, du \\
&= \frac{N_0}{2} \int_{-\infty}^{\infty} \tilde{h}(u)\, h(\tau - u)\, du. \quad\quad (3.80)
\end{aligned}
$$

In other words, $R_Y(\tau) = (N_0/2)\, f(\tau)$, where $f = \tilde{h} * h$, a relationship that can be derived directly from (3.54). So if the input to a time-invariant linear system with impulse response h is white Gaussian noise, then the output $Y(t)$ is a zero-mean Gaussian random process with autocorrelation function

$$
R_Y(\tau) = \frac{N_0}{2}\, f(\tau),
$$

where $f = \tilde{h} * h$. The actual calculation of $R_Y(\tau)$ is usually easier if we employ the fact that $R_Y(\tau) = R_Y(-\tau)$ to observe from (3.80) that

$$
R_Y(\tau) = \frac{N_0}{2} \int_{-\infty}^{\infty} h(-u)\, h(-\tau - u)\, du = \frac{N_0}{2} \int_{-\infty}^{\infty} h(t)\, h(t - \tau)\, dt. \quad\quad (3.81)
$$

Exercise 3-15
Suppose that white Gaussian noise with spectral density $N_0/2$ is the input to a time-invariant linear filter with impulse response $2 \exp(-4t)\, u(t)$, where $u(t)$ is the unit step function. The output process, denoted by $Y(t)$, is sampled at times $t = 0$ and $t = 1$. The samples are added to give the random variable $W = Y(0) + Y(1)$. Find the probability that the random variable W is greater than 5.

Solution.   First, observe that $Y(t)$ is a Gaussian, wide-sense stationary random process, because the input is white Gaussian noise. Therefore, the distribution function for W is $F_W(u) = \Phi[(u - \mu)/\sigma]$, where μ is the mean and σ^2 is the variance of W. The probability that we are asked to determine is therefore given by $P(W > 5) = 1 - \Phi[(5 - \mu)/\sigma]$. Next notice that $\mu = 0$, because $Y(t)$ has zero mean. It follows from this that σ^2 is just the second moment of W, which is given by $E\{W^2\} = 2R_Y(0) + 2R_Y(1)$. The solution to Exercise 3-7 ($\beta = 2$ and $\alpha = 4$) shows that $R_Y(0) = N_0/4$ and $R_Y(1) = N_0\, e^{-4}/4$; therefore, $\sigma^2 = E\{W^2\} = N_0(1 + e^{-4})/2$. Combining all of these facts, we see that $P(W > 5) = 1 - \Phi\left[5\sqrt{2/\{N_0[1 + \exp(-4)]\}}\right]$. ∎

The next example further illustrates the analysis of white noise in linear systems, and it also introduces a new random process.

Example 3-5   The Wiener–Lévy Process

Suppose the input to the system described in the solution to Exercise 3-10 is white Gaussian noise. That is, the random process $Y(t)$, $t \geq 0$, is defined by

$$Y(t) = \int_0^t X(\tau) \, d\tau,$$

where $X(t)$ is a white Gaussian noise process. In the alternative approach given as part of the solution to that exercise, we found that

$$R_Y(t, s) = \int_0^t \int_0^s R_X(\tau - \lambda) \, d\lambda \, d\tau.$$

Thus, if $X(t)$ is white noise with spectral density $N_0/2$, then

$$R_Y(t, s) = \frac{N_0}{2} \int_0^t \int_0^s \delta(\tau - \lambda) \, d\lambda \, d\tau.$$

If $0 \leq t \leq s$, then, for each τ in the range $0 \leq \tau \leq t$, there is a value of λ in the range $0 \leq \lambda \leq s$ for which $\lambda = \tau$. It follows that,

$$\int_0^s \delta(\tau - \lambda) \, d\lambda = 1,$$

and thus,

$$R_Y(t, s) = \frac{N_0}{2} \int_0^t 1 \, d\tau = \frac{N_0 t}{2}.$$

On the other hand, for $0 \leq s < t$, there are values of τ in $[0, t]$ for which there is no λ in $[0, s]$ such that $\lambda = \tau$ (namely, those τ's in the interval (s, t)). Thus, for $t > s$, we consider the reverse order of integration to obtain

$$R_Y(t, s) = \frac{N_0}{2} \int_0^s \int_0^t \delta(\tau - \lambda) \, d\tau \, d\lambda,$$

to which the argument in the preceding paragraph applies. (Just reverse the roles of t and s.) Hence, for $t > s$,

$$\int_0^t \delta(\tau - \lambda) \, d\tau = 1$$

and

$$R_Y(t, s) = \frac{N_0}{2} \int_0^s 1 \, d\lambda = \frac{N_0 s}{2}.$$

The conclusion is that $Y(t)$, $t \geq 0$, is a zero-mean Gaussian random process with autocorrelation function

$$R_Y(t, s) = \begin{cases} N_0 t/2, & t \leq s, \\ N_0 s/2, & t > s. \end{cases}$$

This autocorrelation function can be written in the more compact form

$$R_Y(t, s) = \frac{N_0}{2} \min(t, s)$$

for all $t \geq 0$ and $s \geq 0$. Notice that $\min(t, s)$ does not depend on $t - s$ only. Notice also that

$$\text{Var}\{Y(t)\} = R_Y(t, t) = \frac{N_0 t}{2},$$

so that the variance of $Y(t)$ depends on t. The random process $Y(t)$, $t \geq 0$, obtained in this manner is the *Wiener–Lévy process* (often called the Brownian motion process). It is a *nonstationary*, second-order Gaussian process.   ■

3.6   Determination of Correlation Functions

In all of the preceding examples and exercises that involve either autocorrelation functions or crosscorrelation functions, we have assumed that the autocorrelation function of at least one of the random processes (e.g., the input process) is known. If $X(t)$ is the random process with known autocorrelation function, the problem is to find the autocorrelation function for a related random process $Y(t)$ or to find the crosscorrelation function for the two random processes $X(t)$ and $Y(t)$. Also, for problems in which $X(t)$ and $Y(t)$ are the system input and output, respectively, we have assumed that the impulse response of the system is known.

Actually, these assumptions are quite often valid in practice. For instance, it is frequently true in a satellite communication system that the input at some point in the system is white noise (described in Section 2.7), which implies that the autocorrelation function of this process is a delta function. The impulse responses for the linear filters in such a system are known quite accurately in practice, so autocorrelation functions and crosscorrelation functions can be determined for the processes at the outputs of these filters.

However, for certain engineering design and analysis problems, such correlation functions may be unknown and not directly computable (by using only the methods of this chapter) from the information that is known. If this is the case, the first step in the solution to the problem is to determine the correlation functions for the processes of interest. Usually, this is accomplished in one of two ways: If enough information is known about a random process, its autocorrelation function can be determined analytically; otherwise, the autocorrelation function is determined experimentally. Each of these approaches is illustrated in this section. First, we give an example in which the analytical approach is used, and then we describe a general technique for the experimental determination of correlation functions for wide-sense stationary processes.

3.6.1   The Autocorrelation Function for a Poisson Process

Consider for the moment the counting process described in Example 2-3. Suppose we are modeling a particular random process as a counting process for which we require the number of events that occur in one time interval to be independent of the number of events that occur in all other nonoverlapping time intervals. For instance, in most situations, the number of phone calls received at a switchboard on Wednesday can be modeled as a random variable that is independent of the number of phone calls received on Tuesday. Secondly, we require that the probability distribution for the number of events occurring in the time interval $(t, t + \tau)$ be the same for all $t \geq 0$. That is, the probability $p_n(\tau)$ of exactly n events occurring in a time interval of length τ is the same, regardless of the time at which the interval starts.

Mathematically, we describe such a process as follows: Let $t_0 > 0$ be arbitrary, and let $t_1, t_2, \ldots, t_{2K}$ be such that $0 \leq t_i \leq t_j$ whenever $1 \leq i \leq j \leq 2K$. Consider the random vector $\mathbf{Z} = (Z_1, Z_2, \ldots, Z_K)$ for which $Z_k = X(t_{2k} + t_0) - X(t_{2k-1} + t_0)$ for $1 \leq k \leq K$. If $X(t)$ is the counting process just described, then the components $Z_1, Z_2, \ldots, Z_K$ are mutually independent, and the distribution function for $\mathbf{Z}$ does not depend on t_0.

For further consideration of this random process, it is helpful to rely on some standard terminology. First, for $s < t$, the quantity $X(t) - X(s)$ is called an *increment* of the random process $X(t)$. If, for each positive integer K and each choice of $t_0, t_1, \ldots, t_{2K}$, the distribution function for the random vector $\mathbf{Z}$ does not depend on t_0, the process $X(t)$ is said to have *stationary increments*. If, for each choice of K and $t_0, t_1, \ldots, t_{2K}$, the random variables $Z_1, Z_2, \ldots, Z_K$ are mutually independent, the process $X(t)$ is said to have *independent increments*. Notice that if $X(t)$ has independent increments, then, for $0 < s < t$, the random variables $Z_1 = X(s) - X(0)$ and $Z_2 = X(t) - X(s)$ are independent, which implies that they are also *uncorrelated*. Thus,

$$
\begin{aligned}
E\{Z_1 Z_2\} &= E\{[X(s) - X(0)][X(t) - X(s)]\} \\
&= E\{X(s) - X(0)\} E\{X(t) - X(s)\},
\end{aligned}
\tag{3.82}
$$

a fact that will be useful later. A random process with independent increments also has *uncorrelated increments*, but, in general, the reverse implication is not true.

Recall that in Example 2-3, we require that all jumps or steps in the process must be of size one; that is, no more than one event can take place at a time. Stated more precisely, this requirement is that the probability of two or more events occurring at exactly the same time is zero. Whenever we use the term *counting process* for a random process $X(t)$, we assume that $X(0) = 0$, $X(t) \leq X(t + \tau)$ for all $t \geq 0$ and all $\tau \geq 0$, and at most one event can take place at a time. These assumptions are made in Example 2-3.

It is known that for any counting process that has stationary, independent increments, there is a constant $\nu > 0$ for which the probability of exactly n events occurring in a time interval of length τ is given by

$$
p_n(\tau) = (\nu \tau)^n \, e^{-\nu \tau} / n!
$$

for all $\tau \geq 0$ and all nonnegative integers n. In other words, for each $t \geq 0$, $X(t)$ has the Poisson distribution with parameter $\lambda = \nu t$. The process $X(t)$ is called a *Poisson process*.

Our real interest at the moment is in the autocorrelation function for such a process. First, recall that a random variable that has the Poisson distribution with parameter λ

has mean λ and variance λ. Thus, if $X(t)$ is a Poisson process, then

$$E\{X(t)\} = \nu t$$

and

$$\mathrm{Var}\{X(t)\} = \nu t$$

for any $t > 0$. These relationships imply that

$$E\{[X(t)]^2\} = \nu t + (\nu t)^2 = \nu t(1 + \nu t). \tag{3.83}$$

Next, we assume that $0 < t_1 < t_2$ and evaluate

$$\begin{aligned}
R_X(t_1, t_2) &= E\{X(t_1) X(t_2)\} \\
&= E\{X(t_1) [X(t_2) - X(t_1) + X(t_1)]\} \\
&= E\{X(t_1) [X(t_2) - X(t_1)]\} + E\{[X(t_1)]^2\}. \tag{3.84}
\end{aligned}$$

Now, from (3.83), we have

$$E\{[X(t_1)]^2\} = \nu t_1(1 + \nu t_1),$$

and because $X(0) = 0$,

$$E\{X(t_1)[X(t_2) - X(t_1)]\} = E\{[X(t_1) - X(0)][X(t_2) - X(t_1)]\}.$$

But the process $X(t)$ has independent increments, so, from (3.82), we have

$$E\{X(t_1)[X(t_2) - X(t_1)]\} = E\{X(t_1)\}E\{X(t_2) - X(t_1)\} = \nu t_1(\nu t_2 - \nu t_1),$$

where we have used the fact that $E\{X(t_i)\} = \nu t_i$. Consequently, (3.84) implies that

$$R_X(t_1, t_2) = \nu t_1(\nu t_2 - \nu t_1) + \nu t_1(1 + \nu t_1) = \nu t_1(1 + \nu t_2).$$

Similarly (just interchange the roles of t_1 and t_2), we find that for $0 < t_2 < t_1$,

$$R_X(t_1, t_2) = \nu t_2(1 + \nu t_1).$$

These two cases can be combined to give

$$R_X(t_1, t_2) = \nu \min(t_1, t_2)[1 + \nu \max(t_1, t_2)], \tag{3.85}$$

which is valid for all $t_1 \geq 0$ and $t_2 \geq 0$. (Observe that (3.83) takes care of $t_1 = t_2$.) For a process with nonzero mean, such as the Poisson process, it is often easier to work with the autocovariance function $C_X(t_1, t_2)$. From (3.85), it is easy to show that

$$C_X(t_1, t_2) = \nu \min(t_1, t_2).$$

This relationship follows from $E\{X(t_i)\} = \nu t_i$,

$$R_X(t_1, t_2) = C_X(t_1, t_2) + E\{X(t_1)\} E\{X(t_2)\},$$

and the fact that

$$\min(t_1, t_2) \max(t_1, t_2) = t_1 t_2.$$

Notice that the Poisson process has the same autocovariance function as the Wiener–Lévy process.

The main point is that the autocorrelation and autocovariance functions are obtained *analytically* on the basis of some physical properties of the process. In fact, many of the key properties that we used (e.g., stationarity and independence) are *qualitative* rather than *quantitative*. This is in contrast to the experimental approach described next, wherein we seek to obtain directly some quantitative information about the correlation function (e.g., the value of $R_X(2,0)$). The analytical approach, whenever it can be employed, typically gives solutions to a wider class of problems. In the foregoing analysis, we obtained the autocorrelation function for the class of all Poisson processes. Furthermore, along the way, we established some results that are valid for any independent increment process (e.g., (3.82)). In fact, a review of the derivation will reveal that we really used only the property that the increments are *uncorrelated*, a property which is considerably weaker than the independence of the increments.

It is not correct to infer from the preceding discussion that experimental methods are without merit or that they are unnecessary for engineering problems involving random processes. Experimental results are extremely useful in many such problems. Even in the example of the Poisson process, notice that we have not yet completely solved the problem, because our solution is in terms of the parameter ν. In practice, experimentation may be needed to determine the correct value of ν for a given application. However, it is generally more efficient to estimate only a single parameter than to attempt to estimate $R_X(t_1, t_2)$ for each different value of t_1 and t_2. A simple estimate for ν is $\hat{\nu} = x(T_0)/T_0$, which is based on observing the sample function $x(t)$, $0 \leq t \leq T_0$, from the process $X(t)$. Notice that this is just the average number of jumps per unit time in the sample function $x(t)$.

3.6.2   Experimental Determination of Means and Correlation Functions for Ergodic Random Processes

In this section, we restrict attention to random processes that are *ergodic*, which, for our purposes, means that almost all sample functions for the random process are typical of the entire process. For such processes, certain time averages converge to probabilistic averages as the time interval for the averaging becomes large. Actually, in what follows, we really require only a considerably weaker property, which is often referred to as *weak ergodicity*. A stationary Gaussian random process, for example, is weakly ergodic if its autocovariance function is absolutely integrable. In many situations, even if we do not know the autocovariance function precisely, we may know that it is absolutely integrable (e.g., it may be enough to have a bound on the autocovariance function).

Suppose we can devise an experiment that allows us to observe a sample function of an ergodic, wide-sense stationary, random process $X(t)$ for an arbitrarily long period of time. On the basis of this observation, we can estimate the autocorrelation function of $X(t)$ with any specified degree of accuracy. More generally, given sample functions

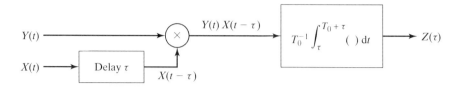

Figure 3-8: Estimating a crosscorrelation function.

from a jointly ergodic, jointly wide-sense stationary pair of processes $X(t)$ and $Y(t)$, we can estimate the crosscorrelation function of the two processes. Since the latter problem includes the former as a special case (simply let $X(t) = Y(t)$), we examine the estimation of the crosscorrelation function.

Consider the system shown in Figure 3-8. The output of the integrator is given by

$$Z(\tau) = T_0^{-1} \int_\tau^{T_0+\tau} Y(t) \, X(t-\tau) \, dt = T_0^{-1} \int_0^{T_0} Y(t+\tau) \, X(t) \, dt, \qquad (3.86)$$

where τ is arbitrary, but held constant for the moment. For sufficiently large values of T_0, $Z(\tau)$ is a good approximation to $R_{Y,X}(\tau)$ for most of the jointly wide-sense stationary random processes that are encountered in engineering (i.e., those which have the joint ergodicity property). Notice, for instance, that

$$\begin{aligned}
E\{Z(\tau)\} &= T_0^{-1} \int_0^{T_0} E\{Y(t+\tau)\, X(t)\} \, dt \\
&= R_{Y,X}(\tau)
\end{aligned}$$

for any $T_0 > 0$.

A commonly used measure of accuracy for estimating $R_{Y,X}(\tau)$ by $Z(\tau)$ is the mean-squared error $E\{[Z(\tau) - R_{Y,X}(\tau)]^2\}$. For ergodic processes, the mean-squared error will converge to zero as $T_0 \to \infty$. In particular, the mean-squared error will converge to zero for any stationary jointly Gaussian random processes $X(t)$ and $Y(t)$ for which

$$\int_{-\infty}^\infty |C_X(\tau)| \, d\tau < \infty \qquad (3.87)$$

and

$$\int_{-\infty}^\infty |C_Y(\tau)| \, d\tau < \infty. \qquad (3.88)$$

A related, but somewhat simpler, problem is to find the *mean* of a wide-sense stationary random process $X(t)$. In this case, we simply use the integrator of Figure 3-8 with $X(t)$ alone as its input. The output is the random variable

$$W = T_0^{-1} \int_0^{T_0} X(t) \, dt,$$

which has the property

$$E\{W\} = E\{X(t)\} = \mu_X.$$

The mean-squared error for this estimate depends on the value of T_0 and is given by

$$
\begin{aligned}
\varepsilon(T_0) &= E\{[W - \mu_X]^2\} \\
&= E\left\{\left[T_0^{-1}\int_0^{T_0} X(t)\,dt - \mu_X\right]^2\right\} \\
&= E\left\{\left[T_0^{-1}\int_0^{T_0} (X(t) - \mu_X)\,dt\right]^2\right\} \\
&= E\left\{\left[T_0^{-1}\int_0^{T_0} Y(t)\,dt\right]^2\right\},
\end{aligned}
$$

where $Y(t) = X(t) - \mu_X$. The problem of finding the second moment of the integral of a random process is a problem that we have already solved. In the solution to Exercise 3-10, we found that

$$
\begin{aligned}
E\left\{\left[\int_0^{T_0} Y(t)\,dt\right]^2\right\} &= \int_0^{T_0}\int_0^{T_0} R_Y(t-s)\,dt\,ds \\
&= 2\int_0^{T_0} (T_0 - \lambda)\,R_Y(\lambda)\,d\lambda.
\end{aligned}
$$

(See (3.56) for instance.) Next, we notice that $R_Y(\lambda) = C_X(\lambda)$ and that, for $0 \le \lambda \le T_0$,

$$0 \le (T_0 - \lambda) \le T_0.$$

Since $E\{[W - \mu_X]^2\} \ge 0$, we have

$$
\begin{aligned}
0 \le \varepsilon(T_0) &= 2T_0^{-2}\int_0^{T_0} (T_0 - \lambda)\,C_X(\lambda)\,d\lambda \\
&\le 2T_0^{-2}\int_0^{T_0} (T_0 - \lambda)|C_X(\lambda)|\,d\lambda \\
&\le 2T_0^{-1}\int_0^{T_0} |C_X(\lambda)|\,d\lambda.
\end{aligned}
\tag{3.89}
$$

If the process $X(t)$ satisfies (3.87), then

$$
\begin{aligned}
\lim_{T_0\to\infty} \varepsilon(T_0) &= \left(\lim_{T_0\to\infty} T_0^{-1}\right)\left(\lim_{T_0\to\infty} 2\int_0^{T_0} |C_X(\lambda)|\,d\lambda\right) \\
&= \left(\lim_{T_0\to\infty} T_0^{-1}\right)\int_{-\infty}^{\infty} |C_X(\lambda)|\,d\lambda = 0.
\end{aligned}
$$

Hence, (3.87) and (3.89) imply that the mean-squared error converges to zero as $T_0 \to \infty$. Furthermore, for any finite T_0, (3.89) provides the following upper bound on the mean-squared error:

$$\varepsilon(T_0) \le 2T_0^{-1} \int_0^{T_0} |C_X(\lambda)| \, d\lambda$$

$$\le T_0^{-1} \int_{-\infty}^{\infty} |C_X(\lambda)| \, d\lambda = T_0^{-1} \Gamma_X.$$

In the rightmost term,

$$\Gamma_X = \int_{-\infty}^{\infty} |C_X(\lambda)| \, d\lambda < \infty.$$

Hence, if $C_X(\tau)$ is known, then Γ_X can be determined, and a prespecified mean-square error $\varepsilon_{\mathrm{ms}}$ can be guaranteed by selecting T_0 to satisfy $T_0 \ge \Gamma_X/\varepsilon_{\mathrm{ms}}$.

3.7   Applications

The material presented in this chapter is applicable to such fields as communications, control, radar, signal processing, and system design. Unfortunately, for many of the key applications that arise in these disciplines, further, more specialized training in the particular field is required in order to understand and appreciate the applications. The next three examples provide a limited insight based on the results obtained thus far.

Example 3-6   System Identification
Suppose that the impulse response of a continuous-time, time-invariant linear system is unknown. Theoretically, we could use a delta function as an input and observe the response. Letting $x(\tau) = \delta(\tau)$ in (3.1), we find that $y(t) = h(t)$. The problem is that delta functions are impossible to generate in the laboratory (as are ideal step functions and related signals), and even if one could be generated, its effect on the system would be disastrous. An alternative approach is to let the input to the system be white Gaussian noise (see Section 3.5) or at least noise whose spectrum is flat over a band wider than the bandwidth of the system we are trying to identify; usually, a rough estimate of the system's bandwidth will suffice. White Gaussian noise in the form of thermal noise is relatively easy to generate: Any resistive component generates thermal noise as a result of the random motion of its conduction electrons. (See Example 2-2.)

   If the input to the system is white noise, the autocorrelation function (not the signal itself) is a delta function. Thus, the same effect as that obtained by a delta-function input can be produced without actually generating a delta-function signal. To determine the impulse response h in this manner, we need to know only the crosscorrelation function between the input and output. This fact follows from (3.48) with $R_X(\tau) = \delta(\tau)$, which implies that $R_{Y,X}(\tau) = h(\tau)$. But $R_{Y,X}(\tau)$ can be determined as described in Section 3.6.2. Thus, $h(\tau)$ can be determined by the system shown in Figure 3-8 if $X(t)$ is the input to the linear time-invariant system and $Y(t)$ is the output. ∎

Example 3-7   Prediction of Gaussian Processes

Consider the problem of predicting the value of a discrete-time Gaussian random process $X(t)$ at some *future* time $t = k$, based on the observation of the value of $X(t)$ at the *present* time $t = m$. (Clearly, $m < k$.) Suppose we wish to require the prediction $\hat{X}(k)$ to have the minimum possible mean-squared error

$$\varepsilon_{\mathrm{ms}} = E\{[X(k) - \hat{X}(k)]^2\}$$

for any prediction that is based only upon the observation of $X(m)$. The optimum predictor turns out to be the conditional expected value of $X(k)$ given $X(m)$; that is,

$$\hat{X}(k) = E\{X(k)|X(m)\}.$$

This is not surprising from an intuitive point of view, since we are simply predicting that $X(k)$ will take on its "average" value, and the "average" should be based on all of the information that is available (namely, the value of $X(m)$). Formally, this "average" is the conditional expectation given the value of $X(m)$.

The conditional expectation is actually the optimum predictor even if the random process is not Gaussian. However, when the process is Gaussian, the prediction $\hat{X}(k)$ has a very simple form. Recall that if Y_1 and Y_2 are jointly Gaussian with means μ_1 and μ_2, variances σ_1^2 and σ_2^2, and correlation coefficient ρ, then

$$E\{Y_1|Y_2\} = \mu_1 + \rho\,\sigma_1\,\sigma_2^{-1}(Y_2 - \mu_2).$$

Since the random process is Gaussian, $X(m)$ and $X(k)$ are jointly Gaussian, so that

$$\begin{aligned}
\hat{X}(k) &= E\{X(k)|X(m)\} \\
&= \mu_X(k) + \rho_X(k, m)\,\sigma_X(k)[\sigma_X(m)]^{-1}[X(m) - \mu_X(m)], \qquad (3.90)
\end{aligned}$$

where

$$\rho_X(k, m) = [\sigma_X(k)\,\sigma_X(m)]^{-1}C_X(k, m).$$

That is, given that we observe $X(m) = x$, the minimum mean-squared-error (MMSE) prediction of $X(k)$ is

$$\begin{aligned}
\hat{X}(k) &= E\{X(k)|X(m) = x\} \\
&= \mu_X(k) + [\mathrm{Var}\{X(m)\}]^{-1}C_X(k, m)[x - \mu_X(m)] \\
&= [\mathrm{Var}\{X(m)\}]^{-1}C_X(k, m)\,x + \{\mu_X(k) - \mu_X(m)[\mathrm{Var}\{X(m)\}]^{-1}C_X(k, m)\}.
\end{aligned}$$

$$(3.91)$$

If the random process is Gaussian *and stationary*, the MMSE prediction based on the observation $X(m) = x$ is given by the simpler expression

$$\begin{aligned}
\hat{X}(k) &= \mu_X + [\sigma_X^2]^{-1}C_X(k - m)[x - \mu_X] \\
&= \mu_X\{1 - \sigma_X^{-2}C_X(k - m)\} + \sigma_X^{-2}C_X(k - m)\,x.
\end{aligned}$$

Finally, if the stationary Gaussian random process $X(t)$ has *zero mean*, then

$$\hat{X}(k) = \sigma_X^{-2} R_X(k-m)\, x.$$

Even if the random process is not Gaussian, the predictor given by (3.90) is the MMSE *linear* predictor. (Strictly speaking, what we mean is *affine*, which is "linear plus a constant," but the convention is to call this a linear predictor.) The precise statement of this fact is as follows: Of all predictors of the form

$$\hat{X}(k) = a\, X(m) + b,$$

where a and b are constants, the one that minimizes the mean-squared error

$$\varepsilon_{ms} = E\{[X(k) - \hat{X}(k)]^2\} = E\{[X(k) - a\, X(m) - b]^2\}$$

is obtained by setting $a = C_X(k, m)/\mathrm{Var}\{X(m)\}$ and

$$b = \mu_X(k) - \mu_X(m) \left\{ \frac{C_X(k, m)}{\mathrm{Var}\{X(m)\}} \right\} = \mu_X(k) - a\, \mu_X(m),$$

as in (3.91).   ■

Example 3-8   Digital Communication System

Consider a baseband communication system in which binary digits are transmitted as rectangular pulses of duration T. A binary zero is transmitted as a positive rectangular pulse $s_0(t) = A\, p_T(t)$, and a binary one is transmitted as a negative rectangular pulse $s_1(t) = -A\, p_T(t)$, where A is a deterministic, positive constant that represents the amplitude of the signal and

$$p_T(t) = \begin{cases} 1, & 0 \le t < T, \\ 0, & \text{otherwise.} \end{cases}$$

These pulses are transmitted over an additive white Gaussian noise channel, and the signal that is received is

$$Y_i(t) = s_i(t) + X(t),$$

where $i = 0$ or 1, depending on whether the binary digit that is transmitted is 0 or 1, and $X(t)$ is a white Gaussian noise process with spectral density $N_0/2$.

The first receiver consists of a linear time-invariant filter followed by a sampler. The filter is a simple RC filter with $RC = T/2$. Its impulse response is

$$h(t) = (RC)^{-1}\, e^{-t/RC} u(t),$$

where

$$u(t) = \begin{cases} 1, & t \ge 0, \\ 0, & t < 0, \end{cases}$$

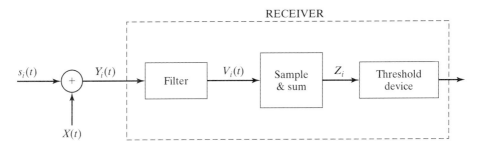

Figure 3-9: A communication system with a suboptimum receiver.

is the unit step function. The output $V_i(t)$ of the RC filter is sampled at times $t_1 = T/2$, $t_2 = 3T/4$, and $t_3 = T$. The sample values are then summed to give

$$Z_i = \sum_{k=1}^{3} V_i(t_k).$$

The receiver decides that 0 was sent if the sum of the sample values exceeds zero and it decides that 1 was sent if the sum is less than or equal to zero. A block diagram of the receiver is shown in Figure 3-9.

An alternative receiver for the same transmitted signals $s_0(t)$ and $s_1(t)$ and the same additive white Gaussian noise channel is shown in Figure 3-10. This receiver is optimum for the given signals and channel in the sense that it gives the minimum possible error probability of any communication receiver.

To clarify what is meant by *minimum possible error probability*, we let $P_{e,i}$ denote the probability of error for a given receiver when the symbol i is sent (i is either 0 or 1). That is, $P_{e,0}$ is the probability that the receiver decides that 1 was sent when in fact 0 was sent, and $P_{e,1}$ is the probability that the receiver decides that 0 was sent when 1 was actually sent. We then define

$$P_{e,m} = \max\{P_{e,0},\, P_{e,1}\}$$

and

$$\overline{P}_e = \frac{P_{e,0} + P_{e,1}}{2}.$$

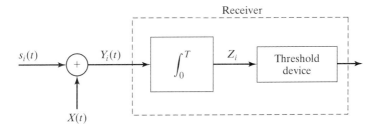

Figure 3-10: An optimum receiver.

The receiver of Figure 3-10 gives the smallest possible value of $P_{e,m}$ that can be obtained with any receiver. It also gives the smallest possible value of $\overline{P_e}$.

To illustrate the general analytical methods applicable to this type of problem, and, in particular, to demonstrate the role played by the autocorrelation functions in such methods, we derive expressions for $P_{e,0}$ and $P_{e,1}$ for the receivers of Figures 3-9 and 3-10. This derivation requires the evaluation of

$$P_{e,0} = P(Z_0 \le 0)$$

and

$$P_{e,1} = P(Z_1 > 0)$$

for each of the two receivers.

For the suboptimum receiver of Figure 3-9, Z_i is a Gaussian random variable with mean

$$\mu_i = E\left\{\sum_{k=1}^{3} V_i(t_k)\right\} = \sum_{k=1}^{3} v_i(t_k)$$

and variance

$$\begin{aligned}
\sigma_i^2 &= \mathrm{Var}\left\{\sum_{k=1}^{3} V_i(t_k)\right\} \\
&= E\left\{\left[\sum_{k=1}^{3}\{V_i(t_k) - v_i(t_k)\}\right]^2\right\},
\end{aligned} \tag{3.92}$$

where

$$v_i(t_k) = E\{V_i(t_k)\}.$$

Now, $V_i(t)$ consists of the sum of a signal component

$$\hat{s}_i(t) = \int_{-\infty}^{\infty} h(\tau)\, s_i(t-\tau)\, d\tau$$

and a noise component

$$\hat{X}(t) = \int_{-\infty}^{\infty} h(\tau)\, X(t-\tau)\, d\tau.$$

That is, $V_i(t) = \hat{s}_i(t) + \hat{X}(t)$, where $\hat{s}_i = s_i * h$ is the output of the filter due to the signal alone and $\hat{X}(t)$ is the output of the filter due to noise alone. Since the noise $X(t)$ is white, the output process $\hat{X}(t)$ has zero mean and autocorrelation function

$$R_{\hat{X}}(\tau) = \frac{N_0}{2}\int_{-\infty}^{\infty} h(u)\, h(u-\tau)\, du.$$

Because $\hat{X}(t)$ has zero mean,

$$
\begin{aligned}
v_i(t) &= E\{V_i(t)\} = E\{\hat{s}_i(t) + \hat{X}(t)\} \\
&= \hat{s}_i(t) + E\{\hat{X}(t)\} = \hat{s}_i(t).
\end{aligned}
$$

Thus, the mean of $V_i(t)$ is just its signal component. Consequently, from (3.92) we see that

$$
\begin{aligned}
\sigma_i^2 &= E\left\{\left[\sum_{k=1}^{3} \hat{X}(t_k)\right]^2\right\} \\
&= E\left\{\sum_{k=1}^{3}\sum_{j=1}^{3} \hat{X}(t_k)\,\hat{X}(t_j)\right\} \\
&= \sum_{k=1}^{3}\sum_{j=1}^{3} R_{\hat{X}}(t_k - t_j).
\end{aligned}
$$

For the given values of t_1, t_2, and t_3,

$$
|t_1 - t_2| = |t_2 - t_3| = \frac{T}{4}
$$

and

$$
|t_1 - t_3| = \frac{T}{2}.
$$

Thus,

$$
\sigma_i^2 = 3R_{\hat{X}}(0) + 4R_{\hat{X}}(T/4) + 2R_{\hat{X}}(T/2). \tag{3.93}
$$

Notice, in particular, that σ_i^2 does not depend on i (i.e., $\sigma_1^2 = \sigma_0^2$). For simplicity, we drop the subscripts and denote the common value of σ_1^2 and σ_0^2 by σ^2.

Next, we must find $v_i(t)$ and $R_{\hat{X}}(\tau)$. The first step is to evaluate

$$
\begin{aligned}
v_i(t) &= \hat{s}_i(t) = \int_{-\infty}^{\infty} h(\tau)\, s_i(t - \tau)\, d\tau \\
&= (-1)^i \frac{A}{RC} \int_{0}^{\infty} e^{-\tau/RC}\, p_T(t - \tau)\, d\tau.
\end{aligned}
$$

Since $p_T(t - \tau) = 0$ for $t < 0$, it follows that $v_i(t) = 0$ for $t < 0$. For $0 \le t < T$,

$$
\begin{aligned}
v_i(t) &= (-1)^i \frac{A}{RC} \int_{0}^{t} e^{-\tau/RC}\, d\tau \\
&= (-1)^i A[1 - \exp(-t/RC)],
\end{aligned}
$$

and for $t \ge T$,

$$
\begin{aligned}
v_i(t) &= (-1)^i \frac{A}{RC} \int_{t-T}^{t} \exp(-\tau/RC)\, d\tau \\
&= (-1)^i A[\exp(T/RC) - 1]\exp(-t/RC).
\end{aligned}
$$

Thus, the mean of Z_i is

$$\mu_i = (-1)^i A[3 - e^{-T/2RC} - e^{-3T/4RC} - e^{-T/RC}]$$
$$= (-1)^i A[3 - e^{-1} - e^{-3/2} - e^{-2}],$$

where, in the last step, we have used the fact that $RC = T/2$. The autocorrelation function for $\hat{X}(t)$ is given by

$$R_{\hat{X}}(\tau) = \frac{N_0}{2}(RC)^{-2} \int_{-\infty}^{\infty} e^{-u/RC} e^{-(u+\tau)/RC} \, du$$

for $\tau \geq 0$. For $\tau < 0$, we can use the fact that $R_{\hat{X}}(\tau) = R_{\hat{X}}(-\tau)$.

Evaluating the integral in the expression for the autocorrelation function, we find that for $\tau \geq 0$,

$$R_{\hat{X}}(\tau) = \frac{N_0}{4RC} e^{-\tau/RC} = \frac{N_0}{2T} e^{-2\tau/T}.$$

Therefore, since $R_{\hat{X}}(\tau) = R_{\hat{X}}(-\tau)$, it follows that

$$R_{\hat{X}}(\tau) = \frac{N_0}{2T} \exp(-2|\tau|/T)$$

for all τ. From (3.93), we see that the variance of Z_i, which does not depend on i, is given by

$$\sigma^2 = \frac{N_0}{2T}[3 + 4e^{-1/2} + 2e^{-1}].$$

The error probability $P_{e,0}$ is given by

$$P_{e,0} = P(Z_0 \leq 0) = \Phi\left(\frac{0 - \mu_0}{\sigma}\right) = 1 - \Phi\left(\frac{\mu_0}{\sigma}\right),$$

and the error probability $P_{e,1}$ is given by

$$P_{e,1} = P(Z_1 > 0) = 1 - P(Z_1 \leq 0) = 1 - \Phi\left(\frac{0 - \mu_1}{\sigma}\right) = 1 - \Phi\left(\frac{-\mu_1}{\sigma}\right).$$

Evaluating the expressions for μ_i and σ, we find that

$$\mu_i \approx (2.2737)(-1)^i A$$

and

$$\sigma \approx 2.4823 \sqrt{\frac{N_0}{2T}}.$$

Hence,

$$\frac{\mu_i}{\sigma} \approx (-1)^i 0.9159 \sqrt{\frac{2A^2 T}{N_0}}$$

$$= (-1)^i 0.9159 \sqrt{\frac{2\mathcal{E}_b}{N_0}},$$

where $\mathcal{E}_b = A^2 T$ is the energy per bit; that is,

$$\mathcal{E}_b = \int_{-\infty}^{\infty} [s_i(t)]^2 \, dt = A^2 T.$$

Next, we notice that

$$P_{e,0} = P_{e,1} = 1 - \Phi \left(\frac{|\mu_i|}{N_0} \right)$$

$$\approx 1 - \Phi \left[(0.9159) \sqrt{\frac{2\,\mathcal{E}_b}{N_0}} \right].$$

If $\mathcal{E}_b/N_0 = 5$, for instance, then

$$P_{e,0} = P_{e,1} \approx 1 - \Phi \left[0.9159\sqrt{10} \right]$$

$$\approx 1 - \Phi(2.896)$$

$$\approx 0.00189 = 1.89 \times 10^{-3}.$$

The optimum receiver is actually much simpler to analyze. It is shown in Chapter 5 that

$$P_{e,0} = P_{e,1} = 1 - \Phi \left(\sqrt{\frac{2A^2 T}{N_0}} \right)$$

$$= 1 - \Phi \left(\sqrt{\frac{2\,\mathcal{E}_b}{N_0}} \right)$$

for the optimum receiver. For the same numerical example as before (i.e., $\mathcal{E}_b/N_0 = 5$), the error probabilities for the optimum receiver are

$$P_{e,0} = P_{e,1} = 1 - \Phi(\sqrt{10})$$

$$\approx 1 - \Phi(3.162)$$

$$\approx 0.000783 = 7.83 \times 10^{-4}. \qquad \blacksquare$$

Problems

3.1 A random process is defined by $X(t) = V p_T(t)$, where V is a random variable that is uniformly distributed on $[0, 1]$ and $p_T(t)$ is the rectangular pulse defined in Exercise 3-5. The random process is the input to a linear filter with impulse response $h(t) = p_{T_0}(t)$, and $Y(t)$ is the corresponding output.

   (a) Sketch some typical sample functions for the random process $X(t)$.

   (b) Sketch some typical sample functions for the output random process $Y(t)$ if $T_0 = T$.

   (c) Is the input random process wide-sense stationary?

(d) Find the mean of $Y(t)$ if $T_0 > T$.

(e) Find the autocorrelation function for the input random process $X(t)$.

(f) Find $R_Y(t, t)$ if $T_0 = T$.

3.2 The zero-mean random process $Z(t)$ is Gaussian with autocorrelation function $R_Z(\tau) = \exp(-\tau^2)$. Find the probability that the sum $Z(0) + Z(1) + Z(2)$ exceeds the threshold γ. Express your answer in terms of γ, the exponential function, and the standard Gaussian distribution function Φ.

3.3 A discrete-time linear time-invariant filter has pulse response given by $h(n) = 1$ for $n = 0, 1$, and 2, and $h(n) = 0$ otherwise. The input is a discrete-time random process $X(n)$, $n \in \mathbb{Z}$, that has mean $\mu_X(n)$ and autocorrelation function $R_X(n, k) = \exp\{-(n - k)^2\}$. The corresponding output is the random process $Y(n)$, $n \in \mathbb{Z}$.

(a) Find the mean $\mu_Y(n)$ of the output in terms of the mean of the input.

(b) Assume that the input mean is identically zero, and find the crosscorrelation function $R_{X,Y}(i, j)$.

(c) Assume that the input mean is identically zero, and find the output autocorrelation function $R_Y(i, j)$.

(d) If the input mean is identically zero, are the processes $X(n)$ and $Y(n)$ jointly wide-sense stationary? Explain carefully why or why not.

3.4 The zero-mean random process $X(t)$ has autocorrelation function $R_X(\tau) = \beta \delta(\tau)$. $X(t)$ is the input to a time-invariant linear filter with impulse response given by

$$h(t) = \begin{cases} 1, & t_1 \le t < t_2, \\ 0, & \text{otherwise.} \end{cases}$$

where $0 < t_1 < t_2$. Let $Y(t)$ be the corresponding output random process.

(a) Are the input and output processes jointly wide-sense stationary?

(b) Find the crosscorrelation function $R_{X,Y}(t, s)$ for the random processes $X(t)$ and $Y(t)$.

(c) Find the autocorrelation function $R_Y(t, s)$ for the random process $Y(t)$.

3.5 Suppose $X(t)$ is a zero-mean, wide-sense stationary, continuous-time process with autocorrelation function $R_X(\tau) = \eta \exp\{-\gamma|\tau|\}$, $-\infty < \tau < \infty$. Let $X(t)$ be the input to the linear time-invariant filter with impulse response $h(t) = p_T(t)$. (Notice that this is the filter considered in Exercise 3-5.) Find the expected value of the instantaneous power in the output process at time t.

3.6 The random process $X(t)$ described in Problem 3.5 is the input to a simple RC filter with impulse response $h(t) = b \exp(-bt)$ for $t \ge 0$, where $b = 1/RC$. Assume that $1/RC \ne \gamma$. The corresponding output is $Y(t)$. Explain why the random processes $X(t)$ and $Y(t)$ are automatically jointly wide-sense stationary. Find the crosscorrelation function $R_{X,Y}(\tau)$.

3.7 Consider the same input random process and linear filter as in Problem 3.6. Find the output autocorrelation function $R_Y(\tau)$. The solution to Exercise 3-7 may be of help.

3.8 A continuous-time linear filter has impulse response $h(t) = p_\lambda(t)$. The input random process has the triangular autocorrelation function given as the second entry in Table 2.1 and illustrated in Figure 2-11(b). The output random process is $Y(t)$. Find the expected value of the instantaneous power in the output at time t_0. Consider three cases: $\lambda < T$, $\lambda = T$, and $\lambda > T$.

3.9 Suppose $X_1, X_2, X_3, \ldots$ is a sequence of independent, identically distributed random variables with distribution specified by

$$P(X_k = 0) = p, \quad P(X_k = 1) = q, \quad P(X_k = 2) = 1 - p - q,$$

and

$$P(X_k = n) = 0, \text{ for } n < 0 \text{ and for } n > 2.$$

Assume that p and q are positive numbers that satisfy the relation $p + q < 1$. From the sequence $X_1, X_2, X_3, \ldots$, we obtain another sequence $Z_1, Z_2, Z_3, \ldots$ by letting $Z_k = X_k + X_{k+1} + X_{k+2}$ for each positive integer k.

    (a) Find $E\{Z_k\}$ and $E\{(Z_k)^2\}$ for each positive integer k.

    (b) Find the autocorrelation $R_Z(k, k + 1) = E\{Z_k Z_{k+1}\}$ for each positive integer k.

    (c) Find the distribution for Z_k; that is, specify $P(Z_k = n)$ for each value of n and each positive integer k.

    (d) Is the sequence $Z_1, Z_2, Z_3, \ldots$ a sequence of independent random variables? Answer this question by direct application of the definition of independent random variables.

3.10 The input to a linear filter with impulse response $p_T(t)$ is the sum of a deterministic signal $s(t) = \alpha\, p_T(t)$ and noise $X(t)$. Assume that the noise is wide-sense stationary with zero mean and autocorrelation function $R_X(\tau) = \beta \exp(-\tau^2/2)$. The output consists of the sum of $\hat{s}(t)$, the filtered version of the signal, and $\hat{X}(t)$, the filtered version of the noise. Suppose this output is sampled at time t_0, and the output signal-to-noise ratio is defined as $\mathrm{SNR}(t_0) = \hat{s}(t_0)/\sigma$, where σ is the standard deviation of the noise $\hat{X}(t_0)$.

    (a) Find the output signal-to-noise ratio as a function of t_0 (for $-\infty < t_0 < \infty$).

    (b) Find the value of t_0 that maximizes the output signal-to-noise ratio, and find the resulting maximum value of $\mathrm{SNR}(t_0)$.

    (c) The input is sampled at time τ_0 ($0 < \tau_0 < T$). Give the corresponding definition for the input signal-to-noise ratio.

    (d) For the definition of the signal-to-noise ratio obtained in part **(c)**, find the input signal-to-noise ratio.

    (e) Compare the input signal-to-noise ratio and the maximum output signal-to-noise ratio from part **(b)**, and show that for large values of T (i.e., $T \gg 1$), the filter improves the signal-to-noise ratio.

3.11 A discrete-time feedback control system has the property that its output voltage $X(k+1)$ at time $k+1$ is a linear combination of the output voltage $X(k)$ at time k and a random error $Y(k+1)$ that is independent of past outputs. The equation that governs the system is

$$X(k+1) = \alpha\, X(k) + Y(k+1),$$

for each integer k, where $|\alpha| < 1$. Assume that the random process $Y(n)$, $n \in \mathbb{Z}$, is a sequence of independent, identically distributed random variables with mean zero and standard deviation β. Assume also that the random process $X(n)$, $n \in \mathbb{Z}$, has zero mean. The statement that the random error $Y(k+1)$ is independent of past outputs means that $Y(k+1)$ is independent of $\{X(n) : n \le k\}$.

(a) Sketch a block diagram for this system.

(b) Is the output random process $X(n)$, $n \in \mathbb{Z}$, wide-sense stationary? Explain.

(c) Find the crosscorrelation function $R_{X,Y}(i, j)$ for all i and j.

(d) Find the autocorrelation $R_X(k, k+1)$ for arbitrary k.

(e) Find the complete autocorrelation function for $X(n)$, $n \in \mathbb{Z}$. (It may be easier to first find $R_X(k, k+2)$, $R_X(k, k+3)$, etc., and observe the pattern.)

3.12 A time-invariant linear filter has impulse response given by $h(t) = t/T$ for $0 \le t < T$ and $h(t) = 0$ otherwise, as shown in the following diagram:

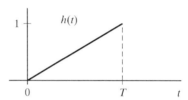

The input to this filter is a continuous-time random process $X(t)$, and the corresponding output is the random process $Y(t)$. If the input random process is white noise with autocorrelation function $R_X(\tau) = \beta\,\delta(\tau)$, find the autocorrelation function for the output process.

3.13 Consider the following system:

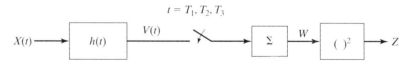

The input random process $X(t)$ is a Gaussian white-noise process that has spectral density $N_0/2$. The filter is a linear, time-invariant, causal filter with impulse response $h(t) = \exp(-t/T)$ for $0 \le t < \infty$. The filter output is $V(t)$ when $X(t)$ is the input. The random variable W is given by

$$W = V(T_1) + V(T_2) + V(T_3),$$

and the sampling times are $T_1 = T$, $T_2 = 2T$, and $T_3 = 4T$. The system output is $Z = W^2$.

(a) Is $V(t)$ wide-sense stationary? Explain why or why not.

(b) Find the autocorrelation function for $V(t)$. Express your answer in terms of the parameters N_0 and T only. Give a mathematical expression for the autocorrelation function, and also draw a sketch of its graph.

(c) Find $\mu = E\{Z\}$. Express your answer in terms of the parameters N_0 and T only.

(d) Find $P(Z \leq \gamma)$ for each choice of the real number γ. Express your answer in terms of γ, the function Φ, and the parameter μ of part (c), not in terms of N_0 or T.

3.14 The continuous-time random process $X(t)$, $-\infty < t < \infty$, is wide-sense stationary and has autocorrelation function $R_X(\tau)$. The random process $Y(t)$, $-\infty < t < \infty$, is the output of a continuous-time, linear, time-invariant filter when $X(t)$ is the input.

(a) What facts about the filter and the random processes permit you to conclude that the input and output processes are *jointly* wide-sense stationary?

(b) Give the definition for the crosscorrelation function $R_{Y,X}(\tau)$ in terms of the expectation operator $E\{\ \}$ and the random processes $X(t)$, $-\infty < t < \infty$, and $Y(t)$, $-\infty < t < \infty$.

(c) Suppose the linear time-invariant filter has impulse response

$$h(t) = \begin{cases} 3, & t_1 \leq t < t_2, \\ 0, & \text{otherwise}, \end{cases}$$

where $0 < t_1 < t_2$. Give an integral expression for $R_{Y,X}(\tau)$ with finite limits on the integral. Your answer should be in terms of τ, the parameters t_1 and t_2, and the autocorrelation function for the process $X(t)$, not in terms of the function h.

(d) Now suppose that the autocorrelation function for the process $X(t)$ is

$$R_X(\tau) = 7\exp(-5|\tau|), \quad -\infty < \tau < \infty.$$

Give a simple expression (e.g., evaluate all integrals) for $R_{Y,X}(\tau)$ for $\tau < 0$ only. Give your answer in terms of the parameters t_1 and t_2. A sketch of the functions involved may be of help.

(e) Is the correct answer to (d) also valid for some positive values of τ?

(f) If you said yes in part (e), give the range of positive values of τ for which the correct answer to (d) is valid. If you said no in part (e), explain in words or with a sketch why the correct answer to (d) is invalid for all $\tau > 0$.

3.15 The input to a time-invariant linear system with impulse response $h(t)$ is a white Gaussian noise process $X(t)$ with two-sided spectral density $N_0/2$. The output is the random process $Y(t)$. The filter's impulse response is given in terms of $p_T(t)$, the rectangular pulse of duration T, by $h(t) = \sin(2\pi t/T)\, p_T(t)$.

(a) Find the crosscorrelation function $R_{X,Y}(\tau)$, $-\infty < \tau < \infty$.

(b) Find $R_Y(0)$.

3.16 The continuous-time, wide-sense stationary random process $X(t)$, $-\infty < t < \infty$, has autocorrelation function given by

$$R_X(\tau) = \begin{cases} 3(T - |\tau|)/T, & -T < \tau < T, \\ 0, & \text{otherwise}, \end{cases}$$

for $T > 0$. The random process $Y(t)$, $-\infty < t < \infty$, is the output of a continuous-time, time-invariant linear filter when $X(t)$ is the input. The impulse response of the filter is $h(t) = 5\, p_T(t)$. For each of the following, express your answer in terms of the parameter T:

(a) Give an expression for $R_{Y,X}(T/2)$.

(b) Give an expression for $R_{X,Y}(T/2)$.

(c) Give an expression for $E\{[Y(t)]^2\}$.

3.17 Consider the suboptimum receiver described in Example 3-8 and illustrated in Figure 3-9. The filter is the simple RC filter defined in Example 3-8, except that $RC = T$. The channel noise $X(t)$ is white Gaussian noise with spectral density $N_0/2$. Suppose that the output of the filter is sampled twice, with sample times $t_1 = T/3$ and $t_2 = 2T/3$, and let $Z_i = V_i(t_1) + V_i(t_2)$. The variance of Z_i is denoted by σ_i^2, as in Example 3-8. Give an exact expression for σ_i^2 for each i. Express your answers in terms of the parameters N_0 and T.

3.18 Consider the receiver illustrated in Figure 3-10, and assume the random process $X(t)$ is wide-sense stationary and its mean is zero. The autocorrelation function for $X(t)$ is given by $R_X(\tau) = \beta \exp(-|\tau|)$ for $-\infty < \tau < \infty$. Suppose that $s_i(t) = 0$ for all t, and let $\sigma_i^2 = \text{Var}\{Z_i\}$ for each i. Give an expression for σ_i^2 for each i. Express your answers in terms of the parameters β and T.

Chapter 4

Frequency-Domain Analysis of Random Processes in Linear Systems

4.1 The Use of Fourier Transform Techniques

Just as they do in the analysis of deterministic signals in linear systems, Fourier transform techniques offer an alternative approach to the time-domain analysis of random processes in linear systems. The application of Fourier transform methods frequently leads to considerable computational savings and a better physical understanding of the problem at hand. The computational advantage stems from the relative ease of multiplying two Fourier transforms, compared with evaluating the convolution of two time-domain functions. The physical insight is derived from the ability to investigate the frequency distribution of the power in the random process, and such an investigation often provides important guidance in the design of filters to smooth the noise or improve the signal-to-noise ratio of a system.

There is one very important difference between the frequency-domain analysis of deterministic signals and the frequency-domain analysis of random processes: Just as in the time-domain analysis of random processes, we do not (and, in fact, cannot) apply Fourier transform methods to the individual sample functions of a random process. Instead, Fourier analysis techniques are applied primarily to the autocorrelation functions of the wide-sense stationary random processes involved. Although Fourier techniques can obviously be employed to evaluate the mean of the output process of a linear time-invariant system (i.e., to evaluate the convolution integral in equation (3.31)), this is not a very important application of Fourier techniques to the analysis of random processes. The most important applications deal with the Fourier transforms of the autocorrelation functions of wide-sense stationary random processes.

In this chapter, we introduce the spectral density function for wide-sense stationary random processes and present some of the properties of that function. We then develop the spectral analysis of random processes in linear systems and close the chapter with discussions of the spectral density of a modulated signal and the representations of bandpass random processes and bandpass filters.

4.2   The Spectral Density of a Wide-Sense Stationary Random Process

The (power) *spectral density function* for a wide-sense stationary random process $X(t)$ with autocorrelation function R_X is defined by

$$S_X(\omega) = \int_{-\infty}^{\infty} R_X(\tau)\, e^{-j\omega\tau}\, d\tau, \tag{4.1}$$

whenever the integral exists. That is, *the spectral density function is the Fourier transform of the autocorrelation function.*

A few remarks about Fourier transforms are in order before we discuss the properties of spectral densities. In general, if g is a real-valued function of a real variable (i.e., $g : \mathbb{R} \to \mathbb{R}$), the Fourier transform of the function g is another function $G = \mathcal{F}\{g\}$ that is a complex-valued function of a real variable (i.e., $G : \mathbb{R} \to \mathbb{C}$ where $\mathbb{C}$ denotes the set of all complex numbers). The function $G = \mathcal{F}\{g\}$ is defined by

$$G(\omega) = \int_{-\infty}^{\infty} g(t)\, e^{-j\omega t}\, dt, \quad -\infty < \omega < \infty, \tag{4.2}$$

again, provided that the integral exists. The integral does exist whenever

$$\int_{-\infty}^{\infty} |g(t)|\, dt < \infty,$$

for instance, but this is not a necessary condition for defining the Fourier transform of the function g. We can consider Fourier transforms of functions such as $g(t) = \sin(2\pi t)$ by the use of generalized functions in the frequency domain (e.g., delta functions).

The preceding comments imply that the integral in (4.1) exists and the spectral density function S_X is well defined if the autocorrelation function satisfies

$$\int_{-\infty}^{\infty} |R_X(\tau)|\, d\tau < \infty, \tag{4.3}$$

even though this condition is not an absolute requirement. In applications to communications system design and analysis, it is often necessary to work with the spectral density function for a wide-sense stationary random process $X(t)$ whose autocorrelation function does *not* satisfy (4.3). There are many important autocorrelation functions, such as $R_X(\tau) = \cos(2\pi f_0 \tau)$, that do not satisfy (4.3), but that can be handled by the use of generalized functions. In fact, the spectral density that corresponds to this autocorrelation function consists of a pair of delta functions in the frequency domain. (See Table 4.1 at the end of the section.)

The variable ω in (4.1) and (4.2) is usually referred to as the *radian frequency* and is measured in radians per second. The frequency $\omega/2\pi$ is the "usual" frequency—the number of cycles per second, measured in hertz. A standard abuse of notation is to write $S_X(f)$ for the spectral density as a function of frequency in hertz. A more precise notation is $S_X(2\pi f)$, but the notation $S_X(f)$ is more compact and more common in the literature. There is one important fact to remember in the use of the notation $S_X(f)$ and in the expression of spectral densities in terms of the frequency in hertz:

> To convert an expression for the spectral density as a function of ω to an expression for the same spectral density as a function of f, we must replace ω by $2\pi f$, except in the argument of the function S_X itself.

For example, if $S_X(\omega) = (1 + \omega^2)^{-1}$ for all ω, then $S_X(f) = (1 + 4\pi^2 f^2)^{-1}$ for all f. If $S_X(\omega) = N_0/2$ for all ω, then $S_X(f) = N_0/2$ for all f.

Formally, the spectral density as a function of frequency in hertz is defined as

$$S_X(f) = \int_{-\infty}^{\infty} R_X(\tau) e^{-j2\pi f\tau} \, d\tau.$$

A comparison of this expression with (4.1) shows that ω has been replaced by $2\pi f$ in the integrand, which is the basis for the conversion described in the previous paragraph.

Physically, $S_X(\omega)$ represents the density of power at frequency ω radians/sec, and $S_X(f)$ plays the same role for frequencies in hertz. The power in any frequency band is obtained by integrating the spectral density over the range of frequencies that make up the band. To develop the relationship between power and the integral of the spectral density function, the approach used here is to show that the inverse transform of the spectral density, when evaluated at the origin, gives the expected value of the instantaneous power in the random process.

Because $S_X = \mathcal{F}\{R_X\}$, the *inverse transform* of the spectral density is the auto-correlation function: $R_X = \mathcal{F}^{-1}\{S_X\}$. In integral form, this inverse transform can be written in either of the following two ways:

$$R_X(\tau) = \int_{-\infty}^{\infty} S_X(\omega) e^{+j\omega\tau} \, d\omega/2\pi$$

$$= \int_{-\infty}^{\infty} S_X(f) e^{+j2\pi f\tau} \, df.$$

In particular, the power in the random process $X(t)$ is

$$E\{[X(t)]^2\} = R_X(0) = \frac{1}{2\pi} \int_{-\infty}^{\infty} S_X(\omega) e^0 \, d\omega$$

$$= \frac{1}{2\pi} \int_{-\infty}^{\infty} S_X(\omega) \, d\omega = \int_{-\infty}^{\infty} S_X(f) \, df.$$

This relationship shows that the power in the random process is the total area under the spectral density $S_X(f)$. Careful attention should be paid to the presence of the factor 2π in the integral of $S_X(\omega)$ and the absence of this factor in the integral of $S_X(f)$. The issue has a lot to do with the preference among many authors for working with $S_X(f)$ rather than $S_X(\omega)$. In general, the power in a frequency band from f_1 Hz to f_2 Hz is given by

$$\frac{1}{2\pi} \int_{2\pi f_1}^{2\pi f_2} S_X(\omega)\, d\omega = \int_{f_1}^{f_2} S_X(f)\, df.$$

This equation is a special case of some results developed in Section 4.3.

The properties of the spectral density function for a wide-sense stationary random process $X(t)$ that are most useful in frequency-domain analysis are as follows:

Property 1: $S_X(\omega) = S_X(-\omega)$, for all ω.

Property 2: $S_X(\omega) = [S_X(\omega)]^*$, for all ω.

Property 3: $S_X(\omega) \geq 0$, for all ω.

Property 4: If $\int_{-\infty}^{\infty} |R_X(\tau)|\, d\tau < \infty$, then $S_X(\omega)$ is a continuous function of ω.

In property 2, $[S_X(\omega)]^*$ denotes the *complex conjugate* of $S_X(\omega)$. The first three properties can be summarized by saying that a spectral density function is (1) even, (2) real, and (3) nonnegative. The first two of these properties are easily derived; the third and fourth are somewhat more difficult, and their derivations are postponed until Section 4.4.

To derive the formula $S_X(\omega) = S_X(-\omega)$, we make a change of variable $(\lambda = -\tau)$ and use the fact that $R_X(\lambda) = R_X(-\lambda)$. This procedure yields the following sequence of identities:

$$S_X(\omega) = \int_{-\infty}^{\infty} R_X(\tau)\, e^{-j\omega\tau}\, d\tau = \int_{-\infty}^{\infty} R_X(-\lambda)\, e^{+j\omega\lambda}\, d\lambda$$

$$= \int_{-\infty}^{\infty} R_X(\lambda)\, e^{-j(-\omega)\lambda}\, d\lambda$$

$$= S_X(-\omega).$$

To prove that $S_X(\omega) = [S_X(\omega)]^*$, we can simply use the fact that $S_X(-\omega) = [S_X(\omega)]^*$ and then apply property 1. To see that $S_X(-\omega) = [S_X(\omega)]^*$, notice that

$$[S_X(\omega)]^* = \left\{ \int_{-\infty}^{\infty} R_X(\tau)\exp(-j\omega\tau)\, d\tau \right\}^* = \int_{-\infty}^{\infty} R_X(\tau)[\exp(-j\omega\tau)]^*\, d\tau$$

$$= \int_{-\infty}^{\infty} R_X(\tau)\exp(+j\omega\tau)\, d\tau = \int_{-\infty}^{\infty} R_X(\tau)\, e^{-j(-\omega)\tau}\, d\tau$$

$$= S_X(-\omega).$$

The three exercises that follow illustrate the conversion between autocorrelation functions and spectral density functions.

Exercise 4-1

Suppose the continuous-time random process $X(t)$ has autocorrelation function

$$R_X(\tau) = 1 + e^{-\alpha|\tau|}, \quad -\infty < \tau < \infty,$$

where $\alpha > 0$. Find the spectral density function for $X(t)$.

Solution.   The Fourier transform of the constant 1 is $2\pi\,\delta(\omega)$. The Fourier transform of $e^{-\alpha|\tau|}$ is $2\alpha/(\alpha^2 + \omega^2)$. Hence,

$$S_X(\omega) = 2\pi\,\delta(\omega) + \frac{2\alpha}{\alpha^2 + \omega^2}.$$

Exercise 4-2

A zero-mean wide-sense stationary random process $X(t)$, $-\infty < t < \infty$, has power spectral density

$$S_X(\omega) = \frac{1}{1 + \omega^2}, \quad -\infty < \omega < \infty.$$

Find the mean and variance of the random process defined by

$$Y(t) = \sum_{k=0}^{2} X(t + k).$$

Solution.   The inverse transform of $S_X(\omega)$ is

$$R_X(\tau) = \tfrac{1}{2}\,e^{-|\tau|}.$$

Because $E\{Y(t)\} = 0$,

$$
\begin{aligned}
\mathrm{Var}\{Y(t)\} &= E\{[Y(t)]^2\}\\
&= E\left\{\sum_{k=0}^{2}\sum_{i=0}^{2} X(t+k)\,X(t+i)\right\}\\
&= \sum_{k=0}^{2}\sum_{i=0}^{2} R_X(k-i)\\
&= 3R_X(0) + 2R_X(2) + 4R_X(1)\\
&= \left(\tfrac{3}{2}\right) + 2\,e^{-1} + e^{-2}.
\end{aligned}
$$

The solution to this exercise requires an understanding of *both* time- and frequency-domain descriptions, which is often the case for engineering problems that deal with wide-sense stationary random processes.

Exercise 4-3
Given a nonnegative, even function S, find a random process $X(t)$ that has S as its power spectral density.

Solution.   Because S is nonnegative, it can be normalized to give a probability density function g. If P is the total power in a random process with spectral density S, then

$$P = \frac{1}{2\pi} \int_{-\infty}^{\infty} S(\omega)\, d\omega.$$

Since we want the function g to have unit area, let

$$g(u) = \frac{S(u)}{2\pi P}, \qquad -\infty < u < \infty, \tag{4.4}$$

which is then a valid probability density function. Let W be a random variable that has probability density function g, and let Θ be uniformly distributed on the interval $[0, 2\pi]$ and independent of W. It follows that

$$E\{\cos(\Theta)\} = E\{\sin(\Theta)\} = E\{\cos(2\,\Theta)\} = E\{\sin(2\,\Theta)\} = 0.$$

Define the random process $X(t)$ by

$$X(t) = \sqrt{2\,P}\,\cos(Wt + \Theta). \tag{4.5}$$

Using the independence of W and Θ, it is easy to show that $E\{X(t)\} = 0$ and

$$R_X(t + \tau, t) = P\,E\{\cos(W\tau)\}$$

for each t. It follows that $X(t)$ is wide-sense stationary. Next, observe that because g is the density function for the random variable W,

$$E\{\cos(W\tau)\} = \int_{-\infty}^{\infty} \cos(u\tau)\,g(u)\, du.$$

Thus,

$$R_X(\tau) = P \int_{-\infty}^{\infty} g(\omega) \cos(\omega\,\tau)\, d\omega. \tag{4.6}$$

Because g is even and $j\sin(\omega\,\tau) = -j\sin(\omega(-\tau))$ for all τ, (4.6) is equivalent to

$$R_X(\tau) = P \int_{-\infty}^{\infty} g(\omega) \exp(j\,\omega\,\tau)\, d\omega. \tag{4.7}$$

Substituting from (4.4) into (4.7) we see that R_X is the inverse Fourier transform of S, which implies that S is the spectral density function for the random process $X(t)$ that is defined by (4.5).   ∎

Recall that white noise has an autocorrelation function that is proportional to a delta function. Specifically, if $X(t)$ is a continuous-time white-noise process, its auto-correlation function is $R_X(\tau) = (N_0/2)\,\delta(\tau)$. If we substitute this expression into (4.1), we find that *the spectral density function for the white-noise process* is

$$S_X(\omega) = \frac{N_0}{2}, \quad -\infty < \omega < \infty.$$

Thus, white noise has a constant power spectral density over the entire frequency band. This spectral density function can also be written as

$$S_X(f) = \frac{N_0}{2}, \quad -\infty < f < \infty.$$

It is common to refer to $N_0/2$ as the *two-sided spectral density* and N_0 is sometimes called the *one-sided spectral density*.

Table 4.1 gives a list of useful autocorrelation functions and their spectral density functions that arise frequently in engineering problems. The parameters α, T, and W that appear in the table are arbitrary positive constants. The list can be expanded easily by the use of simple identities such as the fact that, for an arbitrary positive constant c, the spectral density corresponding to $c\,R_X(\tau)$ is just $c\,S_X(\omega)$. Also, if $R_Z(\tau) = R_X(\tau) + R_Y(\tau)$, then $S_Z(\omega) = S_X(\omega) + S_Y(\omega)$.

Recall that multiplication in the time domain corresponds to convolution in the frequency domain. That is, if

$$g(t) = g_1(t)\,g_2(t),$$

and if $G_1 = \mathcal{F}\{g_1\}$, $G_2 = \mathcal{F}\{g_2\}$, and $G = \mathcal{F}\{g\}$, then

$$G(\omega) = \frac{1}{2\pi} \int_{-\infty}^{\infty} G_1(\alpha)\,G_2(\omega - \alpha)\,d\alpha, \quad -\infty < \omega < \infty, \tag{4.8a}$$

and

$$G(f) = \int_{-\infty}^{\infty} G_1(\beta)\,G_2(f - \beta)\,d\beta, \quad -\infty < f < \infty. \tag{4.8b}$$

It follows that if autocorrelation functions are related by

$$R_Z(\tau) = R_X(\tau)\,R_Y(\tau),$$

then the corresponding spectral densities are related by

$$S_Z(\omega) = \frac{1}{2\pi}(S_X * S_Y)(\omega) \tag{4.9a}$$

and

$$S_Z(f) = (S_X * S_Y)(f). \tag{4.9b}$$

For example, we can use (4.9a) to obtain entry (6) in Table 4.1 from entries (4) and (5). It is also possible to expand the list by using the fact that convolution in the time domain corresponds to multiplication in the frequency domain.

Table 4.1: Some Common Autocorrelation-Function/Spectral-Density Pairs[1]

	$R_X(\tau)$	$S_X(\omega)$
(1)	$\begin{cases}(T-|\tau|)/T, & |\tau| < T, \\ 0, & \text{otherwise}\end{cases}$	$T\{\sin(\omega T/2)/(\omega T/2)\}^2 = T\,\text{sinc}^2(fT)$
(2)	1	$2\pi\,\delta(\omega)$
(3)	$\delta(\tau)$	1
(4)	$\exp(-\alpha|\tau|)$	$2\alpha/(\alpha^2 + \omega^2)$
(5)	$\cos(\omega_0\tau)$	$\pi\,\delta(\omega - \omega_0) + \pi\,\delta(\omega + \omega_0)$
(6)	$\exp(-\alpha|\tau|)\cos(\omega_0\tau)$	$\{\alpha/[\alpha^2 + (\omega - \omega_0)^2]\} + \{\alpha/[\alpha^2 + (\omega + \omega_0)^2]\}$
(7) $2W\,\text{sinc}(2W\tau) = \sin(2\pi W\tau)/\pi\tau$		$\begin{cases}1, & |\omega| \leq 2\pi W \\ 0, & \text{otherwise}\end{cases}$

[1] Note the definition of $\text{sinc}(u) = \sin(\pi u)/\pi u$, $-\infty < u < \infty$.

4.3   Spectral Analysis of Random Processes in Linear Systems

The main problem that we are concerned with in this section is the determination of the spectral density $S_Y(\omega)$ of the output $Y(t)$ of a time-invariant linear system for which the input is $X(t)$, a wide-sense stationary random process having spectral density $S_X(\omega)$. We let $h(t)$ be the impulse response of the system and $H(\omega)$ be the Fourier transform of $h(t)$ (i.e., $H = \mathcal{F}\{h\}$). The function $H(\omega)$ is called the *transfer function* of the system. Often, it is $H(\omega)$, rather than $h(t)$, that is specified in practice.

The first step in determining the spectral density is to notice that, if $\tilde{h}$ is the "time-reverse impulse response" (defined in Chapter 3), the Fourier transform of $\tilde{h}$ is

$$\tilde{H}(\omega) = \int_{-\infty}^{\infty} h(-t)\,e^{-j\omega t}\,dt = \int_{-\infty}^{\infty} h(u)\,e^{+j\omega u}\,du$$
$$= \left[\int_{-\infty}^{\infty} h(u)\,e^{-j\omega u}\,du\right]^*$$
$$= [H(\omega)]^*,$$

where, as before, z^* denotes the complex conjugate of the complex number z.

Thus, $H(\omega)\tilde{H}(\omega) = H(\omega)[H(\omega)]^* = |H(\omega)|^2$. If we now use the fact that convolution in the time domain corresponds to multiplication in the frequency domain, we see that (3.51) corresponds to

$$S_Y(\omega) = H(\omega)\tilde{H}(\omega)\,S_X(\omega) = |H(\omega)|^2 S_X(\omega). \tag{4.10}$$

This is the key result for the spectral analysis of wide-sense stationary random processes in time-invariant linear systems. Several important facts about spectral densities can be deduced from (4.10), including the fact that the power (i.e., the expected value of the instantaneous power) in the output process is

$$E\{Y^2(t)\} = R_Y(0) = \int_{-\infty}^{\infty} |H(\omega)|^2 S_X(\omega)\, d\omega/2\pi. \qquad (4.11)$$

This equation can also be written as

$$R_Y(0) = \int_{-\infty}^{\infty} |H(f)|^2 S_X(f)\, df. \qquad (4.12)$$

Because of (4.11) and (4.12), $|H(\omega)|^2$ (or $|H(f)|^2$) is sometimes called the *power transfer function* for the linear filter. It is often easier to measure $|H(\omega)|^2$ than it is to measure $H(\omega)$. (The latter requires both phase and amplitude measurements.) If $H(\omega)$ or $|H(\omega)|^2$ is either specified or easily determined from the information that is given, then (4.10) is usually easier to work with than the corresponding time-domain expression, which is (3.51).

Exercise 4-4

Suppose the random process $X(t)$ of Exercise 4-1 is the voltage across a series RLC circuit. Let $Y(t)$ be the voltage across the capacitor. Find the spectral density function for the process $Y(t)$.

Solution.   Since $Y(t)$ is the output of a linear system for which $X(t)$ is the input, it follows that

$$S_Y(\omega) = |H(\omega)|^2 S_X(\omega).$$

For the given system,

$$H(\omega) = [j\omega RC - \omega^2 LC + 1]^{-1}$$

and

$$|H(\omega)|^2 = [\omega^2(RC)^2 + (1 - \omega^2 LC)^2]^{-1}.$$

Thus,

$$
\begin{aligned}
S_Y(\omega) &= [\omega^2(RC)^2 + (1 - \omega^2 LC)^2]^{-1}\{2\pi\, \delta(\omega) + 2\alpha(\omega^2 + \alpha^2)^{-1}\} \\
&= 2\pi\, \delta(\omega) + 2\alpha\{(\omega^2 + \alpha^2)\{(LC)^2\omega^4 + [(RC)^2 - 2LC]\omega^2 + 1\}\}^{-1}.
\end{aligned}
$$

∎

If the input random process $X(t)$ is white noise with spectral density $N_0/2$, its spectral density function is defined by $S_X(\omega) = N_0/2$, $-\infty < \omega < \infty$. As a result, (4.10) simplifies to

$$S_Y(\omega) = \frac{N_0}{2}|H(\omega)|^2.$$

Similarly, if the input process is white noise, (4.11) and (4.12) simplify to

$$R_Y(0) = \frac{N_0}{2} \int_{-\infty}^{\infty} |H(\omega)|^2 \, d\omega/2\pi \tag{4.13}$$

and

$$R_Y(0) = \frac{N_0}{2} \int_{-\infty}^{\infty} |H(f)|^2 \, df, \tag{4.14}$$

respectively. From Chapter 3, we know that if $R_X(\tau) = (N_0/2)\,\delta(\tau)$, then

$$R_Y(0) = \frac{N_0}{2} \int_{-\infty}^{\infty} [h(t)]^2 \, dt. \tag{4.15}$$

(Equation (4.15) can be obtained by setting $\tau = 0$ in (3.81), for instance.) It follows from (4.13)–(4.15) that if the input to a time-invariant linear system is white noise, the power in the output process can be obtained by integrating either the square of the *magnitude* of the transfer function in the frequency domain or the square of the impulse response in the time domain. Some readers may notice that the equality between (4.14) and (4.15) can also be deduced from Parseval's relation.

The choice of which equation to use, (4.14) or (4.15), depends on the linear system under consideration. This is true in general in the choice between frequency-domain methods and time-domain methods. *Usually*, if $H(\omega)$ or $|H(\omega)|^2$ is either specified or easily determined, frequency-domain methods are easier than time-domain methods. On the other hand, if the time-domain functions are simple to describe mathematically, such as a rectangular or triangular pulse, time-domain methods are nearly always easier than frequency-domain methods. In particular, time-limited functions lend themselves to time-domain methods, while band-limited functions are usually easier to handle in the frequency domain.

4.4   Derivation of Properties 3 and 4

In this section, we derive the following two properties of the spectral density function for a wide-sense stationary random process $X(t)$:

Property 3: $S_X(\omega) \geq 0$, for all ω.

Property 4: If $\displaystyle\int_{-\infty}^{\infty} |R_X(\tau)| \, d\tau < \infty$, then $S_X(\omega)$ is a continuous function of ω.

There are two different approaches to deriving property 3. We include both derivations here, since both are instructive and illustrate two different analysis techniques.

For the first approach, we start with the observation that for $T > 0$,

$$0 \leq E\left\{\left|\int_0^T X(t)\,e^{-j\omega t}\,dt\right|^2\right\} = E\left\{\left[\int_0^T X(t)\,e^{-j\omega t}\,dt\right]\left[\int_0^T X(s)\,e^{-j\omega s}\,ds\right]^*\right\}$$

$$= E\left\{\int_0^T X(t)\,e^{-j\omega t}\,dt\int_0^T X(s)\,e^{+j\omega s}\,ds\right\}$$

$$= E\left\{\int_0^T \int_0^T X(t)\,X(s)\,e^{-j\omega(t-s)}\,dt\,ds\right\}$$

$$= \int_0^T \int_0^T R_X(t-s)\,e^{-j\omega(t-s)}\,dt\,ds.$$

Thus, this last double integral is nonnegative. Because $T > 0$, T^{-1} is also nonnegative, and we conclude that

$$T^{-1}\int_0^T \int_0^T R_X(t-s)\,e^{-j\omega(t-s)}\,dt\,ds \geq 0$$

for all positive values of T. Property 3 then follows from the fact that, for each ω and each $T > 0$,

$$\int_{-T}^T T^{-1}(T-|\tau|)\,R_X(\tau)\,e^{-j\omega\tau}\,d\tau = T^{-1}\int_0^T \int_0^T R_X(t-s)\,e^{-j\omega(t-s)}\,dt\,ds \quad (4.16)$$

and

$$\lim_{T\to\infty}\int_{-T}^T T^{-1}(T-|\tau|)\,R_X(\tau)\,e^{-j\omega\tau}\,d\tau = S_X(\omega). \quad (4.17)$$

Note that (4.16) and (4.17) establish that, for each ω, $S_X(\omega)$ is the limit of a sequence of nonnegative numbers, and the limit of a sequence of nonnegative numbers cannot be negative. Thus, property 3 is verified once we derive (4.16) and (4.17).

To derive (4.16), we use a change-of-variable transformation $F(t,s) = (t, t-s)$ and then substitute τ for $t-s$. The transformation F maps the original region of integration $\{(t,s) : 0 \leq t \leq T, 0 \leq s \leq T\}$ onto the region $G = \{(t,\tau) : t-T \leq \tau \leq t, 0 \leq t \leq T\}$. The original region is a square with vertices at the points $(0,0)$, $(0,T)$, (T,T), and $(T,0)$; the region G is a parallelogram with corner points $(0,0)$, $(0,-T)$, $(T,0)$, and (T,T). As illustrated in Figure 4-1, the region G can be written as

$$G = G_1 \cup G_2,$$

where

$$G_1 = \{(t,\tau) : \tau \leq t \leq T, 0 \leq \tau \leq T\}$$

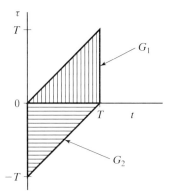

Figure 4-1: The region G.

and

$$G_2 = \{(t, \tau) : 0 \leq t \leq T + \tau, -T \leq \tau \leq 0\}.$$

From the decomposition of G into G_1 and G_2, it should be clear that the change-of-variable transformation permits the integral to be written as the sum of an integral over the region G_1 and an integral over the region G_2 as follows:

$$\int_0^T \int_0^T R_X(t - s) \, e^{-j\omega(t-s)} \, dt \, ds$$

$$= \int_0^T \left[\int_\tau^T R_X(\tau) \, e^{-j\omega\tau} \, dt \right] d\tau + \int_{-T}^0 \left[\int_0^{T+\tau} R_X(\tau) \, e^{-j\omega\tau} \, dt \right] d\tau$$

$$= \int_0^T (T - \tau) \, R_X(\tau) \, e^{-j\omega\tau} \, d\tau + \int_{-T}^0 (T + \tau) \, R_X(\tau) \, e^{-j\omega\tau} \, d\tau$$

$$= T \int_{-T}^T \frac{(T - |\tau|)}{T} R_X(\tau) \, e^{-j\omega\tau} \, d\tau.$$

This completes the derivation of (4.16).

To derive (4.17), note that if we define the function $\hat{R}_T$ by

$$\hat{R}_T(\tau) = \left[1 - \frac{|\tau|}{T} \right] R_X(\tau), \quad \text{for} \quad |\tau| \leq T$$

and $\hat{R}_T(\tau) = 0$ for $|\tau| > T$, then, for each fixed τ,

$$\lim_{T \to \infty} \hat{R}_T(\tau) = R_X(\tau).$$

Therefore,

$$\lim_{T \to \infty} \int_{-T}^{T} \left(1 - \frac{|\tau|}{T} \right) R_X(\tau) \, e^{-j\omega\tau} \, d\tau = \lim_{T \to \infty} \int_{-\infty}^{\infty} \hat{R}_T(\tau) \, e^{-j\omega\tau} \, d\tau$$

$$= \int_{-\infty}^{\infty} \lim_{T \to \infty} \hat{R}_T(\tau) \, e^{-j\omega\tau} \, d\tau$$

$$= \int_{-\infty}^{\infty} R_X(\tau) \, e^{-j\omega\tau} \, d\tau = S_X(\omega).$$

This derivation of (4.17) assumes that the order of the limit and the integration can be interchanged as in the foregoing sequence of steps. Because $|\hat{R}_T(\tau)| \leq |R_X(\tau)|$ for $T > 0$ and $-\infty < \tau < \infty$, this interchange is valid, for instance, if

$$\int_{-\infty}^{\infty} |R_X(\tau)| \, d\tau < \infty,$$

which is (4.3). The proof of the validity of the interchange follows from the dominated convergence theorem, which is stated and used shortly in the derivation of property 4. Before moving on to property 4, however, it is worth making two further observations about property 3. First, property 3 is actually valid for all processes of interest in engineering, even if (4.3) is not satisfied. However, the derivation of property 3 in the more general situation is more difficult than that just given. Second, we will show that property 3 also follows from (4.3) and property 4 (which is derived later).

The proof that (4.3) and property 4 imply property 3 is by contradiction. We suppose that (4.3) and property 4 hold. We then assert that property 3 is false and show that this leads to a contradiction. Because of our assertion that property 3 is false, there is at least one value of ω_0 for which $S_X(\omega_0) < 0$. By property 1, $S_X(-\omega_0) < 0$ also, so we can assume that $\omega_0 > 0$ if we wish. Since (4.3) is satisfied, property 4 implies that $S_X(\omega)$ is *continuous*, so there must exist an ω_L and an ω_U such that $\omega_L < \omega_0 < \omega_U$ and $S_X(\omega) < 0$ for all ω in the interval (ω_L, ω_U). Let $Y(t)$ be the output of a bandpass filter $H(\omega)$ defined by letting $H(\omega) = 1$ for $|\omega|$ in the interval (ω_L, ω_U) and letting $H(\omega) = 0$, otherwise. Then

$$2\pi R_Y(0) = \int_{-\infty}^{\infty} S_Y(\omega) \, d\omega < 0,$$

which is a *contradiction* because $R_Y(0) \geq 0$ for any wide-sense stationary random process $Y(t)$. Hence, there cannot exist such an ω_0 as we have hypothesized; that is, it must be that $S_X(\omega) \geq 0$ for *all* ω. This proves that property 3 follows from (4.3) and property 4.

Notice that an *ideal* bandpass filter is not required in the preceding derivation. Any sufficiently close approximation suffices. All that is actually required of $H(\omega)$ is that

$$\int_{\omega_L}^{\omega_U} |S_X(\omega)| \, |H(\omega)|^2 \, d\omega > \int_{0}^{\omega_L} S_X(\omega) \, |H(\omega)|^2 \, d\omega + \int_{\omega_U}^{\infty} S_X(\omega) \, |H(\omega)|^2 \, d\omega.$$

The next step is to derive property 4. We make use of a result in mathematics known as the *dominated convergence theorem*, which gives one set of conditions under which it is valid to interchange the order of a limit and an integral. We have already used this theorem implicitly in establishing other results (e.g., (4.17)).

Theorem 4-1   Suppose that $f(x, y)$ is an integrable function of x for each y, and suppose also that

$$\lim_{y \to y_0} f(x, y) = f(x, y_0)$$

for each x. If there exists a function g such that

$$|f(x, y)| \le g(x)$$

for each x and y, and if g satisfies

$$\int_{-\infty}^{\infty} g(x) \, dx < \infty,$$

then

$$\lim_{y \to y_0} \int_{-\infty}^{\infty} f(x, y) \, dx = \int_{-\infty}^{\infty} \lim_{y \to y_0} f(x, y) \, dx = \int_{-\infty}^{\infty} f(x, y_0) \, dx. \qquad \blacksquare$$

This theorem is applied to our problem by letting

$$f(\tau, \omega) = R_X(\tau) \, e^{-j\omega\tau}$$

and noticing that if $g(\tau) = |R_X(\tau)|$, then $|f(\tau, \omega)| \le g(\tau)$ for each ω. Also, $f(\tau, \omega)$ is a continuous function of ω for each τ, so

$$\lim_{\omega \to \omega_0} f(\tau, \omega) = f(\tau, \omega_0) = R_X(\tau) \, e^{-j\omega_0\tau}.$$

Therefore,

$$\lim_{\omega \to \omega_0} S_X(\omega) = \lim_{\omega \to \omega_0} \int_{-\infty}^{\infty} R_X(\tau) \, e^{-j\omega\tau} \, d\tau = \lim_{\omega \to \omega_0} \int_{-\infty}^{\infty} f(\tau, \omega) \, d\tau.$$

An application of the dominated convergence theorem to the last expression establishes that

$$\lim_{\omega \to \omega_0} S_X(\omega) = \int_{-\infty}^{\infty} f(\tau, \omega_0) \, d\tau = \int_{-\infty}^{\infty} R_X(\tau) \, e^{-j\omega_0\tau} \, d\tau = S_X(\omega_0).$$

Thus, we have shown that, for any ω_0,

$$\lim_{\omega \to \omega_0} S_X(\omega) = S_X(\omega_0),$$

which is just the definition of continuity for the function S_X at the point ω_0.
    Now for a word of warning: Although the requirement

$$\int_{-\infty}^{\infty} |R_X(\tau)| \, d\tau < \infty$$

is not essential for several of our results (e.g., property 3), it is critical to property 4. In fact, it is easy to find examples of random processes for which

$$\int_{-\infty}^{\infty} |R_X(\tau)| \, d\tau = \infty$$

and the corresponding spectral densities have discontinuities. For instance, if

$$R_X(\tau) = 2 W \left[\frac{\sin 2\pi W\tau}{2\pi W\tau} \right],$$

then $S_X(\omega) = 1$ for $-2\pi W \leq \omega \leq 2\pi W$ and $S_X(\omega) = 0$, otherwise. This spectral density function has discontinuities at $-2\pi W$ and $+2\pi W$. The example $R_X(\tau) = \cos \tau$ is even worse: *Its spectral density consists of delta functions.*

4.5   Spectrum of Amplitude-Modulated Signals

Consider a signal of the form

$$Y(t) = \sqrt{2}\, A(t) \cos(\omega_c t + \Theta), \tag{4.18}$$

where $A(t)$ is a random process representing the amplitude modulation and Θ is a random variable representing the phase of the sinusoidal carrier upon which $A(t)$ is modulated. The carrier signal is $\sqrt{2}\cos(\omega_c t + \Theta)$. For analog communications, $A(t)$ may represent a speech signal. Such a signal is very noiselike, so for many applications it is best modeled as a random process. In a digital communication system, $A(t)$ is typically a continuous-time waveform that represents a sequence of data pulses, and there is randomness in the data sequence represented by the sequence of amplitudes of these pulses. In addition, there may be a random time shift in the sequence of pulses, which represents the delay in the communication channel or the random starting time of the transmission.

In nearly all systems, the phase angle Θ and the information-bearing signal $A(t)$ originate from different physical mechanisms, and it is appropriate to model them as statistically independent. In all that follows, we assume that Θ and $A(t)$ are independent, which implies that

$$E\{Y(t)\} = \sqrt{2}\, E\{A(t)\}\, E\{\cos(\omega_c t + \Theta)\}$$

and

$$E\{Y(t+\tau)\, Y(t)\} = 2\, E\{A(t+\tau)\, A(t)\}\, E\{\cos[\omega_c(t+\tau) + \Theta]\cos(\omega_c t + \Theta)\}.$$

We also assume that $A(t)$ is a wide-sense stationary process with autocorrelation function

$$R_A(\tau) = E\{A(t+\tau)\, A(t)\}.$$

By the trigonometric identity for the product of the cosines, we see that

$$E\{Y(t+\tau)\, Y(t)\} = R_A(\tau)\, E\{\cos(\omega_c \tau) + \cos[\omega_c(2t+\tau) + 2\Theta]\}. \tag{4.19}$$

As is always true for phase angles, the random variable Θ appears inside a trigonometric function, so we can subtract multiples of 2π from Θ without changing its effective value. Thus, there is no harm in limiting the range of Θ to the interval $[0,2\pi]$. This is accomplished formally by considering the phase angle modulo 2π: If φ' takes values in the interval $-\infty < \varphi' < \infty$ and φ takes values in the interval $0 \leq \varphi < 2\pi$, then φ' is equal to φ modulo 2π if $\varphi = \varphi' + 2\pi n$ for some choice of the integer n.

For most applications, the distribution for the random variable Θ should be modeled as the uniform distribution on the interval $[0,2\pi]$. If Θ is so distributed, then

$$E\{\cos(\omega_c t + \Theta)\} = 0 \tag{4.20a}$$

and

$$E\{\cos[\omega_c(2t + \tau) + 2\Theta]\} = 0. \tag{4.20b}$$

Equation (4.20a) implies that $E\{Y(t)\} = 0$ for all t. Because the term $\cos(\omega_c\tau)$ is deterministic, it follows from (4.19) and (4.20b) that

$$E\{Y(t+\tau)Y(t)\} = R_A(\tau)\cos(\omega_c\tau).$$

We conclude that if $A(t)$ is wide-sense stationary, then $Y(t)$ is a zero-mean, wide-sense stationary random process with autocorrelation function

$$R_Y(\tau) = R_A(\tau)\cos(\omega_c\tau). \tag{4.21}$$

The *modulation theorem of Fourier transforms* states that if $v(t)$ has Fourier transform $V(f)$, then the Fourier transform of the signal

$$w(t) = v(t)\cos(\omega_c t) \tag{4.22a}$$

is

$$W(f) = \frac{V(f - f_c) + V(f + f_c)}{2}, \tag{4.22b}$$

where $f_c = \omega_c/2\pi$. The theorem is easily proved by using the fact that multiplication in the time domain corresponds to convolution in the frequency domain (e.g., see (4.8)). The proof is completed by noting that the Fourier transform of $\cos(2\pi f_c t)$ is

$$\frac{\delta(f - f_c) + \delta(f + f_c)}{2}$$

and that the convolution of $\delta(f \pm f_c)$ with $V(f)$ is $V(f \pm f_c)$. An alternative proof is outlined in Problem 4.6.

If the modulation theorem is applied to (4.21), we find that the spectral density for $Y(t)$ is given by

$$S_Y(f) = \frac{S_A(f - f_c) + S_A(f + f_c)}{2}, \tag{4.23}$$

where $S_A(f)$ denotes the spectral density for $A(t)$.

Example 4-1   A Random Data Signal with Rectangular Pulses

We consider one model for a random sequence of rectangular pulses in which each pulse has unit amplitude and duration T. Thus, the basic waveform is the rectangular pulse defined by

$$p_T(t) = \begin{cases} 1, & 0 \le t < T, \\ 0, & \text{otherwise.} \end{cases}$$

Define a random process by

$$D(t) = \sum_{n=-\infty}^{\infty} A_n \, p_T(t - nT),$$

where A_n is a random variable that represents the amplitude of the nth pulse. Note that the pulses do not overlap: The nth pulse is nonzero only in the interval $nT \le t < (n+1)T$. Even if the random variables A_n, $-\infty < n < \infty$, are independent and identically distributed, the random process $D(t)$ is still not wide-sense stationary. To see this, simply compare $E\{D(T/4)\,D(3T/4)\}$ with $E\{D(3T/4)\,D(5T/4)\}$. The former is $E\{A_0^2\}$ while the latter is $E\{A_0\,A_1\} = E\{A_0\}\,E\{A_1\}$. Suppose, for example, that for each n, the mean of A_n is zero and the variance of A_n is unity. Then

$$E\{D(T/4)\,D(3T/4)\} = 1,$$

but

$$E\{D(3T/4)\,D(5T/4)\} = 0.$$

Similarly, $E\{D(T/4)\,D(T/2)\} = 1$, while $E\{D(-T/8)\,D(T/8)\} = 0$.   ∎

Example 4-1 illustrates the need for a random time offset between the data signal and the carrier if it is desired to model the amplitude modulation signal as a wide-sense stationary random process. Let the data pulse waveform be represented by the function ζ. For instance, it might be that $\zeta(t) = p_T(t)$, the rectangular pulse of Example 4-1. Suppose that

$$D(t) = \sum_{n=-\infty}^{\infty} A_n \, \zeta(t - nT), \tag{4.24}$$

as in that example, and the signal

$$A(t) = D(t - U), \quad -\infty < t < \infty,$$

where U is a random variable that is uniformly distributed on the interval $[0, T]$. The variable U is a random time offset, which may be used to account for a random starting time or propagation delay between the transmitter and receiver. As long as the random variables A_n, $-\infty < n < \infty$, are independent and identically distributed, the range of U can be restricted to the interval $[0, T]$ for essentially the same reason that phase angles can be restricted to the interval $[0, 2\pi]$. Moreover, only minor changes are required in the development that follows if U is uniformly distributed on any interval of the form $[n_1 T, n_2 T]$ for arbitrary integers n_1 and n_2 (with $n_1 < n_2$, of course).

Suppose that the data sequence A_n, $-\infty < n < \infty$, is a sequence of independent random variables with $E\{A_n\} = 0$ and $E\{A_n^2\} = \alpha^2$ for each n. First, notice that $E\{A_n\} = 0$ for each n implies that

$$E\{A(t)\} = E\{D(t)\} = 0$$

for each t. In particular, the random process $A(t)$ has a constant mean. Next, notice that because the random variables A_n, $-\infty < n < \infty$, are independent, they are uncorrelated. Also, because the mean of each A_n is zero,

$$E\{A_n A_k\} = 0$$

for $n \neq k$.

The next step is to evaluate $E\{A(t + \tau) A(t)\}$ and show that it does not depend on t. This, together with the fact that $A(t)$ has a constant mean, implies that $A(t)$ is a wide-sense stationary random process, so that we can consider its spectral density.

For any t and τ,

$E\{A(t + \tau) A(t)\}$

$$= E\left\{ \sum_{n=-\infty}^{\infty} A_n \zeta(t + \tau - n T - U) \sum_{k=-\infty}^{\infty} A_k \zeta(t - k T - U) \right\}$$

$$= \sum_{n=-\infty}^{\infty} \sum_{k=-\infty}^{\infty} E\{A_n A_k\} E\{\zeta(t + \tau - n T - U) \zeta(t - k T - U)\}$$

$$= \sum_{n=-\infty}^{\infty} \alpha^2 E\{\zeta(t + \tau - n T - U) \zeta(t - n T - U)\}. \qquad (4.25)$$

The last step follows from the fact that $E\{A_n A_k\} = 0$ for $n \neq k$. Because the random variable U is uniformly distributed on the interval $[0, T]$,

$E\{\zeta(t + \tau - n T - U) \zeta(t - n T - U)\}$

$$= T^{-1} \int_0^T \zeta(t + \tau - n T - u) \zeta(t - n T - u) \, du$$

$$= T^{-1} \int_{n T}^{(n+1)T} \zeta(t + \tau - v) \zeta(t - v) \, dv.$$

The last step follows from the change of variable $v = u + n T$. Substituting into (4.25) gives

$$
\begin{aligned}
E\{A(t+\tau)\,A(t)\} &= \sum_{n=-\infty}^{\infty} \alpha^2 T^{-1} \int_{nT}^{(n+1)T} \zeta(t+\tau-v)\,\zeta(t-v)\,dv \\
&= \alpha^2 T^{-1} \sum_{n=-\infty}^{\infty} \int_{nT}^{(n+1)T} \zeta(t+\tau-v)\,\zeta(t-v)\,dv \\
&= \alpha^2 T^{-1} \int_{-\infty}^{\infty} \zeta(t+\tau-v)\,\zeta(t-v)\,dv \\
&= \alpha^2 T^{-1} \int_{-\infty}^{\infty} \zeta(u)\,\zeta(u-\tau)\,du.
\end{aligned}
\tag{4.26}
$$

The last step is just the change of variable $u = t + \tau - v$. Note in particular that

$$
E\{A^2(t)\} = \alpha^2 T^{-1} \int_{-\infty}^{\infty} \zeta^2(u)\,du.
$$

For $\zeta(t) = p_T(t)$, this gives $E\{A^2(t)\} = \alpha^2$.

It follows from (4.26) that $E\{A(t+\tau)\,A(t)\}$ does not depend on t, so the random process $A(t)$ is a zero-mean, wide-sense stationary random process with autocorrelation function

$$
R_A(\tau) = \alpha^2 T^{-1} \int_{-\infty}^{\infty} \zeta(u)\,\zeta(u-\tau)\,du.
\tag{4.27}
$$

This expression can also be written as

$$
R_A(\tau) = \alpha^2 T^{-1}(\zeta * \tilde{\zeta})(\tau),
\tag{4.28}
$$

where $\tilde{\zeta}(t) = \zeta(-t)$, $-\infty < t < \infty$. That is, the function $\tilde{\zeta}$ is the time reverse of the pulse function ζ, and R_A is proportional to the convolution of the pulse waveform and its time reverse. Note the similarity between (4.28) and the results for time-invariant linear filtering of random processes (e.g., (3.53)).

Let Z denote the Fourier transform of ζ. We know that the Fourier transform of $\tilde{\zeta}$ is Z^*, the complex conjugate of Z. (This property is derived in Section 4.3.) In a manner analogous to the relationship between (3.51) and (4.10), it follows from (4.28) that the spectral density function for the random process $A(t)$ is given by

$$
S_A(f) = \alpha^2 T^{-1}|Z(f)|^2.
\tag{4.29}
$$

This result can be employed in (4.23) to show that the spectral density function for the amplitude-modulated signal

$$
Y(t) = \sqrt{2}\,A(t)\cos(\omega_c t + \Theta)
$$

is

$$
S_Y(f) = \frac{\alpha^2 \left\{ |Z(f-f_c)|^2 + |Z(f+f_c)|^2 \right\}}{2T}.
\tag{4.30}
$$

Example 4-2   A Random Data Signal with Rectangular Pulses (Revisited)

Consider a signal of the form

$$Y(t) = \sqrt{2}\, D(t - U) \cos(\omega_c t + \Theta), \qquad (4.31)$$

where

$$D(t) = \sum_{n=-\infty}^{\infty} A_n\, p_T(t - nT), \qquad (4.32)$$

as in Example 4-1. Notice that $D(t)$ is just a sequence of rectangular pulses of duration T. The amplitudes of these pulses are determined by the sequence A_n, $-\infty < n < \infty$. If we ignore the random time delay for the moment (i.e., temporarily set $U = 0$), then, for each value of t in the range $nT \le t < (n+1)T$,

$$Y(t) = \sqrt{2}\, A_n \cos(\omega_c t + \Theta).$$

Information can be conveyed in the sequence of amplitudes of this signal, which is a common method in digital communications. We are interested in determining the spectral density of $Y(t)$ if U is modeled as a random variable that is uniformly distributed on $[0, T]$, Θ is uniformly distributed on $[0, 2\pi]$, and the sequence of amplitudes is a sequence of independent random variables satisfying $E\{A_n\} = 0$ and $E\{A_n^2\} = \alpha^2$ for each n. The random variables U, Θ, and A_n, $-\infty < n < \infty$, are assumed to be mutually independent. The determination of the spectral density for $Y(t)$ is accomplished by an application of (4.30) with $\zeta(t) = p_T(t)$.

There are two ways to proceed. We can use the fact that $P_T(f)$, the Fourier transform of the unit-amplitude rectangular pulse of duration T, satisfies

$$|P_T(f)| = |(\pi f)^{-1} \sin(\pi f T)|, \qquad (4.33)$$

so that for $\zeta(t) = p_T(t)$,

$$|Z(f)|^2 = (\pi f)^{-2} \sin^2(\pi f T).$$

The second approach is to use the fact that if $\zeta(t) = p_T(t)$, the function $\zeta * \tilde{\zeta}$ is a triangular function centered at the origin with base $2T$ and height T. In fact, $(\zeta * \tilde{\zeta})(\tau) = f(\tau)$, the function derived in Exercise 3-5 and illustrated in Figure 3-4. We can then use the fact that the Fourier transform of such a triangular pulse is $(\pi f)^{-2} \sin^2(\pi f T)$, the same result as that obtained in the first approach. Using the fact that $\operatorname{sinc}(x) = (\pi x)^{-1} \sin(\pi x)$, we see that $|Z(f)|^2$ can be written more compactly as

$$|Z(f)|^2 = T^2 \operatorname{sinc}^2(fT). \qquad (4.34)$$

It follows from (4.29) that the spectral density for $A(t)$ is

$$S_A(f) = \alpha^2 T \operatorname{sinc}^2(fT),$$

and it follows from (4.30) that the spectral density for $Y(t)$ is

$$S_Y(f) = \frac{\alpha^2 T \{\operatorname{sinc}^2[(f - f_c)T] + \operatorname{sinc}^2[(f + f_c)T]\}}{2}. \qquad (4.35)$$

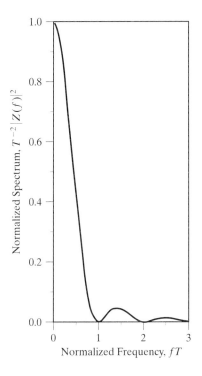

Figure 4-2: Spectral density for a random data signal with a rectangular pulse waveform.

The spectrum of (4.35) can best be illustrated by taking advantage of the fact that its shape is derived from the spectrum

$$|Z(f)|^2 = T^2 \operatorname{sinc}^2(fT).$$

It is convenient to normalize $|Z(f)|^2$ by dividing it by T^2 and then plot the normalized spectrum as a function of fT. Because the units of f are inverse seconds (hertz) and the units of T are seconds, fT is a dimensionless parameter. A graph of $T^{-2}|Z(f)|^2$ as a function of the normalized frequency is shown in Figure 4-2 for $0 \leq fT \leq 3$. This spectrum is an even function of fT, so there is no need to show the graph for negative frequencies.

It is interesting to compare the results we have obtained by considering the communication signal as a random process with the corresponding results that are obtained if the signal is modeled as a deterministic signal. The key step in the latter approach is to use the energy spectrum for the deterministic signal. If the Fourier transform of the deterministic signal $f(t)$ is denoted by $F(\omega)$, then the *energy spectrum* of the signal is defined to be $|F(\omega)|^2$. Thus, the energy spectrum for the pulse waveform $\zeta(t)$ is $|Z(f)|^2$. Notice that the shape of the spectral density given in (4.30) depends only on the energy spectrum of the pulse waveform $\zeta(t)$ and the carrier frequency ω_c. This relationship is pursued further in Problem 4.8.

4.6   Bandpass Frequency Functions

Recall that if V is the Fourier transform of v, then

$$V(\omega) = \int_{-\infty}^{\infty} v(t)\, e^{-j\omega t}\, dt \tag{4.36a}$$

and

$$v(t) = \int_{-\infty}^{\infty} V(\omega)\, e^{+j\omega t}\, \frac{d\omega}{2\pi}, \tag{4.36b}$$

where the variable ω represents frequency in radians per second. In terms of the frequency variable f in hertz, these equations become

$$V(f) = \int_{-\infty}^{\infty} v(t)\, e^{-j2\pi f t}\, dt \tag{4.37a}$$

and

$$v(t) = \int_{-\infty}^{\infty} V(f)\, e^{+j2\pi f t}\, df. \tag{4.37b}$$

In this section, much of the development is in terms of inverse Fourier transforms, for which it is convenient to use transform expressions (4.37) in order to avoid the need to carry along the factor 2π. Thus, all Fourier transforms in this section are given in terms of the variable f, the frequency in hertz.

The first result that we need follows easily from (4.37a): If $v(t)$ is real for each t, and V is the Fourier transform of v, then

$$\begin{aligned}
V^*(f) &= \left\{ \int_{-\infty}^{\infty} v(t)\, e^{-j2\pi f t}\, dt \right\}^* \\
&= \left\{ \int_{-\infty}^{\infty} v(t)\, e^{+j2\pi f t}\, dt \right\} = V(-f). \tag{4.38}
\end{aligned}$$

Additional steps are given for this derivation in Section 4.2, where it is shown that $[S_X(\omega)]^* = S_X(-\omega)$ for each ω. Recall that S_X is the Fourier transform of the function R_X, and $R_X(\tau)$ is real for each τ. We refer to the property wherein $V^*(f) = V(-f)$ as *conjugate symmetry*, so (4.38) is summarized by saying that the Fourier transform of a real function has conjugate symmetry. If a time function is real, its Fourier transform has conjugate symmetry.

4.6.1   Alternative Definitions of the Bandwidth for Frequency Functions

The time-domain functions of interest in this section not only are real-valued functions, but also have Fourier transforms that are band limited in a special way. Suppose that W is a function that represents a frequency-domain mathematical description of some entity such as a signal, filter, or noise process. For example, W might be the transfer function for a time-invariant linear system, or it might be the spectral density for a wide-sense stationary random process. Because the function W characterizes the entity in the frequency domain, we refer to W as a *frequency function*. Its inverse transform is a real function that characterizes the entity in the time domain. Examples include the impulse response for a time-invariant linear filter and the autocorrelation function for a random process.

Although all of the illustrations in this section are for real frequency functions, all of the analysis is for complex-valued functions. Of course, this analysis applies to real frequency functions as a special case. Spectral densities are real functions, but transfer functions for filters of interest in applications are usually complex valued. A complex-valued frequency function can always be written as

$$W(f) = A(f) \exp \{ j \Psi(f) \},$$

where A and Ψ are real frequency functions that represent the amplitude and phase, respectively, of the frequency function W. Notice that W has conjugate symmetry if and only if $A(f) = A(-f)$ and $\Psi(f) = -\Psi(-f)$ for all f.

A complete illustration of a complex-valued frequency function requires the display of both its amplitude and phase functions. Often, the amplitude function is of greater interest than the phase function and is the only one of the two that is displayed. One of the features that we wish to examine is the bandwidth of the frequency function, a parameter that depends only on the amplitude function. Our illustrations of real frequency functions can be viewed as illustrations of the amplitude functions for complex-valued frequency functions.

If there are two frequencies f' and f'' such that $0 < f' < f''$ and for which $W(f) \approx 0$ for all f outside the intervals $[-f'', -f']$ and $[f', f'']$, the frequency function W is referred to as a *bandpass frequency function*. If $W(f) = 0$ for all frequencies f outside the intervals $[-f'', -f']$ and $[f', f'']$, W is referred to as an *ideal bandpass frequency function*. The frequencies f' and f'' are not unique, because it is always possible to increase f'' and decrease f'. For the ideal bandpass frequency functions that are considered in this book, it is possible to let f_1 be the largest value of f' and f_2 the smallest value of f'' for which $W(f) = 0$ for all frequencies f outside the intervals $[-f'', -f']$ and $[f', f'']$. The resulting interval $[f_1, f_2]$ is referred to as the *frequency support* for W. An example of a bandpass frequency function is illustrated in Figure 4-3.

If the ideal bandpass frequency function W is a continuous function, the frequencies f_1 and f_2 can be defined mathematically by

$$f_1 = \max\{u : W(f) = 0 \text{ for } 0 \le f \le u\} \tag{4.39a}$$

and

$$f_2 = \min\{v : W(f) = 0 \text{ for } f \ge v\}. \tag{4.39b}$$

Continuity guarantees that the maximum and minimum in (4.39) exist. The frequencies f_1 and f_2 represent the lower and upper *cutoff frequencies*, respectively, for the ideal

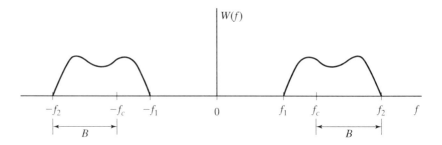

Figure 4-3: An ideal bandpass frequency function.

bandpass frequency function W. The difference $f_2 - f_1$ is referred to as the *absolute bandwidth* or *cutoff bandwidth* of the bandpass frequency function. Thus, the absolute bandwidth is just the width of the frequency support for W. Notice that because W has conjugate symmetry, its cutoff frequencies and its bandwidth can be defined by considering positive frequencies only.

Other measures of bandwidth are useful in characterizing the spectral occupancy of signals and random processes. We mention here only two of the many alternative definitions. The first is the half-power bandwidth for a filter with transfer function W. Let W_m be the maximum value of $|W(f)|$, $-\infty < f < \infty$; this represents the maximum gain of the filter across the frequency band. The transfer function W has *half-power bandwidth* $f_4 - f_3$ if $0 < f_3 < f_4$ and if each of the following is true:

$$|W(f_3)| = |W(f_4)| = \frac{W_m}{\sqrt{2}},$$

$$|W(f)| > \frac{W_m}{\sqrt{2}} \quad \text{for } f_3 < f < f_4,$$

and

$$|W(f)| < \frac{W_m}{\sqrt{2}} \quad \text{for } 0 < f < f_3 \text{ and } f_4 < f < \infty.$$

Notice that a transfer function may not have a well-defined half-power bandwidth. (See Problem 4.11.) If the half-power bandwidth exists, then the frequencies f_3 and f_4 are such that the power transfer function $|W(f)|^2$ satisfies

$$|W(f_3)|^2 = |W(f_4)|^2 = \frac{(W_m)^2}{2}.$$

If W represents a power spectral density, the conditions for f_3 and f_4 become

$$W(f_3) = W(f_4) = \frac{W_m}{2}.$$

Because $10 \log_{10}(1/2) \approx -3$ dB, the half-power bandwidth is often referred to as the 3-dB bandwidth.

Another bandwidth of interest is the *null-to-null bandwidth*. Under the following conditions, $f_6 - f_5$ is the null-to-null bandwidth of the frequency function W: f_5 and f_6 are such that $0 < f_5 < f_6$, the maximum value of $W(f)$ occurs between frequencies f_5 and f_6, $W(f) \neq 0$ for $f_5 < f < f_6$, and $W(f_5) = W(f_6) = 0$. For the frequency function illustrated in Figure 4-3, the null-to-null bandwidth is equal to $f_2 - f_1$, the same as the absolute bandwidth. As another example, it is easy to show that the null-to-null bandwidth of the spectral density

$$S_Y(f) = \frac{\alpha^2 T \{ \operatorname{sinc}^2[(f - f_c)T] + \operatorname{sinc}^2[(f + f_c)T] \}}{2}$$

is $2/T$. This spectral density is derived in Example 4-2.

4.6.2   Time-Domain Descriptions of Ideal Bandpass Frequency Functions

In this section, we seek the time-domain characterization of an ideal bandpass frequency function W that has conjugate symmetry (i.e., W is the Fourier transform of a real function). It is shown that if f_c is a suitable frequency, then the inverse Fourier transform of W can be written in the form

$$w(t) = v_1(t) \cos(2\pi f_c t) - v_2(t) \sin(2\pi f_c t),$$

where v_1 and v_2 are baseband functions. A convenient representation for $w(t)$ is

$$w(t) = \operatorname{Re}\{\tilde{v}(t) \exp(j 2\pi f_c t)\},$$

where $\tilde{v}(t) = v_1(t) + j v_2(t)$. The complex representation of signals and linear systems is described in Appendix C. This representation is the primary reason for writing $w(t)$ as a difference of two signals rather than as a sum; however, both conventions are common in the literature.

The choice of the frequency f_c depends on the application. If $w(t)$ represents a carrier-modulated communication signal, f_c is almost always the carrier frequency of the signal. If $w(t)$ is the impulse response of a bandpass filter, f_c might be selected to be the center frequency for this filter, or it might be the carrier frequency of a signal that is the input to the filter. Under normal circumstances, f_c is located somewhere near the middle of the frequency support of the frequency function W, but this is not a requirement. All that is required is that f_c be within the frequency support of W; that is, if W has lower and upper cutoff frequencies f_1 and f_2, respectively, then $f_1 < f_c < f_2$. In this section, it is shown that for such a choice of f_c, if the absolute bandwidth of W is not greater than $2B$, then v_1 and v_2 have Fourier transforms that are identically zero for $|f| \geq B$.

If W is an ideal bandpass frequency function with lower and upper cutoff frequencies f_1 and f_2, respectively, and if f_c is between f_1 and f_2, then B can be defined by

$$B = \max\{f_2 - f_c, f_c - f_1\}.$$

That is, B is the frequency separation between f_c and the cutoff frequency that is farthest from f_c. It follows that $W(f) = 0$ for all f outside the intervals $[-f_c - B, -f_c + B]$

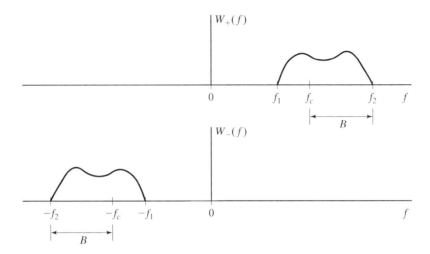

Figure 4-4: The positive and negative parts of W.

and $[f_c - B, f_c + B]$, so the absolute bandwidth is not greater than $2B$. The relationship that exists among f_1, f_2, f_c, and B is illustrated in Figure 4-3.

For the mathematical development that follows, we concentrate on ideal bandpass frequency functions. In practice, however, it is usually sufficient if $W(f)$ is *approximately* zero outside the intervals $[-f_2, -f_1]$ and $[f_1, f_2]$ in order that the time and frequency functions have the key properties derived in this section. Under this weaker condition, v_1 and v_2 have Fourier transforms that are approximately zero for $|f| \geq B$.

Let W_+ be the *positive part* of the bandpass frequency function W; that is,

$$W_+(f) = \begin{cases} W(f), & f > 0, \\ 0, & f \leq 0. \end{cases} \qquad (4.40)$$

The positive part of W is illustrated in Figure 4-4. Notice that the frequency support for the positive part of W is a subset of the interval $(0, \infty)$.

Similarly, let W_- be the *negative part* of W; that is, $W_-(f) = W(f)$ for $f < 0$ and $W_-(f) = 0$ for $f \geq 0$. The negative part of W is also illustrated in Figure 4-4. The frequency support for the negative part of W is a subset of the interval $(-\infty, 0)$. The bandpass frequency function W can be written as

$$W(f) = W_+(f) + W_-(f).$$

The frequency support for W_+ and the frequency support for W_- are disjoint for a bandpass frequency function. That is, for each frequency f, either $W_+(f) = 0$ or $W_-(f) = 0$. This can be expressed concisely in terms of the product of the two functions:

$$W_+(f)\, W_-(f) = 0, \quad -\infty < f < \infty.$$

Because $W(f)$ is the Fourier transform of a real function, it has conjugate symmetry: $W^*(f) = W(-f)$. But this implies that $[W_+(f)]^* = W_-(-f)$, which guarantees that the inverse Fourier transform of W can be written in terms of the function W_+ only. The inverse transform of W is given by

$$
\begin{aligned}
w(t) &= \int_{-\infty}^{\infty} W(f)\,e^{+j2\pi f t}\,\mathrm{d}f \\
&= \int_{-\infty}^{\infty} W_+(f)\,e^{+j2\pi f t}\,\mathrm{d}f + \int_{-\infty}^{\infty} W_-(f)\,e^{+j2\pi f t}\,\mathrm{d}f.
\end{aligned}
\tag{4.41}
$$

But

$$
\begin{aligned}
\int_{-\infty}^{\infty} W_-(f)\,e^{+j2\pi f t}\,\mathrm{d}f &= \int_{-\infty}^{\infty} W_-(-u)\,e^{-j2\pi u t}\,\mathrm{d}u \\
&= \int_{-\infty}^{\infty} [W_+(u)]^*\,e^{-j2\pi u t}\,\mathrm{d}u.
\end{aligned}
$$

Therefore,

$$
\int_{-\infty}^{\infty} W_-(f)\,e^{+j2\pi f t}\,\mathrm{d}f = \left\{ \int_{-\infty}^{\infty} W_+(u)\,e^{+j2\pi u t}\,\mathrm{d}u \right\}^*.
\tag{4.42}
$$

Because $z + z^* = 2\,\mathrm{Re}\{z\}$ for an arbitrary complex number z, (4.41) and (4.42) imply that

$$
w(t) = 2\,\mathrm{Re}\left\{ \int_{-\infty}^{\infty} W_+(u)\,e^{+j2\pi u t}\,\mathrm{d}u \right\}.
\tag{4.43}
$$

Now define the frequency function Y by $Y(f) = 2\,W_+(f + f_c)$. For the frequency function W illustrated in Figures 4-3 and 4-4, the corresponding frequency function Y is as shown in Figure 4-5. It should be clear from the definition of Y that $Y(f) = 0$ for $|f| \geq B$ and $2\,W_+(u) = Y(u - f_c)$ for $-\infty < u < \infty$.

It follows from (4.43) that

$$
w(t) = \mathrm{Re}\left\{ \int_{-\infty}^{\infty} Y(u - f_c)\,e^{+j2\pi u t}\,\mathrm{d}u \right\}.
\tag{4.44}
$$

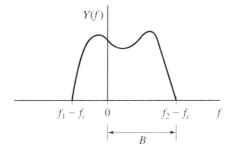

Figure 4-5: The frequency function $Y(f) = 2\,W_+(f + f_c)$.

By substituting f for $u - f_c$, we can rewrite (4.44) as

$$w(t) = \text{Re}\left\{ \int_{-\infty}^{\infty} Y(f) \exp\{j2\pi(f + f_c)t\} \, df \right\}.$$

The term $\exp\{j2\pi f_c t\}$ does not depend on f, so

$$w(t) = \text{Re}\left\{ \int_{-\infty}^{\infty} Y(f) \exp\{j2\pi f t\} \, df \exp\{j2\pi f_c t\} \right\}. \tag{4.45}$$

But the integral that appears in (4.45) is just the inverse Fourier transform integral for the frequency function Y. Thus, if y is the inverse transform of Y, then

$$w(t) = \text{Re}\{y(t) \exp\{j\,2\pi f_c\,t\}\}. \tag{4.46}$$

From (4.46), we obtain

$$w(t) = [\text{Re}\{y(t)\}] \cos(2\pi f_c t) - [\text{Im}\{y(t)\}] \sin(2\pi f_c t). \tag{4.47}$$

The expression we seek for the time-domain representation for the frequency function W follows from (4.47).

Recall that our goal is to express $w(t)$ in the form

$$w(t) = v_1(t) \cos(2\pi f_c t) - v_2(t) \sin(2\pi f_c t), \tag{4.48}$$

where v_1 and v_2 are baseband functions with Fourier transforms that are identically zero for $|f| \geq B$. The functions v_1 and v_2 needed for this representation can be obtained from the real and imaginary parts of $y(t)$. If

$$v_1(t) = \text{Re}\{y(t)\} \tag{4.49a}$$

and

$$v_2(t) = \text{Im}\{y(t)\}, \tag{4.49b}$$

then (4.47) is equivalent to (4.48). It remains to show that v_1 and v_2 are baseband functions with the desired bandwidth.

Before dealing with the bandwidth issue for v_1 and v_2, it is worthwhile to discuss the function y that is used to obtain those functions. It may seem at first glance that y ought to be a real function. That this is not true in general is illustrated by Figure 4-5. Notice that $Y(f)$ does not have conjugate symmetry in this example, so its inverse Fourier transform cannot be a real function. Even though $W(-f) = [W(f)]^*$ in this example, it is not true that $Y(-f) = [Y(f)]^*$. For some bandpass frequency functions, such as the one in the next example, no choice of the frequency f_c gives $Y(-f) = [Y(f)]^*$. For certain other bandpass frequency functions, f_c can be selected to give $Y(-f) = [Y(f)]^*$. (See Problem 4.12.)

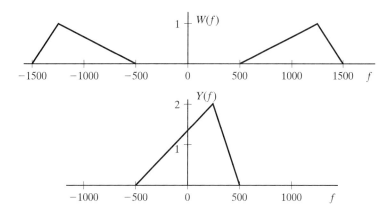

Figure 4-6: An example of a bandpass frequency function.

Example 4-3   A Bandpass Frequency Function

Consider the ideal bandpass frequency function defined by

$$W(f) = \begin{cases} (|f| - 500)/750, & 500 < |f| < 1250, \\ -(|f| - 1500)/250, & 1250 < |f| < 1500, \\ 0, & \text{otherwise.} \end{cases}$$

This frequency function is illustrated in Figure 4-6, along with the corresponding frequency function Y. Clearly, no choice of f_c leads to a function Y that has conjugate symmetry. ∎

The next step is to show that the Fourier transforms of v_1 and v_2 are identically zero for $|f| \geq B$. Recall from (4.49a) that v_1 is obtained from y by extracting the real part:

$$v_1(t) = \text{Re}\{y(t)\}.$$

Because $z + z^* = 2 \, \text{Re}\{z\}$ for an arbitrary complex number z,

$$v_1(t) = \frac{\{y(t) + y^*(t)\}}{2}. \tag{4.50}$$

But

$$y(t) = \int_{-\infty}^{\infty} Y(f) \, e^{+j2\pi f t} \, df \tag{4.51}$$

and

$$y^*(t) = \left[\int_{-\infty}^{\infty} Y(f) \, e^{+j2\pi f t} \, df \right]^*$$

$$= \int_{-\infty}^{\infty} Y^*(f) \, e^{-j2\pi f t} \, df.$$

It follows from the change of variable $u = -f$ that

$$y^*(t) = \int_{-\infty}^{\infty} Y^*(-u)\, e^{+j2\pi ut}\, du. \tag{4.52}$$

From (4.50)–(4.52), it follows that

$$v_1(t) = \tfrac{1}{2} \int_{-\infty}^{\infty} \{Y(f) + Y^*(-f)\}\, e^{+j2\pi ft}\, df. \tag{4.53}$$

Equation (4.53) implies that the Fourier transform of v_1 is given by

$$V_1(f) = \frac{\{Y(f) + Y^*(-f)\}}{2}. \tag{4.54a}$$

But $Y(f) = 0$ for $|f| \geq B$, so $Y^*(-f) = 0$ for $|f| \geq B$ as well. It then follows from (4.54a) that $V_1(f) = 0$ for $|f| \geq B$.

A similar development shows that the Fourier transform of v_2 is

$$V_2(f) = \frac{-j\{Y(f) - Y^*(-f)\}}{2}, \tag{4.54b}$$

which implies that $V_2(f) = 0$ for $|f| \geq B$. Thus, the functions v_1 and v_2 have Fourier transforms that are identically zero for $|f| \geq B$, as we set out to show.

Nearly all applications of bandpass frequency functions satisfy $B < f_c$. In fact, for most applications, $B \ll f_c$. For example, a typical VHF radio transmission system for digitized voice signals might have a bandwidth $(2B)$ of 25 kHz and employ a carrier frequency of 60 MHz, in which case B would be more than three orders of magnitude smaller than f_c.

Suppose that v is an ideal bandpass signal with absolute bandwidth not greater than $2B$. We have shown that such a signal can be expressed as

$$v(t) = v_1(t) \cos(2\pi f_c t) - v_2(t) \sin(2\pi f_c t),$$

where v_1 and v_2 are baseband functions whose Fourier transforms are identically zero for $|f| \geq B$. So far we have given only mathematical descriptions of v_1 and v_2. However, it is easy to show that if $B < f_c$, then v_1 and v_2 are baseband signals that can be generated by the system illustrated in Figure 4-7. The system consists of two multipliers and two lowpass filters, and the only need for the filters is to remove double-frequency components at the outputs of the multipliers. (See Problem 4.14 for further details.)

Recall that for a bandpass frequency function W that has conjugate symmetry, the frequency function Y is defined by $Y(f) = 2\, W_+(f + f_c)$, $-\infty < f < \infty$. There are bandpass frequency functions for which Y has conjugate symmetry for some choice of the frequency f_c. For Y to have such conjugate symmetry, the bandpass frequency function W must have an additional property known as local symmetry. A bandpass frequency function W is said to have *local symmetry* about f_c if, for each real number δ,

$$[W_+(f_c + \delta)]^* = W_+(f_c - \delta), \quad -\infty < f < \infty.$$

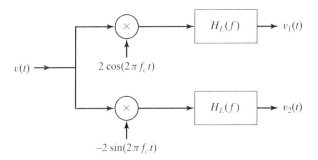

Figure 4-7: Generation of the baseband signals v_1 and v_2.

This condition is illustrated in Figure 4-8. It is easy to show that it is equivalent to $[W_-(-f_c + \delta)]^* = W_-(-f_c - \delta)$, $-\infty < f < \infty$. Because $Y(f) = 2\,W_+(f + f_c)$, local symmetry of W about f_c implies that $[Y(f)]^* = Y(-f)$ for all f. Thus, if the bandpass frequency function W has both local symmetry about f_c and conjugate symmetry, the function Y has conjugate symmetry.

If W has local symmetry about f_c, the time-domain representation of W is simpler than it is for a general bandpass frequency function. The local symmetry of W guarantees that Y has conjugate symmetry, and this in turn guarantees that y, the inverse Fourier transform of Y, is a real function. It follows that $\mathrm{Im}\{y(t)\}$ is identically zero, so (4.48) reduces to

$$w(t) = v_1(t)\cos(2\pi f_c t).$$

Because $y(t)$ is real,

$$v_1(t) = \mathrm{Re}\{y(t)\} = y(t), \quad -\infty < t < \infty.$$

As a result,

$$w(t) = y(t)\cos(2\pi f_c t). \tag{4.55}$$

The conclusion is that if W is a locally symmetric ideal bandpass frequency function, the function y is real and the representation of w is given by (4.55).

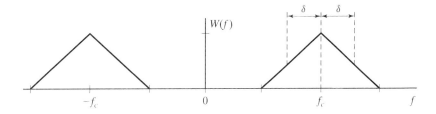

Figure 4-8: An ideal bandpass frequency function with local symmetry.

If the locally symmetric bandpass frequency function is the *transfer function* $H(f)$ for a linear time-invariant filter, we follow the convention that the impulse response of the filter is written as

$$h(t) = 2\,g(t)\cos(2\pi f_c t). \qquad (4.56)$$

The inclusion of the factor 2 in (4.56) leads to several notational advantages in the results that follow. If $h(t)$ and $g(t)$ are related as in that equation and h is the impulse response of a locally symmetric bandpass filter, the filter with impulse response g is referred to as the *baseband equivalent* of the bandpass filter. It follows from the modulation theorem of Fourier transforms that

$$H(f) = G(f - f_c) + G(f + f_c), \quad -\infty < f < \infty, \qquad (4.57)$$

and it follows from (4.57) that

$$G(f) = H_+(f + f_c), \quad -\infty < f < \infty.$$

Consider the problem of evaluating the output of an ideal bandpass filter if the input is a carrier-modulated signal. The evaluation is simplified greatly if the signal is an ideal bandpass signal and the transfer function of the filter has local symmetry. Suppose that the input signal is

$$w(t) = v_1(t)\cos(2\pi f_c t) - v_2(t)\sin(2\pi f_c t).$$

It should be clear from the definition of an ideal bandpass signal that, regardless of the transfer function of the filter, if the input to the filter is an ideal bandpass signal, the output must also be an ideal bandpass signal. It follows that the output can be written as

$$\hat{w}(t) = w_1(t)\cos(2\pi f_c t) - w_2(t)\sin(2\pi f_c t), \qquad (4.58)$$

for some baseband functions w_1 and w_2. If the filter impulse response is given by (4.56), these baseband functions are just the results of passing the signals v_1 and v_2 through the baseband equivalent filter. That is, $w_1 = v_1 * g$ and $w_2 = v_2 * g$. As a result, the output of a locally symmetric bandpass filter can be determined by working with the baseband functions v_1, v_2, and g. The reader may wish to write out the integral expression for convolving w with h in order to gain an appreciation for the amount of labor that is saved by the use of (4.58) together with $w_1 = v_1 * g$ and $w_2 = v_2 * g$.

To derive (4.58), first observe that it follows from the modulation theorem of Fourier transforms that the bandpass signal

$$w(t) = v_1(t)\cos(2\pi f_c t) - v_2(t)\sin(2\pi f_c t)$$

has Fourier transform

$$W(f) = \frac{V_1(f + f_c) + V_1(f - f_c)}{2} - \frac{j\{V_2(f + f_c) - V_2(f - f_c)\}}{2}. \qquad (4.59)$$

The Fourier transform of the output of the filter is $W(f)\,H(f)$, where

$$H(f) = G(f + f_c) + G(f - f_c)$$

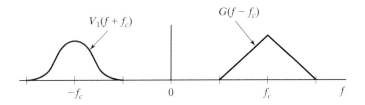

Figure 4-9: Illustration of the fact that $V_1(f + f_c)\,G(f - f_c) = 0$.

is the transfer function for the filter. Therefore, the output of the filter has terms of the form

$$\{V_i(f + f_c) \pm V_i(f - f_c)\}H(f)$$
$$= \{V_i(f + f_c) \pm V_i(f - f_c)\}\{G(f + f_c) + G(f - f_c)\}, \quad (4.60a)$$

or, equivalently,

$$\{V_i(f + f_c) \pm V_i(f - f_c)\}H(f)$$
$$= V_i(f + f_c)\,G(f + f_c) \pm V_i(f - f_c)\,G(f - f_c)\}, \quad (4.60b)$$

which implies that

$$\{V_i(f + f_c) \pm V_i(f - f_c)\}H(f) = W_i(f + f_c) \pm W_i(f - f_c), \quad (4.60c)$$

where $W_i(f) = V_i(f)\,G(f)$ for $-\infty < f < \infty$ and for $i = 1$ and $i = 2$. To see that (4.60b) is true, notice that if the right-hand side of (4.60a) is expanded, it has four terms. An examination of these terms shows that two of them are identically zero. For example, $V_1(f + f_c)\,G(f - f_c) = 0$ for all f, as illustrated in Figure 4-9.

More generally, for each value of i,

$$V_i(f + f_c)\,G(f - f_c) = 0, \quad -\infty < f < \infty, \quad (4.61a)$$

and

$$V_i(f - f_c)\,G(f + f_c) = 0, \quad -\infty < f < \infty. \quad (4.61b)$$

It follows that

$$W(f)\,H(f) = \frac{W_1(f + f_c) + W_1(f - f_c)}{2} - \frac{j\{W_2(f + f_c) - W_2(f - f_c)\}}{2}.$$

Thus, the output signal $\hat{w}(t)$ has Fourier transform

$$\widehat{W}(f) = W(f)\,H(f)$$
$$= \frac{W_1(f + f_c) + W_1(f - f_c)}{2} - \frac{j\{W_2(f + f_c) - W_2(f - f_c)\}}{2}, \quad (4.62)$$

where $W_i(f) = V_i(f)\,G(f)$ for $-\infty < f < \infty$.

Now let w_i be the inverse Fourier transform of W_i. From the modulation theorem of Fourier transforms, we know that, because of (4.62), $\widehat{W}(f)$ corresponds to the time-domain signal

$$\hat{w}(t) = w_1(t) \cos(2\pi f_c t) - w_2(t) \sin(2\pi f_c t).$$

The frequency-domain representation

$$W_i(f) = V_i(f) G(f)$$

corresponds to $w_i = v_i * g$ in the time domain. Hence, we have shown that

$$\hat{w}(t) = w_1(t) \cos(2\pi f_c t) - w_2(t) \sin(2\pi f_c t),$$

where $w_i = v_i * g$ for $i = 1$ and $i = 2$.

Even if the ideal bandpass filter does not have local symmetry about f_c, (4.58) is still valid, and it can be shown that four convolutions involving baseband functions give the necessary results for determining w_1 and w_2. The derivation of the result for locally symmetric bandpass filters can be generalized easily to provide the more general result for ideal bandpass filters that do not have local symmetry. As demonstrated in Appendix C, the derivation of the general result is even simpler if complex representations are used for the signals and impulse responses.

4.7   Bandpass Random Processes

One of the applications of the time-domain representations derived in the previous section is to the representation of the autocorrelation function for a random process whose spectral density is a bandpass frequency function. Let S_X be such a spectral density function for a zero-mean, wide-sense stationary random process $X(t)$, and apply the results of Section 4.6 by letting

$$W(f) = S_X(f), \quad -\infty < f < \infty.$$

Thus, we can write $S_X(f) = W_+(f) + W_-(f)$, where $W_+(f)$ is the positive part of the spectral density and $W_-(f)$ is the negative part. Assume, as illustrated in Figure 4-10, that the absolute bandwidth of the spectral density function is not greater than $2B$. Throughout this section, it is also assumed that $B \le f_c$.

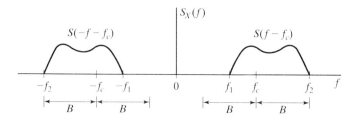

Figure 4-10: A bandpass spectral density function.

The situation is somewhat simpler than the general application of the results of Section 4.6, because the spectral density function must be real and even; that is, $S_X(f)$ is real for each f, and $S_X(f) = S_X(-f)$, $-\infty < f < \infty$. It follows that the positive and negative parts of W are real functions that are related by $W_+(f) = W_-(-f)$ for $-\infty < f < \infty$. Next, define $S(f) = W_+(f + f_c)$ for $-\infty < f < \infty$, and observe that this implies that $S(f - f_c) = W_+(f)$. But since $W_+(f) = W_-(-f)$, we have $W_-(f) = S(-f - f_c)$. As a result, the bandpass spectral density can be written as

$$S_X(f) = S(f - f_c) + S(-f - f_c), \tag{4.63}$$

where the frequency function S satisfies $S(f) = 0$, for $|f| \geq B$. This situation is illustrated in Figure 4-10. It is not true in general that $S(f) = S(-f)$. Of course, if the spectral density S_X has *local symmetry* about f_c, then S is an even function.

The spectral density function S_X is an ideal bandpass frequency function, so the results of Section 4.6 imply that the inverse Fourier transform of S_X—the autocorrelation function for $X(t)$—can be expressed as

$$R_X(\tau) = f_1(\tau) \cos(2\pi f_c \tau) - f_2(\tau) \sin(2\pi f_c \tau), \tag{4.64}$$

where f_1 and f_2 have Fourier transforms that are zero for $|f| \geq B$.

The function Y defined in Section 4.6 is given by

$$Y(f) = 2 W_+(f + f_c) = 2 S(f),$$

so, from (4.45),

$$R_X(\tau) = 2 \operatorname{Re} \left\{ \left[\int_{-\infty}^{\infty} S(f) \exp\{j\, 2\pi f \tau\} \, df \right] \exp\{j\, 2\pi f_c \tau\} \right\}. \tag{4.65}$$

It follows that

$$f_1(\tau) = 2 \int_{-\infty}^{\infty} S(f) \cos(2\pi f \tau) \, df \tag{4.66a}$$

and

$$f_2(\tau) = 2 \int_{-\infty}^{\infty} S(f) \sin(2\pi f \tau) \, df. \tag{4.66b}$$

If S_X has local symmetry about f_c, then $S(f) = S(-f)$ for all f, and it is clear from (4.66b) that $f_2(\tau)$ is identically zero. In this special case, (4.64) simplifies to

$$R_X(\tau) = f_1(\tau) \cos(2\pi f_c \tau).$$

Of much greater interest for applications in communications systems analysis is the fact that the random process itself can be expressed in the form of the signal representation of (4.48). It can be shown that if the zero-mean random process $X(t)$ is wide-sense stationary and has an ideal bandpass spectral density, it can be expressed as

$$X(t) = X_1(t) \cos(2\pi f_c t) - X_2(t) \sin(2\pi f_c t), \tag{4.67}$$

where $X_1(t)$ and $X_2(t)$ are zero-mean, jointly wide-sense stationary random processes with spectral densities that are identically zero for $|f| \geq B$. Some consequences of this representation in terms of jointly wide-sense stationary random processes are given in the next exercise. Let R_i be the autocorrelation function $X_i(t)$, and let $R_{i,k}$ be the crosscorrelation function for $X_i(t)$ and $X_k(t)$ if $i \neq k$.

Exercise 4-5

Show that (4.67) implies that

$$R_1(\tau) = R_2(\tau), \quad -\infty < \tau < \infty \tag{4.68a}$$

and

$$R_{2,1}(\tau) = -R_{1,2}(\tau), \quad -\infty < \tau < \infty. \tag{4.68b}$$

Solution.   The autocorrelation function for $X(t)$ is given by

$$R_X(\tau) = E\{X(t + \tau) X(t)\}.$$

Substituting from (4.67), expanding the product, taking the expectation term by term, and employing standard trigonometric identities, we obtain

$$
\begin{aligned}
R_X(\tau) = {} & \left\{ \frac{R_1(\tau) + R_2(\tau)}{2} \right\} \cos(2\pi f_c \tau) \\
& + \left\{ \frac{R_{1,2}(\tau) - R_{2,1}(\tau)}{2} \right\} \sin(2\pi f_c \tau) \\
& + \left\{ \frac{R_1(\tau) - R_2(\tau)}{2} \right\} \cos(4\pi f_c t + 2\pi f_c \tau) \\
& - \left\{ \frac{R_{1,2}(\tau) + R_{2,1}(\tau)}{2} \right\} \sin(4\pi f_c t + 2\pi f_c \tau),
\end{aligned}
\tag{4.69}
$$

where

$$R_i(\tau) = E\{X_i(t + \tau) X_i(t)\}$$

and

$$R_{i,k}(\tau) = E\{X_i(t + \tau) X_k(t)\}$$

for $i = 1, 2$ and $k = 1, 2$. Because $X(t)$ is wide-sense stationary, the last two terms of (4.69) must be constant. (There can be no dependence on t.) The only way for the third term to be constant is for $R_1(\tau) - R_2(\tau)$ to be zero for all τ, and the only way for the fourth term to be constant is for $R_{2,1}(\tau) + R_{1,2}(\tau)$ to be zero for all τ. But these observations imply that (4.68) must hold.   ■

Notice that it follows from (4.68) and (4.69) that

$$R_X(\tau) = R_1(\tau) \cos(2\pi f_c \tau) - R_{2,1}(\tau) \sin(2\pi f_c \tau), \tag{4.70}$$

from which the functions f_1 and f_2 of (4.64) can be identified, namely, $f_1 = R_1$ and $f_2 = R_{2.1}$. Thus, we can write (4.66a) as

$$R_1(\tau) = 2 \int_{-\infty}^{\infty} S(f) \cos(2\pi f \tau) \, df \tag{4.71a}$$

and (4.66b) as

$$R_{2.1}(\tau) = 2 \int_{-\infty}^{\infty} S(f) \sin(2\pi f \tau) \, df. \tag{4.71b}$$

One important observation that follows from (4.70) is that the function f_1 in (4.64) is an autocorrelation function ($f_1 = R_1$), but the function f_2 is not. We see that $f_2 = R_{2.1}$, which is a crosscorrelation function. Hence, there is no guarantee that $f_2(-\tau) = f_2(\tau)$, because crosscorrelation functions need not have such symmetry. In fact, it turns out that $f_2(-\tau) = f_2(\tau)$ only in the trivial situation in which $f_2(\tau) = 0$ for all τ.

To see this, observe from the definitions of the two crosscorrelation functions that

$$R_{2.1}(-\tau) = R_{1.2}(\tau).$$

Together with (4.68b), this equation implies that

$$R_{2.1}(-\tau) = -R_{2.1}(\tau). \tag{4.72}$$

Equation (4.72) can also be derived from (4.71b) by using the fact that

$$\sin(-2\pi f \tau) = -\sin(2\pi f \tau).$$

As a result of (4.72), the only way for $R_{2.1}$ to satisfy $R_{2.1}(-\tau_0) = R_{2.1}(\tau_0)$ for a particular τ_0 is if $-R_{2.1}(\tau_0) = R_{2.1}(\tau_0)$, and this is true only if $R_{2.1}(\tau_0) = 0$. Therefore, $R_{2.1}(-\tau_0) = R_{2.1}(\tau_0)$ only if the random processes $X_1(t)$ and $X_2(t)$ are such that

$$E\{X_1(t + \tau_0) X_2(t)\} = 0$$

for all t. Furthermore,

$$R_{2.1}(-\tau) = R_{2.1}(\tau). \quad \infty < \tau < \infty,$$

only if

$$E\{X_1(t + \tau) X_2(t)\} = 0$$

for all t and all τ (i.e., the two zero-mean random processes are uncorrelated).

For the special case in which $X_1(t)$ and $X_2(t)$ are uncorrelated random processes, $R_{2.1}(\tau) = 0$ for all τ, and the autocorrelation for $X(t)$ can be written as

$$R_X(\tau) = R_1(\tau) \cos(2\pi f_c \tau). \tag{4.73}$$

From the modulation theorem of Fourier transforms, it follows that the spectral density for $X(t)$ is given by

$$S_X(f) = S(f - f_c) + S(f + f_c), \tag{4.74}$$

where S is one-half the Fourier transform of R_1. As the Fourier transform of a real function, S must be an even function, so S_X has local symmetry about f_c. Also, since S is an even function, $S(f + f_c) = S(-f - f_c)$. It follows that if $X_1(t)$ and $X_2(t)$ are uncorrelated, (4.74) agrees with (4.63) and $S = \mathcal{F}\{R_1\}/2$. Note, however, that (4.63) is valid for any bandpass random process, but (4.74) is valid only if $X_1(t)$ and $X_2(t)$ are uncorrelated random processes. Also, (4.70) and (4.71) are valid for any bandpass random process, but $S = \mathcal{F}\{R_1\}/2$ only if $X_1(t)$ and $X_2(t)$ are uncorrelated random processes.

Although it is not true in general that $R_{2,1}(\tau) = 0$ for all τ, it is true in general that $R_{2,1}(0) = 0$. This follows from (4.71b) by setting $\tau = 0$. The implication is that, for any choice of t_0, $E\{X_2(t_0) X_1(t_0)\} = 0$. Thus, although the two random *processes* are not uncorrelated, the two random *variables* $X_1(t_0)$ and $X_2(t_0)$ *are* uncorrelated for any choice of the sampling time t_0.

Because R_1 is the autocorrelation function for the random process $X_1(t)$, the Fourier transform of R_1, denoted by S_1, is the spectral density function for $X_1(t)$. Because $R_2 = R_1$, then S_1 is also the spectral density function for the random process $X_2(t)$. Let $S_{2,1}$ be the Fourier transform of the crosscorrelation function $R_{2,1}$; that is,

$$S_{2,1}(f) = \int_{-\infty}^{\infty} R_{2,1}(\tau)\, e^{-j2\pi f \tau}\, d\tau. \tag{4.75}$$

The function $S_{2,1}$ is known as the *cross-spectral density function* for the two random processes $X_1(t)$ and $X_2(t)$. As shown in the next exercise, the spectral density for $X(t)$ can be expressed in terms of the functions S_1 and $S_{2,1}$.

Exercise 4-6

Let $X(t)$ be a zero-mean, wide-sense stationary random process with ideal bandpass spectral density. Use the representation given by (4.70) and the modulation theorem of Fourier transforms to find the spectral density for $X(t)$ in terms of the spectral density for the random process $X_1(t)$ and the cross-spectral density for the random processes $X_1(t)$ and $X_2(t)$.

Solution.   The general form of the modulation theorem of Fourier transforms, which is developed in Problem 4.6, gives the Fourier transforms of $v(t)\cos(2\pi f_c t)$ and $v(t)\sin(2\pi f_c t)$ in terms of the Fourier transform of $v(t)$. Applying these results shows that the Fourier transform of $R_1(\tau)\cos(2\pi f_c t)$ is

$$\frac{S_1(f - f_c) + S_1(f + f_c)}{2}$$

and the Fourier transform of $R_{2,1}(\tau)\sin(2\pi f_c t)$ is

$$-\frac{j[S_{2,1}(f - f_c) - S_{2,1}(f + f_c)]}{2}.$$

Combining the two terms, we find that the spectral density for $X(t)$ is given by

$$S_X(f) = \frac{S_1(f - f_c) + S_1(f + f_c)}{2} - \frac{j[S_{2,1}(f - f_c) - S_{2,1}(f + f_c)]}{2}. \qquad \blacksquare$$

We know from previous sections in this chapter that S_1 and $S_{2,1}$ have conjugate symmetry, because they are Fourier transforms of the real functions R_1 and $R_{2,1}$, respectively. In addition, because $R_1(-\tau) = R_1(\tau)$ for all τ, S_1 is an even function. However, according to (4.72),

$$R_{2,1}(-\tau) = -R_{2,1}(\tau)$$

for all τ. The complex conjugate of the cross-spectral density is given by

$$[S_{2,1}(f)]^* = \int_{-\infty}^{\infty} R_{2,1}(\tau)\, e^{+j2\pi f\tau}\, d\tau = \int_{-\infty}^{\infty} R_{2,1}(-u)\, e^{-j2\pi fu}\, du$$

$$= \int_{-\infty}^{\infty} -R_{2,1}(u)\, e^{-j2\pi fu}\, du = -S_{2,1}(f). \qquad (4.76)$$

Just as $z^* = z$ implies that the complex number z is real, $z^* = -z$ implies that z is purely imaginary. As demonstrated in (4.76), the cross-spectral density for the random processes $X_1(t)$ and $X_2(t)$ is purely imaginary. As mentioned in the first sentence of this paragraph, $S_{2,1}$ has conjugate symmetry; that is,

$$[S_{2,1}(f)]^* = S_{2,1}(-f), \quad -\infty < f < \infty.$$

This fact, together with (4.76), implies that

$$S_{2,1}(-f) = -S_{2,1}(f)$$

for each f. It follows that $j\, S_{2,1}(f)$ is real and is an odd function of f; that is,

$$j\, S_{2,1}(-f) = -j\, S_{2,1}(f), \quad -\infty < f < \infty,$$

as illustrated in Figure 4-11.

The random processes $X_1(t)$ and $X_2(t)$ can be generated from $X(t)$ by the use of the system shown in Figure 4-12. The system is identical to the one employed for deterministic signals in Section 4.6.2. (See Figure 4-7.) This approach is not used in the analytical development of the properties of the correlation functions and spectral densities, largely because the system of Figure 4-12 is time varying. In particular, the inputs to the low-pass filters are not wide-sense stationary random processes. As a result, spectral analysis methods cannot be applied to the filtering operations performed in that system.

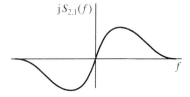

Figure 4-11: Cross-spectral density function $S_{2,1}$.

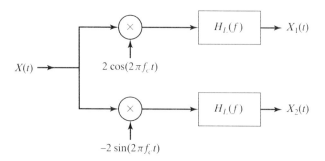

Figure 4-12: Decomposition of the ideal bandpass random process $X(t)$.

The derivation of the representation of a bandpass random process is typically given in terms of the Hilbert transform (e.g., see [4.1] and [4.2]). An alternative approach is given in [4.3]. Even though the system of Figure 4-12 may be difficult to analyze, it generates random processes $X_1(t)$ and $X_2(t)$ that are equivalent to those obtained by the use of the Hilbert transform. (See [4.1] or [4.4].) The outputs $X_1(t)$ and $X_2(t)$ are wide-sense stationary, even if the intermediate processes generated within the system are not; furthermore, $X_1(t)$ and $X_2(t)$ have the correlation functions and spectral densities described in this section.

One property that is obvious from Figure 4-12 is very important: If $X(t)$ is a Gaussian random process, then $X_1(t)$ and $X_2(t)$ are jointly Gaussian. Even though the system shown is not time invariant, it is linear, and that is all that is required to guarantee that the output processes are jointly Gaussian. If the spectral density S_X for the Gaussian random process $X(t)$ is such that $R_{2.1}(\tau) = 0$ for all τ, then the output processes are uncorrelated. Because they are jointly Gaussian, the output processes are also independent. In general, however, $R_{2.1}(\tau)$ is not zero for all τ, and the random processes $X_1(t)$ and $X_2(t)$ are statistically dependent.

References and Suggestions for Further Reading

[4.1] A. Papoulis and S. U. Pillai, *Probability, Random Variables, and Stochastic Processes*, 4th ed., New York: McGraw-Hill, 2002.

[4.2] B. P. Lathi, *An Introduction to Random Signals and Communication Theory*, Scranton, PA: International, 1968.

[4.3] A. J. Viterbi, *Principles of Coherent Communication*, New York: McGraw-Hill, 1966.

[4.4] H. Stark and J. W. Woods, *Probability and Random Processes with Applications to Signal Processing*, 3rd ed., Upper Saddle River, NJ: Prentice Hall, 2002.

Problems

4.1 Derive the spectral density in item (6) of Table 4.1 in two ways. First use items (4) and (5) from the table and the fact that multiplication in the time domain corresponds to convolution in the frequency domain. (Watch out for the constants.) Second, use item (4) and the modulation theorem of Fourier transforms, which states that if $v(t)$ has Fourier transform $V(\omega)$ then the Fourier transform of the signal

$$W(t) = v(t)\cos(\omega_c t)$$

is

$$W(\omega) = \frac{V(\omega - \omega_c) + V(\omega + \omega_c)}{2}.$$

The proof of the modulation theorem is outlined in Problem 4.6.

4.2 Consider the signal $v(t) = 1/\sqrt{T}$ for $-T/2 < t < T/2$ and $v(t) = 0$, otherwise. The Fourier transform of this signal is

$$V(f) = \sqrt{T}\,\mathrm{sinc}(fT).$$

Show how this fact can be used to derive item (1) in Table 4.1. (*Hint:* The convolution of $v(t)$ with itself is a triangle, and convolution in the time domain corresponds to multiplication in the frequency domain.)

4.3 A linear time-invariant filter has transfer function $H(f) = 1$ for $|f| < B$ and $H(f) = 0$, otherwise. The input random process $X(t)$ is a zero-mean, wide-sense stationary, Gaussian random process with autocorrelation function $R_X(\tau) = \beta \exp(-\alpha|\tau|)$. The corresponding output is $Y(t)$.

    (a) Find $E\{[Y(t)]^2\}$.

    (b) Find $P[|Y(t_0)| \le \gamma]$ for an arbitrary positive value of γ.

4.4 The wide-sense stationary random process $X(t)$ has spectral density $S_X(f) = 1$ for $|f| < B$ and $S_X(f) = 0$, otherwise. Let $Y(t)$ be the output when $X(t)$ is the input to a linear time-invariant filter with transfer function $H(f) = \cos(\pi f/2W)$ for $|f| < W$ and $H(f) = 0$ otherwise. Find the output spectral density $S_Y(f)$, and give the value of $E\{[Y(t)]^2\}$. Consider the two cases $B \ge W$ and $B < W$.

4.5 A time-invariant linear filter has transfer function

$$H(\omega) = \exp(-\omega^2), \quad -\infty < \omega < \infty.$$

The input to this filter is a white-noise process $X(t)$ with power spectral density

$$S_X(\omega) = \frac{N_0}{2}, \quad -\infty < \omega < \infty,$$

and the corresponding output is the random process $Y(t)$. Find the expected value of the instantaneous power in the output process. (*Hint:* To help evaluate the resulting integral, notice that the transfer function has a Gaussian shape. Can you relate it to a probability density function? How might this help solve the problem?)

4.6 **(a)** Prove the modulation theorem of Fourier transforms by using the following steps: First, note that the signal $v(t)$ has Fourier transform given by

$$V(\omega) = \int_{-\infty}^{\infty} v(t) e^{-j\omega t} \, dt.$$

Next, use the fact that

$$\cos(\omega_c t) = \tfrac{1}{2}\{\exp(+j\omega_c t) + \exp(-j\omega_c t)\}$$

to infer that the Fourier transform of $w_1(t) = v(t)\cos(\omega_c t)$ can be written as

$$W_1(\omega) = \tfrac{1}{2}\int_{-\infty}^{\infty} v(t)\exp[-j(\omega - \omega_c)t]\,dt + \tfrac{1}{2}\int_{-\infty}^{\infty} v(t)\exp[-j(\omega + \omega_c)t]\,dt.$$

By comparing these two integrals with the Fourier transform integral, show that the expression for W_1 is equivalent to

$$W_1(\omega) = \tfrac{1}{2}[V(\omega - \omega_c) + V(\omega + \omega_c)].$$

(b) Repeat the steps in part **(a)** for $w_2(t) = v(t)\cos(\omega_c t + \theta)$, using the identity

$$\cos(\omega_c t + \theta) = \tfrac{1}{2}\{\exp[+j(\omega_c t + \theta)] + \exp[-j(\omega_c t + \theta)]\}.$$

Show that

$$W_2(\omega) = \tfrac{1}{2}[V(\omega - \omega_c)e^{+j\theta} + V(\omega + \omega_c)e^{-j\theta}].$$

Notice that this last result reduces to the result in part **(a)** if $\theta = 0$.

(c) Show that the result in part **(b)** applied for the special case $\theta = -\pi/2$ proves that the Fourier transform of $v(t)\sin(\omega_c t)$ is $j[V(\omega + \omega_c) - V(\omega - \omega_c)]/2$.

4.7 Suppose that a random process is defined by

$$Z(t) = A_1(t)\cos(\omega_c t + \Theta) - A_2(t)\sin(\omega_c t + \Theta).$$

Assume that Θ is uniformly distributed on $[0, 2\pi]$ and independent of $A_1(t)$ and $A_2(t)$. Suppose further that $A_1(t)$ and $A_2(t)$ are each zero-mean, wide-sense stationary random processes with autocorrelation function $R_A(\tau)$. Suppose also that $A_1(t)$ and $A_2(t)$ are uncorrelated. Find the spectral density for $Z(t)$ in terms of the spectral density $S_A(\omega)$ for $A_1(t)$ and $A_2(t)$. Compare this result with (4.23).

4.8 Let $\zeta(t)$ represent the waveform for a data pulse, as in Section 4.5.

(a) Suppose that a communication signal is given by

$$x(t) = \sqrt{2}\,\zeta(t)\cos(\omega_c t).$$

What is the energy spectrum $|X(\omega)|^2$? Give sufficient conditions for the shape of this energy spectrum to be identical to the shape of the spectral density $S_Y(\omega)$ given in (4.30). That is, give sufficient conditions for

$$|X(\omega)|^2 = \text{constant} \times S_Y(\omega).$$

(b) Consider the signal defined by

$$y(t) = \sqrt{2}\,\zeta(t)\cos(\omega_c t + \theta).$$

The parameter θ is a deterministic constant. What is the energy spectrum for $y(t)$? (*Hint*: Judicious use of the result of Problem 4.6(b) may be helpful.)

4.9 Consider the random signal given by $A(t) = D(t - U)$, $-\infty < t < \infty$, where U is a random variable that is uniformly distributed on the interval $[0, T]$ and the random signal $D(t)$ is given by (4.24). Suppose that the random sequence (A_n) is wide-sense stationary, is independent of U, and has autocorrelation function $\rho_k = E\{A_0 A_k\}$, $-\infty < k < \infty$.

(a) Show that $A(t)$ is wide-sense stationary and has autocorrelation function given by

$$R_A(\tau) = \sum_{k=-\infty}^{\infty} \rho_k\, r_\zeta(\tau - kT),$$

where

$$r_\zeta(\tau) = T^{-1} \int_{-\infty}^{\infty} \zeta(t)\,\zeta(t + \tau)\,dt.$$

(b) Apply the result obtained in part **(a)** to find the autocorrelation and spectral density functions for the random signal $A(t)$ if $\zeta(t) = p_T(t)$, $\rho_0 = \alpha^2$, $\rho_1 = \rho_{-1} = \beta$, and $\rho_k = 0$ for $|k| \geq 2$.

(c) For $\alpha = 1$ and $\beta = 1/2$, compare a plot of the spectral density obtained in part **(b)** with the spectral density that is obtained if the sequence (A_n) is a sequence of independent zero-mean random variables with $E\{A_n^2\} = 1$.

4.10 Which of the following frequency functions are ideal bandpass frequency functions as defined in Section 4.6?

(a) $W(f) = \begin{cases} j\sin(\pi f), & 10 < |f| < 15, \\ 0, & \text{otherwise.} \end{cases}$

(b) $W(f) = \exp\{-(f - 100)^2\}, \quad -\infty < f < \infty.$

(c) $W(f) = \begin{cases} \exp\{j\pi f/50\}, & 1000 < |f| < 2000, \\ 0, & \text{otherwise.} \end{cases}$

Find the cutoff bandwidth for each of the ideal bandpass frequency functions.

4.11 Suppose the functions in Problem 4.10 are transfer functions for time-invariant linear filters. Which of them have well-defined half-power bandwidths? For each that does have a well-defined half-power bandwidth, determine the maximum gain W_m and the half-power bandwidth.

4.12 Consider the ideal bandpass frequency function defined by

$$W(f) = \begin{cases} (|f| - 500)/500, & 500 < |f| < 1000, \\ -(|f| - 1500)/500, & 1000 < |f| < 1500, \\ 0, & \text{otherwise.} \end{cases}$$

This frequency function is illustrated in the following diagram:

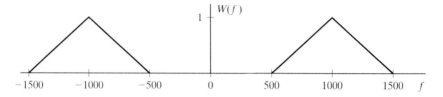

Consider the function y, defined as the inverse transform of

$$Y(f) = 2 W_+(f + f_c), \quad -\infty < f < \infty.$$

(a) What choice of f_c makes y a real function?

(b) Suppose that $f_c = 750$. Specify $Y(f)$ for $-\infty < f < \infty$ and $y(t)$ for $-\infty < t < \infty$.

4.13 **(a)** Following the development of the results for V_1 in (4.50) through (4.54a), begin with the fact that $v_2(t) = \mathrm{Im}\{y(t)\}$, and show that

$$V_2(f) = \frac{-j\{Y(f) - Y^*(-f)\}}{2}.$$

(b) Use your result from part **(a)** together with (4.54a) to show that, for both $i = 1$ and $i = 2$,

$$V_i(-f) = V_i^*(f), \quad -\infty < f < \infty.$$

4.14 Let v be an ideal bandpass signal with absolute bandwidth $2B$. Then v can be expressed as in (4.48), where v_1 and v_2 are baseband functions with Fourier transforms that are identically zero for $|f| \geq B$. Assume that $B < f_c$. Consider the system shown in the following diagram:

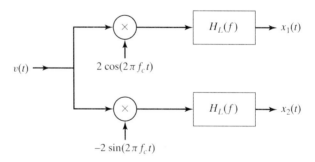

The linear filters in this system are defined by

$$H_L(f) = \begin{cases} 1, & -B \leq f \leq +B, \\ 0, & \text{otherwise.} \end{cases}$$

Show that the system generates the signals v_1 and v_2 by demonstrating that $x_1(t) = v_1(t)$ and $x_2(t) = v_2(t)$ for $-\infty < t < \infty$. (*Hint:* It suffices to show that the Fourier transform of x_i is V_i for each i, and (4.54) may be of help in doing so.)

4.15 Let $X(t)$ be an ideal bandpass process that is wide-sense stationary and has auto-correlation function given by

$$R_X(\tau) = f_1(\tau)\cos(2\pi f_c\tau) - f_2(\tau)\sin(2\pi f_c\tau), \quad -\infty < \tau < \infty.$$

Show that if f_1 is an autocorrelation function, f_2 cannot be an autocorrelation function. (*Hint:* Notice that R_X, f_1, and cosine are even functions, but sine is an odd function.)

4.16 A wide-sense stationary, zero-mean, Gaussian random process $X(t)$ is the input to the following system:

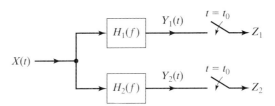

The spectral density for $X(t)$ is given by $S(f) = \alpha \exp(-\beta f^2)$, $-\infty < f < \infty$. The outputs $Y_1(t)$ and $Y_2(t)$ of the filters are sampled at time t_0 to give the random variables Z_1 and Z_2. The linear time-invariant filters $H_1(f)$ and $H_2(f)$ are ideal bandpass filters with center frequencies f_1 and f_2, respectively, and each has bandwidth $2W$. Assume that $H_1(f) = 1$ for frequencies in the passband and that $0 < W < f_1 < f_2 - 2W$, so that the passbands of the filters do not overlap. In parts **(a)–(e)** and **(g)**, evaluate all integrals, and give your answers in terms of the function Φ and the parameters α, β, t_0, W, f_1, and f_2 only.

(a) What constraints must be placed on α and β to guarantee that $S(f)$ is a valid spectral density for a wide-sense stationary random process with power P if $0 < P < \infty$? Assume that α and β satisfy these constraints in parts **(b)–(h)**.

(b) Give an expression for P, the power in the random process $X(t)$.

(c) Give an expression for $S_1(f)$, the spectral density for $Y_1(t)$.

(d) Find $m_1 = E\{[Y_1(t_0)]^2\}$ and $m_2 = E\{[Y_2(t_0)]^2\}$.

(e) Give expressions for $\mu_1 = E\{Z_1\}$ and $\mu_2 = E\{Z_2\}$.

(f) Give an integral expression (with a single or double integral) for the cross-correlation function $R_{1,2}(\tau) = E\{Y_1(t+\tau)Y_2(t)\}$ in terms of the impulse responses of the two filters and the autocorrelation function for the random process $X(t)$ (i.e., your answer should be an integral expression involving the functions h_1, h_2, and R_X). Give an equivalent expression in terms of one or two convolutions.

(g) Give an expression for $E\{Z_1 Z_2\}$. (*Hint:* It may be best to first relate $E\{Z_1 Z_2\}$ to the crosscorrelation function $R_{1,2}$ defined in part **(f)** and then use frequency-domain methods to evaluate the result.)

(h) Find $P(Z_2 > Z_1 - 1)$. Express your answer in terms of Φ and the parameters m_1, m_2, μ_1, and μ_2 only. Explain how the solution to part **(g)** is used to obtain your answer.

Chapter 5

Baseband Transmission of Binary Data

Many of the important concepts in digital communications are more easily learned in the context of a binary baseband data transmission system. In particular, it is beneficial to introduce certain performance measures and decision criteria in the restricted setting of a baseband system. The correlator, matched filter, and whitening filter are described for baseband receivers before we deal with the additional complexity required in radio-frequency (RF) communication systems. The term *baseband* refers to the portion of the frequency spectrum below some frequency W. The actual value of this upper limit W depends on the data rate of the system, but common values range from a few hundred hertz for low-data-rate systems to a few hundred megahertz for high-speed transmission.

If a digital communication system is not a baseband system, then it is a carrier-modulated communication system. A *carrier* is a pure sinusoidal signal, and its frequency, known as the *carrier frequency*, is usually denoted by f_c. A carrier-modulated system utilizes a frequency band from $f_c - W$ to $f_c + W$. The carrier frequency is always greater than, and is often much greater than, W. Carrier-modulated communication systems include RF communication systems, which typically utilize carrier frequencies in the range from 30 Hz to 300 GHz. At considerably higher frequencies, one enters the realm of optical communication, including infrared and laser communications.

As far as we are concerned in this book, the main feature of digital RF and optical communication systems is that the communication signals are obtained by modulating a baseband signal onto a carrier. The demodulation of the carrier-modulated signal typically requires the receiver to know or estimate the value of the carrier frequency, and it may require determination of the phase of the carrier as well. For our purposes, it is this property of carrier-modulated transmission, and not the spectral occupancy, that distinguishes it from baseband transmission. The issue of estimating carrier frequency or phase deals with an aspect of digital communications that we prefer to postpone until several basic concepts have been introduced, which is the reason for devoting this chapter to baseband transmission exclusively. Carrier-modulated communication systems are discussed in Chapters 6 and 7. Even for carrier-modulated systems, a substantial amount of the processing in the receiver is performed after the signal has been converted to a baseband signal. Thus, the concepts introduced here are applicable to carrier-modulated systems as well as to baseband systems.

5.1   Signal Sets for Binary Data Transmission

In an M-ary data transmission system there is a collection $\{s_i : 0 \leq i \leq M - 1\}$ of M signals, which are also referred to as waveforms. Information is conveyed to the receiver by transmitting signals from this collection. For a binary data transmission system, the collection has only two signals. For a *binary baseband communication system*, each of the two signals has its energy concentrated at low frequencies (below some frequency W). In order to send a sequence of binary digits, a corresponding sequence of waveforms is transmitted to the receiver. If the binary digits are generated at the rate of one digit every T units of time, the waveforms must be transmitted at a rate of $1/T$. If the waveforms are time limited and of duration T, the transmitted signal consists of a sequence of nonoverlapping waveforms in consecutive time intervals of duration T.

The collection of waveforms that is available to the transmitter is known as the *signal set*. A binary signal set consists of two waveforms s_0 and s_1. If these waveforms have duration T, then for both $i = 0$ and $i = 1$, $s_i(t) = 0$ for $t < 0$ and $t \geq T$. Let $b_0, b_1, \ldots, b_{N-1}$ be a sequence of binary digits (i.e., for each n, either $b_n = 0$ or $b_n = 1$). In order to send this sequence of binary digits to a receiver, a sequence of waveforms is transmitted. The composite signal $v(t)$ formed by the sequence of waveforms can be written as

$$v(t) = \sum_{n=0}^{N-1} s_{b_n}(t - nT); \tag{5.1}$$

that is, b_n is sent by transmitting the waveform $s_{b_n}(t - nT)$.

Unless stated otherwise, the waveforms in each signal set considered in the remainder of this section are time limited to the interval 0 to T. For such waveforms, the composite signal $v(t)$ given in (5.1) satisfies

$$v(t) = s_{b_n}(t - nT), \quad nT \leq t < (n+1)T. \tag{5.2}$$

If, for example, the binary sequence 010 is to be sent, then $v(t) = s_0(t)$ for $0 \leq t < T$, $v(t) = s_1(t)$ for $T \leq t < 2T$, and $v(t) = s_0(t)$ for $2T \leq t < 3T$. If the receiver can distinguish between the two waveforms in each of the three time intervals, it can determine the binary sequence from $v(t)$. Unfortunately, in an actual system the receiver does not see $v(t)$; instead, it has available only a corrupted version of $v(t)$. The extraction of the original information from a corrupted version of a transmitted signal is the principal topic of the book.

An example of a binary signal set is illustrated in Figure 5-1(a). There is only one nonzero waveform, a rectangular pulse of duration T. The signal set is defined by

$$s_0(t) = \begin{cases} 1, & 0 \leq t < T, \\ 0, & \text{otherwise,} \end{cases}$$

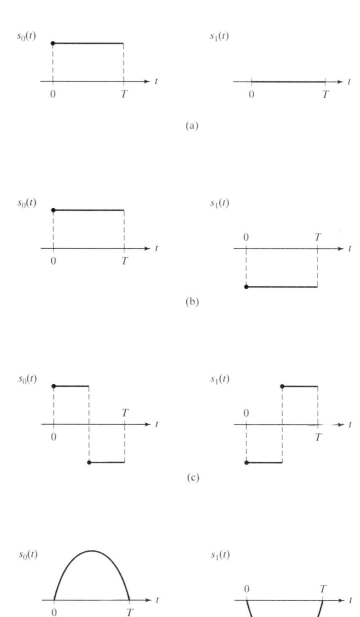

Figure 5-1: Four examples of baseband signal sets.

and $s_1(t) = 0$ for all t. This signal set is one form of *on–off signaling*: The data bit 0 is represented by the presence of the pulse (on), and the data bit 1 is represented by its absence (off). The signal set is also an example of an *orthogonal signal set*. We say that signals s_0 and s_1 are orthogonal on the interval from 0 to T if

$$\int_0^T s_0(t)\, s_1(t)\, dt = 0. \tag{5.3}$$

The waveform s_0 in this example arises so frequently in digital communications that we give it the special notation p_I. Thus, we define the *rectangular pulse of duration T* by

$$p_I(t) = \begin{cases} 1, & 0 \le t < T, \\ 0, & \text{otherwise.} \end{cases} \tag{5.4}$$

Another simple example is obtained by letting $s_0(t) = p_I(t)$ and $s_1(t) = -p_I(t)$, as illustrated in Figure 5-1(b). In this example, a positive pulse represents a 0 and a negative pulse represents a 1. If the pulse shapes are identical and the pulses have opposite polarity, as in this example, the resulting signal set is an antipodal signal set. A signal set $\{s_0, s_1\}$ is an *antipodal signal set* and the two signals are said to be *antipodal* if $s_1(t) = -s_0(t)$ for all t. Many binary data transmission systems, both baseband and RF, employ antipodal signals. Under certain conditions, the optimum binary signal set is an antipodal signal set. If it is known that a signal set is antipodal, the set is defined completely by specifying $s_0(t)$ for all t. Obviously, s_1 can be determined from $s_1(t) = -s_0(t)$.

For antipodal signals, it is fairly common in the digital communications literature that the data bit 0 is represented by a positive waveform and 1 is represented by a negative waveform. We follow this convention for binary signal sets that have clearly defined positive and negative waveforms. As an example of a signal set that does not have clearly defined positive and negative waveforms, consider the antipodal signals defined by

$$s_0(t) = \begin{cases} 1, & 0 \le t < T/2, \\ -1, & T/2 \le t < T, \\ 0, & \text{otherwise.} \end{cases} \tag{5.5}$$

For this signal set, which is illustrated in Figure 5-1(c), it is impossible to say which waveform is negative and which is positive.

As a final example, consider the antipodal signal set $\{s_0, s_1\}$ illustrated in Figure 5-1(d) and defined by

$$s_0(t) = \sin(\pi t / T)\, p_I(t). \tag{5.6}$$

The pulse in (5.6) is referred to as the *sine pulse* or, sometimes, the *half-wave sine pulse*. It represents half a period of a full sine wave. The positive pulse is $s_0(t)$, and the negative pulse is $s_1(t)$.

The four binary baseband signal sets shown in Figure 5-1 are used in several examples throughout the chapter, and they arise again in carrier-modulated communication systems in the chapters that follow. For instance, the antipodal signal set based on the rectangular pulse is employed in phase-shift-keyed (PSK) communication systems, and the set based on the sine pulse is used in minimum-shift-keyed (MSK) communication systems. Among the techniques for carrier-modulated digital communications, PSK and MSK are two of the most popular.

All signal sets illustrated in Figure 5-1 employ time-limited signals that start at $t = 0$ and have duration T. We refer to such a signal as being *time limited to the interval* $[0, T]$. As is true of s_1 in Figure 5-1(a), it may be that such a signal satisfies $s_i(t) = 0$ for one or more values of t in the range $0 \leq t < T$; however, a signal that is time limited to the interval $[0, T]$ cannot be nonzero outside this range. Also, notice that each of the signals in the four sets satisfies

$$\int_{-\infty}^{\infty} [s_i(t)]^2 \, dt < \infty, \tag{5.7}$$

and this constraint is imposed on all signal sets that we consider. This condition is just a requirement that s_i have finite energy, which is, of course, a necessary condition for implementation.

5.2   Analysis of Linear Receivers

As mentioned in the previous section, the receiver in a communication system cannot observe the transmitted signal. Instead, it observes a signal that is only statistically related to the transmitted signal. Based on this observation, the receiver makes inferences about the binary sequence that was sent. At present, we consider *binary decisions* only. For each binary digit that is sent, the receiver must decide whether the digit is a 0 or a 1. Such a receiver is often referred to as a hard-decision receiver. Other types of decisions are possible, such as including the option of allowing the receiver to declare it does not know the value of a digit, but such an option is not allowed for the performance analysis given in this section.

For binary decisions, there are only two possible outcomes: The receiver's decision is correct or it is wrong. If the receiver makes the wrong decision, we say it has made an error. Our goal is to design the receiver in a way that minimizes the probability that it makes an error.

A linear baseband receiver consists of one or more linear filters followed by one or more samplers and a decision device. A sampler and a decision device are used in all receivers. The restriction that is imposed by the requirement for linearity involves the processing of the received signal. For the receiver structure under consideration, only *linear filters* are allowed. For baseband communications over channels in which the only disturbance is additive Gaussian noise, the optimum receiver is linear. Therefore, for the communication channels considered in this chapter, the best linear receiver is in fact the optimum receiver of any type.

In this section, expressions for the probability of error are obtained for linear receivers that make binary decisions. These expressions are derived for an arbitrary

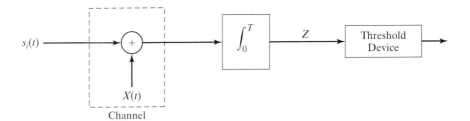

Figure 5-2: A binary data transmission system model.

signal set, an arbitrary linear filter, and an arbitrary sampling time. The decision is based on the comparison of a decision statistic with a threshold. No constraints are imposed on the value of the threshold in the decision device.

5.2.1   A Special Case

We begin by considering a special case. The model for the communication system under consideration is shown in Figure 5-2. The signals s_0 and s_1 are given by

$$s_0(t) = A \, p_T(t) \tag{5.8a}$$

and

$$s_1(t) = -A \, p_T(t), \tag{5.8b}$$

where p_T is the rectangular pulse of duration T defined in (5.4). The noise $X(t)$ is a white Gaussian noise process with spectral density $N_0/2$, and the noise process is independent of the channel input.

The output of the channel is the random process $Y(t)$, which is the sum of the channel input and the noise $X(t)$. The input to the threshold device is the random variable

$$Z = \int_0^T Y(t) \, dt. \tag{5.9}$$

In this receiver, the sampler is part of the integrate-and-dump filter, as discussed in Example 3-2. The decision device is a threshold device: If Z exceeds the threshold γ, the device decides that the transmitted signal is s_0; if Z does not exceed γ, the device decides that s_1 was sent. In this example, we let $\gamma = 0$, so the decision is based on whether Z is positive or negative.

Suppose that a 0 is sent; that is, the signal s_0 is the input to the channel. The output of the channel is $Y(t) = s_0(t) + X(t)$, and the input to the threshold device is $Z = Z_0$, where

$$Z_0 = \int_0^T s_0(t) \, dt + \int_0^T X(t) \, dt$$

$$= AT + \int_0^T X(t) \, dt. \tag{5.10}$$

From the definition of white Gaussian noise (Section 3.5), we know that Z_0 is Gaussian and

$$E\left\{\int_0^T X(t)\,dt\right\} = 0. \tag{5.11}$$

Thus, if s_0 is the input to the channel, the input to the threshold device is a Gaussian random variable Z_0 that has mean $\mu_0 = AT$. The variance of Z_0 is given by

$$\sigma_0^2 = \text{Var}\{Z_0\} = E\{[Z_0 - AT\,]^2\}. \tag{5.12}$$

From (5.10) and (5.12), we see that this variance can be written as

$$\sigma_0^2 = E\left\{\left[\int_0^T X(t)\,dt\right]^2\right\}. \tag{5.13}$$

Next, we must recall some facts from Chapter 3 (e.g., let $t = T$ in Example 3-5). Because $X(t)$ is a white-noise process with spectral density $N_0/2$,

$$E\left\{\left[\int_0^T X(t)\,dt\right]^2\right\} = \int_0^T \int_0^T \tfrac{1}{2}N_0\,\delta(t - s)\,ds\,dt. \tag{5.14}$$

Consider the inner integral

$$\int_0^T \tfrac{1}{2}N_0\,\delta(t - s)\,ds = \tfrac{1}{2}N_0 \int_0^T \delta(t - s)\,ds.$$

Notice that if t is in the range $0 < t < T$, then

$$\int_0^T \delta(t - s)\,ds = 1. \tag{5.15}$$

This is a consequence of two facts: (1) The area under the impulse is 1 and (2) the integral on the left-hand side of (5.15) is equal to the area under the impulse if the location of the impulse is within the interval $(0, T)$. But from (5.14), we see that t is in the range $0 < t < T$. Moreover, the impulse $\delta(t - s)$ is located at $s = t$, so it is in the interval $(0, T)$.

Returning to (5.14), we conclude that, because of (5.15),

$$E\left\{\left[\int_0^T X(t)\,dt\right]^2\right\} = \int_0^T \tfrac{1}{2}N_0\,dt. \tag{5.16}$$

Finally, the desired result follows from (5.13) and (5.16). Combining these two expressions we conclude that if Z_0 is given by (5.10), it has variance

$$\sigma_0^2 = \tfrac{1}{2}N_0 T. \tag{5.17}$$

We have already noted that Z_0 is Gaussian and its mean is $\mu_0 = AT$. It follows that the distribution function for Z_0 is given by

$$
\begin{aligned}
F_0(z) &= \Phi([z - \mu_0]/\sigma_0) \\
&= \Phi\left([z - AT]/\sqrt{N_0T/2}\right).
\end{aligned}
\tag{5.18}
$$

The distribution function F_0 has a dual role. It is, as we have just shown, the distribution function for the random variable Z_0. However, it is also the *conditional* distribution function for Z given that the signal s_0 is transmitted. That is,

$$
P(Z \le z \mid s_0) = P(Z_0 \le z) = F_0(z),
\tag{5.19}
$$

where $P(E|s_0)$ denotes the conditional probability of the event E given that the transmitted signal is s_0.

We are now ready to examine the likelihood that an error is made by the receiver of Figure 5-2 if the transmitted signal is s_0. Let $P_{e,0}$ denote the error probability that results when s_0 is transmitted. Notice that, for the communication system under consideration, an error occurs when s_0 is transmitted if and only if $Z \le 0$. Consequently, the probability of error when s_0 is transmitted is given by

$$
P_{e,0} = P(Z \le 0 \mid s_0) = P(Z_0 \le 0).
\tag{5.20}
$$

From (5.18)–(5.20), we see that

$$
\begin{aligned}
P_{e,0} = F_0(0) &= \Phi([0 - \mu_0]/\sigma_0) \\
&= \Phi\left(-AT/\sqrt{N_0T/2}\right).
\end{aligned}
\tag{5.21}
$$

Recall that

$$
\Phi(x) = \int_{-\infty}^{x} \exp(-u^2/2)\, du/\sqrt{2\pi}
$$

so that $\Phi(-x) = 1 - \Phi(x)$. Also, since

$$
Q(x) = \int_{x}^{\infty} \exp(-u^2/2)\, du/\sqrt{2\pi},
$$

it follows that $Q(x) = 1 - \Phi(x) = \Phi(-x)$. Consequently, (5.21) has the alternative equivalent forms

$$
P_{e,0} = 1 - \Phi\left(AT/\sqrt{N_0T/2}\right)
\tag{5.22}
$$

and

$$
P_{e,0} = Q\left(AT/\sqrt{N_0T/2}\right).
\tag{5.23}
$$

The notation in (5.23) is simpler than in (5.21) and (5.22), and the use of the function Q is consistent with most of the current literature in digital communications, so the final expressions for error probabilities are given in terms of the function Q in this book.

Notice that (5.23) can be written as

$$P_{e,0} = Q\left(\sqrt{2A^2T/N_0}\right).$$

The amount of energy in the waveform $s_0(t) = A\,p_T(t)$ is

$$\mathcal{E}_0 = \int_0^T [s_0(t)]^2\,dt = A^2T.$$

Therefore, the probability of error when s_0 is transmitted can be written as

$$P_{e,0} = Q\left(\sqrt{2\mathcal{E}_0/N_0}\right). \tag{5.24}$$

The form of the right-hand side appears many times in the sections that follow.

Next, we consider the probability that an error is made when the transmitted signal is s_1. Let $P_{e,1}$ denote the conditional probability of error given that s_1 is transmitted. Because an error occurs when s_1 is transmitted if and only if $Z > 0$, this error probability is

$$P_{e,1} = P(Z > 0 \mid s_1). \tag{5.25}$$

If the signal s_1 is transmitted, the random variable Z is equal to the random variable Z_1 that is defined by the following expression (cf. (5.10)):

$$Z_1 = \int_0^T s_1(t)\,dt + \int_0^T X(t)\,dt$$

$$= -AT + \int_0^T X(t)\,dt. \tag{5.26}$$

Consequently, Z_1 is Gaussian with mean $\mu_1 = -AT$ and variance (cf. (5.12)–(5.17))

$$\sigma_1^2 = E\left\{[Z_1 + AT]^2\right\} = E\left\{\left[\int_0^T X(t)\,dt\right]^2\right\}$$

$$= N_0T/2. \tag{5.27}$$

It is important to notice that $\sigma_0^2 = \sigma_1^2$. This observation can be stated as follows: The conditional variance of Z given that the transmitted signal is s_i does not depend on i. Thus, we can write

$$\text{Var}\{Z \mid s_i\} = \sigma_i^2 = \sigma^2 \tag{5.28}$$

for both $i = 0$ and $i = 1$, and σ^2 does not depend on i. For the system of Figure 5-2, we have shown that the value of σ^2 is $N_0T/2$, which depends on the noise spectral density and the integration time only.

Returning to (5.25) and making use of our conclusion that Z_1 is Gaussian with mean $\mu_1 = -AT$ and variance $\sigma_1^2 = N_0 T/2$, we see that

$$
\begin{aligned}
P_{e,1} &= P(Z > 0 \mid s_1) = P(Z_1 > 0) \\
&= 1 - P(Z_1 \le 0) = 1 - F_1(0) \\
&= 1 - \Phi([0 - \mu_1]/\sigma_1) \\
&= 1 - \Phi\left(AT/\sqrt{N_0 T/2}\right),
\end{aligned}
\tag{5.29}
$$

where F_1 is the distribution function for Z_1.

A comparison of (5.22) and (5.29) shows that $P_{e,0} = P_{e,1}$; that is, the error probability $P_{e,i}$ that results when s_i is transmitted does not depend on i. Whenever $P_{e,i}$ does not depend on i, we denote the common value of $P_{e,0}$ and $P_{e,1}$ by P_e. For the system of Figure 5-2, the resulting probability of error is given by the expression

$$
P_e = Q\left(\sqrt{2\,\mathcal{E}/N_0}\right).
\tag{5.30}
$$

In writing (5.30), we have made use of the fact that, for the signal set specified by (5.8), the amount of energy in $s_0(t)$ is the same as the energy in $s_1(t)$. We denote this common value by $\mathcal{E}$ and call it the energy per pulse. In general, we define

$$
\mathcal{E}_i = \int_0^T [s_i(t)]^2 \, dt
\tag{5.31}
$$

for $i = 0$ and $i = 1$, but the signals under consideration are such that $\mathcal{E}_0 = \mathcal{E}_1$, so we may omit the subscripts and denote the common value by $\mathcal{E}$.

Notice from (5.20) and (5.29) that a necessary and sufficient condition for $P_{e,1}$ to be the same as $P_{e,0}$ is

$$
P(Z_1 > 0) = P(Z_0 \le 0),
\tag{5.32}
$$

a condition that is satisfied in many current binary communication systems. In general, if Z_i is Gaussian for each i, then (5.32) is equivalent to

$$
1 - \Phi(-\mu_1/\sigma_1) = \Phi(-\mu_0/\sigma_0),
\tag{5.33}
$$

where μ_i and σ_i^2 are the mean and variance, respectively, of Z_i. In obtaining (5.33), we have used the fact that

$$
P(Z_1 > 0) = 1 - P(Z_1 \le 0)
$$

and, because Z_i is Gaussian,

$$
P(Z_i \le 0) = \Phi([0 - \mu_i]/\sigma_i).
$$

Now (5.33) can be written in a variety of ways by using the fact that

$$
1 - \Phi(x) = \Phi(-x) = Q(x).
$$

For example, notice that

$$1 - \Phi(-\mu_1/\sigma_1) = \Phi(\mu_1/\sigma_1),$$

so (5.33) is equivalent to

$$\Phi(\mu_1/\sigma_1) = \Phi(-\mu_0/\sigma_0). \qquad (5.34)$$

Because $\Phi(x)$ is an increasing continuous function of x, we conclude that if Z_1 and Z_0 are Gaussian and the system employs the decision device under consideration (with $\gamma = 0$), the error probabilities $P_{e,1}$ and $P_{e,0}$ are equal if and only if

$$\mu_1/\sigma_1 = -\mu_0/\sigma_0. \qquad (5.35)$$

If $\sigma_1 = \sigma_0$, as we have shown is true for the system of Figure 5-2, then $\mu_1 = -\mu_0$ is a necessary and sufficient condition for the two error probabilities to be equal. For the system of Figure 5-2,

$$\mu_i = \int_0^T s_i(t)\,dt.$$

From this result, it is clear that any antipodal signal set gives $\mu_1 = -\mu_0$.

The expression $Q\left(\sqrt{2\mathcal{E}/N_0}\right)$ arises frequently in the study of digital communications. Consequently, it is very beneficial for our discussions in subsequent sections to have a table of numerical values for this quantity. Communications engineers usually express the ratio $\mathcal{E}/N_0$, and other ratios of this type, in decibels (dB). The ratio $\mathcal{E}/N_0$ expressed in dB is defined as

$$(\mathcal{E}/N_0)_{\mathrm{dB}} = 10\log_{10}(\mathcal{E}/N_0).$$

Values of $Q\left(\sqrt{2\mathcal{E}/N_0}\right)$ are given in Table 5.1 for several values of $(\mathcal{E}/N_0)_{\mathrm{dB}}$ that are in the range of interest for most practical digital communication systems. The corresponding values of $\mathcal{E}/N_0$ are included in Table 5.1 for convenience, and a graph of $Q\left(\sqrt{2\mathcal{E}/N_0}\right)$ is given in Figure 5-3.

5.2.2   The General Model

The signals considered in this section are given by

$$s_0(t) = A\,\psi_0(t)$$

and

$$s_1(t) = A\,\psi_1(t),$$

where $\{\psi_0, \psi_1\}$ is a binary signal set of the type described in Section 5.1. Thus, ψ_0 and ψ_1 are finite-energy, time-limited signals of duration T. If $A < \infty$, which is assumed to

Table 5.1: Values of the Error Probability $Q\left(\sqrt{2\,\mathcal{E}/N_0}\right)$

$(\mathcal{E}/N_0)_{\mathrm{dB}}$	$\mathcal{E}/N_0$	$Q\left(\sqrt{2\,\mathcal{E}/N_0}\right)$
5.0	3.16	5.95×10^{-3}
5.5	3.55	3.86×10^{-3}
6.0	3.98	2.39×10^{-3}
6.5	4.47	1.40×10^{-3}
7.0	5.01	7.73×10^{-4}
7.5	5.62	3.99×10^{-4}
8.0	6.31	1.91×10^{-4}
8.5	7.08	8.40×10^{-5}
9.0	7.94	3.36×10^{-5}
9.5	8.91	1.21×10^{-5}
10.0	10.00	3.87×10^{-6}
10.5	11.22	1.08×10^{-6}
11.0	12.59	2.61×10^{-7}
11.5	14.13	5.33×10^{-8}
12.0	15.85	9.01×10^{-9}

be true in all that follows, then the signals s_0 and s_1 are also finite-energy, time-limited signals of duration T. The energy $\mathcal{E}_i$ in the signal s_i can be written as

$$
\mathcal{E}_i = \int_{-\infty}^{\infty} [s_i(t)]^2 \, dt = \int_0^T [s_i(t)]^2 \, dt
$$

$$
= A^2 \int_0^T [\psi_i(t)]^2 \, dt. \tag{5.36}
$$

The general model for the binary baseband data transmission system is shown in Figure 5-4. The received signal $Y(t)$ is the sum of the noise process $X(t)$ and the signal $s_i(t)$, where $i = 0$ if the binary digit 0 is sent and $i = 1$ if the binary digit 1 is sent. The filter shown in the figure is a time-invariant linear filter with impulse response $h(t)$. The output of this filter, which is denoted by $Z(t)$, is sampled at time T_0. The output $Z(T_0)$ of the sampler is then compared with a threshold γ in order to make a decision between the two alternatives 0 and 1. A detailed description of the general model follows.

First, consider the channel noise process. This random process $X(t)$ is a stationary, zero-mean, Gaussian process that does not depend on the transmitted signal. The spectral density of $X(t)$ is denoted by $S_X(\omega)$. In this section, $X(t)$ need not be a white-noise process; on the contrary, the shape of the spectral density is arbitrary for the results obtained.

A communication channel that has the properties given in the preceding paragraph is called an *additive Gaussian noise (AGN) channel*. The essential features of the additive Gaussian noise channel model are as follows. The output of the channel is equal to the sum of the input signal and the channel noise. The channel noise is stationary and

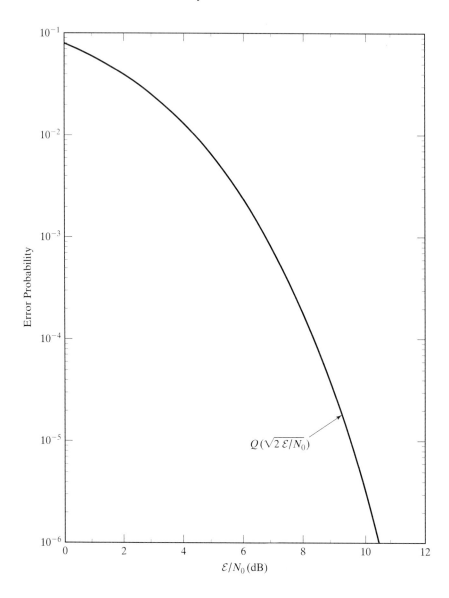

Figure 5-3: The error probability $Q\left(\sqrt{2\,\mathcal{E}/N_0}\right)$ as a function of $\mathcal{E}/N_0$ in dB.

Gaussian, has zero mean, and is independent of the input to the channel. If the noise is also white, the channel is called an *additive white Gaussian noise (AWGN) channel*.

Although we speak of "channel" noise, it must be remembered that much of the noise originates in the receiver itself. This noise is the thermal noise due to random motion of the conduction electrons in the resistive components of the receiver. Our channel model includes the RF portion of the receiver, or receiver front end as it is sometimes called; thus, the thermal noise is viewed as part of the channel. Our concern in

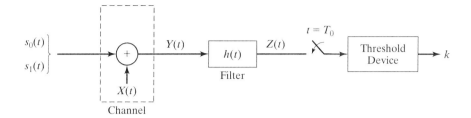

Figure 5-4: General model for binary data transmission.

the design of the communication system is with the demodulation of the received signal, and this demodulation takes place after the signal is converted to a lower frequency. Consequently, the received signal in our model is not the RF signal. In this chapter, it is, in fact, the baseband signal that is considered to be the received signal.

Next consider the filter. Its input is the random process

$$Y(t) = s_i(t) + X(t). \tag{5.37}$$

This process has a signal component $s_i(t)$ and a noise component $X(t)$. Since the filter is linear, its output $Z(t)$ can also be written as the sum of a signal component and a noise component. The signal component of $Z(t)$ is just the convolution of the signal s_i with the impulse response h. If $\hat{s}_i$ denotes the signal component of the output of the filter, then

$$\hat{s}_i(t) = \int_{-\infty}^{\infty} h(t - \tau)\, s_i(\tau)\, d\tau, \tag{5.38}$$

which is denoted in compact form by $\hat{s}_i = s_i * h$. Of course, $s_i * h = h * s_i$; thus, $\hat{s}_i(t)$ can also be written as

$$\hat{s}_i(t) = \int_{-\infty}^{\infty} s_i(t - \tau)\, h(\tau)\, d\tau. \tag{5.39}$$

The noise component $X(t)$ of the input gives rise to a noise component $\hat{X}(t)$ at the output of the filter. The properties of the random process $\hat{X}(t)$ are obtained by applying the methods of Chapters 3 and 4. Since $X(t)$ is a wide-sense stationary Gaussian random process and the filter is linear and time invariant, the output process $\hat{X}(t)$ is also a wide-sense stationary Gaussian random process. Thus, for our purposes, $\hat{X}(t)$ is completely specified by its mean $\mu_{\hat{X}} = E\{\hat{X}(t)\}$, which is zero because $X(t)$ is a zero-mean random process, and its autocorrelation function

$$R_{\hat{X}}(\tau) = E\left\{\hat{X}(t)\,\hat{X}(t + \tau)\right\}.$$

In order to evaluate the performance of the communication system of Figure 5-4, it is necessary to determine only $R_{\hat{X}}(0)$, as is shown later in this section.

The filter is followed by a sampler. The sampler is basically a switch that closes briefly at time T_0. The value of T_0 is arbitrary for now, but we will eventually be concerned with the determination of the optimum sampling time. The output of the sampler is the random variable $Z(T_0)$, which is the input to the threshold device as illustrated in Figure 5-4.

In the threshold device, $Z(T_0)$ is compared with a threshold level γ. The output k of the threshold device is determined by whether $Z(T_0) > \gamma$ or $Z(T_0) \leq \gamma$. The decision that the binary digit 0 was transmitted corresponds to $k = 0$ as the output of the threshold device, and the decision that 1 was sent corresponds to $k = 1$. For a fixed value of γ, there are only two nontrivial decision rules that are possible with this type of threshold device. One is to decide 0 was sent if $Z(T_0) > \gamma$ and 1 was sent if $Z(T_0) \leq \gamma$. The other is just the opposite: Decide 1 was sent if $Z(T_0) > \gamma$ and 0 was sent if $Z(T_0) \leq \gamma$.

The choice as to which of these two rules should be used depends on the relative values of $\hat{s}_0(T_0)$ and $\hat{s}_1(T_0)$. Since the signal $\hat{s}_i$ is the convolution of the signal s_i with the impulse response h, the value of $\hat{s}_i(T_0)$ depends on s_i, h, and the sampling time T_0. If s_0, s_1, h, and T_0 are such that

$$\hat{s}_0(T_0) > \hat{s}_1(T_0), \tag{5.40}$$

then the decision rule that should be used is to decide 0 was sent (i.e., $k = 0$) if $Z(T_0) > \gamma$ and 1 was sent (i.e., $k = 1$) if $Z(T_0) < \gamma$. It is assumed in all that follows that (5.40) holds. This assumption is made without any loss of generality, because, for a given impulse response and sampling instant, we can always reverse the indices on s_0 and s_1, if necessary, in order to ensure that (5.40) is true. That is, if $\hat{s}_1(T_0) > \hat{s}_0(T_0)$, then we simply interchange the roles of s_0 and s_1 so that the inequality is reversed. Alternatively, for a given signal set $\{s_0, s_1\}$ and sampling time T_0, we can replace the impulse response h by its negative $-h$ if we wish to reverse the inequality.

This concludes the description of the components of the communication system illustrated by Figure 5-4. The next step is to analyze its performance. The performance parameters of interest are the two probabilities $P_{e,0}$ and $P_{e,1}$. As in Section 5.2.1, the symbol $P_{e,i}$ denotes the probability of error when the binary digit i is sent. In order to evaluate $P_{e,i}$, it is necessary to examine the output of the filter in greater detail.

5.2.2.1   The Filter Output

The output of the filter is the random process $Z(t)$, which can be written as

$$Z(t) = \hat{s}_i(t) + \hat{X}(t). \tag{5.41}$$

The index i denotes the binary digit that is sent. The signal component $\hat{s}_i(t)$ of the output of the filter is easily evaluated from the convolution integral

$$\hat{s}_i(t) = \int_{-\infty}^{\infty} s_i(t - \tau) h(\tau) \, d\tau. \tag{5.42}$$

The autocorrelation function for the process $\hat{X}(t)$ can be obtained by the methods of Section 3.4.2. In particular, (3.48)–(3.54) apply because the channel noise is wide-sense stationary and the linear filter is time invariant. Thus, we have

$$R_{\hat{X}}(\tau) = \int_{-\infty}^{\infty} f(\tau - \lambda) R_X(\lambda) \, d\lambda, \tag{5.43}$$

where the function f is defined by

$$f(t) = \int_{-\infty}^{\infty} h(u-t) h(u) \, du. \tag{5.44}$$

Expressions (5.43) and (5.44) can be combined and written in more compact form as

$$R_{\hat{X}} = \tilde{h} * h * R_X, \tag{5.45}$$

where $\tilde{h}$ is the time reverse of h (i.e., $\tilde{h}(t) = h(-t)$). Notice that (5.45) is just (3.51) and that $f = \tilde{h} * h$.

The additive white Gaussian noise (AWGN) channel is an extremely important special case of the additive Gaussian noise channel. For the AWGN channel, $X(t)$ is a white Gaussian noise process (described in Section 3.5). It follows that for the AWGN channel $R_X(\tau) = (N_0/2)\,\delta(\tau)$, as discussed in Section 3.5. Consequently, the autocorrelation function for the output process is given by

$$R_{\hat{X}}(\tau) = \tfrac{1}{2} N_0 f(\tau) = \tfrac{1}{2} N_0 \int_{-\infty}^{\infty} h(u-\tau) h(u) \, du. \tag{5.46}$$

This result is obtained by substituting $\tfrac{1}{2} N_0 \delta(\lambda)$ for $R_X(\lambda)$ in (5.43) to obtain the expression

$$R_{\hat{X}}(\tau) = \tfrac{1}{2} N_0 \int_{-\infty}^{\infty} f(\tau - \lambda)\,\delta(\lambda) \, d\lambda. \tag{5.47}$$

For any sufficiently well-behaved function f, the integral in (5.47) is just $f(\tau)$. Equation (5.46) follows from this fact and (5.44).

Equations (5.43)–(5.45) provide a time-domain approach to the problem of evaluating the autocorrelation function for the noise process $\hat{X}(t)$. For the system under consideration, $\hat{X}(t)$ is wide-sense stationary; thus, there is an alternative frequency-domain approach in terms of spectral densities. By applying the results of Chapter 4 to (5.43)–(5.45), we obtain

$$S_{\hat{X}}(\omega) = |H(\omega)|^2 S_X(\omega), \tag{5.48}$$

where $H(\omega)$ is the transfer function for the linear system (H is the Fourier transform of h), $S_X(\omega)$ is the spectral density for the channel noise process, and $S_{\hat{X}}(\omega)$ is the spectral density for the output process $\hat{X}(t)$. Of course, the autocorrelation function for $\hat{X}(t)$, if needed, can be computed from its spectral density by taking the inverse Fourier transform, but in many problems this is not required.

As with the time-domain approach, there is a significant simplification in the frequency-domain expression if the channel noise process is white. The spectral density for the random process $\hat{X}(t)$ is

$$S_{\hat{X}}(\omega) = \tfrac{1}{2} N_0 |H(\omega)|^2,$$

where $N_0/2$ is the spectral density for the white noise process $X(t)$.

Since the noise process $\hat{X}(t)$ is Gaussian, it is completely characterized by its mean (which is zero) and its autocorrelation function. Moreover, since the output of the filter $Z(t)$ is the sum of the signal $\hat{s}_i(t)$ and the zero-mean Gaussian noise $\hat{X}(t)$, the filter output is completely characterized by (5.42) together with the autocorrelation function for $\hat{X}(t)$, which is given by (5.45).

Example 5-1   A Simple Binary Communication System

Consider the binary communication system model of Figure 5-4. Suppose that the signals are $s_0(t) = A\,p_I(t)$ and $s_1(t) = -A\,p_I(t)$ for $A > 0$. The filter impulse response is $h(t) = p_I(t)$, and the channel is an AWGN channel. The output signal $\hat{s}_0(t)$ can be evaluated from (5.42) with $i = 0$, which is

$$\hat{s}_0(t) = \int_{-\infty}^{\infty} s_0(t - \tau)\,h(\tau)\,d\tau. \tag{5.49}$$

Alternatively, $\hat{s}_0(t)$ can be determined from the observation that the convolution of two rectangular pulses, each of which has duration T, is a triangular pulse of duration $2T$. Taking the latter approach, we only need to determine the amplitude and location of the triangular pulse. In order to do this, first notice from the form of the functions s_0 and h in this example that (5.49) reduces to

$$\hat{s}_0(t) = \int_{-\infty}^{\infty} A\,p_I(t - \tau)\,p_I(\tau)\,d\tau$$

$$= A \int_{t-T}^{t} p_I(\tau)\,d\tau.$$

Since $p_I(\tau) = 1$ for $0 < \tau < T$ and $p_I(\tau) = 0$ otherwise, the integral is maximized when the interval of integration $(t - T, t)$ is the interval $(0, T)$. This occurs if and only if $t = T$; thus, the maximum of $\hat{s}_0(t)$ occurs for $t = T$. The resulting maximum value is

$$\hat{s}_0(T) = A \int_{0}^{T} p_I(\tau)\,d\tau = AT.$$

Thus, $\hat{s}_0(t)$ is a triangular pulse of the form

$$\hat{s}_0(t) = \begin{cases} At, & 0 \le t \le T, \\ A(2T - t), & T < t \le 2T, \\ 0, & \text{otherwise.} \end{cases}$$

Since $s_1(t) = -s_0(t)$ then $\hat{s}_1(t) = -\hat{s}_0(t)$. The autocorrelation function for the noise process can be found from (5.46) by a calculation that is essentially the same as the calculation that led to the preceding expression for $\hat{s}_0(t)$. From (5.46),

$$R_{\hat{X}}(\tau) = \tfrac{1}{2}N_0 \int_{-\infty}^{\infty} p_I(u - \tau)\,p_I(u)\,du$$

$$= \tfrac{1}{2}N_0 \int_{\tau}^{\tau+T} p_I(u)\,du$$

from which it follows easily that

$$R_{\hat{X}}(\tau) = \begin{cases} \frac{1}{2}N_0(T - |\tau|), & |\tau| \leq T, \\ 0, & |\tau| > T. \end{cases}$$

In order to determine the error probabilities, we really don't need a complete characterization of the random process $Z(t)$. All that matters is the value of this process at the sampling time T_0, and what is actually required is a characterization of the random variable $Z(T_0)$. The decision made by the receiver is based solely on the value of $Z(T_0)$, since this is the only input to the threshold device.

5.2.2.2   The Input to the Threshold Device

The input to the threshold device is the random variable $Z(T_0)$, which is the sum of the deterministic quantity $\hat{s}_i(T_0)$ and the random variable $\hat{X}(T_0)$. Because the receiver's decision is based entirely on the random variable $Z(T_0)$, this random variable is called the *decision statistic*. The decision statistic contains all of the information from the receiver input $Y(t)$ that is actually used in making the decision.

Clearly, for any system of interest, $\hat{X}(T_0)$ does not depend on which signal is sent. This follows from the definition of the additive Gaussian noise channel: The process $X(t)$ is independent of the channel input for such a channel; consequently, the process $\hat{X}(t)$ is independent of the transmitted signal. It follows that the random variable $\hat{X}(T_0)$ does not depend on i.

Suppose that 0 is sent. The decision statistic is then given by

$$Z(T_0) = \hat{s}_0(T_0) + \hat{X}(T_0).$$

For convenience, we define a random process

$$Z_0(t) = \hat{s}_0(t) + \hat{X}(t). \tag{5.50}$$

When 0 is sent, $Z(t)$ is equal to $Z_0(t)$ for all t (in particular for $t = T_0$, the sampling time). The random variable $Z_0(T_0)$ is Gaussian because it is the sum of a deterministic quantity $\hat{s}_0(T_0)$ and a Gaussian random variable $\hat{X}(T_0)$. The mean of $Z_0(T_0)$ is

$$\mu_0(T_0) = E\{Z_0(T_0)\} = \hat{s}_0(T_0) + E\left\{\hat{X}(T_0)\right\}, \tag{5.51}$$

which can be expressed in terms of the impulse response of the filter by employing (5.42) and using the fact that $\hat{X}(t)$ is a zero-mean process. The result is

$$\begin{aligned} \mu_0(T_0) &= \hat{s}_0(T_0) \\ &= \int_{-\infty}^{\infty} s_0(T_0 - \tau)\,h(\tau)\,d\tau. \end{aligned} \tag{5.52}$$

The variance of $Z_0(T_0)$ is given by

$$
\begin{aligned}
\mathrm{Var}\{Z_0(T_0)\} &= E\left\{[Z_0(T_0) - \mu_0(T_0)]^2\right\} \\
&= E\left\{[Z_0(T_0) - \hat{s}_0(T_0)]^2\right\} \\
&= E\left\{[\hat{X}(T_0)]^2\right\} \\
&= R_{\hat{X}}(0).
\end{aligned}
\tag{5.53}
$$

This parameter can be expressed in terms of the impulse response of the filter by employing (5.43)–(5.45). The result is

$$
\begin{aligned}
\mathrm{Var}\{Z_0(T_0)\} &= (\tilde{h} * h * R_X)(0) \\
&= \int_{-\infty}^{\infty} f(\lambda) R_X(\lambda)\, d\lambda \\
&= \int_{-\infty}^{\infty} \left[\int_{-\infty}^{\infty} h(u - \lambda) h(u)\, du\right] R_X(\lambda)\, d\lambda.
\end{aligned}
\tag{5.54}
$$

Notice that it is not necessary to evaluate the entire autocorrelation function $R_{\hat{X}}(\tau)$, $-\infty < \tau < \infty$. It suffices to compute $R_{\hat{X}}(0)$ only.

Next, suppose that 1 is sent. The decision statistic is then the random variable

$$
Z(T_0) = Z_1(T_0),
$$

where $Z_1(t)$ is the random process defined by

$$
Z_1(t) = \hat{s}_1(t) + \hat{X}(t).
\tag{5.55}
$$

The random variable $Z_1(T_0)$ is Gaussian with mean $\mu_1(T_0)$ given by

$$
\begin{aligned}
\mu_1(T_0) &= E\{Z_1(T_0)\} = \hat{s}_1(T_0) + E\left\{\hat{X}(T_0)\right\} \\
&= \hat{s}_1(T_0) \\
&= \int_{-\infty}^{\infty} s_1(T_0 - \tau) h(\tau)\, d\tau.
\end{aligned}
\tag{5.56}
$$

The variance of $Z_1(T_0)$ is given by (5.53) with $Z_0(T_0)$ and $\hat{s}_0(T_0)$ replaced by $Z_1(T_0)$ and $\hat{s}_1(T_0)$ respectively. Hence, we have

$$
\begin{aligned}
\mathrm{Var}\{Z_1(T_0)\} &= E\left\{[\hat{X}(T_0)]^2\right\} \\
&= R_{\hat{X}}(0) \\
&= \mathrm{Var}\{Z_0(T_0)\}.
\end{aligned}
\tag{5.57}
$$

As a result, if σ is defined by $\sigma^2 = R_{\hat{X}}(0)$, it follows that

$$
\sigma^2 = \int_{-\infty}^{\infty} \left[\int_{-\infty}^{\infty} h(u - \lambda) h(u)\, du\right] R_X(\lambda)\, d\lambda
\tag{5.58}
$$

(cf. (5.54)) and

$$
\mathrm{Var}\{Z_i(T_0)\} = \sigma^2
$$

for both $i = 0$ and $i = 1$.

One conclusion that can be drawn from the preceding development is that the only difference between $Z_0(T_0)$ and $Z_1(T_0)$ is the mean. Both of these are Gaussian random variables; hence, they are completely characterized by their means and variances. However, they have the same variance σ^2. Thus, the decision device at the output of the sampler must discriminate between two random variables that differ only in their mean values. If 0 is sent, the mean is $\mu_0(T_0) = \hat{s}_0(T_0)$, but if 1 is sent, the mean is $\mu_1(T_0) = \hat{s}_1(T_0)$. Intuitively, then, we expect that the ability of the decision device to discriminate between these two cases should depend on the difference of the means $\mu_0(T_0) - \mu_1(T_0)$. Notice that

$$
\begin{aligned}
\mu_0(T_0) - \mu_1(T_0) &= \hat{s}_0(T_0) - \hat{s}_1(T_0) \\
&= \int_{-\infty}^{\infty} [s_0(T_0 - \tau) - s_1(T_0 - \tau)]\, h(\tau)\, d\tau \\
&= \int_{-\infty}^{\infty} [s_0(\tau) - s_1(\tau)]\, h(T_0 - \tau)\, d\tau.
\end{aligned}
\tag{5.59}
$$

Thus, the difference of the means is a function of the impulse response of the filter and the difference of the two signals s_0 and s_1. The performance of the communication system is optimized by use of signals that are as different as possible and a filter with an impulse response that accentuates this difference as much as possible. The selection of the signals and the filter to optimize performance is discussed later in this chapter.

Example 5-2   A Simple Binary Communication System Revisited

This is a continuation of Example 5-1 for the system of Figure 5-4 with $s_0(t) = -s_1(t) = A\, p_T(t)$ and $h(t) = p_T(t)$. If 0 is sent, the input to the threshold device is the random variable

$$
Z(T_0) = Z_0(T_0) = \hat{s}_0(T_0) + \hat{X}(T_0),
$$

which is Gaussian with mean $\mu_0(T_0) = \hat{s}_0(T_0)$ and variance $\sigma^2 = R_{\hat{X}}(0) = \frac{1}{2} N_0 T$. If 1 is sent, the input to the threshold device is

$$
Z(T_0) = Z_1(T_0) = \hat{s}_1(T_0) + \hat{X}(T_0),
$$

which is a Gaussian random variable with mean $\mu_1(T_0) = \hat{s}_1(T_0)$ and variance $\sigma^2 = R_{\hat{X}}(0) = \frac{1}{2} N_0 T$. The actual expression for $\mu_i(T_0)$ can be obtained from the expression for $\hat{s}_i(t)$ given in Example 5-1. Notice that the difference in the means is

$$
\begin{aligned}
\mu_0(T_0) - \mu_1(T_0) &= \hat{s}_0(T_0) - \hat{s}_1(T_0) \\
&= \hat{s}_0(T_0) - [-\hat{s}_0(T_0)] \\
&= 2\hat{s}_0(T_0).
\end{aligned}
$$

For the signals and filter under consideration,

$$
\mu_0(T_0) - \mu_1(T_0) = \begin{cases} 2AT_0, & 0 \le T_0 \le T, \\ 2A(2T - T_0), & T < T_0 \le 2T, \\ 0, & \text{otherwise.} \end{cases}
$$

The difference is maximized if the sampling time T_0 is equal to T. ∎

For certain filters, the transfer function $H(\omega)$ is easier to work with than the impulse response. This is true for an ideal lowpass filter, for instance. Since convolution of time-domain signals corresponds to multiplication of their Fourier transforms, expressions such as (5.52), (5.56), and (5.58) can be evaluated by computing the inverse Fourier transform of the product of the transforms. This is particularly useful in the evaluation of the variance σ^2. Notice that (5.58) is just the statement that

$$\sigma^2 = R_{\tilde{X}}(0) = (h * \tilde{h} * R_X)(0),$$

which is equivalent to

$$\sigma^2 = \frac{1}{2\pi} \int_{-\infty}^{\infty} S_{\tilde{X}}(\omega) \, d\omega = \frac{1}{2\pi} \int_{-\infty}^{\infty} |H(\omega)|^2 S_X(\omega) \, d\omega. \tag{5.60}$$

For the AWGN channel with spectral density $S_X(\omega) = N_0/2$, the expression for σ^2 in (5.60) simplifies to

$$\sigma^2 = \frac{N_0}{4\pi} \int_{-\infty}^{\infty} |H(\omega)|^2 \, d\omega. \tag{5.61a}$$

Letting $f = \omega/2\pi$ denote the frequency in hertz, this becomes

$$\sigma^2 = \tfrac{1}{2} N_0 \int_{-\infty}^{\infty} |H(f)|^2 \, df. \tag{5.61b}$$

By applying Parseval's theorem or by using (5.58) with $R_X(\lambda) = \tfrac{1}{2} N_0 \, \delta(\lambda)$, we see that

$$\sigma^2 = \tfrac{1}{2} N_0 \int_{-\infty}^{\infty} [h(t)]^2 \, dt. \tag{5.61c}$$

Thus, for the AWGN channel, σ^2 can be evaluated by integrating either the square of the magnitude of the transfer function or the square of the impulse response.

Example 5-3   An Ideal Lowpass Filter
Suppose the filter is an ideal lowpass filter with transfer function

$$H(\omega) = \begin{cases} \beta, & |\omega| \le 2\pi W, \\ 0, & |\omega| > 2\pi W. \end{cases}$$

If the channel noise is a white Gaussian random process with spectral density $N_0/2$, the output process has spectral density $S_{\tilde{X}}(\omega) = \tfrac{1}{2} N_0 |H(\omega)|^2$. Applying (5.60), we find via an easy calculation that the variance is $\sigma^2 = \beta^2 N_0 W$. The evaluation of σ^2 from (5.58) is considerably more difficult for this example.   ∎

5.2.2.3   The Error Probabilities

Having characterized the decision statistic $Z(T_0)$, we can now give analytical expressions for the error probabilities $P_{e,0}$ and $P_{e,1}$ for the communication system of Figure 5-4. The probability $P_{e,0}$ is the probability that the decision made by the receiver is wrong when 0 is sent (i.e., when the signal s_0 is transmitted). The receiver is wrong when 0 is sent if and only if the input to the threshold device is less than or equal to the threshold γ. Under the condition that 0 is sent, the input $Z(T_0)$ to the threshold device is the random variable $Z_0(T_0)$. Thus,

$$P_{e,0} = P[Z(T_0) \leq \gamma \mid s_0] = P[Z_0(T_0) \leq \gamma]. \tag{5.62}$$

We use $P[A \mid s_i]$ to denote probability of the event A when the transmitted signal is s_i (for $i = 0$ or $i = 1$). The random variable $Z_0(T_0)$ is Gaussian with mean $\mu_0(T_0)$ and variance σ^2. Consequently,

$$P[Z_0(T_0) \leq \gamma] = \Phi([\gamma - \mu_0(T_0)]/\sigma).$$

The probability of error when 0 is sent is therefore given by

$$P_{e,0} = \Phi([\gamma - \mu_0(T_0)]/\sigma), \tag{5.63a}$$

where $\mu_0(T_0)$ is given by (5.52) and σ can be determined from (5.58) or (5.60). Because of the fact that $\Phi(x) = 1 - \Phi(-x) = Q(-x)$, the error probability $P_{e,0}$ can be written as

$$P_{e,0} = 1 - \Phi([\mu_0(T_0) - \gamma]/\sigma) \tag{5.63b}$$

or

$$P_{e,0} = Q([\mu_0(T_0) - \gamma]/\sigma). \tag{5.63c}$$

The probability of error when 1 is sent is given by (cf. (5.62) and (5.63a))

$$\begin{aligned} P_{e,1} &= P[Z(T_0) > \gamma \mid s_1] = P[Z_1(T_0) > \gamma] \\ &= 1 - P[Z_1(T_0) \leq \gamma] \\ &= 1 - \Phi([\gamma - \mu_1(T_0)]/\sigma), \end{aligned} \tag{5.64a}$$

where $\mu_1(T_0)$ is given by (5.56). This probability has the equivalent forms

$$P_{e,1} = \Phi([\mu_1(T_0) - \gamma]/\sigma) \tag{5.64b}$$

and

$$P_{e,1} = Q([\gamma - \mu_1(T_0)]/\sigma). \tag{5.64c}$$

By taking advantage of the fact that $(-1)^i$ is equal to $+1$ for $i = 0$ and -1 for $i = 1$, we can summarize the key results as follows. If the transmitted signals are $s_0(t)$ and $s_1(t)$, and if $\mu_i(t)$ is defined by

$$\mu_i(t) = \hat{s}_i(t) = \int_{-\infty}^{\infty} h(t - \tau)\, s_i(\tau)\, d\tau \tag{5.65}$$

for $i = 0$ and $i = 1$, then the probability of error given the ith signal is transmitted is

$$P_{e,i} = Q[(-1)^i \{\mu_i(T_0) - \gamma\}/\sigma], \qquad (5.66)$$

where γ is the threshold, T_0 is the sampling time, and σ is the standard deviation of the output of the filter. The value of $\mu_i(T_0)$ is found by letting $t = T_0$ in (5.65), and the value of σ is obtained from

$$\sigma^2 = (h * \tilde{h} * R_X)(0)$$
$$= \int_{-\infty}^{\infty} \int_{-\infty}^{\infty} h(u - \lambda) h(u) \, du \, R_X(\lambda) \, d\lambda, \qquad (5.67a)$$

or

$$\sigma^2 = \frac{1}{2\pi} \int_{-\infty}^{\infty} |H(\omega)|^2 S_X(\omega) \, d\omega. \qquad (5.67b)$$

For an AWGN channel, equations (5.67) reduce to

$$\sigma^2 = \tfrac{1}{2}N_0(h * \tilde{h})(0) = \tfrac{1}{2}N_0 \int_{-\infty}^{\infty} h^2(t) \, dt \qquad (5.68a)$$

and

$$\sigma^2 = \frac{N_0}{4\pi} \int_{-\infty}^{\infty} |H(\omega)|^2 \, d\omega. \qquad (5.68b)$$

The choice between (5.67a) versus (5.67b) or (5.68a) versus (5.68b) is a matter of convenience: Use the form that is easier to evaluate for the given filter and (for nonwhite noise) the given spectral density or autocorrelation function.

Example 5-4   On–Off Binary Signals

Consider a binary communications system with the signals $s_0(t) = A \, p_T(t)$ for $A > 0$ and $s_1(t) = 0$. The receiver filter has impulse response $h(t) = p_T(t)$ and the sampling time is $T/2$. The output signal $\hat{s}_0(t)$ is given by (see Example 5-1).

$$\hat{s}_0(t) = \begin{cases} At, & 0 \le t \le T, \\ A(2T - t), & T < t \le 2T, \\ 0, & \text{otherwise,} \end{cases}$$

and the output signal $\hat{s}_1(t)$ is identically zero. It follows that $\mu_0(T/2) = \tfrac{1}{2}AT$ and $\mu_1(T/2) = 0$. For an additive Gaussian noise channel and threshold level γ, the error probabilities obtained by applying (5.66) are

$$P_{e,0} = Q([(AT/2) - \gamma]/\sigma)$$

and

$$P_{e,1} = Q(\gamma/\sigma).$$

For a given spectral density or autocorrelation function for the channel noise, the standard deviation σ can be evaluated from (5.67a) or (5.67b). If the noise is white with spectral density $N_0/2$, then either (5.68a) or (5.68b) is used to evaluate σ^2. For the impulse response $h(t) = p_T(t)$, (5.68a) is clearly simpler than (5.68b). For an AWGN channel, the value of σ^2 is $N_0 T/2$, and the error probability expressions become

$$P_{e,0} = Q\left(\left[\tfrac{1}{2} AT - \gamma \right] / \sqrt{\tfrac{1}{2} N_0 T} \right)$$

and

$$P_{e,1} = Q\left(\gamma / \sqrt{\tfrac{1}{2} N_0 T} \right).$$

Notice that $\gamma = 0$ gives $P_{e,1} = Q(0) = 1/2$, so that this threshold setting is not a good choice for the vast majority of practical applications. More generally, $\gamma \leq 0$ gives $P_{e,1} \geq 1/2$ and $\gamma \geq AT/2$ gives $P_{e,0} \geq 1/2$, so we would typically select the threshold to be in the range $0 < \gamma < AT/2$.　■

The results summarized in equations (5.65), (5.67), and (5.68) are formulas for evaluation of the conditional means and the variance of the decision statistic for the system shown in Figure 5-4. For certain problems, it may be easier to evaluate these quantities directly. The following is an example of such a problem.

Example 5-5   A Sampling Receiver
The signal set is $s_0(t) = p_T(t)$ and $s_1(t) = -p_T(t)$, and the channel noise is a zero-mean Gaussian random process with autocorrelation function

$$R_X(\tau) = \begin{cases} \lambda(T - |\tau|), & |\tau| \leq T, \\ 0, & |\tau| > T. \end{cases}$$

The input to the receiver is sampled at times $T/3$ and $2T/3$, and these two samples are added together to give the decision statistic Z, which is in turn the input to the threshold device.

This problem can be solved via (5.65)–(5.67) by observing that the receiver fits the model of Figure 5-4 if we let

$$h(t) = \delta(t - 2T/3) + \delta(t - T/3)$$

and $T_0 = T$. This can be seen by noting that for an arbitrary input $y(t)$ the output at time T is

$$\begin{aligned} \hat{y}(T) &= \int_{-\infty}^{\infty} y(T - \tau) h(\tau) \, d\tau \\ &= \int_{-\infty}^{\infty} y(T - \tau) [\delta(\tau - 2T/3) + \delta(\tau - T/3)] \, d\tau \\ &= y(T - 2T/3) + y(T - T/3) \\ &= y(T/3) + y(2T/3). \end{aligned}$$

which is the desired response. An application of (5.65), for instance, shows that $\mu_i(T) = s_i(T/3) + s_i(2T/3)$, which is equal to $+2$ for $i = 0$ and -2 for $i = 1$. But this could have been obtained directly from the description of the receiver. The receiver input $Y(t)$ is sampled and summed to give the decision statistic $Z = Y(T/3) + Y(2T/3)$. But $Y(t) = s_i(t) + X(t)$ and the noise $X(t)$ has a zero mean, so the conditional mean of the input to the receiver is just the signal $s_i(t)$. It follows that the conditional mean of Z is just the sum of the two samples, $s_i(T/3)$ and $s_i(2T/3)$, of the transmitted signal. Similarly, the noise component of Z is $X(T/3) + X(2T/3)$, and this has variance

$$
\begin{aligned}
\sigma^2 &= E\left\{[X(T/3) + X(2T/3)]^2\right\} \\
&= 2R_X(0) + 2R_X(T/3) \\
&= 2\lambda\{T + (T - T/3)\} = 10\lambda T/3.
\end{aligned}
$$

The evaluation of σ^2 via (5.67a) is more complicated than the evaluation just given because of the products of δ-functions that appear in the integrand of (5.67a).   ∎

5.3   Optimization of the Threshold

Example 5-4 illustrates that some threshold choices are better than others, and some choices can give very poor results. In order to pursue this further, we must consider criteria for comparing the performance of receivers with different threshold settings. The presentation here is limited to two of the many possible criteria.

Before examining the specific criteria, it is helpful to consider the behavior of the error probabilities as the threshold is changed. Observe that if all of the parameters of the signals and the noise are fixed, the error probabilities $P_{e,0}$ and $P_{e,1}$ are functions of the threshold γ only. This can be seen from the expression

$$
P_{e,i} = Q\left[(-1)^i\{\mu_i(T_0) - \gamma\}/\sigma\right].
$$

When it is desired to emphasize the dependence of the error probabilities on the threshold, we display it explicitly by using the notation

$$
P_{e,i}(\gamma) = Q\left[(-1)^i\{\mu_i(T_0) - \gamma\}/\sigma\right].
$$

From the individual expressions for $P_{e,0}(\gamma)$ and $P_{e,1}(\gamma)$, it is clear that one of them is an increasing function of γ and the other is a decreasing function of γ. The increasing function is

$$
P_{e,0}(\gamma) = Q([\mu_0(T_0) - \gamma]/\sigma), \tag{5.69a}
$$

and the decreasing function is

$$
P_{e,1}(\gamma) = Q([\gamma - \mu_1(T_0)]/\sigma). \tag{5.69b}
$$

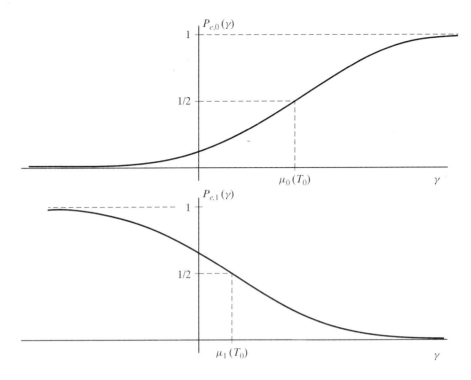

Figure 5-5: The error probabilities as a function of the threshold γ.

Recall that $\Phi(x)$, a distribution function, is an increasing function of x, and so $Q(x) = 1 - \Phi(x) = \Phi(-x)$ is a decreasing function of x. Graphs of $P_{e,0}(\gamma)$ and $P_{e,1}(\gamma)$ are shown in Figure 5-5.

Suppose that the threshold is selected to minimize $P_{e,0}$. Because $P_{e,0}(\gamma)$ is an increasing function of γ, the optimum threshold is $\gamma = -\infty$. This gives $P_{e,0} = 0$! The receiver never makes an error when the binary digit 0 is sent. Why is this true? The reason is clear from the decision rule, which is to decide that 0 was sent if $Z(T_0) > \gamma$. If $\gamma = -\infty$, the decision statistic is always greater than γ, so the receiver always decides 0 was sent, regardless of the input to the receiver.

The fact that the receiver always decides 0 was sent has the unfortunate consequence that it is always wrong when 1 is sent. Indeed, $P_{e,1} = 1$ for $\gamma = -\infty$. Thus, while the minimization of $P_{e,0}$ produces the desirable result $P_{e,0} = 0$, it also produces the undesirable result $P_{e,1} = 1$.

Another way of viewing the situation for $\gamma = -\infty$ is based on an examination of the resulting decision process. As noted, the receiver always decides that 0 was sent, regardless of the input to the receiver. Now this saves the expense of building a communication system—it is not needed if we always decide 0 was sent—but no information is being communicated from the source to the destination. Clearly, minimization of $P_{e,0}$ leads to a degenerate system that is of no use.

The dual of the case $P_{e,0} = 0$ and $P_{e,1} = 1$ results from the other extreme. If the criterion is to minimize $P_{e,1}$, the trivial solution that results is to always decide 1 was sent. This gives $P_{e,1} = 0$ and it corresponds to $\gamma = +\infty$. If $\gamma = +\infty$ and the decision rule is to decide 0 was sent if $Z(T_0) > \gamma$, then the receiver never decides 0 was sent. Hence, it is always wrong when in fact 0 was sent; that is, $P_{e,0} = 1$. Indeed, equations (5.69) show that $P_{e,1} = 0$ and $P_{e,0} = 1$ for the threshold $\gamma = +\infty$.

It is clear that neither extreme $\gamma = +\infty$ or $\gamma = -\infty$ is of interest. It should also be clear that we must minimize some function of *both* error probabilities; considering them individually leads to trivial solutions. The threshold should be selected to minimize $f[P_{e,0}(\gamma), P_{e,1}(\gamma)]$ for *appropriate* choices of the function f. Notice that these trivial solutions $\gamma = -\infty$ and $\gamma = +\infty$ are based on functions $f(x, y) = x$ and $f(x, y) = y$, respectively.

5.3.1   The Minimax Criterion

The first criterion that we consider is known as the *minimax criterion*. According to this criterion, the function f is defined by

$$f(x, y) = \max\{x, y\}.$$

That is, the threshold is selected to give the *min*imum possible value of the *max*imum of the two error probabilities, hence the term "minimax." The solution, denoted by γ_m, satisfies the equation

$$\max\{P_{e,0}(\gamma_m), P_{e,1}(\gamma_m)\} = \min_{\gamma} \left[\max\{P_{e,0}(\gamma), P_{e,1}(\gamma)\}\right].$$

The minimax criterion focuses on the largest of the two error probabilities for each choice of the threshold γ. For a given γ, let the largest of the two error probabilities be denoted by $P_{e,m}(\gamma)$. That is, define

$$P_{e,m}(\gamma) = \max\{P_{e,0}(\gamma), P_{e,1}(\gamma)\}$$

for each value of γ in the range $-\infty < \gamma < +\infty$.

In the selection of the threshold using the minimax criterion, we consider the worst case between "0 sent" versus "1 sent" for each value of γ, and then we choose γ in a way that makes this worst case as favorable as possible. Clearly, the trivial case $\gamma = -\infty$ leads to the very unfavorable worst-case situation: The decision is always wrong when 1 is sent. Similarly, $\gamma = +\infty$ gives an equally poor worst-case situation: The decision is always wrong when 0 is sent. These two choices of γ give $P_{e,m}(\gamma) = 1$, which, of course, is the largest possible value of the error probability $P_{e,m}(\gamma)$.

The proper choice of the threshold is obvious from Figure 5-6, in which graphs of $P_{e,0}(\gamma)$ and $P_{e,1}(\gamma)$ are shown on the same axes. Suppose that we select the threshold to be γ_1 as illustrated in the figure. The resulting value of the maximum error probability is

$$P_{e,m}(\gamma_1) = \max\{P_{e,0}(\gamma_1), P_{e,1}(\gamma_1)\} = P_{e,1}(\gamma_1).$$

Notice that this value would decrease if γ_1 were increased. Hence, γ_1 is too small to be the optimum threshold. Next suppose we select γ_2. The resulting value of the maximum

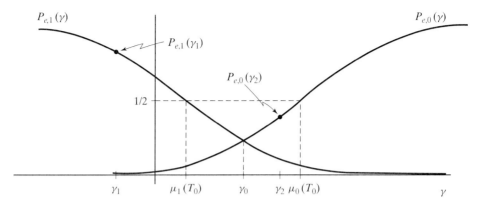

Figure 5-6: Three possible solutions for the threshold. The threshold γ_0 is the equalizer threshold: $P_{e,0}(\gamma_0) = P_{e,1}(\gamma_0)$.

error probability is

$$P_{e,m}(\gamma_2) = \max\{P_{e,0}(\gamma_2), P_{e,1}(\gamma_2)\} = P_{e,0}(\gamma_2),$$

and this time we observe that the maximum error probability would decrease if γ_2 were decreased. In fact, $P_{e,m}(\gamma_1)$ decreases as γ_1 increases up to γ_0, and $P_{e,m}(\gamma_2)$ decreases as γ_2 decreases down to γ_0. The minimax threshold is γ_0, which is the *unique* solution to the equation

$$P_{e,0}(\gamma_0) = P_{e,1}(\gamma_0).$$

The solution γ_0 is also called the *equalizer threshold*, because it equalizes the two error probabilities. It is clear from Figure 5-6 that the equalizer threshold also minimizes $P_{e,m}(\gamma)$; that is, $\gamma_m = \gamma_0$.

A slightly different illustration is shown in Figure 5-7. The solid line is a graph of $P_{e,m}(\gamma)$. The minimax threshold is, by definition, the threshold that corresponds to

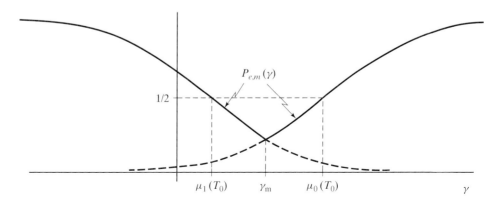

Figure 5-7: The error probability $P_{e,m}(\gamma) = \max\{P_{e,0}(\gamma), P_{e,1}(\gamma)\}$.

the minimum value of this solid line. The dotted lines are included to show that the minimax threshold γ_m is also the (unique) solution to $P_{e,0}(\gamma) = P_{e,1}(\gamma)$.

The next step is to obtain an analytical expression for γ_m in terms of the parameters of the communication system. In order to have $P_{e,0}(\gamma) = P_{e,1}(\gamma)$, we must have

$$Q([\mu_0(T_0) - \gamma]/\sigma) = Q([\gamma - \mu_1(T_0)]/\sigma). \tag{5.70}$$

But since $Q(x)$ is a continuous, strictly decreasing function of x, (5.70) has one and only one solution. This solution is obtained by equating the arguments of the functions in (5.70). That is, the minimax threshold satisfies

$$\frac{\mu_0(T_0) - \gamma_m}{\sigma} = \frac{\gamma_m - \mu_1(T_0)}{\sigma}.$$

But this reduces to

$$\gamma_m = [\mu_0(T_0) + \mu_1(T_0)]/2, \tag{5.71}$$

which says that the minimax threshold is midway between the means $\mu_0(T_0)$ and $\mu_1(T_0)$.

This can be seen graphically from Figure 5-8. Starting at the two means, $\mu_1(T_0)$ as the argument of $P_{e,1}$ and $\mu_0(T_0)$ as the argument of $P_{e,0}$, increase the argument of $P_{e,1}$ by an amount δ and decrease the argument of $P_{e,0}$ by the same amount. From the graph, it appears that we should have

$$P_{e,1}(\mu_1(T_0) + \delta) = P_{e,0}(\mu_0(T_0) - \delta)$$

for any choice of δ. That this is indeed true is proved analytically from equations (5.69): Equation (5.69b) gives

$$P_{e,1}(\mu_1(T_0) + \delta) = Q(\delta/\sigma)$$

and (5.69a) gives

$$P_{e,0}(\mu_0(T_0) - \delta) = Q(\delta/\sigma)$$

for any choice of δ. In particular, for the choice $\delta = \delta_m$, where δ_m is given by

$$\delta_m = [\mu_0(T_0) - \mu_1(T_0)]/2,$$

observe that

$$\begin{aligned} P_{e,1}(\mu_1(T_0) + \delta_m) &= P_{e,1}([\mu_0(T_0) + \mu_1(T_0)]/2) \\ &= Q([\mu_0(T_0) - \mu_1(T_0)]/2\sigma) \\ &= P_{e,0}([\mu_0(T_0) + \mu_1(T_0)]/2) = P_{e,0}(\mu_0(T_0) - \delta_m). \end{aligned}$$

This choice for δ corresponds to the point midway between $\mu_1(T_0)$ and $\mu_0(T_0)$, and it satisfies $P_{e,0}(\gamma_m) = P_{e,1}(\gamma_m)$ with

$$\gamma_m = \mu_1(T_0) + \delta_m = \mu_0(T_0) - \delta_m = [\mu_0(T_0) + \mu_1(T_0)]/2.$$

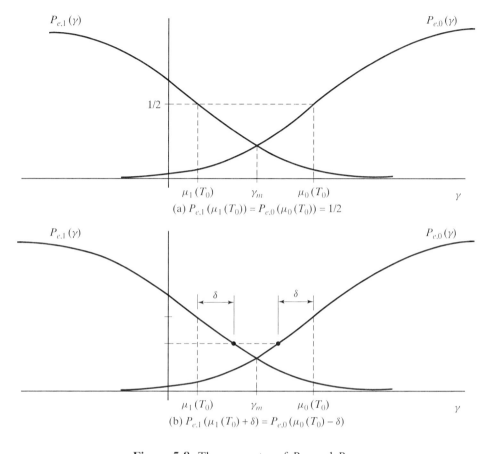

$P_{e.1}(\gamma)$ ... $P_{e.0}(\gamma)$

1/2

$\mu_1(T_0)$ ... γ_m ... $\mu_0(T_0)$ ... γ

(a) $P_{e.1}(\mu_1(T_0)) = P_{e.0}(\mu_0(T_0)) = 1/2$

$P_{e.1}(\gamma)$ ... $P_{e.0}(\gamma)$

δ ... δ

$\mu_1(T_0)$ ... γ_m ... $\mu_0(T_0)$ ... γ

(b) $P_{e.1}(\mu_1(T_0) + \delta) = P_{e.0}(\mu_0(T_0) - \delta)$

Figure 5-8: The symmetry of $P_{e.1}$ and $P_{e.0}$.

Notice that the following fact is obtained as an intermediate result in the preceding development. If the minimax threshold γ_m is employed, the resulting error probabilities are given by

$$P_{e.0} = P_{e.1} = P_{e.m} = Q([\mu_0(T_0) - \mu_1(T_0)]/2\sigma).$$

This result permits the evaluation of the minimax error probability without the need to evaluate the minimax threshold first.

Methods of calculus were not employed in the minimization of $P_{e.m}(\gamma)$. The reason for this is clear from Figure 5-7. Notice that the only values of γ for which

$$\frac{d\left[P_{e.m}(\gamma)\right]}{d\gamma} = 0$$

are $\gamma = -\infty$ and $\gamma = +\infty$. As we have already seen, these correspond to maxima rather than minima. From the figure, we see that, in fact, the derivative of $P_{e.m}(\gamma)$ is not even defined at the point that corresponds to the minimum (i.e., at $\gamma = \gamma_m$). For $\gamma < \gamma_m$, the derivative is strictly negative; and for $\gamma > \gamma_m$, it is strictly positive. But there is a discontinuity in the derivative at $\gamma = \gamma_m$, so the classical approach of setting the derivative equal to zero and solving for γ does not work. There are other optimization problems that are encountered in subsequent sections for which the traditional methods of calculus do not apply (or at least do not give the best approach).

Summary of Results (Minimax)

It is helpful to denote the minimax error probability by $P_{e,m}^*$ in all that follows. As a brief review, recall that $P_{e,m}$ is the maximum of the two error probabilities $P_{e,0}$ and $P_{e,1}$. We display the dependence of these probabilities on the threshold γ by writing

$$P_{e,m}(\gamma) = \max\{P_{e,0}(\gamma), P_{e,1}(\gamma)\}.$$

The *minimax error probability* is defined by

$$P_{e,m}^* = \min\left\{P_{e,m}(\gamma) : -\infty < \gamma < \infty\right\}.$$

It follows from the preceding results that

$$P_{e,m}^* = P_{e,m}(\gamma_m) = P_{e,0}(\gamma_m) = P_{e,1}(\gamma_m),$$

where γ_m is the minimax threshold. For the general additive Gaussian noise channel, the minimax threshold is given by

$$\gamma_m = [\mu_0(T_0) + \mu_1(T_0)]/2,$$

and the minimax error probability is

$$P_{e,m}^* = Q\{[\mu_0(T_0) - \mu_1(T_0)]/2\sigma\}. \tag{5.72}$$

Example 5-6   On-Off Binary Signals Revisited

Consider again the binary communication system described in Example 5-4. The signal set is composed of $s_0(t) = A\,p_T(t)$ for $A > 0$ and $s_1(t) = 0$. The filter impulse response is $h(t) = p_T(t)$, and the sampling time is $T_0 = T/2$. Assume that the channel is an AWGN channel with spectral density $N_0/2$. It is shown in Example 5-4 that $\hat{s}_0(T/2) = AT/2$ and, of course, $\hat{s}_1(T/2) = 0$. Thus, for $T_0 = T/2$, the minimax threshold is

$$
\begin{aligned}
\gamma_m &= [\mu_0(T/2) + \mu_1(T/2)]/2 \\
&= [\hat{s}_0(T/2) + \hat{s}_1(T/2)]/2 \\
&= AT/4.
\end{aligned}
$$

This is halfway between the means $\mu_1(T/2) = 0$ and $\mu_0(T/2) = AT/2$. The resulting minimax error probability is

$$
\begin{aligned}
P_{e,m}^* &= P_{e,0}(\gamma_m) = Q([\mu_0(T/2) - \gamma_m]/\sigma) = Q\left(\left[\tfrac{1}{2}AT - \gamma_m\right]/\sigma\right) \\
&= Q(AT/4\sigma).
\end{aligned}
$$

The value of σ is $\sqrt{N_0 T/2}$, so

$$
P_{e,m}^* = Q\left(\sqrt{A^2 T/8 N_0}\right).
$$

It should be clear by now that $P_{e,m}^*$ can be computed from either $P_{e,0}$ or $P_{e,1}$ with $\gamma = \gamma_m$. ($\gamma = \gamma_m$ implies $P_{e,m} = P_{e,0} = P_{e,1}$.) Working with $P_{e,1}$ gives

$$
\begin{aligned}
P_{e,m}^* &= P_{e,1}(\gamma_m) = Q([\gamma_m - \mu_1(T_0)]/\sigma) \\
&= Q(\gamma_m/\sigma) = Q(AT/4\sigma).
\end{aligned}
$$

Alternatively, if all that is required is to determine $P_{e,m}^*$, equation (5.72) can be used to evaluate $P_{e,m}^*$ directly by using the fact that

$$
\begin{aligned}
\mu_0(T/2) - \mu_1(T/2) &= \mu_0(T/2) = \hat{s}_0(T/2) \\
&= AT/2.
\end{aligned}
$$

Thus, (5.72) implies that $P_{e,m}^* = Q(AT/4\sigma)$.   ■

As an exercise, rework Example 5-6 for a different sampling time. Show that if $T_0 = T$, the resulting minimax threshold and minimax error probability are

$$
\gamma_m = AT/2
$$

and

$$
P_{e,m}^* = Q(AT/2\sigma).
$$

Notice that the minimax error probability is smaller for $T_0 = T$ than for $T_0 = T/2$. Is $T_0 = T$ the best choice for the sampling time if the minimax threshold is used?

Equation (5.72) is not only of value for computing the minimax error probability, it is also the starting point for further minimization of $P_{e,m}^*$. Recall that the filter impulse response, the sampling time, and the signal set are arbitrary, subject only to the condition $\hat{s}_0(T_0) > \hat{s}_1(T_0)$. An optimum system (from the minimax point of view) is obtained if $P_{e,m}^*$ is minimized with respect to these elements of the communication system.

Because Q is a decreasing function, $P_{e,m}^*$ is minimized by maximizing the quantity $[\mu_0(T_0) - \mu_1(T_0)]/2\sigma$. Because $\mu_0(T_0) - \mu_1(T_0)$ depends on the signals, but not on the noise, and σ depends on the noise spectral density, but not on the signals, the quantity of interest can be thought of as a signal-to-noise ratio. It is convenient to define

$$
\text{SNR} = [\mu_0(T_0) - \mu_1(T_0)]/2\sigma, \tag{5.73a}
$$

so that (5.72) becomes $P_{e,m}^* = Q(\text{SNR})$. Keep in mind that SNR depends on the filter impulse response h, the sampling time T_0, and the signal set $\{s_0, s_1\}$. As a review, recall that the relevant equations are

$$\mu_i(T_0) = \hat{s}_i(T_0) = \int_{-\infty}^{\infty} s_i(T_0 - \tau) h(\tau) \, d\tau \qquad (5.73b)$$

(for $i = 0$ and $i = 1$) and

$$\sigma^2 = \int_{-\infty}^{\infty} \int_{-\infty}^{\infty} h(u - \lambda) h(u) R_X(\lambda) \, du \, d\lambda$$

$$= \frac{1}{2\pi} \int_{-\infty}^{\infty} |H(\omega)|^2 S_X(\omega) \, d\omega. \qquad (5.73c)$$

Because the noise variance σ^2 does not depend on the sampling time T_0, the maximization of SNR with respect to T_0 is easily accomplished for most filters and signal sets of interest. The signal-to-noise ratio is maximized by choosing T_0 to maximize the quantity $\Delta_\mu(T_0)$ defined by

$$\Delta_\mu(T_0) = \mu_0(T_0) - \mu_1(T_0) = \int_{-\infty}^{\infty} [s_0(T_0 - \tau) - s_1(T_0 - \tau)] h(\tau) \, d\tau. \qquad (5.74)$$

This is usually accomplished by finding an analytical expression for $\Delta_\mu(T_0)$ and then maximizing with respect to T_0 by inspection or, if necessary, by methods of calculus.

Example 5-7   Determination of the Optimum Sampling Time
Suppose the signals are given by

$$s_0(t) = a_0 \, p_I(t)$$

and

$$s_1(t) = -a_1 \sin(\pi t / T) \, p_I(t).$$

and the filter impulse response is $h(t) = p_I(t)$. Assume that a_0 and a_1 are positive. It follows from (5.73b) that $\mu_0(T_0) \geq 0$ (note that $s_0(t) \geq 0$ and $h(t) \geq 0$ for all t) and $\mu_1(T_0) \leq 0$ (note that $s_1(t) \leq 0$ and $h(t) \geq 0$ for all t). In this particular example, there is a value of T_0 that simultaneously *maximizes* $\mu_0(T_0)$ and *minimizes* $\mu_1(T_0)$; hence, it maximizes $\Delta_\mu(T_0) = \mu_0(T_0) - \mu_1(T_0)$.

To find the optimum sampling time, first observe that

$$\mu_0(T_0) = a_0 \int_{-\infty}^{\infty} p_I(T_0 - \tau) \, p_I(\tau) \, d\tau.$$

But $p_I(T_0 - \tau) = p_I[\tau - (T_0 - T)]$, as can be seen by drawing a graph of the functions or checking the range of their nonzero values. It follows that

$$\mu_0(T_0) = a_0 \int_{-\infty}^{\infty} p_I[\tau - (T_0 - T)] \, p_I(\tau) \, d\tau.$$

This integral is maximized by choosing T_0 to maximize the amount of overlap of the two pulses that appear in the integrand. Clearly, the overlap interval is no larger than $[0, T]$, the interval of nonzero values of $p_T(\tau)$. Moreover, this overlap can be achieved by making the two pulses identical, which is accomplished by letting $T_0 = T$. The resulting maximum value of $\mu_0(T_0)$ is

$$\mu_0(T) = a_0 \int_{-\infty}^{\infty} [p_T(\tau)]^2 \, d\tau = a_0 T.$$

The expression for the conditional mean given that signal s_1 is transmitted is

$$\mu_1(T_0) = \int_{-\infty}^{\infty} h(T_0 - \tau) s_1(\tau) \, d\tau$$

$$= -a_1 \int_{-\infty}^{\infty} p_T(T_0 - \tau) \sin(\pi\tau/T) \, p_T(\tau) \, d\tau.$$

Because the integrand is nonnegative, the same argument as the one used for maximizing $\mu_0(T_0)$ works here as well. The integral is maximized (hence, $\mu_1(T_0)$ is minimized) by letting $T_0 = T$. The resulting value of $\mu_1(T_0)$ is

$$\mu_1(T) = -a_1 \int_{-\infty}^{\infty} \sin(\pi\tau/T)[p_T(\tau)]^2 \, d\tau = -2a_1 T/\pi.$$

This establishes that the maximum value of $\Delta_\mu(T_0)$ for this example is $\Delta_\mu(T)$ $= [a_0 + (2a_1/\pi)]T$.

This answer can be derived more formally as follows: First, observe that

$$\Delta_\mu(T_0) = \int_{-\infty}^{\infty} p_T(T_0 - \tau)[a_0 + a_1 \sin(\pi\tau/T)] \, p_T(\tau) \, d\tau$$

$$= \int_{0}^{T} p_T(T_0 - \tau)[a_0 + a_1 \sin(\pi\tau/T)] \, d\tau$$

$$\leq \int_{0}^{T} [a_0 + a_1 \sin(\pi\tau/T)] \, d\tau.$$

The inequality in the last step follows from the fact that $p_T(u) \leq 1$ for all values of u. Next, observe that the upper bound on $\Delta_\mu(T_0)$ is just $\Delta_\mu(T)$. That is,

$$\Delta_\mu(T) = \int_{-\infty}^{\infty} p_T(T - \tau)[a_0 + a_1 \sin(\pi\tau/T)] \, p_T(\tau) \, d\tau$$

$$= \int_{0}^{T} [a_0 + a_1 \sin(\pi\tau/T)] \, d\tau,$$

because $p_T(T - \tau) = p_T(\tau)$. Thus, we have shown that $\Delta_\mu(T_0) \leq \Delta_\mu(T)$, which proves that

$$\max\{\Delta_\mu(T_0) : -\infty < T_0 < \infty\} = \Delta_\mu(T).$$

That this maximum is unique can also be proved by observing that

$$\int_0^T p_T(T_0 - \tau)[a_0 + a_1 \sin(\pi \tau / T)] \, d\tau \; < \; \int_0^T [a_0 + a_1 \sin(\pi \tau / T)] \, d\tau$$

if $T_0 \neq T$.

We conclude that the maximum signal-to-noise ratio is

$$\text{SNR} = \Delta_\mu(T)/2\sigma = [a_0 + (2a_1/\pi)]T/2\sigma.$$

If the noise is white, then $\sigma^2 = \frac{1}{2} N_0 T$ and the maximum signal-to-noise ratio is

$$\text{SNR} = [a_0 + (2a_1/\pi)]\sqrt{T/2N_0}.$$

The minimax error probability that results when the optimum sampling time is used is

$$P_{e,m}^* = Q\left\{[a_0 + (2a_1/\pi)]\sqrt{T/2N_0}\right\}.$$

■

5.3.1.1   Antipodal Signals

The minimax threshold for *antipodal signals* is especially simple, and so is the resulting expression for the minimax error probability. Recall that the defining property for antipodal signals is $s_1(t) = -s_0(t)$ for all t. For convenience, let $s(t) = s_0(t)$ so that $s_1(t) = -s(t)$.

First, observe that

$$\begin{aligned}
\mu_1(T_0) &= (s_1 * h)(T_0) = -(s * h)(T_0) \\
&= -(s_0 * h)(T_0) = -\mu_0(T_0).
\end{aligned}$$

Next, recall that the minimax threshold for an arbitrary signal set is halfway between the means:

$$\gamma_m = [\mu_0(T_0) + \mu_1(T_0)]/2.$$

We conclude that the minimax threshold for antipodal signals is $\gamma_m = 0$.

The fact that $\gamma_m = 0$ for antipodal signals seems reasonable. For such signals, there is no bias in the signal set. An antipodal signal set is symmetric about zero in the sense that $s_0(t) = -s_1(t)$. Because the channel noise has a zero mean, the channel adds no bias. Finally, the linear filter does not add a bias, so the decision statistic is itself unbiased (i.e., symmetric about zero). This intuitive argument dealing with symmetry and the lack of bias is made mathematically precise by looking at the conditional density functions for the decision statistic $Z(T_0)$.

The conditional density for $Z(T_0)$ given that s_i is transmitted is

$$f_i(z) = \frac{1}{\sqrt{2\pi}\,\sigma} e^{-[z-\mu_i(T_0)]^2/2\sigma^2}.$$

As we have shown, $\mu_0(T_0) = -\mu_1(T_0)$ for antipodal signals. Therefore,

$$
\begin{aligned}
\sqrt{2\pi}\,\sigma f_1(-z) &= \exp\left\{-[-z - \mu_1(T_0)]^2/2\sigma^2\right\} \\
&= \exp\left\{-[z + \mu_1(T_0)]^2/2\sigma^2\right\} \\
&= \exp\left\{-[z - \mu_0(T_0)]^2/2\sigma^2\right\} \\
&= \sqrt{2\pi}\,\sigma f_0(z).
\end{aligned}
$$

That is, the conditional densities are the duals of each other in the sense that $f_1(-z) = f_0(z)$ for all z. A given range of positive values is just as likely under the hypothesis that s_0 was sent as the corresponding range of negative values under the hypothesis that s_1 was sent. For this reason, it seems natural, under the minimax criterion, not to bias the threshold toward either positive or negative values.

The signal-to-noise ratio for antipodal signals and the minimax threshold is

$$
\mathrm{SNR} = [\mu_0(T_0) - \mu_1(T_0)]/2\sigma = \mu_0(T_0)/\sigma.
$$

The minimax error probability is therefore

$$
P_{e,m}^* = Q(\mu_0(T_0)/\sigma),
$$

where

$$
\mu_0(T_0) = (s_0 * h)(T_0).
$$

An important consideration for implementation concerns the dependence of the threshold on the received signal levels (i.e., the peak voltages or average powers in the received signals $s_0(t)$ and $s_1(t)$). In general, one must know these signal levels in order to set the threshold to its minimax value. This is illustrated by Example 5-6, in which we found that $\gamma_m = AT/4$, where A is the peak amplitude of the signal $s_0(t)$.

The practical problem is that the received signal levels may not be known accurately. Moreover, these levels may change due to relative motion of the transmitter and receiver (changes in the communication range), fluctuations in the propagation loss due to atmospheric conditions, or variations in the output power levels for the various transmitters that are sending data to the receiver in question. For optimum performance, these changes must be tracked and the threshold must be adjusted to compensate for the changes.

A significant advantage of antipodal signals is that the minimax threshold does not depend on the received signal levels. The optimum threshold from the minimax point of view is zero, regardless of transmitter power levels, propagation losses, etc. This is a very important feature of antipodal signals.

5.3.1.2   The Conditional Densities and the Minimax Threshold

An interesting byproduct of our consideration of antipodal signals is the fact that $f_1(-z) = f_0(z)$ for all z. This implies that, for antipodal signals, *a solution* of the equation

$$
f_1(\gamma) = f_0(\gamma)
$$

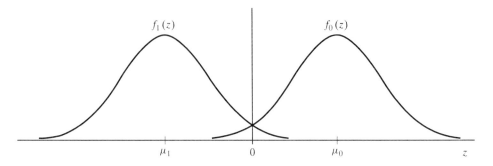

Figure 5-9: The conditional density functions for the decision statistic (antipodal signals).

is the minimax threshold $\gamma = \gamma_m = 0$. That this is in fact the *only solution* is illustrated in Figure 5-9. For convenience, $\mu_i(T_0)$ is denoted by μ_i for the present discussion, because the sampling time is fixed throughout.

Is it true that the minimax threshold is the solution to $f_1(\gamma) = f_0(\gamma)$ for more general signal sets than antipodal? In order to examine this question, we begin with the fact that the minimax threshold is given by

$$\gamma_m = (\mu_0 + \mu_1)/2$$

and that the densities are

$$f_i(z) = \frac{1}{\sqrt{2\pi}\,\sigma}\, \exp\left\{-(z - \mu_i)^2/2\sigma^2\right\}.$$

Consider $f_0(z)$ and $f_1(z)$ for $z = \gamma_m$. First, observe that

$$(\gamma_m - \mu_1)^2 = [(\mu_0 - \mu_1)/2]^2$$

and

$$(\gamma_m - \mu_0)^2 = [(\mu_1 - \mu_0)/2]^2.$$

It follows that

$$f_1(\gamma_m) = \frac{1}{\sqrt{2\pi}\,\sigma}\, \exp\left\{-(\mu_0 - \mu_1)^2/8\sigma^2\right\} = f_0(\gamma_m).$$

Because no restrictions were placed on the relationship between μ_0 and μ_1, this demonstrates that the minimax threshold satisfies

$$f_1(\gamma_m) = f_0(\gamma_m)$$

for an arbitrary signal set $\{s_0, s_1\}$. An illustration is given in Figure 5-10. The fact that $f_1(\gamma_m) = f_0(\gamma_m)$ is just a consequence of (1) the location of the minimax threshold, which is midway between the means μ_1 and μ_0, and (2) the symmetry of the Gaussian density about its mean value.

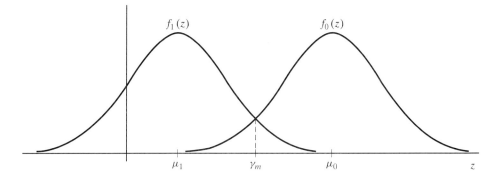

Figure 5-10: The conditional densities for the decision statistics (general signals).

The following interpretation of the role of the threshold device is suggested by Figure 5-10. Given that the decision statistic is $Z = z$, the decision is that 0 was sent if $f_0(z) > f_1(z)$ and that 1 was sent if $f_1(z) \geq f_0(z)$. It turns out that this type of decision rule arises in more general circumstances, as discussed in Section 5.4.

5.3.2   The Bayes Criterion

The minimax criterion does not attempt to give preference to one symbol over the other. In fact, it does not even distinguish between the symbols 0 and 1. Suppose that it is known in advance that a given data source will emit a much larger number of 0s than 1s, or suppose that it is felt that it is more important to make a correct decision when 0 is sent than when 1 is sent. In such cases, it may be preferred to bias the decision in favor of the symbol 0. Because the threshold device decides that 0 was sent when the decision statistic exceeds the threshold, the threshold should be decreased in order to bias the decision in favor of 0. To bias it in favor of the symbol 1, the threshold should be increased.

A decision criterion that incorporates the ability to adjust the threshold according to the transmission probabilities for the 0s and 1s is the *Bayes criterion*. Formally, the Bayes criterion fits our model, in which we minimize $f[P_{e,0}(\gamma), P_{e,1}(\gamma)]$, if the function f is defined by $f(x, y) = \pi_0 x + \pi_1 y$ for some choice of $\pi_0 > 0$ and $\pi_1 > 0$. That is, γ is selected to minimize a (positive) linear combination of $P_{e,0}(\gamma)$ and $P_{e,1}(\gamma)$. In practice, the coefficients π_0 and π_1 are nearly always defined to be the probabilities that 0 and 1 are sent, respectively. Thus, in addition to the constraints $\pi_0 > 0$ and $\pi_1 > 0$, we also require $\pi_0 + \pi_1 = 1$. In any event, there is no harm in imposing this last constraint, because it is just a normalization that does not affect the minimization of $\pi_0 P_{e,0}(\gamma) + \pi_1 P_{e,1}(\gamma)$. In all that follows, π_i represents the probability that i is sent (for $i = 0$ and $i = 1$).

We can look at this from an alternative point of view by considering the *average* probability of error $\overline{P}_e$ for transmission of binary digits. Let E denote the event that an error occurs at the receiver, and let H_i denote the event that the signal s_i is transmitted (representing the binary digit i). Bayes' rule gives an expression for the probability of error in terms of the conditional probabilities:

$$P(E) = P(E|H_0)P(H_0) + P(E|H_1)P(H_1)$$
$$= P_{e,0}\pi_0 + P_{e,1}\pi_1.$$

Thus, $P(E)$ is just the weighted *average* of the two conditional error probabilities. In this setting, $P(E)$ is referred to as the *average probability of error*, and it is denoted by $\overline{P_e}$.

The *Bayes threshold* is selected to minimize the average probability of error. In order to emphasize the dependence on the threshold, the average probability of error is written as

$$\overline{P_e}(\gamma) = P_{e,0}(\gamma)\,\pi_0 + P_{e,1}(\gamma)\,\pi_1.$$

The dependence of the average probability of error on γ is governed by the fact that

$$P_{e,0}(\gamma) = Q([\mu_0(T_0) - \gamma]/\sigma)$$

is an increasing function of γ and

$$P_{e,1}(\gamma) = Q([\gamma - \mu_1(T_0)]/\sigma)$$

is a decreasing function of γ. Therefore, minimization of $\overline{P_e}(\gamma)$ involves a tradeoff between these two conditional probabilities, much like the minimization of the maximum error probability $P_{e,m}(\gamma)$ in the previous subsection.

An important difference, however, is that $\overline{P_e}(\gamma)$ has a well-defined derivative over the entire range $-\infty < \gamma < \infty$. Thus, unlike the solution to the minimax problem, the determination of the Bayes threshold follows from standard methods of calculus. The procedure is to find solutions to the equation $\overline{P_e}'(\gamma) = 0$, where $\overline{P_e}'$ denotes the derivative of the function $\overline{P_e}$.

The derivative of $\overline{P_e}(\gamma)$ with respect to the threshold γ is

$$\overline{P_e}'(\gamma) = P_{e,0}'(\gamma)\,\pi_0 + P_{e,1}'(\gamma)\,\pi_1.$$

Now, $P_{e,0}$ is an increasing function and $P_{e,1}$ is a decreasing function, so $P_{e,0}'(\gamma) > 0$ and $P_{e,1}'(\gamma) < 0$. We solve $\overline{P_e}'(\gamma) = 0$ by solving the equation

$$P_{e,0}'(\gamma)\,\pi_0 = -P_{e,1}'(\gamma)\,\pi_1,$$

each side of which is positive for all γ.

Recall that

$$P_{e,0}(\gamma) = P[Z_0(T_0) \le \gamma].$$

This is just equation (5.62). Therefore, $P_{e,0}(\gamma)$ is the distribution function for $Z_0(T_0)$ evaluated at γ. That is, $P_{e,0}(\gamma) = F_0(\gamma)$, where F_0 can be thought of as either the distribution function for $Z_0(T_0)$ or the conditional distribution function for $Z(T_0)$ given that s_0 is transmitted. It follows that

$$P_{e,0}'(\gamma) = f_0(\gamma).$$

where f_0 is the density function for $Z_0(T_0)$. But $Z_0(T_0)$ is a Gaussian random variable with mean $\mu_0(T_0)$ and variance σ^2, so

$$P'_{e,0}(\gamma) = \exp\{-[\gamma - \mu_0(T_0)]^2/2\sigma^2\}/\sqrt{2\pi}\,\sigma. \qquad (5.75a)$$

Similarly,

$$\begin{aligned} P_{e,1}(\gamma) &= P[Z_1(T_0) > \gamma] \\ &= 1 - P[Z_1(T_0) \le \gamma] \\ &= 1 - F_1(\gamma), \end{aligned}$$

where F_1 is the distribution function for $Z_1(T_0)$. Proceeding as before, we see that

$$P'_{e,1}(\gamma) = -f_1(\gamma),$$

and therefore,

$$P'_{e,1}(\gamma) = -\exp\{-[\gamma - \mu_1(T_0)]^2/2\sigma^2\}/\sqrt{2\pi}\,\sigma. \qquad (5.75b)$$

Exercise 5-1
Establish (5.75a) and (5.75b) by the following alternative method: Define functions g_0 and g_1 by

$$g_0(x) = [\mu_0(T_0) - x]/\sigma$$

and

$$g_1(x) = [x - \mu_1(T_0)]/\sigma$$

for real numbers x. Use the chain rule to differentiate $Q[g_i(\gamma)]$, and show that the results give (5.75a) for $i = 0$ and (5.75b) for $i = 1$.

Solution.   First, observe that

$$P_{e,i}(\gamma) = Q[g_i(\gamma)],$$

which is the basis for this approach. The chain rule applied to the right-hand side gives

$$P'_{e,i}(\gamma) = Q'[g_i(\gamma)]\,g'_i(\gamma).$$

But

$$Q'(y) = -\Phi'(y) = -\exp\{-y^2/2\}/\sqrt{2\pi}$$

for all y. It follows from the definitions of the functions g_0 and g_1 that

$$Q'[g_i(\gamma)] = -\exp\{-[\gamma - \mu_i(T_0)]^2/2\sigma^2\}/\sqrt{2\pi}.$$

Next, observe that $g_0'(x) = -1/\sigma$ and $g_1'(x) = 1/\sigma$ for all x. Combining these results gives

$$P_{e,0}'(\gamma) = \exp\{-[\gamma - \mu_0(T_0)]^2/2\sigma^2\}/\sqrt{2\pi}\,\sigma$$

and

$$P_{e,1}'(\gamma) = -\exp\{-[\gamma - \mu_1(T_0)]^2/2\sigma^2\}/\sqrt{2\pi}\,\sigma.$$

∎

Recall that $\overline{P}_e'(\gamma) = P_{e,0}'(\gamma)\,\pi_0 + P_{e,1}'(\gamma)\,\pi_1$. From (5.75), it is clear that $\overline{P}_e'(\gamma) = 0$ is equivalent to the equation

$$\pi_0 \exp\{-[\gamma - \mu_0(T_0)]^2/2\sigma^2\}/\sqrt{2\pi}\,\sigma = \pi_1 \exp\{-[\gamma - \mu_1(T_0)]^2/2\sigma^2\}/\sqrt{2\pi}\,\sigma.$$

Multiply through by $\sqrt{2\pi}\,\sigma$, and take logarithms (the natural logarithm, denoted by ln, is used). This gives

$$\ln \pi_0 - \{[\gamma - \mu_0(T_0)]^2/2\sigma^2\} = \ln \pi_1 - \{[\gamma - \mu_1(T_0)]^2/2\sigma^2\}.$$

It is assumed that π_0 and π_1 are both nonzero (the trivial cases $\pi_0 = 0$ and $\pi_1 = 0$ are discussed later), so the last equation can be written as

$$\{[\gamma - \mu_1(T_0)]^2 - [\gamma - \mu_0(T_0)]^2\}/2\sigma^2 = \ln(\pi_1/\pi_0).$$

Multiply both sides by $2\sigma^2$, expand the quadratic terms on the left-hand side, and simplify to obtain

$$2\gamma[\mu_0(T_0) - \mu_1(T_0)] - [\mu_0^2(T_0) - \mu_1^2(T_0)] = 2\sigma^2 \ln(\pi_1/\pi_0).$$

Our convention that $\hat{s}_0(T_0) > \hat{s}_1(T_0)$ implies $\mu_0(T_0) - \mu_1(T_0) > 0$, so this equation can be written as

$$2\gamma - [\mu_0(T_0) + \mu_1(T_0)] = 2\sigma^2 \ln(\pi_1/\pi_0)/[\mu_0(T_0) - \mu_1(T_0)].$$

This gives the solution $\gamma = \bar{\gamma}$, where

$$\bar{\gamma} = \tfrac{1}{2}[\mu_0(T_0) + \mu_1(T_0)] + \frac{\sigma^2 \ln(\pi_1/\pi_0)}{[\mu_0(T_0) - \mu_1(T_0)]}. \tag{5.76}$$

There are some matters to address before we can claim that the solution $\bar{\gamma}$ is the Bayes threshold (i.e., that $\bar{\gamma}$ minimizes $\overline{P}_e(\gamma)$ over all γ). First, it was assumed that π_0 and π_1 are nonzero. This is easily handled by noting that if $\pi_0 = 0$, then $\overline{P}_e = P_{e,1}$ so the optimum threshold is $\gamma = +\infty$ (always decide 1 was sent). This gives $\overline{P}_e = 0$. Similarly, if $\pi_1 = 0$, then $\overline{P}_e = P_{e,0}$ and $\gamma = -\infty$ is the optimum threshold (always decide 0 was sent). This also gives $\overline{P}_e = 0$. Of course, these are trivial cases because the receiver knows which symbol is sent before any transmission takes place, so no information is communicated. Notice also that if we are willing to interpret $\ln(0/x) =$

$-\infty$ and $\ln(x/0) = +\infty$ for $x > 0$, then these trivial cases are covered by (5.76) anyway: $\pi_0 = 0$ gives $\bar{\gamma} = +\infty$ and $\pi_1 = 0$ gives $\bar{\gamma} = -\infty$.

Another matter concerns the proof that $\bar{\gamma}$ is in fact a *minimum*. All that has been shown so far is that $\overline{P}_e'(\gamma) = 0$, which is not enough (e.g., $\bar{\gamma}$ could be a maximum for $\overline{P}_e$). The formal proof that $\bar{\gamma}$ is a minimum is given in the following subsection. The method of proof is to show that $\overline{P}_e'(\gamma) > 0$ for $\gamma > \bar{\gamma}$ and $\overline{P}_e'(\gamma) < 0$ for $\gamma < \bar{\gamma}$. This is equivalent to the statement that $\overline{P}_e(\gamma)$ is a decreasing function of γ for $-\infty < \gamma < \bar{\gamma}$ and an increasing function of γ for $\bar{\gamma} < \gamma < \infty$, which proves that $\bar{\gamma}$ is indeed the minimum (and hence it is the Bayes threshold).

5.3.2.1   Proof of the Optimality of $\bar{\gamma}$

The proof is obtained by inverting the sequence of steps that led to (5.76). Only the proof that $\overline{P}_e'(\gamma) > 0$ for $\gamma > \bar{\gamma}$ is given here, because the proof that $\overline{P}_e'(\gamma) < 0$ for $\gamma < \bar{\gamma}$ is the same except that all of the inequalities are reversed. Suppose $\gamma > \bar{\gamma}$. Then, according to (5.76),

$$\gamma > \tfrac{1}{2}[\mu_0(T_0) + \mu_1(T_0)] + \frac{\sigma^2 \ln(\pi_1/\pi_0)}{[\mu_0(T_0) - \mu_1(T_0)]}.$$

Multiply both sides of the inequality by $2[\mu_0(T_0) - \mu_1(T_0)]$ to obtain

$$2\gamma\,\mu_0(T_0) - 2\gamma\,\mu_1(T_0) > \mu_0^2(T_0) - \mu_1^2(T_0) + 2\sigma^2 \ln(\pi_1/\pi_0).$$

Add γ^2 to each side, and manipulate the inequality to show that

$$[\gamma - \mu_1(T_0)]^2 - [\gamma - \mu_0(T_0)]^2 > 2\sigma^2 \ln(\pi_1/\pi_0),$$

and this can be manipulated further to give

$$\ln \pi_0 - \{[\gamma - \mu_0(T_0)]^2/2\sigma^2\} > \ln \pi_1 - \{[\gamma - \mu_1(T_0)]^2/2\sigma^2\}. \tag{5.77}$$

In the next step, the fact that the function $\exp(\cdot)$ is strictly increasing plays a critical role. If f is a strictly increasing function, then $x < y$ if and only if $f(x) < f(y)$. Thus, (5.77) implies (and is implied by) the inequality

$$\pi_0 \exp\{-[\gamma - \mu_0(T_0)]^2/2\sigma^2\} > \pi_1 \exp\{-[\gamma - \mu_1(T_0)]^2/2\sigma^2\}.$$

This is the key inequality. It can be written in more compact form by dividing both sides by $\sqrt{2\pi}\,\sigma$ and using the fact that

$$f_i(\gamma) = \exp\{-[\gamma - \mu_i(T_0)]^2/2\sigma^2\}/\sqrt{2\pi}\,\sigma.$$

The resulting inequality is $\pi_0 f_0(\gamma) > \pi_1 f_1(\gamma)$ for $\gamma > \bar{\gamma}$.

It has already been shown (see (5.75)) that $P_{e,0}'(\gamma) = f_0(\gamma)$ and $P_{e,1}'(\gamma) = -f_1(\gamma)$. From this and the fact that

$$\overline{P}_e'(\gamma) = P_{e,0}'(\gamma)\,\pi_0 + P_{e,1}'(\gamma)\,\pi_1,$$

it follows that

$$\overline{P}_e'(\gamma) = \pi_0 f_0(\gamma) - \pi_1 f_1(\gamma).$$

But, as has been shown, $\gamma > \bar{\gamma}$ implies $\pi_0 f_0(\gamma) > \pi_1 f_1(\gamma)$, and this in turn implies $\pi_0 f_0(\gamma) - \pi_1 f_1(\gamma) > 0$. This completes the proof that $\gamma > \bar{\gamma}$ implies $\overline{P}_e'(\gamma) > 0$. An identical sequence of steps with reversed inequalities proves that $\gamma < \bar{\gamma}$ implies $\overline{P}_e'(\gamma) < 0$, and from these two results it follows that $\bar{\gamma}$ is a minimum. It can also be shown that $\pi_0 f_0(\bar{\gamma}) = \pi_1 f_1(\bar{\gamma})$ so that $\overline{P}_e'(\bar{\gamma}) = 0$.

A summary of the preceding results is as follows:

$$\overline{P}_e'(\gamma) < 0 \quad \text{for } \gamma < \bar{\gamma},$$
$$\overline{P}_e'(\gamma) = 0 \quad \text{for } \gamma = \bar{\gamma},$$

and

$$\overline{P}_e'(\gamma) > 0 \quad \text{for } \gamma > \bar{\gamma}.$$

The important consequence is that

$$\min\{\overline{P}_e(\gamma) : -\infty < \gamma < \infty\} = \overline{P}_e(\bar{\gamma}).$$

Moreover, $\bar{\gamma}$ is the *unique* minimum; any choice of threshold other than $\bar{\gamma}$ gives a larger average probability of error. The Bayes threshold is therefore

$$\bar{\gamma} = \tfrac{1}{2}[\mu_0(T_0) + \mu_1(T_0)] + \frac{\sigma^2 \ln(\pi_1/\pi_0)}{[\mu_0(T_0) - \mu_1(T_0)]}. \tag{5.78a}$$

5.3.2.2   Properties of the Bayes Threshold

Now that the proof is complete, let's examine the result to see if it provides the bias in the decision that we sought. First, observe that the leading term in the expression for the Bayes threshold is the minimax threshold. That is, (5.78a) can be written as

$$\bar{\gamma} = \gamma_m + \{\sigma^2 \ln(\pi_1/\pi_0)/[\mu_0(T_0) - \mu_1(T_0)]\}. \tag{5.78b}$$

The Bayes threshold is the minimax threshold plus a correction term, and the correction term depends on the probabilities π_0 and π_1. Notice that the correction term is zero if $\pi_0 = \pi_1$. *The Bayes threshold is equal to the minimax threshold if the two signals are equally probable.*

Next consider the effects of unequal probabilities. If $\pi_0 > \pi_1$, then $\ln(\pi_1/\pi_0)$ is negative. A negative correction term decreases the Bayes threshold ($\bar{\gamma} < \gamma_m$). Loosely speaking, this makes it more likely that the threshold is exceeded when a given symbol is sent. The receiver decides that 0 was sent if the threshold is exceeded, so the effect of $\pi_0 > \pi_1$ is to cause the Bayes receiver to increase the likelihood of deciding in favor of the symbol 0. On the other hand, if $\pi_1 > \pi_0$, the correction term is positive and increases the threshold ($\bar{\gamma} > \gamma_m$). This decreases the likelihood of deciding in favor of the symbol 0. It is clear that the Bayes receiver does indeed include a bias that reflects the symbol transmission probabilities.

A final observation concerns the fact that the effect of the noise level on the Bayes threshold is via the ratio $\sigma^2/[\mu_0(T_0) - \mu_1(T_0)]$. As we have already mentioned (see (5.73a) and (5.74)), the difference

$$\Delta_\mu(T_0) = \mu_0(T_0) - \mu_1(T_0)$$

should be large for good performance. Clearly, the noise variance should be small. For convenience, let $\beta = \sigma^2/\Delta_\mu(T_0)$. The parameter β is a kind of noise-to-signal ratio at the filter output; it is a measure of the quality or reliability of the decision statistic Z.

Consider that π_0 and π_1 represent information about which signal is being sent. In fact, this is the only such information that we have *prior* to observing the decision statistic $Z(T_0)$. The observation of $Z(T_0)$ provides additional information about which symbol was sent. The combination of the prior information and the observation of $Z(T_0)$ is used to make the decision in the Bayes receiver.

Intuitively, if the channel noise level is so high that the decision statistic is "mostly noise" (and therefore unreliable), then we should de-emphasize the observation and base the decision primarily on the prior information. The Bayes threshold does this for us. As $\beta \to \infty$, $\bar{\gamma} \to \infty$ (always decide 1) if $\pi_1 > \pi_0$ and $\bar{\gamma} \to -\infty$ (always decide 0) if $\pi_0 > \pi_1$. That is, in the absence of a useful observation, the Bayes decision is the symbol that is more likely according to the *prior* information provided by the relative values of π_0 and π_1.

If the channel noise level is very low, then the decision statistic is "mostly signal" (and therefore reliable). In this case, we should base the decision on the observation of Z to a greater extent than on the prior information. Indeed, as $\beta \to 0$, the correction term disappears and the dependence on the prior probabilities π_0 and π_1 disappears along with it. More precisely, as $\beta \to 0$ the Bayes threshold converges to the minimax threshold, and the latter does not depend on π_0 or π_1.

An examination of the expressions

$$\bar{\gamma} = \tfrac{1}{2}[\mu_0(T_0) + \mu_1(T_0)] + \frac{\sigma^2 \ln(\pi_1/\pi_0)}{[\mu_0(T_0) - \mu_1(T_0)]},$$

$$\mu_i(T_0) = \int_{-\infty}^{\infty} s_i(\tau)h(T_0 - \tau)\,d\tau,$$

and

$$\sigma^2 = \frac{1}{2\pi} \int_{-\infty}^{\infty} |H(\omega)|^2 S_X(\omega)\,d\omega,$$

shows that, in general, the Bayes threshold $\bar{\gamma}$ depends on the received signal levels, the noise density level, and the gain of the filter. Multiplication of the signals, the noise, or the filter by a gain constant changes the value of $\bar{\gamma}$. Thus, in order to use the Bayes threshold in a communication system, the received signal level, noise level, and filter gain levels must all be calibrated accurately. This makes it difficult to use the Bayes threshold in certain practical applications.

If antipodal signals are used, then $\gamma_m = 0$ and so (5.78b) implies that

$$\bar{\gamma} = \sigma^2 \ln(\pi_1/\pi_0)/[2\,\mu_0(T_0)].$$

In obtaining this expression, we have also used the fact that

$$\mu_0(T_0) - \mu_1(T_0) = 2\,\mu_0(T_0)$$

for antipodal signals. Unlike the minimax threshold, the Bayes threshold depends on the received signal levels, even if the signals are antipodal. (See Problem 5.7.) Moreover, the Bayes threshold depends on the noise level and the filter gain, even if the signals are antipodal. (See Problem 5.8.)

5.3.2.3   The Error Probability for the Bayes Threshold

The *Bayes error probability* is defined as the minimum of $\overline{P}_e(\gamma)$ for $-\infty < \gamma < \infty$. The value of the Bayes error probability can be determined by finding the Bayes threshold $\bar{\gamma}$ and then evaluating $\overline{P}_e(\bar{\gamma})$. For convenience, we denote the Bayes error probability by $\overline{P}_e^*$. By definition,

$$\overline{P}_e^* = \min\left\{\overline{P}_e(\gamma) : -\infty < \gamma < \infty\right\} = \overline{P}_e(\bar{\gamma}).$$

Analogous to the situation for the minimax error probability, it is not necessary to find the Bayes threshold first in order to evaluate the Bayes error probability. Direct evaluation of the Bayes error probability is possible. Return to the expression for the average probability of error:

$$
\begin{aligned}
\overline{P}_e(\gamma) &= P_{e,0}(\gamma)\,\pi_0 + P_{e,1}(\gamma)\,\pi_1 \\
&= Q([\mu_0(T_0) - \gamma]/\sigma)\,\pi_0 + Q([\gamma - \mu_1(T_0)]/\sigma)\,\pi_1.
\end{aligned}
$$

To find $\overline{P}_e(\gamma)$ for $\gamma = \bar{\gamma}$, first observe that

$$
\begin{aligned}
\mu_0(T_0) - \bar{\gamma} &= \mu_0(T_0) - \gamma_m - [\sigma^2 \ln(\pi_1/\pi_0)/\Delta_\mu(T_0)] \\
&= \tfrac{1}{2}\Delta_\mu(T_0) - [\sigma^2 \ln(\pi_1/\pi_0)/\Delta_\mu(T_0)],
\end{aligned}
$$

and so

$$\frac{\mu_0(T_0) - \bar{\gamma}}{\sigma} = \frac{\Delta_\mu(T_0)}{2\sigma} - \frac{\sigma}{\Delta_\mu(T_0)}\ln(\pi_1/\pi_0).$$

The quantity $\Delta_\mu(T_0)/2\sigma$ is the signal-to-noise ratio SNR defined in (5.73a). Thus, we have shown that

$$[\mu_0(T_0) - \bar{\gamma}]/\sigma = \text{SNR} - \tfrac{1}{2}(\text{SNR})^{-1}\ln(\pi_1/\pi_0).$$

Similarly,

$$
\begin{aligned}
\bar{\gamma} - \mu_1(T_0) &= \gamma_m - \mu_1(T_0) + [\sigma^2 \ln(\pi_1/\pi_0)/\Delta_\mu(T_0)] \\
&= \tfrac{1}{2}\Delta_\mu(T_0) + [\sigma^2 \ln(\pi_1/\pi_0)/\Delta_\mu(T_0)];
\end{aligned}
$$

therefore,

$$[\bar{\gamma} - \mu_1(T_0)]/\sigma = \text{SNR} + \tfrac{1}{2}(\text{SNR})^{-1}\ln(\pi_1/\pi_0).$$

In order to work with more compact expressions, we denote the signal-to-noise ratio SNR by α. The preceding results show that the Bayes error probability is given by

$$\overline{P}_e^* = Q\left[\alpha - \tfrac{1}{2}\alpha^{-1}\ln(\pi_1/\pi_0)\right]\pi_0 + Q\left[\alpha + \tfrac{1}{2}\alpha^{-1}\ln(\pi_1/\pi_0)\right]\pi_1, \qquad (5.79a)$$

where

$$\alpha = \text{SNR} = \Delta_\mu(T_0)/2\sigma = [\mu_0(T_0) - \mu_1(T_0)]/2\sigma. \qquad (5.79b)$$

Exercise 5-2

On the basis of intuition, we should expect that the Bayes error probability decreases as the signal-to-noise ratio increases. It may not be clear from (5.79a) that $\overline{P}_e^*$ is indeed a decreasing function of α. In particular, note that if $\pi_1 > \pi_0$, the term $\tfrac{1}{2}\alpha^{-1}\ln(\pi_1/\pi_0)$ is a decreasing function of α while the term $-\tfrac{1}{2}\alpha^{-1}\ln(\pi_1/\pi_0)$ is an increasing function of α. Prove that $\overline{P}_e$ is in fact a strictly decreasing function of α provided that $0 < \pi_0 < 1$.

Solution.   The approach is to differentiate $\overline{P}_e^*$ with respect to α and show the derivative is negative everywhere. For convenience, let $c = \tfrac{1}{2}\ln(\pi_1/\pi_0)$. Recall that $Q'(x) = -\exp(-x^2/2)/\sqrt{2\pi}$ so that

$$\sqrt{2\pi}\,\frac{d\overline{P}_e^*}{d\alpha} = -\exp\left\{-\tfrac{1}{2}(\alpha - c\alpha^{-1})^2\right\}(1 + c\alpha^{-2})\pi_0$$
$$-\exp\left\{-\tfrac{1}{2}(\alpha + c\alpha^{-1})^2\right\}(1 - c\alpha^{-2})\pi_1.$$

Expand $(\alpha \pm c\alpha^{-1})^2$ and factor out the common terms to obtain

$$\sqrt{2\pi}\,\frac{d\overline{P}_e^*}{d\alpha} = -[\pi_0(1 + c\alpha^{-2})\,e^c + \pi_1(1 - c\alpha^{-2})\,e^{-c}]$$
$$\times \exp\left\{-\tfrac{1}{2}(\alpha^2 + c^2\alpha^{-2})\right\}.$$

The sign of the derivative is the sign of the term in square brackets, because the exponential term is positive for all α. Thus, the derivative is negative if and only if

$$G(\alpha) = -\pi_0(1 + c\alpha^{-2})\,e^c - \pi_1(1 - c\alpha^{-2})\,e^{-c}$$

is negative. Recall that

$$c = \tfrac{1}{2}\ln(\pi_1/\pi_0) = \ln(\sqrt{\pi_1/\pi_0})$$

so that $e^c = \sqrt{\pi_1/\pi_0}$ and $e^{-c} = \sqrt{\pi_0/\pi_1}$. It follows that

$$\pi_0\,e^c = \sqrt{\pi_0\pi_1} = \pi_1\,e^{-c}.$$

Therefore,

$$G(\alpha) = -\sqrt{\pi_0\pi_1}\,(1 + c\alpha^{-2} + 1 - c\alpha^{-2}) = -2\sqrt{\pi_0\pi_1},$$

from which it follows that $G(\alpha) < 0$ for all α.

Summary of Results (Bayes)

The probability $\overline{P}_e$ is the average probability of error, which is the average of the two probabilities $P_{e,0}$ and $P_{e,1}$. We display the dependence of these probabilities on the threshold γ by writing

$$\overline{P}_e(\gamma) = \pi_0 P_{e,0}(\gamma) + \pi_1 P_{e,1}(\gamma).$$

The *Bayes error probability* is defined by

$$\overline{P}_e^* = \min\left\{\overline{P}_e(\gamma) : -\infty < \gamma < \infty\right\}.$$

It follows from preceding results that

$$\overline{P}_e^* = \pi_0 P_{e,0}(\bar{\gamma}) + \pi_1 P_{e,1}(\bar{\gamma}),$$

where $\bar{\gamma}$ is the Bayes threshold. For the general additive Gaussian noise channel, the Bayes threshold is given by

$$\bar{\gamma} = \frac{\mu_0(T_0) + \mu_1(T_0)}{2} + \frac{\sigma^2 \ln(\pi_1/\pi_0)}{\mu_0(T_0) - \mu_1(T_0)}.$$

The Bayes error probability is

$$\begin{aligned}\overline{P}_e^* &= \pi_0\, Q[\text{SNR} - (2\,\text{SNR})^{-1} \ln(\pi_1/\pi_0)]\\ &\quad + \pi_1\, Q[\text{SNR} + (2\,\text{SNR})^{-1} \ln(\pi_1/\pi_0)],\end{aligned}$$

where

$$\text{SNR} = \frac{\mu_0(T_0) - \mu_1(T_0)}{2\sigma}.$$

5.4   General Decision Rules for General Channels

In this section, we take a brief look at more general decision rules than the simple threshold test that we have considered thus far, and we consider general (possibly non-Gaussian) channels. The communication system model shown in Figure 5-11 is the same as the model we have employed up to this point (cf. Figure 5-4), except that the threshold device has been replaced by a general decision device, and we do not require the channel to be an additive Gaussian noise channel.

The most general binary decision device operates as follows. The set Γ of all possible values of the decision statistic $Z(T_0)$ is partitioned into two disjoint subsets, Γ_0 and Γ_1, with the property that the union of Γ_0 and Γ_1 is the entire observation space Γ. If $Z(T_0)$ falls in Γ_0, the decision is that 0 was sent; otherwise, $Z(T_0)$ is in Γ_1 and the decision is that 1 was sent.

Figure 5-11: Model for binary data transmission with a general decision device.

It is easy to see that the threshold device is a special case of this general decision device. The decision regions for the threshold device are intervals. In particular, Γ_0 is the interval (γ, ∞) and Γ_1 is the interval $(-\infty, \gamma]$. If the observation falls in the interval (γ, ∞), the decision is that 0 was sent; otherwise, the decision is that 1 was sent.

Now let's consider the average probability of error for the general binary decision device. As before, let π_i denote the probability that signal s_i is transmitted. The average probability of error is

$$\overline{P}_e = P_{e,0}\pi_0 + P_{e,1}\pi_1,$$

where $P_{e,i}$ is the conditional probability of error given that s_i is transmitted.

First, examine the case $i = 0$. Assuming that s_0 is the transmitted signal (i.e., 0 is sent), an error is made if and only if the decision is that 1 was sent. But this happens if and only if the decision statistic $Z(T_0)$ falls in the set Γ_1. Let f_0 denote the conditional density of $Z(T_0)$ given that 0 was sent. Recall that for the additive Gaussian noise channel, this density is the Gaussian density with mean $\mu_0(T_0)$ and variance σ^2. Now, however, we do not require that f_0 be a Gaussian density; that is, we consider more general channels than the additive Gaussian noise channel.

The probability that a random variable with density f_0 falls in a set Γ_1 is just the integral of that density over the set. Thus, the conditional probability of error given that 0 is sent is

$$P_{e,0} = \int_{\Gamma_1} f_0(z)\,dz.$$

Similarly, the conditional probability of error given that 1 is sent is

$$P_{e,1} = \int_{\Gamma_0} f_1(z)\,dz,$$

where f_1 is the conditional density of $Z(T_0)$ given that 1 is sent.

It follows from these equations that the average probability of error is

$$\overline{P}_e = \pi_0 \int_{\Gamma_1} f_0(z)\,dz + \pi_1 \int_{\Gamma_0} f_1(z)\,dz.$$

Our goal is to choose Γ_1 and Γ_0 to minimize $\overline{P}_e$. This gives the optimum decision (for the Bayes criterion) based on the observation $Z(T_0)$.

Minimization of $\overline{P}_e$ with respect to Γ_0 and Γ_1 may appear to be an intractable problem. The trick—and it is a fairly standard one—is to rewrite the expression for $\overline{P}_e$ so that it is given in terms of one set only. This can be accomplished because

$$P_{e,0} = \int_{\Gamma_1} f_0(z)\,dz = 1 - \int_{\Gamma_0} f_0(z)\,dz = 1 - P_{c,0},$$

where $P_{c,0}$ is the conditional probability that the correct decision is made given that 0 is sent. This relationship can also be deduced from the fact that the union of Γ_0 and Γ_1 is the entire real line $(-\infty, \infty)$. Because Γ_0 and Γ_1 are disjoint,

$$\int_{-\infty}^{\infty} f_0(z)\,dz = \int_{\Gamma_0} f_0(z)\,dz + \int_{\Gamma_1} f_0(z)\,dz.$$

But f_0 is a density, so we have

$$\int_{\Gamma_0} f_0(z)\,dz + \int_{\Gamma_1} f_0(z)\,dz = 1,$$

from which it follows that

$$\int_{\Gamma_1} f_0(z)\,dz = 1 - \int_{\Gamma_0} f_0(z)\,dz.$$

Returning to the expression for the average probability of error, we see that

$$\begin{aligned} \overline{P}_e &= \pi_0 \int_{\Gamma_1} f_0(z)\,dz + \pi_1 \int_{\Gamma_0} f_1(z)\,dz \\ &= \pi_0 \left[1 - \int_{\Gamma_0} f_0(z)\,dz \right] + \pi_1 \int_{\Gamma_0} f_1(z)\,dz. \end{aligned}$$

This can be rewritten as

$$\overline{P}_e = \pi_0 + \int_{\Gamma_0} [\pi_1 f_1(z) - \pi_0 f_0(z)]\,dz. \tag{5.80}$$

Because π_0 is fixed, there is nothing we can do about the first term on the right-hand side of (5.80). We want to choose Γ_0 in a way that makes the second term (the integral) in (5.80) as small as possible.

Notice that the integrand of (5.80) is the difference of two nonnegative terms: $\pi_1 f_1(z)$ and $\pi_0 f_0(z)$. For some values of z, this integrand is positive and for others it is negative. For some, it may be zero. If we include in the set Γ_0 a value of z for which the integrand is positive, this *adds* to the error probability. But, if we include a value of z for which the integrand is negative, this *subtracts* from the error probability. Values of z for which the integrand is zero have no effect at all.

These observations imply that, in order to minimize $\overline{P}_e$, Γ_0 should include *all* of the values of z for which the integrand is negative and *none* of the values of z for which it is positive. This makes the term

$$\int_{\Gamma_0} [\pi_1 f_1(z) - \pi_0 f_0(z)]\,dz$$

as small as possible. Thus, Γ_0 should be the set of all z for which $\pi_1 f_1(z) - \pi_0 f_0(z) < 0$. This is written formally as

$$\Gamma_0 = \{z : \pi_1 f_1(z) - \pi_0 f_0(z) < 0\},$$

which is equivalent to

$$\Gamma_0 = \{z : \pi_0 f_0(z) > \pi_1 f_1(z)\}.$$

Any point not in Γ_0 must be in Γ_1 because the union of the two sets is the entire real line. Thus, Γ_1 is the set of all z for which the integrand is nonnegative, which is written as

$$\Gamma_1 = \{z : \pi_1 f_1(z) - \pi_0 f_0(z) \geq 0\}.$$

This is equivalent to

$$\Gamma_1 = \{z : \pi_0 f_0(z) \leq \pi_1 f_1(z)\}.$$

Notice that, for minimizing $\overline{P}_e$, we really don't care about the set

$$\Gamma' = \{z : \pi_0 f_0(z) = \pi_1 f_1(z)\},$$

because points in this set do not contribute to the error probability. That is,

$$\int_{\Gamma'} [\pi_0 f_0(z) - \pi_1 f_1(z)] \, dz = 0.$$

Notice also that unless Γ' is empty, then there isn't a unique optimum choice of Γ_0 and Γ_1. Points in Γ' can be put in either set with no change in $\overline{P}_e$. The convention that is followed throughout the book is to include all such points in Γ_1 *whenever the Bayes criterion is employed.* This convention cannot always be followed if the minimax criterion is used, as discussed later.

The preceding results are summarized as follows: The general binary decision device is defined by two sets Γ_0 and Γ_1 that represent the decision regions. If the decision statistic falls in the set Γ_0, then decide that 0 was sent; but if the decision statistic falls in the set Γ_1, then decide that 1 was sent. For the Bayes criterion, optimum choices for the decision regions are

$$\Gamma_0 = \{z : \pi_0 f_0(z) > \pi_1 f_1(z)\} \tag{5.81a}$$

and

$$\Gamma_1 = \{z : \pi_0 f_0(z) \leq \pi_1 f_1(z)\}. \tag{5.81b}$$

Use of these decision regions yields the minimum possible value for the average probability of error.

Equations (5.81) actually define a family of sets indexed by π_0. (We can write $\pi_1 = 1 - \pi_0$.) Any decision rule that minimizes the average probability of error for some choice of π_0 is called a *Bayes decision rule*. In this terminology, equations (5.81) define one family of Bayes decision rules. There are others, as we will see later.

Exercise 5-3

Suppose that f_0 and f_1 are Gaussian densities that have equal variances. Show that equations (5.81) correspond to the same decision rule as the threshold device with Bayes threshold $\bar{\gamma}$.

Solution.   The solution presented here is essentially the reverse of the sequence of steps used to prove the optimality of the Bayes threshold $\bar{\gamma}$. A point z is in Γ_0 if and only if $\pi_0 f_0(z) > \pi_1 f_1(z)$. But this inequality is equivalent to

$$\pi_0 \exp\left\{-(z - \mu_0)^2/2\sigma^2\right\} > \pi_1 \exp\left\{-(z - \mu_1)^2/2\sigma^2\right\},$$

where μ_i is the mean corresponding to density f_i. By taking logarithms, we obtain

$$\ln(\pi_0) - (z - \mu_0)^2/2\sigma^2 > \ln(\pi_1) - (z - \mu_1)^2/2\sigma^2.$$

We next expand the quadratic terms and simplify to obtain

$$2\mu_0 z - \mu_0^2 - 2\mu_1 z + \mu_1^2 > 2\sigma^2 \ln(\pi_1/\pi_0),$$

which can be written as

$$z(\mu_0 - \mu_1) > \tfrac{1}{2}(\mu_0^2 - \mu_1^2) + \sigma^2 \ln(\pi_1/\pi_0).$$

The next step depends on the sign of $\mu_0 - \mu_1$. If $\mu_0 - \mu_1 > 0$, the last inequality is equivalent to

$$z > \tfrac{1}{2}(\mu_0 + \mu_1) + \frac{\sigma^2 \ln(\pi_1/\pi_0)}{\mu_0 - \mu_1}.$$

But this is just the inequality $z > \bar{\gamma}$ (see (5.78a)) if $\mu_i = \mu_i(T_0)$, which is the case for the application of (5.81) to the communication system of Figure 5-4. Note that $\mu_0 - \mu_1 > 0$ corresponds to $(s_0 * h)(T_0) > (s_1 * h)(T_0)$. If $\mu_0 - \mu_1 < 0$, the inequality is reversed, because we are multiplying both sides by the negative number $1/(\mu_0 - \mu_1)$. For the system of Figure 5-4, $\mu_0 - \mu_1 < 0$ corresponds to the situation $(s_0 * h)(T_0) < (s_1 * h)(T_0)$, which is not according to our assumption. We have shown that, under the assumption $\mu_0 > \mu_1$, z is in Γ_0 if and only if $z > \bar{\gamma}$. Because Γ_1 is the complement of Γ_0, it follows that z is in Γ_1 if and only if $z \leq \bar{\gamma}$. Hence, equations (5.81) reduce to a threshold test if f_0 and f_1 are Gaussian densities with the same variance. ∎

The solution to Exercise 5-3 proves that the threshold device with threshold $\bar{\gamma}$ is the *optimum* decision device for the AGN channel and the *Bayes criterion* (i.e., it minimizes the average probability of error). If the communication system is such that

$$(s_0 * h)(T_0) > (s_1 * h)(T_0),$$

the optimum decision is to compare the decision statistic $Z(T_0)$ to $\bar{\gamma}$ and decide 0 was sent if $Z(T_0) > \bar{\gamma}$ or 1 was sent if $Z(T_0) \leq \bar{\gamma}$. If

$$(s_1 * h)(T_0) > (s_0 * h)(T_0),$$

then we should decide 1 was sent if $Z(T_0) \geq \bar{\gamma}$ or 0 was sent if $Z(T_0) < \bar{\gamma}$.

It is also true that, under the conditions of Exercise 5-3, the threshold device with the minimax threshold γ_m is the *optimum* decision device for the *minimax criterion*. The proof of this depends on the following result from statistical decision theory [5.1].

Theorem 5-1   If a decision rule is a Bayes rule and also satisfies $P_{e,0} = P_{e,1}$, then it is a minimax decision rule.   ∎

We already know that the threshold test with threshold γ_m is a Bayes decision rule (for $\pi_0 = \pi_1 = 1/2$, see Section 5.3.2.2) and that this threshold test also satisfies $P_{e,0} = P_{e,1}$. Therefore, according to Theorem 5-1, it is a minimax decision rule for the conditions of Exercise 5-3.

There are two warnings that should be issued at this point. First, if f_0 and f_1 are not Gaussian densities with equal variances, a simple threshold test, in which we compare the decision statistic with a single number γ, may not be optimum for either the minimax or Bayes criterion. One such situation is given in Example 5-8, and another is given in Problem 5.12. The second warning is that we have not proved that the receiver of the form shown in Figure 5-4 is optimum. We have, however, proved that the threshold device gives the optimum decision if the decision is based only on the decision statistic $Z(T_0)$ of Figure 5-4. It turns out that the receiver is in fact optimum, but this matter is deferred until Chapter 6 and Appendix D.

Example 5-8   A Minimax Decision Rule from Bayes Decision Rules
Consider the density functions defined by

$$f_0(z) = \begin{cases} 1/4, & 0 \leq z \leq 4, \\ 0, & \text{otherwise} \end{cases}$$

and

$$f_1(z) = \begin{cases} z/4, & 0 \leq z \leq 2, \\ (4-z)/4, & 2 < z \leq 4, \\ 0, & \text{otherwise.} \end{cases}$$

These densities are shown in Figure 5-12. Suppose that $\pi_0 = \pi_1 = 1/2$. The optimum decision regions for minimizing $\overline{P_e}$ are

$$\Gamma_0 = [0, 1) \cup (3, 4]$$

and

$$\Gamma_1 = (1, 3).$$

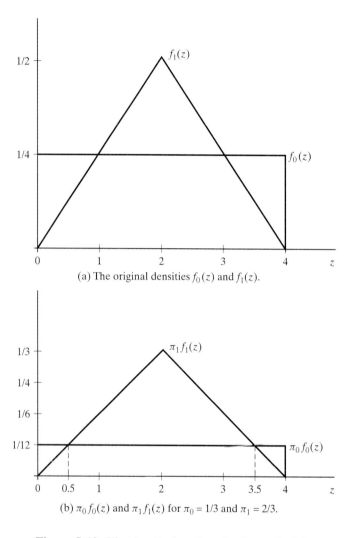

(a) The original densities $f_0(z)$ and $f_1(z)$.

(b) $\pi_0 f_0(z)$ and $\pi_1 f_1(z)$ for $\pi_0 = 1/3$ and $\pi_1 = 2/3$.

Figure 5-12: The density functions for Example 5-8.

We have ignored the intervals $(-\infty, 0)$ and $(4, \infty)$ because

$$\int_{-\infty}^{0} f_i(z)\, dz = \int_{4}^{\infty} f_i(z)\, dz = 0$$

for $i = 0$ and $i = 1$. The error probabilities that result when the preceding decision regions are employed are

$$P_{e,0} = P[1 < Z(T_0) < 3 \mid 0 \text{ sent}] = \tfrac{1}{2}$$

and

$$P_{e,1} = P[0 \le Z(T_0) < 1 \mid 1 \text{ sent}] + P[3 < Z(T_0) \le 4 \mid 1 \text{ sent}] = \tfrac{1}{4}.$$

Clearly, these regions do not correspond to a minimax rule.

A careful examination of Bayes rules for other choices of π_0 and π_1 shows that for $0 < \pi_0 < 2/3$, they all correspond to decision regions of the form

$$\Gamma_0 = [0, \varepsilon) \cup (4 - \varepsilon, 4]$$

and

$$\Gamma_1 = (\varepsilon, 4 - \varepsilon).$$

For instance, if $\pi_0 = 1/2$, we have $\varepsilon = 1$; this is clear from Figure 5-12(a). For the choice $\pi_0 = 1/3$, it is clear from Figure 5-12(b) that the value of ε is $1/2$. In general the value of ε is $\pi_0/(1 - \pi_0)$; this can be shown by solving $\pi_0 f_0(\varepsilon) = (1 - \pi_0) f_1(\varepsilon)$ for ε in terms of π_0. Because we are interested in $\varepsilon < 2$, then we have $f_1(\varepsilon) = \varepsilon/4$. Therefore, we must solve $\pi_0/4 = (1 - \pi_0)\varepsilon/4$, which gives the stated result.

To find a minimax rule, we first determine the functional dependence of the error probabilities on the parameter ε. First, note that for $0 \le \varepsilon \le 2$,

$$P_{e,0} = P[\varepsilon < Z(T_0) < 4 - \varepsilon \mid 0 \text{ sent}]$$

$$= \int_{\varepsilon}^{4-\varepsilon} f_0(z)\,dz = (4 - 2\varepsilon)/4$$

$$= 1 - \frac{\varepsilon}{2}$$

and

$$P_{e,1} = P[0 \le Z(T_0) \le \varepsilon \mid 1 \text{ sent}] + P[4 - \varepsilon \le Z(T_0) \le 4 \mid 1 \text{ sent}]$$

$$= 2\int_0^{\varepsilon} f_1(z)\,dz = \varepsilon^2/4.$$

All choices of ε in the range $0 < \varepsilon < 2$ give Bayes decision rules, so the value of ε for which $P_{e,0} = P_{e,1}$ must be a minimax rule. Thus, we solve $1 - (\varepsilon/2) = \varepsilon^2/4$ to find that $\varepsilon = -1 + \sqrt{5}$, the only solution in the interval $(0, 2)$, gives a minimax decision rule. ∎

Example 5-9   Bayes and Minimax Decision Rules for Two Uniform Densities

Suppose that f_0 is uniform on the interval $(0, 3)$ and that f_1 is uniform on the interval $(1, 2)$. The optimum decision regions for minimizing the average probability of error with prior probabilities $\pi_0 = \pi_1 = 1/2$ can be determined from (5.81). Observe that $f_0(z) = 1/3$ for $0 < z < 3$, while $f_1(z) = 1$ for $1 < z < 2$, but $f_1(z) = 0$ for $0 < z \le 1$ and $2 \le z < 3$. Hence, $f_0(z) > f_1(z)$ for $0 < z \le 1$ and $2 \le z < 3$, but $f_0(z) \le f_1(z)$ for $z \le 0$, $1 < z < 2$, and $z \ge 3$. That is, Γ_0 is the union of the intervals $(0, 1]$ and $[2, 3)$, and Γ_1 is the union of the intervals $(-\infty, 0]$, $(1, 2)$, and $[3, \infty)$. This does not correspond to a simple threshold device.

The intervals $(-\infty, 0)$ and $[3, \infty)$ are not important, because both densities assign zero probability to the event that the decision statistic falls in either of these intervals. Similarly, it doesn't matter whether the endpoints 1 and 2 are included in Γ_0 or

Γ_1. Because the probability distributions are continuous, any finite set of points has zero probability.

This example also serves as a counterexample to any misconception that the minimax decision rule is always a Bayes decision rule for $\pi_0 = \pi_1 = 1/2$. The Bayes rule gives $P_{e,0} = 1/3$ and $P_{e,1} = 0$, so it isn't *guaranteed* to be a minimax decision rule by the preceding theorem. Moreover, it is easy to see that it is in fact not a minimax decision rule. For example, by increasing Γ_0 and reducing Γ_1, we can decrease $P_{e,0}$ at the expense of increasing $P_{e,1}$. This matter is pursued further in Problem 5.14.   ∎

The decision rules considered thus far involve only *deterministic* or *nonrandomized* decisions: The decision is a deterministic function of the decision statistic. In fact, this function is a very simple one, because it depends only on whether the decision statistic is in the set Γ_0 or the set Γ_1. There are situations in which it is better to allow *randomized* decisions. For our purposes, randomized decision rules can be defined in terms of three sets. Two of these sets, Γ_0 and Γ_1, are as we have described; if the decision statistic falls in Γ_i, then the decision is that i was sent ($i = 0$ or $i = 1$). The third set, which we denote by Γ_R, corresponds to a randomized decision; specifically, if the decision statistic falls in Γ_R, then with probability θ we decide that 0 was sent and with probability $1 - \theta$ we decide that 1 was sent, for some choice of θ.

Let us return to our consideration of Bayes decision rules and the set Γ', the set of all z for which $\pi_0 f_0(z) = \pi_1 f_1(z)$. Previously, we arbitrarily included this set in Γ_1, but some or all of the points in Γ' could have been included in Γ_0 without changing $\overline{P_e}$. Alternatively, the points in Γ' could be included in Γ_R, the set that corresponds to a randomized decision. This gives the following decision rule: If the decision statistic $Z(T_0)$ is in the set

$$\Gamma_0 = \{z : \pi_0 f_0(z) > \pi_1 f_1(z)\}, \tag{5.82a}$$

then decide 0 was sent; if it is in the set

$$\Gamma_1 = \{z : \pi_0 f_0(z) < \pi_1 f_1(z)\}, \tag{5.82b}$$

then decide 1 was sent; and if it is in the set

$$\Gamma_R = \{z : \pi_0 f_0(z) = \pi_1 f_1(z)\}, \tag{5.82c}$$

then with probability θ decide that 0 was sent and with probability $1 - \theta$ decide that 1 was sent. Note that θ is a variable ($0 \leq \theta \leq 1$), and so (5.82) is actually a family of decision rules indexed by θ. For $\theta = 0$, (5.82) reduces to the rule defined by (5.81).

Clearly, there is no advantage in using randomized decision rules for the Bayes criterion. The decision rules defined by (5.81) and (5.82) yield the same value of $\overline{P_e}$. However, there is an advantage in using randomized rules for the minimax criterion, as illustrated by the following example and by Problem 5.14.

Example 5-10   A Randomized Minimax Decision Rule
Suppose that the two densities are

$$f_1(z) = \tfrac{1}{2} \exp\{-|z|\}, \quad -\infty < z < \infty,$$

and

$$f_0(z) = \begin{cases} \exp\{-z\}, & 0 \leq z < \infty, \\ 0, & -\infty < z < 0. \end{cases}$$

Bayes rules all have $(-\infty, 0)$ as part of Γ_1, because $\pi_0 f_0(z) < \pi_1 f_1(z)$ for all z if $\pi_1 > 0$. If $\pi_1 = 0$, then we don't care about $(-\infty, 0)$, because $\pi_1 = 0$ implies $Z(T_0) \geq 0$ with probability 1. In order to examine $[0, \infty)$, we look at the inequality

$$\pi_0 f_0(z) > \pi_1 f_1(z)$$

that defines Γ_0. For the densities in this example, $f_0(z) > c f_1(z)$ for $z \geq 0$ if $c < 2$. Thus, for $\pi_1/\pi_0 < 2$, the Bayes rules are such that $\Gamma_1 = (-\infty, 0)$ and $\Gamma_0 = [0, \infty)$. The inequality $\pi_1/\pi_0 < 2$ corresponds to $1/3 < \pi_0 \leq 1$. For such values of π_0, the set Γ_R is empty. (The resulting decision rules are deterministic.) In this case, $P_{e,0} = 0$ and $P_{e,1} = 1/2$; the average error probability is $\pi_1/2$.

For $\pi_1/\pi_0 > 2$ (i.e., $\pi_0 < 1/3$), $\pi_0 f_0(z)$ is never greater than $\pi_1 f_1(z)$, so $\Gamma_1 = (-\infty, \infty)$ and Γ_0 and Γ_R are empty. The optimum decision is to decide that 1 was sent regardless of the value of the decision statistic. In this case, $P_{e,0} = 1$ and $P_{e,1} = 0$. The average error probability is $\pi_0 = 1 - \pi_1$.

Finally, consider the case $\pi_1/\pi_0 = 2$ (i.e., $\pi_0 = 1/3$). Once again, $\pi_0 f_0(z)$ is never greater than $\pi_1 f_1(z)$, so Γ_0 is empty. Now, however, Γ_R is not empty; in fact, $\Gamma_R = [0, \infty)$. To find the probability of error given that 0 was sent, observe that an error is made when 0 is sent if $Z(T_0)$ is in the set Γ_1 or if it is in Γ_R and the random decision is in favor of the decision that 1 was sent. Thus,

$$P_{e,0} = P[Z(T_0) \in \Gamma_1 \mid 0 \text{ sent}] + (1 - \theta) P[Z(T_0) \in \Gamma_R \mid 0 \text{ sent}]$$
$$= 0 + (1 - \theta)(1) = 1 - \theta.$$

Similarly, an error is made when 1 is sent if $Z(T_0)$ is in the set Γ_0 (which is impossible because Γ_0 is empty) or if it is in Γ_R and the random decision is in favor of the decision that 0 was sent. It follows that

$$P_{e,1} = P[Z(T_0) \in \Gamma_0 \mid 1 \text{ sent}] + \theta P[Z(T_0) \in \Gamma_R \mid 1 \text{ sent}] = 0 + \theta\,(1/2) = \theta/2.$$

The case $\pi_0 = 1/3$ points the way to a minimax decision rule. For $\pi_0 = 1/3$, the decision rule corresponding to $\Gamma_1 = (-\infty, 0]$, $\Gamma_0 = \varnothing$, and $\Gamma_R = [0, \infty)$ is a Bayes decision rule for each choice of θ. This rule has error probabilities $P_{e,0} = 1 - \theta$ and $P_{e,1} = \theta/2$. The value $\theta = 2/3$ gives $P_{e,0} = P_{e,1}$; hence, it corresponds to a minimax decision rule according to Theorem 5-1.

In summary, a minimax rule is as follows: If $Z(T_0) < 0$, decide 1 was sent. If $Z(T_0) > 0$, then with probability 2/3 decide that 0 was sent and with probability 1/3 decide that 1 was sent. For this rule, $P_{e,0} = P_{e,1} = 1/3$. ∎

We close this subsection with a brief discussion of the Bayes decision rules for $\pi_0 = \pi_1 = 1/2$. From (5.81), we see that the optimum decision regions are

$$\Gamma_0 = \{z : f_0(z) > f_1(z)\}$$

and

$$\Gamma_1 = \{z : f_0(z) \le f_1(z)\}.$$

We can dispense with those values of z for which $f_0(z) = f_1(z)$, because these points are in the set Γ' that does not contribute to the average probability of error.

The important values of z are those for which $f_0(z) \ne f_1(z)$. Of these, the values of z for which $f_0(z)$ is larger than $f_1(z)$ are the ones that correspond to the decision that 0 was sent, and the values of z for which $f_1(z)$ is larger than $f_0(z)$ are the ones that correspond to the decision that 1 was sent. A decision rule for which

$$f_i(z) = \max\{f_0(z), f_1(z)\}$$

implies that z is in Γ_i is called a *maximum-likelihood* decision rule. The Bayes decision rule for $\pi_0 = \pi_1 = \frac{1}{2}$ is a maximum-likelihood decision rule. Maximum-likelihood decision rules are often used for situations in which the prior probabilities are unknown. For binary communications over an AGN channel, it is shown in Section 5.3.1 that the minimax threshold corresponds to the maximum likelihood decision rule for the system of Figure 5-4. (See especially Figure 5-10 in Section 5.3.1.2.)

The ratio

$$\Lambda(z) = f_1(z)/f_0(z)$$

is called the *likelihood ratio*. A *likelihood ratio test* is a decision rule of the form: Decide that 1 was sent if $\Lambda(z) > \eta$, and decide that 0 was sent if $\Lambda(z) < \eta$. What we do in the case $\Lambda(z) = \eta$ depends on the situation, but the decision can depend on z via $\Lambda(z)$ only. (See Problem 5.14.) If $\Lambda(z) = \eta$, we can use a randomized rule: With probability θ, decide 0 was sent and with probability $1 - \theta$ decide 1 was sent. Alternatively, we can use a deterministic rule (essentially, $\theta = 0$ or $\theta = 1$).

The parameter η depends on the decision criterion. For the optimum Bayes decision, it follows from (5.81) that $\eta = \pi_0/\pi_1$. Note that any likelihood ratio test is a Bayes decision rule: It minimizes $\overline{P_e}$ for some choice of π_0 and π_1. In fact, given η, we can solve $\eta = \pi_0/(1 - \pi_0)$ for π_0 so that we can find the prior probabilities for which a given likelihood ratio test gives the optimum Bayes decision. The solution is

$$\pi_0 = \eta/(1 + \eta)$$

and

$$\pi_1 = 1/(1 + \eta).$$

Note also that the likelihood ratio test with $\eta = 1$ is a maximum-likelihood decision rule.

5.5   The Matched Filter for the AWGN Channel

In Section 5.3, expressions are given for the minimax and Bayes error probabilities. Recall that the minimax error probability is defined by

$$P_{e,m}^* = \min\{P_{e,m}(\gamma) : -\infty < \gamma < \infty\},$$

where $P_{e,m}(\gamma) = \max\{P_{e,0}(\gamma), P_{e,1}(\gamma)\}$. For the general additive Gaussian noise channel, it is shown in Section 5.3.1 that the minimax error probability is given by

$$P_{e,m}^* = Q\{[\mu_0(T_0) - \mu_1(T_0)]/2\sigma\}, \tag{5.83a}$$

where

$$\mu_i(T_0) = (s_i * h)(T_0) \tag{5.83b}$$

and $\sigma^2 = (h * \tilde{h} * R_X)(0)$. For the AWGN channel, the variance σ^2 can be written as

$$\sigma^2 = \frac{N_0}{2} \int_{-\infty}^{\infty} h^2(t)\, dt, \tag{5.83c}$$

where $N_0/2$ is the noise spectral density.

These results are valid for any time-invariant linear filter. The goal of this section is to find the optimum filter; that is, we wish to find the filter that gives the smallest value of the error probability $P_{e,m}^*$. We first consider the minimax error probability, so the threshold is

$$\begin{aligned} \gamma_m &= [\mu_0(T_0) + \mu_1(T_0)]/2 \\ &= [(s_0 * h)(T_0) + (s_1 * h)(T_0)]/2 \\ &= (\{s_0 + s_1\} * h)(T_0)/2. \end{aligned} \tag{5.83d}$$

The threshold depends on the filter h, and it is implicit in the optimization we are carrying out that, as the filter is varied, the threshold is also varied (according to (5.83d)) to give the smallest value of $P_{e,m}(\gamma)$ for each choice of filter.

Next, we return to the expression for the minimax error probability and write it in terms of the signal-to-noise ratio defined in (5.73a) as

$$P_{e,m}^* = Q(\text{SNR}), \tag{5.84a}$$

where

$$\text{SNR} = [\mu_0(T_0) - \mu_1(T_0)]/2\sigma. \tag{5.84b}$$

By definition, we also have

$$P_{e,0} = P_{e,1} = Q(\text{SNR}) \tag{5.84c}$$

if the minimax threshold is used. The signal-to-noise ratio is a function of the filter impulse response, as can be seen from (5.83b) and (5.83c). Applying these expressions to (5.84b), we see that

$$\mathrm{SNR} = \frac{(s_0 * h)(T_0) - (s_1 * h)(T_0)}{\sqrt{2N_0}\,\|h\|}, \tag{5.84d}$$

where $\|h\|$ is defined by

$$\|h\| = \left\{ \int_{-\infty}^{\infty} h^2(t)\,dt \right\}^{1/2}$$

for any function h for which

$$\int_{-\infty}^{\infty} h^2(t)\,dt < \infty.$$

This condition is equivalent to saying that h has finite energy. It is wise to restrict attention to filters satisfying this condition, because

$$\int_{-\infty}^{\infty} h^2(t)\,dt = \infty$$

implies that $\sigma^2 = \infty$, in which case the noise at the filter output has infinite power. As usual, the positive square root is employed in the above definition of $\|h\|$.

The quantity $\|h\|$ is known as the *norm* of the function h; it is a generalization of the notion of the length of a vector $\mathbf{x} = (x_1, x_2, \ldots, x_n)$, which is

$$\|\mathbf{x}\| = \left\{ \sum_{i=1}^{n} x_i^2 \right\}^{1/2}.$$

We also wish to introduce the notion of the inner product of two functions. If f and g are real-valued functions of a real variable, and if each of the functions has finite energy, then the *inner product* of f and g is defined by

$$(f, g) = \int_{-\infty}^{\infty} f(t)\,g(t)\,dt.$$

This is a generalization of the inner product (sometimes called the dot product) of two vectors $\mathbf{x} = (x_1, x_2, \ldots, x_n)$ and $\mathbf{y} = (y_1, y_2, \ldots, y_n)$, which is

$$(\mathbf{x}, \mathbf{y}) = \sum_{i=1}^{n} x_i y_i.$$

Notice that $(f, f) = \|f\|^2$ and that $(f, g) = (g, f)$.

The main reason for introducing the inner product is because the convolutions that appear in (5.84d) can be written as inner products. First, observe that if we define the signal s_{T_0} by

$$s_{T_0}(t) = [s_0(T_0 - t) - s_1(T_0 - t)]/2,$$

then we can express the numerator of (5.84d) as an inner product:

$$
\begin{aligned}
(s_0 * h)(T_0) - (s_1 * h)(T_0) &= \int_{-\infty}^{\infty} s_0(T_0 - \tau)\, h(\tau)\, d\tau - \int_{-\infty}^{\infty} s_1(T_0 - \tau)\, h(\tau)\, d\tau \\
&= \int_{-\infty}^{\infty} [s_0(T_0 - \tau) - s_1(T_0 - \tau)]\, h(\tau)\, d\tau \\
&= 2 \int_{-\infty}^{\infty} s_{T_0}(\tau)\, h(\tau)\, d\tau \\
&= 2(s_{T_0}, h).
\end{aligned}
\tag{5.85}
$$

The signal-to-noise ratio can then be written as

$$
\begin{aligned}
\text{SNR} &= (s_{T_0}, h) / \sqrt{N_0 \|h\|^2 / 2} \\
&= \left\{ 2(s_{T_0}, h)^2 / N_0 \|h\|^2 \right\}^{1/2}
\end{aligned}
\tag{5.86}
$$

In view of the fact that $Q(\text{SNR})$ is a decreasing function of SNR, it is clear that $P_{e,m}^* = Q(\text{SNR})$ is minimized by maximizing SNR. That is, a filter impulse response gives the smallest value of the minimax error probability if and only if the impulse response gives the largest possible value of the signal-to-noise ratio. Hence, our goal, as far as the minimax error probability is concerned, is to maximize SNR over all filters h for which

$$
\int_{-\infty}^{\infty} h^2(t)\, dt < \infty.
$$

Next, we consider the Bayes error probability. In Section 5.3.2, we found that the Bayes error probability is given by

$$
\overline{P}_e^* = Q[\alpha - (2\alpha)^{-1} \ln(\pi_1/\pi_0)]\pi_0 + Q[\alpha + (2\alpha)^{-1} \ln(\pi_1/\pi_0)]\pi_1.
\tag{5.87}
$$

where $\alpha = \text{SNR}$. As shown in Exercise 5-2, $\overline{P}_e^*$ is a decreasing function of SNR. Consequently, the filter impulse response that maximizes SNR also minimizes the Bayes error probability $\overline{P}_e^*$ for any probabilities π_0 and π_1.

This, together with the previous conclusion concerning the minimax error probability, gives the following result:

> A filter that maximizes SNR gives the smallest possible value for the minimax error probability and the Bayes error probability.

Because of this result, we speak of an optimum filter without regard to the criterion (minimax or Bayes) that is being employed.

The problem of finding an optimum filter has just been shown to be equivalent to finding a filter that maximizes the signal-to-noise ratio

$$
\text{SNR} = \left\{ 2(s_{T_0}, h)^2 / N_0 \|h\|^2 \right\}^{1/2}.
\tag{5.88a}
$$

It is more convenient to work with the square of the signal-to-noise ratio, and the maximization of SNR^2 is equivalent to the maximization of SNR. Because N_0 does not depend on h, the goal then is to maximize the quantity

$$\xi = (s_{T_0}, h)^2 / \|h\|^2$$

$$= \left[\int_{-\infty}^{\infty} s_{T_0}(t) \, h(t) \, dt \right]^2 \bigg/ \int_{-\infty}^{\infty} h^2(t) \, dt. \qquad (5.88b)$$

Maximization of ξ is accomplished by appealing to the Schwarz inequality, which is the topic of the next subsection.

5.5.1   The Schwarz Inequality

We consider two functions f and g that have finite energy; that is,

$$\int_{-\infty}^{\infty} f^2(u) \, du < \infty$$

and

$$\int_{-\infty}^{\infty} g^2(u) \, du < \infty.$$

Suppose the function g is given, and it is desired to choose f to maximize the inner product

$$(f, g) = \int_{-\infty}^{\infty} f(u) \, g(u) \, du.$$

The solution is provided by the following theorem.

Theorem 5-2 (Schwarz Inequality)   For any finite-energy functions f and g,

$$\left[\int_{-\infty}^{\infty} f(u) \, g(u) \, du \right]^2 \leq \int_{-\infty}^{\infty} f^2(u) \, du \int_{-\infty}^{\infty} g^2(v) \, dv.$$

■

In the notation of norms and inner products, the Schwarz inequality is expressed as $(f, g)^2 \leq \|f\|^2 \|g\|^2$. This statement of the Schwarz inequality is more general, because it applies to other vector spaces as well (e.g., finite-dimensional spaces of vectors). Furthermore, the notation of norms and inner products is more compact, so it is adopted gradually in the proof of Theorem 5-2. As an aid to the reader, we begin the proof by working with integral expressions and giving their equivalents in terms of norms and inner products. As the proof develops, we omit the integral expressions and work with norms and inner products only.

Proof of Theorem 5-2.   If f and g are functions with finite energy and c is an arbitrary real number, then

$$\|f - c\,g\|^2 = \int_{-\infty}^{\infty} [f(u) - c\,g(u)]^2 \, du \geq 0.$$

By expanding the integrand and then integrating term by term, we see that

$$\| f - c\,g \|^2 = \int_{-\infty}^{\infty} f^2(u)\,du + c^2 \int_{-\infty}^{\infty} g^2(u)\,du - 2\,c \int_{-\infty}^{\infty} f(u)\,g(u)\,du.$$

From this expression and the fact that $\| f - c\,g \|^2 \geq 0$, it follows that

$$2\,c \int_{-\infty}^{\infty} f(u)\,g(u)\,du \leq \int_{-\infty}^{\infty} f^2(u)\,du + c^2 \int_{-\infty}^{\infty} g^2(u)\,du,$$

for any real number c. In terms of norms and inner products, our findings thus far are

$$\| f - c\,g \|^2 = \| f \|^2 - 2\,c\,(f,\,g) + c^2 \| g \|^2$$

and

$$2\,c\,(f,\,g) \leq \| f \|^2 + c^2 \| g \|^2,$$

and each of these two expressions holds for any real number c.

We deal first with the consequences of this last inequality if $\| g \|^2 = 0$. Because $\| f \|^2$ and $\| g \|^2$ are finite and

$$2\,c\,(f,\,g) \leq \| f \|^2 + c^2 \| g \|^2$$

for each real number c, we can conclude that if $\| g \|^2 = 0$, it must be that $(f,\,g) = 0$. This can be established by a two-step procedure. First, suppose that $\| g \|^2 = 0$, but $(f,\,g) > 0$. The inequality becomes

$$2\,c\,(f,\,g) \leq \| f \|^2$$

for each real number c. This inequality cannot be true for all c, because if we let $c \to +\infty$ the left-hand side increases to $+\infty$ but the right-hand side is finite. Thus, we have reached a contradiction, and so if $\| g \|^2 = 0$, it is not possible that $(f,\,g) > 0$. Now suppose that $\| g \|^2 = 0$, but $(f,\,g) < 0$. By letting $c \to -\infty$, we see that the left-hand side again increases to $+\infty$, but the right-hand side is finite. Therefore, it is also not possible that $(f,\,g) < 0$. The only possibility left is $(f,\,g) = 0$.

Rather than using the two-step procedure just described, we can simply let the sign of c be the same as the sign of $(f,\,g)$. If we then let the magnitude of c go to $+\infty$, we show that $\| g \|^2 = 0$ implies $(f,\,g) = 0$. Regardless of which procedure is followed, the conclusion is that if $\| g \|^2 = 0$, then $(f,\,g) = 0$, in which case the inequality that we wish to prove becomes $0 \leq \| f \|^2$, which is obviously a true statement for any function f. Thus, if $\| g \|^2 = 0$, the Schwarz inequality is valid.

Now suppose that $\| g \|^2 \neq 0$ (which, of course, means that $\| g \|^2 > 0$, because $\| g \|^2$ cannot be negative). We have already established that

$$2\,c\,(f,\,g) \leq \| f \|^2 + c^2 \| g \|^2,$$

for any real number c. In particular, the inequality holds for $c = (f,\,g)/\| g \|^2$. Note that it is necessary to have $\| g \|^2 \neq 0$ in order to use this value for c. Substitution for c in the preceding inequality gives

$$2(f,\,g)^2/\| g \|^2 \leq \| f \|^2 + \{(f,\,g)^2/\| g \|^2\},$$

which is equivalent to

$$(f, g)^2 / \|g\|^2 \le \|f\|^2.$$

Multiplication of both sides by $\|g\|^2$ gives the Schwarz inequality, and this completes the proof of Theorem 5-2.   ∎

For our purposes, the statement that "$f(u) = \lambda\, g(u)$ for almost all u" means that

$$\int_{-\infty}^{\infty} [f(u) - \lambda\, g(u)]^2 \, du = 0, \tag{5.89}$$

which is equivalent to $\|f - \lambda\, g\|^2 = 0$. This does not require that $f(u) = \lambda\, g(u)$ for each value of u. It should be obvious to the reader that the value of the integrand can be changed at a finite or countably infinite number of points without changing the value of the integral (e.g., $f(u) - \lambda\, g(u)$ can be changed for each integer value of u, yet the value of the integral in (5.89) is not affected). It follows that even if f and g satisfy (5.89), $f(u)$ may differ from $\lambda\, g(u)$ for a finite or countably infinite number of values of u. Although it is beyond the scope of this book to provide the mathematical background necessary to give a proof, it is true that $f(u)$ can differ from $\lambda\, g(u)$ on certain sets of uncountably many points and still satisfy (5.89).

Before proceeding, it is helpful to record two consequences of Theorem 5-2, which are given in the following corollary. Observe that the requirements on the functions f and h in the corollary can be written concisely as $\|f\|^2 < \infty$, $\|h\|^2 < \infty$, and $\|f - h\|^2 = 0$.

Corollary to Theorem 5-2   Let f and h be functions with finite energy and suppose $f(u) = h(u)$ for almost all u, then

$$\text{(i)}\quad \|f\|^2 = \|h\|^2,$$

and

$$\text{(ii)}\quad (f, g) = (h, g),$$

for any function g that has finite energy.   ∎

Proof of Corollary   To prove (i), observe that

$$0 = \|f - h\|^2 = \|f\|^2 - 2(f, h) + \|h\|^2.$$

Theorem 5-2 implies that $(f, h) \le \|f\|\|h\|$. It follows that $-2(f, h) \ge -2\|f\|\|h\|$. Therefore,

$$0 = \|f\|^2 - 2(f, h) + \|h\|^2 \ge \|f\|^2 - 2\|f\|\|h\| + \|h\|^2 = \{\|f\| - \|h\|\}^2.$$

The conclusion is that $\|f\| = \|h\|$, which proves (i). To prove (ii), we note again that $\|f - h\|^2 = 0$, and apply Theorem 5-2 to conclude that

$$(f - h, g)^2 \le \|f - h\|^2 \|g\|^2 = 0.$$

But $(f - h, g) = (f, g) - (h, g)$, so the conclusion is $(f, g) - (h, g) = 0$, which establishes (ii) and completes the proof of the corollary.   ∎

The Schwarz inequality gives an upper bound on the inner product of two finite-energy functions. For our applications to communication receivers, it is important to know the circumstances under which the upper bound can be achieved by the inner product of two functions. That is, when does equality hold in the Schwarz inequality? The next theorem provides the complete answer to this question.

Theorem 5-3 (Conditions for Equality)   If f and g are finite-energy functions and

$$\int_{-\infty}^{\infty} g^2(v) \, dv > 0,$$

then equality holds in the Schwarz inequality if and only if there is a real number λ such that $f(u) = \lambda \, g(u)$ for almost all u.   ∎

In terms of norms, the statement of Theorem 5-3 is as follows: If $\|g\| > 0$, the necessary and sufficient condition for equality in the Schwarz inequality is that there exists a λ for which $\|f - \lambda g\| = 0$. For our application to the maximization of the signal-to-noise ratio as expressed in (5.88), the theorem guarantees that the solution is "essentially unique" in the sense that if impulse responses h_1 and h_2 each achieve the upper bound on the signal-to-noise ratio, then $h_1(t) = \lambda \, h_2(t)$ for almost all t. In other words, if each of two impulse responses gives the maximum signal-to-noise ratio, one of the impulse responses is essentially a multiple of the other.

Proof of Theorem 5-3.   We show that equality holds in the Schwarz inequality if and only if there exists a λ such that $f(u) = \lambda \, g(u)$ for almost all u (i.e., $\|f - \lambda g\|^2 = 0$). First, suppose there does not exist such a λ; that is, suppose

$$\|f - \lambda g\|^2 > 0$$

for all λ. In particular, for $\lambda = c = (f, g)/\|g\|^2$, as in the proof of Theorem 5-2, we have

$$
\begin{aligned}
0 < \|f - c g\|^2 &= \|f\|^2 - 2c\,(f, g) + c^2 \|g\|^2 \\
&= \|f\|^2 - 2\{(f, g)^2/\|g\|^2\} + \{(f, g)^2/\|g\|^2\} \\
&= \|f\|^2 - \{(f, g)^2/\|g\|^2\}.
\end{aligned}
$$

Multiplying through by $\|g\|^2$, which is valid because $\|g\|^2 \neq 0$, we obtain

$$0 < \|f\|^2 \|g\|^2 - (f, g)^2,$$

which gives the strict inequality

$$(f, g)^2 < \|f\|^2 \|g\|^2.$$

We have proved that if there is no λ for which $f(u) - \lambda \, g(u)$ for almost all u, equality cannot occur in the Schwarz inequality. Thus, the existence of such a λ is a necessary condition for equality in the Schwarz inequality.

It is easy to see that this condition is also sufficient, because if $f(u) = \lambda g(u)$ for almost all u, then we can apply (ii) of the corollary with the function h defined by $h(u) = \lambda g(u)$ for each value of u. This gives

$$(f, g) = (h, g) = (\lambda g, g) = \lambda(g, g) = \lambda \|g\|^2.$$

We next apply (i) to conclude that

$$\|f\|^2 = \|h\|^2 = \|\lambda g\|^2 = \lambda^2 \|g\|^2.$$

Putting these two results together, we see that

$$(f, g)^2 = \lambda^2 \|g\|^4 = \{\lambda^2 \|g\|^2\} \|g\|^2 = \|f\|^2 \|g\|^2.$$

which represents equality in the Schwarz inequality. Thus, the existence of a λ for which $f(u) = \lambda g(u)$ for almost all u is also sufficient to give equality in the Schwarz inequality. This completes the proof of Theorem 5-3. ∎

For convenience in what follows, we often omit the phrase "for almost all u" in statements such as "$f(u) = \lambda g(u)$ for almost all u," and we write simply $f(u) = \lambda g(u)$ with the understanding that "for almost all u" is implied. It is helpful to think of two functions as being essentially the same if they agree for almost all values of their arguments. This is particularly true if the functions represent signal waveforms or impulse responses. In practice, there is usually one of the essentially equivalent functions that is natural and preferred for implementation (e.g., one of them may be continuous). When this is true, there is a clear preference for one particular function from among the set of essentially equivalent functions.

In case it is not obvious, we should mention that the Schwarz inequality actually provides two inequalities. The inner product (f, g) can be either positive or negative, so the inequality

$$(f, g)^2 \leq \|f\|^2 \|g\|^2$$

provides the two useful inequalities

$$-\|f\| \|g\| \leq (f, g) \leq \|f\| \|g\|.$$

Equality occurs in the left-hand inequality if $f(u) = \lambda g(u)$ for $\lambda \leq 0$ and in the right-hand inequality if $f(u) = \lambda g(u)$ for $\lambda \geq 0$.

It may help provide some physical insight into the Schwarz inequality and the conditions for equality if we consider the application to two-dimensional vectors. To avoid trivialities, we assume $\mathbf{x} \neq 0$ for the present discussion. In Figure 5-13, we show vectors $\mathbf{x} = (x_1, x_2)$ and $\mathbf{y} = (y_1, y_2)$ in two-dimensional Euclidean space. The angle between the two vectors is θ. For these vectors, $\|\mathbf{x}\|^2 = x_1^2 + x_2^2$, $\|\mathbf{y}\|^2 = y_1^2 + y_2^2$, and

$$(\mathbf{x}, \mathbf{y}) = x_1 y_1 + x_2 y_2.$$

Another way to express the inner product (also known as the dot product) is

$$(\mathbf{x}, \mathbf{y}) = \|\mathbf{x}\| \|\mathbf{y}\| \cos \theta.$$

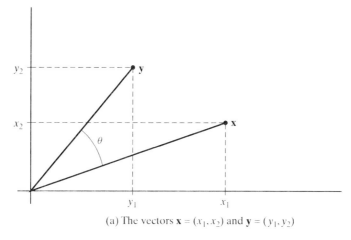

(a) The vectors $\mathbf{x} = (x_1, x_2)$ and $\mathbf{y} = (y_1, y_2)$

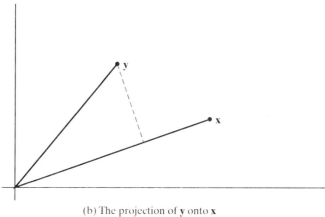

(b) The projection of $\mathbf{y}$ onto $\mathbf{x}$

Figure 5-13: An illustration for the Schwarz inequality in two-dimensional vector space.

Because $|\cos \theta| \le 1$, it follows that

$$(\mathbf{x} . \mathbf{y})^2 \le \|\mathbf{x}\|^2 \|\mathbf{y}\|^2.$$

which is the Schwarz inequality for two-dimensional vectors. If $\|\mathbf{x}\|$ and $\|\mathbf{y}\|$ are not zero, equality holds if and only if $\cos^2 \theta = 1$. But $\cos^2 \theta = 1$ if and only if $\theta = 0$ or $\theta = \pi$. In either case, $\mathbf{x}$ is a scalar multiple of $\mathbf{y}$. If $\theta = 0$, the scalar is positive; if $\theta = \pi$, the scalar is negative.

The validity of the inequality

$$|(\mathbf{x} . \mathbf{y})| \le \|\mathbf{x}\| \|\mathbf{y}\|$$

is also clear from the fact that the magnitude of the inner product $(\mathbf{x} . \mathbf{y})$ is just the length of the projection of $\mathbf{y}$ onto $\mathbf{x}$ times the length of $\mathbf{x}$. The length of the projection of a vector cannot exceed the length of the vector itself. Hence the length of the projection of $\mathbf{y}$ onto $\mathbf{x}$ is equal to the length of $\mathbf{y}$ if and only if $\mathbf{y}$ lies on the straight line that passes through both the origin and the point $\mathbf{x} = (x_1, x_2)$. That is, $\mathbf{y} = \lambda \mathbf{x}$ for some choice of the real number λ.

In the special case $\|\mathbf{y}\| = 0$, $\mathbf{y}$ is a scalar multiple of $\mathbf{x}$ for any choice of $\mathbf{x}$ ($\mathbf{y} = \mathbf{0}$ and $\lambda = 0$ implies $\mathbf{y} = \lambda \mathbf{x}$). In this case, the Schwartz inequality is an equality, and the necessary and sufficient condition $\mathbf{y} = \lambda \mathbf{x}$ is trivially satisfied.

5.5.2 The Matched Filter

The Schwarz inequality is applied to the maximization of the signal-to-noise ratio by letting $g = s_{T_0}$ and $f = h$. This gives, in the notation of inner products,

$$(s_{T_0}, h)^2 \leq \|s_{T_0}\|^2 \|h\|^2$$

with equality if and only if $h(t) = \lambda s_{T_0}(t)$ for almost all t, for some choice of λ. In actual communication systems, it is always true that

$$\int_{-\infty}^{\infty} [s_{T_0}(t)]^2 \, dt > 0,$$

so that the essential uniqueness of the optimizing filter is guaranteed.

As an example, consider an antipodal signal set: $s_0(t) = s(t)$, $s_1(t) = -s(t)$. The quantity $\|s_{T_0}\|^2$ is given by

$$\begin{aligned}
\int_{-\infty}^{\infty} [s_{T_0}(t)]^2 \, dt &= \int_{-\infty}^{\infty} [s_0(T_0 - t) - s_1(T_0 - t)]^2 / 4 \, dt \\
&= \int_{-\infty}^{\infty} [s(T_0 - t)]^2 \, dt \\
&= \int_{-\infty}^{\infty} s^2(t) \, dt = \mathcal{E},
\end{aligned}$$

where $\mathcal{E}$ is the energy in the signal $s(t)$. Clearly, $\mathcal{E}$ is not zero for any system of interest. We conclude that the optimum filter has impulse response

$$\begin{aligned}
h(t) &= \lambda s_{T_0}(t) \\
&= \lambda [s_0(T_0 - t) - s_1(T_0 - t)] / 2 \qquad \text{(5.90a)}
\end{aligned}$$

for some choice of λ. We are not interested in the trivial solution that corresponds to $\lambda = 0$ (i.e., $h(t) = 0$ for all t), because this produces a receiver whose decision does not depend on the receiver input and gives error probabilities $P_{e,0} = P_{e,1} = 1/2$.

The value of the constant λ can be either positive or negative, but only positive values of λ satisfy our convention that $\hat{s}_0(T_0) > \hat{s}_1(T_0)$. To see this, start with

$$\begin{aligned}
\hat{s}_0(T_0) - \hat{s}_1(T_0) &= (s_0 * h)(T_0) - (s_1 * h)(T_0) \\
&= 2(s_{T_0}, h),
\end{aligned}$$

as was shown in (5.85). If $h(t) = \lambda s_{T_0}(t)$, then

$$(s_{T_0}, h) = \lambda(s_{T_0}, s_{T_0}) = \lambda \|s_{T_0}\|^2,$$

and so

$$\hat{s}_0(T_0) - \hat{s}_1(T_0) = 2\lambda \|s_{T_0}\|^2.$$

Thus, in order to have $\hat{s}_0(T_0) > \hat{s}_1(T_0)$, we must have $\lambda > 0$. Again, this is just a matter of convention with regard to the decision made at the output of the threshold device. Since λ is arbitrary, we might as well absorb the factor of 2 that appears in (5.90a) into the constant; that is, replace $\lambda/2$ by the constant c.

The filter with impulse response given by (5.90a) is called the *matched filter*. More precisely, for a particular binary signal set $\{s_0, s_1\}$ and sampling time T_0, we say that a linear time-invariant filter with impulse response $h(t)$ is matched to the signal set and sampling time if

$$h(t) = c \, [s_0(T_0 - t) - s_1(T_0 - t)] \tag{5.90b}$$

for some constant $c \neq 0$. In order to follow the convention $\hat{s}_0(T_0) > \hat{s}_1(T_0)$, the constant c must be positive.

Other than to satisfy the restriction $c > 0$, the value of c is arbitrary. That is, the resulting error probabilities $P_{e,0}$ and $P_{e,1}$ do not depend on the value of the positive constant c. This is very important for implementation, because it means that the hardware designer is free to choose this gain constant.

The fact that the gain constant c is unimportant in terms of the receiver performance is proved by showing that the signal-to-noise ratio does not depend on c. Recall that both the minimax and Bayes error probabilities depend on the signals and the filter via the signal-to-noise ratio only. (See (5.84) and (5.87).) From (5.86), we see that the signal-to-noise ratio can be expressed as

$$\mathrm{SNR} = \frac{(s_{T_0}, h)}{\sqrt{N_0/2} \, \|h\|}.$$

But if the filter impulse response is given by

$$h(t) = c \, [s_0(T_0 - t) - s_1(T_0 - t)],$$

then $h(t) = 2 \, c \, s_{T_0}(t)$. It follows that

$$(s_{T_0}, h) = 2 \, c \, \|s_{T_0}\|^2$$

and

$$\|h\| = 2 \, c \, \|s_{T_0}\|,$$

so that

$$\mathrm{SNR} = \|s_{T_0}\| / \sqrt{N_0/2}, \tag{5.91}$$

which does not depend on the constant c.

We might expect this to be the case, because both the signal and the noise at the filter input are multiplied by the constant c, which does not change the signal-to-noise ratio. This intuitive argument might lead the reader to conjecture that the error probabilities do not depend on the gain constant even if the filter is not the matched filter. This is true only for certain choices of the threshold. The minimax and Bayes thresholds depend on the gain constant, and they are adjusted to compensate for changes in the gain according to (5.71) and (5.78). If the threshold is not adjusted in this way, the error probabilities may depend on the gain constant. (See Problems 5.9–5.11.) Another way of saying this is that for thresholds other than the minimax or Bayes thresholds, the error probabilities $P_{e,0}$ and $P_{e,1}$ may not depend on the signal-to-noise ratio alone. They may depend on other parameters of the signal, noise, and filter such as the signal amplitude or the filter gain constant. (See Problem 5.11.)

For the remainder of this section, assume that a minimax or Bayes threshold is employed, and define the signal $s(t)$ by

$$s(t) = [s_0(t) - s_1(t)]/2.$$

The matched filter is given by

$$h(t) = \lambda s(T_0 - t) = 2 c s(T_0 - t).$$

Because the gain constant is unimportant, we typically let $c = 1/2$ ($\lambda = 1$), so that $h(t) = s(T_0 - t)$. Notice that for antipodal signals, this choice for the constant gives $h(t) = s_0(T_0 - t) = -s_1(T_0 - t)$.

Just as the minimax threshold compensates for the signal amplitude and the filter gain constant, the matched filter, as specified by (5.90), compensates for the sampling time T_0. The filter with impulse response

$$h(t) = c\,[s_0(T_0 - t) - s_1(T_0 - t)]$$

gives the maximum distinguishability between the signals at the sampling time T_0. If the sampling time is increased (i.e., delayed), then the filter impulse response is also delayed. This can be demonstrated by changing the sampling time from T_0 to $T_1 = T_0 + \tau$ for $\tau > 0$. The matched filter for sampling time T_1 has impulse response

$$\begin{aligned}
h_1(t) &= [s_0(T_1 - t) - s_1(T_1 - t)]/2 \\
&= \{s_0[T_0 - (t - \tau)] - s_1[T_0 - (t - \tau)]\}/2 \\
&= h(t - \tau).
\end{aligned}$$

So $h_1(t)$ is just a delayed version of $h(t)$, as illustrated in Figure 5-14.

The signal $s(t)$ is shown in Figure 5-14(a). If the sampling time is T_0, the matched filter for this signal is as shown in Figure 5-14(b). Because the signal satisfies $s(t) = 0$ for $t < 0$, the matched filter satisfies $h(t) = 0$ for $t > T_0$. The signal is time-limited to the interval $[0, T]$, so the filter is time limited to the interval $[T_0 - T, T_0]$.

The effect of delaying the sampling time to T_1, where $T_1 = T_0 + \tau$ for $\tau > 0$ in our previous notation, is illustrated in Figure 5-14(c). Notice that if the sampling time is delayed further, the matched filter can be causal. As illustrated in Figure 5-14(b), the

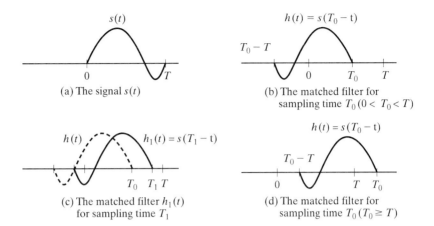

Figure 5-14: The impulse response of the matched filter for various sampling times.

matched filter satisfies $h(t) = 0$ for $t < T_0 - T$. The condition for causality is $h(t) = 0$ for $t < 0$, so the matched filter *is* causal if $T_0 - T > 0$, as illustrated in Figure 5-14(d).

The key feature of the signal that permits the matched filter to be causal is that there exist a time T such that $s(t) = 0$ for $t > T$. The most common signaling waveforms satisfy this constraint and more: They are typically time-limited signals. We summarize our conclusions by giving the following property of the matched filter:

If the signal $s(t) = [s_0(t) - s_1(t)]/2$ satisfies $s(t) = 0$ for $t > T$ and the sampling time satisfies $T_0 \geq T$, then the matched filter is causal.

There is typically no reason to delay the sampling more than necessary, so that in theory we would like to have $T_0 = T$, the earliest sampling time that gives a causal matched filter. There may be constraints in the implementation that give rise to additional delays in the sampling time, so that we may not actually realize this goal. Fortunately, as we show in what follows, the receiver's performance does not depend on the sampling time provided the sampling time is known and is incorporated into the matched filter according to (5.90).*

Recall that for either the minimax or the Bayes threshold, the error probabilities $P_{e,0}$ and $P_{e,1}$ depend on the signals, filter, and sampling time via the signal-to-noise ratio only. The signal-to-noise ratio is in general given by

$$\text{SNR} = \left\{ \frac{2(s_{T_0}, h)^2}{N_0 \|h\|^2} \right\}^{1/2},$$

but for the matched filter the expression was shown (see (5.91)) to reduce to

$$\text{SNR} = \|s_{T_0}\| / \sqrt{N_0/2}.$$

At first glance, this may appear to depend on T_0, but we will see that in fact it does not.

---

*Changes in the sampling time (e.g., due to drift in the receiver's clock) after the filter has been implemented do affect performance. (See Exercise 5-4 and Problem 5.18.)

The signal-to-noise ratio for the matched-filter receiver can be expressed in terms of more fundamental parameters of the signal set $\{s_0, s_1\}$. We start with

$$\text{SNR} = \|s_{T_0}\| / \sqrt{N_0/2}$$

and then use the fact that

$$
\begin{aligned}
\|s_{T_0}\|^2 &= \int_{-\infty}^{\infty} [s_{T_0}(t)]^2 \, dt = \tfrac{1}{4} \int_{-\infty}^{\infty} [s_0(T_0 - t) - s_1(T_0 - t)]^2 \, dt \\
&= \tfrac{1}{4} \left\{ \int_{-\infty}^{\infty} s_0^2(u) + s_1^2(u) \, du \right\} - \tfrac{1}{2} \int_{-\infty}^{\infty} s_0(u) \, s_1(u) \, du \\
&= \tfrac{1}{4}(\mathcal{E}_0 + \mathcal{E}_1) - \tfrac{1}{2}\rho,
\end{aligned}
$$

where ρ is the inner product for signals s_0 and s_1. That is,

$$\rho = \int_{-\infty}^{\infty} s_0(u) \, s_1(u) \, du.$$

Letting $\bar{\mathcal{E}} = (\mathcal{E}_0 + \mathcal{E}_1)/2$ and $r = \rho/\bar{\mathcal{E}}$, we can write $\|s_{T_0}\|^2$ as

$$\|s_{T_0}\|^2 = \bar{\mathcal{E}}(1 - r)/2.$$

The signal-to-noise ratio for a receiver with the matched filter is therefore given by

$$\text{SNR} = \{\bar{\mathcal{E}}(1 - r)/N_0\}^{1/2}. \tag{5.92}$$

The parameter $\bar{\mathcal{E}}$ is called the *average energy* for the signal set, and the parameter r is the *correlation coefficient* (or normalized correlation) for the two signals s_0 and s_1.

The first observation to be made about (5.92) is that the signal-to-noise ratio does not depend on the sampling time T_0 if the matched filter is used. The intuitive reason for this is that the matched filter, as defined by

$$h(t) = c \, [s_0(T_0 - t) - s_1(T_0 - t)],$$

automatically compensates for any changes in the sampling time. It should also be noted that, because the noise is stationary, the variance of the output noise is independent of the sampling time. (This is true for any time-invariant linear filter.)

Another observation to be made about (5.92) is that the signal-to-noise ratio and hence the error probabilities depend on two signal parameters only. These are the average energy in the two signals s_0 and s_1 and the inner product (s_0, s_1). The detailed structure of the signals is unimportant.

Antipodal signals have equal energy, so $\bar{\mathcal{E}} = \mathcal{E}_0 = \mathcal{E}_1$. Recall that for equal-energy signals, we drop the subscripts and denote the energy by $\mathcal{E}$. The inner product for antipodal signals is

$$
\begin{aligned}
\rho = (s_0, s_1) &= \int_{-\infty}^{\infty} s_0(t) \, s_1(t) \, dt \\
&= -\int_{-\infty}^{\infty} [s_0(t)]^2 \, dt = -\mathcal{E}_0.
\end{aligned}
$$

The correlation coefficient for antipodal signals is therefore $r = -1$. The resulting signal-to-noise ratio is

$$\text{SNR} = \sqrt{2\,\mathcal{E}/N_0}, \tag{5.93a}$$

and the error probabilities for the minimax threshold are given by

$$P_{e,0} = P_{e,1} = Q\left(\sqrt{2\,\mathcal{E}/N_0}\right). \tag{5.93b}$$

Recall that the error probabilities of (5.93b) are obtained in Section 5.2.1 for the antipodal signals $s_0(t) = a\,p_I(t)$ and $s_1(t) = -a\,p_I(t)$. The matched filter for this example has an impulse response that is a constant multiple of $a\,p_I(t)$. Because the constant is unimportant, we can let $h(t) = p_I(T_0 - t)$, so that the decision statistic is

$$Z(T_0) = \int_{-\infty}^{\infty} Y(t)\,h(T_0 - t)\,dt$$

$$= \int_{-\infty}^{\infty} Y(t)\,p_I(t)\,dt.$$

Letting $Z = Z(T_0)$, we see that the decision statistic is given by

$$Z = \int_{-\infty}^{\infty} Y(t)\,p_I(t)\,dt = \int_0^T Y(t)\,dt,$$

which is just the decision statistic defined in (5.9) of Section 5.2.1. Finally, we note that the threshold $\gamma = 0$ used in Section 5.2.1 is the minimax threshold (true for any antipodal signal set). We therefore conclude that the example of Section 5.2.1 is just a special case of a communication system with an antipodal signal set, matched filter, and minimax threshold. The fact that (5.93b) agrees with (5.30) is therefore not surprising.

Exercise 5-4
Consider the signal set defined by

$$s_0(t) = p_I(t)$$

and

$$s_1(t) = -p_I(t).$$

Suppose that the *nominal* sampling time is $T_0 = T$. Assume that the threshold is $\gamma = 0$ and the matched filter impulse response is based on the nominal sampling time. When the receiver is placed in operation, however, the actual sampling time varies because of short-term fluctuations in the clock (timing jitter) and long-term changes in other receiver components due to temperature variations. Examine the effects of these a posteriori variations in sampling time for the (fixed) matched filter.

Solution.   Because the nominal sampling time is T, the impulse response of the matched filter is given by

$$
\begin{aligned}
h(t) &= [s_0(T - t) - s_1(T - t)]/2 \\
     &= p_T(T - t) \\
     &= p_T(t),
\end{aligned}
$$

where we have used (5.90b) with $c = 1/2$. Because the threshold is the minimax threshold for any choice of sampling time, the error probabilities are given by $P_{e,0} = P_{e,1} = Q(\text{SNR})$, where SNR can be found from (5.84d). The parameter T_0 that appears in this equation is the actual sampling time, which we denote by T_a. The signal-to-noise ratio as a function of the actual sampling time is

$$
\text{SNR}(T_a) = \frac{(s_0 * h)(T_a) - (s_1 * h)(T_a)}{\sqrt{2N_0}\,\|h\|}.
$$

For the signals and filter under consideration,

$$
\hat{s}_1(T_a) = (s_1 * h)(T_a) = -(s_0 * h)(T_a) = -\hat{s}_0(T_a)
$$

and

$$
\|h\|^2 = \int_{-\infty}^{\infty} [p_T(t)]^2 \, dt = T.
$$

Moreover,

$$
\hat{s}_0(T_a) = (s_0 * h)(T_a) = \Lambda_T(T_a - T),
$$

where Λ_T is the triangular pulse defined by

$$
\Lambda_T(u) = \begin{cases} T - |u|, & 0 \le |u| \le T, \\ 0, & |u| > T. \end{cases}
$$

The output signals, $\hat{s}_0(T_a)$ and $\hat{s}_1(T_a)$, are shown in Figure 5-15 as a function of the actual sampling time T_a. Notice that the matched filter produces maximum distinguishability (i.e., largest value of $\hat{s}_0(T_a) - \hat{s}_1(T_a)$) at the nominal sampling time T, as expected. It follows that

$$
\text{SNR}(T_a) = \Lambda_T(T_a - T)/\sqrt{N_0 T/2}.
$$

For $T_a = T$, this reduces to

$$
\text{SNR}(T) = (2T/N_0)^{1/2}.
$$

This is consistent with (5.92), because, for the signals under consideration,

$$
\bar{\mathcal{E}} = \mathcal{E}_0 = \int_{-\infty}^{\infty} [s_0(t)]^2 \, dt = \int_{-\infty}^{\infty} p_T(t) \, dt = T
$$

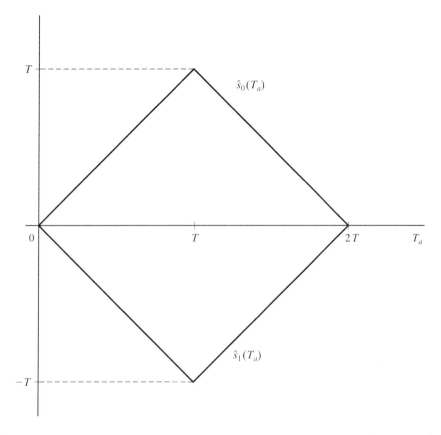

Figure 5-15: The output signals as a function of the actual sampling time (for Exercise 5-4).

and

$$\rho = \int_{-\infty}^{\infty} s_0(t)\, s_1(t)\, \mathrm{d}t = -\int_{-\infty}^{\infty} p_T(t)\, \mathrm{d}t = -T,$$

so that

$$r = \rho/\bar{\mathcal{E}} = -1$$

and

$$\bar{\mathcal{E}}(1-r)/N_0 = 2\mathcal{E}_0/N_0 = 2T/N_0.$$

For T_a in the range $0 \le T_a \le 2T$, the general result is

$$\mathrm{SNR}(T_a) = (T - |T - T_a|)/\sqrt{N_0 T/2},$$

which decreases linearly in the sampling error $T_e = |T_a - T|$.   ∎

Exercise 5-5
Repeat Exercise 5-4 for the signal set

$$s_0(t) = p_T(t)$$

and

$$s_1(t) = 0, \quad -\infty < t < \infty.$$

Assume that the threshold is minimax for the nominal sampling time.

Solution.   This exercise will test your understanding of the material in Sections 5.2 and 5.3 as well as the present section. Equation (5.84d), upon which the solution to Exercise 5-4 is based, does not apply to Exercise 5-5. Why not? Well, consider the fact that the minimax threshold for the nominal sampling time is $\gamma_m = \hat{s}_0(T)/2$, but the minimax threshold for sampling time T_a is

$$\gamma_m(T_a) = \hat{s}_0(T_a)/2.$$

Thus, even if the threshold is a minimax threshold for the nominal sampling time, it may not be a minimax threshold for the actual sampling time. (In fact, it is not minimax for this exercise.) Thus, we must go back to Section 5.2 to find applicable results. In particular, we can apply (5.66) to show that the error probabilities $P_{e,0}$ and $P_{e,1}$, as a function of the threshold and actual sampling time, are given by

$$P_{e,i} = Q[(-1)^i \{\mu_i(T_a) - \gamma\}/\sigma].$$

For the filter

$$
\begin{aligned}
h(t) &= [s_0(T - t) - s_1(T - t)]/2, \\
     &= \tfrac{1}{2} p_T(t),
\end{aligned}
$$

the minimax threshold for the nominal sampling time is

$$\gamma = [\mu_0(T) + \mu_1(T)]/2 = \hat{s}_0(T)/2 = T/4.$$

By contrast, the minimax threshold for the actual sampling time T_a is

$$\gamma(T_a) = [\mu_0(T_a) + \mu_1(T_a)]/2 = \hat{s}_0(T_a)/2 = \{T - |T - T_a|\}/4$$

for $0 \le T_a \le 2T$, and these are equal only if $T_a = T$. It follows that

$$P_{e,0} = Q[\{\mu_0(T_a) - \gamma\}/\sigma]$$

and

$$P_{e,1} = Q[\{\gamma - \mu_1(T_a)\}/\sigma],$$

where

$$\sigma^2 = \tfrac{1}{2} N_0 \int_{-\infty}^{\infty} h^2(t) \, dt = N_0 T/8.$$

Now $\mu_1(T_a) = 0$ for all values of T_a, so

$$P_{e,1} = Q(\gamma/\sigma) = Q\left(\sqrt{T/2N_0}\right),$$

which does not depend on T_a. However,

$$\mu_0(T_a) = \hat{s}_0(T_a) = \{T - |T - T_a|\}/2,$$

so that

$$
\begin{aligned}
P_{e,0} &= Q\left[\{(T - 2|T - T_a|)^2/2N_0 T\}^{1/2}\right] \\
&= Q\left[\sqrt{T/2N_0}\,\{1 - 2\,|1 - (T_a/T)\,|\}\right],
\end{aligned}
$$

which does depend on T_a. In fact, for sufficiently large sampling error (i.e., $|T - T_a| > T/2$) this error probability exceeds $1/2$. This can be predicted by a comparison of $\hat{s}_0(T_a)$ and γ, namely,

$$\hat{s}_0(T_a) = \{T - |T - T_a|\}/2 < \gamma = T/4$$

for $|T - T_a| > T/2$. Thus, both signals s_0 and s_1 give outputs that fall below the threshold γ if the sampling error is larger than $T/2$. Notice also that the "effective signal-to-noise ratio" (i.e., the argument of Q) for evaluating $P_{e,0}$ decreases linearly in the sampling error $T_e = |T_a - T|$, as in Exercise 5-4. Now, however, the slope is twice that in Exercise 5-4. ∎

5.5.3   Frequency-Domain Interpretation of the Matched Filter

There are two features of the frequency-domain analysis of the filter that are explored in this section. First, we examine the transfer function of the matched filter and discuss its relationship to the Fourier transforms of the signals. Second, we demonstrate that the optimum transfer function (i.e., that of the matched filter) can be derived from a frequency-domain analysis using the same methods as in the time-domain analysis of Section 5.5.2.

Recall that if we define

$$s(t) = [s_0(t) - s_1(t)]/2,$$

then the matched filter has impulse response

$$h(t) = h_M(t) = 2\,c\,s(T_0 - t), \tag{5.94a}$$

where T_0 is the sampling time. Let V_i be the Fourier transform of the signal s_i; that is,

$$V_i(\omega) = \int_{-\infty}^{\infty} s_i(t)\,\mathrm{e}^{-j\omega t}\,dt.$$

The Fourier transform of the signal s is then given by

$$V(\omega) = [V_0(\omega) - V_1(\omega)]/2.$$

Now let H_M be the transfer function of the matched filter. We know that

$$H_M(\omega) = 2\,c \int_{-\infty}^{\infty} s(T_0 - t)\, e^{-j\omega t}\, dt.$$

A change of variable ($u = T_0 - t$) gives

$$\begin{aligned}
H_M(\omega) &= 2\,c \int_{-\infty}^{\infty} s(u)\, e^{-j\omega(T_0 - u)}\, du \\
&= 2\,c\, e^{-j\omega T_0} \int_{-\infty}^{\infty} s(u)\, e^{j\omega u}\, du.
\end{aligned}$$

Next, observe that because

$$e^{j\omega u} = [e^{-j\omega u}]^*,$$

where z^* denotes the complex conjugate of the complex number z, then

$$\begin{aligned}
H_M(\omega) &= 2\,c\, e^{-j\omega T_0} \int_{-\infty}^{\infty} s(u)[e^{-j\omega u}]^*\, du \\
&= 2\,c\, e^{-j\omega T_0} \left[\int_{-\infty}^{\infty} s(u)\, e^{-j\omega u}\, du \right]^*.
\end{aligned}$$

The last step follows from the fact that $s(u)$ is real for all u. Consequently,

$$\begin{aligned}
H_M(\omega) &= 2\,c\, e^{-j\omega T_0}\, V^*(\omega) \\
&= c\,[V_0^*(\omega) - V_1^*(\omega)]\, e^{-j\omega T_0}.
\end{aligned}$$

For $c = 1/2$, the transfer function for the matched filter is given by

$$\begin{aligned}
H_M(\omega) &= V^*(\omega)\, e^{-j\omega T_0} \\
&= \tfrac{1}{2}[V_0^*(\omega) - V_1^*(\omega)]\, e^{-j\omega T_0}.
\end{aligned} \tag{5.94b}$$

In particular, notice that the amplitude response for the matched filter is

$$|H_M(\omega)| = |V(\omega)| = \tfrac{1}{2}|V_0(\omega) - V_1(\omega)|.$$

The phase response for the matched filter is the phase of $V_0^*(\omega) - V_1^*(\omega)$ plus a linear term. The only effect of a change in the sampling time T_0 is a corresponding change in the slope of the linear term. In summary, the amplitude response of the matched filter depends only on the magnitude of the Fourier transform of the signal s, and the value of the sampling time affects the phase response only.

The preceding conclusion regarding the amplitude response $|H_M(\omega)|$ of the matched filter is reasonable from an intuitive viewpoint. The filter gain is large over the range of frequencies where the signal energy is large, and the gain is small for frequencies with little signal energy. In particular, if $V(\omega) = 0$, then $H_M(\omega) = 0$; that is, at frequencies for which there is no signal energy, the matched filter gain is zero. Having nonzero gain at such frequencies would increase the noise variance at the filter output without increasing the signal.

Next, we consider the derivation of the matched filter. As before, the goal is to choose the filter in a way that maximizes

$$\mathrm{SNR} = [\mu_0(T_0) - \mu_1(T_0)]/2\sigma.$$

The quantities $\mu_0(T_0)$ and $\mu_1(T_0)$ are just the values of the output signals at the sampling time, and these output signals can be expressed in terms of the transfer function for the filter and the Fourier transforms of the signals. Because

$$\mu_i(T_0) = (s_i * h)(T_0),$$

and because convolution in the time domain corresponds to multiplication in the frequency domain, we can express $\mu_i(T_0)$ as an inverse Fourier transform:

$$\mu_i(T_0) = \frac{1}{2\pi} \int_{-\infty}^{\infty} V_i(\omega) H(\omega) e^{j\omega T_0} \, d\omega.$$

It follows that

$$\mu_0(T_0) - \mu_1(T_0) = \frac{1}{2\pi} \int_{-\infty}^{\infty} 2V(\omega) H(\omega) e^{j\omega T_0} \, d\omega.$$

Recall that

$$\sigma^2 = N_0 \int_{-\infty}^{\infty} |H(\omega)|^2 \, d\omega / 4\pi.$$

It follows that

$$\mathrm{SNR} = \frac{\int_{-\infty}^{\infty} V(\omega) H(\omega) e^{j\omega T_0} \, d\omega}{\left\{ \pi N_0 \int_{-\infty}^{\infty} |H(\omega)|^2 \, d\omega \right\}^{1/2}}. \tag{5.95}$$

In order to proceed further, we require the Schwarz inequality for complex-valued functions. First, however, we need the notions of norm and inner product for such functions. If f is a complex-valued function of a real variable, the norm of f is

$$\|f\| = \left\{ \int_{-\infty}^{\infty} |f(u)|^2 \, du \right\}^{1/2}.$$

and the inner product for two such functions f and g is

$$(f, g) = \int_{-\infty}^{\infty} f(u) g^*(u) \, du.$$

Notice that $(f, f) = \|f\|^2$ and $(f, g) = (g, f)^*$. With these definitions, the inequality

$$|(f, g)|^2 \le \|f\|^2 \|g\|^2$$

is still valid and the proof is nearly the same. The inequality is proved by noting that, for any complex number c,

$$
\begin{aligned}
0 \leq \| f - c\,g \|^2 &= \int_{-\infty}^{\infty} [f(u) - c\,g(u)][f(u) - c\,g(u)]^* \, du \\
&= \int_{-\infty}^{\infty} |f(u)|^2 \, du + |c|^2 \int_{-\infty}^{\infty} |g(u)|^2 \, du \\
&\quad - c^* \int_{-\infty}^{\infty} f(u)\,g^*(u)\, du - c \int_{-\infty}^{\infty} f^*(u)\,g(u)\, du \\
&= \| f \|^2 + |c|^2 \| g \|^2 - 2\,\mathrm{Re}\{c^*(f, g)\}.
\end{aligned}
$$

In particular, if we choose $c = (f, g)/\| g \|^2$, assuming $\| g \| \neq 0$ as before, then the preceding inequality becomes

$$
\begin{aligned}
0 &\leq \| f \|^2 + \{|(f, g)|^2/\| g \|^2\} - 2\{|(f, g)|^2/\| g \|^2\} \\
&= \| f \|^2 - \{|(f, g)|^2/\| g \|^2\},
\end{aligned}
$$

from which it follows that $|(f, g)|^2 \leq \| f \|^2 \| g \|^2$. Clearly, we get equality if there exists a complex number λ such that $f(u) = \lambda\,g(u)$ for almost all u. In this case $\| f \|^2 = |\lambda|^2 \| g \|^2$ and $(f, g) = \lambda \| g \|^2$, so

$$
|(f, g)|^2 = |\lambda|^2 \| g \|^4 = \| f \|^2 \| g \|^2.
$$

Applying the Schwarz inequality for complex-valued functions to (5.95), we let

$$
g(\omega) = [V(\omega)\,e^{j\omega T_0}]^* = V^*(\omega)\,e^{-j\omega T_0}
$$

and

$$
f(\omega) = H(\omega),
$$

so that (5.95) is equivalent to

$$
\begin{aligned}
\mathrm{SNR} &= \frac{\int_{-\infty}^{\infty} f(\omega)\,g^*(\omega)\, d\omega}{\{\pi N_0 \int_{-\infty}^{\infty} |f(\omega)|^2 \, d\omega\}^{1/2}} \\
&= (f, g)/\sqrt{\pi N_0}\, \| f \|.
\end{aligned}
$$

Notice that (f, g) must be real in this application, because SNR and $\| f \|$ are both real numbers. The Schwarz inequality gives

$$
\begin{aligned}
\mathrm{SNR} &\leq \| g \| / \sqrt{\pi N_0} \\
&= \left\{ \int_{-\infty}^{\infty} |V(\omega)|^2 \, d\omega / \pi N_0 \right\}^{1/2},
\end{aligned}
$$

with equality if, for some complex number λ, $f(\omega) = \lambda g(\omega)$ for almost all ω. But this is just the condition

$$
H(\omega) = \lambda V^*(\omega)\,e^{-j\omega T_0},
$$

which, for $\lambda = 1$, is just (5.94b).

Note that, for our application, λ must be a real number, because if

$$H(\omega) = \lambda V^*(\omega)\, e^{-j\omega T_0},$$

then

$$\int_{-\infty}^{\infty} V(\omega)\, H(\omega)\, e^{j\omega T_0}\, d\omega = \lambda \int_{-\infty}^{\infty} |V(\omega)|^2\, d\omega.$$

But this quantity must be real according to (5.95), because SNR is real.

The upper bound on SNR is the quantity

$$\mathrm{SNR}_M = \left\{ \int_{-\infty}^{\infty} |V(\omega)|^2\, d\omega / \pi N_0 \right\}^{1/2}.$$

From Parseval's identity, we know that

$$\frac{1}{2\pi} \int_{-\infty}^{\infty} |V(\omega)|^2\, d\omega = \int_{-\infty}^{\infty} s^2(t)\, dt,$$

and the integral on the right-hand side is shown in Section 5.5.2 to be the quantity $\frac{1}{2}(\bar{\mathcal{E}} - \rho) = \frac{1}{2}\bar{\mathcal{E}}(1 - r)$. As a result, we see that the signal-to-noise ratio for a receiver with the matched filter is

$$\mathrm{SNR} = \left\{ \bar{\mathcal{E}}(1 - r)/N_0 \right\}^{1/2},$$

as obtained in (5.92).

5.5.4   Alternative Implementations for the Matched Filter

Presumably, the matched filter can be implemented by constructing a passive filter that has impulse response $h_M(t)$ or, equivalently, transfer function $H_M(\omega)$, as specified in (5.94). In practice, however, this may not always be the best approach, and there are two alternatives. First, a reasonable approximation to the matched filter impulse response (or transfer function) often suffices, and an appropriately selected approximation can be considerably easier to implement. The performance of some suboptimum filters is considered in Chapter 8.

Here, we consider a second alternative, which is to use a different filter structure that produces the same decision statistic as the matched filter. The resulting performance is identical to that of the matched filter. Hence, the alternative filter structures considered in this section are optimum in the sense defined at the beginning of Section 5.5.

As before, denote the input to the filter by $Y(t)$. For the matched filter, the output is

$$Z(t) = \int_{-\infty}^{\infty} Y(u)\, h_M(t - u)\, du.$$

We are only interested in this output at the sampling time T_0, because $Z(T_0)$ is the statistic used to make the decision. The decision statistic is given by

$$Z(T_0) = \int_{-\infty}^{\infty} Y(u)\, h_M(T_0 - u)\, du.$$

Recall that the matched filter impulse response is

$$h_M(t) = 2\,c\,s(T_0 - t).$$

so that

$$h_M(T_0 - u) = 2\,c\,s[T_0 - (T_0 - u)] = 2\,c\,s(u).$$

Therefore, the decision statistic can be expressed as

$$Z(T_0) = 2\,c \int_{-\infty}^{\infty} Y(u)\,s(u)\,du. \qquad (5.96a)$$

The constant c is arbitrary, as before, so we let $c = 1/2$ or $c = 1$, whichever is more convenient, in what follows. The final observation is that the right-hand side of (5.96a) does not depend on T_0. Therefore, we denote the decision statistic by Z, and we write

$$Z = \int_{-\infty}^{\infty} Y(u)\,s(u)\,du. \qquad (5.96b)$$

which corresponds to $c = 1/2$.

Filters based on (5.96) are referred to as *correlators* or *correlation receivers*. Such a filter correlates the filter input with a replica of the signal $s(t)$. All that is required for implementation is a multiplier, an integrator, and the replica of the signal. The basic correlation receiver is shown in Figure 5-16(a). Notice in particular that there is no sampler.

For the version shown in Figure 5-16(a), the integration time is infinite, which, of course, is not practical. However, if the signal $s(t)$ has finite duration, there is no reason to use an infinite integration interval. In particular, if $s(t) = 0$ for $t < 0$ and for $t > T$, then

$$Z = \int_{-\infty}^{\infty} Y(t)\,s(t)\,dt = \int_0^T Y(t)\,s(t)\,dt.$$

so we can use the version shown in Figure 5-16(b).

Finally, recall that

$$s(t) = [s_0(t) - s_1(t)]/2.$$

so that for $c = 1$, we have

$$Z = \int_{-\infty}^{\infty} Y(t)\,s_0(t)\,dt - \int_{-\infty}^{\infty} Y(t)\,s_1(t)\,dt.$$

The decision statistic can be obtained by subtracting the outputs of two correlators as shown in Figure 5-16(c). This version requires an additional integrator, however, so it would rarely be used in an actual implementation. But it does suggest a nice interpretation of the optimum filter in a receiver with threshold $\gamma = 0$. For this threshold, the operation that consists of subtracting the two outputs and basing the decision on $Z > 0$ vs. $Z \leq 0$ (shown in Figure 5-17(a)) is equivalent to comparing the two outputs and basing the decision on the larger of the two. That is, we decide 0 was sent if $W_0 > W_1$ and 1 was sent if $W_1 \geq W_0$. The interpretation is that the decision is in favor of the signal which has the higher degree of correlation with the receiver input $Y(t)$. The receiver based on this latter approach is shown in Figure 5-17(b).

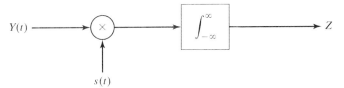

(a) Correlation receiver with infinite integration time

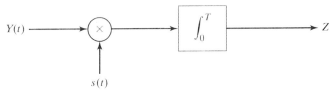

(b) Correlation receiver with finite integration time for signals time limited to $[0, T]$

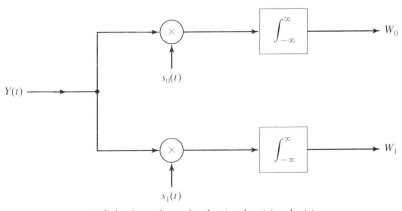

(c) Pair of correlators for the signals $s_0(t)$ and $s_1(t)$

Figure 5-16: Correlation receivers for binary signals $s_0(t)$ and $s_1(t)$ with $s(t) = [s_0(t) - s_1(t)]/2$.

5.5.5 Optimum Thresholds for Receivers with Matched Filters

In this section, we examine the expressions for the minimax and Bayes thresholds for systems in which the matched filter is employed. We begin with the minimax threshold.

Recall that the minimax threshold is given by

$$\gamma_m = [\mu_0(T_0) + \mu_1(T_0)]/2$$
$$= [\hat{s}_0(T_0) + \hat{s}_1(T_0)]/2.$$

where $\mu_i = \hat{s}_i = s_i * h$. This expression for γ_m can be written as

$$\gamma_m = \int_{-\infty}^{\infty} [s_0(u) + s_1(u)] h(T_0 - u) \, du/2.$$

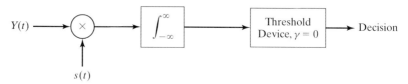

(a) Single correlator with threshold device

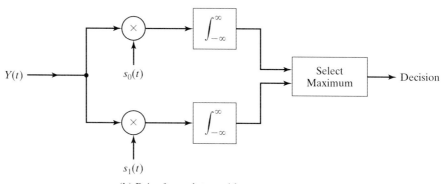

(b) Pair of correlators with comparator

Figure 5-17: Correlation receivers in systems with threshold equal to zero.

But the matched filter has impulse response

$$h(t) = c\,[s_0(T_0 - t) - s_1(T_0 - t)],$$

and thus

$$h(T_0 - u) = c\,[s_0(u) - s_1(u)].$$

It follows that

$$
\begin{aligned}
\gamma_m &= c \int_{-\infty}^{\infty} [s_0(u) + s_1(u)][s_0(u) - s_1(u)]\,du/2 \\
&= c \int_{-\infty}^{\infty} \left\{ [s_0(u)]^2 - [s_1(u)]^2 \right\}\,du/2 \\
&= c\,(\mathcal{E}_0 - \mathcal{E}_1)/2.
\end{aligned}
\tag{5.97}
$$

Recall that the constant c is arbitrary. If we choose $c = 1$, then

$$h(t) = s_0(T_0 - t) - s_1(T_0 - t)$$

and

$$\gamma_m = (\mathcal{E}_0 - \mathcal{E}_1)/2.$$

These results show that for a receiver that uses the matched filter, the minimax threshold depends only on the signal energies $\mathcal{E}_0$ and $\mathcal{E}_1$ and the filter gain constant.

Previously, we observed that, for any choice of filter, $\gamma_m = 0$ for antipodal signals. Now we see that *if the matched filter is used, $\gamma_m = 0$ for all equal-energy signal sets* (those for which $\mathcal{E}_0 = \mathcal{E}_1$). All antipodal signal sets are equal-energy signal sets, but, of course, equal-energy signals need not be antipodal.

Next, we consider the Bayes threshold. The general expression is given in (5.78b) in terms of the minimax threshold γ_m. Based on the preceding results for γ_m, we can rewrite (5.78b) as

$$\bar{\gamma} = \frac{c}{2}\{\mathcal{E}_0 - \mathcal{E}_1\} + \frac{\sigma^2 \ln(\pi_1/\pi_0)}{[\mu_0(T_0) - \mu_1(T_0)]}.$$

But, because the matched filter is employed,

$$\sigma^2 = \tfrac{1}{2}N_0 \int_{-\infty}^{\infty} c^2[s_0(T_0 - t) - s_1(T_0 - t)]^2 \, dt$$
$$= c^2 N_0(\mathcal{E}_0 + \mathcal{E}_1 - 2\rho)/2$$
$$= c^2 N_0 \bar{\mathcal{E}}(1 - r)$$

and

$$\mu_0(T_0) - \mu_1(T_0) = c \int_{-\infty}^{\infty} [s_0(u) - s_1(u)][s_0(u) - s_1(u)] \, du$$
$$= c\,(\mathcal{E}_0 + \mathcal{E}_1 - 2\rho)$$
$$= 2c\bar{\mathcal{E}}(1 - r).$$

It follows that the Bayes threshold for a receiver with the matched filter is given by

$$\bar{\gamma} = \frac{c}{2}(\mathcal{E}_0 - \mathcal{E}_1) + \frac{c}{2}N_0 \ln(\pi_1/\pi_0)$$
$$= c\,[\mathcal{E}_0 - \mathcal{E}_1 + N_0 \ln(\pi_1/\pi_0)]/2. \tag{5.98}$$

Note that this can also be written as

$$\bar{\gamma} = \gamma_m + \tfrac{1}{2}N_0 c \ln(\pi_1/\pi_0).$$

We have already observed that $\gamma_m = 0$ for a communication system with equal-energy signals and the matched filter. From the foregoing expression, we see that for such signals $\bar{\gamma} = \tfrac{1}{2}N_0 c \ln(\pi_1/\pi_0)$. Notice that even if equal-energy signals are employed, the Bayes threshold depends on the filter gain constant. In the special case of signals that are equally probable ($\pi_0 = \pi_1$), the Bayes threshold is equal to the minimax threshold even if the matched filter is not used. For equal-energy signals that are equally probable, $\bar{\gamma} = 0$ if the matched filter is used.

5.5.6 Signal Design

Consider a binary baseband communication system with an additive white Gaussian noise channel. Suppose that the matched filter is employed in the receiver. We have seen in Section 5.5 (see especially (5.83)–(5.88)) that the resulting error probability is

$$P_e = Q(\text{SNR}), \tag{5.99a}$$

where, as we showed in Section 5.5.2, the signal-to-noise ratio is given by

$$\text{SNR} = [\bar{\mathcal{E}}(1-r)/N_0]^{1/2}, \tag{5.99b}$$

provided that the matched filter is used. The probability P_e is the minimax error probability, and it is also the Bayes error probability for $\pi_0 = \pi_1$. Observe that the error probability P_e decreases as SNR increases. The more general expression for the Bayes error probability is given by (5.79a), which is also a decreasing function of SNR. (See Exercise 5-2.)

The results in (5.99) are valid for any binary signal set $\{s_0, s_1\}$. We now wish to examine the signal design problem, which is to select the signal set to maximize SNR. Because the error probability is a decreasing function of SNR, this signal set gives the smallest probability of error achievable by any binary signal set when employed in a communication system with an additive white Gaussian noise channel, matched filter, and optimum threshold.

It should be emphasized that we are not considering a fixed filter and threshold as we optimize the signal set. As the signal set is varied in the optimization process, the filter impulse response must change according to

$$h(t) = c\,[s_0(T_0 - t) - s_1(T_0 - t)],$$

and the threshold must change according to

$$\gamma_m = c\,(\mathcal{E}_0 - \mathcal{E}_1)/2$$

or

$$\bar{\gamma} = c\,[\mathcal{E}_0 - \mathcal{E}_1 + N_0 \ln(\pi_1/\pi_0)]/2$$

for the minimax and Bayes thresholds, respectively.

Returning to the task at hand, we wish to maximize SNR over the class of binary signal sets. This is a constrained maximization, however, because we have to account for the cost of using a particular signal set. The primary component of this cost is the transmitter power required, and this is constrained by constraining the energy used by the signal set. The approach that we follow is to maximize SNR over the class of all binary signal sets that have average energy $\bar{\mathcal{E}}$, and the only restrictions on $\bar{\mathcal{E}}$ are that it be fixed throughout the discussion and that it be finite. Recall that $\bar{\mathcal{E}} = (\mathcal{E}_0 + \mathcal{E}_1)/2$, so $\bar{\mathcal{E}} < \infty$ is just the condition that s_0 and s_1 have finite energy.

Maximizing SNR is equivalent to maximizing SNR^2, which is given by

$$\text{SNR}^2 = \bar{\mathcal{E}}(1-r)/N_0.$$

But $\bar{\mathcal{E}}$ is fixed and N_0 does not depend on the signal set, so maximizing SNR subject to the constraint that the average energy is $\bar{\mathcal{E}}$ is equivalent to maximizing $1 - r$ subject to this constraint. It follows that we should minimize

$$r = \rho/\bar{\mathcal{E}} = \int_{-\infty}^{\infty} s_0(u)\, s_1(u)\, du/\bar{\mathcal{E}}.$$

Again using the fact that $\bar{\mathcal{E}}$ is fixed, we see that the goal is to minimize

$$\rho = (s_0, s_1) = \int_{-\infty}^{\infty} s_0(u)\, s_1(u)\, du.$$

The minimization or maximization of an inner product can be accomplished by an application of the Schwarz inequality. The signals are assumed to have finite energy, so we may apply Theorem 5-2 with $f = s_0$ and $g = s_1$ to conclude that

$$\left\{ \int_{-\infty}^{\infty} s_0(u)\, s_1(u)\, du \right\}^2 \leq \int_{-\infty}^{\infty} [s_0(u)]^2\, du \int_{-\infty}^{\infty} [s_1(u)]^2\, du,$$

which can also be written as

$$(s_0, s_1)^2 \leq \|s_0\|^2 \|s_1\|^2.$$

But this is equivalent to

$$-\|s_0\| \|s_1\| \leq (s_0, s_1) \leq +\|s_0\| \|s_1\|,$$

from which we deduce that ρ can be no smaller than $-\|s_0\| \|s_1\|$.

Next, recall that

$$\|s_i\|^2 = \int_{-\infty}^{\infty} [s_i(u)]^2\, du = \mathcal{E}_i,$$

so that we can bound ρ by

$$-\sqrt{\mathcal{E}_0 \mathcal{E}_1} \leq \rho \leq \sqrt{\mathcal{E}_0 \mathcal{E}_1}.$$

Clearly, $\mathcal{E}_0 = 0$ or $\mathcal{E}_1 = 0$ does not give the smallest value of the bound $-\sqrt{\mathcal{E}_0 \mathcal{E}_1}$ so we may assume that the signal energies are nonzero. In fact, $\mathcal{E}_i = 0$ for either value of i gives $\rho = 0$, whereas we know that we can make $\rho < 0$ (e.g., by using antipodal signals). Because $\mathcal{E}_1 > 0$, we can assert that

$$|\rho| = \sqrt{\mathcal{E}_0 \mathcal{E}_1}$$

if and only if there is a real number λ such that $s_0(u) = \lambda\, s_1(u)$. In particular, we see that

$$\rho \geq -\sqrt{\mathcal{E}_0 \mathcal{E}_1},$$

with equality if and only if $s_0(u) = \lambda\, s_1(u)$. But this latter condition implies that

$$\rho = (\lambda s_1, s_1) = \lambda(s_1, s_1) = \lambda \|s_1\|^2 = \lambda \mathcal{E}_1.$$

Hence, if we want $\rho = -\sqrt{\mathcal{E}_0 \mathcal{E}_1}$, which is negative, then we must have $\lambda < 0$. We conclude that

$$\rho \geq -\sqrt{\mathcal{E}_0 \mathcal{E}_1},$$

with equality if and only if $s_0(u) = \lambda\, s_1(u)$ for some negative real number λ.

How small can we make $-\sqrt{\mathcal{E}_0 \mathcal{E}_1}$ subject to the constraint that $(\mathcal{E}_0 + \mathcal{E}_1)/2 = \bar{\mathcal{E}}$, which is fixed? Because $\mathcal{E}_0 = 2\bar{\mathcal{E}} - \mathcal{E}_1$, we have

$$-\sqrt{\mathcal{E}_0 \mathcal{E}_1} = -\sqrt{(2\bar{\mathcal{E}} - \mathcal{E}_1)\mathcal{E}_1}.$$

Minimization of $-\sqrt{\mathcal{E}_0 \mathcal{E}_1}$ is therefore equivalent to maximization of $(2\bar{\mathcal{E}} - \mathcal{E}_1)\mathcal{E}_1$. If it is not already clear to the reader that the maximum is achieved when

$$\mathcal{E}_1 = \bar{\mathcal{E}}.$$

then the next exercise should take care of any doubt. The result of the exercise can be applied by letting $x = \mathcal{E}_1$ and $c = \bar{\mathcal{E}}$.

Exercise 5-6
Suppose that $0 \leq x \leq 2c$ and

$$f(x) = (2c - x)x.$$

Show that the maximum value of $f(x)$ for x in this range is c^2 and that it occurs at $x = c$.

Solution.   The first solution requires the reader to be aware of the following fact for $x > 0$ and $y > 0$: If $x + y$ is held constant, the product xy is maximized when $x = y$. This fact is applied by letting $y = 2c - x$ and observing that $x = y$ is equivalent to $x = c$ and $f(c) = c^2$.

The second solution (for readers not previously aware of the preceding fact) is to observe that $f'(x) = 2c - 2x$ and $f''(x) = -2$. It follows that $f'(x) = 0$ gives a maximum, and $f'(x) = 0$ implies that $x = c$.   ■

Our conclusion is that $-\sqrt{\mathcal{E}_0 \mathcal{E}_1}$ is minimized subject to the constraint $(\mathcal{E}_0 + \mathcal{E}_1)/2 = \bar{\mathcal{E}}$ by $\mathcal{E}_1 = \bar{\mathcal{E}}$. But $\mathcal{E}_1 = \bar{\mathcal{E}}$ implies that $\mathcal{E}_0 = \mathcal{E}_1$, so we have

$$\rho \geq -\sqrt{\mathcal{E}_0 \mathcal{E}_1} \geq -\bar{\mathcal{E}}.$$

Equality occurs in the left-hand inequality if and only if $s_0(t) = \lambda s_1(t)$ for some negative value of λ, and equality occurs in the right-hand inequality if and only if $\mathcal{E}_0 = \mathcal{E}_1$. Thus, $\rho \geq -\bar{\mathcal{E}}$ with equality if and only if both conditions are met. But $s_0(t) = \lambda s_1(t)$ implies that

$$\int_{-\infty}^{\infty} [s_0(t)]^2 \, dt = \lambda^2 \int_{-\infty}^{\infty} [s_1(t)]^2 \, dt;$$

that is,

$$\mathcal{E}_0 = \lambda^2 \mathcal{E}_1.$$

This, together with the condition $\mathcal{E}_0 = \mathcal{E}_1$, implies that $\lambda^2 = 1$. Because λ must be negative, we conclude that $\lambda = -1$ and, therefore,

$$\rho \geq -\bar{\mathcal{E}}$$

with equality if and only if $s_0(t) = -s_1(t)$ for almost all t. Thus, *antipodal signals give the smallest possible value of ρ.*

Recall that $r = \rho/\bar{\mathcal{E}}$, so the preceding result can be stated as $r \geq -1$ with equality if and only if the signals are antipodal. By comparison, note that orthogonal signals have

$$\int_{-\infty}^{\infty} s_0(t)\, s_1(t)\, dt = 0;$$

hence, $r = 0$ for orthogonal signals. It is easy to establish that the largest possible value of r is $+1$ and that this occurs if and only if $s_0(t) = s_1(t)$: The two signals are the same. The latter signal set obviously does not give a useful communication system.

Returning to our expression for the signal-to-noise ratio, we see that the preceding results imply that

$$\text{SNR} = \sqrt{\bar{\mathcal{E}}(1 - r)/N_0} \leq \sqrt{2\,\bar{\mathcal{E}}/N_0},$$

with equality if and only if the signals are antipodal. We have shown that if the matched filter is employed, antipodal signals give the largest possible value for the signal-to-noise ratio. Because the minimax and Bayes error probabilities are decreasing functions of SNR, such signals also give the smallest possible error probability if either the minimax or Bayes threshold is used. Note that the shape of the signal waveforms is unimportant. (The matched filter compensates for the shape.) The only requirement is that one signal be the negative of the other.

If the minimax threshold is used, the resulting error probability for antipodal signals and a matched-filter receiver is

$$P_c = Q\left(\sqrt{2\,\mathcal{E}/N_0}\right).$$

Because $\mathcal{E}_0 = \mathcal{E}_1 = \bar{\mathcal{E}}$ for antipodal signals, we can omit the subscripts on $\mathcal{E}$. (All energies of interest are the same.) Recall that the minimax threshold is $\gamma_m = 0$ in this case. (See (5.97).) Recall also that values for $Q\left(\sqrt{2\,\mathcal{E}/N_0}\right)$ are given in Table 5.1 and Figure 5-3. We now see that these are the best possible values for the error probability P_c for any receiver of the form shown in Figure 5-4 if the noise is white Gaussian noise with spectral density $N_0/2$ and the signals are constrained to have average energy not exceeding $\mathcal{E}$.

5.6   The Optimum Filter for the AGN Channel

Rather than consider the most general additive Gaussian noise (AGN) channel, we limit our discussion to the most practical model for nonwhite noise. Receivers always generate some thermal noise, and this is accurately modeled as white Gaussian noise for frequencies in the range of interest for the systems under consideration in this book (well below 10^{13} Hz). Consequently, it is reasonable to assume that the channel noise spectral density is nonzero for all frequencies. This is always true if the channel noise has a white-noise component. In this case the channel noise spectral density can be written as $S_X(\omega) = S_N(\omega) + (N_0/2)$, where $S_N(\omega)$ is an arbitrary spectral density and $N_0/2$ is the spectral density of the white-noise component. The weaker requirement $S_X(\omega) > 0$

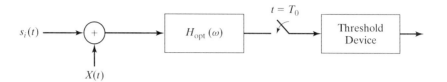

Figure 5-18: Optimum receiver for the general AGN channel.

for $-\infty < \omega < \infty$ is all that is needed in this section. We wish to find the optimum filter for the receiver shown in Figure 5-18 (optimum in the sense of Section 5.5) for spectral densities that are not necessarily constant (i.e., a general AGN channel).

The solution of the problem of finding the optimum filter for the AGN channel is based on the fact that we know how to find the matched filter for the AWGN channel. If we whiten the noise, then we can use the filter derived in the previous section. Toward this end, suppose we pass the received signal plus noise through a time-invariant linear filter as illustrated in Figure 5-19(a). Suppose also that the filter has transfer function $H_W(\omega)$ and that

$$|H_W(\omega)|^2 = \frac{1}{S_X(\omega)}. \tag{5.100}$$

The output consists of a filtered version of the signal, which we denote by $u_i(t)$, plus Gaussian noise $W(t)$ that has spectral density

$$\begin{aligned} S_W(\omega) &= S_X(\omega)|H_W(\omega)|^2 \\ &= 1. \end{aligned}$$

The output noise is white Gaussian noise with spectral density 1. Notice from (5.100) that it is critical that $S_X(\omega) > 0$ for all ω in order that the whitening filter have a finite transfer function.

The problem is now an AWGN problem. Namely, we have the binary signal set $\{u_0, u_1\}$ and an AWGN channel, as illustrated in Figure 5-19(b). The optimum filter for this system is the matched filter for the signal set $\{u_0, u_1\}$, as derived in Section 5.5. Let the transfer function for the filter be denoted by $H_U(\omega)$. This filter follows the whitening filter as shown in Figure 5-19(c).

Let the impulse response corresponding to the transfer function $H_U(\omega)$ be denoted by $h_U(t)$. From the results in Section 5.5, we know that

$$h_U(t) = [u_0(T_0 - t) - u_1(T_0 - t)]/2. \tag{5.101}$$

Thus, we have now specified the optimum filter for the AGN channel; it has transfer function given by

$$H_{\text{opt}}(\omega) = H_W(\omega) H_U(\omega), \tag{5.102}$$

where $H_W(\omega)$ is defined by (5.100). The result is illustrated in Figure 5-19(c).

How do we know that the cascaded filter is optimum? Well, observe that the effects of the first filter are reversible; that is, the second filter could be used to recover the

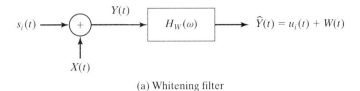

(a) Whitening filter

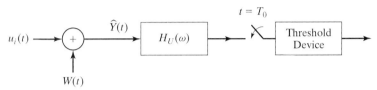

(b) Matched filter for whitened signals and noise

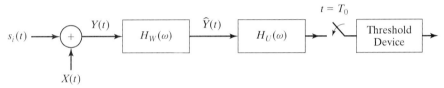

(c) Optimum receiver with cascade of whitening filter and matched filter

Figure 5-19: Optimum filter with the whitening filter for the general AGN channel.

original process $Y(t)$ completely. Therefore, we are not losing anything by passing $Y(t)$ through the whitening filter. Anything done by the whitening filter can be undone by a filter that follows it.

This argument can be made more precise as follows: Suppose $G(\omega)$ is the transfer function for the optimum linear filter for the original problem (i.e., signal set $\{s_0, s_1\}$, AGN channel with nonwhite noise). The transfer function can be written as

$$
\begin{aligned}
G(\omega) &= H_W(\omega)\{G(\omega)/H_W(\omega)\} \\
&= H_W(\omega)\{H_W^{-1}(\omega)G(\omega)\}.
\end{aligned}
$$

Thus, the systems of Figures 5-20(a) and 5-20(b) are equivalent. If $G(\omega)$ is optimum, then the filter with transfer function $H_W^{-1}(\omega)G(\omega)$ must be optimum for the problem of detecting $u_0(t)$ vs. $u_1(t)$ in the presence of AWGN. But we know that the optimum filter for this latter problem is the filter with transfer function $H_U(\omega)$, which is the filter that is matched to the signal set $\{u_0, u_1\}$. Thus, it must be that

$$
H_U(\omega) = H_W^{-1}(\omega)G(\omega).
$$

This is illustrated by comparing Figure 5-20(b) with Figure 5-19(c). It follows that

$$
G(\omega) = H_U(\omega)\,H_W(\omega) = H_{\text{opt}}(\omega).
$$

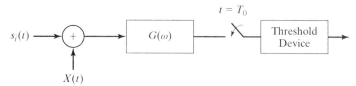

(a) Single-filter representation

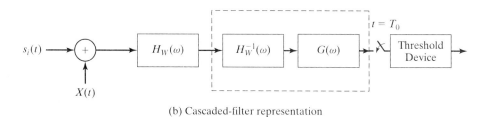

(b) Cascaded-filter representation

Figure 5-20: Two representations for the optimum receiver for the general AGN channel.

Next we return to (5.101) and recall some facts from Section 5.5.3. If U_i is the Fourier transform of u_i, the transfer function for the filter matched to $\{u_0, u_1\}$ is

$$H_U(\omega) = \tfrac{1}{2}[U_0^*(\omega) - U_1^*(\omega)]\,\mathrm{e}^{-j\omega T_0}. \qquad (5.103)$$

This is just (5.94b) of Section 5.5.3 with U_i in place of V_i and H_U in place of H_M. Recall from Section 5.5.3 that V_i is the Fourier transform of s_i.

Now consider the fact that u_i is a filtered version of s_i, and the filter in question has transfer function H_W. Therefore, if h_W is the inverse Fourier transform of H_W, then $u_i = s_i * h_W$. But this is equivalent to

$$U_i(\omega) = V_i(\omega)\,H_W(\omega),$$

which implies that

$$U_i^*(\omega) = V_i^*(\omega)\,H_W^*(\omega).$$

It follows from this result and (5.103) that

$$H_U(\omega) = \tfrac{1}{2}[V_0^*(\omega) - V_1^*(\omega)]\,H_W^*(\omega)\,\mathrm{e}^{-j\omega T_0}.$$

The next step is to apply (5.103) to (5.102), which yields

$$\begin{aligned}
H_{\mathrm{opt}}(\omega) &= \tfrac{1}{2}H_W(\omega)[V_0^*(\omega) - V_1^*(\omega)]\,H_W^*(\omega)\,\mathrm{e}^{-j\omega T_0} \\
&= \tfrac{1}{2}|H_W(\omega)|^2[V_0^*(\omega) - V_1^*(\omega)]\,\mathrm{e}^{-j\omega T_0}.
\end{aligned}$$

Finally, we use (5.100) and (5.94b) to write this as

$$H_{\mathrm{opt}}(\omega) = H_M(\omega)/S_X(\omega); \qquad (5.104)$$

that is, the optimum filter has a transfer function that is the transfer function for the matched filter (matched to the original signal set $\{s_0, s_1\}$) divided by the spectral density of the noise.

As a check on this result, observe that if $S_X(\omega)$ is constant (i.e., white noise) the optimum filter has the shape of the matched filter. Thus, (5.104) reduces to the result obtained in Section 5.5. (Recall that gain constants are unimportant.) For nonwhite noise, (5.104) has the intuitively satisfying feature that the gain of the optimum filter is small for frequencies at which the noise spectral density is large.

5.7   Signal-Space Concepts for Binary Communication on the AWGN Channel

In this section, the notion of a vector representation of signals is introduced for binary baseband data transmission over the AWGN channel. This vector representation of signals is then applied to provide an alternative interpretation of the expression for the probability of error when the matched filter is employed. The geometric interpretation developed here is extended to more complicated binary signals and to nonbinary signals in subsequent chapters.

5.7.1   Geometric Representation of Binary Signals

Consider a binary signal set $\{s_0, s_1\}$. As usual, let $\mathcal{E}_i$ denote the energy in the signal $s_i(t)$, so that $\mathcal{E}_i = \|s_i\|^2$. To avoid trivialities, assume that $\mathcal{E}_0 \neq 0$. Clearly, one of the two signals must have nonzero energy in order for the signal set to be of interest, so let s_0 be such a signal. Define the signal

$$\psi_0(t) = s_0(t)/\sqrt{\mathcal{E}_0}, \tag{5.105}$$

for $-\infty < t < \infty$, and observe that

$$\|\psi_0\|^2 = \int_{-\infty}^{\infty} [\psi_0(t)]^2 \, dt = 1.$$

Next, define

$$\beta_0 = (s_1, \psi_0) = \int_{-\infty}^{\infty} s_1(t) \, \psi_0(t) \, dt. \tag{5.106}$$

If we think of s_1 and ψ_0 as vectors, then β_0 is the length of the projection of s_1 onto ψ_0. In some sense, $\beta_0\psi_0(t)$ is that part of $s_1(t)$ that can be represented by ψ_0. The remaining part of $s_1(t)$ is given by

$$v_1(t) = s_1(t) - \beta_0\psi_0(t). \tag{5.107}$$

If it turns out that $v_1(t) = 0$ for almost all t (i.e., $\|v_1\| = 0$), then we have represented the original binary signal set in terms of the single function $\psi_0(t)$. This is true only if one of the original signals is a multiple of the other: There is some number c for which

$s_1(t) = c\, s_0(t)$ for almost all t. (The value $c = 0$ is allowed.) A pair of binary antipodal signals is one of the most important examples of a signal set that can be represented in terms of a single normalized function $\psi_0(t)$. Each of the signal sets illustrated in Figure 5-1 can be represented in terms of a single function.

If $v_1(t)$ is not equal to zero for almost all t (i.e., $\|v_1\| > 0$), then we proceed as follows: The next step is to observe that $v_1(t)$ and $\psi_0(t)$ are orthogonal. This can be demonstrated by substituting for $v_1(t)$ into the defining integral for (v_1, ψ_0), which gives

$$(v_1, \psi_0) = \int_{-\infty}^{\infty} [s_1(t) - \beta_0 \psi_0(t)]\, \psi_0(t)\, dt$$

$$= (s_1, \psi_0) - \beta_0 \|\psi_0\|^2 = (s_1, \psi_0) - \beta_0.$$

So (5.106) implies $(v_1, \psi_0) = 0$. Next define $\psi_1(t)$ to be a normalized version of $v_1(t)$; that is, let

$$\psi_1(t) = v_1(t)/\|v_1\|.$$

so that $\|\psi_1\|^2 = 1$. Define $\beta_1 = \|v_1\|$, and observe that

$$\psi_1(t) = v_1(t)/\beta_1. \tag{5.108}$$

From (5.108), we know that $\psi_1(t)$ is just a multiple of $v_1(t)$, and because $(v_1, \psi_0) = 0$, we can conclude that $(\psi_1, \psi_0) = 0$. Also observe that $(\psi_1, \psi_1) = \|\psi_1\|^2 = 1$, and (5.108) implies $(\psi_1, \psi_1) = (v_1, \psi_1)/\beta_1$, so it must be that $\beta_1 = (v_1, \psi_1)$. But, from (5.107),

$$(v_1, \psi_1) = (s_1, \psi_1) - \beta_0(\psi_0, \psi_1) = (s_1, \psi_1).$$

so

$$\beta_1 = (s_1, \psi_1).$$

Thus far, we have obtained a binary signal set $\{\psi_1, \psi_0\}$ with the property that each signal has unit energy (normalized) and the two signals are orthogonal. We refer to such a signal set as an *orthonormal signal set*. Our two original signals $s_0(t)$ and $s_1(t)$ can be expressed in terms of this orthonormal signal set as follows: From (5.105), we have

$$s_0(t) = \alpha_0 \psi_0(t). \tag{5.109}$$

where $\alpha_0 = \|s_0\| = \sqrt{\mathcal{E}_0}$. Note also that $\alpha_0 = (s_0, \psi_0)$. From (5.107), it is clear that

$$s_1(t) = v_1(t) + \beta_0 \psi_0(t).$$

and (5.108) implies that $v_1(t) = \beta_1 \psi_1(t)$, so that

$$s_1(t) = \beta_0 \psi_0(t) + \beta_1 \psi_1(t). \tag{5.110}$$

The coefficients β_0 and β_1 that appear in (5.110) were shown earlier to satisfy $\beta_i = (s_1, \psi_i)$ for $i = 0$ and $i = 1$. Equations (5.109) and (5.110) display the signals as linear

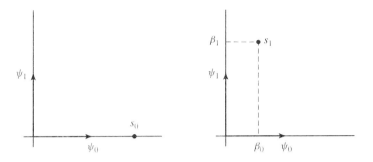

Figure 5-21: Geometric representation of binary signals s_0 and s_1.

combinations of the orthonormal functions ψ_0 and ψ_1. We can think of ψ_0 and ψ_1 as orthogonal vectors of unit length and then illustrate the signals s_0 and s_1 as points in the two-dimensional space spanned by ψ_0 and ψ_1, as shown in Figure 5-21.

This development provides a formal procedure that allows us to begin with a signal set $\{s_0, s_1\}$ and obtain a pair of orthonormal signals $\{\psi_0, \psi_1\}$ with the property that $s_0(t)$ and $s_1(t)$ can be written as linear combinations of $\psi_0(t)$ and $\psi_1(t)$. This procedure is a special case of the Gram–Schmidt procedure, which can be applied to any finite set of signals. For information on the application of the Gram–Schmidt procedure for nonbinary signal sets, see [5.4].

In summary, given signals $s_0(t)$ and $s_1(t)$, we can let

$$\psi_0(t) = s_0(t)/\sqrt{\mathcal{E}_0}$$

and

$$v_1(t) = [s_1(t) - \beta_0 \psi_0(t)].$$

The signal $v_1(t)$ is then normalized to give $\psi_1(t)$. If $\alpha_0 = (s_0, \psi_0)$, $\beta_0 = (s_1, \psi_0)$, and $\beta_1 = (s_1, \psi_1)$, then

$$s_0(t) = \alpha_0 \psi_0(t)$$

and

$$s_1(t) = \beta_0 \psi_0(t) + \beta_1 \psi_1(t).$$

In general, the choice of the functions ψ_0 and ψ_1 need not be unique; in fact, if $\mathcal{E}_1 \neq 0$, we could just as well start with $s_1(t)$ and define $\psi_1(t) = s_1(t)/\sqrt{\mathcal{E}_1}$. This approach produces a different set of orthonormal functions unless the original signals are orthogonal. However, it is not necessary to use the foregoing procedure at all if a set of orthonormal functions can be obtained by other means. In many examples, the orthonormal functions ψ_0 and ψ_1 can be determined by inspection of the signals s_0 and s_1 without the need to go through the formal procedure.

Example 5-11   Orthonormal Functions for a Binary Signal Set

Suppose that the signals are given by

$$s_0(t) = A_0 \cos(2\pi t/T)\, p_T(t)$$

and

$$s_1(t) = [A_1 \cos(2\pi t/T) + B_1 \sin(2\pi t/T)]\, p_T(t).$$

It should be clear that the orthonormal functions can be defined as

$$\psi_0(t) = \sqrt{2/T}\,\cos(2\pi t/T)\, p_T(t)$$

and

$$\psi_1(t) = \sqrt{2/T}\,\sin(2\pi t/T)\, p_T(t).$$

It is easy to verify that these are orthogonal and that each has unit energy (normal). Furthermore, it is easy to see that

$$s_0(t) = \sqrt{A_0^2 T/2}\,\psi_0(t)$$

and

$$s_1(t) = \sqrt{A_1^2 T/2}\,\psi_0(t) + \sqrt{B_1^2 T/2}\,\psi_1(t).$$

Notice that if $A_1 = 0$, the original signals s_0 and s_1 are orthogonal, so the orthonormal functions are just normalized versions of s_0 and s_1. ∎

5.7.2   The Minimum Probability of Error Revisited

If the minimax threshold and matched filter are used, the probability of error is given by

$$P_e = Q(\text{SNR}), \tag{5.111a}$$

and the signal-to-noise ratio is given by

$$\text{SNR} = \{\bar{\mathcal{E}}(1 - r)/N_0\}^{1/2}. \tag{5.111b}$$

There is an alternative expression for SNR that ties in nicely with our geometric interpretation.

We begin by defining the *distance* between signals s_0 and s_1 as

$$d = \| s_0 - s_1 \|.$$

If we represent the signals as points in a two-dimensional space, then d is just the distance between these points, as illustrated in Figure 5-22. Note that this distance is independent of the choice of the orthonormal functions, as illustrated in Figure 5-23.

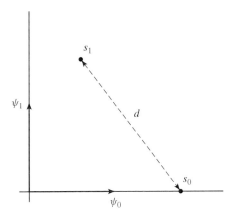

Figure 5-22: Illustration of the distance between signals s_0 and s_1.

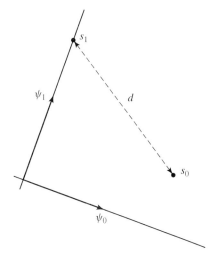

Figure 5-23: Illustration of the distance between signals s_0 and s_1 for a different choice of orthonormal functions.

From the definition of the norm, it is clear that the squared distance is given by

$$d^2 = \int_{-\infty}^{\infty} [s_0(t) - s_1(t)]^2 \, dt. \tag{5.112}$$

Expanding the integrand and integrating term by term, we find that

$$d^2 = \mathcal{E}_0 + \mathcal{E}_1 - 2(s_0, s_1).$$

In the notation of Section 5.5.2, the squared distance can be written as

$$d^2 = \mathcal{E}_0 + \mathcal{E}_1 - 2\rho = 2(\bar{\mathcal{E}} - \rho) = 2\bar{\mathcal{E}}(1 - r).$$

Therefore,

$$\text{SNR} = \{\bar{\mathcal{E}}(1-r)/N_0\}^{1/2} = \{d^2/2N_0\}^{1/2} = d/\sqrt{2N_0}.$$

This development provides an alternative expression for the probability of error for binary signaling on the AWGN channel, the minimax threshold, and the matched filter. The probability of error is given by

$$P_e = Q\left(d/\sqrt{2N_0}\right), \tag{5.113a}$$

where d is the distance between s_0 and s_1, which satisfies

$$d^2 = \int_{-\infty}^{\infty} [s_0(t) - s_1(t)]^2 \, dt. \tag{5.113b}$$

Thus, the performance depends only on the spectral density of the noise on the channel and the distance between the signals. Equations (5.113) not only tie in well with the geometric interpretation of the signals, they also offer an alternative to (5.111) as a means of calculating the probability of error. One should use either (5.111) or (5.113), whichever is easier for the problem at hand.

Example 5-12
The binary signals are as illustrated in Figure 5-24, and the channel is an AWGN channel with spectral density $N_0/2$. It is desired to find the probability of error that results when the optimum minimax receiver is employed. We can use either (5.111) or (5.113). Use of (5.111) requires evaluation of the inner product (s_0, s_1), which appears to be cumbersome. On the other hand, use of (5.113) requires only that we find the distance d between the two signals. This distance is in terms of $s_0(t) - s_1(t)$, which is easily characterized, as illustrated in Figure 5-24. By inspection, we see that $d^2 = T/3$, so that the error probability is $P_e = Q\left(\sqrt{T/6N_0}\right)$. ∎

References and Suggestions for Further Reading

[5.1] T. S. Ferguson, *Mathematical Statistics: A Decision Theoretic Approach*. New York: Academic Press, 1967.

[5.2] E. C. Jordan (ed.), *Reference Data for Engineers: Radio, Electronics, Computer, and Communications*, 7th ed. Indianapolis: Howard Sams, 1985.

[5.3] C. L. Weber, *Elements of Detection and Signal Design*. New York: McGraw-Hill, 1968 [reprinted by Springer-Verlag, New York, 1987].

[5.4] J. M. Wozencraft and I. M. Jacobs, *Principles of Communication Engineering*. New York: Wiley, 1965 [reissued by Waveland Press, 1990].

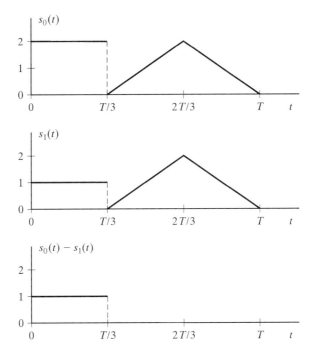

Figure 5-24: The binary signal set for Example 5-12.

Problems

5.1 Binary data must be transmitted over an AWGN channel (noise spectral density $N_0/2$) using the rectangular pulses $s_0(t) = \alpha\, p_T(t)$ and $s_1(t) = -\alpha\, p_T(t)$, where $p_T(t) = 1$ for $0 \le t < T$ and $p_T(t) = 0$ elsewhere. Assume that $\alpha > 0$. The receiver forms an output Z and decides "s_0 was sent" if $Z > 0$ and "s_1 was sent" if $Z \le 0$.

(a) Suppose that the output Z is given by

$$Z = \int_0^T g(t)\, Y(t)\, dt,$$

where $Y(t)$ is the transmitted signal plus noise and $g(t)$ is the triangular waveform

$$g(t) = \begin{cases} t, & 0 \le t < T/2, \\ T - t, & T/2 \le t < T, \\ 0, & \text{otherwise.} \end{cases}$$

Find the error probabilities $P_{e,0}$ and $P_{e,1}$.

(b) Suppose the receiver produces the output $Z = Y(T/2)$; that is, it merely samples the received signal at time $T/2$. Find the error probabilities $P_{e,0}$ and $P_{e,1}$.

5.2 Suppose a binary communications system of the type shown in Figure 5-2 employs the signal set of Figure 5-1(b). The channel is an additive Gaussian noise channel for which the noise process $X(t)$ has zero mean and autocorrelation function

$$R_X(\tau) = \begin{cases} \beta \, [(T - |\tau|)/T], & |\tau| \le T. \\ 0, & |\tau| > T. \end{cases}$$

Find $P_{e,0}$ and $P_{e,1}$ if the threshold is $\gamma = 0$.

5.3 Consider the binary communication system of Figure 5-2. (See Section 5.2.1 for a description of this system.) The signals $s_0(t)$ and $s_1(t)$ are defined by

$$s_0(t) = 4 \, p_T(t)$$

and

$$s_1(t) = 2 \, p_T(t).$$

Suppose $T = 2$ and $N_0 = 4$.

(a) Find the error probabilities $P_{e,0}$ and $P_{e,1}$ for the system with threshold $\gamma = 7$.

(b) What is the minimax threshold for this system?

(c) Find the minimax error probability.

(d) Add a squaring device to the system at the input to the threshold device. The decision statistic is now Z^2 rather than Z. Find expressions in terms of the threshold γ and the function Q for the error probabilities $P_{e,0}$ and $P_{e,1}$ for this new system. (*Warning*: Z^2 is *not* Gaussian.)

5.4 The following expression is often used to find the minimax threshold: $\gamma_m = [\hat{s}_0(T_0) + \hat{s}_1(T_0)]/2$. This expression is valid for a linear receiver and an additive Gaussian noise channel for which the noise has zero mean. Suppose now that the mean $\mu_X(t)$ of the Gaussian noise process $X(t)$ is not identically zero, and find an expression for the minimax threshold.

Hint: Show that the filter inputs are the same for the following two systems: System 1 has signals $s_0(t)$ and $s_1(t)$ and Gaussian channel noise $X(t)$ with mean $\mu_X(t)$ and autocovariance function $C_X(\tau)$. System 2 has signals $s_0'(t) = s_0(t) + \mu_X(t)$ and $s_1'(t) = s_1(t) + \mu_X(t)$ and Gaussian channel noise $W(t)$ with *zero mean* and autocorrelation function $R_W(\tau) = C_X(\tau)$. Because the two systems give equivalent inputs to the filter, the minimax threshold and resulting error probabilities for System 1 are the same as for System 2. The expression for γ_m applies to System 2 (but not to System 1).

5.5 (B. Hajek) For a zero-mean additive Gaussian noise channel, it is intuitively clear that the minimax threshold is zero for a system with an antipodal signal set and a linear receiver. Suppose the signal set is not antipodal. One approach is to "convert" it to an antipodal signal set by subtracting the average value; that is, given signals $s_0(t)$ and $s_1(t)$, form the signals $s_0'(t)$ and $s_1'(t)$ as follows: Let

$$s_a(t) = [s_0(t) + s_1(t)]/2.$$

and define $s_i'(t) = s_i(t) - s_a(t)$ for $i = 0$ and $i = 1$.

(a) Show that the new signal set $\{s_0', s_1'\}$ is antipodal.

(b) Show that both output signals are changed by the same amount by this conversion; that is, show that

$$\hat{s}_1(T_0) - \hat{s}_1'(T_0) = \hat{s}_0(T_0) - \hat{s}_0'(T_0) = \delta_s.$$

Find the value of δ_s in terms of the signal $s_a(t)$, the filter impulse response $h(t)$, and the sampling time T_0.

(c) The minimax threshold for $\{s_0', s_1'\}$ is zero, and the use of the signal set $\{s_0, s_1\}$ increases the filter outputs by an amount δ_s, regardless of whether 0 or 1 is sent. Intuitively, we therefore expect that the minimax threshold for $\{s_0, s_1\}$ is $\gamma_m = \delta_s$. Show that this agrees with our expression (5.71).

(d) Examine the similarities between the arguments used in this problem and those employed in Problem 5.4.

5.6 The received signal in a binary baseband data communications system is given by $Y(t) = s_i(t) + X(t)$, where $X(t)$ is a stationary Gaussian random process with mean 1 and autocovariance function given by

$$C_X(u) = 4\exp(-|u|).$$

The signals $s_i(t)$ are such that $s_0(t) = 3$ for $0 < t < T$ and $s_1(t) = 1$ for $0 < t < T$; for t not in the interval $(0, T)$, $s_0(t) = s_1(t) = 0$. The receiver samples the received signal at times $T/4$, $T/2$, and $3T/4$ and sums the sample values; that is, it forms the statistic Z, which is the random variable given by

$$Z = Y(T/4) + Y(T/2) + Y(3T/4).$$

(a) First suppose the threshold is 6. That is, if $Z > 6$, the receiver decides that 0 was sent; however, if $Z < 6$, it decides that 1 was sent. Find the probability of error when 0 is transmitted and the probability of error when 1 is transmitted. Express these two probabilities in terms of the function Q. For one of these, you should be able to obtain a numerical value without the aid of a table or calculator.

(b) Is 6 the optimum threshold in the minimax sense? If not, what is the optimum minimax threshold? (Be careful—the noise has a nonzero mean.)

(c) Check your answer to part (b) by comparing the two error probabilities. For your value of the threshold, is the probability of error when 0 is sent equal to the probability of error when 1 is sent? Should these two probabilities be equal? Explain why or why not.

5.7 Consider the system of Figure 5-4 and the antipodal signal set defined by $s_0(t) = A\,p_1(t)$ and $s_1(t) = -s_0(t)$. Show that the Bayes threshold depends on the signal level A unless $\pi_0 = \pi_1$. Contrast this with the minimax threshold.

5.8 For the same signal set as in Problem 5.7, find the Bayes threshold if the filter has transfer function

$$H(2\pi f) = \lambda T \exp(-j\pi f T)\,\mathrm{sinc}(fT)$$

for some gain constant $\lambda > 0$, where $\mathrm{sinc}(x) = (\pi x)^{-1}\sin(\pi x)$. Assume that the noise is white with spectral density $N_0/2$. If the gain constant λ increases by a factor of 2, by what factor does $\bar{\gamma}$ change? Note that both the mean $\mu_0(T_0)$ and the variance σ^2 depend on λ.

5.9 Consider the communication system of Figure 5-4. Suppose we replace the linear filter with impulse response $h(t)$ by a linear filter with impulse response $\beta h(t)$ for an arbitrary positive constant β. Show that the error probabilities $P_{e,0}$ and $P_{e,1}$ do not change (i.e., they do not depend on β) if either the minimax or Bayes threshold is used.

5.10 Consider the system of Figure 5-4 with the signal set $s_0(t) = a\, p_T(t)$ and $s_1(t) = 0$ for $0 \leq t \leq T$. This is on–off signaling using a rectangular pulse. Suppose that $a = 1$ and $T = 1$. Let the sampling time be $T_0 = 1$, and assume that the filter has impulse response $\beta\, p_T(t)$ for $\beta > 0$. The channel is an additive white Gaussian noise channel with spectral density $N_0/2 = 1$.

   (a) If the threshold is fixed at $\gamma = 1/2$, show that the error probabilities depend on the gain constant β.

   (b) Show that the minimax and Bayes thresholds for this system depend on the gain constant β.

   (c) Find the error probabilities for the minimax and Bayes thresholds for this system. Do these probabilities depend on the gain constant?

5.11 In Problem 5.10, assume that the signal amplitude a is arbitrary, and let the threshold be fixed at $\gamma = 1/2$.

   (a) Show that the error probability $P_{e,0}$ depends on both the amplitude a and the gain constant β.

   (b) Find the signal-to-noise ratio for this problem (i.e., evaluate (5.73a) of Section 5.3.1), and show that $P_{e,0}$ can be written in terms of SNR and β.

5.12 Find the decision regions that give the minimum average probability of error for the probabilities π_0 and π_1 if f_0 and f_1 are zero-mean Gaussian densities with *unequal* variances σ_0^2 and σ_1^2. Express Γ_0 and Γ_1 as unions of intervals.

5.13 Find the maximum-likelihood decision rule for Example 5-10 in Section 5.4.

5.14   **(a)** Using Theorem 5-1 stated in Section 5.4, prove that the following randomized decision rule is a minimax decision rule for the densities f_0 and f_1 of Example 5-9. The decision rule is to observe z and

$$\text{if } f_1(z) \; > \; 3 f_0(z), \text{ decide 1 was sent;}$$
$$\text{if } f_1(z) \; < \; 3 f_0(z), \text{ decide 0 was sent;}$$

and

$$\text{if } f_1(z) \; = \; 3 f_0(z), \text{ make a random choice.}$$

The random choice is as follows: With probability $1/4$, decide 0 was sent, and with probability $3/4$, decide 1 was sent. In order to apply the theorem you must first show the decision rule is a Bayes decision rule. (What are the values of π_0 and π_1?) You must also prove that the decision rule has the property that $P_{e,0} = P_{e,1}$.

   (b) Show that the following choice of Γ_0 and Γ_1 gives a nonrandomized minimax rule for the densities of Example 5-9:

$$\Gamma_0 \; = \; [0, 5/4] \cup [2, 3];$$
$$\Gamma_1 \; = \; (5/4, 2).$$

Is this decision rule a likelihood ratio test?

5.15 A binary baseband data transmission system uses the antipodal signal set defined by

$$s_0(t) = \begin{cases} 2At/T, & 0 \le t < T/2, \\ A(2t - T)/T, & T/2 \le t < T, \\ 0, & \text{otherwise.} \end{cases}$$

The channel is an additive white Gaussian noise channel with noise spectral density $N_0/2$. The *minimax* criterion is to be used.

(a) What is the minimum probability of error for this system? Give your answer in terms of A, T, N_0, and the function Q.

(b) Give the impulse response of the filter that achieves the minimum error probability (i.e., the matched filter). Simplify your answer as much as possible.

(c) Give the optimum sampling time and optimum (minimax) threshold for the receiver that uses the filter of part (b).

(d) Find the variance σ^2 of the output process when the input to the matched filter is a white-noise process with spectral density $N_0/2$.

(e) Suppose that the filter in the receiver is *not* the filter of part (b), but is instead a filter with impulse response

$$h(t) = p_T(t).$$

Give an expression for the output $\hat{s}_0(t)$ of this filter when the input is $s_0(t)$. Give the value of $\hat{s}_0(t)$ for *all* t in the range $-\infty < t < \infty$.

(f) For the filter of part (e), find the maximum value of $\hat{s}(t)$, where the maximization is over all t in the range $-\infty < t < \infty$. (Note: This part can be solved independently of the solution to part (e).)

5.16 Consider the binary signal set

$$s_0(t) = A \, p_T(t)$$

and

$$s_1(t) = (-4At/T) \, p_T(t).$$

Assume that these are to be used to transmit binary data over an additive white Gaussian noise channel with spectral density $N_0/2$.

(a) Find the matched filter for this signal set.

(b) Find the minimax threshold and the resulting minimax error probability.

5.17 For each signal set that follows, find the minimax error probability for binary communication via an AWGN channel (spectral density $= N_0/2$). Assume for each signal set that the receiver consists of an ideal matched filter, a sampler that samples at an optimum time, and a threshold device with a minimax threshold.

(a)

$$s_0(t) = \begin{cases} +A, & 0 \le t < \frac{1}{3}T, \\ -A, & \frac{2}{3}T \le t < T, \\ 0, & \text{otherwise} \end{cases}$$

$$s_1(t) = \begin{cases} -A, & \frac{1}{3}T \le t < \frac{2}{3}T, \\ +A, & \frac{2}{3}T \le t < T, \\ 0, & \text{otherwise} \end{cases}$$

(b)

$$s_0(t) = A \,|\cos \omega_0 t| \, p_T(t)$$
$$s_1(t) = A \,|\sin \omega_0 t| \, p_T(t)$$

(c)

$$s_0(t) = A\,(1 + \cos \omega_0 t)\, p_T(t)$$
$$s_1(t) = A\,(1 + \sin \omega_0 t)\, p_T(t)$$

In parts **(b)** and **(c)**, assume that $\omega_0 T = 2\pi n$ for some positive integer n.

5.18 Consider the communications system shown in Figure 5-4. The noise process $X(t)$ is a white Gaussian random process with spectral density $N_0/2$, and the signals $s_0(t)$ and $s_1(t)$ are given by

$$s_i(t) = (-1)^i A\, p_T(t)$$

for $i = 0$ and $i = 1$. The threshold is $\gamma = 0$ and the sampling time is $T_0 = \alpha T$ for $0 < \alpha < 2$. Investigate the effects of the sampling time by finding the error probabilities $P_{e,0}$ and $P_{e,1}$ in the following two cases:

(a) The filter is a linear time-invariant filter that is matched to the signals; that is, the impulse response is

$$h(\lambda) = s_0(T - \lambda) - s_1(T - \lambda).$$

Give $P_{e,0}$ and $P_{e,1}$ in terms of α, A, T, and N_0.

(b) The filter is an integrate-and-dump filter with output given by

$$Z(T_0) = \int_0^{T_0} Y(t)\, dt.$$

Give expressions for $P_{e,0}$ and $P_{e,1}$ in terms of α, A, T, and N_0.

(c) Express your answers to **(a)** and **(b)** in terms of α, $\mathcal{E}$, and N_0 (where $\mathcal{E}$ is the energy per pulse), and compare them.

5.19 Suppose a binary communications system with an AWGN channel uses signals of duration T given by

$$s_0(t) = -s_1(t) = \begin{cases} 10t/T, & 0 \le t < 0.1T, \\ 1, & 0.1T \le t < 0.9T, \\ 10[(T-t)/T], & 0.9T \le t < T. \end{cases}$$

The receiver consists of a filter followed by a sampler (sampling time T) followed by a minimax threshold device. The noise spectral density is $N_0/2$.

(a) Find the minimax threshold and resulting error probability for an integrate-and-dump filter. Give your answer in terms of T and N_0.

(b) Repeat part **(a)** for a matched filter.

5.20 A binary communications system operates over an AWGN channel with spectral density $N_0/2$. The transmitted signals are given by

$$s_0(t) = A\, p_T(t) - A\, p_T(t - T)$$

and

$$s_1(t) = 0.$$

Notice that the signal duration is $2T$.

(a) Give an expression (in terms of A, T, N_0, and Q) for the *minimum average* error probability when the two signals are transmitted with equal a priori probabilities (i.e., $\pi_0 = \pi_1$).

(b) Derive the optimum correlation receiver, and give a block diagram for it.

(c) Assume that you use the receiver that you obtained in part **(b)**, but the signal $s_0(t)$ is actually given by

$$s_0(t) = cA\, p_T(t) - cA\, p_T(t - T)$$

instead of as given at the beginning of this problem. Assume that $s_1(t) = 0$, as before. Give an expression (in terms of c, A, T, N_0, and Q) for the average error probability if $\pi_0 = \pi_1$.

5.21 The signal set $\{s_0, s_1\}$ is antipodal and $s_0(t) = \exp(-t^2)$ for all t. The channel is an AGN channel with zero-mean, wide-sense stationary noise having spectral density $S_X(\omega)$. The received signal plus noise is filtered with a time-invariant linear filter. The output of the filter is sampled at time T_0 and compared with a threshold $\gamma = 0$ in the decision device. Assume that the noise is white with spectral density $S_X(\omega) = N_0/2$.

(a) What is the impulse response of the optimum filter?

(b) If the optimum filter is employed, what is the probability of error for this system?

(c) For the filter you obtained in part **(a)**, what is the signal component at the output of the filter at the sampling time if $s_0(t)$ was actually transmitted?

5.22 Repeat Problem 5.21**(a)** under the assumptions of that problem, except that the noise now has spectral density

$$S_X(\omega) = 1/(\omega^2 + \alpha^2), \quad -\infty < \omega < \infty,$$

where α is real and positive.

5.23 Consider the following binary baseband communication system:

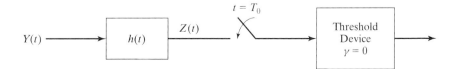

The random process $Y(t)$ is the sum of the signal and a white Gaussian noise process $X(t)$ with spectral density $N_0/2$. The impulse response of the linear filter is given by

$$h(t) = \begin{cases} 3t/T, & 0 \le t < T/3, \\ 1, & T/3 \le t < T, \\ (4T - 3t)/T, & T \le t < 4T/3, \end{cases}$$

and $h(t) = 0$ for $t < 0$ or $t > 4T/3$. A sketch of $h(t)$ is as follows:

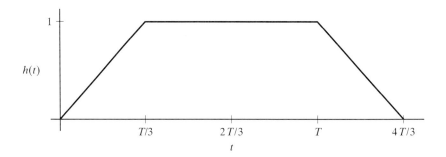

The transmitted signal is $(-1)^i A\, p_T(t)$ for $A > 0$, and the output of the filter is sampled at time T_0. It is not necessary to evaluate the full convolution $p_T * h$ to answer any of the following questions:

(a) What is the value of the output signal at the sampling time if $T_0 = T$?

(b) What is the optimum sampling time? That is, what value of T_0 gives the smallest probability of error?

(c) For the value of T_0 that you selected in part (b), what is the resulting error probability?

5.24 Consider a binary baseband data communication system with an additive Gaussian noise (AGN) channel. The received signal is given by $Y(t) = s_i(t) + X(t)$, where $X(t)$ is a stationary Gaussian random process with mean $E\{X(t)\} = 2$ and autocovariance function $C_X(\tau) = \exp(-\tau^2/T^2)$. The signals are $s_0(t) = 4p_T(t)$ and $s_1(t) = 3p_T(t)$. The receiver samples the received signal at times $T/3$ and $2T/3$, and it forms a decision statistic Z, which is the weighted sum of these sample values:

$$Z = 2Y(T/3) + 3Y(2T/3).$$

This decision statistic is compared with a threshold γ, and the decision is that s_0 was sent if $Z > \gamma$ and that s_1 was sent if $Z < \gamma$.

(a) What range of values for the threshold γ will ensure that the error probabilities $P_{e,0}$ and $P_{e,1}$ are both less than $1/2$? Make sure to specify the *entire* range.

(b) Find the conditional variance of the statistic Z given that s_i is sent for both $i = 0$ and $i = 1$.

(c) Suppose the threshold is $\gamma = 10$. Give expressions for the error probabilities $P_{e,0}$ and $P_{e,1}$.

(d) What is the minimax threshold for this system?

(e) Let f_i be the conditional density function for Z given that s_i was transmitted, and let $\Lambda(z) = f_1(z)/f_0(z)$ be the likelihood ratio. What is the likelihood ratio test that corresponds to the Bayes decision rule for the a priori probabilities $\pi_0 = 1/4$ and $\pi_1 = 3/4$? Describe the test in terms of $\Lambda(z)$.

(f) Give a careful derivation to show that the likelihood ratio test in part (e) reduces to a threshold test.

(g) From part (f), determine the optimum Bayes threshold for the a priori probabilities given in part (e).

5.25 (a) What is the distance d for a binary antipodal signal set? Use the fact that antipodal signals have equal energy.

(b) Compare your answer to part (a) with the distance for a binary orthogonal signal set. Assume that the energy in each signal is the same as for the antipodal signals.

(c) Use these results and (5.113a) to compare the error probabilities for the two signal sets in parts (a) and (b).

Chapter 6

Coherent Communications

Thus far, we have considered baseband communications only. The frequency band occupied by a baseband signal is from 0 Hz up to the cutoff frequency of the transmitter. For binary baseband signaling at R bits per second, this cutoff frequency is typically no more than $2R$ Hz, and it is often much less. For example, if the data rate is 2 kb/s, the baseband signal is usually contained in a band from 0 Hz to 4 kHz; in most cases, this data rate can be accommodated in a much smaller band.

There are numerous communication media (e.g., satellite channels) for which the baseband data signal must be modulated onto a *carrier* in order that the transmitted signal occupy a particular portion of the electromagnetic spectrum other than baseband. For our purposes, a carrier is a sinusoidal signal, and the frequency of this sinusoidal signal is not restricted to any particular band. Many of the modulation techniques and communications principles discussed in this chapter apply equally well to optical communication systems; however, the primary applications of some of the methods and systems are to digital communication systems that operate in the radio-frequency (RF) band, which is the band from 30 Hz to 300 GHz.

There are many reasons why we may prefer (or are required) to operate a data transmission system in a frequency band other than baseband. First, we may be forced to transmit at an assigned frequency, or at least within some assigned band of frequencies, because frequency bands are allocated according to the type of service provided (e.g., radio navigation, broadcasting, or space communication) by organizations such as the U.S. Federal Communication Commission (FCC) and the International Telecommunication Union. (See [6.13].) Second, for the simplest antenna structures, the physical size of the antenna required to radiate a signal is proportional to the wavelength and therefore inversely proportional to the frequency of the signal. Thus, the use of a very low center frequency (large wavelength) requires a very large antenna. Third, the constraints encountered in equipment design usually result in limitations on the *fractional bandwidth* (i.e., the bandwidth of the signal divided by the signal's center frequency). Thus, communication systems with larger bandwidths are feasible at higher carrier frequencies, and the use of larger bandwidths facilitates transmission of higher data rates. Alternatively, the additional bandwidth permits the use of error-control coding, spread-spectrum modulation, and other schemes that provide increased performance at the expense of increased bandwidth.

In some cases, the spectral occupancy of the signal may be restricted by existing equipment. For example, in order to transmit over an existing satellite channel, it is necessary to modulate the baseband signal onto a carrier in a way that produces a carrier-modulated signal whose spectral occupancy matches the passband of the satellite channel. Modulation of different baseband signals onto different carriers also permits simultaneous transmission and reception of several signals, if the carrier frequencies are suitably chosen. Such a scheme is known as frequency-division multiplexing (FDM) or frequency-division multiple access (FDMA). The frequencies of the carriers must be spaced far enough apart to permit separation of the signals (known as demultiplexing) at the receiver.

In this chapter, we consider one class of carrier-modulated communication systems known as *coherent* communication systems. In such systems, the receiver must have a replica of the carrier in order to demodulate the signal. This replica and the actual carrier must be *coherent*; that is, the two carriers must have (approximately) the same frequency and phase. In a few applications, the carrier is supplied to the receiver (e.g., a reference signal is transmitted along with the data signal); however, in most systems, the receiver extracts the carrier from the modulated signal itself.

Much of our discussion of coherent communications is based on the concepts developed for baseband data transmission in Chapter 5. In particular, we need *not* derive the basic results on optimum thresholds, filters, and signals for coherent communications. Our approach permits direct application of the results of Chapter 5 to coherent carrier-modulated communication systems.

6.1   The Optimum Receiver for Binary Coherent Communications

The most general signals that are considered in this section can be represented as a combination of amplitude and phase modulation. The signals $s_0(t)$ and $s_1(t)$ are of the form

$$s_i(t) = \sqrt{2}\, A\, a_i(t) \cos[\omega_c t + \theta_i(t) + \varphi]\, p_T(t), \qquad (6.1)$$

where $a_i(t)$ and $\theta_i(t)$ are baseband signals of the type considered in Chapter 5. For binary signaling, the integer i is either 0 or 1, as usual, and the choice of $a_0(t)$ vs. $a_1(t)$ represents the amplitude modulation while the choice of $\theta_0(t)$ vs. $\theta_1(t)$ is the phase modulation. The phase angle φ represents the phase of the unmodulated carrier at the *receiver* at time $t = 0$; that is, φ includes the phase shifts that arise as the signal propagates from the transmitter to the receiver. Note that the signals are time limited to the interval $[0, T]$.

6.1.1   Binary Phase-Shift Keying (BPSK)

Many binary carrier-modulated signals can be generated in either of two ways: as pure amplitude modulation or as pure phase modulation. This is true for our first example, which is the most important binary signal set for RF coherent communications. *Binary phase-shift keying (BPSK)* is a modulation scheme that can be expressed in the form of (6.1) if the amplitude signals are given by $a_0(t) = a_1(t) = p_T(t)$ (no amplitude modulation) and the phase signals are

$$\theta_0(t) = 0, \ 0 \leq t < T, \tag{6.2a}$$

and

$$\theta_1(t) = \pi, \ 0 \leq t < T. \tag{6.2b}$$

Thus, the signal set for BPSK consists of

$$s_0(t) = \sqrt{2} \, A \, p_T(t) \cos(\omega_c t + \varphi) \tag{6.3a}$$

and

$$\begin{aligned} s_1(t) &= \sqrt{2} \, A \, p_T(t) \cos(\omega_c t + \pi + \varphi) \\ &= -\sqrt{2} \, A \, p_T(t) \cos(\omega_c t + \varphi). \end{aligned} \tag{6.3b}$$

Notice from (6.3) that the signals for BPSK are antipodal.

Next observe that the same signal set can be generated by employing amplitude modulation rather than phase modulation. Let

$$a_0(t) = +p_T(t), \tag{6.4a}$$
$$a_1(t) = -p_T(t), \tag{6.4b}$$

and $\theta_0(t) = \theta_1(t) = 0$ for all t. For these baseband signals, (6.1) gives the carrier-modulated signal set

$$s_0(t) = \sqrt{2} \, A \, p_T(t) \cos(\omega_c t + \varphi) \tag{6.5a}$$

and

$$s_1(t) = -\sqrt{2} \, A \, p_T(t) \cos(\omega_c t + \varphi), \tag{6.5b}$$

which is the same signal set as in (6.3).

Equations (6.1)–(6.5) demonstrate that BPSK can be viewed either as binary phase modulation or as binary amplitude modulation. When considered as amplitude modulation, BPSK is a special case of *amplitude-shift keying (ASK)* in which signals are of the form

$$s_i(t) = \sqrt{2} \, A \, u_i \, \beta(t) \cos(\omega_c t + \varphi) \tag{6.6}$$

and information is conveyed through the choice of u_i. For binary ASK, there are two choices only: u_0 or u_1. A detailed discussion of binary ASK is given in Section 6.3.

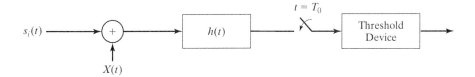

Figure 6-1: A possible receiver for BPSK.

Now that it has been established that the BPSK signal set $\{s_0, s_1\}$ is an antipodal signal set, it follows immediately from the results of Chapter 5 that the minimax threshold γ_m is zero for the system shown in Figure 6-1. If we let $s(t) = s_0(t)$, then $s_1(t) = -s(t)$. This is consistent with our convention that

$$s(t) = [s_0(t) - s_1(t)]/2,$$

because, for antipodal signals,

$$[s_0(t) - s_1(t)]/2 = s_0(t) = -s_1(t).$$

The error probabilities that result when the minimax threshold is used are given by

$$P_{e,0} = P_{e,1} = Q[\hat{s}(T_0)/\sigma]. \tag{6.7}$$

As usual, $\hat{s}$ is the filtered version of the signal s; that is, $\hat{s} = s * h$. The parameter σ is defined by $\sigma^2 = N_0 \|h\|^2/2$, as in Chapter 5.

6.1.2   Matched Filter and Correlation Receiver for BPSK

The next step is to optimize the filter for an additive white Gaussian noise (AWGN) channel. That is, it is desired to choose the impulse response for the filter in Figure 6-1 to minimize the error probability given in (6.7). As in Chapter 5, this error probability is minimized by letting the filter be the matched filter. In terms of the signal $s(t)$, the matched filter impulse response is given by $h_M(t) = c' s(T_0 - t)$, where c' is an arbitrary constant; if the minimax threshold is used, the value of the constant is unimportant. A convenient choice for the constant for the signal set of (6.3) is $c' = A^{-1}$, for which the matched-filter impulse response is

$$h_M(t) = \sqrt{2}\cos[\omega_c(T_0 - t) + \varphi]\, p_T(T_0 - t).$$

Because the BPSK signals under consideration are time limited to $[0, T]$, the most natural choice for the sampling time is $T_0 = T$. For this sampling time, the matched filter can be written as

$$h_M(t) = \sqrt{2}\cos[\omega_c(T - t) + \varphi]\, p_T(t). \tag{6.8}$$

The carrier frequency may be selected to give an integral number of cycles for the time interval $[0, T]$; that is, the carrier frequency satisfies $\omega_c T = 2\pi n$ for some *integer*

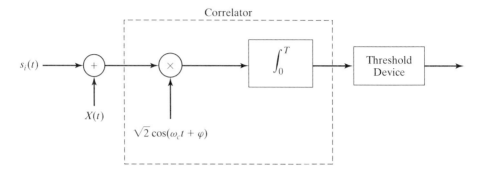

Figure 6-2: Correlation receiver for BPSK.

n. In this case, we say simply that $\omega_c T$ *is a multiple of* 2π. If $\omega_c T$ is a multiple of 2π, and if the sampling time is T, then the matched filter can be written as

$$h_M(t) = \sqrt{2}\cos[-\omega_c t + \varphi]\, p_T(t) = \sqrt{2}\cos[\omega_c t - \varphi]\, p_T(t). \qquad (6.9)$$

Equivalence of the matched-filter receiver and the correlation receiver is demonstrated in Chapter 5 for an arbitrary binary signal set. As a special case, it follows that the matched filter can be replaced by a correlator for the BPSK signal set. If $Y(t)$ is the input to the matched filter, the output, sampled at time T, is given by

$$\hat{Y}(T) = \int_0^T Y(t)\, h_M(T-t)\, \mathrm{d}t = \int_0^T Y(t)\, \sqrt{2}\cos(\omega_c t + \varphi)\, \mathrm{d}t.$$

The last equality is derived from the fact that (6.8) implies

$$h_M(T-t) = \sqrt{2}\cos(\omega_c t + \varphi)\, p_T(t).$$

The correlation receiver for BPSK is illustrated in Figure 6-2.

6.1.3   Probability of Error for BPSK Signals and an AWGN Channel

The probability of error that results when the matched filter (or correlation receiver) is employed in the receiver for BPSK signaling over an AWGN channel follows immediately from results in Chapter 5. For the minimax approach, the minimum possible error probability is

$$P_{e,m}^* = Q\left(\sqrt{2\,\mathcal{E}/N_0}\,\right). \qquad (6.10)$$

Recall that this is the minimum of $P_{e,m} = \max\{P_{e,0}, P_{e,1}\}$ over all choices for the filter impulse response and the threshold. The energy $\mathcal{E}$ is the common value of the energy in the signals s_0 and s_1, which is given by

$$\mathcal{E} = \int_0^T 2\,[A\cos(\omega_c t + \varphi)]^2\, \mathrm{d}t$$

$$= \int_0^T A^2\,[1 + \cos(2\,\omega_c t + 2\,\varphi)]\, \mathrm{d}t.$$

Integrating the constant term in the integrand, we find that

$$\mathcal{E} = A^2 T + \int_0^T A^2 \cos(2\,\omega_c t + 2\,\varphi)\,dt. \tag{6.11}$$

Under the condition that $\omega_c T$ is a multiple of 2π, the integral in (6.11) is zero, and the energy in each of the signals is $\mathcal{E} = A^2 T$.

Even if $\omega_c T$ is not a multiple of 2π, there are other conditions under which the last integral of (6.11) can be neglected. If a sequence of rectangular pulses is transmitted, and each pulse is of duration T seconds, then the transmission rate is T^{-1} pulses per second. This is also a rough measure of the bandwidth (in hertz) of the system. In fact, as we will see later, the first null in the spectrum of the BPSK signal is located T^{-1} Hz away from the carrier frequency. For most applications, the carrier frequency ω_c must be much larger than the data rate. A typical example is a carrier frequency $f_c = \omega_c/2\pi$ of approximately 50 MHz and a transmission rate of 20 kb/s (i.e., an interval of duration $T = 50\ \mu\mathrm{s}$). In this example, it is certainly true that $\omega_c \gg T^{-1}$; in fact, $\omega_c > 10^4 T^{-1}$.

One of the implications of having ω_c much larger than T^{-1} is that the integral in (6.11) can be neglected. Notice from (6.11) that

$$\mathcal{E} = A^2\{T + (2\,\omega_c)^{-1}[\sin(2\,\omega_c T + 2\,\varphi) - \sin(2\,\varphi)]\}.$$

The magnitude of the term $\sin(2\,\omega_c T + 2\,\varphi) - \sin(2\,\varphi)$ can be upper bounded by using the fact that $|\sin(x) - \sin(y)| \le 2$ for all x and y. As a result,

$$(2\,\omega_c)^{-1}|\sin(2\,\omega_c T + 2\,\varphi) - \sin(2\,\varphi)| \le 1/\omega_c.$$

If $\omega_c \gg T^{-1}$, then $1/\omega_c$ is small compared with T; therefore,

$$T + (2\,\omega_c)^{-1}[\sin(2\,\omega_c T + 2\,\varphi) - \sin(2\,\varphi)] \approx T.$$

It follows that $\mathcal{E} \approx A^2 T$. This is an example of a situation in which the double-frequency component can be neglected in the analysis of a carrier-modulated communication system. Further discussion of this issue can be found in Section 6.3.3.

Returning to (6.10), we see that optimum reception of BPSK signaling achieves the smallest probability of error that is possible for the reception of a single pulse on the AWGN channel. Consequently, BPSK is a very popular signaling scheme for satellite communication systems and other applications in which the physical channel is accurately modeled as an AWGN channel. The probability of error specified in (6.10) is valid for the optimum receiver, a receiver that knows the phase φ of the received signal. In Section 6.2, we examine the performance of the receiver if the phase is not known precisely.

6.1.4   On the Optimality of the Correlation Receiver for BPSK

At the beginning of Chapter 5, we focused on a receiver that consists of a time-invariant linear filter, sampler, and threshold device. Although it is stated there that this linear receiver configuration is the optimum receiver configuration for an AWGN channel, no proof of this fact is given in Chapter 5. So far, we have not offered any evidence that some completely different type of system, perhaps utilizing a nonlinear filter, cannot be better than the linear receiver configuration. The development in Chapter 5 is concerned with optimization of the components of the threshold and the filter impulse response within the framework of the linear receiver configuration. In this section and Appendix D, we offer evidence of the optimality of the linear configuration for the AWGN channel.

From Section 5.4, we do know that if a linear filter is used to obtain the decision statistic, the threshold device does indeed produce the optimum decision for binary signaling on the AWGN channel. We also know the optimum impulse response for the linear filter, and we know that the linear filter and sampler can be replaced by an equivalent correlator. In particular, for the transmission of BPSK signals on the AWGN channel, the optimum correlation receiver is as shown in Figure 6-2.

In this section, we demonstrate that the correlation receiver is the optimum filter (linear or otherwise) for BPSK signaling on the AWGN channel; hence, the receiver configuration studied in Chapter 5 is the optimum configuration. The reason for postponing this demonstration until now is that we prefer to develop the result for the special case of BPSK signaling. This is because the proof of the optimality of the correlation receiver for BPSK signaling on the AWGN channel can be based on the Fourier series representation of the BPSK signals, and most electrical engineers are more familiar with Fourier series expansions than general orthogonal expansions. The results can be extended to arbitrary signal sets, both binary and nonbinary, by use of general orthogonal expansions for the signals in the set or by the sampling method of Appendix D. In addition, the optimum receiver can be derived for BPSK signaling over the general AGN channels by using a whitening filter as in Section 5.6.

The signals for BPSK can be expressed as

$$s_i(t) = \sqrt{2}\, A\, u_i \cos(\omega_c t + \varphi)$$

for $0 \le t \le T$. We assume that ω_c is a multiple of $2\pi/T$; that is, $\omega_c = 2\pi m/T$ for some integer m. Then we define ω_0 by $\omega_0 = 2\pi/T$, so that $\omega_c = m\omega_0$. Recalling that $u_0 = +1$ and $u_1 = -1$ for BPSK, we observe that the signals are just scalar multiples of the common signal $\cos(\omega_c t + \varphi) = \cos(m\omega_0 t + \varphi)$. The trigonometric form of the Fourier series expansion for an arbitrary signal $s(t)$ that is defined on the interval $[0, T]$ can be expressed as

$$s(t) = c_0 + \sum_{k=1}^{\infty} \{c_k \cos(k\omega_0 t + \varphi) + d_k \sin(k\omega_0 t + \varphi)\}$$

for $0 \le t \le T$, where

$$c_0 = \int_0^T s(t)\, dt,$$

$$c_k = \int_0^T s(t) \cos(k\omega_0 t + \varphi)\, dt$$

for each positive integer k, and

$$d_k = \int_0^T s(t) \sin(k\omega_0 t + \varphi)\,dt$$

for each positive integer k. Of course, this Fourier series representation can be related to a Fourier series representation in terms of $\cos(k\omega_0 t)$ and $\sin(k\omega_0 t)$ by expanding $\cos(k\omega_0 t + \varphi)$ and $\sin(k\omega_0 t + \varphi)$ as the cosine and sine, respectively, of the sum of two angles.

Certain regularity conditions must be placed on the signal in order to guarantee the Fourier series expansion exists and is equal (in some sense) to the signal $s(t)$, but these regularity conditions are satisfied by any signal of interest for communication systems applications. For example, if we want the Fourier series to be equal to $s(t)$ in a mean-squared sense, the integral of the square of the difference between $s(t)$ and the sum of the Fourier series terms up to $k = N$ must converge to zero as $N \to \infty$. To guarantee that $s(t)$ has a mean-squared-sense Fourier series representation, it suffices that the signal $s(t)$ have finite energy.

For our purposes, it is advantageous to renumber the terms in the Fourier series. Let $\psi_0(t) = \sqrt{1/T}$, $\psi_{2k-1}(t) = \sqrt{2/T}\cos(k\omega_0 t + \varphi)$ for each positive integer k, and $\psi_{2k}(t) = \sqrt{2/T}\sin(k\omega_0 t + \varphi)$ for each positive integer k. Similarly, let $a_0 = c_0\sqrt{T}$; and, for each positive integer k, let $a_{2k-1} = c_k\sqrt{T/2}$ and $a_{2k} = d_k\sqrt{T/2}$. The trigonometric form of the Fourier series is equivalent to

$$s(t) = \sum_{n=0} a_n \psi_n(t),$$

where

$$a_n = \int_0^T s(t)\,\psi_n(t)\,dt$$

for each nonnegative integer n.

It is easy to check that the functions $\{\psi_n : 0 \le n < \infty\}$ are *orthonormal* on the interval $[0, T]$; that is, they are orthogonal and they have been normalized to have unit energy. The functions $\{\psi_n : 0 \le n < \infty\}$ are *orthogonal* on $[0, T]$ if

$$(\psi_k, \psi_n) = \int_0^T \psi_k(t)\,\psi_n(t)\,dt = 0, \ \text{for } k \ne n,$$

and the requirement that they have unit energy is the condition that

$$\|\psi_n\|^2 = \int_0^T [\psi_n(t)]^2\,dt = 1$$

for each n.

We can think of generating the Fourier series coefficients a_n with a bank of correlators as illustrated in Figure 6-3. The nth correlator computes the nth coefficient

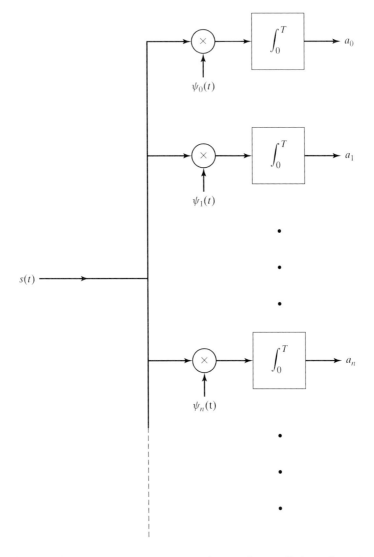

Figure 6-3: Correlators that generate the Fourier coefficients for $s(t)$.

a_n $(0 \leq n < \infty)$, and the bank of correlators computes the infinite-dimensional vector $\mathbf{a} = (a_0, a_1, \ldots, a_n, \ldots)$. The system illustrated in Figure 6-3 does not lose any information contained in the signal, because the signal $s(t)$ can be recovered completely from the vector $\mathbf{a}$. The signal can be reconstructed, if desired, by following the bank of correlators with a system that forms the Fourier series

$$s(t) = \sum_{n=0}^{\infty} a_n \, \psi_n(t).$$

All of the information that we need about $s(t)$ is contained in the vector $\mathbf{a}$.

Suppose the bank of correlators shown in Figure 6-3 is used as the filter in a communication system. Clearly, such a filter preserves all of the information about the signal that is needed to make a decision. In fact, if the input is *any* finite-energy waveform, the output of the bank of correlators completely specifies the input, whether the input is a communication signal, noise, or the sum of the signal and noise.

There is a slight complication that arises with the white Gaussian noise model, because it does not have finite power. This complication can be removed, however, by considering bandlimited white Gaussian noise with as large a bandwidth as desired (e.g., larger than the bandwidth of any hardware implementation of the bank of correlators). Such a noise process has finite power, and yet it is a valid model for thermal noise in electronic systems.

Suppose the input to the bank of correlators is the signal

$$s_i(t) = \sqrt{2}\, A\, u_i \cos(\omega_c t + \varphi).$$

To determine the output vector $\mathbf{a}$ corresponding to this input, first recall that $\omega_c = m\omega_0$, where m is an integer. Therefore,

$$s_i(t) = \sqrt{2}\, A\, u_i \cos(m\omega_0 t + \varphi).$$

But $\psi_{2m-1}(t) = \sqrt{2/T}\, \cos(m\omega_0 t + \varphi)$, so

$$s_i(t) = A\, u_i \sqrt{T}\, \psi_{2m-1}(t).$$

It follows that

$$a_n = \int_0^T s_i(t)\, \psi_n(t)\, dt = A\, u_i \sqrt{T} \int_0^T \psi_{2m-1}(t)\, \psi_n(t)\, dt.$$

Because $\psi_{2m-1}(t)$ and $\psi_n(t)$ are orthogonal for $n \neq 2m - 1$, we conclude that $a_n = 0$ for $n \neq 2m - 1$. Thus, only one of the correlators has a nonzero output, and this correlator is the one that is matched to the signal $s_i(t)$. The output of this correlator is $a_{2m-1} = A\, u_i \sqrt{T}$.

If the bank of correlators is used as the filter in a communication system, the input is actually signal plus noise, as shown in Figure 6-4. Because this filter is linear, the output due to $Y(t) = s_i(t) + X(t)$ is the sum of the output due to $s_i(t)$ and the output due to $X(t)$. We have already determined the output due to the signal component; we must now determine the output of the bank of correlators due to the noise input $X(t)$.

The noise process $X(t)$ is white Gaussian noise with spectral density $N_0/2$. If $X(t)$ is the input to the bank of correlators, then the nth correlator output is a zero-mean, Gaussian random variable defined by

$$X_n = \int_0^T X(t)\, \psi_n(t)\, dt.$$

Thus, when the input is $s_i(t) + X(t)$, the output Y_n of the nth correlator is just the noise component X_n for $n \neq 2m - 1$. For $n = 2m - 1$, the output consists of the

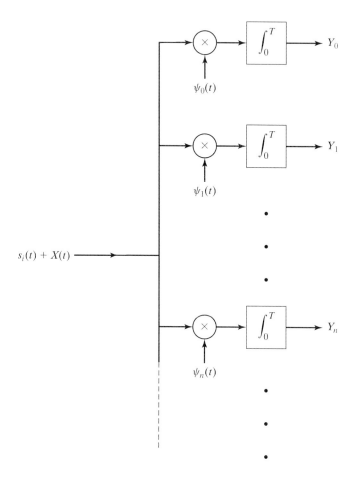

Figure 6-4: Bank of correlators in a communication receiver.

noise component X_{2m-1} plus the signal component, which we have already shown to be $a_{2m-1} = A u_i \sqrt{T}$.

In summary, if the input to the bank of correlators is $Y(t) = s_i(t) + X(t)$, then the outputs are given by

$$Y_n = X_n \text{ for } n \neq 2m - 1$$

and

$$Y_n = A u_i \sqrt{T} + X_n \text{ for } n = 2m - 1.$$

The infinite-length random vector $\mathbf{Y} = (Y_0, Y_1, Y_2, \dots)$ is a *sufficient statistic* for making a decision regarding which signal was sent, because it tells us everything we need to know about $Y(t)$ in order to make the decision. In fact, we can always reconstruct

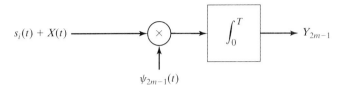

Figure 6-5: Single correlator that produces a sufficient statistic.

$Y(t)$ from **Y**, if we choose to do so, by using the Fourier series

$$Y(t) = \sum_{n=0}^{\infty} Y_n \, \psi_n(t).$$

Notice that the Fourier coefficients are now the random variables Y_n for $0 \le n < \infty$.

The output of the correlator indexed by $2\,m - 1$ is the only one of the correlator outputs that has a signal component. This output is the random variable

$$Y_{2m-1} = A\, u_i \sqrt{T} + X_{2m-1},$$

which is the sum of a signal component and a noise component. The mean of Y_{2m-1} is $A\, u_i \sqrt{T}$, which carries the information that is being sent to the receiver. None of the other correlators produces any information about the transmitted signal. Therefore, we are inclined to simply eliminate the other correlators from the system and use only the correlator that is matched to the signal (i.e., the correlator indexed by $2\,m - 1$). If we can do this, the bank of correlators is replaced by the single correlator shown in Figure 6-5. The optimality of the correlation receiver then follows by observing the equivalence of the correlator in Figure 6-5 and the correlator used in the receiver of Figure 6-2 (i.e., the correlation receiver for BPSK signaling). The two differ by a constant only, and multiplication of the received signal and the noise by the same constant does not affect the optimality of the correlator.

We are tempted, then, to say that Y_{2m-1} is also a sufficient statistic for deciding which signal was sent. Certainly, the other random variables (i.e., Y_n for $n \ne 2\,m - 1$) do not have signal components. However, we must consider one question: Is there any useful information about the *noise* component X_{2m-1} that can be obtained from any of the other correlator outputs? Is it possible that, by observing X_0, for example, we are able to deduce something about X_{2m-1}? If so, we might use this information to reduce the effects of the noise at the output of the correlator indexed by $2\,m - 1$. In the extreme, is it possible that, by observing X_n for all n except $n = 2\,m - 1$, we could learn the value of X_{2m-1} and subtract it from the output of the correlator indexed by $2\,m - 1$?

The answer to each of these questions is *no*. To prove this, we have to show that the noise components at the outputs of different correlators are statistically independent. This would show that the collection $\{X_n : n \ne 2\,m - 1\}$ cannot possibly contain information about X_{2m-1}. As a result, we would know that the random variables $\{Y_n : n \ne 2\,m - 1\}$ have no signal component and that they are independent of the noise component of the only correlator output that does have a signal component. In other words, it would show that the random variables $\{Y_n : n \neq 2m - 1\}$ are *irrelevant* to the decision regarding which signal was sent.

The task that remains can be accomplished by proving that the collection of random variables $\{X_n : 0 \leq n < \infty\}$ is a collection of mutually independent random variables. To do this, we first observe that the system of Figure 6-4 is a single-input linear system with multiple outputs. If the input is the random process $X(t)$, then the output is the infinite-length random vector $\mathbf{X} = (X_0, X_1, X_2, \ldots)$. Because the system is linear, $\mathbf{X}$ is a Gaussian random vector, which means that the random variables $\{X_n : 0 \leq n < \infty\}$ are jointly Gaussian. To show that jointly Gaussian random variables are independent, it suffices to show that each pair of them is uncorrelated. We consider an arbitrary pair of these random variables, X_k and X_n for $k \neq n$, and show that they are uncorrelated. Because the random variables X_k and X_n have zero mean, they are uncorrelated if and only if $E\{X_k X_n\} = 0$. Writing $E\{X_k X_n\}$ as the expected value of the product of two integrals, and then writing the product of two integrals as a double integral, we have

$$E\{X_k X_n\} = E\left\{\int_0^T \int_0^T X(t)\,\psi_k(t) X(s)\,\psi_n(s)\,\mathrm{d}t\,\mathrm{d}s\right\}.$$

Taking the expectation inside the integral, we find that

$$E\{X_k X_n\} = \int_0^T \int_0^T E\{X(t)\,X(s)\}\,\psi_k(t)\,\psi_n(s)\,\mathrm{d}t\,\mathrm{d}s.$$

But $E\{X(t)\,X(s)\} = \frac{1}{2}N_0\,\delta(t - s)$, so

$$E\{X_k X_n\} = \frac{N_0}{2}\int_0^T \psi_k(s)\,\psi_n(s)\,\mathrm{d}s = \frac{N_0}{2}(\psi_k, \psi_n).$$

Because $k \neq n$, the functions ψ_k and ψ_n are orthogonal on the interval $[0, T]$, so $(\psi_k, \psi_n) = 0$. We have therefore shown that $E\{X_k X_n\} = 0$ for $k \neq n$, and this completes the proof that the noise components of the outputs of different correlators are statistically independent.

6.2   Binary Phase-Shift Keying with an Imperfect Phase Reference

It is clear from an examination of the matched-filter receiver (Figure 6-1 with $h(t) = h_M(t)$) and the correlation receiver (Figure 6-2) that knowledge of the phase φ of the received signal is required in the optimum receiver for BPSK signals. In practice, the receiver may not know φ exactly, but it may at least have an estimate $\hat{\varphi}$ of the phase of the received signal. In this section, we consider the error probability that results when there is an error in the estimate of the phase that is used in the correlation receiver for BPSK signaling over an additive white Gaussian noise channel. The analysis of the matched-filter receiver with a phase error is completely analogous.

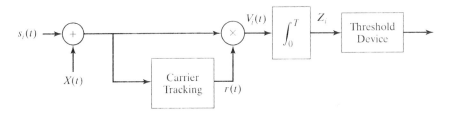

Figure 6-6: Correlation receiver with imperfect phase reference.

6.2.1   Error Probability as a Function of Phase Error

The system under consideration is shown in Figure 6-6. For simplicity, we again consider the minimax threshold for the antipodal BPSK signals ($\gamma_m = 0$), so that gain constants are unimportant. The carrier tracking subsystem produces a sinusoidal reference signal that agrees with the received signal in frequency, but it may differ in phase. The job of the carrier tracking subsystem is to estimate the phase φ of the carrier upon which the data signal is modulated, but it will not be able to obtain a perfect estimate of φ because of the presence of noise at its input. The amplitude constant β for the output of the carrier tracking subsystem may differ from the amplitude of the input signal, but this turns out to be of no consequence (because the minimax threshold is used).

The starting point is to consider the random process $V_i(t)$ at the input to the integrator. This process contains a component due to the signal $s_i(t)$ and a component due to the noise $X(t)$. The signal component is the mean value of the process, and the second moment of the noise component is the variance of the process.

We concentrate first on the signal component of $V_i(t)$, the input to the integrator in the system of Figure 6-6. The reference signal at the output of the carrier tracking loop is

$$r(t) = \sqrt{2}\,\beta \cos(\omega_c t + \hat{\varphi}).$$

and the received signal is either

$$s_0(t) = \sqrt{2}\,A \cos(\omega_c t + \varphi)$$

or

$$s_1(t) = -\sqrt{2}\,A \cos(\omega_c t + \varphi).$$

A convenient illustration of these three signals is the vector diagram of Figure 6-7(a) (also known as a phasor diagram in elementary signal analysis). The situation illustrated is for $A < \beta$ and $\varphi > \hat{\varphi}$, but similar diagrams can be drawn for other cases.

Ignoring the effects of the noise process for the moment, we see that the output of the integrator due to the signal alone is given by

$$z_0 = \int_0^t s_0(t)\, r(t)\, dt$$

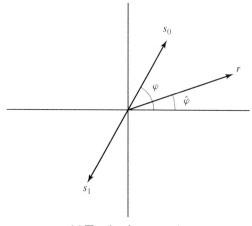

(a) The signals s_0, s_1, and r

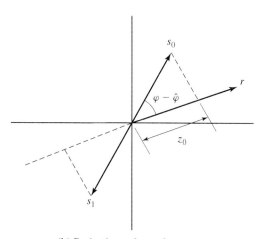

(b) Projections of s_0 and s_1 onto r

Figure 6-7: Vector representations.

if the signal $s_0(t)$ is transmitted. If $s_1(t)$ is transmitted, the integrator output is $z_1 = -z_0$. But these outputs can be expressed in terms of the inner products (s_0, r) and (s_1, r); namely, $z_0 = (s_0, r)$ and $z_1 = (s_1, r)$. In the vector diagram of Figure 6-7(a), these inner products can be thought of as the projections of the vectors s_0 and s_1 onto the vector r. This is illustrated for z_0 in Figure 6-7(b) if $r(t)$ has unit energy on the interval $[0, T]$; that is, $\|r\| = 1$. The reader can verify that $\|r\| = 1$ if $\beta = 1/\sqrt{T}$.

Clearly, the magnitudes are the same for the projections of s_0 and s_1 onto r, and they are proportional to $\cos(\varphi - \hat{\varphi})$. In particular, the largest magnitude occurs for $\hat{\varphi} = \varphi$ (the vectors are co-linear), and the resulting projections are of opposite sign. The smallest magnitude occurs for $\hat{\varphi} = \varphi + \pi/2$ (the vectors are orthogonal), and the resulting projections are indistinguishable.

Because one goal of a communication receiver is to obtain a large signal component (i.e., the magnitudes of z_0 and z_1 should be as large as possible), it seems reasonable that we should strive for $\hat{\varphi} = \varphi$ in the design of the carrier tracking system. An even more important goal is that the two output signals that correspond to the two different transmitted signals should be as "different" as possible. This is accomplished by making $z_0 - z_1$ as large as possible. Clearly, $z_0 = -z_1$ for all values of $\hat{\varphi}$ and φ, but the difference $z_0 - z_1$ is largest for $\hat{\varphi} = \varphi$. A third goal is that the noise component of the output be small, but, as we will see, this is not influenced by the value of $\hat{\varphi}$.

Continuing with the analysis of the signal component, we see that

$$z_0 = \int_0^T \left\{ \sqrt{2}\, A \cos(\omega_c t + \varphi) \right\} \left\{ \sqrt{2}\, \beta \cos(\omega_c t + \hat{\varphi}) \right\} dt,$$

which reduces to

$$z_0 = \int_0^T \beta A[\cos(\varphi - \hat{\varphi}) + \cos(2\,\omega_c t + \varphi + \hat{\varphi})]\, dt.$$

The second term in the integrand is a double-frequency component. As in the previous section, we assume that this term is negligible (e.g., it is zero if $\omega_c T$ is a multiple of 2π), so that

$$z_0 = \beta A T \cos(\varphi - \hat{\varphi}).$$

Similarly,

$$z_1 = -\beta A T \cos(\varphi - \hat{\varphi}).$$

Now, we must consider the effects of the noise process $X(t)$. We focus on the process $V_i(t)$ at the multiplier output and write it as

$$\begin{aligned} V_i(t) &= 2\beta A(-1)^i \cos(\omega_c t + \varphi) \cos(\omega_c t + \hat{\varphi}) + \sqrt{2}X(t)\,\beta \cos(\omega_c t + \hat{\varphi}) \\ &= \beta A(-1)^i [\cos(\varphi - \hat{\varphi}) + \cos(2\,\omega_c t + \varphi + \hat{\varphi})] + \sqrt{2}X(t)\,\beta \cos(\omega_c t + \hat{\varphi}). \end{aligned}$$

The symbol $(-1)^i$ denotes $+1$ for $i = 0$ and -1 for $i = 1$. We have already integrated the signal portion of this expression, and we found that the resulting integral is z_i. Combining this fact with the integral of the noise term, we see that the output of the integrator is the random variable

$$Z_i = z_i + \sqrt{2}\beta \int_0^T X(t) \cos(\omega_c t + \hat{\varphi})\, dt. \qquad (6.12\text{a})$$

Substituting for z_i, we find that the expression for the integrator output is

$$Z_i = \beta A T(-1)^i \cos(\varphi - \hat{\varphi}) + \sqrt{2}\beta \int_0^T X(t) \cos(\omega_c t + \hat{\varphi})\, dt. \qquad (6.12\text{b})$$

Because the integral in (6.12) is a random variable with zero mean, it is clear that the mean value of Z_i is

$$\mu_i = z_i = \beta A T(-1)^i \cos(\varphi - \hat{\varphi}).$$

Let the phase error $\varphi - \hat{\varphi}$ be denoted by θ in what follows. We have shown that the mean value of Z_i, which is the conditional mean of the integrator output given that the signal s_i is sent, is given by

$$\mu_i = \beta A T (-1)^i \cos \theta. \tag{6.13}$$

Again, observe that $\mu_0 = -\mu_1$ and that the difference $\mu_0 - \mu_1$ is proportional to $\cos \theta$. This difference is therefore maximized for $\theta = 0$, which corresponds to the greatest possible distinguishability between the two signals for given values of β, A, and T.

Because the mean is subtracted out in computing the variance of Z_i, this variance is just the second moment of the second term that appears on the right-hand side of (6.12b). That is, the variance of Z_i is given by

$$\sigma^2 = 2\beta^2 \int_0^T \int_0^T E\{X(t)\,X(s)\} \cos(\omega_c t + \hat{\varphi}) \cos(\omega_c s + \hat{\varphi})\, dt\, ds$$

$$= N_0 \beta^2 \int_0^T [\cos(\omega_c t + \hat{\varphi})]^2\, dt,$$

which can be written as

$$\sigma^2 = N_0 \beta^2 \int_0^T [1 + \cos(2\,\omega_c t + 2\,\hat{\varphi})]/2\, dt. \tag{6.14}$$

Assuming that the integral of the double-frequency term is negligible, we see that (6.14) reduces to

$$\sigma^2 = \beta^2 N_0 T / 2. \tag{6.15}$$

Notice that the variance does not depend on i; that is, the variance is the same regardless of which of the two signals is actually transmitted. Clearly, the variance should not depend on the phase angle φ of the received signal, and, indeed, (6.15) shows that it does not. Only the mean of $V_i(t)$ depends on φ, and the mean is subtracted from $V_i(t)$ in the process of evaluating the variance of Z_i.

Notice also that the variance does not depend on the phase $\hat{\varphi}$ of the reference signal from the carrier tracking loop. This is because the double frequency term of (6.14) is negligible, and the baseband term is not a function of $\hat{\varphi}$. We conclude that, while the mean of Z_i depends on φ and $\hat{\varphi}$, the variance does not depend on either of these two phase angles. This fact is important in the analysis that follows, because it implies that the probability of error depends on the phase angles φ and $\hat{\varphi}$ via the mean value of Z_i only. Moreover, as shown earlier, the mean of Z_i depends on φ and $\hat{\varphi}$ via their difference θ only.

The next step is to determine the probability of error. First, consider $P_{e,0}$, the conditional probability of error given that s_0 is the transmitted signal. Because this probability depends on the phase error $\theta = \varphi - \hat{\varphi}$, we write this as $P_{e,0}(\theta)$. The random variable Z_i is Gaussian with mean μ_i and variance σ^2. Hence, the error probability is

$$P_{e,0}(\theta) = P(Z_0 < 0) = \Phi(-\mu_0/\sigma) = Q(\mu_0/\sigma).$$

We now observe that the ratio of μ_0 and σ does not depend on the constant β, and, as mentioned previously, only the mean depends on θ. Thus, the error probability is given by

$$P_{e,0}(\theta) = Q\big(AT\cos\theta/\sqrt{N_0T/2}\,\big) = Q\big(\sqrt{2}\,AT\cos\theta/\sqrt{N_0T}\,\big).$$

Next, we use the fact that $\mathcal{E} = A^2T$ to write

$$P_{e,0}(\theta) = Q\big(\sqrt{2\,\mathcal{E}/N_0}\cos\theta\big).$$

The same sequence of steps applied to the situation in which s_1 is transmitted shows that $P_{e,1}(\theta)$ is identical to $P_{e,0}(\theta)$. Thus, $P_{e,i}(\theta)$ does not depend on i, and so we denote this probability by $P_e(\theta)$.

We conclude that the probability of error for a BPSK communication system in which the receiver's carrier reference has a phase error θ is given by

$$P_e(\theta) = Q\big(\sqrt{2\,\mathcal{E}/N_0}\cos\theta\big). \tag{6.16}$$

As a check, notice that, for $\theta = 0$, equation (6.16) reduces to

$$P_e(0) = Q\big(\sqrt{2\,\mathcal{E}/N_0}\,\big).$$

which is just the expression obtained in Section 6.1 for BPSK with a perfect phase reference. As another check, observe that $\theta = \pi/2$ gives $P_e(\pi/2) = Q(0) = 1/2$. If the phase error is $\pi/2$, the communication receiver can do no better than a random guess (e.g., tossing a coin). For this value of phase error, there is no signal component present at the output of the integrator in Figure 6-6; the decision statistic Z consists of noise only. This phenomenon can be seen geometrically in Figure 6-7(b), in which $\theta = \pi/2$ corresponds to $\varphi - \hat\varphi = \pi/2$, and so $z_0 = 0$.

Notice from (6.16) that the error probability depends on the magnitude of the phase error only: A phase error of θ gives the same error probability as a phase error of $-\theta$. Also notice that, because θ appears as an argument of the cosine function, the error probability depends on the modulo-2π representation of the phase error. That is, if θ and θ' are related by $\theta = \theta' + 2\pi n$ for some integer n, then $P_e(\theta) = P_e(\theta')$. As a result, we might as well consider θ to be a variable in the range from 0 to 2π (reduce all phase angles modulo 2π).

It is also instructive to observe from (6.16) that if the magnitude of the phase error is between 90 degrees and 270 degrees (or, in radians, θ is between $\pi/2$ and $3\pi/2$), then the argument of the function Q is negative, and so the error probability is greater than $1/2$. For such a phase error, the receiver is wrong more often than it is right! In fact, as $N_0 \to 0$, the argument of the function Q in (6.16) is approaching $-\infty$ whenever $\cos\theta$ is negative, and $Q(-\infty) = \Phi(\infty) = 1$. Thus, in the absence of noise, the receiver will always make an error. The problem is, of course, that for phase angles in this range, the receiver has the two signals confused: What appears to the receiver to be $s_0(t)$ is actually $s_1(t)$ and vice versa. Notice that any antipodal signals of the form given by (6.1) are reversed by a phase change of π radians.

For some purposes, it is convenient to write $P_e(\theta)$ in a slightly different form to reflect the degradation in effective signal energy. This is accomplished by taking the $\cos\theta$ term inside the square root to give

$$P_e(\theta) = Q\left(\sqrt{2\,\mathcal{E}(\cos\theta)^2/N_0}\,\right). \tag{6.17}$$

To achieve a given error probability P, suppose that the required energy is $\mathcal{E}'$ if the phase error is 0. For this same error probability, let $\mathcal{E}$ denote the energy that is required to achieve error probability P by a receiver that has phase error θ. From the preceding results, we conclude that

$$P = Q\left(\sqrt{2\,\mathcal{E}'/N_0}\,\right) = Q\left(\sqrt{2\,\mathcal{E}(\cos\theta)^2/N_0}\,\right). \tag{6.18}$$

Because Q is a strictly decreasing, continuous function, we must have $\mathcal{E}' = \mathcal{E}\cos^2\theta$. In addition, since $\cos^2\theta < 1$ for $\theta \neq 0$ (the case $\theta = \pi$ is not of interest, as has been discussed), it follows that $\mathcal{E}' < \mathcal{E}$ if there is a nonzero phase error. That is, more energy is required to achieve a given error probability if there is a phase error in the receiver (which is not surprising). We conclude that, to achieve a given error probability, the amount of energy required by a receiver with phase error θ is

$$\mathcal{E} = \mathcal{E}'/\cos^2\theta,$$

where $\mathcal{E}'$ is the amount of energy required to achieve the same error probability in a receiver with no phase error.

The term $\cos^2\theta$ can be interpreted as the degradation in effective energy due to the phase error. That is, for a receiver with phase error θ, the actual energy is $\mathcal{E}$, but the effective energy is $\mathcal{E}' = \mathcal{E}\cos^2\theta$, which is the term that must be inserted in the expression for the error probability. The degradation can be expressed as a ratio $\delta = \mathcal{E}'/\mathcal{E} = \cos^2\theta$, or it can be expressed in decibels (dB). First, let

$$\mathcal{E}'_{dB} = 10\log_{10}(\mathcal{E}'),$$

and observe that

$$\mathcal{E}'_{dB} = 10\log_{10}(\mathcal{E}\cos^2\theta) = 10\log_{10}(\mathcal{E}) + 10\log_{10}(\cos^2\theta).$$

Letting

$$\mathcal{E}_{dB} = 10\log_{10}(\mathcal{E}),$$

we see that the degradation in effective energy (in decibels) is

$$\mathcal{E}'_{dB} - \mathcal{E}_{dB} = 10\log_{10}(\cos^2\theta) = 20\log_{10}(\cos\theta)$$

for $0 < \theta < \pi/2$. We define $\delta_{dB} = 10\log_{10}\delta$ and observe that

$$\delta_{dB} = 10\log_{10}(\mathcal{E}'/\mathcal{E}) = \mathcal{E}'_{dB} - \mathcal{E}_{dB}.$$

The amount of additional energy, measured in decibels, that is required to compensate for a phase error in the receiver is $\mathcal{E}_{dB} - \mathcal{E}'_{dB} = -\delta_{dB}$.

Table 6.1: Degradation in Effective Energy Due to a Phase Error

Phase Error θ (degrees)	Degradation δ	δ_{dB}
5	0.992	−0.03
7	0.985	−0.07
9	0.976	−0.11
10	0.970	−0.13
15	0.933	−0.30
20	0.883	−0.54
45	0.500	−3.01

Table 6.2: Degradation in Error Probability Due to a Phase Error ($\mathcal{E}/N_0$ is 10.5 dB for all entries)

Phase Error θ (degrees)	Error Probability $P_e(\theta)$
0	1.08×10^{-6}
5	1.19×10^{-6}
10	1.54×10^{-6}
15	2.38×10^{-6}
20	4.27×10^{-6}
45	4.05×10^{-4}

Numerical values for δ and δ_{dB} are given in Table 6.1 for several values of θ. For most applications, degradations of a few tenths of a decibel can be tolerated, and for certain communication systems larger degradations are allowable. According to Table 6.1, phase errors of 10 to 15 degrees are acceptable in such systems. In some communication systems, such as satellite systems, it is often required to keep the degradation due to the phase error to about 0.1 dB or less. For these systems, phase errors should be no more than about 9 degrees for satisfactory performance. Although the values for δ_{dB} are given to the nearest hundredth of a decibel in Table 6.1, it is usually impossible to measure system gains and losses (e.g., antenna gain or propagation loss) to this degree of accuracy. Therefore, for most applications, the numbers should be rounded to the nearest tenth of a decibel.

In Table 6.2, the degradation due to the phase error is illustrated in a different way. For Table 6.1, we hold constant the probability of error and determine how much additional energy is required to compensate for the phase error. For the data in Table 6.2, the energy is fixed and the table entries give the error probability as a function of the phase error. As in Table 6.1, the numerical results presented in Table 6.2 are obtained from equation (6.16). For the data in this table, $(\mathcal{E}/N_0)_{\mathrm{dB}}$ is 10.5 dB, which corresponds to $\mathcal{E}/N_0 \approx 11.22$, so the error probability in the absence of a phase error is $P_e(0) \approx Q(\sqrt{22.44}) \approx 1.08 \times 10^{-6}$. Notice that the error probability increases by more than two orders of magnitude as the phase error increases from 0 degrees to 45 degrees.

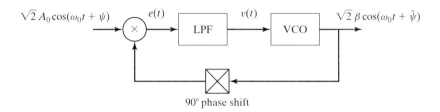

Figure 6-8: A phase-locked loop.

6.2.2   Estimation of the Phase of the BPSK Signal

We begin with the problem of estimating the phase of an unmodulated sinusoidal signal. A commonly used suboptimum system for estimating the phase of a sinusoidal signal is shown in Figure 6-8. This system is known as a *phase-locked loop*.

   If the input to the phase-locked loop is the signal $\sqrt{2}\,A_0\cos(\omega_0 t + \psi)$, it is desired that the output of the loop be a sinusoidal signal whose phase $\hat{\psi}$ is approximately ψ; the amplitude of this signal is unimportant. The components of the phase-locked loop are a lowpass filter (LPF), a voltage-controlled oscillator (VCO), a phase shifter, and an analog multiplier. The output of the VCO is a sinusoidal signal whose phase is proportional to the integral of the input voltage $v(t)$. The phase is the integral of the frequency, so the *frequency* of the VCO output is proportional to the input voltage.

   The operation of the phase-locked loop can be understood by examining the signals at various points in the loop. The output of a VCO with quiescent frequency ω_0 (the frequency that results if the input voltage is zero for a long period of time) is denoted by $\sqrt{2}\,\beta\cos(\omega_0 t + \hat{\psi})$. If this is the input to the phase shifter, the output is

$$\sqrt{2}\,\beta\cos[\omega_0 t + \hat{\psi} + (\pi/2)] = -\sqrt{2}\,\beta\sin(\omega_0 t + \hat{\psi}).$$

This is multiplied by $\sqrt{2}\,A_0\cos(\omega_0 t + \psi)$, and the output of the multiplier is given by

$$e(t) = A_0\,\beta\{\sin(\psi - \hat{\psi}) + \sin(2\,\omega_0 t + \psi + \hat{\psi})\}.$$

Because the lowpass filter removes the double-frequency term, the filter output is

$$v(t) = A_0\,\beta\sin(\psi - \hat{\psi}).$$

   Let us examine the result of an error in the phase estimate. If ψ is greater than $\hat{\psi}$, then the VCO input is positive. A positive input to the VCO results in an increase in the phase of the output, so $\hat{\psi}$ is increased, thereby reducing the phase error $|\psi - \hat{\psi}|$. On the other hand, if ψ is less than $\hat{\psi}$, the VCO input is negative. A negative input to the VCO results in a decrease in the phase of the output, so $\hat{\psi}$ is decreased, thereby reducing the phase error $|\psi - \hat{\psi}|$. In actual applications of the phase-locked loop, there will also be noise at the input, so the phase error will never be reduced to zero.

   Because the BPSK signal is a *modulated* sinusoidal signal, it is not sufficient to use the basic phase-locked loop to estimate the phase of the received signal in a BPSK

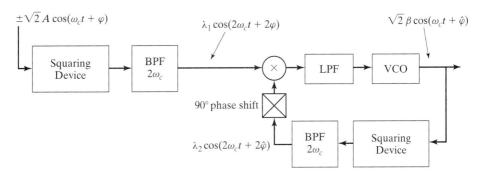

Figure 6-9: The squaring loop for BPSK.

communication system. For this reason, we consider the *squaring loop*, which is one type of loop that can be used to obtain the phase estimate needed for demodulation of BPSK signals. (One alternative is the Costas loop shown in Problem 6.4.) The key idea behind the squaring loop is that squaring the BPSK signal removes its modulation. Recall from Section 6.1 that the BPSK signal is of the form $\pm\sqrt{2}\,A\cos(\omega_c t + \varphi)$. The double-frequency component of the square of this signal is $A^2 \cos(2\,\omega_c t + 2\,\varphi)$, which is a constant-amplitude sinusoidal signal. From this unmodulated sinusoidal signal, a phase-locked loop can obtain an estimate of the phase φ. A system to accomplish this is shown in Figure 6-9. The quiescent frequency of the VCO is ω_c for this system, and the center frequency for the bandpass filters is $2\,\omega_c$.

The purpose of the bandpass filters is to remove the baseband components of the outputs of the squaring devices; it is desired to pass the second harmonic only. The implementation of the squaring device is not too critical. Because its only function is to generate a second harmonic of its input signal, it does not actually have to be a square-law device. The second harmonic can, in principle, be generated by a wide range of nonlinear devices, but some may be more efficient than others in that they may generate a larger second harmonic. Other harmonics generated by the nonlinear device can be removed by the bandpass filters. The constants λ_1 and λ_2 depend on the implementation of the squaring device and the bandpass filters, and they play no essential role. Notice that the loop portion of this system has an input at frequency $2\,\omega_c$ and an output at frequency ω_c, hence, it performs a frequency division (by two) on the input sinusoidal signal.

One difficulty with a squaring loop is that the resulting carrier reference has a phase ambiguity of π radians; that is, the loop is also in lock if the VCO output is $\sqrt{2}\,\beta\cos(\omega_0 t + \hat{\varphi} + \pi)$. The effect of this π-radian phase error is to change the sign of the decision statistic (see Problem 6.3), thereby reversing the decision made by the threshold device. The loop cannot recognize a π-radian phase error. There are several ways to deal with this difficulty, the simplest of which is to begin each segment of a data sequence with a set of symbols that are known to the receiver. If the decision device produces the wrong decisions on these symbols, it is assumed that the carrier reference is π radians out of phase, and the decisions for these and all subsequent symbols are reversed from the normal convention.

Another solution to the phase ambiguity is to employ differential encoding of the data, which is illustrated in Problem 6.5 and described in detail in Section 7.6.1. In this approach, the information is conveyed by agreements or disagreements in the polarities of consecutive pulses, rather than by the polarities themselves. For example, the data symbol 0 can be sent by keeping the polarities the same for two consecutive pulses, and the data symbol 1 can be sent by making the polarities different. The idea is that if a π-radian phase error reverses *all* of the decision statistics corresponding to a sequence of transmitted pulses, then the agreement in the polarities of two consecutive decision statistics is unchanged. The same is true of the disagreement in the polarities of two consecutive decision statistics. One disadvantage of differential encoding is that it increases the probability of error, because an error occurs in making a decision on the data symbol if an error is made in either one (but not both) of the two pulses that represent that data symbol. See Problem 6.5 for details. If the value of $\mathcal{E}/N_0$ is large, the use of differential encoding for coherent reception of BPSK results in an approximate doubling of the probability of error.

For some applications that require fairly small error probabilities, a doubling of the error probability is not too serious, because only a small increase in $\mathcal{E}/N_0$ is required to offset a factor-of-two increase in the probability of error. To determine the amount of increase required, we must evaluate $\mathcal{E}_2/\mathcal{E}_1$ if $\mathcal{E}_2$ and $\mathcal{E}_1$ are related by

$$Q\left(\sqrt{2\mathcal{E}_2/N_0}\right) = \tfrac{1}{2}Q\left(\sqrt{2\mathcal{E}_1/N_0}\right).$$

This relationship implies that coherent reception of BPSK with energy $\mathcal{E}_2$ reduces the probability of error by a factor of two compared with the error probability for coherent reception of BPSK with energy $\mathcal{E}_1$. Let

$$p = Q\left(\sqrt{2\mathcal{E}_2/N_0}\right),$$

and consider $p = 10^{-3}$. At this error probability, calculations give $\mathcal{E}_2/\mathcal{E}_1 \approx 1.15$, which corresponds to an approximate 15% increase in energy to offset a factor-of-two increase in error probability. Because $10\log_{10}(1.15) \approx 0.61$, this is an approximate increase of 0.6 dB, which may be unacceptably large for some applications. However, for $p = 10^{-6}$ the required increase in energy is much smaller. At an error probability of 10^{-6}, $\mathcal{E}_2/\mathcal{E}_1 \approx 1.03$ and $10\log_{10}(\mathcal{E}_2/\mathcal{E}_1) \approx 0.13$, so that the approximate increase is about 0.1 dB only. Such a small increase in the required energy is acceptable for most applications.

6.3   One-Dimensional Signals for Coherent Communications

We begin by reviewing BPSK. In order to introduce certain forms of nonbinary modulation, it is helpful to first examine more general ways to view BPSK signals. The notion of BPSK signaling is then extended to obtain other types of binary signaling. After this, the generalization is carried one step further to obtain a class of *nonbinary* modulation techniques. The emphasis in this section is on a special form of amplitude

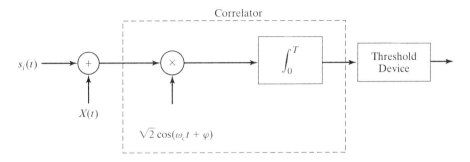

Figure 6-10: Correlation receiver for BPSK.

modulation known as binary amplitude-shift keying. All signals in this section can be viewed as one dimensional in a sense that is made precise in Section 6.3.5.

6.3.1   BPSK Signaling Revisited

As discussed in Section 6.1.1, the BPSK signal for a specific signaling interval (e.g., $[0, T]$) can be viewed as binary phase modulation (by 0 or π radians) or as binary amplitude modulation (by $+1$ or -1). Viewing it as amplitude modulation, we can write the general form of the transmitted signal for BPSK modulation as

$$s(t) = \sqrt{2}\, A\, a(t) \cos(\omega_c t + \varphi). \tag{6.19}$$

For the transmission of a single pulse via BPSK, the baseband signal $a(t)$ is either $+p_T(t)$ or $-p_T(t)$, depending on whether 0 or 1, respectively, is being sent. This can be expressed as

$$a(t) = b\, p_T(t),$$

where b is either $+1$ or -1. If $X(t)$ represents the channel noise, the received signal is given by

$$Y(t) = s(t) + X(t).$$

The ideal correlation receiver, which has a perfect phase reference, multiplies the received signal $Y(t)$ by

$$r(t) = \sqrt{2} \cos(\omega_c t + \varphi)$$

and integrates the result over the interval $[0, T]$, as shown in Figure 6-10.

We can generalize our interpretation of (6.19) by allowing baseband waveforms other than rectangular pulses and amplitudes other than ± 1. For example, a more general binary signal, known as *binary amplitude modulation* (binary AM), is obtained if the baseband signal $a(t)$ of (6.19) can be either of two appropriate baseband waveforms, each of duration T. Binary AM can be described mathematically in terms of such

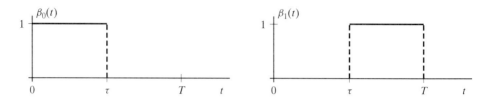

Figure 6-11: Example of a binary AM signal set.

waveforms, $\beta_0(t)$ and $\beta_1(t)$, as follows: Using the signal of (6.19), we let $a(t) = \beta_0(t)$ to send a 0 and $a(t) = \beta_1(t)$ to send a 1.

In a commonly used special form of binary AM, known as *binary amplitude-shift keying* (binary ASK), $\beta_0(t)$ is a real multiple of $\beta_1(t)$. This restriction permits us to express the amplitude modulation $a(t)$ in the form

$$a(t) = b\,\beta(t),$$

where $\beta(t)$ is any baseband waveform of duration T and b takes on one of two possible values, which we denote by u_0 and u_1. The signal $\beta(t)$ determines the basic baseband waveform, but the variable b is used to convey the information: 0 is sent by letting $b = u_0$ and 1 is sent by letting $b = u_1$. Consequently, the information is represented by shifts in the amplitude of a common baseband waveform; hence, the name *amplitude-shift keying*. The signaling scheme known as *on–off signaling* is a special case of binary ASK obtained by letting $u_0 = +1$ (on) and $u_1 = 0$ (off).

This is a good time for a word of warning about the terminology: Some authors do not distinguish between the terms *amplitude-shift keying* and *amplitude modulation*. It is particularly common to refer to binary ASK as binary AM. Of course, binary ASK is a special case of binary AM, so there is no *error* in doing so, but we prefer the more precise terminology for pedagogical reasons.

Example 6-1   A Binary AM Signal Set
As an example of a binary AM signal that is not binary ASK, let $\tau = T/2$ and define $\beta_0(t)$ to be $p_\tau(t)$ and $\beta_1(t)$ to be $p_\tau(t - \tau)$, as shown in Figure 6-11. This signal set is sometimes referred to as binary pulse-position modulation, because the information is conveyed by the position or location of the rectangular pulse. Observe that the two signals are orthogonal and each has energy $T/2$.   ∎

6.3.2   Optimum Receiver for Binary ASK

The principles of Chapter 5 can be applied to determine the optimum receiver for binary AM. In this section, we derive the optimum receiver for binary ASK. The derivation of the optimum receiver for general binary AM is given as Problem 6.6.

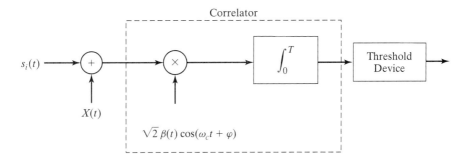

Figure 6-12: Correlation receiver for binary ASK.

The binary signal set $\{s_0, s_1\}$ representing binary ASK is defined by

$$s_i(t) = \sqrt{2}\,A\,u_i\,\beta(t)\cos(\omega_c t + \varphi), \tag{6.20}$$

for $i = 0$ and $i = 1$. As a result,

$$s_0(t) - s_1(t) = \sqrt{2}\,A\,(u_0 - u_1)\,\beta(t)\cos(\omega_c t + \varphi),$$

so the optimum correlation receiver multiplies the receiver input by a signal proportional to $\beta(t)\cos(\omega_c t + \varphi)$ and integrates the result over the duration of the signal. The block diagram for such a receiver is shown in Figure 6-12.

The procedures for determining the optimum thresholds for minimax, Bayes, and maximum-likelihood decisions are exactly as described in Chapter 5. In order to simplify the presentation, we focus on maximum-likelihood decisions in this chapter. Recall from Section 5.3.1.2 (especially Figure 5-10) that the maximum-likelihood threshold is equal to the minimax threshold for binary communications over the AGN channel. Recall also that $\bar{\mathcal{E}} = (\mathcal{E}_0 + \mathcal{E}_1)/2$ is the average energy per bit for the binary signal set. Because there are only two signals, $\bar{\mathcal{E}}$ is also the average energy per signal. Since larger signal sets are considered in this chapter, it is convenient to let $\bar{\mathcal{E}}_s$ denote the average energy per signal in the set, regardless of the number of signals.

Since $P_{e,0} = P_{e,1}$, the subscripts 1 and 0 can be dropped and the probability of error is denoted by P_e. From the results of Chapter 5 for binary signaling over the AWGN channel with the optimum correlator and the maximum-likelihood threshold, we have

$$P_e = Q\left(\sqrt{\bar{\mathcal{E}}_s(1 - r)/N_0}\,\right), \tag{6.21}$$

where r is the correlation coefficient $r = (s_0, s_1)/\bar{\mathcal{E}} = (s_0, s_1)/\bar{\mathcal{E}}_s$. This result is valid for any signal set. In order to apply the result to binary ASK, we must determine the average energy and the inner product (s_0, s_1) for signals of the form

$$s_i(t) = \sqrt{2}\,A\,u_i\,\beta(t)\cos(\omega_c t + \varphi).$$

This is relatively easy, but there is a minor technical problem concerning the double-frequency terms that arise in the analysis. Most of the details in the next section are required only to establish that the double-frequency terms can be neglected in nearly all cases of interest. The reader who is willing to accept this premise may prefer to skip to Section 6.3.4.

6.3.3   Consideration of the Double-Frequency Components

The energy in the signal $s_i(t) = \sqrt{2}\, A\, u_i\, \beta(t) \cos(\omega_c t + \varphi)$ is given by

$$\mathcal{E}_i = \int_0^T [s_i(t)]^2 \, dt = 2A^2 u_i^2 \int_0^T \beta^2(t) \cos^2(\omega_c t + \varphi) \, dt.$$

By expressing the integrand as the sum of a baseband component and a double-frequency component, we have

$$\mathcal{E}_i = A^2 u_i^2 \int_0^T \beta^2(t) \left[1 + \cos(2\,\omega_c t + 2\,\varphi) \right] dt.$$

An approximation based on the leading term of this expression is

$$\mathcal{E}_i \approx A^2 u_i^2 \int_0^T \beta^2(t) \, dt.$$

In other words, $\mathcal{E}_i \approx A^2 u_i^2 \mathcal{E}_\beta$, where $\mathcal{E}_\beta$ is the energy in the waveform $\beta(t)$. This is a good approximation whenever the magnitude of the integral of the double-frequency term is small compared with $\mathcal{E}_\beta$. That is, we need $|I_{df}| \ll \mathcal{E}_\beta$, where

$$I_{df} = \int_0^T \beta^2(t) \cos(2\,\omega_c t + 2\,\varphi) \, dt. \tag{6.22}$$

There are many important signal sets for which the double-frequency term is *identically zero*. For example, suppose the carrier frequency is such that $2 f_c T$ is an integer (i.e., $\omega_c T$ is a multiple of π) and the waveform $\beta(t)$ is constant on intervals of the form $[nT_0, (n+1)T_0]$, where $T_0 = 1/2 f_c$. Of course, this is satisfied in a trivial way if the waveform is a rectangular pulse of width T, or, more generally, if its width is a multiple of T_0. Note that T_0 is the period of a sinusoidal signal of frequency $2 f_c$ (e.g., the signal $\cos(2\,\omega_c t + 2\,\varphi)$). Observe that if $N = (T/T_0) - 1$, then

$$I_{df} = \sum_{n=0}^N \int_{nT_0}^{(n+1)T_0} \beta^2(t) \cos(2\,\omega_c t + 2\,\varphi) \, dt.$$

That is, the integral of the double-frequency component is the sum of integrals over intervals of the form $[nT_0, (n+1)T_0]$. Next, notice that $\beta(t)$ is constant for each of these intervals, so each term in the sum is the integral of a sinusoidal signal over one full period. As a result, each integral is zero and $I_{df} = 0$. In this case,

$$\mathcal{E}_i = A^2 u_i^2 \int_0^T \beta^2(t) \, dt.$$

These conditions are satisfied, for example, if $\beta(t)$ is composed of a sequence of rectangular pulses of possibly different amplitudes and widths. If the width of each pulse is a multiple of τ, then it is sufficient to require τ to be a multiple of T_0. In particular, if all pulses are of width τ, this waveform can be represented as

$$\beta(t) = \sum_{j=0}^{N'-1} a_j \, p_\tau(t - j\tau),$$

where $a_0, a_1, \ldots, a_{N'-1}$ is the sequence of pulse amplitudes and $N' = T/\tau$. Waveforms of this type are used in certain classes of spread-spectrum communication systems.

In many other cases of interest, the integral of the double-frequency term is at least *approximately* zero, such as if the bandwidth of the baseband signal $\beta(t)$ is small compared with the carrier frequency (i.e., the Fourier transform of $\beta(t)$ is approximately zero at and beyond frequency f_c). In this situation, the waveform $\beta(t)$ cannot change much over the interval $[nT_0, (n+1)T_0]$, because to change significantly over an interval of length T_0, the Fourier transform of $\beta(t)$ would have to be nonzero for at least one frequency near or above frequency $1/T_0$. Because $1/T_0 = 2f_c$, this is not possible. Because $\beta(t)$ is nearly constant over intervals of length T_0, each of the integrals in the expression for I_{df} satisfies

$$\int_{nT_0}^{(n+1)T_0} \beta^2(t) \cos(2\omega_c t + 2\varphi)\,dt \approx \beta^2(nT_0) \int_{nT_0}^{(n+1)T_0} \cos(2\omega_c t + 2\varphi)\,dt = 0.$$

For some signal sets, the condition $|I_{df}| \ll \mathcal{E}_\beta$ can be verified directly, as in the following exercise.

Exercise 6-1
Consider the signal set in which

$$s_0(t) = 2\sqrt{2}\,A\sin(\pi t/T)\cos(\omega_c t + \varphi), \quad 0 \le t \le T,$$

and

$$s_1(t) = \sqrt{2}\,A\sin(\pi t/T)\cos(\omega_c t + \varphi), \quad 0 \le t \le T.$$

Assume that $\omega_c \gg T^{-1}$. Verify by evaluation of I_{df} and $\mathcal{E}_\beta$ that $I_{df} \ll \mathcal{E}_\beta$.

Solution.   For this signal set, $\beta(t) = \sin(\pi t/T)$, so

$$\mathcal{E}_\beta = \int_0^T \sin^2(\pi t/T)\,dt = T/2.$$

From (6.22), we see that

$$I_{df} = \int_0^T \sin^2(\pi t/T)\cos(2\omega_c t + 2\varphi)\,dt.$$

Let $\omega_0 = 2\pi/T$, and write $\sin^2(\pi t/T)$ as $[1 - \cos(\omega_0 t)]/2$. Then combine $\cos(\omega_0 t)$ with $\cos(2\omega_c t + 2\varphi)$, using $\cos(x)\cos(y) = [\cos(x-y) + \cos(x+y)]/2$. Evaluate the resulting integrals to obtain

$$\begin{aligned}
I_{df} = {}& (4\,\omega_c)^{-1}\{\sin(2\,\omega_c T + 2\,\varphi) - \sin(2\,\varphi)\} \\
& - [4\,(2\,\omega_c - \omega_0)]^{-1}\{\sin[(2\,\omega_c - \omega_0)T + 2\,\varphi] - \sin(2\,\varphi)\} \\
& - [4\,(2\,\omega_c + \omega_0)]^{-1}\{\sin[(2\,\omega_c + \omega_0)T + 2\,\varphi] - \sin(2\,\varphi)\}.
\end{aligned}$$

Using the fact that $|\sin(u) - \sin(v)| \le 2$, we see that

$$|I_{df}| \le [2\,\omega_c]^{-1} + [2\,(2\,\omega_c - \omega_0)]^{-1} + [2\,(2\,\omega_c + \omega_0)]^{-1}.$$

Because $\omega_c \gg T^{-1}$ and $\omega_0 = 2\pi/T$, it follows that $\omega_c \gg \omega_0$; therefore,

$$\omega_c < (2\,\omega_c - \omega_0) < (2\,\omega_c + \omega_0).$$

From this fact, we see that

$$|I_{df}| < 3\,[2\,\omega_c]^{-1} = 3/(2\,\omega_c).$$

All that remains is to compare this bound with $\mathcal{E}_\beta = T/2$, again using the fact that $\omega_c \gg T^{-1}$, and hence $\omega_c^{-1} \ll T$, to conclude that $|I_{df}| \ll \mathcal{E}_\beta$.   ∎

In view of these considerations, we assume that double-frequency terms are negligible in all that follows, and we do not distinguish between the situations in which this term is *identically* zero and those in which it is only *approximately* zero. In particular, we take $\mathcal{E}_i \approx A^2 u_i^2 \mathcal{E}_\beta$ to be an exact relationship for the signal energy, and so we write $\mathcal{E}_i = A^2 u_i^2 \mathcal{E}_\beta$.

6.3.4   Error Probability for Binary ASK

We are now ready to consider the probability of error for binary ASK signals, an additive white Gaussian noise channel, and the optimum correlation receiver as shown in Figure 6-12. Recall that the signals are of the form

$$s_i(t) = \sqrt{2}\,A\,u_i\,\beta(t)\cos(\omega_c t + \varphi). \tag{6.23}$$

The average energy per signal for this signal set is

$$\bar{\mathcal{E}}_s = (\mathcal{E}_0 + \mathcal{E}_1)/2 = A^2 \mathcal{E}_\beta (u_0^2 + u_1^2)/2. \tag{6.24}$$

For instance, antipodal signals have the property that $u_1 = -u_0$. If $u_0 = 1$, this gives $\mathcal{E}_i = A^2 \mathcal{E}_\beta$ and $\bar{\mathcal{E}}_s = A^2 \mathcal{E}_\beta$.

The inner product for the binary ASK signal set is given by

$$(s_0, s_1) = 2A^2 u_0\,u_1 \int_0^T \beta^2(t)\cos^2(\omega_c t + \varphi)\,dt.$$

As in Section 6.3.3, the term $\cos^2(\omega_c t + \varphi)$ is expressed as the sum of a baseband term and a double-frequency term. Ignoring the double-frequency term (for the same reasons as in Section 6.3.3), we have

$$(s_0, s_1) = A^2 u_0\,u_1 \int_0^T \beta^2(t)\,dt,$$

which is equivalent to $(s_0, s_1) = A^2 u_0\,u_1 \mathcal{E}_\beta$.

The parameter r is the correlation coefficient $r = (s_0, s_1)/\bar{\mathcal{E}}_s$. Because

$$\bar{\mathcal{E}}_s = A^2 \mathcal{E}_\beta (u_0^2 + u_1^2)/2.$$

the correlation coefficient is given by

$$r = 2\,u_0\,u_1/(u_0^2 + u_1^2).\tag{6.25}$$

Notice that $u_0^2 = u_1^2$ for equal-energy signals, so that r is either -1 (antipodal signals) or $+1$ (identical signals) if the binary ASK signals have equal energy. From Chapter 5, we know these values represent the extremes of the range of possible values for r.

The expressions just given for $\bar{\mathcal{E}}_s$ and r can now be substituted into the expression

$$P_e = Q\left(\sqrt{\bar{\mathcal{E}}_s(1 - r)/N_0}\,\right)$$

to find the minimax error probability for binary ASK modulation, an AWGN channel with spectral density $N_0/2$, and the optimum receiver. The result will be in terms of the parameters u_0, u_1, and $\mathcal{E}_\beta$.

Example 6-2   On–Off Signaling
Suppose that the signal set is given by $s_0(t) = \sqrt{2}\,A\cos(\omega_c t + \varphi)$, $0 \le t \le T$, and $s_1(t) = 0$, $0 \le t \le T$. As usual, assume that double-frequency terms can be neglected. From (6.23), we see that $u_0 = +1$, $u_1 = 0$, and $\beta(t) = p_T(t)$. From (6.24) and the fact that $\mathcal{E}_\beta = T$ for this waveform, we conclude that $\bar{\mathcal{E}}_s = A^2 T/2$. From (6.25), we see that $r = 0$, as expected. (This is an orthogonal signal set.) Consequently, the probability of error is given by $P_e = Q(A\sqrt{T/2N_0})$ for the given signal set and the optimum (minimax) receiver. By comparison, *antipodal* signals of the same amplitude and pulse shape as $s_0(t)$ have $\bar{\mathcal{E}}_s = A^2 T$ and $r = -1$, so that they achieve $P_e = Q(A\sqrt{2T/N_0})$. ∎

6.3.5   *M*-ary Amplitude-Shift Keying (*M*-ASK)

The most general form of one-dimensional nonbinary AM is based on a collection $\{\beta_i : 0 \le i \le M - 1\}$ of baseband waveforms ($M > 2$), and the transmitted signal during the interval $[0, T]$ is

$$s(t) = \sqrt{2}\,A\,a(t)\cos(\omega_c t + \varphi),$$

where $a(t) = \beta_i(t)$ for some choice of i. In general, the waveforms are not related to each other except that they are defined on the common time interval $[0, T]$.

Nonbinary ASK is a special case of nonbinary AM in which the transmitted signals have the form

$$s_i(t) = \sqrt{2}\,A\,u_i\,\beta(t)\cos(\omega_c t + \varphi).\tag{6.26}$$

This is the same general form as for binary ASK signals, the only difference is that the data variable u_i is allowed to take on more than two values in nonbinary ASK. It is customary to denote the number of different values for the data variable by M, denote the set of values by $\{u_i : 0 \le i \le M - 1\}$, and refer to the resulting modulation as *M-ary amplitude-shift keying*, abbreviated as *M*-ASK. For instance, if there are four different values, denoted by u_0, u_1, u_2, and u_3, the resulting signal set is a *4-ASK* signal set. Special names are occasionally used for $M = 2$ (binary ASK), $M = 3$ (ternary ASK), and $M = 4$ (quaternary ASK). Because the parameters u_i are sometimes referred to as the "levels" of the signals, M-ASK is also referred to as multilevel ASK in the literature. The term M-ary ASK is also used commonly.

Observe that, in general, the signals obtained in this manner do not all have the same energy; in fact, nonbinary ASK signals cannot all have the same energy. Two parameters that are of interest in characterizing the energy for the signal set are the maximum energy and the average energy. If $\mathcal{E}_i$ denotes the energy in the signal $s_i(t)$, then the *maximum energy* is

$$\mathcal{E}_{\max} = \max\{\mathcal{E}_i : 0 \le i \le M - 1\},$$

and the *average energy per signal* is

$$\bar{\mathcal{E}}_s = (\mathcal{E}_0 + \mathcal{E}_1 + \cdots + \mathcal{E}_{M-1})/M.$$

Because the elements of the signal set are often referred to as symbols, $\bar{\mathcal{E}}_s$ is also known as the *average energy per symbol*. It is desirable, of course, that $\mathcal{E}_{\max}$ and $\bar{\mathcal{E}}_s$ are as small as possible to minimize the required transmitter power, antenna gain, etc.

It is convenient to define the signal $\xi(t)$ by

$$\xi(t) = \sqrt{2}\, A\, \beta(t) \cos(\omega_c t + \varphi).$$

Because

$$s_i(t) = u_i\, \xi(t)$$

for each value of i ($0 \le i \le M-1$), it follows that $\xi(t)$ characterizes the basic waveform for the signal set, just as $\beta(t)$ characterizes the baseband waveform for the set. Each signal in the set has the shape of the waveform $\xi(t)$; hence, the signal set consists of real multiples of $\xi(t)$. For this reason, we refer to $\xi(t)$ as a *basis function* for the signal set defined by (6.26).

To generate the signals for the M-ASK signal set, it suffices to generate the single signal $\xi(t)$ and then multiply it by the various gain constants u_i. Notice that if $\mathcal{E}$ denotes the energy in $\xi(t)$, then the energy in the ith signal is $\mathcal{E}_i = u_i^2 \mathcal{E}$.

Example 6-3   The Energies for Two Possible 4-ASK Signal Sets

It is natural to select the values of the u_i in a way that makes them equally spaced. In this example, we consider two choices for quaternary AM signals:

(a) One choice for the set of values for the u_i is the set $\{-3, -1, +1, +3\}$, which corresponds to $u_i = 2i - 3$ for $0 \le i \le 3$. For this set, the energies for the four signals are $\mathcal{E}_0 = \mathcal{E}_3 = 9\,\mathcal{E}$, corresponding to $u_0 = -3$ and $u_3 = +3$, and $\mathcal{E}_1 = \mathcal{E}_2 = \mathcal{E}$, corresponding to $u_1 = -1$ and $u_2 = +1$. This gives a maximum energy requirement of $9\,\mathcal{E}$ and an average energy requirement of $5\,\mathcal{E}$.

(b) Another set with the same spacing as the set in part (a) is $\{0, 2, 4, 6\}$, which corresponds to $u_i = 2i$. For this set, $\mathcal{E}_0 = 0$, $\mathcal{E}_1 = 4\mathcal{E}$, $\mathcal{E}_2 = 16\mathcal{E}$, and $\mathcal{E}_3 = 36\mathcal{E}$, so the maximum energy is $36\mathcal{E}$, and the average is $14\mathcal{E}$.

Clearly, the second set is inferior to the first in terms of either the maximum or the average energy that is required to use the two signal sets. Because the spacing between the signals is the same for the two sets, it is natural to conjecture that the error probability is also the same. Results developed in this section can be applied to show that this conjecture is correct. The two sets provide the same error probability, but the set defined in part (a) requires less energy in order to do so. As a result, the set defined in part (a) is preferred over the set defined in part (b). ∎

The issues brought out by Example 6-3 can be generalized to sets with an arbitrary number of signals. If M is any even integer, the set

$$S_e = \{-M + 1, -M + 3, \ldots, -1, +1, \ldots, M - 3, M - 1\}$$

provides a set of M signals with equal spacing. For this set, the values of the data variables can be expressed as $u_i = 2i - (M - 1)$ for $0 \le i \le M - 1$, so that the spacing is $u_i - u_{i-1} = 2$ for each value of $i > 0$. On the other hand, if M is odd, the same spacing is obtained by using the set

$$S_o = \{-M + 1, -M + 3, \ldots, -2, 0, +2, \ldots, M - 3, M - 1\},$$

which also corresponds to $u_i = 2i - (M - 1)$. Of course, there are other choices that achieve the same spacing. For instance, regardless of whether M is even or odd, it can be shown that, for an additive white Gaussian noise channel, the set

$$S' = \{0, 2, 4, \ldots, 2(M - 1)\}$$

will result in the same error probabilities $P_{e,i}$ as the appropriate set S_e (for M even) or S_o (for M odd) if the optimum receiver is used for each signal set. However, the maximum energy required by the signals in the set S' is four times that required by the sets S_e or S_o, and the average energy required for S' is at least twice that required by S_e or S_o. Note that in the special case $M = 2$, this is just the observation that the set $\{-1, +1\}$, which is an antipodal signal set, is a better choice than the set $\{0, +2\}$, which is an orthogonal signal set. The error probabilities $P_{e,0}$ and $P_{e,1}$ are the same for the two sets (this is a straightforward exercise based on results from Chapter 5), but the latter set requires four times the maximum energy and twice the average energy of the former.

It is very helpful in the design and analysis of nonbinary signal sets to represent the signals as points in n-dimensional Euclidean space for an appropriate choice of the integer n. This is referred to as a *signal-space representation* of the signal set.

For ASK signal sets, either binary or nonbinary, it is sufficient for the signal space to be one dimensional, because the ASK signal set can be described in terms of a single basis function only, the function $\xi(t)$. Each signal is a real multiple of this basis function, so the ASK signals can be represented as points on the real line, as illustrated in Figure 6-13 for the 4-ASK signals of Example 6-3(a). Recall that the set of data variables

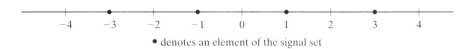

• denotes an element of the signal set

Figure 6-13: 4-ASK signals of Example 6-3(a).

for this example is $\{-3, -1, +1, +3\}$. Because these signals can be represented as points on the real line, ASK signal sets are considered to be *one-dimensional* signal sets. In the next section, we consider signals that require two orthogonal basis functions for their descriptions, and they are referred to as *two-dimensional* signal sets.

Exercise 6-2

Find the ratio of the maximum energy to the average energy for M-ASK signals given by (6.26) with the data variable set $\{u_i : 0 \leq i \leq M - 1\}$ equal to

$$S_e = \{-M + 1, -M + 3, \ldots, -1, +1, \ldots, M - 3, M - 1\}$$

if M is even and

$$S_o = \{-M + 1, -M + 3, \ldots, -2, 0, +2, \ldots, M - 3, M - 1\}$$

if M is odd.

Solution. If $\mathcal{E}$ is the energy in the signal $\xi(t)$ and $\mathcal{E}_i$ is the energy in the signal $s_i(t)$, then $\mathcal{E}_i = u_i^2 \mathcal{E}$ for $0 \leq i \leq M - 1$. Clearly, the maximum energy is $\mathcal{E}_{\max} = (M - 1)^2 \mathcal{E}$. If M is even, the average energy is

$$\bar{\mathcal{E}}_s = 2[(1)^2 + (3)^2 + (5)^2 + \cdots + (M - 1)^2]\mathcal{E}/M$$
$$= 2\mathcal{E}M^{-1}\sum_{k=1}^{M/2}(2k - 1)^2.$$

Using the fact that

$$\sum_{j=1}^{N} j^2 = N(N + 1)(2N + 1)/6$$

and that

$$\sum_{k=1}^{n}(2k - 1)^2 = \sum_{i=1}^{2n} i^2 - \sum_{i=1}^{n}(2i)^2,$$

we conclude that $\bar{\mathcal{E}}_s = (M - 1)(M + 1)\mathcal{E}/3 = (M^2 - 1)\mathcal{E}/3$. It is left to the reader to show that this conclusion is also true if M is odd. It follows that the ratio of the maximum to average energy for the signal set is $\mathcal{E}_{\max}/\bar{\mathcal{E}}_s = 3(M - 1)/(M + 1)$. Notice that $\mathcal{E}_{\max}/\bar{\mathcal{E}}_s \approx 3$ if M is large. ∎

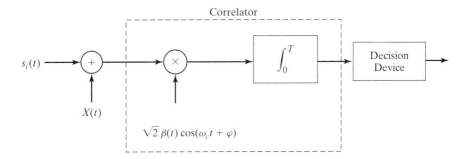

Figure 6-14: Optimum correlation receiver for M-ASK.

6.3.6   Receivers for M-ASK

Recall that the M-ASK signals are of the form

$$s_i(t) = \sqrt{2}\, A\, u_i\, \beta(t) \cos(\omega_c t + \varphi),$$

$0 \le i \le M - 1$, where $\beta(t)$ is time limited to $[0, T]$. These signals can be expressed as $s_i(t) = u_i\, \xi(t)$ where

$$\xi(t) = \sqrt{2}\, A\, \beta(t) \cos(\omega_c t + \varphi).$$

All of the arguments used to prove the optimality of the correlation receiver for BPSK in Section 6.1.4 can be generalized to prove that the optimum receiver for this signal set is matched to $\xi(t)$. In fact, if $\beta(t)$ is the rectangular pulse of duration T, the proof in Section 6.1.4 applies directly to the M-ASK signal set.

For the matched filter formulation, the optimum receiver employs a time-invariant filter with an impulse response that is proportional to $\xi(T - t)$. For the optimum correlation receiver, the received signal is multiplied by the reference signal

$$r(t) = \sqrt{2}\, \beta(t) \cos(\omega_c t + \varphi),$$

and the result is integrated from 0 to T as shown in Figure 6-14. Note that multiplication by $r(t) = \xi(t)/A$ is equivalent to multiplication by $\xi(t)$, provided the constant A is accounted for in the decision device.

In general, the correlator reference may not be optimum for the given signal set. If the reference signal is

$$r(t) = \sqrt{2}\, \zeta(t) \cos(\omega_c t + \hat{\varphi}),$$

the waveform $\zeta(t)$ may not be proportional to $\beta(t)$ or the reference phase $\hat{\varphi}$ may differ from the actual phase φ. The basic analysis is given for the general correlator before the results are specialized to the optimum correlator.

The output of the correlator consists of a term due to the received signal and a term due to the noise. The signal component at the output of the correlator is

$$\hat{u}_i = \int_0^T s_i(t)\, \sqrt{2}\, \zeta(t) \cos(\omega_c t + \hat{\varphi})\, dt.$$

Figure 6-15: Output signal levels for the 4-ASK signals with amplitude levels of Example 6-3(a).

Now, use the fact that $s_i(t) = \sqrt{2}\, A\, u_i\, \beta(t) \cos(\omega_c t + \varphi)$ to obtain

$$\hat{u}_i = 2A\, u_i \int_0^T \beta(t)\, \zeta(t) \cos(\omega_c t + \varphi) \cos(\omega_c t + \hat{\varphi})\, dt.$$

Let $\theta = \varphi - \hat{\varphi}$ denote the phase error, and neglect the double-frequency term to conclude that

$$\hat{u}_i - A\, u_i \cos(\theta) \int_0^T \beta(t)\, \zeta(t)\, dt,$$

which can be written as $\hat{u}_i = u_i A(\beta,\zeta) \cos(\theta)$. If the correlator is optimum (i.e., $\zeta = \beta$ and $\hat{\varphi} = \varphi$), then $\hat{u}_i = u_i A \mathcal{E}_\beta$, where $\mathcal{E}_\beta$ is the energy in the waveform $\beta(t)$. The conclusion is that if $s_i(t)$ is the transmitted signal, the signal component at the output of the correlator is $\hat{u}_i = u_i A\chi$, where $\chi = (\beta,\zeta) \cos(\theta)$ for a general correlator and $\chi = \mathcal{E}_\beta$ for the optimum correlator of Figure 6-14.

For the signal set of Example 6-3(a), which is illustrated in Figure 6-13, $\hat{u}_0 = -3A\chi$, $\hat{u}_1 = -A\chi$, $\hat{u}_2 = +A\chi$, and $\hat{u}_3 = +3A\chi$. Because the output signal levels are proportional to the corresponding data variables u_i, they can be illustrated by a sketch that is essentially the same as Figure 6-13. The output levels are shown as dots in Figure 6-15. The only difference between a signal point in Figure 6-13 and the corresponding output point in Figure 6-15 is the scale factor $A\chi$.

The noise component of the output of the suboptimum correlator is the random variable Y defined by

$$Y = \int_0^T X(t)\, \sqrt{2}\, \zeta(t) \cos(\omega_c t + \hat{\varphi})\, dt,$$

where $X(t)$ is a white Gaussian noise process with spectral density $N_0/2$. The mean of Y is zero. If $\mathcal{E}_\zeta$ denotes the energy in $\zeta(t)$, the variance of Y is given by

$$\sigma^2 = N_0 \int_0^T \zeta^2(t)\, dt/2 = N_0\, \mathcal{E}_\zeta/2.$$

For M-ASK with $M > 2$, the optimum decision device for Figure 6-14 is not the standard threshold device used for binary signaling (cf. Figure 6-12). For M-ASK with a receiver that uses any reasonable linear filter, optimum or not, the decision regions are intervals for the minimax, Bayes, and maximum-likelihood decision criteria. As such, each region is determined completely by its endpoints. The decision device simply determines which interval contains the decision statistic. One possible choice for the decision regions is shown in Figure 6-15 for the signal set of Example 6-3. In Figure 6-15, Γ_i denotes the region that corresponds to the decision that s_i was sent.

6.4   Optimum Decisions in *M*-ary Communication Systems

Thus far, we have examined the optimum decision rules for communication systems with binary signaling only. In order to determine the optimum decision device for such systems as the *M*-ASK communication systems introduced in Section 6.3, it is necessary to generalize our previous decision rules for minimax, Bayes, and maximum-likelihood criteria. This generalization is given in Section 6.4.1, and the results are applied to coherent reception of *M*-ASK in Section 6.4.2. A geometric interpretation of the error-probability expressions for *M*-ASK is given in Section 6.4.3.

6.4.1   Decision Rules for *M*-ary Signaling

First, we must consider the *M*-ary communication scenario. The source wishes to send one of *M* symbols. These symbols are indexed by the integers 0 through $M - 1$. For binary signaling, we have $M = 2$ and we need the indices 0 and 1 only. In an *M*-ary communication system, there are *M* signal waveforms, $s_0, s_1, \ldots, s_{M-1}$, that are available to the transmitter. In order to send symbol i, the signal s_i is transmitted over the channel. The receiver processes the incoming waveform (e.g., using a filter, correlator, or bank of filters or correlators), and it obtains a *decision statistic* upon which the decision is to be based.

An example of a decision statistic is the output of the integrator in Figure 6-14. The decision statistic, which we denote by Z, may be a random variable, as in Section 5.4, or it may be a random vector. For the systems considered in this book, the decision statistic is a continuous random variable or vector. For each i in the range $0 \leq i \leq M - 1$, let f_i denote the conditional density function for Z given that symbol i is sent. For the AGN channel and a linear filter or correlator in the receiver, each of the densities f_i is a Gaussian density.

Our formulation of the decision problem for *M*-ary signaling parallels the approach in Section 5.4. The decision statistic is the input to a decision device, and our goal is to describe the operation of an optimum decision device. The operation of the decision device can be described in terms of the sets $\Gamma_0, \Gamma_1, \ldots, \Gamma_{M-1}$, where Γ_i denotes the region corresponding to the decision that s_i was sent; that is, if it is observed that $Z = z$ and if z is in the set Γ_k, then the decision is that signal s_k was sent. The sets $\Gamma_0, \Gamma_1, \ldots, \Gamma_{M-1}$ are disjoint, and their union is the set of all possible values for the decision statistic.

A *symbol error* is made if the receiver decides that symbol k was sent when in fact symbol i was sent and $i \neq k$. The *symbol error probability* is the probability that the receiver's decision differs from the symbol that was actually sent. More precisely, the *conditional probability of symbol error given that k was sent*, denoted by $P_{e,k}$, is the conditional probability that the decision is some symbol other than k, given that symbol k was actually sent. The conditional probability that the symbol is correct, given that symbol k was sent, is

$$P_{c,k} = P(Z \in \Gamma_k \mid \text{symbol } k \text{ sent}),$$

and, of course, $P_{c,k} = 1 - P_{e,k}$.

Note that unlike for binary signaling and binary decision rules, an M-ary receiver can make an error in more than one way. For example, if the symbol 0 is sent, decisions $1, 2, \ldots,$ and $M - 1$ all result in a symbol error, but they correspond to $M - 1$ different decisions being made by the receiver (each of which is wrong). The symbol error probability, as a performance measure for the system, does not distinguish between the different types of errors. In fact, because the sets $\Gamma_0, \Gamma_1, \ldots, \Gamma_{M-1}$ are disjoint,

$$P_{e,k} = \sum_{i \neq k} P(Z \in \Gamma_i \mid \text{symbol } k \text{ sent}),$$

where the sum is over all i not equal to k.

The goal in designing the receiver, and especially in designing the decision device for the receiver, is to make the symbol error probability small in some sense. The issue of interpreting the phrase "make the error probability small" arises here in the same way as in Chapter 5; namely, in the process of decreasing $P_{e,k}$, we are likely to increase $P_{e,i}$ for one or more values of i other than $i = k$. Hence, we must focus on some specific criterion, such as the minimax, Bayes, or maximum-likelihood criterion, in order to proceed.

In Section 5.4 it is shown that for binary communications the optimum decision regions for the Bayes criterion are

$$\Gamma_0 = \{z : \pi_0 f_0(z) > \pi_1 f_1(z)\}$$

and

$$\Gamma_1 = \{z : \pi_1 f_1(z) > \pi_0 f_0(z)\},$$

where f_i is the conditional density for the decision statistic given that symbol i was sent. Recall that for the Bayes decision rule we do not care about the set $\Gamma' = \{z : \pi_0 f_0(z) = \pi_1 f_1(z)\}$, because it does not contribute to the average probability of error.

For M-ary communication with decision regions $\Gamma_0, \Gamma_1, \ldots, \Gamma_{M-1}$, the probability of making the correct decision given that symbol i was sent is

$$P_{c,i} = \int_{\Gamma_i} f_i(z) \, dz.$$

The average probability of making the correct decision is

$$\overline{P}_c = \pi_0 P_{c,0} + \pi_1 P_{c,1} + \cdots + \pi_{M-1} P_{c,M-1},$$

which can be written in terms of the decision regions Γ_i as

$$\overline{P}_c = \sum_{i=0}^{M-1} \pi_i \int_{\Gamma_i} f_i(z) \, dz.$$

A *Bayes decision rule for M-ary communications* is a decision rule that corresponds to a set of decision regions that gives the maximum possible value of $\overline{P}_c$.

Consider a particular value of z, and suppose that, for this value of z,

$$\pi_k f_k(z) = \max\{\pi_i f_i(z) : 0 \le i \le M - 1\}. \tag{6.27}$$

Clearly, the contribution that such a value of z makes to the quantity $\overline{P}_c$ is as large as possible if we include this value of z in the decision region Γ_k. That is, we maximize the value of $\overline{P}_c$, and therefore minimize $\overline{P}_e = 1 - \overline{P}_c$, by using decision regions for which z is in Γ_k if z satisfies (6.27). There may be ties in selecting the decision region into which a point z should be placed; that is, for some values of z it can be true that (6.27) holds for more than one value of k. For such a value of z, any such value of k will give a Bayes decision for the observation $Z = z$. As a final observation, notice that for the special case $M = 2$ the regions defined by (6.27) reduce to the Bayes decision regions given in Section 5.4 for binary communications.

If the probabilities π_i are all equal, the right-hand side of (6.27) reduces to M^{-1} times the maximum of the conditional densities, and z should be included in Γ_k if

$$M^{-1} f_k(z) = \max\{M^{-1} f_i(z) : 0 \le i \le M - 1\}.$$

Thus, if z is observed, we should decide that signal s_k was sent if

$$f_k(z) = \max\{f_i(z) : 0 \le i \le M - 1\},$$

which is the *maximum-likelihood decision rule for M-ary communications*. This is just the generalization to M-ary signaling of the maximum-likelihood rule defined in Section 5.4 for binary signaling.

Finally, we mention minimax decisions for M-ary communication. A *minimax decision rule* minimizes the quantity

$$P_{e,m} = \max\{P_{e,i} : 0 \le i \le M - 1\}.$$

The method for finding minimax rules is essentially the same as discussed in Section 5.4. One examines the Bayes rules for a given problem and attempts to find an equalizer rule. For one-dimensional signal sets, such as the M-ASK signal sets introduced in Section 6.3.5, this procedure usually corresponds to selecting Bayes thresholds to give equal conditional probabilities of symbol error. This procedure is discussed in the next section for a specific 4-ASK signal set.

If an M-ary communication system is to be used to transmit binary data, a method is required to assign binary digits to the M-ary symbols. For example, if $M = 8$, we can let symbol k represent the 3-bit binary representation for the integer k. That is, symbol 0 represents the bit pattern 000, 1 represents 001, 2 represents 010, etc. It is not always possible to determine the resulting bit error probability directly from the symbol error probability. In fact, the bit error probability depends critically on how the bits are assigned to the symbols, whereas the symbol error probability is not affected by this assignment.

As we have learned, there are different types of symbol errors, and the symbol error probability does not distinguish between these. Different types of symbol errors result in different numbers of bit errors. Consequently, we must distinguish between the different types of symbol errors in computing the bit error probability for a particular signal set and assignment of bits to symbols. For instance, in the above assignment for $M = 8$, the symbol 7 represents 111. If 0 is sent (representing 000) and the decision is symbol 1 (representing 001), then one bit error occurs, but if the decision is symbol 7, three bit errors occur. Further discussion of this issue is best given in the context of a specific signal set, such as the M-ASK signal set considered in the next section.

6.4.2   Applications to M-ASK

Consider the signal set of Example 6-3(a) and a correlation receiver with reference signal $r(t) = \sqrt{2}\,\zeta(t)\cos(\omega_c t + \hat{\varphi})$. If the optimum reference signal is used, $\zeta(t) = \beta(t)$ and $\hat{\varphi} = \varphi$, which gives the receiver of Figure 6-14. If the correlator is optimum (as in Figure 6-14) and the maximum-likelihood decision regions are employed in the decision device, the receiver makes maximum-likelihood decisions on the received waveform $s_i(t) + X(t)$. Even if the correlator is not optimum, the receiver may apply maximum-likelihood decision regions that are designed for the decision statistic at the correlator's output. However, for a suboptimum correlator, the resulting decisions are not maximum-likelihood decisions for the given received waveform, and the error probabilities are larger than for the optimum correlator. A *maximum-likelihood receiver* consists of an optimum correlator and a decision device that uses maximum-likelihood decision regions. The reason there is interest in maximum-likelihood decisions, even for a suboptimum correlator, is to obtain the best possible performance for the given correlator. The correlator may be suboptimum due to implementation imperfections or as the result of a compromise that is made to reduce the cost or complexity of the receiver.

For the general correlator, the density f_i is a Gaussian density with mean $\hat{u}_i$ for $0 \le i \le 3$. The four density functions are illustrated in Figure 6-16. The actual values for the parameters $\hat{u}_i$, which are derived in Section 6.3.6, are not needed at the moment, but we assume that they are ordered as illustrated in Figure 6-16. The locations of the boundaries between the maximum-likelihood decision regions are determined in a straightforward way from two observations. First, the density f_i is symmetrical about $\hat{u}_i$. Second, the density f_1 is just a shifted version of the density f_0; it is shifted by the amount $\hat{u}_1 - \hat{u}_0$. It follows from these two observations that the value of z for which $f_1(z) = f_0(z)$ is $z = (\hat{u}_1 + \hat{u}_0)/2$. Thus, if Γ_0 and Γ_1 are the maximum-likelihood decision regions for deciding s_0 and s_1 were sent, respectively, the boundary between these two regions is midway between $\hat{u}_1$ and $\hat{u}_0$. Similarly, the boundary between Γ_2 and Γ_3 is midway between $\hat{u}_3$ and $\hat{u}_2$. In summary, for the general correlator, the endpoints for the maximum-likelihood regions for the signals of Example 6-3(a) are located halfway between the output signal levels, except for the left-hand end point of Γ_0, which is at $-\infty$, and the right-hand endpoint of Γ_4, which is at $+\infty$. These are also the Bayes regions for $\pi_0 = \pi_1 = \pi_2 = \pi_3$.

It is straightforward to determine the error probabilities that result if the maximum-likelihood decision regions are employed for either a suboptimum or optimum correlator. In Section 6.3.6, it is shown that the values of the parameters $\hat{u}_i$ for Example 6-3(a) are $\hat{u}_0 = -3A\chi$, $\hat{u}_1 = -A\chi$, $\hat{u}_2 = +A\chi$, and $\hat{u}_3 = +3A\chi$, where $\chi = (\beta,\zeta)\cos\theta$. Thus, the boundary between Γ_0 and Γ_1 is at $-2A\chi$, the boundary between Γ_1 and Γ_2 is at 0, and the boundary between Γ_2 and Γ_3 is at $2A\chi$. Then $P_{e,3}$ is just the conditional probability that the decision statistic is less than $2A\chi$ given that symbol 3 was sent.

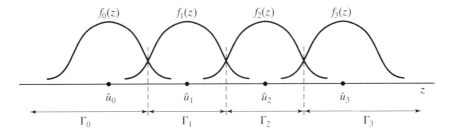

Figure 6-16: Densities and maximum-likelihood decision regions for the 4-ASK signals of Example 6-3(a).

When conditioned on the event that symbol 3 was sent, the decision statistic has mean $\hat{u}_3 = 3A\chi$ and variance $\sigma^2 = N_0\mathcal{E}_\zeta/2$; therefore,

$$
\begin{aligned}
P_{e,3} &= \Phi((2A\chi - \hat{u}_3)/\sigma) = \Phi(-A\chi/\sigma) \\
&= Q\left(\sqrt{2A^2\chi^2/N_0\mathcal{E}_\zeta}\right) = Q\left(\sqrt{2}\,A\chi/\sqrt{N_0\mathcal{E}_\zeta}\right).
\end{aligned}
\tag{6.28a}
$$

Similarly, $P_{e,2}$ is the conditional probability that the decision statistic is outside the interval $\Gamma_2 = (0, 2A\chi)$, and this probability is given by

$$
\begin{aligned}
P_{e,2} &= 1 - P_{c,2} = 1 - \{\Phi((2A\chi - \hat{u}_2)/\sigma) - \Phi((0 - \hat{u}_2)/\sigma)\} \\
&= 2\,Q\left(\sqrt{2}\,A\chi/\sqrt{N_0\mathcal{E}_\zeta}\right).
\end{aligned}
\tag{6.28b}
$$

By symmetry, $P_{e,1} = P_{e,2}$ and $P_{e,0} = P_{e,3}$. If the correlator is optimum, then $\chi = \mathcal{E}_\zeta = \mathcal{E}_\beta$, so $Q\left(\sqrt{2}\,A\chi/\sqrt{N_0\mathcal{E}_\zeta}\right)$ can be replaced by $Q\left(A\sqrt{2\,\mathcal{E}_\beta/N_0}\right)$ in (6.28).

If the minimax criterion is employed, then the decision regions should be selected to minimize

$$
P_{e,m} = \max\{P_{e,i} : 0 \le i \le M - 1\}.
$$

As we found for binary signaling, the minimax regions for M-ary signaling are such that the conditional probability of error that results does not depend on which signal was sent; that is, if the minimax decision regions are used, $P_{e,0} = P_{e,1} = \cdots = P_{e,M-1}$. It is easy to show that the maximum-likelihood regions illustrated in Figure 6-16 do not give a minimax decision rule for the given signal set and receiver. For example, $P_{e,0} < P_{e,1}$ for these decision regions. In order to obtain a minimax rule, $P_{e,1}$ must be decreased. To accomplish this, the set Γ_1 must be enlarged by moving the boundary between Γ_0 and Γ_1 to the left. (By symmetry, we can see that the right-hand endpoint of Γ_1 is correctly placed.) This decreases Γ_0 and therefore increases $P_{e,0}$. The boundary between Γ_0 and Γ_1 should be moved to the left until equality between $P_{e,0}$ and $P_{e,1}$ is achieved. Similarly, the boundary between Γ_2 and Γ_3 should be moved to the right until $P_{e,2} = P_{e,3}$. The final result is illustrated in Figure 6-17.

Expressions involving the Gaussian distribution function Φ (or its complement Q) can be developed easily, and these expressions can be used to relate the endpoints of Γ_1 and Γ_2 to the parameters A, N_0, and either $\mathcal{E}_\beta$ or χ and $\mathcal{E}_\zeta$, depending on whether the

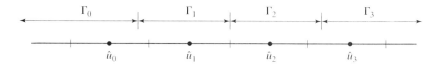

Figure 6-17: Illustration of minimax decision regions.

correlators are optimum. In general, it is not possible to obtain a closed-form expression for the location of each endpoint (see Problem 6.8), but a numerical solution for specific parameter values is easy to obtain.

The ability to reduce significantly the maximum error probability $P_{e,m}$ by adjusting the endpoints of the decision regions decreases greatly as M increases. Moreover, the difficulty in determining the minimax regions increases as M increases. The net result is that the maximum-likelihood regions are typically used, rather than the minimax regions.

Returning to the 4-ary signal set and the maximum-likelihood decision regions, we can evaluate $\overline{P}_e$, the average probability of symbol error, from the expressions given previously for the probabilities $P_{e,i}$ for $0 \leq i \leq 3$. The results are given here for the optimum correlator only, but analogous results for a suboptimum correlator can be obtained by making the obvious substitutions in the arguments of the function Q. (See (6.28).) The average probability of error is given by

$$\overline{P}_e = \pi_1 P_{e,1} + \pi_2 P_{e,2} + \pi_3 P_{e,3} + \pi_4 P_{e,4},$$

so

$$\overline{P}_e = [2(\pi_1 + \pi_2) + \pi_0 + \pi_3] \, Q\left(\sqrt{2A^2 \mathcal{E}_\beta / N_0}\right).$$

In the special case $\pi_i = 1/4$ for each i, this expression reduces to

$$\overline{P}_e = (3/2) \, Q\left(\sqrt{2A^2 \mathcal{E}_\beta / N_0}\right).$$

All of these generalize to arbitrary values of M. For example, the M-ASK signals described in Exercise 6-2 are of the form

$$s_i(t) = u_i \, \xi(t), \quad 0 \leq i \leq M - 1,$$

where

$$\xi(t) = \sqrt{2} \, A \, \beta(t) \cos(\omega_c t + \varphi).$$

The values for the data variables u_i of Exercise 6-2 are specified by

$$u_0 = -M + 1$$

and

$$u_i = u_{i-1} + 2$$

for each $i \geq 1$. If $\pi_i = M^{-1}$ for each i, the average probability of symbol error for this M-ASK signal set is (see Problem 6.10).

$$\overline{P}_e = [2\,(M-1)/M]\,Q\!\left(\sqrt{2A^2\mathcal{E}_\beta/N_0}\,\right).$$

This expression can be given in terms of the average energy transmitted per M-ary symbol by recalling from Exercise 6-2 that $\bar{\mathcal{E}}_s = (M^2-1)\mathcal{E}/3$, where $\mathcal{E}$ is the energy in the signal $\xi(t)$. It is easy to show that $\mathcal{E} = A^2\mathcal{E}_\beta$, so

$$A^2\mathcal{E}_\beta = \mathcal{E} = 3\,\bar{\mathcal{E}}_s/(M^2-1).$$

It follows that

$$\overline{P}_e = [2\,(M-1)/M]\,Q\!\left(\sqrt{6\,\bar{\mathcal{E}}_s/(M^2-1)N_0}\,\right).$$

The average energy per M-ary symbol can be related to $\bar{\mathcal{E}}_b$, the average energy per bit. For convenience, assume that M is a power of two: $M = 2^m$ for some positive integer m. Each of the 2^m different signals can be identified with a unique m-bit sequence; for example, the ith symbol can represent the m-bit binary representation for the integer i, for $0 \leq i \leq M-1$. So the number of bits per symbol is $m = \log_2 M$. Now use the fact that the energy per symbol is the energy per bit times the number of bits per symbol to conclude that

$$\bar{\mathcal{E}}_s = \bar{\mathcal{E}}_b \log_2 M.$$

The resulting expression for the average probability of symbol error is

$$\overline{P}_e = [2\,(M-1)/M]\,Q\!\left(\sqrt{6\,\bar{\mathcal{E}}_b \log_2 M/(M^2-1)N_0}\,\right). \tag{6.29}$$

This expression gives the *average probability of symbol error* in terms of the *average energy per bit* for M-ASK. Keep in mind that because each symbol represents $m = \log_2 M$ bits, a symbol error can result in as few as one or as many as m bit errors.

Example 6-4   Binary Data Transmission Using 4-ASK

Suppose binary data are to be transmitted using the 4-ary signal set of Example 6-3(a). Let the signal s_0 represent the binary digits 00, s_1 the binary digits 01, s_2 the binary digits 10, and s_3 the binary digits 11. If 00 is to be sent, then s_0 is transmitted, and, in the absence of noise, $\hat{u}_0$ is received. Thus, we associate the binary digits 00 with the point $\hat{u}_0$. Similarly, 01 is associated with $\hat{u}_1$, 10 with $\hat{u}_2$, and 11 with $\hat{u}_3$, as illustrated in Figure 6-18. Suppose 00 is to be sent, so s_0 is transmitted. If the receiver decides that s_3 was transmitted, then it concludes that 11 was sent, and so both bits are incorrect. In this case, one symbol error causes two bits to be in error. If, on the other hand, the receiver decides that s_1 was transmitted, which corresponds to 10, there is only one bit error. ∎

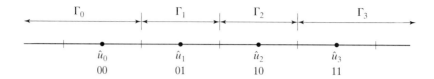

Figure 6-18: One possible assignment of binary digits to the signals.

Assume that M is a power of two, and consider a general M-ary system. If $m = \log_2 M$, each symbol represents a unique sequence of m binary digits. For purposes of illustration, suppose the sequence m zeros was sent. If the binary sequence corresponding to the symbol decision made by the receiver has one or more ones, then the symbol must be incorrect. That is, a bit error in any position implies a symbol error. If $E_{b,j}$ denotes the event that the bit in the jth position is incorrect and E_s denotes the event that the symbol is incorrect, the preceding observation implies that $E_{b,j} \subset E_s$.

An examination of some examples should convince the reader that the bit error probability need not be the same for each of the bit positions. Let $P_{b,j}$ denote the probability of a bit error in the jth position. It follows from $E_{b,j} \subset E_s$ that $P_{b,j} \leq P_e$, where P_e denotes the probability of a symbol error. This inequality holds for all j, and it is valid regardless of which symbol is sent. So, in particular it implies the average bit error probability, averaged over the bit positions, is upper bounded by the symbol error probability, regardless of which symbol is sent. Thus, the inequality is also valid if we average over all the symbols in the signal constellation using any prior distribution whatsoever, and this gives $\overline{P_b} \leq \overline{P_e}$ as one relationship between the average bit error probability and the average symbol error probability.

On the other hand, the symbol is in error if and only if there are errors in one or more bit positions. That is,

$$E_s = \bigcup_{j=1}^{m} E_{b,j}.$$

The union bound then implies that

$$P_e \leq \sum_{j=1}^{m} P_{b,j}.$$

This implies that the symbol error probability does not exceed m times the average of error probabilities for the bits, averaged over the bit positions. It follows by averaging over the symbols in the signal constellation that $\overline{P_e}/m \leq \overline{P_b}$. This and $\overline{P_b} \leq \overline{P_e}$ show that the average probability of bit error is bounded between $\overline{P_e}/m$ and $\overline{P_e}$.

An alternative derivation of $\overline{P_e}/m \leq \overline{P_b} \leq \overline{P_e}$ can be obtained by a relative-frequency argument. Consider a message of N_s symbols, and let N_{se} be the number of errors among these symbols. Such a message consists of $N_b = mN_s$ bits. Let N_{be} denote the number of bit errors in the message. Finally, let N denote the average number of bit errors per symbol error (i.e., $N = N_{be}/N_{se}$). Because each symbol error must result in at least one bit error among the m bits represented by the symbol, $N \geq 1$; because

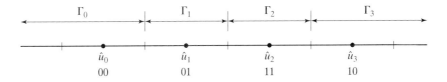

Figure 6-19: A better assignment of binary digits to the signals (Gray code).

each symbol error can produce at most m bit errors, $N \leq m$. The relative frequency of bit errors in the message is N_{be}/N_b, the ratio of the number of bit errors to the total number of bits in the message. By the definition of N, we have $N_{be} = N N_{se}$, so

$$N_{be}/N_b = N N_{se}/N_b = N N_{se}/m N_s. \tag{6.30a}$$

Because $N \leq m$, (6.30a) implies that

$$N_{be}/N_b = N N_{se}/m N_s \leq N_{se}/N_s. \tag{6.30b}$$

Also, (6.30a) and the fact that $N \geq 1$ imply that

$$N_{se}/m N_s = N_{be}/N N_b \leq N_{be}/N_b.$$

This inequality and (6.30b) combine to give

$$N_{se}/m N_s \leq N_{be}/N_b \leq N_{se}/N_s. \tag{6.30c}$$

The argument is completed by matching the bit and symbol error probabilities, $\overline{P_b}$ and $\overline{P_e}$, with the corresponding relative frequencies N_{be}/N_b and N_{se}/N_s in (6.30c) to obtain the bounds

$$\overline{P_e}/m \leq \overline{P_b} \leq \overline{P_e}. \tag{6.31}$$

Because the actual bit error probabilities depend on the assignment of the sequences of bits to the symbols, it makes sense to try to optimize this assignment. Notice in Example 6-4 that if s_1 is actually transmitted, but the receiver decides that s_2 was transmitted, then two bit errors occur. On the other hand, if s_1 is transmitted, but the receiver decides that s_3 was transmitted, then only one bit error occurs. By examining Figure 6-16, we can see that the former error is more likely to occur than the latter, so it is better if we reverse the assignments of bits to signals s_2 and s_3. The resulting assignment scheme, $00 \rightarrow s_0$, $01 \rightarrow s_1$, $11 \rightarrow s_2$, and $10 \rightarrow s_3$, is illustrated in Figure 6-19. This assignment has the property that errors between nearest neighbors cause a single bit error only. This is an example of *Gray coding*, in which nearest neighbors in a signal constellation are assigned bit patterns than differ in exactly one bit position. Under usual conditions, the most probable types of errors correspond to the demodulated signal being one of the nearest neighbors of the transmitted signal. At high signal-to-noise ratio, nearly all symbol errors that occur are of this type. Consequently, signal sets that use Gray coding result in a bit error probability $\overline{P_b} \approx \overline{P_e}/m$ for high signal-to-noise ratio. Notice this approximation is equal to the lower bound for $\overline{P_b}$.

Exercise 6-3
Consider a binary antipodal ASK signal set and a 4-ASK signal set. Each is used for communication over the same AWGN channel, and a maximum-likelihood receiver is used for each. Assume that $\bar{\mathcal{E}}_b/N_0 = 10$ for each system. Which is larger, the average probability of bit error for the binary ASK system or the average probability of bit error for 4-ASK? Does the answer depend on how the binary digits are assigned to the symbols?

Solution.   The best 4-ASK signal set for the criteria of this exercise is the set of Example 6-3(a). Because this 4-ASK signal set is of the type described in Exercise 6-2, its average probability of symbol error is given by (6.29). Let $\bar{P}_c(M)$ and $\bar{P}_b(M)$ denote the average probabilities of symbol error and bit error, respectively, for an M-ASK signal set, and let $\bar{\mathcal{E}}_b/N_0 = \lambda$ for each system. It follows from (6.29) and (6.31) that, for the 4-ASK set of Example 6-3(a),

$$\bar{P}_b(4) \geq \bar{P}_c(4)/2 = 3\,Q\big(\sqrt{4\lambda/5}\,\big)/4,$$

for any assignment of bits to symbols. Binary antipodal ASK has bit error probability $Q(\sqrt{2\lambda})$. If we show that $Q(\sqrt{2\lambda})$ is smaller than the lower bound $3\,Q(\sqrt{4\lambda/5})/4$, it is clear that the 4-ASK signal set has a larger average probability of bit error than the binary signal set. Because 2λ is more than a factor of two larger than $4\lambda/5$, $Q(\sqrt{2\lambda})$ is much smaller than $Q(\sqrt{4\lambda/5})$ for moderate to large values of λ; therefore, the factor of 3/4 is of no consequence. Thus, for such values of λ, the probability of bit error for binary ASK is much smaller than $3\,Q(\sqrt{4\lambda/5})/4$. In particular, for $\lambda = 10$ we see that $Q(\sqrt{4\lambda/5}) \approx 2.3 \times 10^{-3}$ and $Q(\sqrt{2\lambda}) \approx 3.9 \times 10^{-6}$. Note that this value of λ corresponds to $(\bar{\mathcal{E}}_b/N_0)_{dB} = 10$. Next consider small values of λ. If $(\bar{\mathcal{E}}_b/N_0)_{dB} = 0$, then $\lambda = 1$, $Q(\sqrt{4\lambda/5}) \approx 0.19$, and $Q(\sqrt{2\lambda}) = Q(\sqrt{2}) \approx 0.079$. Even if $\lambda = \ln(2)$, which corresponds to $(\bar{\mathcal{E}}_b/N_0)_{dB} \approx -1.6$, we have $Q(\sqrt{4\lambda/5}) \approx 0.23$ and $Q(\sqrt{2\lambda}) \approx 0.12$. Thus, the average bit error probability is larger for 4-ASK than for binary ASK, even for small values of λ. As discussed in Section 6.6.3, $\bar{\mathcal{E}}_b/N_0 = \ln(2)$ is the smallest value of interest.   ∎

6.4.3   A Geometric Representation for M-ASK and the Resulting Symbol Error Probability Expressions

Signal space concepts are introduced for binary baseband communication systems in Section 5.7. These concepts also apply to carrier-modulated signals, and they allow us to give alternative expressions for the symbol error probability in an M-ASK communication system with a maximum-likelihood receiver. The signals are given by

$$s_i(t) = u_i\,\xi(t), \quad 0 \leq i \leq M - 1,$$

where

$$\xi(t) = \sqrt{2}\,A\,\beta(t)\cos(\omega_c t + \varphi).$$

For convenience, assume that $u_0 < u_1 < \cdots < u_{M-1}$.

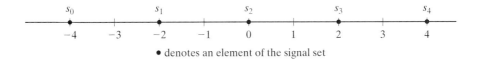

$\bullet$ denotes an element of the signal set

Figure 6-20: M-ASK signal representation ($M = 5$).

Because the M-ASK signal set is a one-dimensional set, the M-ASK signals can be represented as points on a line. Such a representation is illustrated in Figure 6-20 for $M = 5$. The numbers below the axis are the values of the variables u_i. This is an example of an *equally spaced* M-ASK signal set; in particular, $u_i - u_{i-1} = 2$ for each $i \geq 1$. A family of equally spaced M-ASK signal sets is described in Exercise 6-2 of Section 6.3.5. The outputs of the correlation receiver can also be viewed as points on a line (e.g., compare Figures 6-13 and 6-15), and the maximum-likelihood decision regions are just intervals with boundaries that are halfway between adjacent points.

For each i in the range $1 \leq i \leq M - 1$, let d_i be distance between the signals s_i and s_{i-1}. That is, d_i is defined by

$$d_i^2 = \int_{-\infty}^{\infty} [s_i(t) - s_{i-1}(t)]^2 \, dt \tag{6.32}$$

for $1 \leq i \leq M - 1$. Equivalently, the parameter d_i can be written as $d_i = \|s_i - s_{i-1}\|$.

Exercise 6-4

Express d_i in terms of the parameters u_i, u_{i-1}, A, and $\mathcal{E}_\beta$ for the M-ASK signal set.

Solution.   First, observe that

$$d_i^2 = (u_i - u_{i-1})^2 \int_{-\infty}^{\infty} \xi^2(t) \, dt,$$

so that $d_i^2 = (u_i - u_{i-1})^2 \mathcal{E}$. But $\mathcal{E} = A^2 \mathcal{E}_\beta$, so $d_i^2 = A^2 \mathcal{E}_\beta (u_i - u_{i-1})^2$. We conclude that the squared-distance between consecutive signals is proportional to $(u_i - u_{i-1})^2$. If the waveform $\beta(t)$ is normalized to have unit energy, then $d_i = A(u_i - u_{i-1})$.   ∎

For equally spaced signals, the value of d_i does not depend on i, and we denote the common value by d. The energy $\mathcal{E}_i$ in signal s_i from a set of equally spaced M-ASK signals can be related to d. For example, the 5-ASK signal set in Figure 6-20 has $\mathcal{E}_3 = \mathcal{E}_1 = d^2$ and $\mathcal{E}_4 = \mathcal{E}_0 = 4d^2$ ($\mathcal{E}_2 = 0$, of course). The 4-ASK signal set of Example 6-3(a), which is illustrated in Figure 6-13, has $\mathcal{E}_2 = \mathcal{E}_1 = (d/2)^2$ and $\mathcal{E}_3 = \mathcal{E}_0 = (3d/2)^2$.

The quantity $A^2 \mathcal{E}_\beta$ can also be expressed in terms of d. If $u_i - u_{i-1} = 2$ for each i, the solution of Exercise 6-4 implies that $d^2 = 4 \mathcal{E} = 4A^2 \mathcal{E}_\beta$. Recall that the conditional error probabilities $P_{e,i}$ are functions of $A^2 \mathcal{E}_\beta$, so they can be written in terms of d. For

instance, in Section 6.4.2, we found that the error probabilities for the 4-ASK signal set
of Example 6-3(a) are

$$P_{e,0} = P_{e,3} = Q\left(\sqrt{2A^2\mathcal{E}_\beta/N_0}\,\right)$$

and

$$P_{e,1} = P_{e,2} = 2\,Q\left(\sqrt{2A^2\mathcal{E}_\beta/N_0}\,\right).$$

An examination of the derivations of these results reveals that the derivations are valid
for any value of M, and the general results for the equally spaced M-ASK signals of
Exercise 6-2 are

$$P_{e,0} = P_{e,M-1} = Q\left(\sqrt{2A^2\mathcal{E}_\beta/N_0}\,\right)$$

and

$$P_{e,i} = 2\,Q\left(\sqrt{2A^2\mathcal{E}_\beta/N_0}\,\right), \quad 1 \le i \le M-2.$$

As a side note, recall from Section 6.4.2 that

$$A^2\mathcal{E}_\beta = 3\,\bar{\mathcal{E}}_s/(M^2-1) = 3\,\bar{\mathcal{E}}_b(\log_2 M)/(M^2-1),$$

so the error probabilities can also be written in terms of either the average energy per
symbol or the average energy per bit. Returning to our goal of expressing the error
probabilities as functions of d, we use the fact that $d^2 = 4A^2\mathcal{E}_\beta$ to conclude that

$$P_{e,0} = P_{e,M-1} = Q\left(d/\sqrt{2N_0}\,\right) \tag{6.33a}$$

and

$$P_{e,i} = 2\,Q\left(d/\sqrt{2N_0}\,\right), \quad 1 \le i \le M-2. \tag{6.33b}$$

Thus, an alternative method for evaluating the symbol error probabilities for equally
spaced M-ASK signals is to first determine the squared distance from

$$d^2 = \int_{-\infty}^{\infty} [s_i(t) - s_{i-1}(t)]^2 \, dt \tag{6.34}$$

and then substitute for d in (6.33). For the equally spaced signal set under consideration,
any value of i in the range $1 \le i \le M-1$ can be used in (6.34). The use of (6.33)
and (6.34) to evaluate symbol error probabilities gives the same alternative method as
suggested for binary baseband signal sets in Section 5.7.2. In fact, equation (5.113a) of
Section 5.7.2 is identical to (6.33a) for $M = 2$.

   If written in terms of the parameter d, the expression for the conditional probability
of symbol error given that a particular *exterior point* of the M-ASK signal constellation
was transmitted is the same as the error probability expression for binary signaling.

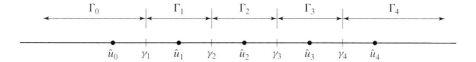

Figure 6-21: Output signal levels for the 5-ASK signal set.

This is because only a single threshold is involved if the transmitted signal is an exterior point in the constellation. This is illustrated in Figure 6-21 for 5-ASK.

If we are concerned only with the symbol error probability, then there is only one way to make an error if one of the exterior points is transmitted. For the signal set illustrated in Figure 6-21, for example, if the signal s_0 is transmitted, an error is made if and only if the decision statistic exceeds the threshold labeled γ_1. Thus, $P_{e,0}$ is the same as if there were only two signals, s_0 and s_1, in the signal constellation (i.e., binary signaling).

On the other hand, the conditional probability of symbol error given one of the *interior points* of the M-ASK signal constellation was transmitted is larger by a factor of two than the error probability expression for binary signaling. This is because there are two thresholds that come into play when a signal corresponding to an interior point is transmitted. Basically, there are two ways to make an error if the transmitted signal corresponds to an interior point. For example, in the situation depicted in Figure 6-21, if the signal s_1 is transmitted, an error is made if the decision statistic is less than γ_1 or if it is greater than γ_2. The probability of each of these individual events is the same as $P_{e,0}$. Because the events are disjoint, $P_{e,1}$ is the sum of the probabilities of the individual events, which is $2P_{e,0}$.

Notice that the symbol error probability $P_{e,1}$ is the same as it would be for the signal set $\{s_0, s_1, s_2\}$. The other signals in the constellation illustrated in Figure 6-21 do not affect the conditional probability of symbol error given that s_1 was transmitted. In fact, any number of signals of the form

$$s_i(t) = \sqrt{2}\, A\, u_i\, \beta(t) \cos(\omega_c t + \varphi)$$

can be added to the signal set, and the value of $P_{e,1}$ will not change provided the additional signals correspond to points that are no closer than distance d to s_1.

Expressions (6.33) and (6.34) are for equally spaced one-dimensional signal constellations. If the signal constellation is one dimensional, but the signals are not equally spaced, these expressions can be generalized (Problem 6.11) and the results are in terms of the parameters d_i defined in (6.32). Notice that the only distances needed to determine symbol error probabilities are the distances between *nearest neighbors* in the signal constellations.

We close this section with a warning. Equations (6.33) are easy to use because they require only the evaluation of the distance parameter d. For example, the mean and variance of the correlator output need not be determined. However, these equations are applicable only for maximum-likelihood receivers. Recall that for the receiver to be a maximum-likelihood receiver, the correlator must be optimum and the decision device must employ maximum-likelihood decision regions. If the correlator is not optimum, the distance parameters are not sufficient to characterize the performance of the receiver even if the maximum-likelihood decision regions are used. In this situation, (6.28) must be used instead of (6.33).

6.5   Two-Dimensional Modulation for Coherent Communications

There are certain advantages to be gained if the signal sets are permitted to have dimension greater than one. For example, two-dimensional signals can be constructed to have the same error probability vs. bit energy to noise density ratio as BPSK and also possess many other desirable characteristics of BPSK, but require only half the bandwidth of BPSK. In this section, we examine a special class of two-dimensional signal sets in which each signal in the set is generated by modulating a pair of orthogonal sinusoidal signals of the same frequency.

6.5.1   Two-Dimensional Signal Sets

The basic idea we have developed for one-dimensional signals can be extended to two-dimensional signals as follows. The general form for the one-dimensional signals is

$$s(t) = \sqrt{2}\, A\, a(t) \cos(\omega_c t + \varphi),$$

and the signal $a(t)$ is a pulse of duration T (or a sequence of such pulses). Now suppose that the transmitted signal is

$$s(t) = A\, a_1(t) \cos(\omega_c t + \varphi) - A\, a_2(t) \sin(\omega_c t + \varphi), \qquad (6.35)$$

and the signals $a_1(t)$ and $a_2(t)$ are pulses of duration T'. The reason for denoting the pulse duration for the two-dimensional signals by T' will become clear later.

The signal $s(t)$ of (6.35) has two components. The first is given by

$$s_I(t) = A\, a_1(t) \cos(\omega_c t + \varphi) \qquad (6.36a)$$

and is called the *inphase* component of the signal $s(t)$. The second component is given by

$$s_Q(t) = -A\, a_2(t) \sin(\omega_c t + \varphi) \qquad (6.36b)$$

and is referred to as the *quadrature* component of $s(t)$. Under certain conditions, which will be described later, the inphase and quadrature components are orthogonal signals. As a result, $s(t)$ is described in terms of two orthogonal basis functions, and so it is a two-dimensional signal. The type of signaling specified by (6.35) is referred to as *quadrature amplitude modulation* (QAM). In the most general form of QAM, the inphase and quadrature signals are nonbinary AM signals of the type discussed in Section 6.3.1.

The presence of the minus sign in (6.35) is strictly a matter of convention. The decision to include the minus sign is motivated by the relationship with certain kinds of phase modulation and the desire to permit interpretation of the two-dimensional signals as phasors. In view of the identity $\cos[\theta + (\pi/2)] = -\sin(\theta)$, we can say that the quadrature carrier

$$c_Q(t) = -\sin(\omega_c t + \varphi)$$

leads the inphase carrier

$$c_I(t) = \cos(\omega_c t + \varphi)$$

by $\pi/2$ radians, and this orientation is used in many of the diagrams that follow. In addition, the minus sign arises naturally in the complex representation of two-dimensional signals, as shown in Appendix C.

The first specific two-dimensional modulation scheme we consider provides a set of four signals. Given a baseband waveform $\alpha(t)$ of duration T', information can be transmitted via the inphase component of $s(t)$ by letting $a_1(t)$ be either $+\alpha(t)$ or $-\alpha(t)$, depending on whether a 0 or 1 is being sent on the inphase component of the signal. Similarly, information is transmitted on the quadrature component by letting $a_2(t)$ be either $+\alpha(t)$ or $-\alpha(t)$, depending on whether a 0 or a 1 is being sent on the quadrature component of the signal. Thus, there are two independent choices for the amplitude of each component, giving a total of *four* different signals in the set. With this signal, a total of two bits of information can be sent during an interval of length T'. One possible choice for the baseband waveform $\alpha(t)$ is $p_{T'}(t)$, the rectangular pulse of duration T'.

This example can be extended to generate larger signal sets if $a_1(t)$ and $a_2(t)$ are not required to be binary. Given sets $\{u_i : 0 \leq i \leq M' - 1\}$ and $\{v_k : 0 \leq k \leq M'' - 1\}$ of real numbers, let the baseband signals $a_1(t)$ and $a_2(t)$ be given by $a_1(t) = u_i\,\alpha(t)$ and $a_2(t) = v_k\,\alpha(t)$ for t in the signaling interval $[0, T']$. These baseband signals are then used in (6.35) to give a signal set of size $M = M'M''$. The resulting signals are given by

$$s(t) = A\,u_i\,\alpha(t)\cos(\omega_c t + \varphi) - A v_k\,\alpha(t)\sin(\omega_c t + \varphi). \tag{6.37}$$

Because each of the signals $s_I(t)$ and $s_Q(t)$ is an ASK signal, the modulation scheme that results is referred to as *M-ary quadrature amplitude-shift keying (M-QASK)*. Just as one-dimensional ASK is a special case of one-dimensional AM, QASK is a special case of QAM.

If the baseband waveform $\alpha(t)$ used to define the QASK signals is a rectangular pulse of duration T', and if the carrier frequency is such that $\omega_c T'$ is a multiple of π, then the inphase and quadrature signals are orthogonal on the interval $[0, T']$. To illustrate this concept, consider the fact that, because $\alpha(t)$ is a rectangular pulse, the product $a_1(t)\,a_2(t)$ is just the constant $u_i v_k$ for $0 \leq t \leq T'$, where u_i is the symbol being sent on the inphase component and v_k is the corresponding symbol for the quadrature component. Thus, the inner product of the inphase and quadrature signals can be written as

$$
\begin{aligned}
(s_I, s_Q) &= \int_0^{T'} s_I(t)\, s_Q(t)\, dt \\
&= -A^2 u_i v_k \int_0^{T'} \cos(\omega_c t + \varphi) \sin(\omega_c t + \varphi)\, dt \\
&= \tfrac{1}{2} A^2 u_i v_k \int_0^{T'} \sin(2\,\omega_c t + 2\,\varphi)\, dt = 0.
\end{aligned}
$$

We have therefore established that if the basic pulse shape $\alpha(t)$ is rectangular and if $\omega_c T'$ is a multiple of π, then the QASK signal is a linear combination of two orthogonal functions. More generally, even if the pulse shape is not rectangular, or the carrier frequency and pulse duration are not related by $\omega_c T' = n\pi$ for some integer n, these two components of the QASK signal are orthogonal on $[0, T']$ if the pulse waveform $\alpha(t)$ and carrier frequency are such that

$$
\int_0^{T'} \alpha^2(t) \sin(2\,\omega_c t + 2\,\varphi)\, dt = 0 \qquad (6.38)
$$

for each value of φ.

If (6.38) is satisfied, it is convenient to picture the resulting QASK signal as a linear combination of two orthogonal vectors. This view can be formalized by defining the orthogonal basis signals

$$
\psi_0(t) = \sqrt{2/\mathcal{E}_\alpha}\, \alpha(t) \cos(\omega_c t + \varphi)
$$

and

$$
\psi_1(t) = -\sqrt{2/\mathcal{E}_\alpha}\, \alpha(t) \sin(\omega_c t + \varphi),
$$

where $\mathcal{E}_\alpha$ is the energy in the waveform $\alpha(t)$. Notice that $\psi_0(t)$ and $\psi_1(t)$ have been normalized to have unit energy (hence, they are *orthonormal* signals). They are the continuous-time equivalents of unit-length orthogonal vectors in the plane. Any M-QASK signal of the form of (6.37) can be represented as a linear combination of $\psi_0(t)$ and $\psi_1(t)$.

For most of our illustrations, it is not necessary to represent the communication signal in terms of unit-energy signals. Typically, it suffices to use the orthogonal signals

$$
A\, \alpha(t) \cos(\omega_c t + \varphi)
$$

and

$$
-A\, \alpha(t) \sin(\omega_c t + \varphi)
$$

in our illustrations.

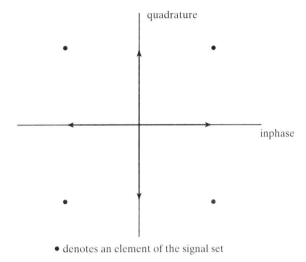

 denotes an element of the signal set

Figure 6-22: Standard 4-QASK signal constellation.

For an example of the representation of communication signals, return to the first QASK signals that were discussed: Those with binary antipodal modulation on each component (i.e., $a_1(t) = \pm\alpha(t)$ and $a_2(t) = \pm\alpha(t)$). The signals in this QASK signal set can be represented as linear combinations of orthogonal vectors, as illustrated in Figure 6-22. In this diagram, the inphase modulation is shown on the horizontal axis, and the quadrature modulation is shown on the vertical axis. The four possible linear combinations of these vectors (shown as dots in Figure 6-22) represent the four possible signals that are obtained if binary antipodal modulation is employed on each component. Mathematically, these four signals correspond to the four possible combinations of signs in the expression

$$\pm A\,\alpha(t)\cos(\omega_c t + \varphi) \pm A\,\alpha(t)\sin(\omega_c t + \varphi).$$

The resulting signal set, illustrated by the dots in Figure 6-22, is referred to as the *signal constellation* for the particular modulation scheme, and in this case it is referred to as *standard 4-QASK*. For this signal set, $M' = M'' = 2$ and each of the sets $\{u_i : 0 \le i \le M' - 1\}$ and $\{v_j : 0 \le j \le M'' - 1\}$ is just the binary set $\{-1, +1\}$. A larger QASK signal set is considered in the next example.

Example 6-5   A Two-Dimensional Signal Set with Sixteen Signals
If each of these two signals $a_1(t)$ and $a_2(t)$ is allowed to be $\pm\alpha(t)$ or $\pm3\alpha(t)$, we obtain a signal set with more than four signals. Because there are four possible amplitudes for the inphase component and four possible amplitudes for the quadrature component of the QASK signal, $M' = M'' = 4$ and the signal constellation has a total of sixteen points. The resulting modulation scheme is *16-ary quadrature amplitude-shift keying (16-QASK)*. For this specific 16-QASK signal set, referred to as *standard 16-QASK*, each of the sets $\{u_i : 0 \le i \le M' - 1\}$ and $\{v_k : 0 \le k \le M'' - 1\}$ is $\{-3, -1, +1, +3\}$. The signal constellation is illustrated in Figure 6-23. ∎

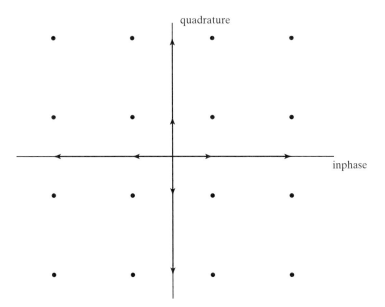

Figure 6-23: Standard 16-QASK signal constellation.

Suppose the elements of each set $\{u_i : 0 \leq i \leq M' - 1\}$ and $\{v_k : 0 \leq k \leq M'' - 1\}$ are indexed in ascending order; that is, $u_i < u_{i+1}$ for $0 \leq i \leq M' - 2$ and $v_k < v_{k+1}$ for $0 \leq k \leq M'' - 2$. A QASK signal constellation is said to be *rectangular* if there are positive numbers ϵ_1 and ϵ_2 for which $u_{i+1} - u_i = \epsilon_1$ for each i and $v_{k+1} - v_k = \epsilon_2$ for each k. A QASK signal constellation is said to be *regular* if it is rectangular and $\epsilon_1 = \epsilon_2$. Thus, the points in a rectangular QASK constellation are aligned in each dimension, but the spacing between neighboring points in one dimension may differ from the spacing between neighboring points in the other dimension. The points in a regular QASK constellation are aligned in each dimension and uniformly spaced in the plane. The QASK constellations illustrated in Figures 6-22 and 6-23 are regular constellations.

6.5.2   Binary ASK vs. QASK: A Comparison of Bandwidth Efficiency

Recall that the general form of binary ASK gives a transmitted signal of the form

$$s(t) = \sqrt{2}\, A\, b\, \beta(t) \cos(\omega_c t + \varphi) \qquad (6.39a)$$

for the interval $[0, T]$, where b is either $+1$ or -1, depending on whether 0 or 1 is sent, and $\beta(t)$ is a baseband waveform of duration T. Of course, the same signal format can be used in subsequent intervals with possibly different choices for the pulse polarities.

We can express this formally by writing the signal as

$$s(t) = \sqrt{2}\, A\, b_n\, \beta(t - nT) \cos(\omega_c t + \varphi), \quad nT \le t < (n+1)T, \qquad (6.39b)$$

where n is an arbitrary integer and b_n represents the polarity of the pulse in the nth signaling interval.

In this manner, the binary ASK signal can be used to convey one bit of information in each of the T-second intervals of the form $[nT, (n+1)T]$. Strictly speaking, the right-hand endpoint should not be included in this interval, unless the signal $s(t)$ is continuous at this point. If the signal is not continuous at $(n+1)T$, the interval should be denoted by $[nT, (n+1)T)$, but we rarely need to worry about this detail. In practice, the signals that we can actually generate and use in a communication system are continuous. It is just that it is both convenient and conventional to permit mathematical signal models that may not be continuous at the endpoints of the signaling intervals. So, engaging in an abuse of notation, we will write all such intervals as closed intervals in this section.

The 4-QASK signal can be written as

$$s(t) = A\, b_1\, \alpha(t) \cos(\omega_c t + \varphi) - A\, b_2\, \alpha(t) \sin(\omega_c t + \varphi) \qquad (6.40a)$$

for the interval $[0, T']$. Because b_1 is either $+1$ or -1 and b_2 is either $+1$ or -1, the 4-QASK signal can be used to convey two bits of information in the interval $[0, T']$. The expression for the 4-QASK signal that is analogous to (6.39b) is

$$s(t) = A\, b_{1,n}\, \alpha(t - nT') \cos(\omega_c t + \varphi) - A\, b_{2,n}\, \alpha(t - nT') \sin(\omega_c t + \varphi),$$
$$nT' \le t < (n+1)T', \quad (6.40b)$$

where $b_{1,n}$ represents the polarity of the pulse on the inphase component during the nth signaling interval and $b_{2,n}$ represents the polarity of the pulse on the quadrature component during the nth signaling interval.

Now suppose that the length of the signaling interval for 4-QASK is twice the length of the signaling interval for binary ASK (i.e., $T' = 2T$). In this case, the 4-QASK signal can be used to represent two bits of information during each $2T$-second interval of the form $[2nT, 2(n+1)T]$. That is, during the nth signaling interval, one bit of information is conveyed by the polarity of $b_{1,n}$, and another bit is conveyed by the polarity of $b_{2,n}$. Notice also that, during this same time interval, two bits of information can be transmitted using binary ASK to send a sequence of two consecutive pulses, each having a duration of T seconds. Thus, both the binary ASK and 4-QASK signals can be used to transmit two bits of information in a time interval of length $2T$, and so the data rate is the same: The QASK signal of (6.40) and the binary ASK signal of (6.39) each have an information rate of $1/T$ bits per second.

We can now point out the advantage of the 4-QASK signal over the binary ASK signal. Suppose the 4-QASK signal uses a pulse of the same *shape* as the binary ASK signal and that $T' = 2T$, as just discussed. It follows that $\alpha(t) = \beta(t/2)$ for $0 \le t \le T'$. However, the 4-QASK signal uses pulses that are twice as long as the binary ASK signal. Because the bandwidth required by these signals depends only on the Fourier transform of the pulse being used (see Problem 6.21), and because an increase in the pulse duration by a factor of two in the time domain corresponds to a decrease by a factor of two in the width of its Fourier transform, the 4-QASK signal requires only *one-half the bandwidth* of the binary ASK signal. Thus, 4-QASK can achieve the same data rate as binary ASK signaling in half the bandwidth required by binary ASK. Even greater decreases in bandwidth are obtainable from higher-order versions of QASK. For example, a similar argument shows that 16-QASK requires only one-fourth the bandwidth of binary ASK for the same information rate.

It is also true that the one-dimensional 4-ASK signal set of Example 6-3 has the same bandwidth advantage over binary ASK. If the one-dimensional 4-ASK signals use a pulse that is twice as long as that used for binary ASK, the two signal sets can provide the same data rate. For example, we can encode pairs of binary digits into the 4-ASK signals according to $00 \to u_0$, $01 \to u_1$, $10 \to u_2$, and $11 \to u_3$, so that each transmitted pulse represents two bits of information. However, the primary disadvantage of one-dimensional 4-ASK is that it requires a larger maximum and average energy for a given probability of error when compared with binary ASK. This is demonstrated in Exercise 6-3 in terms of the average probability of bit error and the average energy per bit. On the other hand, it is shown in Section 6.5.4 that, for a given probability of bit error and a given information rate, the energy requirements for 4-QASK are *identical* to those for binary ASK.

6.5.3   Nonbinary Phase Modulation

A common choice for the waveform $\alpha(t)$ in 4-QASK is the rectangular pulse $p_{T'}(t)$. For this choice of waveform, signals in the standard 4-QASK signal set are of the form

$$s(t) = A\{\pm \cos(\omega_c t + \varphi) \pm \sin(\omega_c t + \varphi)\} \tag{6.41}$$

for $0 \le t \le T'$. Just as for binary ASK with rectangular pulses, the QASK signal given in (6.41) can be expressed as phase-shift keying, except the number of different phases is four for QASK rather than two for binary ASK. Because the phase of this signal takes on one of four possible values in each signaling interval, the resulting signal set is referred to as *quadriphase-shift keying (QPSK)* when it is being viewed as phase modulation. Note that the letter "Q" in QPSK means quadriphase, whereas the letter "Q" in QASK stands for quadrature. Another name for QPSK is 4-PSK.

In order to demonstrate the equivalence between standard 4-QASK signals with rectangular pulses and QPSK signals, we begin with the general form for a carrier-modulated signal with phase modulation $\psi(t)$:

$$s(t) = \sqrt{2}\, A \cos(\omega_c t + \psi(t) + \varphi). \tag{6.42}$$

For QPSK, $\psi(t)$ is constant on intervals of length T' and it takes on one of the four values $\pi/4, 3\pi/4, 5\pi/4$, and $7\pi/4$. In order to show that these four values are sufficient for (6.42) to produce the QASK signal set, it is convenient to expand (6.42) to give

$$s(t) = \sqrt{2}\, A\{\cos(\omega_c t + \varphi) \cos[\psi(t)] - \sin(\omega_c t + \varphi) \sin[\psi(t)]\},$$

and then use the fact that $\cos x$ and $\sin x$ are equal to $\pm 1/\sqrt{2}$ for values of x equal to $(2n+1)\pi/4$ for any integer n. In view of this, it is better to group the factor $\sqrt{2}$ with

Table 6.3: Equivalence of QPSK and Standard QASK

Phase Modulation $\psi(t)$	Equivalent Amplitude Modulation	
	Inphase $a_1(t)$	Quadrature $a_2(t)$
$\pi/4$	$+1$	$+1$
$3\pi/4$	-1	$+1$
$5\pi/4$	-1	-1
$7\pi/4$	$+1$	-1

the terms $\cos[\psi(t)]$ and $\sin[\psi(t)]$ and express the signal as

$$s(t) = A\left\{\sqrt{2}\cos[\psi(t)]\cos(\omega_c t + \varphi) - \sqrt{2}\sin[\psi(t)]\sin(\omega_c t + \varphi)\right\}. \qquad (6.43)$$

If we now let $a_1(t) = \sqrt{2}\cos[\psi(t)]$ and $a_2(t) = \sqrt{2}\sin[\psi(t)]$, then the QPSK signal of (6.42) can be written as

$$s(t) = A\{a_1(t)\cos(\omega_c t + \varphi) - a_2(t)\sin(\omega_c t + \varphi)\}. \qquad (6.44)$$

Equation (6.44) can be used to prove that the QPSK signal set is equivalent to the signal set obtained from standard 4-QASK with rectangular pulses of duration T' (i.e., $\alpha(t) = p_{T'}(t)$ in (6.40a)). Consider the signals $a_i(t)$ for $i = 1$ and $i = 2$, and note that $a_i(t)$ is constant on intervals of length T', because $\psi(t)$ is constant on such intervals. Furthermore, because $\psi(t)$ takes on values of the form $(2n + 1)\pi/4$ only (for integer values of n), then $a_i(t)$ takes on values $+1$ and -1 only. Hence, we can let $a_i(t) = b_i\,\alpha(t)$ and $\alpha(t) = p_{T'}(t)$ to complete the proof. We observe that the mathematical form of (6.42) is such that the values $(2n + 1)\pi/4$ and $(2n' + 1)\pi/4$ for $\psi(t)$ produce the same signals if n' and n are related by $n' = n + 4N$ for some integer N. Hence, (6.42) produces four distinct signals only, and these can be obtained by letting $n = 0, 1, 2$, and 3.

We have shown that phase modulation with $\psi(t) = (2n + 1)\pi/4$ for $n = 0, 1, 2$, or 3 corresponds to amplitude modulation by positive or negative rectangular pulses of unit amplitude on the inphase and quadrature components of the signal. The exact relationship between the four possible phase angles and the amplitudes of the two components is given in Table 6.3. Because QPSK is equivalent to standard 4-QASK with rectangular pulses, and because BPSK is equivalent to binary ASK with rectangular pulses, we conclude that QPSK has the same bandwidth improvement over BPSK as 4-QASK has over binary ASK; namely, if the two signals are used to transmit information at the same rate, the QPSK signal has one-half the bandwidth of the BPSK signal.

One useful way to display the signals of QPSK, as well as other forms of phase modulation, is to represent the signals in the signal set by phasors in two-dimensional diagrams of the type shown in Figure 6-24. The four vectors in the diagrams of Figure 6-24 illustrate the four possible QPSK signals. These four signals are specified mathematically by (6.42) with $\psi(t) = (2n + 1)\pi/4$, $0 \le t \le T'$, for an integer value of n. Note that each of the four vectors is the sum of two orthogonal vectors (shown

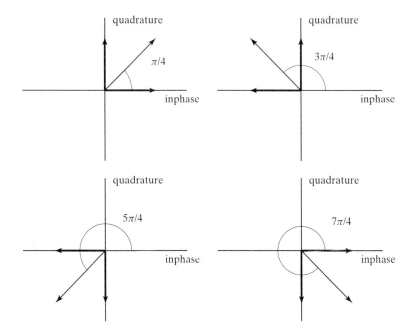

Figure 6-24: Illustration of the QPSK signal set.

as boldface vectors). The orthogonal vectors are the inphase and quadrature components of the QPSK signal. (This is also illustrated by equations (6.42)–(6.44) and by Table 6.3.) For each of the four cases illustrated in Figure 6-24, the boldface vectors can also be thought of as representing the two orthogonal components of the equivalent 4-QASK signal. A comparison of the resultant vectors of Figure 6-24 and the points in the signal constellation for regular 4-QASK shown in Figure 6-22 provides another illustration of the equivalence of QPSK and regular 4-QASK with rectangular pulses.

A set of M signals is obtained by letting $\psi(t) = (2n + 1)\pi/M$, $0 \le t \le T'$, in (6.42). Such a signal set is referred to as M-PSK. Its constellation consists of M points uniformly spaced around a circle centered at the origin. Except for $M = 4$, M-PSK does not correspond to a regular QASK signal constellation. The phase angles used in the modulation for 8-PSK are $(2n + 1)\pi/8$ for $0 \le n \le 7$. If Gray coding is applied to M-PSK, the bit patterns 000, 001, 011, 010, 110, 111, 101, and 100 are assigned to the phase angles in order according to the index n, which corresponds to a counterclockwise rotation around the origin. The patterns for consecutive phase angles differ in a single bit position, including the patterns for $\pi/8$ and $15\pi/8$, which are 000 and 100, respectively. Unlike in M-ASK, the signals for $n = 0$ and $n = M - 1$ are adjacent in M-PSK.

An expression for the average probability of bit error for optimum coherent reception of Gray-coded 8-PSK with maximum-likelihood decision regions is [6.14]

$$\overline{P}_b = 2\left\{Q\left[\sqrt{6\,\mathcal{E}_b/N_0}\,\sin(\pi/8)\right]\right. \\ \left. + Q\left[\sqrt{6\,\mathcal{E}_b/N_0}\,\sin(3\pi/8)\right]\left(1 - Q\left[\sqrt{6\,\mathcal{E}_b/N_0}\,\sin(\pi/8)\right]\right)\right\}/3.$$

Note that $\sin(\pi/8) = 0.5[2 - \sqrt{2}]^{1/2}$ and $\sin(3\pi/8) = 0.5[2 + \sqrt{2}]^{1/2}$. If $\mathcal{E}_b/N_0$ is large, the first term inside the braces is much larger than the remaining term. Thus,

$$\overline{P_b} \approx 2\, Q\left[\sqrt{6\,\mathcal{E}_b/N_0}\,\sin(\pi/8)\right]\big/3$$

if $\mathcal{E}_b/N_0$ is large. For $(\mathcal{E}_b/N_0)_{\mathrm{dB}} = 5$, the first term exceeds the remaining term by three orders of magnitude, the exact expression gives 3.186×10^{-2}, and the approximation gives 3.184×10^{-2}. The bandwidth of M-PSK is $1/\log_2(M)$ times that of BPSK, but M-PSK requires a larger value of $\mathcal{E}_b/N_0$ to achieve the same error probability. For a bit error probability of 10^{-6}, the value of $\mathcal{E}_b/N_0$ needed by BPSK and QPSK is 10.5 dB, but 8-PSK needs 14 dB (more than twice as much energy). However, 8-PSK needs only 1/3 the bandwidth of BPSK and 2/3 the bandwidth of QPSK for the same bit rate.

6.5.4   Bit Error Probability for 4-QASK

Recall that a 4-QASK signal has the form

$$s(t) = A\, u_i\, \alpha(t) \cos(\omega_c t + \varphi) - A v_k\, \alpha(t) \sin(\omega_c t + \varphi), \qquad (6.45)$$

where $\alpha(t)$ is a pulse of duration T'. The inphase component is the binary ASK signal

$$s_I(t) = A\, u_i\, \alpha(t) \cos(\omega_c t + \varphi), \qquad (6.46a)$$

and the quadrature component is the binary ASK signal

$$s_Q(t) = -A v_k\, \alpha(t) \sin(\omega_c t + \varphi). \qquad (6.46b)$$

In this section, the data modulation on the inphase and quadrature signals is binary, so the data variables are u_0 and u_1 for the inphase component and v_0 and v_1 for the quadrature component. The input to the receiver is

$$Y(t) = s(t) + X(t),$$

where $X(t)$ is white Gaussian noise with spectral density $N_0/2$.

Consider the *inphase–quadrature (I–Q) correlation receiver* shown in Figure 6-25. The upper branch of this receiver is called the *inphase branch*, and it is a correlation receiver that is optimum for a system in which the binary ASK signal $s_I(t)$ is transmitted on an AWGN channel. The lower branch is called the *quadrature branch*, and it is an optimum correlation receiver for the binary ASK signal $s_Q(t)$ when transmitted on an AWGN channel. Therefore, the inphase branch is used to make decisions concerning $a_1(t)$, the inphase data signal, and the lower branch is used to make decisions concerning $a_2(t)$, the quadrature data signal.

From the preceding discussion, it is clear that the I–Q correlation receiver shown in Figure 6-25 is the optimum receiver for the 4-QASK signals provided that there is no *crosstalk* between the inphase and quadrature components of the 4-QASK signal. We now wish to obtain a key analytical result for I–Q correlation receivers for general rectangular QASK (i.e., the constellation does not have to be regular and there is no restriction on its size). We know that the inphase branch of the receiver operates

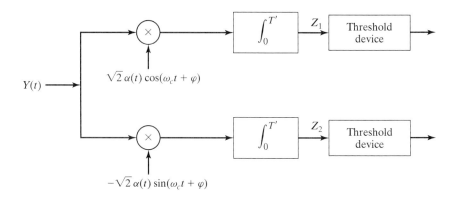

Figure 6-25: I–Q correlation receiver for 4-QASK.

optimally on $s_I(t)$ plus white Gaussian noise and that the quadrature branch operates optimally on $s_Q(t)$ plus white Gaussian noise. The receiver operates optimally on the sum of $s_I(t)$, $s_Q(t)$, and white Gaussian noise if the inphase branch of the receiver does not respond to the quadrature component of the signal and the quadrature branch of the receiver does not respond to the inphase component of the signal. Thus, the condition for optimality is that the output of the receiver's inphase branch should be zero if the input signal is $s_Q(t)$ only, and the output of the quadrature branch should be zero if the input signal is $s_I(t)$ only. If these conditions are satisfied, the receiver shown in Figure 6-25 gives the optimum decisions in each branch.

To examine these conditions further, we begin by considering the response of the inphase (upper) branch to the signal $s_Q(t)$. Notice that

$$\alpha(t) \cos(\omega_c t + \varphi) = (u_i A)^{-1} s_I(t),$$

and observe that, if the input is $s_Q(t)$, the output of the integrator in the inphase branch is

$$\int_0^{T'} s_Q(t) \alpha(t) \cos(\omega_c t + \varphi)\, dt = (u_i A)^{-1} \int_0^{T'} s_Q(t) s_I(t)\, dt.$$

Thus, the condition that the quadrature component of the signal produce no output in the inphase part of the receiver is just the condition that the inphase and quadrature components of the QASK signal be orthogonal; that is, $(s_I, s_Q) = 0$. As discussed in Section 6.5.1 (see (6.38)) this condition reduces to

$$\int_0^{T'} \alpha^2(t) \sin(2\omega_c t + 2\varphi)\, dt = 0, \tag{6.47}$$

which is satisfied in a number of important situations, such as if $\alpha(t)$ is a rectangular pulse and $\omega_c T$ is a multiple of π.

Similarly, if the input is $s_I(t)$, the output of the quadrature branch is

$$-\int_0^{T'} s_I(t) \alpha(t) \sin(\omega_c t + \varphi)\, dt = -(v_k A)^{-1} \int_0^{T'} s_Q(t) s_I(t)\, dt,$$

which is also equal to zero if (6.47) is satisfied. Therefore, under the condition that the inphase and quadrature components of the QASK signal are orthogonal—a condition that is equivalent to (6.47)—optimum decisions are being made in each branch of the I–Q correlation receiver. Observe that the orthogonality of the inphase and quadrature components is established for all forms of QASK that satisfy (6.47).

Returning to the analysis of standard 4-QASK, we wish to find the bit error probability for the I–Q correlation receiver if (6.47) is satisfied. Assume that the minimax criterion, Bayes criterion with equal prior probabilities, or the maximum-likelihood criterion is employed for the binary decision made in each branch of the receiver of Figure 6-25. This implies that the value for the optimum threshold is zero. Because the inphase branch of the I–Q correlation receiver is the optimum receiver for the binary ASK signal $s_I(t)$ on an AWGN channel, and because the presence of the quadrature signal $s_Q(t)$ does not affect the output of the inphase branch of this receiver, the probability of error for the binary decisions made on the inphase data signal is $Q\left(\sqrt{\bar{\mathcal{E}}(1-r)/N_0}\right)$.

If the signal $s_I(t)$ is an antipodal signal (i.e., $u_0 = -u_1$), then $\mathcal{E}_0 = \mathcal{E}_1 = \bar{\mathcal{E}}$ and $r = -1$. Let $\mathcal{E}$ denote the common value of the energy (i.e., $\mathcal{E} = \mathcal{E}_0 = \mathcal{E}_1$), and observe that $\mathcal{E}$ is the amount of energy in $s_I(t)$ during the interval $0 \leq t \leq T'$. Because the signal $s_I(t)$ can convey one bit of information during $0 \leq t \leq T'$, this energy is often referred to as the *energy per bit* for the signal, and it is denoted by $\mathcal{E}_b$. Thus, the error probability is $Q\left(\sqrt{2\,\mathcal{E}_b/N_0}\right)$ for the optimum decisions made in the inphase branch about the binary digits sent on the inphase component of the signal. Analogous conclusions hold for the decisions made on the quadrature signal in the quadrature branch of the receiver if $s_Q(t)$ is antipodal.

6.5.5   Summary: One-Dimensional vs. Two-Dimensional Modulation

We found that the inphase branch of the I–Q correlation receiver is the optimum receiver for the inphase binary ASK signal, and the quadrature branch is optimum for the quadrature binary ASK signal. In the absence of crosstalk, each of these receivers achieves the minimum bit error probability for reception of a single pulse on each of the inphase and quadrature components of the signal. If the inphase and quadrature signals are each antipodal and the minimax criterion, Bayes criterion with equal prior probabilities, or maximum-likelihood criterion is employed, this error probability is $Q\left(\sqrt{2\,\mathcal{E}_b/N_0}\right)$.

We also observed that the binary ASK signal with pulses of duration T and the 4-QASK signal with pulses of duration $T' = 2T$ can be used to transmit data at a rate of $1/T$ bits per second. Recall that $\alpha(t)$, the pulse waveform for 4-QASK, is related to $\beta(t)$, the pulse waveform for binary ASK, by $\alpha(t) = \beta(t/2)$ for $0 \leq t \leq T'$. Thus, the pulse waveform for 4-QASK has the same shape as the pulse waveform for binary ASK, but the 4-QASK waveform has twice the duration of the binary ASK waveform. It follows that the Fourier transform of the 4-QASK waveform has half the width of the Fourier transform of the binary ASK waveform; therefore, 4-QASK requires only half the bandwidth needed by binary ASK to achieve the data rate of $1/T$ bits per second. The relationship $\alpha(t) = \beta(t/2)$, $0 \leq t \leq T'$, also implies that $\mathcal{E}_\alpha = 2\mathcal{E}_\beta$. For binary ASK with $\{u_0, u_1\} = \{-1, +1\}$, the energy per bit is $\mathcal{E}_b = A^2 \mathcal{E}_\beta$ (e.g., see Section 6.3.4).

Table 6.4: Summary of Comparison: Binary ASK vs. 4-QASK

Parameter	Binary ASK	4-QASK
Pulse duration	T	$T' = 2T$
Pulse waveform	$\beta(t)$	$\alpha(t) = \beta(t/2)$
Data rate	$1/T$ bps	$1/T$ bps
Energy per bit	$\mathcal{E}_b = A^2 \mathcal{E}_\beta$	$\mathcal{E}_b = A^2 \mathcal{E}_\beta$
Bit error probability	$Q(\sqrt{2\mathcal{E}_b/N_0})$	$Q(\sqrt{2\mathcal{E}_b/N_0})$
Bandwidth	W	$W/2$

Notes:
1. $\mathcal{E}_\beta$ is the energy in the waveform $\beta(t)$, and W is the bandwidth of $\beta(t)$.
2. If $\beta(t) = p_T(t)$ and $\alpha(t) = p_{2T}(t)$, the power in the binary ASK and 4-QASK signals is $P = A^2$, the energy in the waveform $\beta(t)$ is $\mathcal{E}_\beta = T$, the energy per bit for each signal is $\mathcal{E}_b = A^2 T$, and the null-to-null bandwidth of $\beta(t)$ is $W = 2/T$.

For regular QASK with $\{u_0, u_1\} = \{v_0, v_1\} = \{-1, +1\}$, it easy to show from (6.45) that the energy in the signal for the interval from 0 to T' is $A^2 \mathcal{E}_\alpha$. The QASK signal represents two bits of information during the interval, so the energy per bit for QASK is $\mathcal{E}_b = A^2 \mathcal{E}_\alpha/2$. Since $\mathcal{E}_\alpha = 2\mathcal{E}_\beta$, the energy per bit for QASK is $\mathcal{E}_b = A^2 \mathcal{E}_\beta$, the same as for binary ASK. These and other conclusions are summarized in Table 6.4.

Because BPSK is equivalent to binary ASK with rectangular pulses, and because QPSK is equivalent to standard 4-QASK with rectangular pulses, a comparison of BPSK with QPSK can be obtained from Table 6.4 by letting $\beta(t) = p_T(t)$. The conclusion is that, for the same data rate, BPSK and QPSK have the same performance, except that QPSK requires half the bandwidth of BPSK.

6.5.6   Decision Regions for QASK Receivers

In Section 6.5.4, the performance criterion is the *bit error probability* on the inphase and quadrature branches of the 4-QASK signal. For this criterion, the optimum receiver makes independent decisions on the inphase and quadrature components, and the two-dimensional QASK signal is demodulated as a pair of one-dimensional binary ASK signals. In this section, we view the demodulation process in two-dimensional signal space and reconsider the demodulation of QASK signals. Moreover, consideration is given to the *symbol error probability* as well as the bit error probability.

We begin with 4-QASK, which can be expressed mathematically as

$$s(t) = A\,a_1(t)\cos(\omega_c t + \varphi) - A\,a_2(t)\sin(\omega_c t + \varphi), \qquad (6.48)$$

where $a_1(t) = u_i\,\alpha(t)$ and $a_2(t) = v_k\,\alpha(t)$ for the time interval $0 \le t \le T'$, $\alpha(t)$ is a baseband waveform of duration T', and u_i and v_k are binary data variables. The inphase modulation is $a_1(t)$, and the quadrature modulation is $a_2(t)$. A signal with these characteristics can be used to transmit two bits of information in a time interval of length T', so its data rate is $2/T'$ bits per second. Assume throughout this section that the inphase and quadrature components of $s(t)$ are orthogonal (i.e., (6.47) is satisfied).

For convenience, we focus on standard 4-QASK, for which the inphase and quadrature signals are each antipodal and $u_0 = +1$, $u_1 = -1$, $v_0 = +1$, and $v_1 = -1$. Notice that the binary symbols represented by the data variables u_i and v_k are the indices i and k, respectively. That is, $i = 0$ corresponds to the inphase modulation

$$a_1(t) = u_0 \alpha(t) = +\alpha(t),$$

which is used to send a 0 on the inphase component, and $i = 1$ corresponds to

$$a_1(t) = u_1 \alpha(t) = -\alpha(t),$$

which is used to send a 1 on the inphase component. Similarly, in the quadrature channel, $k = 0$ corresponds to

$$a_2(t) = v_0 \alpha(t) = +\alpha(t),$$

and $k = 1$ corresponds to

$$a_2(t) = v_1 \alpha(t) = -\alpha(t).$$

A particular choice for the pair (i, k) specifies the 4-QASK signal for one time interval. For example, from (6.48) and the definitions of $a_1(t)$ and $a_2(t)$, the choice $i = 0$ and $k = 1$ corresponds to the signal

$$
\begin{aligned}
s(t) &= A\,\alpha(t)\cos(\omega_c t + \varphi) + A\,\alpha(t)\sin(\omega_c t + \varphi) \\
&= \sqrt{2}\,A\,\alpha(t)\cos[\omega_c t + \varphi - (\pi/4)].
\end{aligned}
$$

If numerical values are given for i and k, we often omit the parentheses and the comma. For example, we denote $i = 0$ and $k = 1$ by 01, rather than (0, 1). Thus, we can label the points in the 4-QASK signal constellation by pairs of binary digits as in Figure 6-26. Notice that this labeling has the property that the bit patterns assigned to nearest neighbors differ in a single bit (Gray coding), as discussed in Section 6.4.2.

Let the 4-QASK signals be denoted by s_0, s_1, s_2, and s_3, and encode pairs of binary digits into these signals according to $00 \rightarrow s_0$, $10 \rightarrow s_1$, $11 \rightarrow s_2$, and $01 \rightarrow s_3$, as illustrated in Figure 6-26. For example, from the preceding discussion, we see that the signal that corresponds to $i = 0$ and $k = 1$ is given by

$$s_3(t) = \sqrt{2}\,A\,\alpha(t)\cos[\omega_c t + \varphi - (\pi/4)].$$

This signal is in the fourth quadrant of the diagram in Figure 6-26.

If s(t) is the transmitted signal, the input to the receiver is

$$Y(t) = s(t) + X(t),$$

where $X(t)$ is white Gaussian noise with spectral density $N_0/2$. It should be clear by now (e.g., from the discussion in Section 6.1.4) that the correlators shown in Figure 6-27 extract all of the useful information from $Y(t)$. The decision as to which of the four symbols was transmitted can be based on the outputs Z_1 and Z_2 only, where Z_1 is the

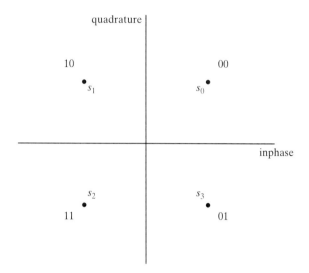

Figure 6-26: A labeled 4-QASK signal constellation.

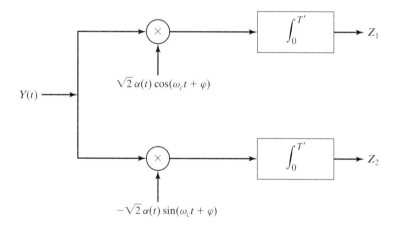

Figure 6-27: I–Q correlators for QASK.

output of the correlator in the inphase branch of the receiver and Z_2 is the output of the correlator in the quadrature branch. These outputs can be used to form a vector decision statistic $\mathbf{Z} = (Z_1, Z_2)$, and the decision can be based on this statistic alone.

Values for the decision statistic $\mathbf{Z}$ can be represented as points in the plane, just as values for the decision statistic for one-dimensional signals can be represented by points on the real line. Furthermore, if the I–Q correlation receiver is used for reception of QASK signals, the decision regions can be represented as subsets of the plane, just as the decision regions for the correlation receiver for one-dimensional signals are intervals of the real line.

The results of Section 6.5.4 imply that in order to minimize the *bit error probability* in each branch, the optimum symbol decision for QASK is based on making independent decisions on the inphase and quadrature components, and the decision procedure for each component is a simple threshold test. As a result, the decision regions that minimize the bit error probability for QASK signaling are *rectangular*; that is, they are of the form

$$\{(z_1, z_2) : a < z_1 < b, c < z_2 < d\}.$$

The parameters a and c can be finite or $-\infty$, and the parameters b and d can be finite or $+\infty$. From the results of Section 6.5.4, it is easy to see that if the bit error probability is the desired performance measure and we employ the minimax criterion, maximum-likelihood criterion, or the Bayes criterion with $\pi_i = 1/4$ for $0 \le i \le 3$, then the optimum decision regions are

$$\Gamma_0 = \{(z_1, z_2) : 0 < z_1 < +\infty, 0 < z_2 < +\infty\},$$
$$\Gamma_1 = \{(z_1, z_2) : -\infty < z_1 < 0, 0 < z_2 < +\infty\},$$
$$\Gamma_2 = \{(z_1, z_2) : -\infty < z_1 < 0, -\infty < z_2 < 0\},$$

and

$$\Gamma_3 = \{(z_1, z_2) : 0 < z_1 < +\infty, -\infty < z_2 < 0\}.$$

The decision regions Γ_0 and Γ_2 are illustrated in Figure 6-28.

Next we examine the symbol error probability. If the symbol error probability is the performance measure of interest, it is not necessary to associate binary digits with the transmitted symbols. A standard 4-QASK signal constellation is illustrated in Figure 6-29 without an assignment of binary digits to the four symbols.

Recall that the QASK signal is given by

$$s(t) = A u_i \alpha(t) \cos(\omega_c t + \varphi) - A v_k \alpha(t) \sin(\omega_c t + \varphi), \tag{6.49}$$

where $\alpha(t)$ is a baseband waveform of duration T' and u_i and v_k are the data variables from the sets $\{u_i : 0 \le i \le M' - 1\}$ and $\{v_k : 0 \le k \le M'' - 1\}$, respectively. This gives a QASK signal set of size $M = M'M''$. The signal set is denoted by

$$\{s_n : 0 \le n \le M - 1\}.$$

Each pair (i, k) is a *symbol* to be sent to the receiver. This is accomplished by transmitting a different signal for each symbol. For each symbol that is sent, the receiver can make $M - 1$ different types of errors, one for each of the $M - 1$ other symbols. Because our interest is in the symbol error probability, the different types of errors all count the same. That is, if symbol (i, k) is sent, but the receiver decides that (i', k') was sent, then we say the receiver makes a *correct symbol decision* if both $i' = i$ and $k' = k$; otherwise, we say that a *symbol error* has occurred. As a result, the symbol error probability is not affected by the way in which the signals are assigned to the different symbols. We can use any assignment of the M signals to the pairs $(i, k), 0 \le i \le M'-1$, $0 \le k \le M'' - 1$, as long as no two pairs are assigned to the same signal. For instance, we can order the M pairs of integers in an arbitrary way and represent the first pair by s_0, the second pair by s_1, etc. If the pair (i, k) is represented by the signal s_n, we write $(i, k) \leftrightarrow n$. An example of the assignment of the pairs of integers to signals is given in Figure 6-26 for $M = 4$ ($M' = M'' = 2$).

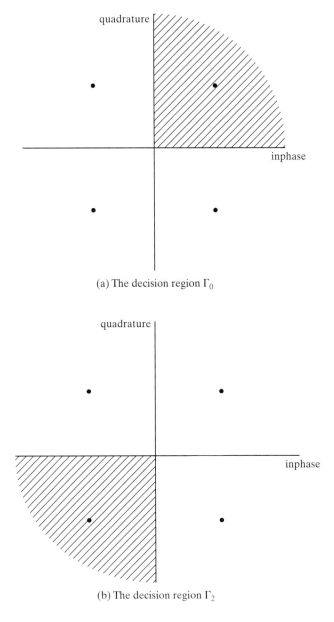

(a) The decision region Γ_0

(b) The decision region Γ_2

Figure 6-28: The decision regions Γ_0 and Γ_2.

If symbol (i, k) is to be sent and $(i, k) \leftrightarrow n$, signal $s_n(t)$ is transmitted over the channel. The receiver bases its decision as to which symbol was sent on the decision statistic $\mathbf{Z} = (Z_1, Z_2)$ that is produced by the correlators of Figure 6-27. In order to determine the probability of symbol error, we first consider the components of Z_1 and

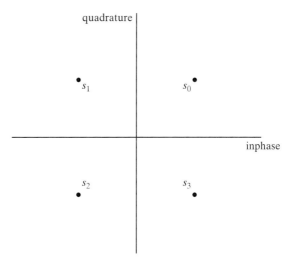

Figure 6-29: Standard 4-QASK signal constellation.

Z_2 that are due to the signal $s_n(t)$ only. This step is analogous to the corresponding development for the one-dimensional M-ASK signals in Section 6.3.6.

In the absence of noise, the output of the inphase (upper) correlator of Figure 6-27 is

$$\hat{u}_i = u_i A \, \mathcal{E}_\alpha / \sqrt{2}, \tag{6.50a}$$

where $\mathcal{E}_\alpha$ is the energy in the waveform $\alpha(t)$. The output of the quadrature (lower) correlator is

$$\hat{v}_k = v_k A \, \mathcal{E}_\alpha / \sqrt{2}. \tag{6.50b}$$

Even if the noise is present, the variables $\hat{u}_i$ and $\hat{v}_k$ represent the mean values of the outputs of the inphase and quadrature correlators, respectively, when the symbol (i, k) is sent. This is because the noise has zero mean.

Just as the outputs for M-ASK can be represented as points on the line (e.g., Figure 6-15 in Section 6.3.6), the outputs for M-QASK can be represented as points in the plane. The coordinates for these points are specified by $(\hat{u}_i, \hat{v}_k)$, as illustrated in Figure 6-30 for a 16-QASK signal set with data symbol alphabets given by

$$\{u_0, u_1, u_2, u_3\} = \{v_0, v_1, v_2, v_3\} = \{-3, -1, +1, +3\}.$$

Geometrical representations such as Figure 6-30 are of use in the determination of the decision regions and in the analysis of the symbol error probability for M-QASK, just as similar representations are of use for one-dimensional signals in Section 6.4.

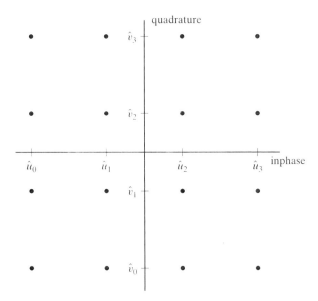

Figure 6-30: Output signal components for 16-QASK.

The foregoing considerations provide a first step in specifying the bivariate density for the vector decision statistic $\mathbf{Z} = (Z_1, Z_2)$. In particular, $\hat{u}_i$ and $\hat{v}_k$ are the conditional means for the components of $\mathbf{Z} = (Z_1, Z_2)$ given that the data variables are u_i and v_k. The random variables Z_1 and Z_2 are jointly Gaussian, and they are uncorrelated because of the orthogonality of the inphase and quadrature reference signals (i.e., the condition expressed in equation (6.47) is satisfied). As a result, the two components, Z_1 and Z_2, are independent, and their joint density is simply the product of their individual marginal densities. But each of these random variables is Gaussian with variance $\sigma^2 = N_0 \mathcal{E}_\alpha / 2$, so their densities are given by

$$f_{Z_1}(x) = \exp\{-(x - \hat{u}_i)^2 / 2\sigma^2\} / \sqrt{2\pi}\sigma$$

and

$$f_{Z_2}(y) = \exp\{-(y - \hat{v}_k)^2 / 2\sigma^2\} / \sqrt{2\pi}\sigma.$$

The optimum decision rules are derived from the conditional densities for the vector decision statistic $\mathbf{Z} = (Z_1, Z_2)$. Let f_n denote the conditional density of $\mathbf{Z} = (Z_1, Z_2)$ given that the symbol (i, k) is sent and $(i, k) \leftrightarrow n$ (i.e., given that the transmitted signal is s_n). Because Z_1 and Z_2 are independent, we can write the density function as

$$f_n(x, y) = \exp\{-[(x - \hat{u}_i)^2 + (y - \hat{v}_k)^2] / 2\sigma^2\} / 2\pi\sigma^2. \qquad (6.51)$$

The density f_n is a two-dimensional Gaussian density with mean $(\hat{u}_i, \hat{v}_k)$, so its maximum value is $f_n(\hat{u}_i, \hat{v}_k)$. Notice that the density has circular symmetry about the point $(\hat{u}_i, \hat{v}_k)$ at which the maximum occurs: The value of $f_n(x, y)$ depends only on the

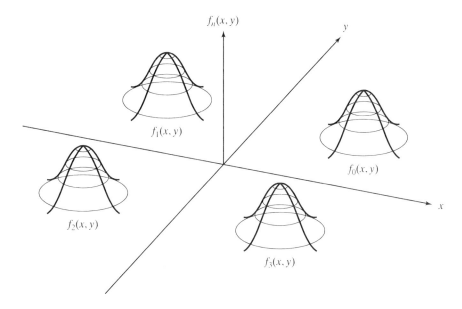

Figure 6-31: The four two-dimensional Gaussian densities $f_n(x, y)$, $0 \leq n \leq 3$.

distance between the mean $(\hat{u}_i, \hat{v}_k)$ and the point (x, y); the direction from $(\hat{u}_i, \hat{v}_k)$ to the point (x, y) is unimportant. Thus, circles centered at $(\hat{u}_i, \hat{v}_k)$ are lines of constant values for the function f_n. Because the Gaussian densities f_n, $0 \leq n \leq M - 1$, have the same variance, they are just shifted versions of each other. In fact, they are all shifted versions of the zero-mean density

$$f(x, y) = \exp\{-(x^2 + y^2)/2\sigma^2\}/2\pi\sigma^2.$$

As an example, consider the 4-QASK signal constellation of Figure 6-29. For this example, the four density functions f_n, $0 \leq n \leq 3$, are symmetrical about points of the form $(\pm c, \pm c)$ for some constant c. These densities are illustrated in Figure 6-31 for $M' = M'' = 2$ ($M = 4$), $u_0 = -u_1 = +1$, $v_0 = -v_1 = +1$ (i.e., the standard 4-QASK signal set). The conditional means for this example are given by

$$\hat{u}_0 = -\hat{u}_1 = \hat{v}_0 = -\hat{v}_1 = A\,\mathcal{E}_\alpha/\sqrt{2}.$$

Thus, the points $(\hat{u}_i, \hat{v}_k)$ are of the form $(\pm c, \pm c)$, and the constant c is $A\,\mathcal{E}_\alpha/\sqrt{2}$.

The maximum-likelihood decision is as follows. If it is observed that $\mathbf{Z} = (x, y)$, the decision is that symbol (i', k') was sent if $(i', k') \leftrightarrow m$ and

$$f_m(x, y) = \max\{f_n(x, y) : 0 \leq n \leq M - 1\}.$$

In other words if, for each n in the range $0 \leq n \leq M - 1$, Γ_n denotes the set of observations for which the maximum-likelihood decision is that s_n was the transmitted signal, then the point (x, y) is included in Γ_m if $f_m(x, y) \geq f_n(x, y)$ for all n. For some values of x and y there will be ties, of course, but any method of resolving such ties gives a maximum-likelihood decision rule.

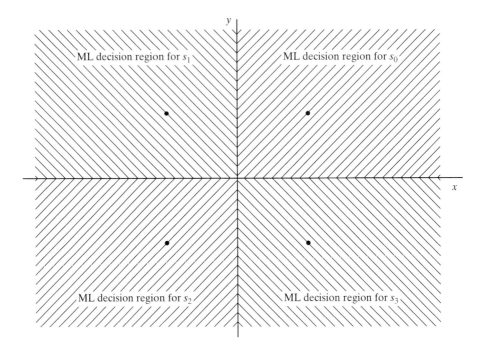

Figure 6-32: The maximum-likelihood (ML) decision regions for standard 4-QASK.

The next step is to characterize the maximum-likelihood decision regions Γ_n. This can be done in two different ways. The simplest approach is to note that the value of $f_n(x, y)$ decreases as the point (x, y) gets farther from the mean of the density. Thus, if f_n and f_m are two-dimensional Gaussian densities with means (μ_n, ν_n) and (μ_m, ν_m), respectively, then $f_m(x, y) > f_n(x, y)$ if and only if the distance between (x, y) and (μ_m, ν_m) is less than the distance between (x, y) and (μ_n, ν_n). As a result, if the observation is that $Z_1 = x$ and $Z_2 = y$, the maximum-likelihood decision is that signal s_m was transmitted if (x, y) is closer to (μ_m, ν_m) than it is to any of the other means (μ_n, ν_n), $n \neq m$. This conclusion is valid for any signal constellation in two dimensions (and it generalizes easily to higher dimensions). When applied to the 4-QASK signals of Figure 6-29, this observation implies that the maximum-likelihood decision regions are as illustrated in Figure 6-32.

Another approach, more algebraic in nature, is to examine the set $\Gamma_{m,n}$ of all (x, y) for which $f_m(x, y) = f_n(x, y)$ if $m \neq n$. If Γ_m and Γ_n intersect, the intersection must be the common boundary of Γ_m and Γ_n, and this common boundary is a subset of $\Gamma_{m,n}$. The key result needed for this approach is contained in the following exercise:

Exercise 6-5

Suppose that g_1 and g_2 are two-dimensional Gaussian densities given by

$$g_1(x, y) = \exp\{-[(x - \mu_1)^2 + (y - \nu_1)^2]/2\sigma^2\}/2\pi\sigma^2$$

and

$$g_2(x, y) = \exp\{-[(x - \mu_2)^2 + (y - \nu_2)^2]/2\sigma^2\}/2\pi\sigma^2.$$

Assume that $\mu_1 \neq \mu_2$ or $\nu_1 \neq \nu_2$ (or both); otherwise, the densities are identical and they do not model a practical communication system. Find the set of all (x, y) for which $g_1(x, y) = g_2(x, y)$. Simplify the answer for the situation shown in Figure 6-31 in which μ_1, μ_2, ν_1, and ν_2 all have the same magnitude. In particular, suppose that either $\mu_1 = -\mu_2$ and $\nu_1 = \nu_2$ or else $\mu_1 = \mu_2$ and $\nu_1 = -\nu_2$.

Solution.   Set $g_1(x, y) = g_2(x, y)$ and solve for x and y. Because the two densities have the same variance, $g_1(x, y) = g_2(x, y)$ if and only if

$$(x - \mu_1)^2 + (y - \nu_1)^2 = (x - \mu_2)^2 + (y - \nu_2)^2.$$

Expanding and canceling common terms, we find that this is equivalent to

$$-2\mu_1 x + \mu_1^2 - 2\nu_1 y + \nu_1^2 = -2\mu_2 x + \mu_2^2 - 2\nu_2 y + \nu_2^2.$$

But this is just the equation for the *straight line*

$$2(\nu_2 - \nu_1)y = 2(\mu_1 - \mu_2)x + (\mu_2^2 + \nu_2^2 - \mu_1^2 - \nu_1^2).$$

If $\nu_1 \neq \nu_2$, this can be written in the more familiar form $y = mx + b$, where

$$m = (\mu_1 - \mu_2)/(\nu_2 - \nu_1)$$

and

$$b = (\mu_2^2 + \nu_2^2 - \mu_1^2 - \nu_1^2)/2(\nu_2 - \nu_1).$$

Or, if $\mu_1 \neq \mu_2$, it can be written as $x = m'y + b'$, where

$$m' = (\nu_2 - \nu_1)/(\mu_1 - \mu_2)$$

and

$$b' = (\mu_1^2 + \nu_1^2 - \mu_2^2 - \nu_2^2)/2(\mu_1 - \mu_2).$$

Now if μ_1, μ_2, ν_1, and ν_2 all have the same magnitude, the straight line has the simpler equation

$$(\nu_2 - \nu_1)y = (\mu_1 - \mu_2)x,$$

which is the equation for a straight line through the origin. If $\mu_1 = -\mu_2$ and $\nu_1 = \nu_2$, as is true for the 4-QASK signal set of Figure 6-29, the above equation reduces to the line $x = 0$, but if $\mu_1 = \mu_2$ and $\nu_1 = -\nu_2$, it is the line $y = 0$. As expected, these two lines are the boundaries of the decision regions of Figure 6-32.   ■

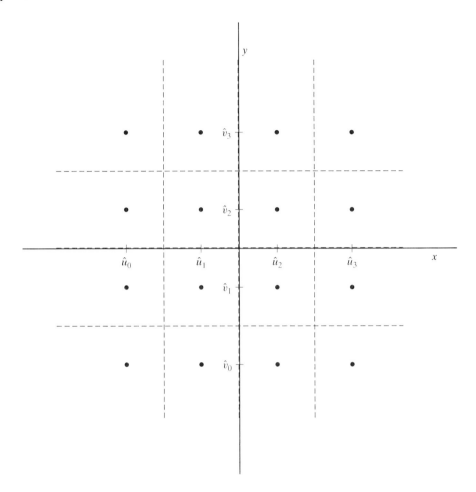

Figure 6-33: The maximum-likelihood decision regions for standard 16-QASK.

Either of the preceding approaches can be applied to any QASK signal constellation to determine the maximum-likelihood decision regions. For example, when applied to the standard 16-QASK signal constellation illustrated in Figure 6-30, the results show that the decision regions are as illustrated in Figure 6-33. For this application, the parameters μ_1 and μ_2 take values among the $\hat{u}_i$, $0 \leq i \leq 3$, and ν_1 and ν_2 are selected from among the values of $\hat{v}_k$, $0 \leq k \leq 3$. As illustrated in this figure, if the two-dimensional densities are Gaussian and have a common variance, the boundary between two adjoining maximum-likelihood decision regions is a line segment or a point. (The latter is just a degenerate form of a line segment.) For this particular signal constellation, the boundaries for the maximum-likelihood decision regions are located midway between the points of the constellation. Notice also, that the decision regions for *exterior* points of the constellation are semi-infinite rectangles, while the decision regions for *interior* points are squares of finite area.

6.5.7   Symbol Error Probabilities for QASK Receivers

The evaluation of the symbol error probabilities depends on the signal constellation, the correlators, and the decision regions. We begin with an arbitrary signal constellation, general correlators, and general rectangular decision regions. Thus, the only initial restriction is that each decision region must be given by

$$\Gamma = \{(x, y) : x_1 < x < x_2, y_1 < y < y_2\} \tag{6.52}$$

for some choice of x_1, x_2, y_1, and y_2. Each of the strict inequalities in (6.52) can be changed to include equality without altering any of the resulting error probabilities. To avoid trivialities, we require $x_1 < x_2$ and $y_1 < y_2$. The values for x_1 and y_1 can be finite or $-\infty$, and the values for x_2 and y_2 can be finite or $+\infty$.

The receiver produces a decision statistic $\mathbf{Z} = (Z_1, Z_2)$. For example, the receiver could be the I–Q correlator of Figure 6-27, but the only important feature is that the decision statistic have the required density. The conditional density function for the observation $\mathbf{Z} = (Z_1, Z_2)$, given that signal s_n was transmitted, is the two-dimensional Gaussian density with mean (μ, ν) and variance σ^2. That is,

$$f_n(x, y) = \exp\{-[(x - \mu)^2 + (y - \nu)^2]/2\sigma^2\}/2\pi\sigma^2. \tag{6.53}$$

Consider the decision region illustrated in Figure 6-34. The conditional probability of a *correct decision* given that signal s_n is transmitted, which is denoted by $P_{c,n}$, is the conditional probability that the decision statistic $\mathbf{Z}$ is inside the set Γ_n given that s_n is the transmitted signal. From Chapter 1, we know that if the random vector $\mathbf{Z}$ has the Gaussian density of (6.53), the probability that $\mathbf{Z}$ is in Γ_n is given by

$$P_{c,n} = \{\Phi[(x_2 - \mu)/\sigma] - \Phi[(x_1 - \mu)/\sigma]\}\{\Phi[(y_2 - \nu)/\sigma] - \Phi[(y_1 - \nu)/\sigma]\}. \tag{6.54}$$

Of course, the conditional probability of error given that signal s_n is transmitted is $P_{e,n} = 1 - P_{c,n}$. As before, x_1 and y_1 are allowed to be $-\infty$ in (6.54), and x_2 and y_2 are permitted to be $+\infty$. Thus, (6.54) can be used to obtain the expression for the symbol error probability for any M-QASK signal set and any choice of rectangular decision regions. Note in particular that the probability of error does not depend on the location in two-dimensional space of the decision rectangle; it depends only on the standard deviation σ and the distances from the mean (μ, ν) to each of the four boundaries. See Problem 6.16 for the basic property that underlies this fact.

Recall from Section 6.5.1 that the points in a regular QASK signal constellation are aligned in each dimension and uniformly spaced, as illustrated in Figure 6-23. For a regular QASK signal constellation and maximum-likelihood decision regions, the boundaries for the decision regions are straight lines that are located midway between pairs of closest points among the $(\hat{u}_i, \hat{v}_k)$, $0 \le i \le M' - 1$, $0 \le k \le M'' - 1$, as illustrated for 16-QASK in Figure 6-33. This figure is reproduced in Figure 6-35(a) with the addition of labels for the points and regions. The terminology illustrated in Figure 6-35(a) is employed in the discussion that follows.

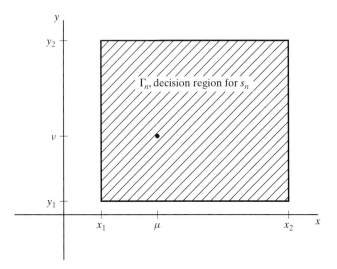

Figure 6-34: The general rectangular decision region Γ_n.

The error probability expressions simplify considerably for maximum-likelihood decisions and a regular QASK signal set. First, we consider the interior points of the signal constellation. For these points, the maximum-likelihood decision regions are squares as shown in Figure 6-35(b). In this figure, the parameter δ is the width of the squares that represent the decision regions for the interior points of the regular QASK signal set.

Because the boundaries of the maximum-likelihood decision regions are straight lines that are located midway between points in the constellation, it is clear that δ also represents the distance between nearest neighbors among the points $(\hat{u}_i, \hat{v}_j)$. That is,

$$\delta = \hat{u}_{i+1} - \hat{u}_i \text{ for } 0 \leq i \leq M' - 2$$

and

$$\delta = \hat{v}_{k+1} - \hat{v}_k \text{ for } 0 \leq k \leq M'' - 2.$$

Thus, for interior points, it follows that (6.54) applies with

$$x_2 - \mu = \mu - x_1 = y_2 - \nu = \nu - y_1 = \delta/2.$$

The probability that the symbol is received correctly is therefore given by

$$
\begin{aligned}
P_{c,n} &= [\Phi(\delta/2\sigma) - \Phi(-\delta/2\sigma)][\Phi(\delta/2\sigma) - \Phi(-\delta/2\sigma)] \\
&= [2\Phi(\delta/2\sigma) - 1]^2 = [1 - 2Q(\delta/2\sigma)]^2 = [2Q(\delta/2\sigma) - 1]^2. \quad (6.55a)
\end{aligned}
$$

For the corner points of a regular QASK signal constellation, either $x_2 = +\infty$ or $x_1 = -\infty$ (but not both) and either $y_2 = +\infty$ or $y_1 = -\infty$ (but not both). For the other exterior points, exactly one of the parameters $x_1, x_2, y_1,$ or y_2 is infinite. Equation (6.54)

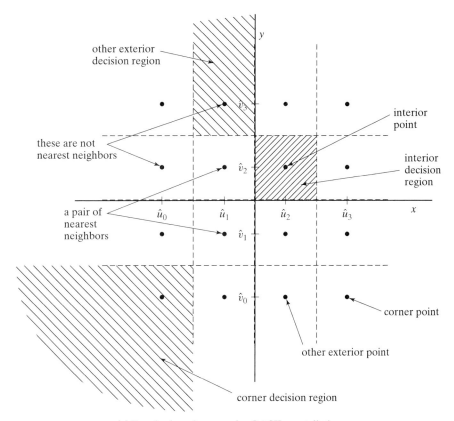

(a) Terminology for a regular QASK constellation
with its maximum-likelihood decision regions

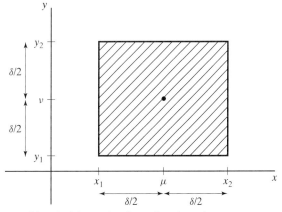

(b) A decision region for an interior point
in a regular QASK signal constellation

Figure 6-35: Decision regions for QASK.

applies for any values of these parameters, and substitution of the appropriate values for the parameters provides the results for the various special cases. For the corner points, the probability that the symbol decision is correct is

$$P_{c,n} = [\Phi(\delta/2\sigma)]^2 = [1 - Q(\delta/2\sigma)]^2. \tag{6.55b}$$

For the other exterior points, the probability of correct reception is

$$P_{c,n} = \Phi(\delta/2\sigma)[2\,\Phi(\delta/2\sigma) - 1] = [1 - Q(\delta/2\sigma)][1 - 2\,Q(\delta/2\sigma)]. \tag{6.55c}$$

From an examination of either Figure 6-35(a) or equations (6.55), it is clear that the smallest of these probabilities corresponds to the interior points (i.e., (6.55a)). Thus, if it is desirable to guarantee that the probability of correct reception is greater than a specified value q, regardless of which symbol is sent, we must have

$$[2\,Q(\delta/2\sigma) - 1]^2 > q.$$

Equivalently, we can consider the *probability of symbol error* (i.e., the probability that the symbol is received incorrectly) for each of the points in the signal constellation. For example, the probability of symbol error for the interior points is

$$P_{e,n} = 1 - [2\,Q(\delta/2\sigma) - 1]^2 = 4\,Q(\delta/2\sigma)[1 - Q(\delta/2\sigma)]. \tag{6.56a}$$

The symbol error probability for the corner points is

$$P_{e,n} = Q(\delta/2\sigma)[2 - Q(\delta/2\sigma)], \tag{6.56b}$$

and the symbol error probability for the other exterior points is

$$P_{e,n} = Q(\delta/2\sigma)[3 - 2\,Q(\delta/2\sigma)]. \tag{6.56c}$$

For the optimum correlators of Figure 6-27 and the maximum-likelihood decision regions (i.e., the maximum-likelihood receiver),

$$\hat{u}_i = u_i A \mathcal{E}_\alpha / \sqrt{2}$$

and

$$\hat{v}_k = v_k A \mathcal{E}_\alpha / \sqrt{2}.$$

If $u_{i+1} - u_i = 2$ for $0 \le i \le M' - 2$ and $v_{k+1} - v_k = 2$ for $0 \le k \le M'' - 2$, then

$$\delta = \hat{u}_{i+1} - \hat{u}_i = \hat{v}_{k+1} - \hat{v}_k = \sqrt{2}\,A\,\mathcal{E}_\alpha$$

for each such i and k. Furthermore, it is easy to show that $\sigma^2 = N_0 \mathcal{E}_\alpha / 2$ for the receiver of Figure 6-27, so

$$\delta/2\sigma = A\sqrt{\mathcal{E}_\alpha / N_0}.$$

Thus, equations (6.56) can be rewritten in terms of the parameters A, $\mathcal{E}_\alpha$, and N_0.

Alternatively, for the maximum-likelihood receiver, we can express (6.56) in terms of the distance between nearest neighbors in the signal constellation to provide expressions analogous to those obtained for one-dimensional modulation (i.e., M-ASK) in Section 6.4.3. Proceeding as in Exercise 6-4 of that section, we let d be the distance between nearest neighbors in a regular QASK signal constellation. That is, d is the minimum distance between any pair of signals in the signal set:

$$d^2 = \min_{m \neq n} \int_{-\infty}^{\infty} [s_n(t) - s_m(t)]^2 \, dt. \tag{6.57}$$

It is easy to show that $d^2 = 2A^2 \mathcal{E}_\alpha$ for the regular QASK signal constellation given by (6.49) with $u_{i+1} - u_i = 2$ for each i and $v_{k+1} - v_k = 2$ for each k. It follows that

$$\delta/2\sigma = d/\sqrt{2N_0}.$$

Substitution into (6.56) gives the desired expressions for the error probabilities. If the transmitted signal s_n corresponds to an interior point of the signal constellation, then

$$P_{e,n} = 4 \, Q\!\left(d/\sqrt{2N_0}\right)\!\left[1 - Q\!\left(d/\sqrt{2N_0}\right)\right]. \tag{6.58a}$$

If the transmitted signal corresponds to a corner point, then

$$P_{e,n} = Q\!\left(d/\sqrt{2N_0}\right)\!\left[2 - Q\!\left(d/\sqrt{2N_0}\right)\right], \tag{6.58b}$$

and if the transmitted signal corresponds to an exterior point other than a corner point, then

$$P_{e,n} = Q\!\left(d/\sqrt{2N_0}\right)\!\left[3 - 2 \, Q\!\left(d/\sqrt{2N_0}\right)\right]. \tag{6.58c}$$

It should be clear from intuitive considerations, and it is easy to prove (see Problem 6.15), that the interior points give the largest probability of error, and the corner points give the smallest. In order to design a communication system that provides a symbol error probability less than a prescribed value regardless of which symbol is transmitted, it is necessary to concentrate on the interior points of the constellation. For such a constraint, the probability

$$P_e = 4 \, Q\!\left(d/\sqrt{2N_0}\right)\!\left[1 - Q\!\left(d/\sqrt{2N_0}\right)\right] \tag{6.59}$$

should be employed as the symbol error probability for a regular QASK signal set with maximum-likelihood reception and $M \geq 16$. Moreover, for $M \geq 16$, there are sufficiently many interior points that they should be given primary consideration for most applications. For smaller values of M, there are few, if any, interior points, so the appropriate results from (6.58) should be used instead of relying on (6.59) only.

Note that if $d/\sqrt{2N_0}$ is large (i.e., the signal-to-noise ratio is large), then

$$Q\!\left(d/\sqrt{2N_0}\right) \ll 1.$$

Thus, (6.58a) implies that the interior points have a symbol error probability that is approximated by

$$P_{e,n} \approx 4 \, Q\!\left(d/\sqrt{2N_0}\right)$$

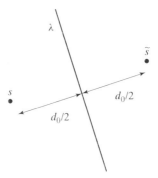

Figure 6-36: Illustration for Exercise 6-6.

for large signal-to-noise ratios. Similar approximations can be obtained for the other points in the constellation. (See Problem 6.17.) Furthermore, other probabilities can be obtained from the notion of distance in signal space, as illustrated in the next exercise.

Exercise 6-6
Consider a signal $s(t)$ that represents data variables u_i and v_k as specified by (6.49). In Figure 6-36, line λ in the two-dimensional signal space is distance $d_0/2$ from signal s. The channel is an additive white Gaussian noise channel with spectral density $N_0/2$, and the receiver employs the optimum correlators of Figure 6-27. Give an expression in terms of d_0, N_0, and the function Q for the probability that the decision statistic $\mathbf{Z} = (Z_1, Z_2)$ falls on the opposite side of the line from $(\hat{u}_i, \hat{v}_k)$.

Solution.   The key results needed for the solution are in Section 5.7.2. Although this exercise is posed in two-dimensional space, it can be solved using the one-dimensional methods of Sections 5.7.2 and 6.4.3. (See Problem 6.16.) Consider a signal $\tilde{s}$ that placed at the unique location that makes line λ the maximum-likelihood decision boundary for binary communication with signals $s_0 = s$ and $s_1 = \tilde{s}$; that is, $\tilde{s}$ is the mirror image of s with respect to line λ, as illustrated in Figure 6-36. The probability in question is equal to the probability of error for signal set $\{s_0, s_1\}$ and maximum-likelihood coherent reception. Since the distance between s and $\tilde{s}$ is d_0, we know from Section 5.7.2 that the probability of error is $Q\left(d_0/\sqrt{2N_0}\right)$.   ∎

By use of methods similar to those employed in Exercise 6-2, we can relate the maximum and average energies for a particular signal constellation to the distance parameter d. In the evaluation of the average energy for a standard M-QASK signal constellation in which M is an even power of 2 (i.e., $M = 2^m$ for an even integer m), it suffices to restrict attention to one quadrant of the constellation because of the symmetry of the signal set. Thus, for the signal constellation of Figure 6-37(a), the average energy for the four signals s_{10}, s_{11}, s_{14}, and s_{15} is the same as the average energy for all 16 signals. Notice also that $\mathcal{E}_{14} = \mathcal{E}_{11}$, so it suffices to find expressions for $\mathcal{E}_{10}$, $\mathcal{E}_{11}$, and $\mathcal{E}_{15}$ in terms of d.

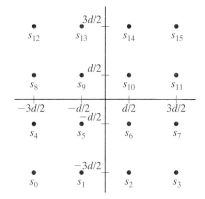

(a) The standard 16-QASK signals

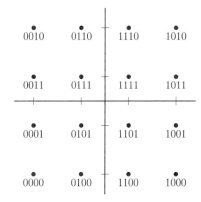

(b) Gray coding for 16-QASK

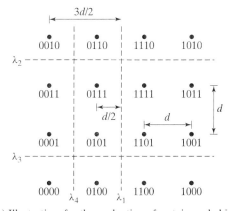

(c) Illustration for the evaluation of certain probabilities

Figure 6-37: Standard 16-QASK signal constellation.

From Figure 6-37(a), we see that $s_{10}(t) = -s_5(t)$ for all t and the distance between s_{10} and s_5 is $d\sqrt{2}$. It follows that $\|s_{10} - s_5\|^2 = \|2s_{10}\|^2 = 2d^2$ and $\mathcal{E}_{10} = d^2/2$. The same conclusion can be obtained from the observation that the square of the distance from the origin to signal s is $\|s\|^2$, the energy in signal s. The distance from the origin to s_{10} is $d/\sqrt{2}$. Similarly, the distance from the origin to s_{15} is $3d/\sqrt{2}$, so $\mathcal{E}_{15} = 9d^2/2$. Finally, the distance from the origin to s_{11} is $\sqrt{10}\,d/2$, so $\mathcal{E}_{11} = 5d^2/2$. Since the average of $d^2/2$ and $9d^2/2$ is also $5d^2/2$, it follows that the average energy for the signals of the regular 16-QASK signal constellation is $\bar{\mathcal{E}}_s = 5d^2/2$, so $d^2 = 2\,\bar{\mathcal{E}}_s/5$. As illustrated in Figure 6-37(b), each of the 16-QASK signals represents four bits of information, so the average energy per bit for the standard 16-QASK signal constellation is $\bar{\mathcal{E}}_b = 5d^2/8$, and $d^2 = 8\,\bar{\mathcal{E}}_b/5$. Next, notice that each of the expressions in (6.58) can be written in terms of the square root of $d^2/2N_0$, and then $d^2/2N_0$ can be replaced by $\bar{\mathcal{E}}_s/5N_0$ or $4\,\bar{\mathcal{E}}_b/5N_0$ to give expressions for the symbol error probabilities for standard 16-QASK in terms of the average energy per symbol or the average energy per bit, respectively. In Problem 6.18, you are asked to apply the methods of Exercise 6-2 to a standard M-ASK signal constellation for which M is an even power of 2 and show that

$$\bar{\mathcal{E}}_s = (M - 1)\,d^2/6,$$
$$\bar{\mathcal{E}}_b = (M - 1)\,d^2/(6 \log_2 M),$$
$$Q\big(d/\sqrt{2N_0}\,\big) = Q\big(\sqrt{3\,\bar{\mathcal{E}}_s/[(M - 1)N_0]}\,\big),$$

and

$$Q\big(d/\sqrt{2N_0}\,\big) = Q\big(\sqrt{3\,\bar{\mathcal{E}}_b \log_2(M)/[(M - 1)N_0]}\,\big).$$

Substitution into (6.58) gives expressions for the symbol error probabilities in terms of the average energy per symbol or the average energy per bit.

6.5.8   Bit Error Probabilities for QASK Receivers

It should be clear from examination of diagrams such as Figure 6-37(a) that certain types of symbol errors are more likely than others. For example, as long as the signal-to-noise ratio is sufficiently large, the most likely symbol errors involve nearest neighbors. That is, if a particular signal is transmitted, the incorrect decision that a particular nearest-neighbor signal was transmitted has a higher probability than an incorrect decision involving a specific signal that is more distant in the constellation. This issue is explored in Problem 6.20 for a large signal constellation; however, the conclusion is true for signal constellations of any size.

Suppose M-QASK signaling is to be employed to transmit binary data. Because different types of symbol errors can result in different numbers of bit errors, care should be taken in assigning sequences of binary digits to the M-ary symbols. The assignment of binary digits to the symbols in the signal set should be done in such a way that the most likely symbol errors correspond to the fewest number of bit errors. If the signal-to-noise ratio is large, then $d/\sqrt{2N_0}$ is large and the most likely symbol errors involve nearest neighbors only. As a result, it is of interest to assign binary digits to the symbols in such a way that the binary sequences assigned to nearest neighbors differ in a single bit only. This can be accomplished by use of Gray coding, the technique discussed for one-dimensional signals in Section 6.4.2. The binary digits are numbered from left to right (e.g., the first bit is in the leftmost position and the last bit is in the rightmost position).

Gray coding in two dimensions is illustrated in Figure 6-37(b) for 16-QASK. The assignment depicted in the figure can be viewed as a separate Gray code for each of the two dimensions if the first two bits represent the vertical dimension and the last two bits represent the horizontal dimension. The Gray code for 4-ASK in Section 6.4.2 (see Figure 6-19) is applied to each dimension of the 16-QASK signal constellation. With this assignment of binary digits to signals, a symbol error that involves nearest neighbors results in a single bit error.

Suppose that $M = 2^m$ for some positive integer m. Then each point in the M-QASK signal set represents m binary digits. Suppose also that each pair of nearest neighbors is assigned a pair of m-bit sequences that differ in exactly one bit (e.g., via Gray coding). If the signal-to-noise ratio is large, the argument of Section 6.4.2 establishes that the average bit error probability is approximately equal to the average symbol error probability divided by m. In the notation of Section 6.4.2, $\overline{P}_b \approx \overline{P}_e/m$, where $\overline{P}_b$ denotes the average bit error probability and $\overline{P}_e$ denotes the average symbol error probability. Of course, the bound $\overline{P}_e/m \leq \overline{P}_b \leq \overline{P}_e$ applies for any assignment of binary sequences to M-QASK signals, so Gray coding achieves the lower bound asymptotically as the signal-to-noise ratio increases.

Recall that nearest neighbors are separated by distance d. The closest pairs of points that are not nearest neighbors, referred to as *next-nearest neighbors*, are at distance $d\sqrt{2}$. The assignment illustrated in Figure 6-37(b) also guarantees that symbol errors involving next-nearest neighbors result in exactly two bit errors.

Exercise 6-7

For the standard 16-QASK signal constellation and bit assignment of Figure 6-37(c), use the result of Exercise 6-6 to give an expression in terms of the function Q and the parameters d and N_0 for the probability p_1, the conditional probability that the first bit is incorrect given that 0000 is sent. The channel has additive white Gaussian noise with spectral density $N_0/2$, and the receiver employs maximum-likelihood decision regions and the optimum correlators of Figure 6-27. Also give expressions for p_2 through p_6 if p_2 is the probability the first bit is incorrect given 0111 is sent, p_3 is the probability the last bit is incorrect given 1001 is sent, p_4 is the probability the last bit is incorrect given 1100 is sent, p_5 is the probability the first and second bits are correct given 0101 is sent, and p_6 is the probability that at least one bit is incorrect given 0000 is sent.

Solution.   Conditioned on 0000 being sent, the first bit is incorrect if and only if the decision statistic is on the right-hand side of line λ_1 in Figure 6-37(c). Because this line is a distance $3d/2$ from the point that corresponds to 0000, the solution of Exercise 6-6 with $d_0 = 3d$ shows that

$$p_1 = Q\big(3d/\sqrt{2N_0}\,\big).$$

Similarly, the distance from the point for 0111 to line λ_1 is $d/2$, so

$$p_2 = Q\big(d/\sqrt{2N_0}\,\big).$$

If 1001 is sent, the last bit is incorrect if and only if the decision statistic is above line λ_2 or below line λ_3. These are mutually exclusive events, so p_3 is the sum of their conditional probabilities. Because line λ_2 is at a distance $3d/2$ and line λ_3 is at distance $d/2$,

$$p_3 = Q(3\,d/\sqrt{2N_0}) + Q(d/\sqrt{2N_0}).$$

Given that 1100 is sent, the last bit is incorrect if and only if the decision statistic is above line λ_3, but not above line λ_2, so

$$p_4 = Q(d/\sqrt{2N_0}) - Q(3d/\sqrt{2N_0}).$$

Alternatively, the conditional probability the last bit is correct given that 1100 is sent is the conditional probability the decision statistic is below line λ_3 or above line λ_2, which are mutually exclusive events. Thus, $1 - p_4$ is the sum of $1 - Q(d/\sqrt{2N_0})$ and $Q(3d/\sqrt{2N_0})$. Given that 0101 is sent, the probability that at least one of the first two bits is incorrect is the probability the decision statistic is to the right of line λ_1 or to the left of line λ_4 (mutually exclusive events). Each line is at distance $d/2$, so each event has probability $Q(d/\sqrt{2N_0})$. Thus, $1 - p_5$ is twice this probability, and the probability that both of the first two bits are correct is

$$p_5 = 1 - 2\,Q(d/\sqrt{2N_0}).$$

Since the probability the first bit is correct is $1 - Q(d/\sqrt{2N_0})$ and the probability the second bit is correct is $1 - Q(3d/\sqrt{2N_0}) - Q(d/\sqrt{2N_0})$, it is clear that p_5 is not equal to the product of the two probabilities. Thus, errors in different bit positions are not independent. Finally, p_6 is equal to the probability the symbol is incorrect given that the corner point representing 0000 is sent. We could obtain p_6 from (6.58c), but instead let's use the method of Exercise 6-6. At least one bit is incorrect if and only if the decision statistic is above line λ_3 or to the right of line λ_4. Since each line is at distance $d/2$, each of the two events has conditional probability $Q(d/\sqrt{2N_0})$, but the events are not mutually exclusive. Since the noise is independent in the two orthogonal directions, the two events are independent and the probability that both events occur is the product of their probabilities. Thus, the probability that at least one bit is incorrect given 0000 is sent is

$$p_6 = 2\,Q(d/\sqrt{2N_0}) - Q^2(d/\sqrt{2N_0}) = Q(d/\sqrt{2N_0})[2 - Q(d/\sqrt{2N_0})].$$

If the signal constellation is given by (6.49) with $u_{i+1} - u_i = v_{k+1} - v_k = 2$ for each i and k, the p_i can be written in terms of A, $\mathcal{E}_\alpha$, and N_0 by substituting $A\sqrt{\mathcal{E}_\alpha/N_0}$ for $d/\sqrt{2N_0}$ in each expression. ∎

6.6   *M*-ary Orthogonal Signaling with Coherent Demodulation

Thus far, we have considered signal sets of dimensions one and two only. In this section, signal sets of arbitrarily large dimension are introduced. These signal sets also have the property that any two signals in the set are orthogonal. The optimum coherent receiver for such signals is given, and the resulting error probabilities are derived.

6.6.1   *M*-ary Orthogonal Signals

The signal sets described in Section 6.5 contain M signals, but span a dimension of two only. That is, for each of these signal sets, there are two orthogonal basis functions ψ_0 and ψ_1 with the property that each of the M signals in the set can be expressed as a linear combination of ψ_0 and ψ_1. An example is the M-QASK signal set, for which the basis functions are defined in Section 6.5.1 as

$$\psi_0(t) = \sqrt{2/\mathcal{E}_\alpha}\,\alpha(t)\cos(\omega_c t + \varphi) \tag{6.60a}$$

and

$$\psi_1(t) = -\sqrt{2/\mathcal{E}_\alpha}\,\alpha(t)\sin(\omega_c t + \varphi). \tag{6.60b}$$

The factor $\sqrt{2/\mathcal{E}_\alpha}$ is included in order to make the functions *orthonormal* as discussed in Section 6.5.1 (i.e., the two functions are orthogonal, and each has unit energy).

The indexing for the basis functions begins with the number 0 rather than 1. This is simply a matter of following a standard convention, in part because it is consistent with the indexing used for M-ary signals. Recall that the notation for a set of M signals is

$$\{s_i : 0 \leq i \leq M - 1\}.$$

A signal set $\{s_i : 0 \leq i \leq M - 1\}$ is an M-ary orthogonal signal set if $(s_i, s_j) = 0$ for all choices of i and j such that $i \neq j$. Given a set of M orthonormal functions

$$\{\psi_i : 0 \leq i \leq M - 1\},$$

any signal set defined by

$$s_i(t) = \sqrt{\mathcal{E}_i}\,\psi_i(t), \quad -\infty < t < \infty, \quad 0 \leq i \leq M - 1,$$

is an M-ary orthogonal signal set. Notice that $\mathcal{E}_i$ is just the energy in signal s_i, because ψ_i has unit energy. Furthermore, given an M-ary orthogonal signal set, each signal in the set can be normalized to have unit energy, thereby providing a set of M orthonormal functions.

An illustration of an M-ary orthogonal signal constellation is given in Figure 6-38 for $M = 3$. The three orthonormal basis functions ψ_0, ψ_1, and ψ_2 are illustrated as unit vectors in each of the three dimensions, and the three orthogonal signals s_0, s_1, and s_2 are just scalar multiples of the three basis functions.

One of the most common M-ary orthogonal signal sets consists of a set of sinusoidal signals with suitable frequencies. If the frequencies are carefully selected, the sinusoidal signals are orthogonal. Normally, these sinusoidal signals also have the same power, which provides an example of an equal-energy, orthogonal signal set. This signal set is the subject of the next example.

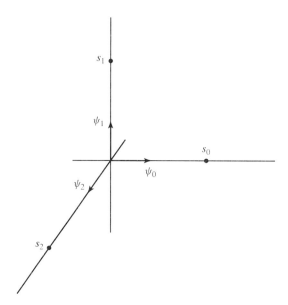

Figure 6-38: 3-ary orthogonal signal constellation.

Example 6-6   The MFSK Signal Set

Suppose that $\omega_0 = 2\pi n/T$ for some integer n, and $\beta = 2\pi m/T$ for some integer m. For most applications, $m = 1$. For each integer i in the range $0 \le i \le M - 1$, define

$$\psi_i(t) = \sqrt{2/T}\,\cos[(\omega_0 + i\beta)t + \varphi_i], \quad 0 \le t \le T.$$

It is easy to verify that the set $\{\psi_i : 0 \le i \le M - 1\}$ is an orthonormal set of functions defined on the interval $[0, T]$ for any choice of the phase angles φ_i, $0 \le i \le M - 1$. Consequently, any signal set $\{s_i : 0 \le i \le M - 1\}$ for which

$$s_i(t) = \sqrt{2}\,A_i \cos[(\omega_0 + i\beta)t + \varphi_i], \quad 0 \le t \le T, \quad 0 \le i \le M - 1,$$

is an M-ary orthogonal signal set for each choice of the parameters A_i, $0 \le i \le M - 1$, and φ_i, $0 \le i \le M - 1$.

For most applications, it is desirable that the signals have the same energy; furthermore, there is typically no advantage in permitting different phase angles among the M signals. If the signals have the same amplitudes and phase angles, they can be represented as

$$s_i(t) = \sqrt{2}\,A \cos[(\omega_0 + i\beta)t + \varphi], \quad 0 \le t \le T, \tag{6.61}$$

and the corresponding orthonormal basis functions are

$$\psi_i(t) = \sqrt{2/T}\,\cos[(\omega_0 + i\beta)t + \varphi], \quad 0 \le t \le T. \tag{6.62}$$

Notice that $\mathcal{E}$, the common value of the energy in the signals s_i of (6.61), is given by $\mathcal{E} = A^2 T$, and the signals can be written as

$$s_i(t) = \sqrt{\mathcal{E}}\,\psi_i(t) = \sqrt{2\mathcal{E}/T}\,\cos[(\omega_0 + i\beta)t + \varphi], \quad 0 \le t \le T. \tag{6.63}$$

The signals defined by (6.61) or (6.63) are referred to as *M-ary frequency-shift-keyed (M-ary FSK)* or *multiple frequency-shift-keyed (MFSK) signals*.   ∎

Other *M*-ary orthogonal signal sets can be constructed by using orthogonal amplitude modulation on a common sinusoidal carrier. Such signals are of the form

$$s_i(t) = \sqrt{2}\, A\, a_i(t) \cos(\omega_c t + \varphi) \tag{6.64}$$

for $0 \leq i \leq M - 1$. The baseband signals a_i, $0 \leq i \leq M - 1$, are orthogonal, but this is not enough to guarantee that the resulting carrier-modulated signals are orthogonal. Generally, they are at least approximately orthogonal if the baseband signals are orthogonal and have bandwidths that are small compared to the carrier frequency. However, there are other conditions under which the carrier-modulated signals are orthogonal. If the signals are orthogonal and are generated by amplitude modulation by a set of orthogonal baseband signals, we refer to the signal set as *M-ary orthogonal amplitude modulation*, or simply *M-ary orthogonal AM*.

It is not necessary for orthogonality of the *M*-ary amplitude-modulated signals that the bandwidth of the baseband signals be small relative to the carrier frequency, as is illustrated in what follows. Suppose that, in addition to being orthogonal, the signals a_i are time limited to the interval $[0, T]$, and they consist of sequences of rectangular pulses of duration T_0, where $T_0 = T/N$ for some positive integer N. Then if $\omega_c = 2\pi m/T_0$ for some integer m, the resulting carrier-modulated signals of the form given in (6.64) are orthogonal. A mathematical representation for a signal of this type is

$$a_i(t) = \sum_{n=0}^{N-1} \alpha_{i,n}\, p_{T_0}(t - nT_0). \tag{6.65}$$

It is easy to show that orthogonality of the signals a_i is guaranteed by orthogonality of the vectors $\alpha_i = (\alpha_{i,0}, \alpha_{i,1}, \ldots, \alpha_{i,N-1})$. This follows from the fact that, for signals given by (6.65),

$$\int_0^T a_i(t)\, a_j(t)\, dt = T_0 \sum_{n=0}^{N-1} \alpha_{i,n}\, \alpha_{j,n}. \tag{6.66}$$

Notice also that the rectangular pulse in (6.65) can be replaced by any other pulse waveform that has duration T_0, and (6.66) is valid if T_0 is replaced by some appropriate constant that depends on the waveform, but not on the vectors α_i and α_j.

A simple example for $N = M$ is to let $\alpha_{i,n} = 1$ for $i = n$ and $\alpha_{i,n} = 0$ for $i \neq n$. The resulting signals a_i, $0 \leq i \leq M - 1$, are illustrated in Figure 6-39 for $M = 3$. The basic signal design philosophy used to construct these signals is that only one signal is "on" at a time. (All others are zero during that time.) Therefore, not only is the inner product of (6.66) equal to zero for any two signals from the set, the product of the signals is zero for each value of t. We refer to signals of this type as *orthogonal on–off baseband signals*. For such signals, the integrand of the left-hand side of (6.66) is identically zero for each choice of i and j $(i \neq j)$.

The main disadvantage of orthogonal on–off signals is that they provide a rather small amount of energy for a given signal amplitude. If there is a limit on the signal

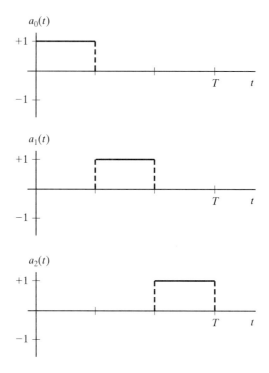

Figure 6-39: Three orthogonal on–off baseband signals.

amplitude, such as the result of a peak-power constraint in the transmitter, this signal design is very poor, because the energy in each signal is small compared with what can be achieved for the given constraint. For example, if the constraint is $|a_i(t)| \leq 1$, the maximum energy in the signal a_i is T, whereas orthogonal on–off signals provide energy T/N.

A better signal design for orthogonal baseband signals is obtained from Hadamard matrices [6.9]. The signal design that is described in this section is derived from the rows of certain Hadamard matrices, but because of the close relationship with Walsh functions [6.11], the resulting signals have also been referred to as Hadamard–Walsh signals [6.22].

The simplest Hadamard matrix is the 2×2 matrix

$$H_1 = \begin{bmatrix} +1 & +1 \\ +1 & -1 \end{bmatrix}$$

for which the row vectors are the orthogonal vectors $(+1, +1)$ and $(+1, -1)$. The 4×4 Hadamard matrix is

$$H_2 = \begin{bmatrix} +1 & +1 & +1 & +1 \\ +1 & -1 & +1 & -1 \\ +1 & +1 & -1 & -1 \\ +1 & -1 & -1 & +1 \end{bmatrix},$$

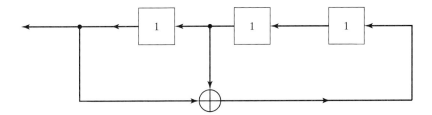

Figure 6-40: A linear feedback shift register with initial loading 111.

which has orthogonal row vectors $(+1, +1, +1, +1)$, $(+1, -1, +1, -1)$, $(+1, +1, -1, -1)$, and $(+1, -1, -1, +1)$. When used as pulse amplitudes in (6.65), the rows of H_2 give a set of four orthogonal signals with $T_0 = T/4$. Notice that H_2 can also be expressed as

$$H_2 = \begin{bmatrix} H_1 & H_1 \\ H_1 & -H_1 \end{bmatrix}$$

in terms of the 2×2 Hadamard matrix H_1. For an arbitrary integer n, the relationship between H_{n+1} and H_n is

$$H_{n+1} = \begin{bmatrix} H_n & H_n \\ H_n & -H_n \end{bmatrix},$$

where H_{n+1} is a $2^{n+1} \times 2^{n+1}$ matrix, and H_n is a $2^n \times 2^n$ matrix. Thus, if N is a power of two, a set of N orthogonal vectors can be obtained from the rows of the Hadamard matrix H_m, where $m = \log_2(N)$. Of course, any smaller number of orthogonal vectors can be obtained by simply deleting rows of the matrix, so it is possible to obtain M orthogonal vectors of length N as long as $M \le N$.

A similar signal design can be obtained from maximal-length linear feedback shift-register sequences, known as *m-sequences*. A discussion of m-sequences is given in Section 9.3. As an example of the construction of orthogonal signals from m-sequences, consider the linear feedback shift register shown in Figure 6-40.

This register has three storage elements and a single modulo-2 adder (also referred to as an exclusive-OR gate). Modulo-2 addition is defined by

$$0 \oplus 0 = 1 \oplus 1 = 0 \quad \text{and} \quad 0 \oplus 1 = 1 \oplus 0 = 1.$$

Suppose that initially each register holds a 1, as illustrated in Figure 6-40. The next element that is fed back and placed into the rightmost register is $1 \oplus 1 = 0$, and the register contents become 110 as the leftmost 1 shifts out of the register. Continuing in this fashion, we see that the sequence that shifts out of the register is

$$1110010111001011100\dots .$$

This is a periodic sequence with period seven, the largest period that can be obtained from a register with three storage elements. That the maximum period is seven follows

Table 6.5: Sequence Sets

Shift-Register Sequences	Orthogonal Sequences
1110010	11100100
1100101	11001010
1001011	10010110
0010111	00101110
0101110	01011100
1011100	10111000
0111001	01110010
0000000	00000000

from the fact that there are only seven different nonzero patterns of three binary digits, and each of these seven patterns can appear in the registers at most once per period. The all-zeros pattern cannot appear, because it would produce zeros from that time on (i.e., if 000 appears in the 3-stage register, the output is the sequence 00000000..., which has period 1).

Next we form the array of all seven-bit sequences that can be generated by the shift register. Each sequence is generated by the initial loading that consists of the first three bits of the sequence. We refer to the resulting sequences as shift-register sequences. The shift-register sequences obtained from the shift register in Figure 6-40 are listed in Table 6.5. The first seven sequences can also be described as left-cyclic-shifts of the initial pattern 1110010. At each step, the pattern is shifted one position to the left, and the leftmost digit is moved to the rightmost position to form the next sequence. The orthogonal sequences are derived from the shift-register sequences by adding a zero to the end of each shift-register sequence. The resulting set is the set of orthogonal sequences listed in Table 6.5.

The next step is to convert the zeros and ones to $+1$s and -1s, respectively. These represent the pulse amplitudes $\alpha_{i,n}$ of the continuous-time signals $a_i(t)$ that are defined by (6.65). For the signals derived from the orthogonal sequences of Table 6.5, $N = 8$ and the signals are given by

$$a_i(t) = \sum_{n=0}^{7} \alpha_{i,n} \, p_{T_0}(t - nT_0)$$

for $T_0 = T/8$. As an illustration, the first three signals in the set are shown in Figure 6-41. It is easy to check that these signals are orthogonal on the interval $[0, T]$. The set of continuous-time signals derived from the shift-register sequences in the same manner forms a simplex set. The waveform for each simplex signal is the same as for its counterpart in the orthogonal set except the last pulse is omitted from the corresponding orthogonal signal.

The distance between any two signals in the orthogonal set is $d = \sqrt{2\mathcal{E}}$. Since the last pulse is identical for each of the orthogonal signals, it contributes nothing to the distance between pairs of signals. Deleting the last pulse reduces the energy by

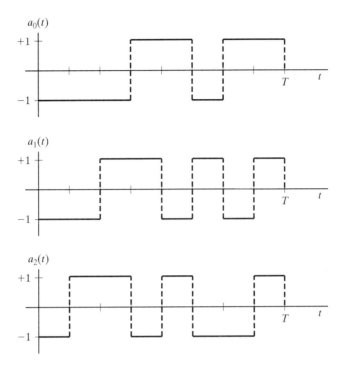

Figure 6-41: Three orthogonal signals derived from the linear feedback shift register of Figure 6-40.

12.5% without changing the distances between signals. It is easy to show that the correlation coefficient is $r = -1/7$ for the simplex signal set derived from the shift-register sequences in Table 6.5, whereas any orthogonal set has $r = 0$. In general, a *simplex signal set of size M* has correlation coefficient $r = -1/(M-1)$. In comparison with an orthogonal signal set of size M, the simplex set can achieve the same distance parameter d with less energy per symbol. The energy for the simplex set is $\mathcal{E}' = (M-1)\mathcal{E}/M$, where $\mathcal{E}$ is the energy per signal for the orthogonal set.

The symbol error probabilities for orthogonal signals are derived in Section 6.6.3. Since the distance parameter is the same for the two signal sets, the symbol error probabilities for maximum-likelihood reception are also the same for the two sets. Thus, for the same probability of symbol error, a set of M simplex signals requires less energy than a set of M orthogonal signals. As illustrated for $M = 8$ in Table 6.5, a set of M simplex signals for $M = 2^m$ can be obtained from an m-sequence generator with m storage elements. See Section 9.3 for more information on m-sequence generators.

6.6.2   The Optimum Coherent Receiver for *M*-ary Orthogonal Signals

The optimum coherent receiver for equal-energy, *M*-ary orthogonal signaling on the additive white Gaussian noise channel can be derived using the type of development given in Section 6.1.4. In fact, for the MFSK signal set, the Fourier series is exploited for the M sinusoidal frequency-shift keyed signals in the same manner as it is for the two sinusoidal PSK signals in Section 6.1.4. The orthonormal basis functions are the sinusoids of the trigonometric Fourier series.

For other orthogonal signal sets, the basis functions may not be sinusoidal, but the basic ideas are the same. A key step is the representation of the receiver's input by a series expansion in terms of orthonormal basis functions that are convenient for describing the signal set under consideration. If the M-ary orthogonal signal set is derived from a set of M orthonormal functions $\{\psi_i : 0 \leq i \leq M - 1\}$ according to

$$s_i(t) = \sqrt{\mathcal{E}}\, \psi_i(t), \quad -\infty < t < \infty, \quad 0 \leq i \leq M - 1,$$

then the infinite set of orthonormal functions should include each ψ_i. For theoretical purposes, it may be necessary to complete the set of orthonormal basis functions to give an infinite set $\{\psi_i : 0 \leq i < \infty\}$ that can be used to represent any finite energy signal on the interval $[0, T]$. For practical purposes, only the subset $\{\psi_i : 0 \leq i \leq M - 1\}$ is needed. We refer to this subset as the *orthonormal basis functions of primary importance*. These basis functions are just constant multiples of the signals in the signal set.

For simplicity, we emphasize equal-energy orthogonal signals. Nearly all M-ary orthogonal signal sets used in practical applications have equal-energy signals. Among the equal-energy orthogonal signal sets, the sets that are most popular for implementation are the MFSK signals and the M-ary orthogonal AM signals. For MFSK signals, the orthonormal basis functions of primary importance are

$$\psi_i(t) = \sqrt{2/T} \cos[(\omega_0 + i\beta)t + \varphi_i], \quad 0 \leq t \leq T, \quad 0 \leq i \leq M - 1.$$

If the orthogonal baseband signals $\{a_i : 0 \leq i \leq M - 1\}$ that are used to generate the M-ary orthogonal AM signals are such that $|a_i(t)| = 1$ for $0 \leq t \leq T$, as is true for the Hadamard–Walsh baseband signals and the signals derived from m-sequences, then the orthonormal basis functions of primary importance are given by

$$\psi_i(t) = \sqrt{2/T}\, a_i(t) \cos(\omega_c t + \varphi), \quad 0 \leq t \leq T, \quad 0 \leq i \leq M - 1.$$

For more general orthogonal baseband signals, the orthonormal basis functions of primary importance are of the form

$$\psi_i(t) = A_i\, a_i(t) \cos(\omega_c t + \varphi), \quad 0 \leq t \leq T, \quad 0 \leq i \leq M - 1,$$

where A_i is a normalization constant whose value may differ from signal to signal.

In Figure 6-42, the transmitted signal $s(t)$ is from the set $\{s_i : 0 \leq i \leq M - 1\}$ of equal-energy, orthogonal signals. The optimum coherent receiver includes a bank of M coherent correlators in which the ith correlator is matched to the basis function ψ_i. Because the signals have equal energy (recall that $s_i = \sqrt{\mathcal{E}}\, \psi_i$ for each i), the correlators can be matched to the signals themselves, as shown in Figure 6-42. The maximum-likelihood decision rule for the set $\{Z_i : 0 \leq i \leq M - 1\}$ of M decision statistics is a natural extension of the maximum-likelihood decision rule for binary signals: If the

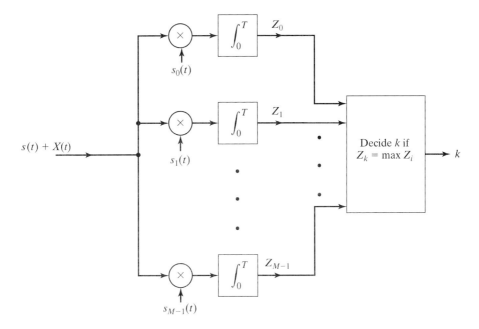

Figure 6-42: Optimum receiver for equal-energy, *M*-ary signals and an additive white Gaussian noise channel.

vector of correlator outputs is $\mathbf{Z} = \mathbf{z}$, then the decision is that s_k is the transmitted signal if

$$f_k(\mathbf{z}) = \max\{f_i(\mathbf{z}) : 0 \leq i \leq M - 1\}.$$

It follows from results in Section 5.4 that the decision is in favor of the correlator that produces the largest output. That is, the maximum-likelihood decision is that the kth signal was transmitted if

$$Z_k = \max\{Z_i : 0 \leq i \leq M - 1\}.$$

This decision rule is also a Bayes rule if $\pi_i = M^{-1}$ for $0 \leq i \leq M - 1$. As shown in the next subsection, this decision rule gives equality among the conditional probabilities $P_{e,i}, 0 \leq i \leq M - 1$, so it is also a minimax decision rule. Thus, the decision device in Figure 6-42 decides on the basis of the largest statistic among the decision statistics Z_i.

Even if the signals are not orthogonal, the receiver of Figure 6-42 is optimum, as demonstrated in Appendix D. Furthermore, if the signals do not have equal energy, the only change needed in Figure 6-42 is that Z_i must be reduced by $\mathcal{E}_i/2$ for each i. That is, $\mathcal{E}_i/2$ is subtracted from the output of the ith correlator to obtain Z_i. A block diagram for the optimum receiver for unequal-energy signals is illustrated in Figure D-2 of Appendix D. Although the same receiver structure is optimum for general signals, the error probabilities are not the same as those obtained in the next subsection for orthogonal signals. In fact, as discussed in Section 6.6.1, the simplex signals can achieve the same probability of symbol error with less energy per symbol. Equivalently, simplex signals can achieve a smaller probability of symbol error with the same energy per symbol.

The receiver structure of Figure 6-42 is optimum, but it may not be the most convenient structure. Alternative structures for the optimum receiver for the M-ary signals derived from the sequences in Table 6.5 are given in Appendix E. In particular, it is shown that for such signals it is not necessary to have M parallel continuous-time correlators.

6.6.3   Error Probabilities for M-ary Orthogonal Signals

In this section, expressions are obtained for the symbol error probabilities that result if coherent maximum-likelihood reception is employed for a system with M-ary orthogonal, equal-energy signals. The channel is an additive white Gaussian noise channel with spectral density $N_0/2$.

The starting point is to observe that if 0 is sent, then the event that a correct decision is made is

$$C_0 = \bigcap_{i=1}^{M-1} \{Z_i < Z_0\}.$$

That is, C_0 is the event that none of the other $M-1$ correlator outputs is greater than the output of the correlator that is matched to the signal s_0. Because the random variables $Z_i, 0 \le i \le M-1$, have a continuous distribution, the event $\{Z_i = Z_0\}$ has probability zero for each i in the range $1 \le i \le M-1$. Consequently, we can ignore such events in the analysis of the receiver.

The event E_0 that an error is made is the complement of C_0; therefore,

$$E_0 = \bigcup_{i=1}^{M-1} \{Z_i \ge Z_0\}.$$

The conditional probability of making an error given that 0 is sent is

$$P_{e,0} = P(E_0 \mid 0 \text{ sent}).$$

From this expression, it is easy to obtain an upper bound on the probability of error. The event E_0 is the union of events of the form $\{Z_i \ge Z_0\}$, and the probability of a union of two or more events is no greater than the sum of the probabilities of the individual events. Therefore, the union bound on the probability of error is

$$P_{e,0} \le \sum_{i=1}^{M-1} P(Z_i \ge Z_0 \mid 0 \text{ sent}).$$

The probability $P(Z_i \ge Z_0 \mid 0 \text{ sent})$ is the probability of error in a binary, coherent, communication system with the orthogonal signals s_i and s_0. In the receiver of Figure 6-42, the two branches that use the functions s_i and s_0 are optimum for this binary signal set, and these produce decision statistics Z_i and Z_0, respectively. Consequently,

$$P(Z_i \ge Z_0 \mid 0 \text{ sent}) = Q\left(\sqrt{\mathcal{E}/N_0}\right).$$

This is the familiar expression from Chapter 5 for the probability of error for binary signaling with equal-energy orthogonal signals and an optimum receiver (e.g., equations (5.111) with $r = 0$). Combining the preceding results, we see that the *union bound* for the probability of error that results in the receiver of Figure 6-42 is

$$P_{e,0} \leq (M-1)Q\left(\sqrt{\mathcal{E}/N_0}\,\right).$$

The union bound is very loose for small values of $\mathcal{E}/N_0$, but it is a reasonably tight bound if $\mathcal{E}/N_0$ is large.

To obtain an expression for the exact probability of error, it is easier to work with the probability of a correct decision. The conditional probability of making a correct decision given that 0 is sent is

$$P_{c,0} = P(C_0 \mid 0 \text{ sent}).$$

This probability can be evaluated by first conditioning on the value of Z_0 and then averaging with respect to the density of Z_0. That is,

$$P_{c,0} = \int_{-\infty}^{\infty} P(C_0 \mid 0 \text{ sent}, Z_0 = u)\, f_{Z_0}(u \mid 0 \text{ sent})\, du. \tag{6.67}$$

To simplify the notation, let

$$g_0(u) = f_{Z_0}(u \mid 0 \text{ sent}).$$

Because we are conditioning on the event that 0 is sent, the random variables Z_i, $0 \leq i \leq M-1$, are independent. This follows from the fact that when conditioned on 0 being sent, Z_i has no signal component for $i \neq 0$, and the noise components of Z_i and Z_j are independent for $i \neq j$. The independence of the noise components is due to the orthogonality of s_i and s_j for $i \neq j$. Because of the conditional independence of the random variables involved, (6.67) can be expressed as

$$P_{c,0} = \int_{-\infty}^{\infty} \prod_{i=1}^{M-1} P(Z_i < Z_0 \mid 0 \text{ sent}, Z_0 = u)\, g_0(u)\, du$$

$$= \int_{-\infty}^{\infty} \prod_{i=1}^{M-1} P(Z_i < u \mid 0 \text{ sent})\, g_0(u)\, du. \tag{6.68}$$

For $i \neq 0$, the conditional distribution of Z_i given that 0 is sent does not depend on i; in fact, for each such i, Z_i is a zero-mean Gaussian random variable with variance given by

$$\sigma^2 = (N_0/2) \int_{-\infty}^{\infty} [s_i(t)]^2\, dt = \mathcal{E}N_0/2.$$

As a result,

$$P(Z_i < u \mid 0 \text{ sent}) = \Phi(u/\sigma), \quad 1 \leq i \leq M-1.$$

Applying this to (6.68), we find that

$$P_{c,0} = \int_{-\infty}^{\infty} [\Phi(u/\sigma)]^{M-1} g_0(u) \, du. \tag{6.69}$$

We have already established that the conditional density for Z_0 is Gaussian with variance $\sigma^2 = \mathcal{E} N_0/2$. Thus,

$$g_0(u) = \exp\{-(u - \mu_0)^2/2\sigma^2\}/\sqrt{2\pi}\sigma, \tag{6.70}$$

where $\mu_0 = E\{Z_0 \mid 0 \text{ sent}\}$. Recall from Section 6.6.2 that

$$s_0(t) = \sqrt{\mathcal{E}} \, \psi_0(t).$$

For this signal and the receiver of Figure 6-42, it is easy to see that $\mu_0 = \mathcal{E}$.

By combining (6.69) and (6.70) and letting $v = (u - \mu_0)/\sigma$, we see that

$$P_{c,0} = \int_{-\infty}^{\infty} [\Phi(v + \sigma^{-1}\mu_0)]^{M-1} \exp(-v^2/2) \, dv/\sqrt{2\pi}.$$

The expression for the probability of a correct decision given that 0 is sent is obtained by observing that $\sigma^2 = \mathcal{E} N_0/2$ and $\mu_0 = \mathcal{E}$ imply that

$$\sigma^{-1}\mu_0 = \sqrt{2\mathcal{E}/N_0},$$

and therefore

$$P_{c,0} = \int_{-\infty}^{\infty} \left[\Phi\left(v + \sqrt{2\mathcal{E}/N_0}\,\right)\right]^{M-1} \exp(-v^2/2) \, dv/\sqrt{2\pi}.$$

From this expression, it is clear that the probability of a correct decision does not depend on which signal is sent. Thus, we can drop the subscript 0 and let P_c denote the probability of a correct decision given that the ith symbol is sent for an arbitrary value of i ($0 \leq i \leq M - 1$). The final expression is then

$$P_c = \int_{-\infty}^{\infty} \left[\Phi\left(v + \sqrt{2\mathcal{E}/N_0}\,\right)\right]^{M-1} \exp(-v^2/2) \, dv/\sqrt{2\pi}. \tag{6.71}$$

Of course the probability of symbol error can be obtained from $P_e = 1 - P_c$. In general, the integral in (6.71) cannot be evaluated in closed form, and numerical integration is required to calculate the probability of error. If the error probability is very small, then P_c is very close to 1 and the use of $P_e = 1 - P_c$ to compute the error probability may lead to computational problems. An alternative expression for P_e is

$$P_e = (M - 1) \int_{-\infty}^{\infty} [\Phi(v)]^{M-2} \Phi\left(v - \sqrt{2\mathcal{E}/N_0}\,\right) \exp(-v^2/2) \, dv/\sqrt{2\pi}, \tag{6.72}$$

which is developed in Problem 6.31. This expression must also be evaluated using numerical integration, but the direct evaluation of P_e is preferred if P_e is very small.

Notice in the special case $M = 2$, (6.71) reduces to

$$P_c = \int_{-\infty}^{\infty} \Phi(v + \alpha) \exp(-v^2/2) \, dv / \sqrt{2\pi},$$

where $\alpha = \sqrt{2\,\mathcal{E}/N_0}$. This integral can be evaluated by writing $\Phi(v+\alpha)$ as an integral and then examining the resulting double integral. The double integral can be written as

$$P_c = \int_{-\infty}^{\infty} \left\{ \int_{-\infty}^{v+\alpha} \exp\{-(u^2 + v^2)/2\} \, du \right\} dv / 2\pi,$$

which is just an expression for the probability that a random variable X does not exceed $Y + \alpha$ if X and Y are independent, zero-mean, unit-variance Gaussian random variables. This is the same as the probability that $X - Y$ does not exceed α, and the random variable $X - Y$ is a zero-mean Gaussian random variable with variance 2. Hence, for $M = 2$,

$$P_c = \Phi(\alpha/\sqrt{2}) = \Phi(\sqrt{\mathcal{E}/N_0}),$$

from which we obtain

$$P_e = Q(\sqrt{\mathcal{E}/N_0}),$$

which is the error probability for optimum reception of binary orthogonal signals.

   Note that for binary signaling, each signal represents one bit, so the energy per bit is the same as the energy $\mathcal{E}$ in the signal waveform. For $M > 2$, each M-ary signal represents more than one bit, so the energy per bit is not the same as the energy in the signal waveform. Suppose $M = 2^m$ for some positive integer m. Then each signal represents m bits, and the energy per bit is given by

$$\mathcal{E}_b = \mathcal{E}/m = \mathcal{E}/\log_2 M. \tag{6.73}$$

The results of a numerical evaluation of (6.72) are shown in Figure 6-43 for $M = 2, 4, 8, 16$, and 32. The symbol error probabilities are shown as a function of $\mathcal{E}_b/N_0$.

   If an M-ary signal set is used to send binary data, the bit error probability is of at least as much interest as the symbol error probability. As we have seen in previous sections, for some signal constellations the bit error depends on how the bits are assigned to the signals in the M-ary signal set. Our first step is to argue that this is not true for orthogonal signals.

   Suppose that symbol 0 was sent (i.e., s_0 is the transmitted signal). There are $M - 1$ different types of symbol errors corresponding to the $M - 1$ possible decisions other than the decision that 0 was sent. For the maximum-likelihood receiver of the form shown in Figure 6-42, the conditional probability that the receiver decides symbol j was sent, given that 0 was actually sent, is the same for all $j \neq 0$. This is because when conditioned on 0 being sent, each of the random variables $Z_j, 1 \leq j \leq M - 1$, has the same probability distribution. From the symmetry of the problem, it is clear that the same is true no matter which symbol is sent: Each of the possible types of symbol errors has the same probability of occurrence. Recall that different types of symbol errors have different probabilities of occurrence for the M-ASK and M-QASK signal

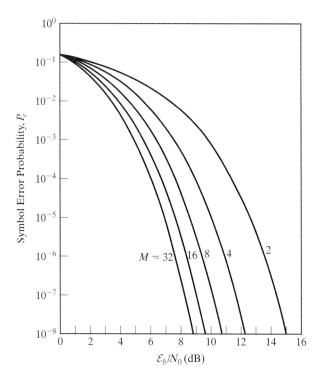

Figure 6-43: Symbol error probabilities for M-ary orthogonal signaling (courtesy of D. L. Noneaker).

sets discussed in Sections 6.4 and 6.5. Such signal sets do not have the symmetry that is displayed by M-ary orthogonal signal sets.

Because the different types of symbol errors are equally likely, it does not matter how we assign binary digits to the M-ary orthogonal signals. We continue to assume that $M = 2^m$ for some positive integer m, which is the normal situation if M-ary signals are used for the transmission of binary data. A common assignment of binary digits to symbols is to let the sequence of binary digits associated with the jth symbol be the m-bit binary representation for the integer j ($0 \leq j \leq M - 1$). The sequence of m zeros is assigned to the symbol 0, the sequence of $m - 1$ zeros followed by a single one is assigned to the symbol 1, etc. This is illustrated in Table 6.6 for $M = 8$.

Different types of symbol errors result in different numbers of bit errors. We are interested in the average probability of bit error, averaged over the $M - 1$ different symbol errors that are possible. Continue to assume that 0 is sent, so that the correct sequence of binary data is the sequence of m zeros. Focus on the ith position within the m bit positions, and count the total number of bit errors that occur in this bit position for the set of $M - 1$ symbol errors. Notice that this is the same as the total number of ones in the ith positions for all M binary sequences, because the inclusion of the all-zeros sequence adds no ones. There are a total of M binary digits in the ith positions of the

Table 6.6: Assignment of Binary Data to *M*-ary Signals for $M = 8$

Symbol	Binary Data
0	000
1	001
2	010
3	011
4	100
5	101
6	110
7	111

M different binary sequences of length *m*, and exactly half of these are ones. Thus the total number of ones in the *i*th positions of the *M* binary sequences is $M/2$. Because there are $M - 1$ different types of symbol errors, and each has the same probability of occurrence, the average number of bit errors in the *i*th position is $M/[2\,(M - 1)]$.

To illustrate this conclusion, consider the bits in the first positions of the binary sequences given in Table 6.6 for $M = 8$ ($m = 3$). If symbol 0 is sent, there are seven possible types of symbol errors, and four of these correspond to a bit error in the first position. Thus, the average number of bit errors in the first position is 4/7. It is easy to verify that this conclusion is true no matter which bit position is considered and no matter which symbol is sent.

The preceding observations establish that P_b, the probability of bit error, is the same for any position in the sequence and the same no matter which sequence is transmitted. It is related to the symbol error probability by

$$P_b = M P_e/[2\,(M - 1)]. \tag{6.74}$$

By combining (6.72)–(6.74), we can determine the bit error probability as a function of $\mathcal{E}_b/N_0$, the ratio of the energy per bit to the noise density.

An important observation is that for large values of *M*, the use of *M*-ary orthogonal signaling and maximum-likelihood coherent demodulation provides lower bit error probabilities for a given value of $\mathcal{E}_b/N_0$ than can be obtained by use of coherent BPSK. A comparison of Figures 6-44 and 5-3 shows that *M*-ary orthogonal signaling for $M \geq 8$ gives a smaller bit error probability than binary antipodal signaling.

This might at first seem to contradict the observations made earlier in the book concerning the optimality of binary antipodal signals. The correct statement is that, for transmission of a *single binary digit* over an additive white Gaussian noise channel with spectral density $N_0/2$, the smallest bit error probability that can be obtained if the energy in the signal is $\mathcal{E}_b$ is

$$P_b = Q\left(\sqrt{2\,\mathcal{E}_b/N_0}\,\right).$$

To achieve this, it is sufficient to make decisions on individual binary digits that are transmitted over the channel. If we use *M*-ary signaling, we are not transmitting a

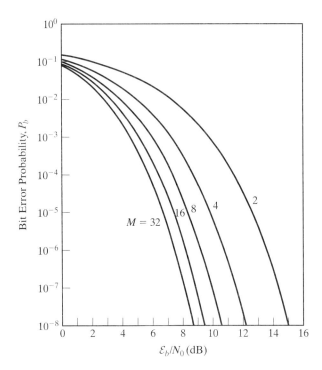

Figure 6-44: Bit error probabilities for M-ary orthogonal signaling (courtesy of D. L. Noneaker).

single bit and making individual binary decisions. We are transmitting a sequence of binary digits and making a decision on the entire sequence. This is often referred to as a *word decision* (as opposed to a *bit decision*). The results of Figure 6-44 show that as the length of the sequence increases (m increases, and therefore M also increases), the bit error probability decreases. This demonstrates that if it is possible to group several bits together and represent them by a signal from a larger signal set, it is possible to obtain a smaller error probability than by use of bit-by-bit transmission with bit-by-bit decisions.

It can be shown from (6.72)–(6.74) that as $M \to \infty$, $P_b \to 0$ for any fixed value of $\mathcal{E}_b/N_0$ that is greater than $\ln(2)$. For $\mathcal{E}_b/N_0 < \ln(2)$, M-ary orthogonal signaling gives $P_b \to 1/2$ as $M \to \infty$. It can be shown that no other signaling scheme gives better performance in the following sense: No signaling method can achieve $P_b \to 0$ for $\mathcal{E}_b/N_0 < \ln(2)$. This is a consequence of a theorem of Shannon that is presented in any textbook on information theory (e.g., see [6.5], [6.7], or [6.8]). For applications of these results, it is convenient to express the limit on $\mathcal{E}_b/N_0$ as

$$(\mathcal{E}_b/N_0)_{\text{dB}} = 10\log_{10}(\mathcal{E}_b/N_0) > 10\log_{10}[\ln(2)] \approx -1.6 \text{ dB}.$$

If $(\mathcal{E}_b/N_0)_{\text{dB}}$ exceeds -1.6 dB, then the bit error probability can be made arbitrarily small by the use of optimum coherent reception of M-ary orthogonal signaling for sufficiently large M.

The major disadvantages of M-ary orthogonal signaling for large values of M are the bandwidth requirements and the complexity of implementation. An examination of the M-ary orthogonal signaling schemes used as examples in this section shows that the signal bandwidth increases linearly in M. Because the number of bits that can be represented increases only as $\log_2(M)$, the bandwidth per binary digit increases as $M/\log_2(M)$. A comparison of $M = 64$ with $M = 2$ indicates that the bandwidth per binary digit for 64-ary orthogonal signaling is more than five times the bandwidth per binary digit for binary orthogonal signaling. Moreover, the implementation illustrated in Figure 6-42 requires 64 correlators in the receiver for 64-ary orthogonal signaling. There are some methods for decreasing the complexity (see Appendix E), but it is still much greater than for binary signaling. As a result, M-ary orthogonal signaling for large values of M is used primarily in applications for which power limitations are more severe than bandwidth or complexity limitations.

6.7   Offset Two-Dimensional Modulation

It is common in the literature to use the *terminology* that corresponds to phase modulation rather than amplitude modulation (e.g., BPSK instead of binary ASK, QPSK instead of 4-QASK). On the other hand, it is more convenient to express the signals mathematically in terms of their amplitude-modulation representations. As a result, in what follows, all modulation schemes are described mathematically as a form of quadrature amplitude modulation. Keep in mind that each has a mathematically equivalent description as phase modulation. In comparisons with QPSK, we assume that *rectangular pulses* are used for 4-QASK, unless stated otherwise.

First, recall that the mathematical representation for BPSK is

$$s(t) = \sqrt{2}\, A\, a(t) \cos(\omega_c t + \varphi),$$

where the amplitude modulation for the nth signaling interval is

$$a(t) = b_n, \quad nT \le t < (n+1)T$$

and $b_n = \pm 1$. The analogous expression for 4-QASK and QPSK is

$$s(t) = A\{a_1(t) \cos(\omega_c t + \varphi) - a_2(t) \sin(\omega_c t + \varphi)\},$$

where

$$a_1(t) = b_{1,n}, \quad nT' \le t < (n+1)T',$$
$$a_2(t) = b_{2,n}, \quad nT' \le t < (n+1)T',$$

and $b_{1,n} = \pm 1$ and $b_{2,n} = \pm 1$. It is convenient to introduce the baseband pulse waveform $\alpha(t)$ and to express the signals in terms of this waveform. For QPSK, this is accomplished by letting $\alpha(t) = p_{T'}(t)$, the rectangular pulse of duration T'. For $nT' \le t < (n+1)T'$, $s(t)$ can be expressed as

$$s(t) = A\{b_{1,n}\, \alpha(t - nT') \cos(\omega_c t + \varphi) - b_{2,n}\, \alpha(t - nT') \sin(\omega_c t + \varphi)\}.$$

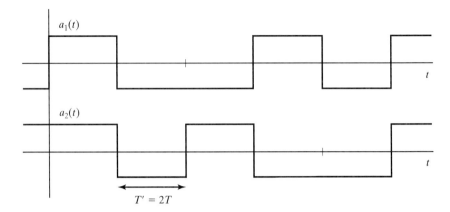

Figure 6-45: QASK or QPSK.

Next, observe that if $s(t)$ conveys a finite-length message sequence, it can be written as

$$s(t) = A \sum_{n=n_1}^{n_2} \{b_{1,n}\, \alpha(t - nT') \cos(\omega_c t + \varphi) - b_{2,n}\, \alpha(t - nT') \sin(\omega_c t + \varphi)\}, \quad (6.75)$$

where n_1 is the index corresponding to the first interval in which data is sent and n_2 is the index corresponding to the last. That is, the transmission begins at time $n_1 T'$ and ends at time $(n_2 + 1)T'$. If the message is sufficiently long (e.g., consider even a few minutes of data transmitted at 100 Mbps), it is often convenient to think of it as an *infinite* sequence and set $n_1 = -\infty$ and $n_2 = +\infty$.

Recall that the baseband pulse waveform $\alpha(t)$ used in the preceding expressions is the rectangular pulse: $\alpha(t) = p_{T'}(t)$. A comparison of these expressions shows that the baseband signals on the inphase and quadrature components are given by

$$a_1(t) = \sum_{n=n_1}^{n_2} b_{1,n}\, \alpha(t - nT') \quad \text{and} \quad a_2(t) = \sum_{n=n_1}^{n_2} b_{2,n}\, \alpha(t - nT').$$

Examination of these expressions should make it clear that both the inphase and quadrature components of the 4-QASK and QPSK signals can switch at time nT'; that is, at each of the times $t = nT'$, it is possible for both $a_1(t)$ and $a_2(t)$ to change sign, as illustrated in Figure 6-45. If both do change at time nT', this means that $s(t)$ itself changes sign at that time. If the modulation is viewed as amplitude modulation, such a sign change can be thought of as an instantaneous change in the polarity of the signal amplitude. Viewed as phase modulation, it corresponds to an instantaneous phase shift of $180°$. Of course, an instantaneous amplitude or phase change is impossible in an actual system, because this would require infinite bandwidth.

For a number of reasons, it is of interest to have modulation schemes that do not permit both components to change signs at the same time. To explore this consideration further, it is necessary to define the envelope of a signal of the form

$$s(t) = x(t)\cos(\omega_c t + \varphi) - y(t)\sin(\omega_c t + \varphi),$$

where $x(t)$ and $y(t)$ are signals that have bandwidths much smaller than the carrier frequency ω_c. A typical example is a 20-Mbps 4-QASK signal transmitted at a center frequency of 2 GHz. If the 4-QASK signal uses rectangular pulses for both the inphase and quadrature components, the pulse duration is $0.1\,\mu s$ and the null-to-null bandwidth is 20 MHz, which is only one percent of the carrier frequency. For such a signal, the *envelope* of the signal is defined by

$$e(t) = \sqrt{x^2(t) + y^2(t)}. \tag{6.76}$$

The original signal $s(t)$ can be written in terms of its envelope as

$$s(t) = e(t)\cos(\omega_c t + \psi(t) + \varphi), \tag{6.77}$$

where $\psi(t)$ is determined by $x(t)$ and $y(t)$; specifically, $\psi(t) = \tan^{-1}(y(t)/x(t))$.

Because the positive square root is assumed in the defining expression for the envelope, we might as well work with $e^2(t)$, the *square* of the envelope; the resulting expressions are somewhat simpler. Another reason for doing so is that $e^2(t)/2$ is the instantaneous power in the signal $s(t)$. The squared envelope of the signal

$$s(t) = A\{a_1(t)\cos(\omega_c t + \varphi) - a_2(t)\sin(\omega_c t + \varphi)\} \tag{6.78}$$

is

$$e^2(t) = A^2\{[a_1(t)]^2 + [a_2(t)]^2\}. \tag{6.79}$$

The signals $a_1(t)$ and $a_2(t)$ are sequences of rectangular pulses of amplitudes ± 1, so $[a_1(t)]^2 = [a_2(t)]^2 = 1$ for all t. It follows that $e^2(t) = 2A^2$ for all t (i.e., the instantaneous power is A^2). Thus, the *ideal* 4-QASK and QPSK signals have constant envelopes, but this observation depends critically on the fact that the pulses are rectangular. Unfortunately, if the signal is filtered or in some other way has its bandwidth restricted (e.g., due to the bandwidth limitations of the transmitter or the effects of the transmission medium), then the pulses do not retain their perfect rectangular shapes.

In general, a filtered 4-QASK or QPSK signal can be represented as

$$s(t) = A\{a_1(t)\cos(\omega_c t + \varphi) - a_2(t)\sin(\omega_c t + \varphi)\}, \tag{6.80}$$

where

$$a_1(t) = \sum_{n=n_1}^{n_2} b_{1,n}\,\alpha(t - nT') \quad \text{and} \quad a_2(t) = \sum_{n=n_1}^{n_2} b_{2,n}\,\alpha(t - nT'), \tag{6.81}$$

but the baseband waveform $\alpha(t)$ now represents a *filtered* rectangular pulse. As such, $\alpha(t)$ is a continuous waveform with a nonzero rise time. In particular, it does not produce inphase and quadrature signals for which $[a_1(t)]^2 = [a_2(t)]^2 = 1$ for all t. A simple example of this can be obtained by letting $\alpha(t)$ be the output of a single-pole RC filter that has a rectangular pulse as its input. In practice, of course, the communication signals do not even start out with perfect rectangular pulses, because we cannot generate such pulses in the first place. In particular, a rectangular pulse has discontinuities at the beginning and the end of the pulse. The pulses that are actually generated are continuous, and they can be thought of as the output of a finite-bandwidth linear filter with a rectangular pulse input.

With filtered versions of a rectangular pulse as the baseband waveform, the envelope of 4-QASK and QPSK is not constant. In fact, at a time when both the inphase and quadrature components of the signal switch signs, the squared envelope

$$e^2(t) = A^2\{[a_1(t)]^2 + [a_2(t)]^2\}$$

typically goes from approximately $2A^2$ to zero and back to $2A^2$ in a short time period. This represents a large excursion in signal power, which can produce unwanted transients in power amplifiers and other nonlinear devices. It imposes very large dynamic range requirements on system components. One way to avoid these large excursions in signal power is to prevent both the inphase and quadrature components from switching at the same time.

For certain variations of QASK and QPSK, called *offset QASK* (OQASK) and *offset QPSK* (OQPSK), respectively, at most one component of the signal can switch at a given time. The term *staggered* is sometimes used in place of the term *offset* (e.g., SQPSK in place of OQPSK), but the signal is the same. OQASK is a QASK-type signal with the same inphase and quadrature components as QASK. The difference is that in OQASK there is a time offset of $T = T'/2$ in the inphase component relative to the quadrature component, as displayed in

$$s(t) = A\{a_1(t + T)\cos(\omega_c t + \varphi) - a_2(t)\sin(\omega_c t + \varphi)\}, \qquad (6.82)$$

where $a_1(t)$ and $a_2(t)$ are the same baseband signals as for QASK. Because of this time offset, the inphase component can switch only at times $(2j+1)T$, where j is an arbitrary integer, as illustrated in Figure 6-46. The quadrature component, on the other hand, can switch only at times $2jT$. For filtered 4-ary OQASK (4-OQASK), the corresponding squared envelope

$$e^2(t) = A^2\{[a_1(t + T)]^2 + [a_2(t)]^2\}$$

is still not constant, but it typically decreases from approximately $2A^2$ to approximately A^2 (as opposed to zero) and back to $2A^2$ in a short time period. This is only a 3-dB variation in signal power. When compared with the large variations that result in binary ASK and QASK (theoretically infinite when expressed in dB), the 3-dB variation that occurs in OQPSK results in far fewer problems as the signal passes through power amplifiers and other nonlinear components of a communication system.

If we consider the OQASK signal with rectangular pulses in its phase-modulation form, we have OQPSK. The result of staggering the inphase and quadrature components of the signal is a limitation in the possible phase transitions that can take place in the

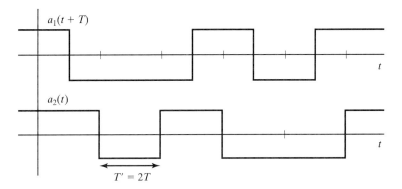

Figure 6-46: OQASK or OQPSK.

OQPSK signal. Since only one of the components can switch at a time, it is impossible to have a 180° phase transition in OQPSK. (See Figure 6-24.) Only 90° or 270° phase transitions are possible, and the envelope of a bandlimited OQPSK signal does not pass through zero with either 90° or 270° phase transitions.

Offsetting the inphase and quadrature components does not solve all the problems with the QPSK signal, however. In particular, OQPSK is still based on the rectangular pulse, and it has the same problems with the discontinuities in this pulse waveform as QPSK.

Minimum-shift key (MSK) modulation is offset QASK with a sine pulse rather than a rectangular pulse; that is,

$$s(t) = A\{a_1(t+T)\cos(\omega_c t + \varphi) - a_2(t)\sin(\omega_c t + \varphi)\}, \qquad (6.83a)$$

where

$$a_1(t) = b_{1,n}\sqrt{2}\sin\{\pi(t-nT')/T'\}, \quad nT' \le t < (n+1)T', \quad b_{1,n} = \pm 1, \quad (6.83b)$$

and

$$a_2(t) = b_{2,n}\sqrt{2}\sin\{\pi(t-nT')/T'\}, \quad nT' \le t < (n+1)T', \quad b_{2,n} = \pm 1. \quad (6.83c)$$

Unlike the rectangular pulse, the sine pulse has no discontinuity, as illustrated in Figure 6-47. The squared envelope for this signal is

$$e^2(t) = A^2\{[a_1(t+T)]^2 + [a_2(t)]^2\}.$$

The time offset of $T = T'/2$ in the inphase channel results in the baseband signal

$$a_1(t+T) = b_{1,n}\sqrt{2}\sin\{\pi(t+\tfrac{1}{2}T' - nT')/T'\}, \quad (n-\tfrac{1}{2})T' \le t < (n+\tfrac{1}{2})T',$$

and since $\sin(\theta + \tfrac{1}{2}\pi) = \cos(\theta)$, this is equivalent to

$$a_1(t+T) = b_{1,n}\sqrt{2}\cos\{\pi(t-nT')/T'\}, \quad (n-\tfrac{1}{2})T' \le t < (n+\tfrac{1}{2})T'.$$

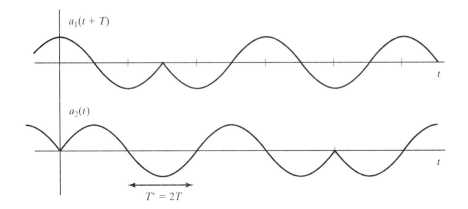

Figure 6-47: MSK.

Thus, the squared envelope is given by

$$e^2(t) = A^2\{2[\cos\{\pi(t-nT')/T'\}]^2 + 2[\sin\{\pi(t-nT')/T'\}]^2\} = 2A^2.$$

The MSK signal achieves a constant envelope without the use of discontinuous baseband waveforms.

For some purposes, it is convenient to define the pulse $\alpha(t)$ by

$$\alpha(t) = \sqrt{2}\sin(\pi t/T'), \quad 0 \le t \le T',$$

and then express $a_1(t)$ and $a_2(t)$ in terms of this pulse:

$$\begin{aligned}
a_1(t) &= b_{1,n}\sqrt{2}\sin\{\pi(t-nT')/T'\} = b_{1,n}\,\alpha(t-nT'), \quad nT' \le t < (n+1)T', \\
a_2(t) &= b_{2,n}\sqrt{2}\sin\{\pi(t-nT')/T'\} = b_{2,n}\,\alpha(t-nT'), \quad nT' \le t < (n+1)T'.
\end{aligned}$$

Also, observe that

$$\begin{aligned}
\alpha(t-nT') &= \sqrt{2}\sin\{\pi(t-nT')/T'\} = \sqrt{2}\sin\{(\pi t/T')-n\pi\} \\
&= (-1)^n\sqrt{2}\sin\{\pi t/T'\},
\end{aligned}$$

which is $+\sqrt{2}\sin\{\pi t/T'\}$ if n is even and $-\sqrt{2}\sin\{\pi t/T'\}$ if n is odd. The term $(-1)^n$ can then be combined with $b_{2,n}$ to give $\hat{b}_{2,n} = (-1)^n b_{2,n}$, and $a_2(t)$ can be written as

$$a_2(t) = b_{2,n}\sqrt{2}\sin\{\pi(t-nT')/T'\} = \hat{b}_{2,n}\sqrt{2}\sin\{\pi t/T'\}, \quad nT' \le t < (n+1)T'.$$

The possible values for $\hat{b}_{2,n}$ are ± 1, the same two values as for $b_{2,n}$. Next, we consider $a_1(t+T)$ and use the fact that, for $T = T'/2$,

$$\begin{aligned}
\sin\{\pi(t+T-nT')/T'\} &= \cos\{\pi(t-nT')/T'\} = \cos\{(\pi t/T')-n\pi\} \\
&= (-1)^n\cos\{\pi t/T'\}.
\end{aligned}$$

Table 6.7: Summary: BPSK, QPSK, and MSK

Parameter	BPSK	QPSK	MSK
Pulse duration	T	$T' = 2T$	$T' = 2T$
Pulse waveform	$p_T(t)$	$p_{T'}(t)$	$\sqrt{2}\sin(\pi t/T')\,p_{T'}(t)$
Data rate	$1/T$ bps	$1/T$ bps	$1/T$ bps
Envelope	$e^2(t) = 2A^2$	$e^2(t) = 2A^2$	$e^2(t) = 2A^2$
Power	$P = A^2$	$P = A^2$	$P = A^2$
Energy per bit	$\mathcal{E}_b = A^2 T$	$\mathcal{E}_b = A^2 T$	$\mathcal{E}_b = A^2 T$
Bit error probability	$Q(\sqrt{2\mathcal{E}_b/N_0}\,)$	$Q(\sqrt{2\mathcal{E}_b/N_0}\,)$	$Q(\sqrt{2\mathcal{E}_b/N_0}\,)$

It follows that if $\hat{b}_{1,n} = (-1)^n\, b_{1,n}$, then

$$
\begin{aligned}
a_1(t+T) &= b_{1,n}\sqrt{2}\sin\{\pi(t+T-nT')/T'\} \\
&= \hat{b}_{1,n}\sqrt{2}\cos\{\pi t/T'\}, \quad (n-\tfrac{1}{2})T' \le t < (n+\tfrac{1}{2})T'.
\end{aligned}
$$

The MSK signal can then be expressed as

$$
s(t) = \sqrt{2}\,A\{\hat{b}_{1,n}\cos(\pi t/T')\cos(\omega_c t + \varphi) - \hat{b}_{2,n}\sin(\pi t/T')\sin(\omega_c t + \varphi)\} \quad (6.84)
$$

for t in the appropriate interval. The mapping of the data variables $b_{1,n}$ and $b_{2,n}$ into the new data variables $\hat{b}_{1,n}$ and $\hat{b}_{2,n}$ is a form of *precoding*. This particular precoding operation is a one-to-one mapping that can be reversed at the demodulator. A summary of the features of BPSK, QPSK, and MSK is given in Table 6.7.

6.8   Bandwidth Comparisons for Two-Dimensional Modulation

Methods for the evaluation of the spectral density of communication signals are described in Section 4.5 and Problems 4.4.7–4.4.9. Application of these methods provides the spectral density for various two-dimensional signal sets including binary ASK, QASK, BPSK, QPSK, OQPSK, and MSK. In this section, we fix the data rate of each signal to be $1/T$ bits per second, and we compare the spectral densities for BPSK, QPSK, OQPSK, and MSK. Binary ASK with rectangular pulses is equivalent to BPSK, and 4-QASK with rectangular pulses is equivalent to QPSK; therefore, the spectral densities for these two forms of ASK are the same as their PSK counterparts.

Recall that the basic one-dimensional signal format is

$$
Y_1(t) = \sqrt{2}\,D(t-U)\cos(\omega_c t + \Theta), \quad\quad (6.85)
$$

where

$$
D(t) = \sum_{n=-\infty}^{\infty} A_n\, p_T(t - nT), \quad\quad (6.86)
$$

as in Example 4-1. The amplitudes A_n, $-\infty < n < \infty$, for different data symbols are assumed to be statistically independent and identically distributed. Extensions to correlated data symbols are possible, as outlined in Problem 4.4.9. The random time delay U and phase angle Θ are independent and uniformly distributed on the appropriate intervals, and they are independent of the pulse amplitudes (e.g., see Example 4-2 in Section 4.5). The random variables that represent the pulse amplitudes are assumed to satisfy

$$P(A_n = +A) = P(A_n = -A) = 1/2$$

for each n, where A is a fixed positive number. Then $E\{A_n\} = 0$ and $E\{A_n^2\} = A^2$ for each n. Thus, we can set $\alpha = A$ in (4.35) to conclude that the spectral density for BPSK is

$$S_1(f) = A^2 T \{ \text{sinc}^2[(f - f_c)T] + \text{sinc}^2[(f + f_c)T] \}/2. \tag{6.87}$$

The QPSK signal can be written as

$$Y_2(t) = D_1(t - U) \cos(\omega_c t + \Theta) - D_2(t - U) \sin(\omega_c t + \Theta), \tag{6.88}$$

where $D_1(t)$ and $D_2(t)$ are independent data signals of the type defined in (6.86), except that the rectangular pulses are of duration $2T$ and the random time delay is uniformly distributed on $[0, 2T]$. These data signals can be expressed as

$$D_i(t) = \sum_{n=-\infty}^{\infty} A_{i,n}\, p_{2T}(t - 2nT) \tag{6.89}$$

for $i = 1$ and $i = 2$, where the random variables $A_{1,n}$, $-\infty < n < \infty$, and $A_{2,n}$, $-\infty < n < \infty$, are mutually independent and identically distributed. As for BPSK, we assume that $A_{i,n}$ takes on the values $+A$ and $-A$ with equal probability. Notice that one data bit can be sent on each of the two components of the QPSK signal during each interval of length $2T$, so the total data rate for the two components of the QPSK signal is $1/T$ bits per second. The spectral density for the QPSK signal can be obtained by comparing (6.88)–(6.89) with (6.85)–(6.86) and then applying Problem 4.4.7. The resulting spectral density is

$$S_2(f) = A^2 T \{ \text{sinc}^2[2\,(f - f_c)T] + \text{sinc}^2[2\,(f + f_c)T] \}. \tag{6.90}$$

At this point, it is convenient to introduce the function G_1 that is defined by

$$G_1(x) = \text{sinc}^2(x) = [\sin(\pi x)/(\pi x)]^2, \quad -\infty < x < \infty.$$

In terms of this function, the spectral density of BPSK is given by

$$S_1(f) = A^2 T \{ G_1[(f - f_c)T] + G_1[(f + f_c)T] \}/2. \tag{6.91}$$

The spectral density for QPSK (and OQPSK) is given by

$$S_2(f) = A^2 T \{ G_2[(f - f_c)T] + G_2[(f + f_c)T] \}/2, \tag{6.92}$$

where $G_2(x) = 2 G_1(2x)$ for each x. The introduction of the functions G_1 and G_2 gives a convenient way for comparing the spectra of the BPSK and QPSK signals. From (6.91) and (6.92), it is clear that the only difference between these spectra is the difference between the functions G_1 and G_2. In particular, if $f_c T \gg 1$, the positive and negative components of the spectra (i.e., those centered at f_c and $-f_c$, respectively) do not overlap significantly. Mathematically, this means that, for $i = 1$ and $i = 2$,

$$G_i[(f - f_c)T] \gg G_i[(f + f_c)T], \quad f > 0,$$

and

$$G_i[(f - f_c)T] \ll G_i[(f + f_c)T], \quad f < 0.$$

It follows that if $f_c T \gg 1$, then

$$S_i(f) \approx G_i[(f - f_c)T], \quad f > 0,$$

and

$$S_i(f) \approx G_i[(f + f_c)T], \quad f < 0.$$

As a result, the shape of the spectral density $S_i(f)$ for $f > f_c$ is just the shape of $G_i(x)$ for $x > 0$. Because $G_i(-x) = G_i(x)$, the function G_i specifies the shape of the spectral density for other ranges of f as well. These spectral densities are similar to the bandpass frequency functions that have local symmetry. (See Section 4.6, especially Figure 4-8.) The only difference is that we do not require the positive parts of the spectral densities to be identically zero for negative frequencies. It is sufficient that they be approximately zero. Similarly, we do not require the negative parts of the spectral densities to be identically zero for positive frequencies.

As suggested by the definition of OQPSK in Section 6.7 (see (6.82)), the signal format for OQPSK is just a minor modification of (6.88): replace $D_1(t - U)$ by $D_1(t + T - U)$. In the notation of Section 6.7, the offset parameter t_0 is equal to T. Again we use Problem 4.4.7, this time to observe that a time offset does not change the spectral density. The conclusion is that the spectral density for OQPSK is also given by (6.90).

Finally, we consider MSK. The MSK signal can also be represented by (6.88) if $D_1(t - U)$ is replaced by $D_1(t + T - U)$ and if the pulse waveform in (6.89) is changed from the rectangular pulse of duration $2T$ to the sine pulse of the same duration, as discussed in Section 6.7. The spectral density is obtained by the same method as for QPSK, the only difference is that the Fourier transform of the rectangular pulse is replaced by the Fourier transform of the sine pulse. The resulting spectral density is

$$S_3(f) = A^2 T \{G_3[(f - f_c)T] + G_3[(f + f_c)T]\}/2, \qquad (6.93)$$

where the function G_3 is defined by [6.16]

$$G_3(x) = \{[(4/\pi) \cos(2\pi x)]/(1 - 16x^2)\}^2. \qquad (6.94)$$

The spectral densities for BPSK, QPSK, OQPSK, and MSK are compared in Figures 6-48 and 6-49. Because of the symmetry involved, the best way to illus-

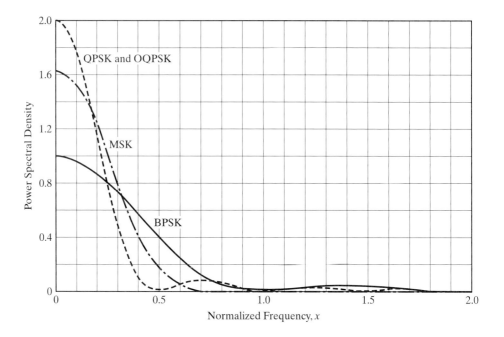

Figure 6-48: Power spectral densities for BPSK, QPSK, OQPSK, and MSK.

trate the spectral densities of these signals is by means of graphs of the functions G_i for $i = 1, 2$, and 3. In Figure 6-48, graphs of $G_1(x)$, $G_2(x)$, and $G_3(x)$ are given for $x > 0$. The parameter x is a normalized measure of frequency. In particular, x represents $(f - f_c)T$, where T is the symbol duration for BPSK. The symbol duration is $2T$ for QPSK, OQPSK, and MSK, but two symbols are sent simultaneously on the inphase and quadrature components of each of these signals. Therefore, $R_b = 1/T$ is the data rate in bits per second for each of the four modulation formats. Clearly, $f - f_c$ is the separation between the frequency of interest and the carrier frequency, and

$$x = (f - f_c)T = (f - f_c)/R_b$$

is the frequency separation divided by the data rate. Typically, the frequency separation is measured in hertz and the data rate is in bits per second, so the parameter x is dimensionless. Use of the normalized frequency facilitates the comparison of spectral densities in a way that is independent of the carrier frequency and data rate of the signals.

Some aspects of the spectral densities are more clearly displayed if they are given in decibels (dB), and because $G_i(x)$ represents a power spectral density, the representation in dB is

$$[G_i(x)]_{dB} = 10 \log_{10}[G_i(x)].$$

In Figure 6-49, graphs of $G_1(x)$, $G_2(x)$, and $G_3(x)$ are given in dB for $x > 0$. Many of the common specifications of bandwidth and frequency occupancy are defined in terms of the spectral densities given in dB, so Figure 6-49 is a more convenient way of displaying the spectral occupancy of BPSK, QPSK, OQPSK, and MSK.

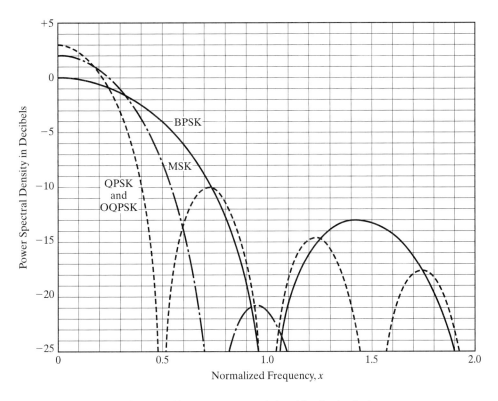

Figure 6-49: Power spectral densities in decibels.

For the types of signals under consideration, the *3-dB bandwidth* for the spectral density $S_i(f)$ is the width of the positive band of frequencies outside of which $S_i(f)$ is at least 3 dB below $S_i(f_c)$, the value of the spectral density at the carrier frequency. If f_{3dB} is the largest frequency for which the locally symmetrical function S_i satisfies

$$[S_i(f)]_{dB} \leq [S_i(f_c)]_{dB} - 3, \quad f > f_{3dB},$$

then the 3-dB bandwidth is $2f_{3dB}$. As discussed in Section 4.6.1, the 3-dB bandwidth is also known as the *half-power bandwidth*. (Recall that $10\log_{10}(1/2)$ is approximately -3 dB.)

Similar definitions are made for other power levels (e.g., 10-dB, 35-dB, and 50-dB bandwidths), and extensions of these definitions to spectral densities that do not have local symmetry about the carrier frequency are straightforward. Basically, they involve the determination of two frequencies, one that is above and one that is below the carrier frequency, and the bandwidth is the difference between these two frequencies. For locally symmetrical spectral densities, the carrier is exactly half way between these two frequencies, so only the upper frequency need be determined. Another common measure of bandwidth, the *null-to-null bandwidth*, is defined and discussed in Section 4.6.1. This measure of bandwidth is just the width of the main lobe of the spectrum, and it gives no information for the other lobes (known as *side lobes*).

Several different definitions of bandwidth are described and compared in [6.1], and numerical values are given for the bandwidths of BPSK, QPSK, MSK, and other modulation schemes. The null-to-null bandwidths, which are easily determined from (6.91)–(6.94) are 2.0 for BPSK, 1.0 for QPSK and OQPSK, and 1.5 for MSK. The 3-dB bandwidths are 0.88 for BPSK, 0.44 for QPSK and OQPSK, and 0.59 for MSK. Notice that the null-to-null and 3-dB bandwidths are larger for MSK than for QPSK.

To see the real advantage of MSK, we must consider measures of bandwidth that reflect the rate of decay of the peaks of the side lobes of the spectral densities. For example, the 35-dB bandwidths are 35.12 for BPSK, 17.56 for QPSK and OQPSK, and 3.24 for MSK [6.1], which demonstrate that MSK has a much smaller spectral density at high frequencies than BPSK and QPSK. This can be seen by the comparison of $G_1(x)$, $G_2(x)$, and $G_3(x)$ for large values of x. The fact that $G_1(x)$ and $G_2(x)$ decrease as $1/x^2$ as x increases implies that the peaks of the side lobes of BPSK and QPSK decrease as the inverse of the second power of $f - f_c$ as f increases. It follows from (6.94) that for large values of x, $G_3(x)$ decreases as $1/x^4$ as x increases, which implies that for large values of f the peaks of the side lobes of MSK decrease as the inverse of the fourth power of $f - f_c$ as f increases.

Another important measure of bandwidth is the bandwidth that contains 99% of the total power in the signal. This measure of bandwidth involves the integral of the spectral density between two positive limits, one below the carrier and one above. The limits are adjusted to make the integral equal to 99% of the entire area under the spectral density function for positive frequencies. If the spectral density has local symmetry about its carrier frequency, then it is sufficient to make the integral of $G_i(x)$ from 0 to x_0 equal to 0.99 times the integral of $G_i(x)$ from 0 to ∞. The 99%-power bandwidth is then $2x_0$. The 99%-power bandwidths are 20.56 for BPSK, 10.28 for QPSK and OQPSK, and 1.18 for MSK [6.1], again demonstrating the spectral efficiency of MSK as compared with the PSK forms of modulation.

6.9   Continuous-Phase Frequency Modulation

Recall from Section 6.5.3 that the general form for carrier-modulated signals with digital phase modulation $\psi(t)$ is

$$s(t) = \sqrt{2}\, A \cos(2\pi f_c t + \psi(t) + \varphi). \tag{6.95}$$

The unmodulated carrier signal is

$$c(t) = \sqrt{2}\, A \cos(2\pi f_c t + \varphi). \tag{6.96}$$

The signal in (6.95) can also be viewed as a frequency-modulated signal for which the frequency modulation is $\psi'(t)/2\pi$, where $\psi'(t)$ denotes the derivative of the phase modulation. The instantaneous phase of the signal in (6.95) is $2\pi f_c t + \psi(t) + \varphi$, which is a continuous function of t if and only if $\psi(t)$ is continuous. The general form of modulation in which the phase is continuous in a signal of the form given in (6.95) is referred to as continuous-phase frequency modulation or simply *continuous-phase modulation* (CPM).

At time τ, the instantaneous frequency of the signal in (6.95) is $f_c + [\psi'(\tau)/2\pi]$. The instantaneous phase of the carrier signal in (6.96) is $2\pi f_c \tau + \varphi$. If the instantaneous phase of carrier is subtracted from the instantaneous phase of the modulated signal, the result is referred to as the *excess phase* of the modulated signal. The excess phase at time τ for the signal of (6.95) is $\psi(\tau)$.

We begin the discussion of CPM with one type of binary frequency-shift key (BFSK) modulation that does not have a continuous phase and hence is not a form of CPM. Nevertheless, it is a good starting point. Two sinusoidal signals are employed, each of which has an arbitrary constant phase angle as one component of the phase modulation. In each symbol interval, the phase modulation takes one of two forms. If 0 is sent in the nth interval, then

$$\psi(t) = 2\pi f_d t + \varphi_0, \quad nT \le t < (n+1)T, \tag{6.97a}$$

but if 1 is sent, then

$$\psi(t) = -2\pi f_d t + \varphi_1, \quad nT \le t < (n+1)T. \tag{6.97b}$$

The parameter f_d is referred to as the *frequency deviation*. The radian-frequency deviation is $2\pi f_d$. The frequency modulation corresponding to (6.97) can be written as

$$\psi'(t) = 2\pi b_n f_d, \tag{6.98}$$

where $b_n = +1$ if a 0 is sent in the nth interval and $b_n = -1$ if a 1 is sent.

We can think of such a BFSK signal as the result of switching between two oscillators, one of which has frequency $f_c + f_d$ and phase $\varphi_0 + \varphi$ and the other has frequency $f_c - f_d$ and phase $\varphi_1 + \varphi$. The phase angles φ_0 and φ_1, which represent the excess phases at $t = 0$, need not be related to each other. The signals at the outputs of the oscillators are often referred to as *tones*. The signal with frequency $f_c + f_d$ is called the *space*, and the signal with frequency $f_c - f_d$ is called the *mark*. Thus, the binary digit 0 is sent by transmitting a space, and the binary digit 1 is sent by transmitting a mark. Without additional restrictions, the phase need not be continuous for BFSK.

It is easy to show that if $2f_c T$ and $2f_d T$ are integers, the mark and space signals are orthogonal on each interval of duration T, regardless of the values of φ_0, φ_1, and φ. If $f_c T \gg 1$, the signals are approximately orthogonal provided that $2f_d T = m$ for some integer m. The *modulation index* for a BFSK signal is defined by $h = 2f_d T$. In terms of the modulation index, the condition for orthogonality in signals with arbitrary phase angles is $h = m$ for some integer m. If $2f_c T$ and $2f_d T$ are integers or if $(f_c - f_d)T \gg 1$, the energy in each signal is $\mathcal{E} = A^2 T$. From Section 6.6.3, we know that the probability of error for binary orthogonal signals, an additive white Gaussian noise channel with spectral density $N_0/2$, and an optimum coherent receiver is

$$P_c = Q\left(\sqrt{\mathcal{E}/N_0}\right).$$

Results for orthogonal BFSK in systems with noncoherent receivers are given in Chapter 7. It is important to remember that $2f_d T = m$ is required for orthogonality if the phase angles φ_0 and φ_1 are arbitrary.

For some applications, the phase modulation includes a phase angle that depends on the interval in which the signal is transmitted. In such an application, the transmitted signal for the nth interval is

$$s(t) = \sqrt{2}\,A\cos(2\pi f_c t + 2\pi b_n f_d t + \theta_n + \varphi), \quad nT \le t < (n+1)T. \quad (6.99)$$

Such a generalization is needed for BFSK signals in frequency-hop spread spectrum (see Chapter 9), and it is also required for continuous-phase frequency-shift key (CPFSK) modulation.

6.9.1   Continuous-Phase Frequency-Shift-Key Modulation

Phase continuity can be obtained by controlling the phase angle θ_n in a signal of the form given by (6.99). For any such signal, the excess phase in the nth interval is

$$\psi(t) = 2\pi b_n f_d t + \theta_n, \quad nT \le t < (n+1)T. \quad (6.100)$$

It is clear that $\psi(t)$ is continuous for t in the interval $(n-1)T < t < nT$; however, we must ensure the phase is also continuous at the endpoints of the interval. To provide phase continuity at $t = nT$, θ_n must satisfy

$$2\pi b_{n-1} f_d nT + \theta_{n-1} = 2\pi b_n f_d nT + \theta_n.$$

Solving for θ_n and rearranging terms, we see that the condition for phase continuity is

$$\theta_n = \theta_{n-1} + 2\pi n f_d T[b_{n-1} - b_n]. \quad (6.101)$$

Thus, if the two successive data symbols b_{n-1} and b_n are the same, then $\theta_n = \theta_{n-1}$ regardless of the frequency deviation. If the two successive symbols differ, then $\theta_n = \theta_{n-1} \pm 4\pi n f_d T$, in which case θ_n depends on the frequency deviation. The condition for phase continuity is equivalent to

$$\theta_n = \theta_{n-1} + \pi n h[b_{n-1} - b_n], \quad (6.102)$$

for each integer n. If $b_{n-1} \ne b_n$ then $\theta_n = \theta_{n-1} \pm 2\pi n h$, in which case the relationship between θ_n and θ_{n-1} depends on the modulation index.

For the simplest CPFSK signals, the modulation index is an integer, so $2\pi n h$ is a multiple of 2π. Phase angles that differ by multiples of 2π are equivalent, so the requirement for phase continuity in a BFSK system with an integer-valued modulation index is $\theta_n = \theta_{n-1}$ for each n. Thus, we can let $\theta_n = \theta$, a constant, for CPFSK modulation in which h is an integer.

Because of the presence of the fixed phase angle φ in (6.99), there is no loss of generality if we let $\theta_0 = 0$, which implies that $\psi(0) = 0$. Let $\psi(nT)$ be denoted by ψ_n for each n, and notice that (6.100) implies that

$$\psi_n = 2\pi b_n f_d nT + \theta_n.$$

Solving for θ_n, we find that

$$\theta_n = -2\pi b_n f_d nT + \psi_n.$$

Substituting for θ_n in (6.100) gives

$$\psi(t) = 2\pi b_n f_d(t - nT) + \psi_n, \quad nT \le t < (n+1)T. \tag{6.103}$$

The resulting CPFSK signal is

$$s(t) = \sqrt{2}\,A\cos[2\pi f_c t + 2\pi b_n f_d(t - nT) + \psi_n + \varphi], \quad nT \le t < (n+1)T. \tag{6.104}$$

Notice that (6.103) implies that ψ_{n+1} and ψ_n are related by

$$\psi_{n+1} = 2\pi b_n f_d T + \psi_n \tag{6.105}$$

for each n. Thus, ψ_1 depends on b_1 only ($\psi_0 = 0$); ψ_2 depends on b_2 and (through ψ_1) b_1; ψ_3 depends on b_3, b_2, and b_1; etc. Hence, ψ_{n+1} depends not only on b_n, but also on b_k for all $k < n$, so CPFSK modulation has memory. (The signal in the present interval depends on the signal in past intervals.) As a result, optimum detection of the data symbol b_n cannot be accomplished by processing the signal for the interval $nT \le t < (n+1)T$ only. Instead, some form of sequence detection must be employed, perhaps using the Viterbi algorithm. Methods for demodulating such signals are described in [6.2], [6.17], and [6.20].

If the modulation index is $h = 1/2$, then $f_d T = 1/4$ and the excess phase for the nth interval is

$$\psi(t) = \frac{\pi}{2T} b_n(t - nT) + \psi_n, \quad nT \le t < (n+1)T. \tag{6.106}$$

The resulting CPFSK signal for $h = 1/2$ is

$$s(t) = \sqrt{2}\,A\cos\left[2\pi f_c t + \frac{\pi}{2T} b_n(t - nT) + \psi_n + \varphi\right], \quad nT \le t < (n+1)T. \tag{6.107}$$

It follows from (6.106) that if $h = 1/2$, then

$$\psi_{n+1} = \frac{\pi}{2} b_n + \psi_n \tag{6.108}$$

for each integer n. Because $\psi_0 = 0$ and $b_n = \pm 1$, (6.108) implies that ψ_n is a multiple of $\pi/2$ for each n. The *phase states* are the possible values for ψ_n. For $h = 1/2$, the phase states are $0, \pi/2, \pi$, and $3\pi/2$. The allowed transitions between consecutive phase states for $h = 1/2$ are shown in the trellis diagram of Figure 6-50. The phase angle ψ_n is either $\pi/2$ or $3\pi/2$ if n is odd, and it is either 0 or π if n is even. (Recall that $\psi_0 = 0$.) From the trellis diagram, we see that

$$\cos(\psi_{2k-1}) = 0 \tag{6.109}$$

and

$$\sin(\psi_{2k}) = 0 \tag{6.110}$$

for each integer k. Furthermore, $\sin(\psi_{2k-1}) = \pm 1$ and $\cos(\psi_{2k}) = \pm 1$ for each integer k.

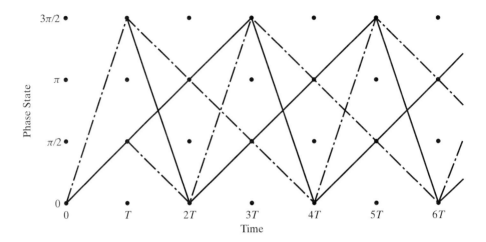

Figure 6-50: Phase trellis for CPFSK with $h = 1/2$ (dashed lines correspond to $b_n = -1$, solid lines correspond to $b_n = +1$).

6.9.2   Minimum-Shift-Key Modulation Revisited

In Section 6.7, MSK modulation is described as offset QASK in which the pulse waveform is the sine pulse. The sine pulse, which is the positive half of a full sine wave, is given by

$$\alpha(t) = \sqrt{2} \sin\left(\frac{\pi t}{2T}\right), \quad 0 \le t < 2T. \tag{6.111}$$

From (6.83) with $T' = 2T$, we know that the MSK signal can be expressed as

$$s(t) = A\{a_1(t + T)\cos(\omega_c t + \varphi) - a_2(t)\sin(\omega_c t + \varphi)\}, \tag{6.112}$$

where

$$a_1(t) = b_{1,k}\,\alpha(t - 2kT), \quad 2kT \le t < (2k+2)T, \tag{6.113}$$

and

$$a_2(t) = b_{2,k}\,\alpha(t - 2kT), \quad 2kT \le t < (2k+2)T, \tag{6.114}$$

for each integer k. Notice that (6.113) implies that

$$a_1(t + T) = b_{1,k}\,\alpha(t - (2k-1)T), \quad (2k-1)T \le t < (2k+1)T.$$

Combine the data sequences $(b_{1,k})$ and $(b_{2,k})$ into a single data sequence by defining $\beta_{2k-1} = b_{1,k}$ and $\beta_{2k} = b_{2,k}$ for each k. In many applications, the data sequences $(b_{1,k})$ and $(b_{2,k})$ are derived from a single sequence using the reverse procedure, so the sequence (β_n) is the original data sequence. It follows from (6.113) and (6.114) that

$$a_1(t + T) = \beta_{2k-1}\,\alpha(t - (2k-1)T), \quad (2k-1)T \le t < (2k+1)T, \tag{6.115a}$$

and

$$a_2(t) = \beta_{2k}\,\alpha(t - 2kT), \quad 2kT \leq t < (2k+2)T, \tag{6.115b}$$

for each integer k. In the remainder of the subsection, it is shown that the MSK signal represented by (6.112) and (6.108) is a form of CPFSK with a modulation index of $h = 1/2$.

The first step is to expand the right-hand side of (6.95) to obtain

$$s(t) = A\{\sqrt{2}\cos[\psi(t)]\cos(2\pi f_c t + \varphi) - \sqrt{2}\sin[\psi(t)]\sin(2\pi f_c t + \varphi)\}. \tag{6.116}$$

In order to equate this with the MSK signal in (6.112), we must show that

$$\sqrt{2}\cos[\psi(t)] = a_1(t + T), \quad (2k-1)T \leq t < (2k+1)T, \tag{6.117a}$$

and

$$\sqrt{2}\sin[\psi(t)] = a_2(t), \quad 2kT \leq t < (2k+2)T. \tag{6.117b}$$

Observe that (6.106) implies that, for $nT \leq t < (n+1)T$,

$$\cos[\psi(t)] = \cos\left[\frac{\pi}{2T}b_n(t - nT)\right]\cos(\psi_n) - \sin\left[\frac{\pi}{2T}b_n(t - nT)\right]\sin(\psi_n) \tag{6.118}$$

and

$$\sin[\psi(t)] = \sin\left[\frac{\pi}{2T}b_n(t - nT)\right]\cos(\psi_n) + \cos\left[\frac{\pi}{2T}b_n(t - nT)\right]\sin(\psi_n). \tag{6.119}$$

We can simplify (6.118) and (6.119) if we use the fact that $b_n = \pm 1$ implies $\sin(b_n x) = b_n \sin(x)$ and $\cos(b_n x) = \cos(x)$ for each real number x.

From (6.118) and (6.109), it follows that

$$\cos[\psi(t)] = -b_{2k-1}\sin(\psi_{2k-1})\sin\left[\frac{\pi}{2T}(t - (2k-1)T)\right], \quad (2k-1)T \leq t < 2kT, \tag{6.120}$$

and from (6.118) and (6.110), we obtain

$$\begin{aligned}\cos[\psi(t)] &= \cos(\psi_{2k})\cos\left[\frac{\pi}{2T}(t - 2kT)\right] \\ &= \cos(\psi_{2k})\sin\left[\frac{\pi}{2T}(t - (2k-1)T)\right], \quad 2kT \leq t < (2k+1)T.\end{aligned} \tag{6.121}$$

From (6.120), we see that the continuity of $\psi(t)$ requires that

$$\cos[\psi(2kT)] = -b_{2k-1}\sin(\psi_{2k-1}),$$

so

$$\cos(\psi_{2k}) = -b_{2k-1}\sin(\psi_{2k-1}). \tag{6.122}$$

Thus, (6.120)–(6.122) yield

$$\sqrt{2}\cos[\psi(t)] = -b_{2k-1}\sin(\psi_{2k-1})\sqrt{2}\sin\left[\frac{\pi}{2T}(t-(2k-1)T)\right] \qquad (6.123)$$

for $(2k-1)T \leq t < (2k+1)T$. It follows from (6.111) that

$$\sqrt{2}\sin\left[\frac{\pi}{2T}(t-(2k-1)T)\right] = \alpha(t-(2k-1)T),$$

so

$$\sqrt{2}\cos[\psi(t)] = -b_{2k-1}\sin(\psi_{2k-1})\alpha(t-(2k-1)T), \qquad (2k-1)T \leq t < (2k+1)T. \qquad (6.124)$$

Since $\sin(\psi_{2k-1}) = \pm 1$, we can define the data variable β_{2k-1} by

$$\beta_{2k-1} = -b_{2k-1}\sin(\psi_{2k-1})$$

for each k, which is just a form of precoding that is similar to that described in Section 6.7. It follows from (6.124) and (6.115a) that

$$\sqrt{2}\cos[\psi(t)] = \beta_{2k-1}\alpha(t-(2k-1)T) = a_1(t+T) \qquad (6.125)$$

for $(2k-1)T \leq t < (2k+1)T$, which establishes (6.117a).

From (6.119) and (6.110), it follows that

$$\sin[\psi(t)] = b_{2k}\cos(\psi_{2k})\sin\left[\frac{\pi}{2T}(t-2kT)\right], \qquad 2kT \leq t < (2k+1)T, \qquad (6.126)$$

and from (6.119) and (6.109), we get

$$\begin{aligned}
\sin[\psi(t)] &= \sin(\psi_{2k+1})\cos\left[\frac{\pi}{2T}(t-(2k+1)T)\right] \\
&= \sin(\psi_{2k+1})\sin\left[\frac{\pi}{2T}(t-2kT)\right], \qquad (2k+1)T \leq t < (2k+2)T.
\end{aligned}$$

$$(6.127)$$

From (6.126) and the continuity of $\psi(t)$, we see that

$$\sin(\psi_{2k+1}) = b_{2k}\cos(\psi_{2k}),$$

Thus, from (6.126) and (6.127), we have

$$\sqrt{2}\sin[\psi(t)] = b_{2k}\cos(\psi_{2k})\alpha(t-2kT), \qquad 2kT \leq t < (2k+2)T. \qquad (6.128)$$

Since $\cos(\psi_{2k}) = \pm 1$, we can let $\beta_{2k} = b_{2k}\cos(\psi_{2k})$ for each k. It follows from (6.128) and (6.115b) that

$$\sqrt{2}\sin[\psi(t)] = \beta_{2k}\alpha(t-2kT) = a_2(t) \qquad (6.129)$$

for $2kT \leq t < (2k+2)T$, which establishes (6.117b). Thus, the MSK signal

$$s(t) = A\{a_1(t+T)\cos(\omega_c t + \varphi) - a_2(t)\sin(\omega_c t + \varphi)\}$$

is equivalent to the CPFSK signal with $h = 1/2$ that is given by

$$s(t) = \sqrt{2}A\cos\left(2\pi f_c t + \frac{\pi}{2T}b_n(t-nT) + \psi_n + \varphi\right), \qquad nT \leq t < (n+1)T,$$

for each integer n.

6.9.3   Continuous-Phase Modulation with a General Frequency Pulse

According to (6.103), the excess phase for CPFSK with modulation index $h = 2f_d T$ is given by

$$\psi(t) = 2\pi\, b_n\, f_d (t - nT) + \psi_n \qquad (6.130)$$

for $nT \leq t < (n+1)T$. Since $\psi_0 = 0$, the excess phase for $0 \leq t < T$ can be written as $\psi(t) = 2\pi f_d t b_0$. An equivalent representation is

$$\psi(t) = 2\pi f_d b_0 \int_{-\infty}^{t} p_T(\lambda)\, d\lambda, \qquad 0 \leq t < T. \qquad (6.131)$$

Because $\psi(t)$ is continuous at $t = T$,

$$\psi_1 = \psi(T) = 2\pi f_d b_0 \int_{-\infty}^{T} p_T(\lambda)\, d\lambda,$$

and because $p_T(\lambda) = 0$ for $\lambda > T$,

$$\psi_1 = 2\pi f_d b_0 \int_{-\infty}^{t} p_T(\lambda)\, d\lambda, \qquad t \geq T. \qquad (6.132)$$

In the second interval, the excess phase is given by

$$\psi(t) = 2\pi f_d b_1 \int_{-\infty}^{t} p_T(\lambda - T)\, d\lambda + \psi_1, \qquad T \leq t < 2T. \qquad (6.133)$$

Substituting for ψ_1 from (6.132), we find that, for $T \leq t < 2T$,

$$\psi(t) = 2\pi f_d \left[b_1 \int_{-\infty}^{t} p_T(\lambda - T)\, d\lambda + b_0 \int_{-\infty}^{t} p_T(\lambda)\, d\lambda \right]. \qquad (6.134)$$

Observe that the first integral is zero for t between 0 and T, so (6.131)–(6.133) imply that (6.134) is valid for the full range $0 \leq t < 2T$. Thus, we can write (6.134) as

$$\psi(t) = 2\pi f_d \int_{-\infty}^{t} \sum_{n=0}^{1} b_n\, p_T(\lambda - nT)\, d\lambda, \qquad 0 \leq t < 2T. \qquad (6.135)$$

The same derivation can be applied for an arbitrary number of intervals of length T to show that, for the data sequence $b_0, b_1, \ldots, b_{N-1}$, the excess phase is given by

$$\psi(t) = 2\pi f_d \int_{-\infty}^{t} \sum_{n=0}^{N-1} b_n\, p_T(\lambda - nT)\, d\lambda, \qquad 0 \leq t < NT. \qquad (6.136)$$

At this stage, it is beneficial to introduce some of the standard terminology for CPM. The pulse $p_T(\lambda)$ in CPFSK modulates the frequency of the carrier (its integral is proportional to the phase), so we refer to it as the *frequency pulse* for the CPFSK signal.

The abbreviation $LREC$ is often used for CPM modulation in which the frequency pulse is a rectangular pulse of duration LT. Thus, CPFSK modulation is also referred to as 1REC modulation. If the frequency pulse has duration T, the CPM is said to be *full response*. CPFSK is an example of full-response CPM. For full-response CPM, the data symbol b_n affects the instantaneous frequency for only the single interval $nT \leq t < (n+1)T$. If the frequency pulse has length in excess of T, the CPM is referred to as *partial response*. In partial-response CPM, such as $LREC$ for $L \geq 2$, each data symbol may influence the instantaneous frequency over multiple intervals.

In descriptions of CPFSK, most authors prefer to replace $p_T(t)$ by the frequency pulse

$$v_T(t) = p_T(t)/2T$$

and consider an arbitrary (perhaps infinite) length for the data sequence (b_n). We define

$$v(\lambda) = \sum_n b_n \, v_T(\lambda - nT), \qquad (6.137)$$

with the understanding that the sum is over all n for which corresponding data symbols are sent. Alternatively, we can consider the sum to be over all integers and let $b_n = 0$ if no symbol is sent in the nth interval. The excess phase can now be written as

$$\psi(t) = 4\pi f_d T \int_{-\infty}^{t} v(\lambda) \, d\lambda. \qquad (6.138)$$

The *phase pulse*, which is sometimes referred to as the *phase response*, is the integral of the frequency pulse. For the frequency pulse function v_T, the corresponding phase pulse function q_T is given by

$$q_T(t) = \int_{-\infty}^{t} v_T(\lambda) \, d\lambda,$$

which gives

$$q_T(t) = \begin{cases} 1/2, & t > T, \\ t/(2T), & 0 \leq t < T, \\ 0, & t < 0. \end{cases} \qquad (6.139)$$

The excess phase can now be written as

$$\psi(t) = 4\pi f_d T \sum_n b_n \, q_T(t - nT). \qquad (6.140)$$

Note that if $NT \leq t < (N+1)T$, then $q(t - nT) = 0$ for $n > N$.

We can generalize (6.137) by considering an arbitrary frequency pulse $w(t)$ and defining the excess phase by

$$\psi(t) = 4\pi f_d T \int_{-\infty}^{t} v(\lambda) \, d\lambda, \qquad (6.141)$$

where

$$v(\lambda) = \sum_n b_n \, w(\lambda - nT). \tag{6.142}$$

A very popular form of CPM is obtained if $h = 1/2$ and the frequency pulse $w(t)$ is the output of a filter with a Gaussian impulse response when the input is a rectangular pulse of duration T. If $h = 1/2$, but the frequency pulse is an unfiltered rectangular pulse, the result is MSK, as shown in Section 6.9.2. If the frequency pulse is a Gaussian-filtered rectangular pulse, the resulting CPM is referred to as *Gaussian MSK* (GMSK). GMSK has been employed in many digital communications systems, including the digital cellular communication system now known as Global System for Mobile Communication or GSM (originally named for *Groupe Spéciale Mobile*, the group that developed the standard for GSM) [6.18]. The sidelobes of the spectrum for GMSK fall off much faster than for MSK, a consequence of the smoothing action of the Gaussian filter.

The transfer function of a Gaussian filter is given by

$$H(f) = \exp(-\alpha f^2), \quad -\infty < f < \infty. \tag{6.143}$$

The corresponding impulse response is

$$h(t) = \sqrt{\pi/\alpha} \exp(-\pi^2 t^2/\alpha), \quad -\infty < t < \infty. \tag{6.144}$$

Notice that the transfer function is real, the filter is not causal, and the impulse response has infinite duration. In practice, a causal, finite-duration, approximation to the Gaussian filter is employed (e.g., by introducing a sufficient delay and truncating the impulse response). The Gaussian filter is usually characterized by its half-power bandwidth B, which is defined as the frequency for which the square of the transfer function is equal to one-half of its maximum value. That is, B is the solution to the equation $H^2(B) = H^2(0)/2$, since the maximum value of $H(f)$ is at $f = 0$. The result of solving this equation is

$$\alpha = \frac{\ln 2}{2B^2}.$$

If the input to the filter is

$$v_T(t) = p_T(t)/2T,$$

the output is

$$w(t) = \frac{1}{2T} \int_{t-T}^{T} h(\tau) \, d\tau. \tag{6.145}$$

The impulse response can be written as

$$h(\tau) = \frac{1}{\sqrt{2\pi}\sigma} \exp\left\{-\frac{\tau^2}{2\sigma^2}\right\},$$

where

$$\sigma^2 = \frac{\alpha}{2\pi^2} = \frac{\ln 2}{(2\pi B)^2}.$$

It follows that the frequency pulse for GMSK is given by

$$w(t) = \frac{1}{2T}\left[\Phi\left(\frac{2\pi Bt}{\sqrt{\ln 2}}\right) - \Phi\left(\frac{2\pi B(t-T)}{\sqrt{\ln 2}}\right)\right] \quad (6.146)$$

for $-\infty < t < \infty$. Of course, (6.146) can also be written in terms of the function Q by using the fact that $\Phi(x) = 1 - Q(x)$ for each real number x. As given in (6.146), the frequency pulse for GMSK has infinite duration. In practice, the pulse is truncated to have duration equal to LT for a fairly small integer L (e.g., 3 or 4). Typical values for B for GMSK are in the range $0.25/T$ to $0.3/T$ [6.20]. Notice that GMSK is a form of partial-response CPM.

References and Suggestions for Further Reading

[6.1] F. Amoroso, "The bandwidth of digital data signals," *IEEE Communications Magazine*, November 1980, pp. 13–24.

[6.2] J. B. Anderson, T. Aulin, and C.-E. Sundberg, *Digital Phase Modulation*, New York: Plenum, 1986.

[6.3] S. Benedetto, E. Biglieri, and V. Castellani, *Digital Transmission Theory*, Englewood Cliffs, NJ: Prentice-Hall, 1987.

[6.4] E. Biglieri, D. Divsalar, P. J. McLane, and M. K. Simon, *Introduction to Trellis-Coded Modulation with Applications*, New York: Macmillan, 1991.

[6.5] R. E. Blahut, *Principles and Practice of Information Theory*, Reading, MA: Addison-Wesley, 1987.

[6.6] R. E. Blahut, *Digital Transmission of Information*, Reading, MA: Addison-Wesley, 1990.

[6.7] T. M. Cover and J. A. Thomas, *Elements of Information Theory*, New York: Wiley, 1991.

[6.8] R. G. Gallager, *Information Theory and Reliable Communication*, New York: Wiley, 1968.

[6.9] S. W. Golomb (ed.), *Digital Communications with Space Applications*, Englewood Cliffs, NJ: Prentice-Hall, 1964.

[6.10] S. W. Golomb, *Shift Register Sequences*, San Francisco: Holden-Day, 1967.

[6.11] K. G. Beauchamp, *Applications of Walsh and Related Functions*, London: Academic Press, 1984.

[6.12] M. I. Irshid and I. S. Salous, "Bit error probability for coherent M-ary PSK systems," *IEEE Transactions on Communications*, vol. 39, no. 3, March 1991, pp. 349–352.

[6.13] E. C. Jordan (ed.), *Reference Data for Engineers: Radio, Electronics, Computer, and Communications*, 7th ed., Indianapolis: Howard Sams, 1985.

[6.14] P. J. Lee, "Computation of the bit error rate of coherent M-ary PSK with Gray code bit mapping," *IEEE Transactions on Communications*, vol. COM-34, no. 5, May 1986, pp. 488–491.

[6.15] M. K. Simon, S. M. Hinedi, and W. C. Lindsey, *Digital Communication Techniques*, Upper Saddle River, NJ: Prentice Hall, 1995.

[6.16] S. Pasupathy, "Minimum shift keying: A spectrally efficient modulation," *IEEE Communications Magazine*, July 1979, pp. 14–22.

[6.17] J. G. Proakis, *Digital Communications*, 4th ed., New York: McGraw-Hill, 2001.

[6.18] S. K. Redl, M. K. Weber, and M. W. Oliphant. *An Introduction to GSM*, Boston: Artech House, 1995.

[6.19] D. V. Sarwate and M. B. Pursley, "Crosscorrelation properties of pseudorandom and related sequences," *Proceedings of the IEEE*, vol. 68, May 1980, pp. 593–619.

[6.20] M. K. Simon, *Bandwidth-Efficient Digital Modulation with Application to Deep-Space Communications*, Monograph 3, Deep-Space Communications and Navigation Series (JPL Publication 00-17), June 2001.

[6.21] J. J. Stiffler, *Theory of Synchronous Communications*, Englewood Cliffs, NJ: Prentice-Hall, 1971.

[6.22] A. J. Viterbi, *CDMA: Principles of Spread Spectrum Communication*, Reading, MA: Addison-Wesley, 1995.

[6.23] C. L. Weber, *Elements of Detection and Signal Design*, New York: McGraw-Hill, 1968 [reprinted by Springer-Verlag, New York, 1987].

[6.24] J. M. Wozencraft and I. M. Jacobs, *Principles of Communication Engineering*, New York: Wiley, 1965 [reissued by Waveland Press, 1990].

Problems

6.1 Suppose that $a_0(t) = a_1(t) = p_T(t)$ as in Section 6.1.1, but (6.2) is replaced by

$$\theta_0(t) = \pi/2, \quad 0 \le t < T$$

and

$$\theta_1(t) = -\pi/2, \quad 0 \le t < T.$$

(a) Show that the signals described by (6.1) for $i = 0$ and $i = 1$ with the amplitude and phase modulation as defined are antipodal signals.

(b) Show that the signals of part (a) are BPSK signals. That is, show they are of the form given by (6.3) with φ replaced by some phase angle φ' that does not depend on t or on the value of i.

(c) Consider the signals described by (6.1) for $i = 0$ and $i = 1$ with

$$a_0(t) = a_1(t) = p_T(t).$$

Show that a necessary and sufficient condition for the phase modulation $\theta_0(t)$ and $\theta_1(t)$ to provide an antipodal signal set is

$$\theta_0(t) - \theta_1(t) = \pi \ (\text{mod } 2\pi), \quad 0 \le t < T.$$

That is, for each t, $\theta_0(t) - \theta_1(t) = \pi + 2\pi n$ for some integer n.

6.2 In Section 5.5, the notion of a vector space of functions is introduced, and the *norm* and *inner product* are defined for functions in this space. The waveforms that arise in the binary communication system can then be illustrated by two-dimensional sketches as in Figures 6-7(a) and 6-7(b), where we use this geometric approach to illustrate the projections of the signals $s_0(t)$ and $s_1(t)$ onto the reference signal $r(t)$. The geometric approach can also be applied to obtain an intuitive description of the operation performed by a correlation receiver for binary signaling on the AWGN channel. Let the sum of the received signal and the noise be $Y(t)$, and consider the receiver shown in Figure 5-16(c). In this receiver, the two correlators are used to form the statistics

$$W_0 = \int_{-\infty}^{\infty} Y(t)\, s_0(t)\, dt \quad \text{and} \quad W_1 = \int_{-\infty}^{\infty} Y(t)\, s_1(t)\, dt.$$

The optimum decision rule, as described in Section 5.5.4, is to decide that 0 was sent if $W_0 > W_1$, and to decide that 1 was sent if $W_1 > W_0$. (The event $W_1 = W_0$ has probability zero.) This rule decides in favor of the signal that has the larger correlation with the received waveform $Y(t)$. Written in vector notation, the decision rule is to decide that 0 was sent if $(Y, s_0) > (Y, s_1)$ and decide that 1 was sent if $(Y, s_1) > (Y, s_0)$. In parts **(a)** and **(b)**, assume that s_0 and s_1 have the same energy.

(a) Using the definition for the norm and inner product, show than an equivalent decision rule is to decide 0 was sent if $\|Y - s_0\| < \|Y - s_1\|$ and decide 1 was sent if $\|Y - s_0\| > \|Y - s_1\|$. Illustrate this fact geometrically. The *distance* between signals v_1 and v_2 is just $\|v_1 - v_2\|$, the norm of the difference between the two signals, so the conclusion is that the signal that has the largest correlation with the received waveform is the same as the signal that is closest to the received waveform.

(b) Apply the geometric viewpoint to the BPSK signal set defined by equations (6.3), and argue that an equivalent decision rule for this signal set is to determine the "phase" of the decision statistic and decide that 0 was sent if this phase is between $\varphi - (\pi/2)$ and $\varphi + (\pi/2)$ and 1 was sent if the phase is not in this range. In order to make this precise, how must the "phase" of the decision statistic be defined? *Hint:* The signals can be expressed as linear combinations of $\cos(\omega_c t)$ and $\sin(\omega_c t)$, and the operation of a correlation receiver matched to $\cos(\omega_c t + \varphi)$ can be expressed in terms of these two orthogonal functions. (What are the coefficients?) Assume that $\omega_c T$ is a multiple of 2π.

6.3 This problem deals with the phase ambiguity in the squaring loop described in Section 6.2.2.

(a) Ignore the effects of noise at the input to the squaring loop. Show that if the output of the VCO is

$$r(t) = \sqrt{2}\,\beta \cos(\omega_c t + \hat{\varphi})$$

and either $\hat{\varphi} = \varphi$ or $\hat{\varphi} = \varphi + \pi$, the loop is locked (i.e., the input to the VCO in the loop is zero). Thus, the loop "cannot tell" if there is a π-radian phase error in the output of the VCO.

(b) Suppose $r(t)$ is employed as a reference signal in a correlation receiver and the input to the receiver is $s_i(t) + X(t)$, as shown in Figure 6-6. Let Z be the decision statistic that results if $\hat{\varphi} = \varphi$. Show the decision statistic that results if $\hat{\varphi} = \varphi + \pi$ is $-Z$, so all decisions made by comparison with a zero threshold are reversed.

6.4 The block diagram of another type of tracking loop for BPSK demodulation is shown in Figure 6-51. Assume that the baseband signal $a(t)$ is a sequence of rectangular pulses, each of unit amplitude and each with either positive or negative polarity; that is,

$$a(t) = d_0 \, p_T(t) + d_1 \, p_T(t - T) + \cdots + d_N \, p_T(t - NT),$$

where, for each i, d_i is $+1$ or -1. As in the consideration of the squaring loop, ignore the effects of noise. Notice that $a^2(t) = 1$ for all t in the range $0 \le t < (N+1)T$.

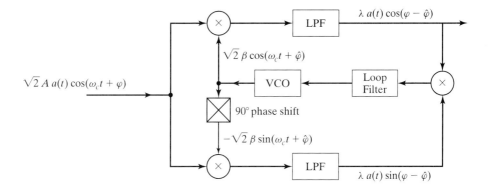

Figure 6-51: Costas loop for BPSK signals.

(a) Show that the Costas loop accomplishes the same objectives as the squaring loop. In particular, show that the input to the VCO is proportional to $\sin[2(\varphi - \hat{\varphi})]$. Thus, if $\hat{\varphi} > \varphi$, the phase of the output of the VCO will be decreased, and if $\hat{\varphi} < \varphi$, the phase of the output of the VCO will be increased.

(b) Show that regardless of the value of the phase error $\varphi - \hat{\varphi}$, the loop output is a binary, antipodal, baseband signal.

(c) Does the Costas loop also have a phase ambiguity?

6.5 This problem deals with the approximate doubling of the error probability at high signal-to-noise ratios for a certain approach to the coherent demodulation of differentially encoded BPSK. The transmitted signal during the interval $(-T, 0)$ is

$$\sqrt{2} \, d_{-1} A \cos(\omega_c t + \varphi),$$

and the transmitted signal during the interval $(0, T)$ is

$$\sqrt{2} \, d_0 A \cos(\omega_c t + \varphi).$$

The data variable d_i is either -1 or $+1$. Assume that $\omega_c T$ is a multiple of 2π, the channel is an additive white Gaussian noise channel, and a perfect phase reference is available. The differential encoding scheme used in this problem is as follows. If 0 is to be sent, $d_0 = +d_{-1}$, and if 1 is to be sent, $d_0 = -d_{-1}$. Assume that both d_{-1} and d_0 are unknown to the receiver. Suppose the optimum correlation receiver is used in each of the intervals $(-T, 0)$ and $(0, T)$ to make decisions on the values of d_{-1} and d_0. This amounts to optimum coherent demodulation of BPSK in each of the individual intervals, so the probability of error in each interval is

$$p = Q\left(\sqrt{2\,\mathcal{E}/N_0}\,\right),$$

where $\mathcal{E} = A^2 T$ is the signal energy in each of the two intervals. The receiver makes decisions $\hat{d}_{-1}$ and $\hat{d}_0$ in the first and second intervals, and it decides that 0 was sent if $\hat{d}_0 = \hat{d}_{-1}$ and that 1 was sent if $\hat{d}_0 = -\hat{d}_{-1}$.

(a) Suppose that 0 is sent, and show that the probability of error is $P_{e,0} = 2p(1 - p)$.

(b) Show that the error probability is the same if 1 is sent.

(c) Show that if $\mathcal{E}/N_0$ is large, the error probability in **(a)** is approximately $2p$. This represents an approximate doubling of the error probability as compared with coherent reception of a BPSK signal that does not use differential encoding.

6.6 Give a block diagram for the optimum coherent correlation receiver for binary AM. Use the minimax criterion, and specify completely all components of this receiver. A good starting point is to observe that the signal set $\{s_0, s_1\}$ for binary AM is defined by

$$s_i(t) = \sqrt{2}\,A\,\beta_i(t)\cos(\omega_c t + \varphi),$$

for $i = 0$ and $i = 1$, so that

$$s_0(t) - s_1(t) = \sqrt{2}\,A[\beta_0(t) - \beta_1(t)]\cos(\omega_c t + \varphi).$$

Specialize your answer for the signal set of Figure 6-11.

6.7 Suppose the threshold for the receiver shown in Figure 6-12 is denoted by γ. The channel is an AWGN channel with spectral density $N_0/2$. Give expressions for $P_{e,0}$ and $P_{e,1}$ if the binary ASK signal set is

$$s_i(t) = \sqrt{2}\,A\,u_i\,p_T(t)\cos(\omega_c t + \varphi)$$

for $i = 0$ and $i = 1$ (i.e., $\beta(t) = p_T(t)$). Assume that $\omega_c \gg T^{-1}$ so that the double-frequency terms can be neglected. Express your answers in terms of the complementary Gaussian distribution function Q and the parameters u_0, u_1, γ, A, T, and N_0.

6.8 Consider the 4-ASK signal set of Example 6-3(a). Assume that the receiver is the optimum correlation receiver (shown in Figure 6-14), and the channel is an AWGN channel. Find expressions that define the *minimax* decision regions. You may not be able to obtain *closed-form* expressions for all of the parameters of these regions, but you should at least be able to obtain expressions in terms of tabulated functions that define these parameters implicitly.

6.9 The signal set for this problem is the 4-ASK signal set of Example 6-3(b).

  (a) Give expressions for the symbol error probabilities in terms of the function Q and the parameters $\mathcal{E}$ and N_0 if the maximum-likelihood receiver is employed.

  (b) Show that these expressions are the same as those for the corresponding symbol error probabilities for the signal set of Example 6-3(a) with its maximum-likelihood receiver.

  (c) Now express your answers to part **(a)** in terms of the average energy $\bar{\mathcal{E}}_s$ and the maximum energy $\mathcal{E}_{max}$, and show that the resulting symbol error probabilities for the signal set of Example 6-3(b) are larger, for a given value of N_0 and for the same values of $\bar{\mathcal{E}}_s$ and $\mathcal{E}_{max}$, than the corresponding symbol error probabilities for the signal set of Example 6-3(a). Assume that the maximum-likelihood receiver is employed for each signal set. This will show that, if the constraint is on either the average energy or the maximum energy, the signal set of Example 6-3(a) is superior to that of Example 6-3(b) when each is used in conjunction with a maximum-likelihood receiver.

6.10 Suppose that the receiver of Figure 6-14 is used for the M-ASK signal set of Exercise 6-2, and the maximum-likelihood decision regions are employed. Show that

$$P_{e,0} = P_{e,M-1} = Q\left(\sqrt{2A^2\mathcal{E}_\beta/N_0}\right)$$

and that

$$P_{e,i} = 2\,Q\left(\sqrt{2A^2\mathcal{E}_\beta/N_0}\right)$$

for $1 \le i \le M - 2$. Find the average probability of symbol error that results if $\pi_i = 1/M$ for each i.

6.11 Consider M-ASK signals that are not necessarily equally spaced. That is, the signals are of the form

$$s_i(t) = \sqrt{2}\,A\,u_i\,\beta(t)\cos(\omega_c t + \varphi),$$

with $u_0 < u_1 < \ldots < u_{M-1}$, and the differences $\delta_i = u_i - u_{i-1}, 1 \le i \le M - 1$, are not necessarily all equal. Give an expression for $P_{e,i}$, the conditional probability of symbol error, in terms of the parameters $d_0, d_1, \ldots, d_{M-1}$ that are defined by (6.32) in Section 6.4.3.

6.12 A digital communication system uses QASK with ternary modulation on the inphase and quadrature components. Specifically, the transmitted signal is of the form

$$s(t) = A\{a_1(t)\cos(\omega_c t + \varphi) + a_2(t)\sin(\omega_c t + \varphi)\},$$

where $a_1(t)$ and $a_2(t)$ are sequences of rectangular pulses, each of duration T. The possible pulse amplitudes for each of the signals $a_1(t)$ and $a_2(t)$ are $-1, 0$, and $+1$. The parameters A and φ are fixed quantities that are known to the receiver, and the channel is an additive white Gaussian noise (AWGN) channel.

  (a) What data rate is possible with this signaling scheme?

  (b) Give a block diagram for an optimum coherent receiver for this signal and channel.

(c) Give *expressions* that specify the minimax decision regions or thresholds for the coherent receiver in part (b). The expressions should be written in terms of the standard Gaussian distribution function Φ and simplified as much as possible, but it is not necessary to obtain explicit solutions.

6.13 The transmitted signal in a 16-QASK communication system is given by

$$s(t) = A\,a_1(t)\cos(\omega_c t + \varphi) - A\,a_2(t)\sin(\omega_c t + \varphi).$$

The signals $a_1(t)$ and $a_2(t)$ are pulses that have duration T and amplitudes from the set $\{-3, -1, +1, +3\}$; that is, $a_1(t) = u_i\,p_T(t)$ for some i, $0 \leq i \leq 3$ and $a_2(t) = v_j\,p_T(t)$ for some j, $0 \leq j \leq 3$, where

$$\{u_i : 0 \leq i \leq 3\} = \{v_j : 0 \leq j \leq 3\} = \{-3, -1, +1, +3\}.$$

The channel is an additive white Gaussian noise channel with spectral density $N_0/2$.

(a) Give a block diagram for the maximum-likelihood receiver for this system.

(b) Specify all parameters of the maximum-likelihood receiver (sampling times, integration intervals, gain constants, thresholds, decision regions, etc.).

(c) Describe in detail the decision device in the maximum-likelihood receiver.

(d) Give expressions for the symbol error probabilities $P_{e,i}$ for $0 \leq i \leq 15$. Does the probability of symbol error depend on which symbol was actually transmitted?

(e) What can you say about the *bit error probability* that results if this system and the maximum-likelihood receiver are used to send binary information? Discuss bounds and asymptotic results (large signal-to-noise ratio) for the bit error probability. Is there a preferred assignment of binary sequences to the 16-ary symbols? If so, what is it?

6.14 A 4-ary signal set is defined as follows:

$$
\begin{aligned}
s_0(t) &= 0, & 0 \leq t \leq T; \\
s_1(t) &= A\sin(\omega_c t + \varphi), & 0 \leq t \leq T; \\
s_2(t) &= A\cos(\omega_c t + \varphi), & 0 \leq t \leq T; \\
s_3(t) &= A\cos(\omega_c t + \varphi) + A\sin(\omega_c t + \varphi), & 0 \leq t \leq T.
\end{aligned}
$$

All four signals are identically zero outside the interval $0 \leq t \leq T$. Assume that $\omega_c T = 2\pi n$ for some integer n. The channel is an additive white Gaussian noise channel with spectral density $N_0/2$. The receiver is a maximum-likelihood receiver for this signal set and channel.

(a) Give a sketch of a geometric representation for this signal set (i.e., the signal constellation).

(b) Find the symbol error probabilities for this system; that is, for each value of i, find $P_{e,i}$, the conditional probability of symbol error given that signal $s_i(t)$ is transmitted. Express your answers in terms of the function Q and the parameters A, T, and N_0.

(c) A binary data sequence is to be transmitted using this signal set, and the assignment of bits to signals is as follows: 00 is represented by $s_0(t)$, 01 by $s_1(t)$, 10 by $s_2(t)$, and 11 by $s_3(t)$. What is the resulting *bit* error probability? Express your answers in terms of the function Q and the parameters A, T, and N_0.

(d) Does the bit error probability depend on which data bits are transmitted? Explain why or why not.

(e) If the assignment of bits to the four signals is changed, do the bit error probabilities change? Explain why or why not.

6.15 Consider the symbol error probabilities for regular QASK signal constellations. Show that if $\mathcal{P}_i$ denotes the probability of symbol error for each interior point, $\mathcal{P}_c$ the probability of symbol error for each corner point, and $\mathcal{P}_o$ the probability of symbol error for other exterior points, then $\mathcal{P}_i > \mathcal{P}_o > \mathcal{P}_c$.

6.16 Suppose the density for the Gaussian random vector $\mathbf{Z} = (Z_1, Z_2)$ is given by

$$f_{Z_1.Z_2}(x, y) = \exp\{-[(x - \mu)^2 + (y - \nu)^2]/2\sigma^2\}/2\pi\sigma^2,$$

$$-\infty < x < \infty, \quad -\infty < y < \infty.$$

Consider an infinite line in two-dimensional space that is distance ε from the point (μ, ν) as illustrated in Figure 6-52. The set Γ is defined to be the set of all points that are on the same side of the line as the point (μ, ν). Find the probability that the random vector $\mathbf{Z}$ is in the set Γ. Notice that the probability in question does not change if the line is rotated about the point (μ, ν), because the given Gaussian density is invariant with respect to such a rotation. (It has circular symmetry about the point (μ, ν).)

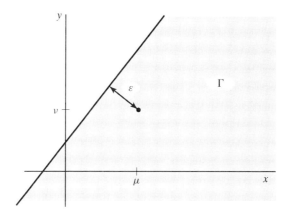

Figure 6-52: Illustration for Problem 6.16.

6.17 The symbol error probabilities are given in equations (6.56) for regular QASK signal constellations.

(a) Show that if $\delta/2\sigma$ is sufficiently large, the approximate probability of symbol error is

$$P_{e.n} \approx 4\, Q(\delta/2\sigma)$$

for the interior points in the signal constellation,

$$P_{e.n} \approx 3\, Q(\delta/2\sigma)$$

for all exterior points except the corner points, and

$$P_{e,n} \approx 2\,Q(\delta/2\sigma)$$

for the corner points.

(b) Give an intuitive explanation for these expressions by examining the decision regions for a regular QASK signal constellation (e.g., Figure 6-33).

(c) Refine your argument in part **(b)** to explain the exact expressions given in equations (6.56).

6.18 Apply the methods of Exercise 6-2 to a standard M-QASK signal constellation for which M is an even power of 2.

(a) Show that the average energy per symbol is

$$\bar{\mathcal{E}}_s = (M-1)d^2/6$$

and the average energy per bit is

$$\bar{\mathcal{E}}_b = (M-1)d^2/(6\log_2 M).$$

(b) Show that the probability $Q\!\left(d/\sqrt{2N_0}\,\right)$ can be written as

$$
\begin{aligned}
Q\!\left(d/\sqrt{2N_0}\,\right) &= Q\!\left(\sqrt{3\,\bar{\mathcal{E}}_s/[(M-1)N_0]}\,\right) \\
&= Q\!\left(\sqrt{3\,\bar{\mathcal{E}}_b\log_2(M)/[(M-1)N_0]}\,\right).
\end{aligned}
$$

(c) Give an expression for $\mathcal{E}_{\max}$, the maximum energy for the signals in the standard M-QASK signal constellation, and give an expression for the ratio $\mathcal{E}_{\max}/\bar{\mathcal{E}}_s$. Show that the ratio can be written as $c_1(\sqrt{M}+c_2)/(\sqrt{M}+c_3)$, and fill in the values for c_1, c_2, and c_3. Compare with the result obtained for M-ASK in Exercise 6-2. What is the limiting value of $\mathcal{E}_{\max}/\bar{\mathcal{E}}_s$ for M-QASK as $M \to \infty$?

6.19 Consider Figure 6-34, and suppose that the point (μ, ν) is not in the set Γ_n. Is (6.54) still valid? Explain how this observation can be used to compute the conditional probability that the receiver decides signal s_n is transmitted given that signal s_m was transmitted and $n \neq m$.

6.20 Consider an interior point of a regular M-QASK signal constellation for a very large value of M. Let this interior point correspond to a particular signal s_n. Suppose that the nearest neighbors of the point are also interior points. In parts **(b)–(d)**, express your answers in terms of the function Q and the parameters d and N_0.

(a) If the distance between nearest neighbors is d, show that the distance between the point that corresponds to s_n and the closest exterior point is at least $3d$.

(b) Suppose the point that corresponds to signal s_m is a nearest neighbor of the point that corresponds to signal s_n. Use the idea conveyed in Problem 6.19 to find the conditional probability that the maximum-likelihood receiver decides that signal s_m was transmitted, given that signal s_n was actually transmitted.

(c) Continue this approach to find the probability that the receiver decides in favor of one of the nearest neighbors, and compare this with the expression for the symbol error probability for the interior point.

(d) Show that for large signal-to-noise ratio, the most likely error event is that the receiver decides in favor of a nearest neighbor rather than one of the other points in the signal constellation.

6.21 Let $A(\omega)$ be the Fourier transform of $\alpha(t)$. Assume that the bandwidth of $A(\omega)$ is small compared with ω_c, so that $A(\omega - \omega_c) A(\omega + \omega_c) = 0$ for all ω. Use the modulation theorem of Fourier transforms to show that the bandwidth of the signal $w_1(t) = u \, \alpha(t) \cos(\omega_c t) - v \, \alpha(t) \sin(\omega_c t)$ depends on $A(\omega)$, but not on the parameters u, v, and ω_c. Let $w_2(t) = \alpha(t) \cos(\omega_c t)$ and show that $w_1(t)$ and $w_2(t)$ have the same bandwidth for any choice of u and v, and explain why this conclusion does not depend on how the bandwidth is defined (e.g., half-power bandwidth, rms bandwidth, equivalent noise bandwidth, etc.).

6.22 Communication systems A and B each employ a signal set that has four signals. Each system transmits data over an additive white Gaussian noise channel with two-sided spectral density $N_0/2$, and coherent maximum-likelihood reception is employed. System A employs 4-ASK modulation with signals of the form

$$s(t) = \sqrt{2} \, A_1 u \, p_{T_1}(t) \cos(\omega_c t + \varphi).$$

The data variable u takes values in the set $\{-3, -1, +1, +3\}$. System B employs 4-QASK with signals of the form

$$s(t) = A_2 u \, p_{T_2}(t) \cos(\omega_c t + \varphi) - A_2 v \, p_{T_2}(t) \sin(\omega_c t + \varphi).$$

Each of the variables u and v takes values in the set $\{-1, +1\}$. The parameters T_1 and T_2 are selected so that each system has the same data rate when measured in bits per second. The amplitudes A_1 and A_2 are selected so that the signals for each system have the same *average* energy per bit. Note that the four signals used in System A do not all have the same energy.

(a) For each of the two signal sets, give a block diagram of the optimum coherent receiver for maximum-likelihood decisions. Specify all components carefully.

(b) If the two systems are used to transmit data at the same information rate, what is the relationship between T_1 and T_2? What are the relative bandwidth requirements for the systems?

(c) If the systems have the same data rate, what is the relationship between A_1 and A_2 that guarantees the two signals have the same average energy per bit?

(d) Determine the symbol error probabilities for each of the four signals in each of the systems if the two systems have the same data rate and the same average energy per bit. Express your answers in terms of the function Q and the parameters A_1, T_1, and N_0.

6.23 This is a classical problem that arises in coherent frequency-shift-key (FSK) communication systems. The binary signals are given by

$$s_0(t) = \sqrt{2} \, A \cos(\omega_0 t + \varphi), \quad 0 \le t \le T,$$

and

$$s_1(t) = \sqrt{2}\,A\cos(\omega_1 t + \varphi), \quad 0 \le t \le T,$$

and the receiver employs maximum-likelihood coherent demodulation. Assume that $\omega_1 > \omega_0$, and ignore the effects of any high-frequency signal components (e.g., ignore the effects of any components at frequency $\omega_1 + \omega_0$). What is the optimum value for the frequency separation $\omega_1 - \omega_0$? *Hint:* For some choices of $\omega_1 - \omega_0$, the two signals are orthogonal (i.e., $r = 0$), but it is possible to obtain a better correlation coefficient for other choices of $\omega_1 - \omega_0$.

6.24 A binary signal set is defined by

$$s_0(t) = A\sin(\omega_0 t + \theta_0 + \theta)\,p_T(t)$$

and

$$s_1(t) = A\sin(\omega_0 t + \pi + \theta_1 + \theta)\,p_T(t),$$

where θ is an arbitrary phase angle that is *known* to the receiver. Note the absence of the normalization factor $\sqrt{2}$ in the expressions for the signals. Assume that $\omega_0 T$ is a multiple of 2π. The receiver is as shown in Figure 6-1. Assume that the sampling time is $T_0 = T$ and the threshold is the minimax threshold for the given signal set. The noise $X(t)$ is white Gaussian noise with spectral density $N_0/2$.

The phase angles θ_0 and θ_1 represent phase shifts that result from imperfections in the oscillators and the communication medium. Under ideal conditions, each of these two phase angles is 0, but because of the imperfections, all we can say is that $|\theta_i| < \pi/2$ for each value of i ($i = 0, 1$). Find the error probabilities $P_{e,0}$ and $P_{e,1}$ for each of the following situations:

(a) The phase angles θ_0 and θ_1 are *known* to the receiver, and the filter is matched to the signal set $\{s_0, s_1\}$ defined at the beginning of this problem.

(b) The phase angles θ_0 and θ_1 are *not* known to the receiver, and so the filter is matched to the signal set $\{v_0, v_1\}$ defined by

$$v_0(t) = A\sin(\omega_0 t + \theta)\,p_T(t)$$

and

$$v_1(t) = A\sin(\omega_0 t + \pi + \theta)\,p_T(t).$$

If we think of θ_0 and θ_1 as each having a nominal value of 0, the filter in part **(b)** corresponds to setting the unknown phase angles to their nominal values and matching the receiver to the resultant signal set. Unfortunately, the phase angles do not always take on their nominal values, and so the receiver in part **(b)** is not always matched to the signals s_0 and s_1 that are actually transmitted. As a result, the answer to part **(b)** is a function of the actual values θ_0 and θ_1.

6.25 Consider the BPSK signals defined by equations (6.5). The receiver is the correlation receiver of Figure 6-6 with an imperfect phase reference $\hat{\varphi}$. Let $\theta = \varphi - \hat{\varphi}$ be the phase error for this receiver. The channel noise is white Gaussian noise with spectral density $N_0/2$, and the decision threshold is $\gamma = 0$. It is known that there is a noncoherent receiver that provides error probability $P_e = 0.5 \exp(-\mathcal{E}/N_0)$, where $\mathcal{E}$ is the energy in each of the two signals. The required error probability for the system is 10^{-5}. Over what range of values of the phase error θ does the correlation receiver give better performance (i.e., permit a smaller value of $\mathcal{E}/N_0$) than this noncoherent receiver? You may wish to use the fact that $Q^{-1}(10^{-5}) \approx 4.27$.

6.26 The BPSK system model shown in Figure 6-53 is for a specular multipath channel that has two propagation paths from the transmitter to the receiver. The propagation loss on the second path is a random variable R, which has density function

$$f_R(r) = (r/\lambda^2)\exp\{-r^2/2\lambda^2\}, \quad r \geq 0,$$

and $f_R(r) = 0$ for $r < 0$. (This is the Rayleigh density with parameter λ.) The phase shift introduced by the second path is a random variable Θ that is independent of R and uniformly distributed on the interval $[0, 2\pi]$.

If the received signal on the first path is

$$s(t) = A b\, p_T(t) \cos(\omega_c t + \varphi),$$

then the received signal on the second path is given by

$$\rho(t) = R A b\, p_T(t) \cos(\omega_c t + \varphi + \Theta).$$

The symbol b represents binary data, so b is $+1$ or -1, depending on whether a 0 or 1 is sent. The noise $X(t)$ is white Gaussian noise with spectral density $N_0/2$, and it is independent of R and Θ.

- **(a)** Give an expression for $P_e(r, \theta)$, the conditional probability of error given $R = r$ and $\Theta = \theta$. Your answer should be in terms of r, θ, A, T, λ, N_0, and the function Q.

- **(b)** Give an integral expression for the average probability of error $\overline{P}_e = E\{P_e(R,\Theta)\}$.

- **(c)** Evaluate your expression in part **(b)** to find the average probability of error in terms of A, T, λ, N_0, and the function Q. It may be beneficial to change from polar coordinates to rectangular coordinates in order to evaluate the integrals. Interpretations of some of the integrals in terms of probabilities may facilitate their evaluations.

- **(d)** Show that the same value of $\overline{P}_e$ results if the signal on the second path is replaced by white Gaussian noise with an appropriate spectral density. Find the value of this spectral density.

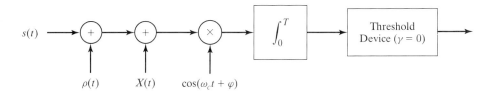

Figure 6-53: System model for Problem 6.26.

6.27 Consider seven signals constructed from the following sequences:

$$1110010$$
$$1100101$$
$$1001011$$
$$0010111$$
$$0101110$$
$$1011100$$
$$0111001$$

These sequences are generated by the shift register of Figure 6-40. The continuous-time signals are obtained from the sequences according to (6.65) of Section 6.6.1. Refer to these seven signals as $s_0, s_1, \ldots, s_6$, where s_0 is the signal corresponding to the first sequence in the list, s_1 to the second sequence, etc. Each signal has duration T, and each consists of a sequence of chips of duration T_c. The signals form a 7-ary signal set for communication over an AWGN channel with spectral density $N_0/2$. The receiver consists of a filter with impulse response $h(t)$, a sampler which samples the filter output at seven different times, and a decision device that bases decisions on the largest sample.

Construct the impulse response for a continuous-time filter that is matched to the sequence 1110010111001 . First obtain the continuous-time signal $s(t)$ from the given sequence as described in Section 6.6.1, and then let the impulse response of the filter be $h(t) = s(T_0 - t)$, $-\infty < t < +\infty$, where T_0 is an arbitrary time reference. Each of the chips for this filter has duration T_c, so $h(t)$ has duration $13T_c = 13T/7$. The value of T_0 is not critical: $T_0 = 0$ or $T_0 = 13T_c$ are acceptable choices. ($T_0 = 13T_c$ provides a causal filter.)

(a) Find the output of the filter for each of the seven signals $s_0, s_1, \ldots, s_6$.

(b) Find the optimum sampling time for each of the seven signals.

(c) How might you use the features demonstrated in part (b) to make symbol decisions?

(d) Consider binary signaling using s_0 and s_1 only. The receiver samples the filter output at the optimum times for signals s_0 and s_1 (from part (b)). Find $P_{e,i}$ for $i = 0, 1$, and compare with the error probabilities for the optimum receiver.

(e) Consider binary signaling using s_0 and s_2 only. Find $P_{e,0}$ and $P_{e,2}$ and compare your answers with the error probabilities for the optimum receiver.

6.28 The transmitted signal in a 20-QASK communication system is given by

$$s(t) = A u_i \cos(\omega_c t + \varphi) - A v_j \sin(\omega_c t + \varphi), \quad 0 \le t \le T.$$

for $0 \le i \le 3$ and $0 \le j \le 4$. The data variables for the inphase component of the signal are from the set $\{-3, -1, +1, +3\}$, and the data variables for the quadrature component are from the set $\{-2, -1, 0, +1, +2\}$; that is,

$$\{u_i : 0 \le i \le 3\} = \{-3, -1, +1, +3\}$$

and

$$\{v_j : 0 \le j \le 4\} = \{-2, -1, 0, +1, +2\}.$$

The channel is an additive white Gaussian noise channel with spectral density $N_0/2$, and the maximum-likelihood receiver is employed.

(a) Provide a sketch similar to Figure 6-23 that illustrates the two-dimensional signal constellation for this 20-QASK signal set.

(b) Consider the probability of symbol error for each of the 20 symbols. How many different values are there for the symbol error probability among these 20 symbols? That is, how many different values are in the set $\{P_{e,n} : 0 \le n \le 19\}$?

(c) Label the points in your sketch of the signal constellation according to their symbol error probabilities. For example, label all those signals with the smallest symbol error probability with an A, those with the next smallest symbol error probability with a B, etc.

(d) Give an expression for the probability of symbol error for each of the 20 symbols. Give your answer in terms of the classification suggested in part (c); that is, give an expression for the symbol error probability for all signals in class A, class B, etc. Express your answers in terms of the parameters A, T, and N_0.

6.29 An 8-ASK signal set is employed for coherent communications over an AWGN channel with spectral density $N_0/2$. The eight signals are given by

$$s_i(t) = \sqrt{2}\, A\, u_i\, p_T(t) \cos(2\pi f_c t + \varphi)$$

for $u_0 = -7$, $u_1 = -5$, $u_2 = -3$, $u_3 = -1$, $u_4 = +1$, $u_5 = +3$, $u_6 = +5$, and $u_7 = +7$. Ignore double-frequency terms, and assume that the maximum-likelihood coherent receiver is employed.

(a) How many distinct values are there in the set $\{P_{e,i} : 0 \le i \le 7\}$?

(b) List all of the values of j for which $P_{e,j} = \max\{P_{e,i} : 0 \le i \le 7\}$.

(c) List all of the values of j for which $P_{e,j} = \min\{P_{e,i} : 0 \le i \le 7\}$.

(d) Suppose $10 \log_{10}(A^2 T/N_0) = 10.0$. Use the following table to determine the value of $P_{e,0}$, the conditional probability of symbol error given that s_0 is transmitted:

$(\mathcal{E}/N_0)_{\mathrm{dB}}$	$Q(\sqrt{2\,\mathcal{E}/N_0}\,)$	$Q(\sqrt{\mathcal{E}/N_0}\,)$	$\frac{1}{2}\exp(-\mathcal{E}/2N_0)$
5.0	5.95×10^{-3}	3.77×10^{-2}	1.03×10^{-1}
6.0	2.39×10^{-3}	2.30×10^{-2}	6.83×10^{-2}
7.0	7.73×10^{-4}	1.26×10^{-2}	4.08×10^{-2}
8.0	1.91×10^{-4}	6.00×10^{-3}	2.13×10^{-2}
9.0	3.36×10^{-5}	2.41×10^{-3}	9.42×10^{-3}
10.0	3.87×10^{-6}	7.83×10^{-4}	3.37×10^{-3}
11.0	2.61×10^{-7}	1.94×10^{-4}	9.23×10^{-4}
12.0	9.01×10^{-9}	3.43×10^{-5}	1.81×10^{-4}

(e) Suppose $10 \log_{10}(A^2 T/N_0) = 8.0$. Use the table to determine the value of $P_{e,4}$, the conditional probability of symbol error given that s_4 is transmitted.

(f) If this 8-ASK signal set is used to transmit binary data at an information rate of 24 Mb/s and the carrier frequency is 950 MHz, what is the null-to-null bandwidth of the transmitted signal? Draw a sketch of the spectrum of the signal and label the important points on the frequency axis.

6.30 A QPSK signal set is employed for coherent communications over an AWGN channel with spectral density $N_0/2$. The QPSK signals are given by

$$s_i(t) = \sqrt{2}\, A\, p_T(t) \cos(2\pi f_c t + \theta_i + \varphi), \quad 0 \le i \le 3,$$

for $\theta_0 = \pi/4$, $\theta_1 = 3\pi/4$, $\theta_2 = 5\pi/4$, and $\theta_3 = 7\pi/4$. Ignore double-frequency terms, and assume that the maximum-likelihood coherent receiver is employed. This QPSK signal set is used to transmit binary data, and the assignment of binary digits to the different signals is as follows: $00 \to s_0$, $10 \to s_1$, $11 \to s_2$, and $01 \to s_3$.

    (a) Suppose $10 \log_{10}[A^2 T/(2N_0)] = 11.0$. Use the table for Problem 6.29 to evaluate the probability of bit error for this QPSK signal set.

    (b) Suppose $10 \log_{10}[A^2 T/(2N_0)] = 6.0$. Use the table for Problem 6.29 to give a numerical expression for the symbol error probabilities $P_{e,i}$, $0 \le i \le 3$, for this QPSK signal set. You need not evaluate the numerical expression, but you must substitute numerical values for any parameters in your expression (i.e., there should be no symbols or variables in the expression, just numbers).

6.31 Several of the expressions for the probability of a correct decision in the reception of M-ary orthogonal signals can be expressed in the form

$$P_c = \int_{-\infty}^{\infty} [F_1(x)]^{M-1} f_2(x)\, dx,$$

where F_1 is a distribution function and f_2 is a density function. Among the expressions that are in this form are (6.69) and (6.71).

    (a) Show that the preceding equation for P_c is equivalent to

$$P_c = 1 - (M-1) \int_{-\infty}^{\infty} [F_1(x)]^{M-2} F_2(x) f_1(x)\, dx,$$

    where f_1 is the density function that corresponds to the distribution function F_1 and F_2 is the distribution function that corresponds to the density function f_2. *Hint*: Use integration by parts.

    (b) Apply this result to (6.71), and obtain an expression for the probability of error. Make a change in the variable of integration, if necessary, to show this expression is equivalent to (6.72).

6.32 Consider the two signal sets illustrated in Figure 6-54. Assume that a maximum-likelihood receiver is used for each signal set, and the channel is an additive white Gaussian noise channel with two-sided spectral density $N_0/2$. The distance between pairs of signals is d for each set. The signals in Set 1 are equal-energy, orthogonal, three-dimensional signals. The signals in Set 2 are equal-energy, two-dimensional signals. Let $\mathcal{E}_1$ denote the energy for each signal in Set 1, and let $\mathcal{E}_2$ denote the energy for each signal in Set 2.

Fact: Since the distance between signals is the same for the two sets, the symbol error probability for maximum-likelihood reception is also the same for the two sets.

    (a) Give an expression for d in terms of $\mathcal{E}_1$.

(b) For Set 1, use (6.72) in Section 6.6.3 and your result in part **(a)** to obtain an expression for the probability of symbol error in terms of d.

(c) Use simple trigonometry to find an expression for $\mathcal{E}_2$ in terms of d. *Hint*: d is the length of each side of the equilateral triangle whose vertices are the points in the constellation for Set 2. From this hint, it is easy to determine various angles that may be useful in the evaluation of $\mathcal{E}_2$.

(d) Use the fact in the problem statement and your results from parts **(b)** and **(c)** to obtain an expression for the probability of symbol error in terms of $\mathcal{E}_2$ for Set 2. For a given symbol error probability, which set requires less energy?

(e) Give a numerical value for the correlation coefficient r for pairs of signals in Set 2. What type of signal set is Set 2?

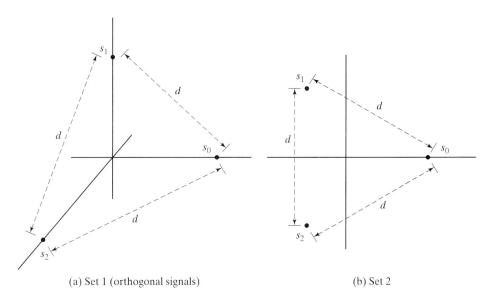

(a) Set 1 (orthogonal signals)   (b) Set 2

Figure 6-54: Signal sets for Problem 6.32.

Chapter 7

Noncoherent Communications

It is possible to communicate reliably using a carrier-modulated signal even if the receiver cannot estimate the carrier's phase with enough accuracy to permit use of the demodulation schemes discussed in Chapter 6. In this chapter, consideration is given to receivers that do not attempt to estimate the phase of the received signal.

The election to use noncoherent communications in a given application may result from a desire to avoid the need for a special subsystem, such as a squaring loop, to estimate the phase of the carrier. In general, noncoherent receivers can be less complex and less expensive than coherent receivers, and so the desire for a simpler, cheaper receiver often leads to the use of noncoherent demodulation. For some applications, it may be that the carrier for the received signal is not sufficiently stable to permit reliable estimation of its phase. The latter situation can arise when communicating on channels with severe fading or when there is excessive phase noise in the transmitter's oscillator. An extreme example of a system for which accurate estimation of the phase of the carrier is difficult (in fact, virtually impossible) is a frequency-hop communication system in which the frequency and phase of the carrier are changed intentionally at the beginning of each data symbol interval.

We begin by presenting the optimum receiver for general binary noncoherent communications. Attention is then focused on binary frequency-shift keying, a particularly important signal set for binary noncoherent communications, and the optimum receiver is developed for this signal set. Consideration is then given to suboptimum filters for binary noncoherent communications. The optimum decision device is derived for both optimum and suboptimum filters, and the resulting error probabilities are determined. Alternative implementations that use envelope detectors are described. We analyze the performance of systems that use nonbinary orthogonal signal sets for noncoherent communications, and evaluate the effects of nonselective fading.

7.1 The Optimum Receiver for Binary Noncoherent Communications

In one sense, there is a similarity between the signals considered for coherent communications in Section 6.1 and the signals considered for noncoherent communications in this section. In both coherent and noncoherent communications, the most general signals that we consider can be represented as a combination of amplitude and phase modulation. There is one important difference, however, and this difference is due to the fact that in a noncoherent communication system, the phase of the received signal is not known to the communications receiver, and no attempt is made to estimate this phase.

Certain signals that are quite good for coherent communication are totally unacceptable for noncoherent communications. For example, the two signals in a BPSK signal set are identical except for their phases: The phase of one signal differs from the phase of the other by π radians. As a result, a noncoherent receiver, which knows nothing about the phase of the received signal, cannot distinguish between the two signals in the BPSK signal set. More generally, any antipodal signal set is inappropriate for noncoherent communications for the same reason: A phase change of π radians in the carrier changes one of the antipodal signals into the other.

For binary noncoherent communications, the two signals $s_0(t)$ and $s_1(t)$ are of the form

$$s_i(t) = \sqrt{2}\, A\, a_i(t) \cos[\omega_c t + \theta_i(t) + \varphi_i]\, p_T(t), \tag{7.1}$$

where $a_i(t)$ and $\theta_i(t)$ are baseband signals of the type considered in Chapter 5, and φ_i is a phase angle that is not known to the receiver. The integer i in (7.1) is either 0 or 1, and the choice of $a_0(t)$ vs. $a_1(t)$ represents the amplitude modulation while the choice of $\theta_0(t)$ vs. $\theta_1(t)$ represents the phase modulation. The phase angle φ_i is the phase of the unmodulated carrier for the ith signal at the *receiver* at time $t = 0$. In many applications, the phase φ_i actually changes from one signaling interval to the next. In order to be able to include such applications in our model, we assume that the noncoherent receiver has no information at all about the value of φ_i. The decision statistics developed in receivers of interest for noncoherent communications do not depend on the value of φ_i.

Because the phase angles always appear inside trigonometric functions, they can be replaced by equivalent phase angles that are in the interval $[0, 2\pi)$. Thus, without loss of generality, we may assume that $0 \le \varphi_i < 2\pi$.

It is usually desirable and often necessary to model the phase angles φ_0 and φ_1 as random variables. In order to reflect complete lack of information about these phase angles, the appropriate density function to use for them is the uniform density on the interval $[0, 2\pi]$. It turns out that the uniform distribution is a good model for phase angles that arise in communication systems. This is due, in part, to the fact (see Problem 7.1) that if a random phase angle φ is the modulo-2π sum of one or more independent random phase components, and if at least one of these phase components has a uniform distribution on $[0, 2\pi]$, then the phase angle φ has a uniform distribution on $[0, 2\pi]$. Typical components of the phase angle at the receiver are the initial phase at the modulator, phase shifts in mixers and other subsystems in the transmitter and receiver, phase rotation encountered in propagating from the transmitter to the receiver, and phase shifts due to reflections off objects in the propagation path.

Example 7-1   Binary Frequency-Shift Key (BFSK)

The most important binary signal set for noncoherent communications is the BFSK signal set. The BFSK signals are defined as a special case of (7.1) by letting $a_0(t) = a_1(t) = 1$ for all t (no amplitude modulation) and

$$\theta_i(t) = (\omega_i - \omega_c)t \tag{7.2}$$

for all t and for both $i = 0$ and $i = 1$. Substituting these choices for the amplitude and phase modulation into (7.1) gives a signal of the form

$$s_i(t) = \sqrt{2}\, A \cos(\omega_i t + \varphi_i)\, p_T(t). \tag{7.3}$$

Thus, we see that the BFSK signals are sinusoidal pulses, sometimes referred to as tones, at frequencies ω_0 and ω_1. Usually, $\omega_i T$ is a multiple* of 2π and $|\omega_i - \omega_c| \ll \omega_c$ for each value of i.   ■

Some of the properties of BFSK signals for which the tone frequencies are multiples of $2\pi/T$ are obtained in the following exercise.

Exercise 7-1

Show that if $\omega_i T$ is a multiple of 2π and $|\omega_i - \omega_c| \ll \omega_c$ for each value of i, then the BFSK signals are *orthogonal* and have *equal energy*.

Solution.   If, for each value of i, $\omega_i = 2\pi n_i/T$ for some integer n_i, then

$$(\omega_1 \pm \omega_0)T = 2\pi(n_1 \pm n_0),$$

so that the sum and difference frequencies are also multiples of $2\pi/T$. This implies that the BFSK signals satisfy

$$(s_0, s_1) = \int_0^T 2A^2 \cos(\omega_0 t + \varphi_0) \cos(\omega_1 t + \varphi_1)\, dt = 0. \tag{7.4}$$

To see that the integral in (7.4) is zero, simply expand the product of $\cos(\omega_0 t + \varphi_0)$ and $\cos(\omega_1 t + \varphi_1)$ into two terms consisting of sinusoidal signals whose frequencies are the sum and difference frequencies $\omega_1 \pm \omega_0$. Because each of these sinusoidal signals is periodic with a period that is a divisor of T, the integral from 0 to T of each signal is zero. Note that (7.4) guarantees that the two signals are orthogonal regardless of the values of the phase angles φ_0 and φ_1.

A similar argument applied to the integral of the square of $s_i(t)$ shows that the energy in each signal is given by

$$\mathcal{E}_i = \int_0^T 2A^2 \cos^2(\omega_i t + \varphi_i)\, dt = A^2 T. \tag{7.5}$$

The expansion of $\cos^2(\omega_i t + \varphi_i)$ gives a baseband term and a double-frequency term. The fact that $\omega_i T$ is a multiple of 2π guarantees that the double-frequency term integrates to zero. As a result, the energy in each signal is $A^2 T$, regardless of the value of the phase angle φ_i.   ■

---

*Recall that in this book "x is a multiple of y" means $x = ny$ for some integer $n \neq 0$.

As discussed in Section 6.3.3, even if $\omega_i T$ is not a multiple of 2π, there are several different conditions that can be imposed to ensure that the integral of the double-frequency term of (7.5) is approximately zero. Similarly, even if ω_0 and ω_1 are not multiples of $2\pi/T$, there are alternative conditions that guarantee the signals are approximately orthogonal. The basic ideas behind these conditions are similar to those that are used in Section 6.3.3. In particular, whenever the double-frequency components are negligible, (7.1) implies that

$$(s_0, s_1) = \int_0^T A^2 \, a_0(t) \, a_1(t) \cos[\theta_0(t) - \theta_1(t) + \varphi_0 - \varphi_1] \, dt. \tag{7.6}$$

Thus, $\{s_0, s_1\}$ is an orthogonal signal set whenever the integral in (7.6) is zero. For BFSK, this integral reduces to

$$(s_0, s_1) = \int_0^T A^2 \cos[(\omega_0 - \omega_1)t + \varphi_0 - \varphi_1] \, dt. \tag{7.7}$$

As we have already pointed out, the integral in (7.7) is zero if $\omega_1 - \omega_0$ is a multiple of $2\pi/T$. It is also approximately zero, in the sense that it is very small compared to the signal energy, if $|\omega_1 - \omega_0| \gg T^{-1}$.

Although BFSK signals are the most commonly used orthogonal signals for noncoherent communications, there is also interest in signals that use amplitude modulation (AM) to achieve orthogonality. For AM signals, $\theta_0(t) = \theta_1(t) = 0$ for all t, so (7.6) implies that the resulting signals are orthogonal whenever the baseband signals $a_0(t)$ and $a_1(t)$ are orthogonal. One example of binary orthogonal AM signals is given in Example 6-1 of Section 6.3.1. In the example, $a_0(t)$ and $a_1(t)$ are nonoverlapping unit-amplitude rectangular pulses of duration $T/2$. The resulting signals also have equal energy. A more efficient signal set, in the sense that it has a larger signal energy for the same peak power, is given in the following example.

Example 7-2   Constant-Power Orthogonal AM

This signal set is given by (7.1) with $\theta_0(t) = \theta_1(t) = 0$ and with $a_0(t)$ and $a_1(t)$ as illustrated in Figure 7-1. Notice that $|a_0(t)| = |a_1(t)| = 1$ for all t, so the resulting signal set has constant power over its duration, unlike the signals of Example 6-1. Because the signals $a_0(t)$ and $a_1(t)$ are orthogonal, the signals $s_0(t)$ and $s_1(t)$ are also orthogonal. ∎

We now introduce the optimum noncoherent receiver for the transmission of signals of the form given by (7.1) over an additive white Gaussian noise channel. In discussing optimality, we must consider a specific criterion, such as the Bayes criterion, the minimax criterion, or the maximum-likelihood criterion. As we found for baseband receivers in Chapter 5 and coherent receivers in Chapter 6, the basic structure of the optimum noncoherent receiver and the optimum choice of filters do not depend on which of these criteria is used, but the optimum decision device does depend on the criterion that is adopted.

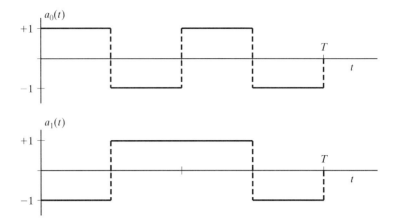

Figure 7-1: Baseband signals for orthogonal AM.

If the minimax criterion is employed, the optimum noncoherent receiver for this signal set when used on an AWGN channel is as illustrated in Figure 7-2. This receiver is optimum for the Bayes criterion if $\pi_0 = \pi_1$ and for the maximum-likelihood criterion. The type of receiver shown in Figure 7-2 is referred to as a *noncoherent correlation receiver*, because it utilizes a correlator in a way that is suitable for noncoherent communications. There is one correlator each for the inphase and quadrature components of $s_0(t)$ and one each for the inphase and quadrature components of $s_1(t)$. The upper two branches of the noncoherent correlation receiver are inphase and quadrature correlators for $s_0(t)$, and the lower two branches are the corresponding correlators for $s_1(t)$.

The decision that is made is that 0 was sent if $R_0^2 > R_1^2$ and that 1 was sent if $R_1^2 > R_0^2$. For an additive Gaussian noise channel, it does not matter what the decision is if $R_0^2 = R_1^2$, because this event has zero probability of occurrence regardless of which signal was actually sent.

If the signals $s_0(t)$ and $s_1(t)$ are orthogonal on the interval $[0, T]$, if each of these signals has energy $\mathcal{E}$, and if the two-sided spectral density of the noise is $N_0/2$, then the minimum possible error probability is given by

$$P_{e,i} = \tfrac{1}{2} \exp(-\mathcal{E}/2N_0) \tag{7.8}$$

for both $i = 0$ and $i = 1$. The error probability in (7.8) is achieved by the receiver of Figure 7-2. Equation (7.8) is derived later in the chapter for certain orthogonal signal sets that are of interest in practical applications, but it is valid for any orthogonal signal set of the form given by (7.1). Although the form of the receiver shown in Figure 7-2 is optimum for any binary signal set that can be described by (7.1), the error probability of (7.8) is for equal-energy, orthogonal signals only. The optimality of the receiver of Figure 7-2 is discussed in Section 7.2 for the important example in which the two signals are sinusoidal signals of different frequencies.

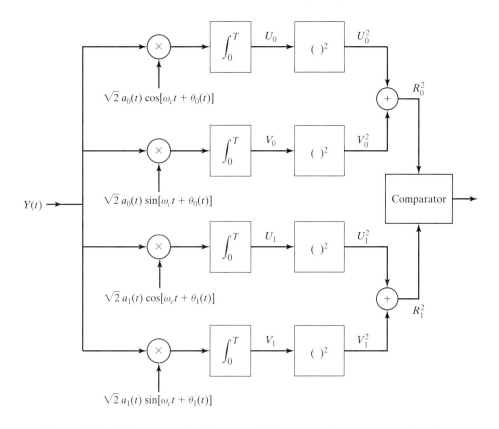

Figure 7-2: Optimum receiver for general binary noncoherent communications.

As should be expected, we must sacrifice something in exchange for not having to estimate the phase of the received signal. The error probabilities given by (7.8) are larger than those achievable with *coherent* communications. Recall that for coherent communications and orthogonal signaling, the optimum receiver achieves an error probability of $Q(\sqrt{\mathcal{E}/N_0})$. For coherent communication, antipodal signals can be used to achieve $Q(\sqrt{2\mathcal{E}/N_0})$, an even smaller probability of error. As discussed earlier, antipodal signals cannot be demodulated noncoherently.

To facilitate comparisons among different systems, the error probabilities for binary coherent communication with antipodal signals, binary coherent communication with orthogonal signals, and binary noncoherent communication with orthogonal signals are listed in Table 7.1 for some typical values of $\mathcal{E}/N_0$ (expressed in dB). The same three probabilities are illustrated graphically in Figure 7-3.

7.2   Binary Frequency-Shift Key (BFSK)

As for coherent communications in the previous chapter, we prefer to discuss the optimality of the noncoherent receiver of Figure 7-2 for sinusoidal signals only. Fortunately,

Table 7.1: Values of the Error Probabilities for Three Binary Communication Systems

$(\mathcal{E}/N_0)_{dB}$	$Q(\sqrt{2\mathcal{E}/N_0})$	$Q(\sqrt{\mathcal{E}/N_0})$	$\frac{1}{2}\exp(-\mathcal{E}/2N_0)$
5.0	5.95×10^{-3}	3.77×10^{-2}	1.03×10^{-1}
6.0	2.39×10^{-3}	2.30×10^{-2}	6.83×10^{-2}
7.0	7.73×10^{-4}	1.26×10^{-2}	4.08×10^{-2}
8.0	1.91×10^{-4}	6.00×10^{-3}	2.13×10^{-2}
9.0	3.36×10^{-5}	2.41×10^{-3}	9.42×10^{-3}
10.0	3.87×10^{-6}	7.83×10^{-4}	3.37×10^{-3}
11.0	2.61×10^{-7}	1.94×10^{-4}	9.23×10^{-4}
12.0	9.01×10^{-9}	3.43×10^{-5}	1.81×10^{-4}

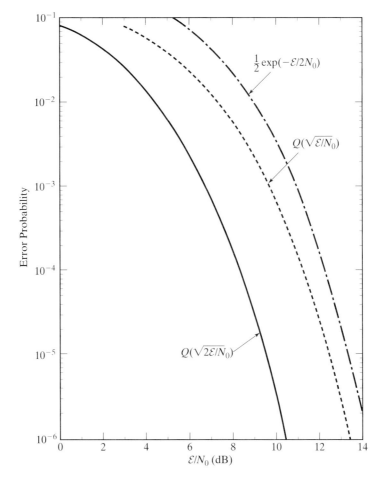

Figure 7-3: Error probabilities for three binary communication systems.

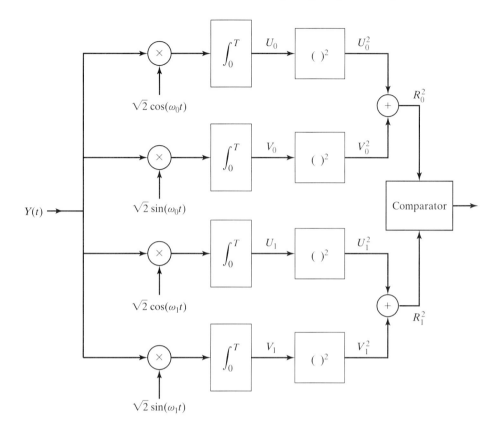

Figure 7-4: Optimum receiver for noncoherent BFSK communications.

as is true for coherent communications, this special case is the most commonly used binary signal set for noncoherent communication. As a consequence, the following discussion of the optimum receiver for the BFSK signal set not only illustrates the principles involved in establishing the optimality of the more general receiver of Figure 7-2, it is also of tremendous importance in its own right. Recall from Example 7-1 of the previous section that the BFSK signals are given by

$$s_i(t) = \sqrt{2}\, A\cos(\omega_i t + \varphi_i)\, p_T(t). \tag{7.9}$$

We will assume that $\omega_i T$ is a multiple of 2π throughout this section.

The optimum noncoherent receiver for BFSK is shown in Figure 7-4. Notice that this is just the receiver of Figure 7-2 after it has been specialized to constant amplitude modulation and to phase modulation of the form $\theta_i(t) = (\omega_i - \omega_c)t$. The receiver in Figure 7-4 is referred to as the *noncoherent correlation receiver* for BFSK. The correlators in Figure 7-4 that use $\cos(\omega_i t)$ are referred to as the *inphase correlators*, and the correlators that use $\sin(\omega_i t)$ are called the *quadrature correlators*.

There are two steps in the demonstration that the receiver of Figure 7-4 is optimum for noncoherent reception of BFSK signals on the additive white Gaussian noise channel. First, it must be shown an optimum decision can be based on U_0, V_0, U_1, and V_1 only; that is, no other statistics that can be extracted from $Y(t)$ help in making a decision as to which signal was sent. In the language of Section 6.1.4, this is just the statement that the random vector (U_0, V_0, U_1, V_1) is a *sufficient statistic* for deciding which signal was sent. The first part of the demonstration of the optimality of the receiver of Figure 7-4 is given in this section.

In the second step, it must be shown that the outputs of these correlators should be processed by summing their squares and making comparisons, as illustrated in Figure 7-4. This is proved as a special case of a conclusion obtained in Section 7.4. In Section 7.4, the optimum decision rule is derived for certain decision statistics that are not necessarily obtained from the correlators of Figure 7-4, but these statistics have certain key properties that are possessed by the outputs of the correlators. The properties in question are described completely in Section 7.4.

The first step, to establish that the random vector (U_0, V_0, U_1, V_1) is a sufficient statistic for deciding which signal was sent, is a straightforward extension of the corresponding discussion on *coherent* communications in Section 6.1.4. The two features of the present model that differ from the model for coherent communications in the previous chapter are that there are now two *orthogonal* signals in the signal set and the receiver does *not* know the phase of the received signal. Nevertheless, we begin at the same point as for coherent communication with PSK signals; that is, we consider the Fourier series representation of the signal set.

The signals have duration T, so let $\omega' = 2\pi/T$. Then define the signals $\psi_n(t)$, for all nonnegative integers n by $\psi_0(t) = \sqrt{1/T}$, $\psi_{2k-1}(t) = \sqrt{2/T}\cos(k\omega't + \varphi)$ for each positive integer k, and $\psi_{2k}(t) = \sqrt{2/T}\sin(k\omega't + \varphi)$ for each positive integer k. For coherent communication, the phase angle φ is the phase of the received signal. However, for noncoherent communications, the phase of the received signal is not known. Because we have no information about the phase of the received signal, the phase angle used in defining the signals $\psi_n(t)$ is unimportant. Therefore, we might as well set $\varphi = 0$ in defining the signals to be used in the Fourier series expansion. The Fourier series representation of $s(t)$ can be written as

$$s(t) = \sum_{n=0}^{\infty} a_n \psi_n(t), \tag{7.10}$$

where

$$a_n = \int_0^T s(t)\,\psi_n(t)\,\mathrm{d}t, \tag{7.11}$$

for each nonnegative integer n.

The Fourier series coefficients a_n, which completely describe the signal $s(t)$, can be generated with a bank of correlators as illustrated in Figure 7-5. The nth correlator computes the nth coefficient a_n $(0 \leq n < \infty)$, and the bank of correlators computes the infinite-dimensional vector $\mathbf{a} = (a_0, a_1, \ldots, a_n, \ldots)$. The system shown in Figure 7-5

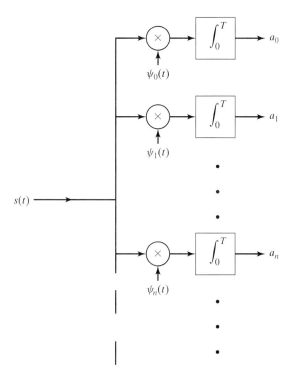

Figure 7-5: Correlators that generate the Fourier coefficients for both $s_0(t)$ and $s_1(t)$.

does not lose any information contained in the signal, because $s(t)$ can be recovered completely from the vector **a**. The signal can be reconstructed from the output of the bank of correlators by a system that forms the Fourier series

$$s(t) = \sum_{n=0}^{\infty} a_n \, \psi_n(t).$$

Thus, all of the information that we need about $s(t)$ is contained in the vector **a**.

So far, the development is the same as for coherent communications. Suppose that the bank of correlators shown in Figure 7-5 is used as the filter in a noncoherent communication system. Clearly, such a filter preserves all of the information about the signal that is needed to make a decision. In fact, if the input is *any* finite-energy waveform, the output of the bank of correlators completely specifies the input, whether the input is a communication signal, noise, or the sum of the signal and noise.

Suppose the input to the bank of correlators in Figure 7-5 is the signal

$$s_i(t) = \sqrt{2} \, A \cos(\omega_i t + \varphi_i).$$

To determine the output vector **a** corresponding to this input, first recall that ω_i is a multiple of $2\pi/T$ for each i. Thus, there are integers m and m' such that $\omega_0 = m\omega'$ and $\omega_1 = m'\omega'$. Therefore,

$$s_0(t) = \sqrt{2} \, A \cos(m\omega't + \varphi_0) = \sqrt{2} \, A\{\cos(\varphi_0) \cos(m\omega't) - \sin(\varphi_0) \sin(m\omega't)\}$$

and

$$s_1(t) = \sqrt{2}\,A\cos(m'\omega't + \varphi_1) = \sqrt{2}\,A\{\cos(\varphi_1)\cos(m'\omega't) - \sin(\varphi_1)\sin(m'\omega't)\}.$$

But $\psi_{2m-1}(t) = \sqrt{2/T}\cos(m\omega't)$ and $\psi_{2m}(t) = \sqrt{2/T}\sin(m\omega't)$, so

$$s_0(t) = A\sqrt{T}\,\{\cos(\varphi_0)\,\psi_{2m-1}(t) - \sin(\varphi_0)\,\psi_{2m}(t)\}. \tag{7.12}$$

Similarly,

$$s_1(t) = A\sqrt{T}\,\{\cos(\varphi_1)\,\psi_{2m'-1}(t) - \sin(\varphi_1)\,\psi_{2m'}(t)\}. \tag{7.13}$$

Thus, each of the signals in the signal set is a linear combination of two of the orthogonal functions $\psi_n(t)$. In particular, $s_0(t)$ is a linear combination of $\psi_{2m-1}(t)$ and $\psi_{2m}(t)$. Because $\psi_{2m-1}(t)$ and $\psi_{2m}(t)$ are orthogonal to all of the functions $\psi_n(t)$ for which $n \neq 2m-1$ and $n \neq 2m$, it follows from (7.12) that $s_0(t)$ is orthogonal to all such functions as well. As a consequence,

$$a_n = \int_0^T s_0(t)\,\psi_n(t)\,dt = 0$$

for all values of n except $2m-1$ and $2m$. Thus, when $s_0(t)$ is the input to the bank of correlators, at most two of the correlators have nonzero outputs. The outputs of these correlators are $a_{2m-1} = A\sqrt{T}\cos(\varphi_0)$ and $a_{2m} = -A\sqrt{T}\sin(\varphi_0)$, as can be seen by comparing (7.12) with (7.10) or by applying (7.11) with

$$s(t) = s_0(t) = \sqrt{2}\,A\cos(\omega_0 t + \varphi_0).$$

From (7.12), it is clear that these two correlators capture the signal $s_0(t)$ completely in the sense that if $s_0(t)$ is the input to the pair of correlators, then $s_0(t)$ can be reconstructed completely from the outputs of these two correlators:

$$s_0(t) = a_{2m-1}\,\psi_{2m-1}(t) + a_{2m}\,\psi_{2m}(t).$$

Notice that for some values of the phase angle φ_0, one of the two correlator outputs is zero. Because φ_0 is not known to the receiver, it is not possible to know in advance which of these correlators might have a zero output, so both correlators are required to represent the signal in a noncoherent receiver.

All of these steps can be repeated to determine the output that results if $s_1(t)$ is the input to the bank of correlators shown in Figure 7-5. The conclusion reached in doing so is that if $s_1(t)$ is the input, then $a_n = 0$ for all values of n except $2m'-1$ and $2m'$, while $a_{2m'-1} = A\sqrt{T}\cos(\varphi_1)$ and $a_{2m'} = -A\sqrt{T}\sin(\varphi_1)$. Furthermore,

$$s_1(t) = a_{2m'-1}\psi_{2m'-1}(t) + a_{2m'}\psi_{2m'}(t).$$

Thus, no more than two correlators are needed to capture the signal $s_1(t)$. In total, at most four correlators are needed to capture whichever signal might be transmitted, and none of the other correlators shown in Figure 7-5 is needed. This establishes that the four correlators illustrated in Figure 7-4 are sufficient to capture whichever signal is transmitted no matter what the values for the phase angles φ_0 and φ_1.

The only difference between the four correlators shown in Figure 7-4 and their counterparts in Figure 7-5 is the normalization factor of $\sqrt{T}$ that appears in the definition of the orthonormal functions $\psi_n(t)$. This normalization factor is unimportant; we can always multiply the outputs of all four correlators in Figure 7-4 by $1/\sqrt{T}$ if we want to match up the constants for the two systems. It should be clear from Figure 7-4 that such gain constants are unimportant as long as the same gain is applied to each of the four correlators. The value of the gain will not affect the comparison that is used in making the decision.

As in Chapter 6 (Section 6.1.4), the argument is completed by considering the output of the bank of correlators in Figure 7-5 when the input is white Gaussian noise only. Again the key issue is the independence of the outputs of different correlators, and this independence follows from the fact that such outputs are jointly Gaussian and uncorrelated. In particular, when the input to the bank of correlators is white Gaussian noise, the outputs of the four correlators indexed by $2\,m-1, 2\,m, 2\,m'-1$, and $2\,m'$ are statistically independent of the collection of outputs of all other correlators. As a result, the outputs of these other correlators cannot help in making a decision as to which of the two signals was transmitted.

The optimum receiver can therefore be based completely on the four correlators shown in Figure 7-4, and the optimum decision rule depends only on the four random variables U_0, V_0, U_1, and V_1 that represent the outputs of these correlators. This completes the first step in the demonstration of the optimality of the receiver of Figure 7-4. Before proceeding, however, it will be helpful to determine some important properties of the random variables U_0, V_0, U_1, and V_1.

When 0 is sent (i.e., $s_0(t)$ is transmitted), U_1 and V_1 have no signal components. In fact, when 0 is sent, $U_1 = X_1$ and $V_1 = Y_1$, where X_1 and Y_1 are defined by

$$X_1 = \int_0^T X(t)\,\sqrt{2}\cos(\omega_1 t)\,dt \tag{7.14a}$$

and

$$Y_1 = \int_0^T X(t)\,\sqrt{2}\sin(\omega_1 t)\,dt, \tag{7.14b}$$

respectively. These integrals represent the outputs of the corresponding correlators when the input is noise only. As a result, X_1 and Y_1 are zero-mean Gaussian random variables. Because $\omega_1 T$ is a multiple of 2π, the signals $\cos(\omega_1 t)$ and $\sin(\omega_1 t)$ are orthogonal on the interval $[0, T]$. In fact, recall that $\psi_k(t) = \sqrt{2/T}\cos(\omega_1 t)$ for some choice of k and $\psi_n(t) = \sqrt{2/T}\sin(\omega_1 t)$ for some choice of n. The development in the last paragraph of Section 6.1.4 shows that $T^{-1}E\{X_1 Y_1\} = \frac{1}{2}N_0(\psi_k, \psi_n) = 0$; and thus X_1 and Y_1 are uncorrelated. This argument will be applied also to $E\{X_0 Y_0\}$, $E\{X_0 Y_1\}$, $E\{X_1 Y_0\}$, $E\{X_0 X_1\}$, and $E\{Y_0 Y_1\}$ in what follows.

Because X_1 and Y_1 are uncorrelated and jointly Gaussian, they are also independent. It is easy to show that their common variance is $\sigma^2 = N_0 T/2$. It follows that U_1 and V_1 are independent, and that each has mean zero and variance σ^2.

When 0 is sent, U_0 and V_0 do have signal components, and these components depend on φ_0. These signal components are proportional to the Fourier series coefficients that we have already determined. The results are that U_0 has signal component $u_0 = AT \cos(\varphi_0)$ and V_0 has signal component $v_0 = -AT \sin(\varphi_0)$. We can then write $U_0 = u_0 + X_0$ and $V_0 = v_0 + Y_0$, where

$$X_0 = \int_0^T X(t)\,\sqrt{2}\cos(\omega_0 t)\,dt \tag{7.15a}$$

and

$$Y_0 = \int_0^T X(t)\,\sqrt{2}\sin(\omega_0 t)\,dt. \tag{7.15b}$$

Clearly X_0 and Y_0 also have mean zero and variance σ^2.

From the fact that $\omega_0 T$ is a multiple of 2π, it follows that $\cos(\omega_0 t)$ and $\sin(\omega_0 t)$ are orthogonal on the interval $[0, T]$, which in turn implies that X_0 and Y_0 are uncorrelated. From the fact that sinusoidal signals at frequency ω_0 and ω_1 are orthogonal on the interval $[0, T]$, it follows that X_0 and X_1 are uncorrelated, X_0 and Y_1 are uncorrelated, Y_0 and X_1 are uncorrelated, and Y_0 and Y_1 are uncorrelated. Because X_0, Y_0, X_1, and Y_1 are jointly Gaussian, they are mutually independent. From this, it follows that X_0, Y_0, U_1, and V_1 are jointly Gaussian, and they are conditionally mutually independent given that 0 is sent.

Now, it might be tempting to conclude that U_0, V_0, U_1, and V_1 are also mutually independent given that 0 is sent, but this is not true. There is a statistical "connection" between U_0 and V_0 due to the phase angle φ_0: Both U_0 and V_0 depend on φ_0. The fact that U_0 and V_0 are not independent can be seen intuitively by considering a special case. Suppose that N_0 is very small, so that X_0 and Y_0 are negligible. By this we mean that with high probability both X_0 and Y_0 are approximately zero. Then with high probability $U_0^2 + V_0^2 \approx u_0^2 + v_0^2 = A^2 T^2$, so $V_0^2 \approx A^2 T^2 - U_0^2$. Thus, U_0 and V_0 are certainly not independent when N_0 is very small. In fact, if N_0 is very small and we observe $U_0 \approx AT$, we know from the relationship $U_0 \approx AT \cos(\varphi_0)$ that either $\varphi_0 \approx 0$ or $\varphi_0 \approx \pi$. For either value of φ_0, $V_0 \approx -AT \sin(\varphi_0) \approx 0$, so the value of U_0 can provide a great deal of information about V_0 through the phase angle φ_0.

Even if the noise is not negligible, U_0 provides information about φ_0, which in turn provides information about V_0, so U_0 and V_0 cannot be statistically independent. On the other hand, if we already know the value of φ_0 before observing U_0, then knowledge of U_0 does not tell anything about V_0. This fact is the key to understanding the relationship between the random variables in question. The correct independence statement regarding U_0, V_0, U_1, and V_1 is that they are *conditionally* independent given that 0 is sent and given the value of φ_0. This will be exploited in the evaluation of the density $f_0(u, v, x, y)$.

In summary, the decision statistics U_0, V_0, U_1, and V_1 have the following properties: When conditioned on 0 being sent and on $\varphi_0 = \varphi$, the random variables U_0, V_0, U_1, and V_1 are conditionally mutually independent and have the form

$$U_0 = \alpha \cos(\varphi) + X_0,$$
$$V_0 = -\alpha \sin(\varphi) + Y_0,$$
$$U_1 = X_1,$$

and

$$V_1 = Y_1.$$

Furthermore, the random variables X_0, Y_0, X_1, and Y_1 are mutually independent, zero-mean, Gaussian random variables that have the same variance σ^2.

A corresponding analysis for the situation when 1 is sent (i.e., $s_1(t)$ is transmitted) shows that when conditioned on 1 being sent and conditioned on $\varphi_1 = \varphi$, the random variables U_0, V_0, U_1, and V_1 are conditionally mutually independent and are given by

$$U_0 = X_0,$$
$$V_0 = Y_0,$$
$$U_1 = \alpha \cos(\varphi) + X_1,$$

and

$$V_1 = -\alpha \sin(\varphi) + Y_1.$$

For the BFSK signal set considered in this section and the correlators of Figure 7-4, α and σ are related to the parameters of the signal set and noise process by $\alpha = AT$ and $\sigma^2 = N_0 T / 2$.

7.3   Suboptimum Correlators in Binary Noncoherent Correlation Receivers

Recall that the general form for the signals used in binary noncoherent communications is

$$s_i(t) = \sqrt{2}\, A\, a_i(t) \cos[\omega_c t + \theta_i(t) + \varphi_i]\, p_T(t).$$

For convenience, let

$$m_i(t) = A\, a_i(t)\, p_T(t),$$

so that the signals can be written as

$$s_i(t) = \sqrt{2}\, m_i(t) \cos[\omega_c t + \theta_i(t) + \varphi_i]. \qquad (7.16)$$

The energy in the ith signal is just the integral of the square of $s_i(t)$, which can be written as

$$\mathcal{E}_i = \int_0^T [m_i(t)]^2 \{1 + \cos[2\,\omega_c t + 2\,\theta_i(t) + 2\,\varphi_i]\}\, dt.$$

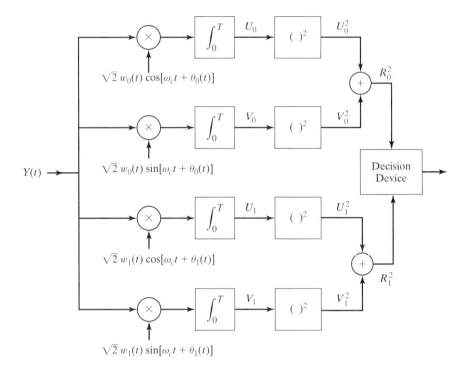

Figure 7-6: Suboptimum receiver for binary noncoherent communications.

Because the double-frequency terms are negligible, the signal energy is given by

$$\mathcal{E}_i = \int_0^T [m_i(t)]^2 \, dt.$$

The most important special case, which should be kept in mind as an illustration throughout this section, is the BFSK signal set described in Example 7-1 and discussed further in Section 7.2. Recall that the BFSK signals are obtained from (7.16) if we let $a_i(t) = 1$ and $\theta_i(t) = (\omega_i - \omega_c)t$ for all t and for both $i = 0$ and $i = 1$. For these signals,

$$m_0(t) = m_1(t) = A \, p_T(t),$$

so

$$s_i(t) = \sqrt{2} \, A \cos[\omega_i t + \varphi_i] \, p_T(t),$$

and $\mathcal{E}_i = A^2 T$ (cf. equation (7.5)).

In this section, we consider the receiver that is illustrated in Figure 7-6. The signals used in the four correlation receivers are

$$f_i(t) = \sqrt{2} \, w_i(t) \cos[\omega_c t + \theta_i(t)] \qquad (7.17)$$

and

$$g_i(t) = \sqrt{2}\, w_i(t) \sin[\omega_c t + \theta_i(t)], \tag{7.18}$$

for both $i = 0$ and $i = 1$. The resulting products of the received waveform and each of these four signals are the inputs to the four integrators shown in Figure 7-6.

The receiver of Figure 7-6 is a generalization of the optimum receiver. By comparing Figures 7-2 and 7-6, we can see the correct correlator *structure* is used in Figure 7-6, but if $w_i(t)$ is not proportional to $m_i(t)$, the correlators of Figure 7-6 are not optimum (i.e., they are not matched to the signal waveform).

It is of interest to consider such suboptimum correlators, because implementation considerations may dictate their use in the receiver. For example, uncertainties or variations in the characteristics of the filtering that occurs in the transmitter or the propagation medium may result in the receiver's lack of knowledge of the waveform $m_i(t)$. In this situation, it may be that the shape of $w_i(t)$ is only an approximation to the shape of $m_i(t)$. Alternatively, in a performance analysis of a noncoherent correlation receiver with imperfect integrators, we might select the waveform $w_i(t)$ to model imperfections in the receiver's integrators. Because of the generality of the model, the analysis that follows can be used to determine the performance of certain suboptimum receivers and permit tradeoffs between performance and complexity in the design of the correlators in the receiver.

The results of this section are applied to optimum receivers by letting $w_i(t)$ be proportional to $m_i(t)$ in the final results. In particular,

$$w_i(t) = a_i(t) = 1, \quad 0 \le t \le T$$

and

$$\theta_i(t) = (\omega_i - \omega_c)\, t, \quad 0 \le t \le T$$

for $i = 0$ and $i = 1$, correspond to the noncoherent correlation receiver that gives the minimum probability of error for noncoherent reception of BFSK.

Similarly, for the signal set of Example 7-2, the optimum receiver and the resulting probability of error are obtained by letting $\theta_0(t) = \theta_1(t) = 0$, $w_0(t) = a_0(t)$, and $w_1(t) = a_1(t)$ for all t. More generally, $w_i(t)$ should be proportional to $a_i(t)$, with the same constant of proportionality for both values of i.

Notice that $|w_0(t)| = |w_1(t)|$ for each t in these two examples. This arises frequently in applications. This condition implies that $\|w_0\| = \|w_1\|$, but this latter condition on the norms is much weaker than the requirement that $|w_0(t)| = |w_1(t)|$ for each t. Although our primary interest is in receivers for which $\|w_0\| = \|w_1\|$, the analysis in this section is carried out without this restriction. This condition on the norms is introduced only at the very end as a special case of the general results.

The analysis that follows is valid only for certain classes of the signals $s_i(t)$, $f_i(t)$, and $g_i(t)$ defined in (7.16)–(7.18). These signals must have the following properties:

(1) The waveforms $m_i(t)$, $w_i(t)$, and $\theta_i(t)$ have bandwidths much smaller than the carrier frequency.

(2) All double-frequency components are negligible.

(3) The signals $w_0(t)\cos[\omega_c t+\theta_0(t)+\varphi]$ and $w_1(t)\cos[\omega_c t+\theta_1(t)+\theta]$ are orthogonal on the interval $[0, T]$ for all phase angles φ and θ.

(4) The signals $m_i(t)\cos[\omega_c t+\theta_i(t)+\varphi]$ and $w_j(t)\cos[\omega_c t+\theta_j(t)+\theta]$ are orthogonal on the interval $[0, T]$ for all phase angles φ and θ and for all pairs i and j such that $i \neq j$.

If the signals $s_0(t)$ and $s_1(t)$ are orthogonal on the interval $[0, T]$ for all phase angles φ_0 and φ_1, and if the optimum correlators are used, then both (3) and (4) follow from the orthogonality of the signals. To see this, just observe that

$$s_i(t) = \sqrt{2}\, m_i(t)\cos[\omega_c t + \theta_i(t) + \varphi_i]$$

and that $w_i(t)$ is proportional to $m_i(t)$ for optimum correlators.

For specific signals and receivers, it may be that not all four of these properties are actually required. The following two examples illustrate this fact.

Example 7-3   Optimum Reception of BFSK
Consider BFSK with its optimum receiver. It turns out that (1) is not necessary for the analysis to be valid. However, the remaining conditions are needed, and it is easy to show that they are satisfied if $\omega_0 \neq \omega_1$ and both $\omega_0 T$ and $\omega_1 T$ are multiples of 2π.   ■

Example 7-4   Spread-Spectrum Signals
The analysis that follows reveals that (1) is not necessary if $m_i(t)$, $w_i(t)$, and $\theta_i(t)$ are all constant on certain common intervals. For instance, if $\omega_c T = 2\pi N$ for some integer N, and $m_i(t)$, $w_i(t)$, and $\theta_i(t)$ are constant on each interval of the form $n\tau < t < (n+1)\tau$, where $\tau = 2\pi/\omega_c$ and $0 \leq n \leq N-1$, then (1) is not required. In practice, signals that are at least approximately constant on the intervals of length τ are sufficient, so perfect rectangular pulses are not actually required. A common example in which the modulation is constant on such intervals is obtained if the amplitude and phase modulation are sequences of rectangular pulses and each pulse has duration $k\tau$ for some integer k. Example 7-2 gives such a signal set if $\omega_c T = 2\pi N$ for some integer N that is a multiple of 4. If each pulse has duration $k\tau$ and k is much smaller than N, then the resulting signals have a bandwidth much larger than $1/T$. If k is very small, the bandwidth is not much smaller than the carrier frequency; therefore, property (1) is not satisfied. Yet the analysis that follows is valid. An important signal set is obtained if $\theta_i(t) = 0$ for $0 \leq t \leq T$ and the amplitude modulation, $m_i(t) = A\, a_i(t)\, p_T(t)$, is such that the signals $a_0(t)$ and $a_1(t)$ are sequences of positive and negative, unit-amplitude rectangular pulses of duration $T_c = k\tau$. For this example, $|a_0(t)| = |a_1(t)| = 1$ for each t. It follows that $|m_0(t)| = |m_1(t)|$ for each t. The signals just described are an example of one form of direct-sequence spread-spectrum signals; they are described in detail in Chapter 9. Notice that the optimum receiver has $|w_0(t)| = |w_1(t)|$ for each t.   ■

As usual, we begin the analysis by assuming that $s_0(t)$ is sent and that $\varphi_0 = \varphi$, so that

$$Y(t) = s_0(t) + X(t) = \sqrt{2}\,m_0(t)\cos[\omega_c t + \theta_0(t) + \varphi] + X(t).$$

It follows that $U_i = u_i + X_i$ and $V_i = v_i + Y_i$ for $i = 0$ and $i = 1$, where

$$u_i = \int_0^T 2\,m_0(t)\,w_i(t)\cos[\omega_c t + \theta_0(t) + \varphi]\cos[\omega_c t + \theta_i(t)]\,dt, \qquad (7.19)$$

$$v_i = \int_0^T 2\,m_0(t)\,w_i(t)\cos[\omega_c t + \theta_0(t) + \varphi]\sin[\omega_c t + \theta_i(t)]\,dt, \qquad (7.20)$$

$$X_i = \int_0^T X(t)\,\sqrt{2}\,w_i(t)\cos[\omega_c t + \theta_i(t)]\,dt, \qquad (7.21)$$

and

$$Y_i = \int_0^T X(t)\,\sqrt{2}\,w_i(t)\sin[\omega_c t + \theta_i(t)]\,dt. \qquad (7.22)$$

Notice that for $i = 0$, (7.19) can be written as

$$u_0 = \int_0^T m_0(t)\,w_0(t)\,\{\cos(\varphi) + \cos[2\,\omega_c t + 2\,\theta_0(t) + \varphi]\}\,dt.$$

Because of property (1), the double-frequency term can be neglected; therefore,

$$u_0 = \int_0^T m_0(t)\,w_0(t)\,dt\,\cos(\varphi). \qquad (7.23)$$

The validity of (7.23) is also guaranteed for the signals of Examples 7-3 and 7-4. For these examples, the product $m_0(t)\,w_0(t)$ is constant on the intervals of length τ, so the integral of the double-frequency term over each such interval is zero. It follows that

$$\int_{n\tau}^{(n+1)\tau} m_0(t)\,w_0(t)\cos[2\,\omega_c t + 2\,\theta_0(t) + \varphi]\,dt = 0$$

for each n. Summing this from $n = 0$ to $n = N - 1$ shows that

$$\int_0^T m_0(t)\,w_0(t)\cos[2\,\omega_c t + 2\,\theta_0(t) + \varphi]\,dt = 0.$$

If we define α_0 by

$$\alpha_0 = (m_0, w_0) = \int_0^T m_0(t)\,w_0(t)\,dt,$$

then our conclusion is that $u_0 = \alpha_0 \cos(\varphi)$. A similar analysis applied to (7.20) for $i = 0$ shows that $v_0 = -\alpha_0 \sin(\varphi)$.

Next, let $i = 1$, and observe that (7.19) and property (4) imply that

$$u_1 = \int_0^T 2\, m_0(t)\, w_1(t) \cos[\omega_c t + \theta_0(t) + \varphi] \cos[\omega_c t + \theta_1(t)]\, dt = 0.$$

Similarly, (7.20) and property (4) imply that $v_1 = 0$.

Consider the noise terms X_0, X_1, Y_0, and Y_1. It is clear from (7.21) and (7.22) that $E\{X_i\} = E\{Y_i\} = 0$ for both values of i. The second moment of X_i is obtained by finding the expectation of the square of the integral in (7.21). The method is the same as employed in Section 6.2.1 (see the analysis leading up to (6.14)), and the result is

$$E\{X_i^2\} = \tfrac{1}{2} N_0 \int_0^T [w_i(t)]^2\, dt,$$

where double-frequency terms have been neglected. This result can be written as $E\{X_i^2\} = N_0 \|w_i\|^2 / 2$. A similar analysis applied to (7.22) shows that $E\{Y_i^2\} = N_0 \|w_i\|^2 / 2$.

We now turn our attention to the correlations among the random variables X_0, X_1, Y_0, and Y_1. As an illustration, consider $E\{X_i Y_i\}$. From (7.21) and (7.22), we see that

$$E\{X_i Y_i\} = \int_0^T N_0 [w_i(t)]^2 \cos[\omega_c t + \theta_i(t)] \sin[\omega_c t + \theta_i(t)]\, dt,$$

which can be written as

$$E\{X_i Y_i\} = \int_0^T \tfrac{1}{2} N_0 [w_i(t)]^2 \sin[2\,\omega_c t + 2\,\theta_i(t)]\, dt.$$

Therefore, $E\{X_i Y_i\} = 0$ by property (1). This result is also valid for Examples 7-3 and 7-4.

As another illustration, consider $E\{X_0 Y_1\}$. From (7.21) and (7.22), we see that

$$E\{X_0 Y_1\} = \int_0^T N_0\, w_0(t)\, w_1(t) \cos[\omega_c t + \theta_0(t)] \sin[\omega_c t + \theta_1(t)]\, dt.$$

It follows from property (3) that $E\{X_0 Y_1\} = 0$. By similar arguments, it can be shown that $E\{X_1 Y_0\} = E\{X_0 X_1\} = E\{Y_0 Y_1\} = 0$.

We conclude that the random variables X_0, X_1, Y_0, and Y_1 are uncorrelated, just as for BFSK in Section 7.2. Because the random variables are also jointly Gaussian, they are mutually independent.

In summary, we find that when conditioned on 0 being sent and $\varphi_0 = \varphi$, the random variables U_0, U_1, V_0, and V_1 are conditionally mutually independent and are given by

$$U_0 = \alpha_0 \cos(\varphi) + X_0,$$
$$V_0 = -\alpha_0 \sin(\varphi) + Y_0,$$
$$U_1 = X_1,$$

and

$$V_1 = Y_1,$$

where $\alpha_0 = (m_0, w_0)$. Furthermore, the random variables X_0, X_1, Y_0, and Y_1 are mutually independent, zero-mean, Gaussian random variables. The random variables X_0 and Y_0 each have variance $\sigma_0^2 = N_0 \|w_0\|^2 / 2$, and X_1 and Y_1 each have variance $\sigma_1^2 = N_0 \|w_1\|^2 / 2$.

A corresponding analysis for the situation when 1 is sent and $\varphi_1 = \varphi$ shows that the random variables U_0, U_1, V_0, and V_1 are conditionally mutually independent and are given by

$$U_0 = X_0,$$
$$V_0 = Y_0,$$
$$U_1 = \alpha_1 \cos(\varphi) + X_1,$$

and

$$V_1 = -\alpha_1 \sin(\varphi) + Y_1,$$

where $\alpha_1 = (m_1, w_1)$. The random variables X_0, X_1, Y_0, and Y_1 have the same properties as in the previous paragraph.

At this point, we make some observations that are used in Section 7.5. For *equal-energy* orthogonal signals of the form

$$s_i(t) = \sqrt{2}\, m_i(t) \cos[\omega_c t + \theta_i(t) + \varphi_i]$$

with $m_i(t) = A\, a_i(t)\, p_T(t)$, the correlators are optimum if for some constant c, $w_0(t) = c\, m_0(t)$ and $w_1(t) = c\, m_1(t)$ for $0 \leq t \leq T$. Because the signals have equal energy, $\|m_0\|^2 = \|m_1\|^2$ and $\|a_0\|^2 = \|a_1\|^2$. It follows that $\|w_0\|^2 = \|w_1\|^2$ for the optimum correlators for equal-energy signals. In general, regardless of whether the correlators are optimum, if $\|w_0\|^2 = \|w_1\|^2$ then the random variables X_0, X_1, Y_0, and Y_1 have common variance

$$\sigma^2 = \tfrac{1}{2} N_0 \|w_0\|^2 = \tfrac{1}{2} N_0 \|w_1\|^2.$$

If the correlators are optimum and the signals have equal energy $\mathcal{E}$, then

$$\alpha_i = (m_i, w_i) = (m_i, c\, m_i) = c \|m_i\|^2 = c\, \mathcal{E}. \tag{7.24}$$

Thus, $\alpha_0 = \alpha_1 = c\, \mathcal{E}$. Because α_i does not depend on i in this situation, we drop the subscript and let α be the common value (i.e., $\alpha = \alpha_0 = \alpha_1 = c\, \mathcal{E}$). But we also have

$\|w_i\|^2 = c^2 \|m_i\|^2 = c^2 \mathcal{E}$. Thus, for the optimum correlators and equal-energy signals,

$$\sigma_i^2 = \tfrac{1}{2} N_0 \|w_i\|^2 = \tfrac{1}{2} N_0 c^2 \mathcal{E}. \tag{7.25}$$

As with the parameter α, we drop the subscript on σ_i and let σ denote the common value. It follows from these observations that for equal-energy signals and correlators that satisfy $w_0(t) = c\, m_0(t)$ and $w_1(t) = c\, m_1(t)$ for $0 \le t \le T$,

$$\alpha^2/\sigma^2 = 2\,\mathcal{E}/N_0. \tag{7.26}$$

The ratio α^2/σ^2 is a signal-to-noise ratio for the output of the correlators. As shown in Section 7.5, the probability of error for the optimum receiver depends on the parameters of the signal and the noise through the ratio α^2/σ^2 only.

7.4   Derivation of the Optimum Decision Rule for Binary Noncoherent Communications

In this section, the decision rules discussed in Section 5.4 are employed to find the optimum decision based on the observations U_0, V_0, U_1, and V_1 when these random variables have certain properties that are satisfied by the outputs of the optimum and suboptimum correlators considered in the two previous sections. For the optimum correlators, the results of this section answer the following question: Now that we have determined that the optimum decision is based on U_0, V_0, U_1, and V_1 exclusively, how do we process these four random variables in order to make an optimum decision? Even if the optimum correlators are not used, it is of interest to make the best decision possible based on the outputs of the suboptimum correlators.

The model described here is more general than is needed for BFSK signaling and optimum noncoherent reception. This permits inclusion of binary orthogonal signal sets that are not based on frequency separation. A prime example is a signal set that uses a pair of orthogonal binary baseband signals to modulate the amplitude of the carrier (e.g., Example 7-2). Similarly, the model for the receiver permits consideration of suboptimum filters in the noncoherent receiver. The derivation given in this section requires the following assumptions only:

(1) When 0 is sent and φ is the phase angle for $s_0(t)$, the decision statistics are given by $U_0 = \alpha_0 \cos(\varphi) + X_0$, $V_0 = -\alpha_0 \sin(\varphi) + Y_0$, $U_1 = X_1$, and $V_1 = Y_1$.

(2) When 1 is sent and φ is the phase angle for $s_1(t)$, these random variables are given by $U_0 = X_0$, $V_0 = Y_0$, $U_1 = \alpha_1 \cos(\varphi) + X_1$, and $V_1 = -\alpha_1 \sin(\varphi) + Y_1$.

(3) The random variables X_0, Y_0, X_1, and Y_1 in (1) and (2) are mutually independent, zero-mean, Gaussian random variables. X_0 and Y_0 have variance σ_0^2, and X_1 and Y_1 have variance σ_1^2.

As shown in Section 7.2, conditions (1)–(3) are satisfied for a communication system with BFSK signals and the optimum correlators of Figure 7-4. For BFSK signals with these correlators, $\alpha_0 = \alpha_1 = \alpha = AT$. The noise variance is given by $\sigma_0^2 = \sigma_1^2 = \sigma^2 = N_0 T/2$.

As shown in Section 7.3, conditions (1)–(3) are also satisfied for more general binary orthogonal signal sets and the receiver of Figure 7-6 with suboptimum correlators. For the latter signals and receiver, we found that $\alpha_0 = (m_0, w_0)$, $\alpha_1 = (m_1, w_1)$, $\sigma_0^2 = N_0 \|w_0\|^2/2$, and $\sigma_1^2 = N_0 \|w_1\|^2/2$. As shown in the previous section, for equal-energy signals and optimum correlators, $\alpha_i = c\,\mathcal{E}$ and $\sigma_i^2 = N_0\,c^2\mathcal{E}/2$, so that $\alpha_i^2/\sigma_i^2 = 2\,\mathcal{E}/N_0$ for both $i = 0$ and $i = 1$.

The claim that we have made is that, for both the Bayes criterion for $\pi_0 = \pi_1$ and the minimax criterion, the receiver of Figure 7-4 is the optimum receiver for noncoherent reception of BFSK signals that are transmitted over an additive white Gaussian noise channel. The receiver of Figure 7-4 uses the decision rule in which $U_0^2 + V_0^2$ is compared with $U_1^2 + V_1^2$, and the decision is made on the basis of which is larger. In this section, we show that this decision rule is optimum for BFSK signaling and the correlators of Figure 7-4. We also derive the optimum decision rule for any signal set and receiver filtering that produces decision statistics U_0, V_0, U_1, and V_1 with properties (1)–(3). This includes receivers in which the filters are not matched to the signal set.

It will be shown that the decision rule used for the BFSK receiver in Figure 7-4 is also optimum for general signals and suboptimum correlators that satisfy (1)–(3) provided that $\alpha_0^2 = \alpha_1^2$ and $\sigma_0^2 = \sigma_1^2$. In other words, even if the decision statistics are not sufficient statistics for making the optimum decision, as long as they satisfy (1)–(3), $\alpha_0^2 = \alpha_1^2$, and $\sigma_0^2 = \sigma_1^2$, the best decision rule to use *for the given decision statistics* is to compare $U_0^2 + V_0^2$ with $U_1^2 + V_1^2$ and make the decision on the basis of which is larger.

We start with arbitrary values for the probabilities π_0 and π_1, so that our results will also show how to generalize the receiver of Figure 7-4 to give the Bayes receiver if $\pi_0 \neq \pi_1$. According to Theorem 5-1, the minimax decision rule can be found by finding a Bayes rule that is also an equalizer rule. So, it suffices to focus attention on the Bayes rules as long as the probabilities π_0 and π_1 are arbitrary. We also know from Section 5.4 that a Bayes decision rule can be written as a likelihood ratio test with the parameter η set equal to the ratio π_0/π_1. So, we cast the problem in the form of a likelihood ratio test.

The observation can be considered to be a vector with U_0, V_0, U_1, and V_1 as its components:

$$\mathbf{Z} = (U_0,\, V_0,\, U_1,\, V_1).$$

The decision-theoretic problem is that we observe that $\mathbf{Z}$ takes on the value $\mathbf{z}$, and we want to find a decision rule that makes the optimum decision for this observation. In order to derive the likelihood ratio test, the starting point is to consider the likelihood ratio $\Lambda(\mathbf{z}) = f_1(\mathbf{z})/f_0(\mathbf{z})$, where f_0 is the density for $\mathbf{Z}$ given that 0 is sent and f_1 is the density for $\mathbf{Z}$ given that 1 is sent. Because $\mathbf{Z}$ is a vector with components U_0, V_0, U_1, and V_1, the densities f_0 and f_1 are actually four-dimensional densities. Letting $\mathbf{z} = (u, v, x, y)$, we can write

$$\Lambda(u, v, x, y) = f_1(u, v, x, y)/f_0(u, v, x, y).$$

The density f_i can be thought of as the conditional joint density function for U_0, V_0, U_1, and V_1 given that i is sent, or it can be viewed as the density for the vector $\mathbf{Z}$ given that i is sent. These interpretations are equivalent.

Recall that the phase angle φ_0 is modeled as a random variable with a uniform distribution on the interval $[0, 2\pi]$. If $\hat{f}_0(u, v, x, y|\varphi)$ is the conditional density for U_0, V_0, U_1, and V_1 given that 0 is sent and given that $\varphi_0 = \varphi$, then

$$f_0(u, v, x, y) = \int_0^{2\pi} \hat{f}_0(u, v, x, y|\varphi) \, d\varphi/2\pi. \tag{7.27}$$

Because of the conditional independence,

$$\hat{f}_0(u, v, x, y|\varphi) = \hat{f}_{U_0}(u|0, \varphi) \hat{f}_{V_0}(v|0, \varphi) \hat{f}_{U_1}(x|0, \varphi) \hat{f}_{V_1}(y|0, \varphi), \tag{7.28}$$

where $\hat{f}_X(\cdot|0, \varphi)$ denotes the conditional density function for an arbitrary random variable X given that 0 is sent and that $\varphi_0 = \varphi$. Let $f_X(\cdot|0)$ denote the conditional density function for X given that 0 is sent (but not conditioned on φ_0).

When conditioned on 0 being sent and $\varphi_0 = \varphi$, each of the random variables U_0, V_0, U_1, and V_1 is Gaussian. With this conditioning, U_0 has mean $u_0 = \alpha_0 \cos(\varphi)$ and V_0 has mean $v_0 - -\alpha_0 \sin(\varphi)$, while both U_1 and V_1 have mean zero. Also, U_0 and V_0 have conditional variance σ_0^2, the same as the variance of the random variables X_0 and Y_0 that represent the noise components of U_0 and V_0, respectively. Similarly, U_1 and V_1 have conditional variance σ_1^2, the same as the variance of the random variables X_1 and Y_1. It follows from the conclusions about U_0 and V_0 that

$$\hat{f}_{U_0}(u|0, \varphi) \hat{f}_{V_0}(v|0, \varphi) = \exp\{-[u - u_0]^2/2\sigma_0^2\} \exp\{-[v - v_0]^2/2\sigma_0^2\}/2\pi\sigma_0^2, \tag{7.29}$$

where $u_0 = \alpha_0 \cos(\varphi)$ and $v_0 = -\alpha_0 \sin(\varphi)$.

Because U_1 and V_1 do not depend on φ_0, $\hat{f}_{U_1}(x|0, \varphi)$ and $\hat{f}_{V_1}(y|0, \varphi)$ do not depend on φ, and $\hat{f}_{U_1}(x|0, \varphi) = f_{U_1}(x|0)$ and $\hat{f}_{V_1}(y|0, \varphi) = f_{V_1}(y|0)$ for all φ. As a result, these densities can be moved outside the integral in (7.27), which gives

$$f_0(u, v, x, y) = f_{U_1}(x|0) \, f_{V_1}(y|0) \int_0^{2\pi} \hat{f}_{U_0}(u|0, \varphi) \hat{f}_{V_0}(v|0, \varphi) \, d\varphi/2\pi. \tag{7.30}$$

But $U_1 = X_1$ and $V_1 = Y_1$, so that

$$f_0(u, v, x, y) = f_{X_1}(x|0) \, f_{Y_1}(y|0) \int_0^{2\pi} \hat{f}_{U_0}(u|0, \varphi) \hat{f}_{V_0}(v|0, \varphi) \, d\varphi/2\pi. \tag{7.31}$$

The random variables X_1 and Y_1 are functions of the noise only, so they do not depend on the fact that 0 is sent. Therefore,

$$f_{X_1}(x|0) = f_{X_1}(x) = \exp\{-x^2/2\sigma_1^2\}/\sqrt{2\pi}\,\sigma_1$$

and

$$f_{Y_1}(y|0) = f_{Y_1}(y) = \exp\{-y^2/2\sigma_1^2\}/\sqrt{2\pi}\,\sigma_1.$$

For the present, it is better to leave the expression for $f_0(u, v, x, y)$ in terms of $f_{X_1}(x)$ and $f_{Y_1}(y)$. Thus, we express (7.31) as

$$f_0(u, v, x, y) = f_{X_1}(x) \, f_{Y_1}(y) \int_0^{2\pi} \hat{f}_{U_0}(u|0, \varphi) \hat{f}_{V_0}(v|0, \varphi) \, d\varphi/2\pi. \tag{7.32}$$

Denote the integral in (7.32) by $\mathcal{K}(u, v)$. In order to evaluate $\mathcal{K}(u, v)$, we substitute for the conditional densities for U_0 and V_0 using (7.29). Multiplying out the quadratic terms in the exponents and moving terms that do not depend on φ outside the integral, we obtain

$$\mathcal{K}(u, v) = (2\pi\sigma_0^2)^{-1} \exp\{-(u^2 + v^2)/2\sigma_0^2\}$$
$$\times \int_0^{2\pi} \exp\{[2\,uu_0 + 2\,vv_0 - u_0^2 - v_0^2]/2\sigma_0^2\}\,d\varphi/2\pi. \qquad (7.33)$$

Notice that

$$(2\pi\sigma_0^2)^{-1} \exp\{-(u^2 + v^2)/2\sigma_0^2\} = f_{X_0}(u)\,f_{Y_0}(v),$$

and recall that $u_0 = \alpha_0 \cos(\varphi)$ and $v_0 = -\alpha_0 \sin(\varphi)$. It follows that $u_0^2 + v_0^2 = \alpha_0^2$, which does not depend on φ. Using these observations, we can rewrite (7.33) as

$$\mathcal{K}(u, v) = f_{X_0}(u)\,f_{Y_0}(v) \exp\{-\alpha_0^2/2\sigma_0^2\}\mathcal{H}(u, v), \qquad (7.34)$$

where

$$\mathcal{H}(u, v) = \int_0^{2\pi} \exp\{(2\,uu_0 + 2\,vv_0)/2\sigma_0^2\}\,d\varphi/2\pi. \qquad (7.35)$$

Substituting for u_0 and v_0, we find that

$$\mathcal{H}(u, v) = \int_0^{2\pi} \exp\{\alpha_0[u\cos(\varphi) - v\sin(\varphi)]/\sigma_0^2\}\,d\varphi/2\pi. \qquad (7.36)$$

This integral cannot be written in closed form, but it can be expressed in terms of the tabulated integral

$$I_0(z) = \int_0^{2\pi} \exp\{z\cos(\varphi)\}\,d\varphi/2\pi,$$

which is the modified Bessel function of the first kind of order zero [7.1]. Note that $I_0(z) = I_0(-z)$. An important property of this function is that it is a strictly increasing function of the magnitude of its argument (i.e., $|z_1| < |z_2|$ implies $I_0(z_1) < I_0(z_2)$). It is left as an exercise (see Problem 7.2) to show that (7.36) reduces to

$$\mathcal{H}(u, v) = I_0\left(\alpha_0\sqrt{u^2 + v^2}/\sigma_0^2\right). \qquad (7.37)$$

Finally, we combine (7.32), (7.34), and (7.37) to conclude that

$$f_0(u, v, x, y) = f_{X_1}(x)\,f_{Y_1}(y)\,f_{X_0}(u)\,f_{Y_0}(v) \exp\{-\alpha_0^2/2\sigma_0^2\}I_0\left(\alpha_0\sqrt{u^2 + v^2}/\sigma_0^2\right). \qquad (7.38)$$

A repeat of the preceding steps under the condition that 1 is sent instead of 0 being sent shows that

$$f_1(u, v, x, y) = f_{X_0}(u)\,f_{Y_0}(v)\,f_{X_1}(x)\,f_{Y_1}(y) \exp\{-\alpha_1^2/2\sigma_1^2\}I_0\left(\alpha_1\sqrt{x^2 + y^2}/\sigma_1^2\right). \qquad (7.39)$$

Now consider the likelihood ratio test: If the observed values of U_0, V_0, U_1, and V_1 are u, v, x, and y, respectively, then, for the likelihood ratio, we have

$$\Lambda(u, v, x, y) = f_1(u, v, x, y)/f_0(u, v, x, y).$$

Decide that 1 was sent if $\Lambda(u, v, x, y) > \eta$ and that 0 was sent if $\Lambda(u, v, x, y) < \eta$. Because the random variables in question are continuous, equality occurs with probability zero in the comparison of the likelihood ratio with the parameter η, and we need not concern ourselves with events of probability zero.

Notice from (7.38) and (7.39) that both $f_0(u, v, x, y)$ and $f_1(u, v, x, y)$ have the product term

$$f_{X_0}(u)\, f_{Y_0}(v)\, f_{X_1}(x)\, f_{Y_1}(y)$$

in common. Canceling this common term in the likelihood ratio, we are left with

$$\Lambda(u, v, x, y) = \frac{I_0\left(\alpha_1\sqrt{x^2+y^2}/\sigma_1^2\right)\exp\{-\alpha_1^2/2\sigma_1^2\}}{I_0\left(\alpha_0\sqrt{u^2+v^2}/\sigma_0^2\right)\exp\{-\alpha_0^2/2\sigma_0^2\}}.$$

Thus, we should decide that 1 was sent if

$$I_0\left(\alpha_1\sqrt{x^2+y^2}/\sigma_1^2\right)\Big/ I_0\left(\alpha_0\sqrt{u^2+v^2}/\sigma_0^2\right) > \eta' \qquad (7.40a)$$

and decide that 0 was sent if

$$I_0\left(\alpha_1\sqrt{x^2+y^2}/\sigma_1^2\right)\Big/ I_0\left(\alpha_0\sqrt{u^2+v^2}/\sigma_0^2\right) < \eta'. \qquad (7.40b)$$

The parameter η' in (7.40) is given by

$$\eta' = \eta\exp\{-\alpha_0^2/2\sigma_0^2\}\big/\exp\{-\alpha_1^2/2\sigma_1^2\},$$

which can be written as

$$\eta' = \eta\exp\{[\alpha_1^2\sigma_0^2 - \alpha_0^2\sigma_1^2]/2\sigma_0^2\sigma_1^2\}.$$

To determine the Bayes decision rule, we simply use the likelihood ratio test with $\eta = \pi_0/\pi_1$. The minimax decision rule is obtained from the set of Bayes rules (for all possible values of π_0 and π_1) by equating $P_{e,0}$ and $P_{e,1}$ to find an equalizer decision rule.

Suppose that $\alpha_0 = \alpha_1$ and $\sigma_0 = \sigma_1$, as is true for BFSK with the receiver of Figure 7-4. Based on the symmetry of the situation, one would expect that the minimax decision rule corresponds to $\pi_0 = \pi_1$ (hence, $\eta = 1$). This is shown to be true in the next subsection, by showing that $\pi_0 = \pi_1$ gives $P_{e,0} = P_{e,1}$. Thus, for $\alpha_0 = \alpha_1$ and $\sigma_0 = \sigma_1$, we see that $\eta = \eta'$, and we conclude that the minimax decision rule, the Bayes decision rule, and the maximum-likelihood decision rule are obtained for $\eta = 1$. But, for this value of η, the decision rule is equivalent to the following: Decide that 1 was sent if

$$I_0\left(\alpha\sqrt{x^2+y^2}/\sigma^2\right) > I_0\left(\alpha\sqrt{u^2+v^2}/\sigma^2\right),$$

and decide that 0 was sent if

$$I_0\left(\alpha\sqrt{x^2+y^2}\big/\sigma^2\right) < I_0\left(\alpha\sqrt{u^2+v^2}\big/\sigma^2\right),$$

where according to the convention adopted in Section 5.4, we should agree to decide that 1 was sent in case

$$I_0\left(\alpha\sqrt{x^2+y^2}\big/\sigma^2\right) = I_0\left(\alpha\sqrt{u^2+v^2}\big/\sigma^2\right),$$

which occurs with probability zero. In these expressions, $\alpha = \alpha_0 = \alpha_1$ and $\sigma = \sigma_0 = \sigma_1$.

Because $I_0(z)$ is a strictly increasing function of $|z|$, this decision rule is equivalent to deciding 0 was sent if $\sqrt{u^2+v^2} > \sqrt{x^2+y^2}$, and deciding 1 was sent if $\sqrt{x^2+y^2} \geq \sqrt{u^2+v^2}$. Recall that the variables u, v, x, and y are the values taken on by the random variables U_0, V_0, U_1, and V_1, respectively. Thus, deciding that 0 was sent if $\sqrt{u^2+v^2} > \sqrt{x^2+y^2}$ corresponds to observing the values of $U_0^2 + V_0^2$ and $U_1^2 + V_1^2$, and making the decision on the basis of whichever is larger. In particular, decide that 0 was sent if $U_0^2 + V_0^2 > U_1^2 + V_1^2$. Our conclusion is summarized in the next statement, which is valid not only for BFSK with optimum correlators but also for any system that satisfies properties (1)–(3) with $\alpha_0 = \alpha_1$ and $\sigma_0 = \sigma_1$.

Maximum-Likelihood Decision for BFSK with Optimum Correlators:
If the observed value of $U_0^2 + V_0^2$ exceeds the observed value of $U_1^2 + V_1^2$, then decide that 0 was sent; otherwise, decide that 1 was sent.

Notice that this decision rule is exactly what is shown in Figure 7-4 for BFSK signaling. Thus, as a special case of the preceding decision rule, we have established that the receiver of Figure 7-4 is indeed the optimum receiver for BFSK signals under the minimax criterion, the Bayes criterion with $\pi_0 = \pi_1$, and the maximum-likelihood criterion.

The decision criteria that are of greatest interest are the minimax criterion, the Bayes criterion with $\pi_0 = \pi_1$, and the maximum-likelihood criterion, all of which have $\eta = 1$. However, we can determine the optimum receiver for the Bayes criterion with $\pi_0 \neq \pi_1$ by making a minor modification to the decision rule just obtained. The same decision statistics $R_0^2 = U_0^2 + V_0^2$ and $R_1^2 = U_1^2 + V_1^2$ are used, but if we observe $R_0 = r_0$ and $R_1 = r_1$, then we must compare the ratio

$$I_0(\alpha r_1/\sigma^2)\big/I_0(\alpha r_0/\sigma^2)$$

with $\eta = \pi_0/\pi_1$ in order to make the optimum decision. From a practical point of view, one of the main complications that results if $\eta \neq 1$ is that the ratio in question depends on the signal amplitude and the noise density. The receiver must know or be able to measure these parameters in order to make the optimum decision. In addition, the receiver must be able to compute values for the modified Bessel function $I_0(\cdot)$. Fortunately, most applications call for $\eta = 1$ to be used, so these difficulties do not actually arise often in practice.

We have now established the optimality of the receiver of Figure 7-4 for noncoherent reception of BFSK signals on an additive white Gaussian noise channel. In the next section, the probability of error that results from using this receiver is derived. In fact, the analysis presented in the next section applies to the more general signals and receivers discussed in Section 7.1, and the results on the error probability for BFSK are obtained as a special case.

7.5   Error Probabilities for the LS Decision Rule

We return to the system architecture illustrated in Figure 7-6. Recall that there is no requirement that the correlators in this system be optimum. In Section 7.4, we determined the decision rule that is optimum in the sense of maximum likelihood, minimax, and Bayes for $\pi_0 = \pi_1$. We found that if $\alpha_0 = \alpha_1$ and $\sigma_0 = \sigma_1$, then the two decision statistics R_0^2 and R_1^2 should be compared and the decision should be made on the basis of which is larger: If R_0^2 is larger than R_1^2, decide that 0 was sent; otherwise, decide that 1 was sent. We refer to such a decision as a *largest statistic* (LS) decision. Because LS decision rules are particularly easy to implement, they are used in some situations even if they are not quite optimum.

The goal of the present section is to determine the probability of error for the system of Figure 7-6 if the decision device uses the LS decision rule applied to the decision statistics R_0^2 and R_1^2. Our investigation considers the performance of this decision rule even if the conditions on the α's and σ's are not met. The reason for interest in this is twofold. First, as mentioned already, the simplicity of the LS decision rule makes it attractive. Second, in actual systems it may be that according to the design values of all of the system and channel parameters, $\alpha_0 = \alpha_1$ and $\sigma_0 = \sigma_1$, but in reality one or both of these equalities do not hold. This can occur because of gain variations in the branches of the receiver, imperfections in the integrators, frequency selectivity in the channel, or differences in the transmitted energies for $s_0(t)$ and $s_1(t)$. Note that if the only departure from ideal is that the received signal energies $\mathcal{E}_0$ and $\mathcal{E}_1$ are unequal, then $\sigma_0 = \sigma_1$ but $\alpha_0 \neq \alpha_1$.

A complete derivation of the probability of error is given for the LS decision rule applied to systems with $\sigma_0 = \sigma_1$. In this derivation, the values of α_0 and α_1 are arbitrary. The expressions that are obtained for the error probabilities are applied to BFSK with its optimum noncoherent receiver and to more general orthogonal signaling with optimum and suboptimum correlators. For systems with $\sigma_1 \neq \sigma_2$, only the final result is given, and its derivation is left as an exercise.

7.5.1   Derivation of the Error Probabilities

We now consider the LS decision rule applied to the decision statistics U_0, V_0, U_1, and V_1. All that is assumed is that properties (1)–(3) of Section 7.4 are satisfied and that $\sigma_0 = \sigma_1$. For convenience, these properties are restated here in a way that accounts for the fact that $\sigma_0 = \sigma_1$ (we denote the common value of $\sigma_0 = \sigma_1$ by σ):

(1) When 0 is sent and φ is the phase angle for $s_0(t)$, the decision statistics are given by $U_0 = \alpha_0 \cos(\varphi) + X_0$, $V_0 = -\alpha_0 \sin(\varphi) + Y_0$, $U_1 = X_1$, and $V_1 = Y_1$.

(2) When 1 is sent and φ is the phase angle for $s_1(t)$, these random variables are given by $U_0 = X_0$, $V_0 = Y_0$, $U_1 = \alpha_1 \cos(\varphi) + X_1$, and $V_1 = -\alpha_1 \sin(\varphi) + Y_1$.

(3) The random variables X_0, Y_0, X_1, and Y_1 in (1) and (2) are mutually independent, zero-mean, Gaussian random variables with common variance σ^2.

Because of the symmetry in the problem, it suffices to determine $P_{e,0}$, the probability of error given that 0 is sent. ($P_{e,1}$ can be determined from the results that we obtain for $P_{e,0}$.) Thus, we begin by assuming that 0 is sent and that φ is the value of the phase angle for $s_0(t)$. It follows from properties (1)–(3) that U_0, V_0, U_1, and V_1 are conditionally mutually independent given that 0 is sent and φ is the phase angle for $s_0(t)$.

A few words about conventions and notation are in order at this time. Because all probabilities in what follows are conditional probabilities given that 0 is sent and φ is the phase angle for $s_0(t)$, we do not display this conditioning explicitly in the expressions that follow, which simplifies the notation greatly. For the probabilities that appear in the expressions, precise statements that display the conditioning can be obtained by replacing $P(\cdot)$ with $P(\cdot|0, \varphi)$ in the appropriate expressions and corresponding substitutions can be made for the density functions. Similarly, all statements regarding conditional independence are assumed to be conditioned on the fact that 0 is sent and φ is the phase angle for $s_0(t)$. Finally, all square roots are defined as positive square roots of their arguments.

Because the notation is simpler, the setting that is employed in the analysis is slightly more general than that given in properties (1) and (2). Let U_0, V_0, U_1, and V_1 be mutually independent, Gaussian random variables with common variance σ^2. Suppose U_0 has mean u_0, V_0 has mean v_0, and U_1 and V_1 have mean 0. We wish to determine certain probabilities involving the random variables $U_0^2 + V_0^2$ and $U_1^2 + V_1^2$. In order to apply the results obtained in this setting to the original problem, we let $u_0 = \alpha_0 \cos(\varphi)$ and $v_0 = -\alpha_0 \sin(\varphi)$ later in the derivation.

Recall that the optimum decision rule is based on $R_0^2 = U_0^2 + V_0^2$ and $R_1^2 = U_1^2 + V_1^2$. If $U_1^2 + V_1^2 < U_0^2 + V_0^2$, the decision is that 0 was sent; otherwise, the decision is that 1 was sent. When 0 is in fact the signal that is sent, the correct decision will be made if and only if $U_1^2 + V_1^2 < U_0^2 + V_0^2$. Thus, the probability that we seek is the probability that the sum of the squares of the two zero-mean Gaussian random variables exceeds the sum of the squares of the two nonzero-mean Gaussian random variables when the four random variables are independent and have common variance σ^2.

Let $R_1 = \sqrt{U_1^2 + V_1^2}$, and observe that the probability of a correct decision is given by

$$P_{c,0} = P\left(R_1 < \sqrt{U_0^2 + V_0^2}\right).$$

Of course, the probability of error can be obtained from $P_{e,0} = 1 - P_{c,0}$. The probability of a correct decision can then be written as

$$P_{c,0} = \int_{-\infty}^{\infty} \int_{-\infty}^{\infty} P\left(R_1 < \sqrt{u^2 + v^2} \,\middle|\, U_0 = u, V_0 = v\right) f_{U_0}(u) f_{V_0}(v)\, du\, dv.$$

In this integral, we are first conditioning on (U_0, V_0) and then averaging out this conditioning by multiplying by the density function $f_{U_0, V_0}(u, v)$ and integrating over the

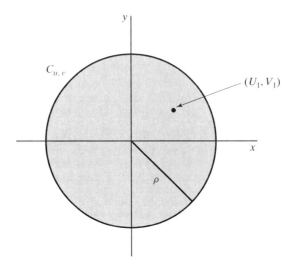

Figure 7-7: The set $C_{u,v}$.

entire range of u and v. We also use the fact that U_0 and V_0 are conditionally independent, which permits replacement of the joint density by the product of the marginal densities for U_0 and V_0.

The conditional mutual independence of U_0, V_0, U_1, and V_1 implies that the pair (U_1, V_1) is conditionally independent of the pair (U_0, V_0). Because R_1 is a function of U_1 and V_1 only, it follows that R_1 is conditionally independent of the pair (U_0, V_0). Therefore,

$$P\left(R_1 < \sqrt{u^2 + v^2} \,\middle|\, U_0 = u, V_0 = v\right) = P\left(R_1 < \sqrt{u^2 + v^2}\right)$$

for all u and v. Making this substitution into the preceding integral, we have

$$P_{c,0} = \int_{-\infty}^{\infty} \int_{-\infty}^{\infty} P\left(R_1 < \sqrt{u^2 + v^2}\right) f_{U_0}(u)\, f_{V_0}(v)\, \mathrm{d}u\, \mathrm{d}v. \tag{7.41}$$

Because $R_1 = \sqrt{U_1^2 + V_1^2}$, the probability $P\left(R_1 < \sqrt{u^2 + v^2}\right)$ is just the probability that the pair (U_1, V_1) falls inside the circle of radius $\rho = \sqrt{u^2 + v^2}$ that is centered at the origin, as illustrated in Figure 7-7. That is,

$$P\left(R_1 < \sqrt{u^2 + v^2}\right) = P[(U_1, V_1) \in C_{u,v}], \tag{7.42}$$

where $C_{u,v} = \left\{(x, y) : \sqrt{x^2 + y^2} < \sqrt{u^2 + v^2}\right\}$.

Next, use the fact that the probability that (U_1, V_1) falls in some set is just the integral of the joint density function for U_1 and V_1 over that set. Because U_1 and V_1 are conditionally independent, this integral can be written as

$$P[(U_1, V_1) \in C_{u,v}] = \iint_{C_{u,v}} f_{U_1}(x)\, f_{V_1}(y)\, \mathrm{d}x\, \mathrm{d}y. \tag{7.43}$$

Because both U_1 and V_1 are Gaussian with mean zero and variance σ^2,

$$
\begin{aligned}
f_{U_1}(x)\, f_{V_1}(y) &= \exp\{-x^2/2\sigma^2\} \exp\{-y^2/2\sigma^2\}/2\pi\sigma^2 \\
&= \exp\{-(x^2+y^2)/2\sigma^2\}/2\pi\sigma^2.
\end{aligned}
\tag{7.44}
$$

By combining (7.42), (7.43), and (7.44), we find that

$$
P\left(R_1 < \sqrt{u^2+v^2}\right) = \iint\limits_{C_{u,v}} \exp\{-(x^2+y^2)/2\sigma^2\}\, dx\, dy/2\pi\sigma^2.
\tag{7.45}
$$

Now, simplify the notation in (7.45) by substituting ρ for $\sqrt{u^2+v^2}$ and letting

$$
C_\rho = \left\{ (x, y) : \sqrt{x^2+y^2} < \rho \right\}.
$$

Notice that $\rho \geq 0$. This substitution in (7.45) gives

$$
P(R_1 < \rho) = \iint\limits_{C_\rho} \exp\left\{ -(x^2+y^2)/2\sigma^2 \right\} dx\, dy/2\pi\sigma^2.
\tag{7.46}
$$

To evaluate the integral in (7.46), change from rectangular coordinates (x, y) to polar coordinates (r, θ) using the usual transformation, which is defined by $x = r\cos(\theta)$ and $y = r\sin(\theta)$. This gives $r^2 = x^2 + y^2$ and $dx\, dy = r\, dr\, d\theta$. The resulting integral in polar coordinates is

$$
P(R_1 < \rho) = (2\pi\sigma^2)^{-1} \int_0^\rho \left\{ \int_0^{2\pi} r\exp\{-r^2/2\sigma^2\}\, d\theta \right\} dr.
$$

The integrand does not depend on θ, so the inner integral is evaluated easily to give

$$
P(R_1 < \rho) = \int_0^\rho (r/\sigma^2)\exp\{-r^2/2\sigma^2\}\, dr
\tag{7.47}
$$

$$
= 1 - \exp\{-\rho^2/2\sigma^2\}.
\tag{7.48}
$$

Exercise 7-2

From these results, find the density and distribution functions for the random variable R_1.

Solution.   Because R_1 is a continuous random variable, its distribution function satisfies $F_{R_1}(r) = P(R_1 \leq r) = P(R_1 < r)$, and thus the integral in (7.47) is equal to $F_{R_1}(\rho)$ for $\rho \geq 0$. Because R_1 is a nonnegative random variable, $F_{R_1}(\rho) = 0$ for $\rho < 0$. It follows that for $r \geq 0$ the integrand of (7.47) is the density function for R_1, and for negative values of its argument the density function for R_1 is zero. That is,

$$
f_{R_1}(r) = \begin{cases} (r/\sigma^2)\exp\{-r^2/2\sigma^2\}, & r \geq 0, \\ 0, & r < 0. \end{cases}
\tag{7.49}
$$

For nonnegative values of its argument, the corresponding distribution function is obtained from (7.48). For negative values of its argument, the distribution function is zero. The complete distribution function for the random variable R_1 is therefore given by

$$F_{R_1}(r) = \begin{cases} 1 - \exp\{-r^2/2\sigma^2\}, & r \geq 0, \\ 0, & r < 0. \end{cases} \tag{7.50}$$

■

The density function given by (7.49) is known as the *Rayleigh density*. The conclusion to be drawn from this exercise is that the random variable $R_1 = \sqrt{U_1^2 + V_1^2}$ has the Rayleigh distribution as specified by the density function in (7.49), or equivalently by the distribution function in (7.50).

Returning to the task at hand, we can now substitute $\sqrt{u^2 + v^2}$ for ρ in (7.48), to conclude that

$$P\left(R_1 < \sqrt{u^2 + v^2}\right) = 1 - \exp\{-(u^2 + v^2)/2\sigma^2\}.$$

This expression is then substituted into (7.41) to give

$$\begin{aligned} P_{c,0} &= \int_{-\infty}^{\infty} \int_{-\infty}^{\infty} [1 - \exp\{-(u^2 + v^2)/2\sigma^2\}] \, f_{U_0}(u) \, f_{V_0}(v) \, du \, dv \\ &= 1 - \int_{-\infty}^{\infty} \int_{-\infty}^{\infty} \exp\{-(u^2 + v^2)/2\sigma^2\} f_{U_0}(u) \, f_{V_0}(v) \, du \, dv. \end{aligned} \tag{7.51}$$

Notice that the integrand in (7.51) factors as

$$\exp\{-u^2/2\sigma^2\} f_{U_0}(u) \times \exp\{-v^2/2\sigma^2\} f_{V_0}(v).$$

Also, the form of $f_{U_0}(u)$ is the same as the form of $f_{V_0}(v)$; that is,

$$f_{U_0}(u) = \exp\{-(u - u_0)^2/2\sigma^2\}/\sqrt{2\pi}\,\sigma$$

and

$$f_{V_0}(v) = \exp\{-(v - v_0)^2/2\sigma^2\}/\sqrt{2\pi}\,\sigma.$$

Thus, to evaluate the double integral that appears in (7.51), we can evaluate the single integral

$$\mathcal{I}(\beta) = \int_{-\infty}^{\infty} \exp\{-x^2/2\sigma^2\} \exp\{-(x - \beta)^2/2\sigma^2\} \, dx/\sqrt{2\pi}\,\sigma, \tag{7.52}$$

where β is an arbitrary real number. We can then use the fact that

$$P_{c,0} = 1 - \mathcal{I}(u_0)\,\mathcal{I}(v_0)$$

in order to obtain an expression for the probability of a correct decision given that 0 is sent.

In order to evaluate (7.52), we write the integrand in a more convenient form by collecting terms inside a single exponential and making the substitution $\lambda = \sigma/\sqrt{2}$. This gives

$$\exp\{-u^2/2\sigma^2\}\exp\{-(u-\beta)^2/2\sigma^2\} = \exp\{-(u^2 - \beta u + \tfrac{1}{2}\beta^2)/2\lambda^2\}.$$

We can now complete the square by adding and subtracting $(\beta/2)^2$ to give

$$\exp\{-(u^2 - \beta u + \tfrac{1}{2}\beta^2)/2\lambda^2\} = \exp\{-[u^2 - 2(\tfrac{1}{2}\beta)u + (\tfrac{1}{2}\beta)^2 - (\tfrac{1}{2}\beta)^2 + \tfrac{1}{2}\beta^2]/2\lambda^2\}$$
$$= \exp\{-(u - \tfrac{1}{2}\beta)^2/2\lambda^2\}\exp\{-\tfrac{1}{4}\beta^2/2\lambda^2\}.$$

It follows that

$$\mathcal{I}(\beta) = (\lambda/\sigma)\exp\{-\tfrac{1}{4}\beta^2/2\lambda^2\}\int_{-\infty}^{\infty}\exp\{-(u - \tfrac{1}{2}\beta)^2/2\lambda^2\}\,du/\sqrt{2\pi}\lambda.$$

But

$$\int_{-\infty}^{\infty}\exp\{-(u - \tfrac{1}{2}\beta)^2/2\lambda^2\}\,du/\sqrt{2\pi}\lambda = 1,$$

because this is just the integral of the Gaussian density with mean $\beta/2$ and variance λ^2. As a result, we see that

$$\mathcal{I}(\beta) = (\lambda/\sigma)\exp\{-\tfrac{1}{4}\beta^2/2\lambda^2\}.$$

After substituting in this expression for λ, by using the fact that $\lambda = \sigma/\sqrt{2}$, we find that

$$\mathcal{I}(\beta) = \exp\{-\beta^2/4\sigma^2\}/\sqrt{2}.$$

Finally, we are ready to conclude that

$$P_{c,0} = 1 - \mathcal{I}(u_0)\mathcal{I}(v_0) = 1 - \tfrac{1}{2}\exp\{-(u_0^2 + v_0^2)/4\sigma^2\}.$$

It follows that the error probability is given by

$$P_{e,0} = \tfrac{1}{2}\exp\{-(u_0^2 + v_0^2)/4\sigma^2\}. \tag{7.53}$$

Recall that for the binary noncoherent communication systems characterized by properties (1)–(3), the parameters u_0 and v_0 are given by $u_0 = \alpha_0\cos(\varphi)$ and $v_0 = -\alpha_0\sin(\varphi)$. Substituting for u_0 and v_0 into (7.53), we see that the error probability is given by

$$P_{e,0} = \tfrac{1}{2}\exp\{-\alpha_0^2/4\sigma^2\}. \tag{7.54}$$

We now interchange the roles of (U_0, V_0) and (U_1, V_1) by letting $U_1 = \alpha_1 \cos(\varphi) + X_1$, $V_1 = -\alpha_1 \sin(\varphi) + Y_1$, $U_0 = X_0$, and $V_0 = Y_0$, so that U_0, V_0, U_1, and V_1 are conditionally mutually independent given that 1 is sent and φ is the phase angle for $s_1(t)$. We can then repeat these steps (or use the symmetry of the problem) to conclude that $P_{e,1}$ is given by

$$P_{e,1} = \tfrac{1}{2} \exp\{-\alpha_1^2/4\sigma^2\}. \tag{7.55}$$

Notice that if $\alpha_0 = \alpha_1$, such as for equal-energy signals and optimum correlators, the error probabilities specified by (7.54) and (7.55) do not depend on which signal is transmitted, nor do they depend on the phase angle of this signal. Letting α denote the common value of α_0 and α_1, we can say simply that the probability of error is

$$P_e = \tfrac{1}{2} \exp\{-\alpha^2/4\sigma^2\} \tag{7.56}$$

for the LS decision rule.

7.5.2   Applications to Specific Systems

For receivers of the form illustrated in Figure 7-6, the preceding results can be combined with those of Section 7.3 to obtain expressions for the error probabilities $P_{e,0}$ and $P_{e,1}$ in terms of the signals and the correlators that are employed in the system. Recall that the transmitted signals are given by

$$s_i(t) = \sqrt{2} m_i(t) \cos[\omega_c t + \theta_i(t) + \varphi],$$

where $m_i(t) = A a_i(t) p_T(t)$. From Section 7.3, we have that $\alpha_0 = (m_0, w_0)$, $\alpha_1 = (m_1, w_1)$, $\sigma_0^2 = N_0 \|w_0\|^2/2$, and $\sigma_1^2 = N_0 \|w_1\|^2/2$. For a receiver with correlators that satisfy $\|w_0\|^2 = \|w_1\|^2$, we can compute σ^2 from $\sigma^2 = N_0 \|w_0\|^2/2$ or $\sigma^2 = N_0 \|w_1\|^2/2$, whichever is more convenient. Also, for such a receiver, (7.54) and (7.55) imply that

$$P_{e,0} = \tfrac{1}{2} \exp\{-(m_0, w_0)^2/2N_0\|w_0\|^2\} \tag{7.57a}$$

and

$$P_{e,1} = \tfrac{1}{2} \exp\{-(m_1, w_1)^2/2N_0\|w_1\|^2\}. \tag{7.57b}$$

Equations (7.57) provide the desired results for a receiver with suboptimum correlators and an LS decision device. Given the modulation waveforms $m_0(t)$ and $m_1(t)$ and the correlator waveforms $w_0(t)$ and $w_1(t)$, these equations require only the evaluations of the inner products and norms in order to determine the error probabilities for the suboptimum receivers under consideration.

From (7.57a) it is clear that $P_{e,0}$ is a decreasing function of $(m_0, w_0)^2/\|w_0\|^2$. Consequently, the Schwarz inequality can be applied, as in Section 5.5, to show that the optimum choice of $w_0(t)$ is a constant multiple of $m_0(t)$, say, $w_0(t) = \lambda_0 m_0(t)$

for $0 \leq t \leq T$. Similarly, (7.57b) implies that the optimum choice of $w_1(t)$ is $w_1(t) = \lambda_1 m_1(t)$ for $0 \leq t \leq T$. If, as suggested by these observations, each correlator waveform has the same shape as the corresponding modulation waveform, equations (7.57) reduce to much simpler expressions.

Suppose the correlator waveforms satisfy $\|w_0\|^2 = \|w_1\|^2$, so that equations (7.57) apply, and suppose the correlator waveforms have the same shapes as the corresponding modulation waveforms. It follows that

$$(m_i, w_i)^2 / 2N_0 \|w_i\|^2 = \lambda_i^2 \|m_i\|^4 / 2N_0 \|\lambda_i m_i\|^2 = \|m_i\|^2 / 2N_0.$$

In particular, observe that $(m_i, w_i)^2 / 2N_0 \|w_i\|^2$ does not depend on λ_i. Next, we use the fact that $\mathcal{E}_i = \|m_i\|^2$, so that equations (7.57) become

$$P_{e,0} = \tfrac{1}{2} \exp\{-\mathcal{E}_0 / 2N_0\} \tag{7.58a}$$

and

$$P_{e,1} = \tfrac{1}{2} \exp\{-\mathcal{E}_1 / 2N_0\}. \tag{7.58b}$$

These are the error probabilities for binary orthogonal signals that do not necessarily have the same energy. The receiver employs correlator waveforms that are matched to the shape of the modulation waveforms, and the LS decision rule is used. If the signals have unequal energies, then it is clear that this receiver is not minimax ($P_{e,0} \neq P_{e,1}$). However, it may be that the signal energies are unknown, so implementation of the minimax receiver is not practical. In such a situation, it is common to design the correlator waveforms to match the shape of the modulation waveforms and satisfy $\|w_0\|^2 = \|w_1\|^2$, even though it may be that $\|m_0\|^2 \neq \|m_1\|^2$.

If the signals have the same energy $\mathcal{E}$ (i.e., if $\mathcal{E} = \mathcal{E}_0 = \mathcal{E}_1$), then expressions (7.58) reduce to

$$P_{e,0} = P_{e,1} = \tfrac{1}{2} \exp\{-\mathcal{E} / 2N_0\}. \tag{7.59}$$

This is the result promised in Section 7.1. (See equation (7.8).) It applies to systems with equal-energy signals and optimum receivers.

Equation (7.59) can also be obtained in a slightly different way. Because the signals have equal energy, it must be that $\|m_0\|^2 = \|m_1\|^2$. Therefore, $\|w_0\|^2 = \|w_1\|^2$ if and only if $\lambda_0 = \lambda_1$. It follows that $\alpha_0 = \alpha_1$ and that

$$\alpha^2 / \sigma^2 = 2(m_0, w_0)^2 / N_0 \|w_0\|^2 = 2\,\mathcal{E} / N_0.$$

Note that this fact is also pointed out in Section 7.3, equation (7.26). Thus, the term $\alpha^2 / 4\sigma^2$ that appears in (7.56) is given by $\alpha^2 / 4\sigma^2 = \mathcal{E} / 2N_0$. With this substitution for $\alpha^2 / 4\sigma^2$, (7.56) gives (7.59).

The expression in (7.59) is the error probability for the general, equal-energy, orthogonal signals considered in this section. In particular, (7.59) is valid for the signal set of Example 7-2 and for certain direct-sequence spread spectrum signals with noncoherent demodulation (Example 7-4).

As an illustration of the application of (7.56) and (7.57), consider BFSK with its optimum receiver. The modulation waveforms are given by $m_i(t) = A\, p_T(t)$ for all t, and the optimum receiver is shown in Section 7.2 to have $w_i(t) = 1$ for all t. As a result, $(m_i, w_i) = AT$ and $\|w_i\|^2 = T$. It follows from (7.57) that

$$P_{e,0} = P_{e,1} = \tfrac{1}{2}\exp\{-A^2T/2N_0\}. \tag{7.60}$$

Alternatively, it is pointed out in Section 7.2 that $\alpha = AT$ and $\sigma^2 = N_0T/2$ for BFSK with the correlators of Figure 7-4. These observations can be used in (7.56) to obtain (7.60) without evaluating the norms and inner products of (7.57). Because $\mathcal{E}_0 = \mathcal{E}_1 = A^2T$ for BFSK signals, we see also that (7.59) is valid for BFSK with its optimum receiver.

7.5.3   Summary of Results

The results of this section are summarized as follows: Suppose four jointly Gaussian random variables U_0, V_0, U_1, and V_1 are mutually independent and have common variance σ^2. The random variables U_0 and V_0 have mean values u_0 and v_0, respectively. The mean values of U_1 and V_1 are each zero. The probability that the sum of the squares of U_1 and V_1 exceeds the sum of the squares of U_0 and V_0 is

$$P(U_1^2 + V_1^2 > U_0^2 + V_0^2) = \tfrac{1}{2}\exp\{-(u_0^2 + v_0^2)/4\sigma^2\}. \tag{7.61}$$

The square root of the sum of the squares of two zero-mean, independent, Gaussian random variables having common variance σ^2 is a random variable with a Rayleigh density (given by (7.49)). It can also be shown (Problem 7.3), that if the means are not zero, the resulting random variable has a Rician density (defined in Problem 7.3).

More generally, it may be that U_0 and V_0 have variance σ_0^2, U_1 and V_1 have variance σ_1^2, and $\sigma_0 \neq \sigma_1$, in which case, equation (7.61) is not applicable. However, it can be shown that in general

$$P(U_1^2 + V_1^2 > U_0^2 + V_0^2) = [\sigma_1^2/(\sigma_0^2 + \sigma_1^2)]\exp\{-(u_0^2 + v_0^2)/2(\sigma_0^2 + \sigma_1^2)\}. \tag{7.62}$$

Equation (7.62) can be used to obtain results for suboptimum correlators in which $\|w_0\|^2 \neq \|w_1\|^2$. The derivation of (7.62) is left as an exercise, but some limiting cases are easy to check. First, notice that if $\sigma_0 = \sigma_1$, then (7.62) reduces to (7.61). Next observe that if $\sigma_1 \to 0$, the probability in (7.62) converges to 0. This should be expected, since for $\sigma_1 = 0$, both the noise and signal components are zero at the output of each of the lower two correlators of Figure 7-6. In essence,

$$P(U_1^2 + V_1^2 > U_0^2 + V_0^2) = P(0 > U_0^2 + V_0^2)$$

if $\sigma_1 = 0$. Finally, if σ_1 is held constant, $0 < \sigma_1 < \infty$, and $\sigma_0 \to \infty$, then the probability in (7.62) converges to 0.

The error probabilities that result when certain signals and correlators are used in receivers that employ the LS decision rule are given in Section 7.5.2. For equal-energy signals and correlators that satisfy the stated restriction, it is shown that the probability of error is given by

$$P_e = \tfrac{1}{2} \exp\{-\mathcal{E}/2N_0\},$$

regardless of which signal is sent and regardless of the value of the phase angle of the received signal. In particular, optimum noncoherent receivers for BFSK achieve a probability of error of $\exp\{-\mathcal{E}/2N_0\}/2$. Numerical values for this error probability are given in Table 7.1 and Figure 7-3.

7.6   Differentially Coherent BPSK

In Chapter 6 and thus far in Chapter 7, we have considered coherent reception and noncoherent reception of various carrier-modulated signals. For phase-shift keying (PSK), there is an alternative that can be viewed as falling between these two extremes. It is known as *differentially coherent demodulation* of PSK. This alternative comes from the realization that it is possible to demodulate a particular pulse by using the previous pulse to obtain a phase reference. In effect, the demodulation is based on the *phase difference* between the present pulse and the previous pulse. Because this previous pulse is corrupted by noise, this approach does not give a perfect phase reference; hence, we would not expect to obtain an error probability as small as can be achieved with perfect coherent reception of PSK. On the other hand, perhaps we can do better than with noncoherent reception of orthogonal signals, in part because differentially coherent demodulation permits the use of a better signal set than the traditional orthogonal signal set.

In the previous sections of this chapter, it is assumed that the phase angle φ is constant for the duration of each individual pulse; that is, φ must be constant for $nT \leq t < (n+1)T$. Now, we must make the stronger assumption that φ does not change very much from one interval to the next (i.e., φ is slowly varying relative to the transmission rate). Ideally, φ should be constant for all time, but this is a stronger condition than necessary to obtain the performance predicted by the analysis that is presented in this section.

7.6.1   Differentially Encoded BPSK Signal Format

We begin by considering a BPSK signal for two consecutive pulses. The pulse intervals are from $-T$ to 0 and from 0 to T. For these intervals, the signal is given by

$$s(t) = \sqrt{2}\, A\, d_{-1}\, p_T(t+T) \cos(\omega_c t + \varphi) + \sqrt{2}\, A\, d_0\, p_T(t) \cos(\omega_c t + \varphi), \qquad (7.63)$$

where d_{-1} is either $+1$ or -1 and represents the polarity of the baseband pulse transmitted during the first interval, and d_0 is either $+1$ or -1 and represents the polarity of the baseband pulse transmitted during the second interval. For all of the usual reasons, we assume that $\omega_c T$ is a multiple of 2π throughout this section.

The receiver does not know φ, and in this chapter we are considering systems and channels for which it is not possible (or at least not feasible) for the receiver to obtain a reliable estimate of φ. But the point of this section is that it is not necessary to know φ

in order to make a reliable decision as to whether d_0 is the same as or different from d_{-1}. So, if the information is conveyed by the agreements and disagreements in consecutive pulse polarities, then it is possible to demodulate the resulting PSK-like signal without knowledge of the phase angle φ. Based on this formulation, it is also clear that the general approach is not limited to BPSK. The same method can be applied to binary antipodal ASK, and the approach generalizes to certain nonbinary modulation formats such as QPSK and some forms of 4-QASK.

Let us first examine the signal set that results if $s(t)$ is as given by (7.63) and the information is conveyed in the product $d_{-1} d_0$. Notice that because $d_i = \pm 1$ for each i, it must be that $d_{-1} d_0 = +1$ if $d_{-1} = d_0$ (agreement in polarity for consecutive pulses) and $d_{-1} d_0 = -1$ if $d_{-1} \neq d_0$ (disagreement in polarity for consecutive pulses). We can rewrite (7.63) as

$$s(t) = \sqrt{2} \, A \, d_{-1} \{ p_T(t+T) + d_{-1} \, d_0 \, p_T(t) \} \cos(\omega_c t + \varphi). \qquad (7.64)$$

In order to obtain (7.64), we have used the fact that $(d_{-1})^2 = +1$ for either value of d_{-1}. This equation clearly displays the dependence on $d_{-1} d_0$, and it also points out that the first pulse can be considered a reference pulse. In order to determine the value of $d_{-1} d_0$, we only need to determine whether the polarity of the second pulse is the same as the first (in which case $d_{-1} d_0 = +1$) or different from the first (in which case $d_{-1} d_0 = -1$). The next step is to examine the two different signals on the interval $[-T, T]$ that result from these two possibilities.

Let $s_0(t)$ be the signal that corresponds to two consecutive pulses having the same polarity. This signal is given by (7.64) with $d_{-1} d_0 = +1$; that is,

$$s_0(t) = \sqrt{2} \, A \, d_{-1} \{ p_T(t+T) + p_T(t) \} \cos(\omega_c t + \varphi). \qquad (7.65a)$$

Let $s_1(t)$ be the signal that corresponds to a difference in polarity between the two pulses, which is the signal given by (7.64) with $d_{-1} d_0 = -1$. Thus, we have

$$s_1(t) = \sqrt{2} \, A \, d_{-1} \{ p_T(t+T) - p_T(t) \} \cos(\omega_c t + \varphi). \qquad (7.65b)$$

The way that these two waveforms are used to convey information is as follows: In order to send a 0 during the interval from 0 to T, the signal $s_0(t)$ is transmitted, and in order to send a 1 during the interval from 0 to T, the signal $s_1(t)$ is transmitted. Note that the information bit does not affect the waveform that is transmitted during the interval from $-T$ to 0; for both 0 and 1, this waveform is

$$\sqrt{2} \, A \, d_{-1} \, p_T(t+T) \cos(\omega_c t + \varphi).$$

Thus, a fixed reference waveform is transmitted in the interval $[-T, 0]$, and this is followed by a waveform in the interval $[0, T]$ that depends on whether a 0 or a 1 is to be sent.

Notice that the value of d_0 is determined by two elements: the value of d_{-1} and the symbol that is to be sent. Let β_0 denote the symbol that is to be sent in the interval $[0, T]$; that is, the variable β_0 represents the information bit the transmitter wishes to convey to the receiver during this time interval. Define b_0 by $b_0 = +1$ if $\beta_0 = 0$ and

$b_0 = -1$ if $\beta_0 = 1$. Then d_0 is given by $d_0 = d_{-1}b_0$. Notice that $d_0 = d_{-1}$ if $\beta_0 = 0$ ($b_0 = +1$) and $d_0 = -d_{-1}$ if $\beta_0 = 1$ ($b_0 = -1$). Observe also that $d_0 = d_{-1}b_0$ implies $d_{-1}d_0 = (d_{-1})^2 b_0 = b_0$, so the transmitted signal given by (7.64) can be written as

$$s(t) = \sqrt{2}\, A\, d_{-1}\{p_T(t + T) + b_0\, p_T(t)\} \cos(\omega_c t + \varphi). \qquad (7.66)$$

The signal that results from the modulation process just described is referred to as a *differentially encoded signal*, because the information is conveyed by polarity differences for consecutive pulses rather than by the absolute polarity of the individual pulses. Since a polarity difference in the amplitude is equivalent to a *phase difference* of π radians and a polarity agreement is equivalent to *no phase difference* (i.e., 0 radians), the form of signaling represented by (7.65) is known as *differentially encoded* BPSK (DBPSK). The equivalent representation of DBPSK in terms of phase modulation can be expressed mathematically by writing the signal that is transmitted in the interval from nT to $(n + 1)T$ as $\sqrt{2}\, A \cos(\omega_c t + \psi_n + \varphi)$, for $n = -1$ and $n = 0$. The equivalence to the previous description of DBPSK is obtained if $\psi_0 = \psi_{-1}$ for $\beta_0 = 0$ and $\psi_0 = \psi_{-1} + \pi$ for $\beta_0 = 1$, which is expressed concisely as $\psi_0 = \psi_{-1} + \pi\beta_0$. (In each such expression that involves phase angles, the sum is to be interpreted as addition modulo 2π.) If we insist that the phase-modulated signal match the reference pulse in the interval from $-T$ to 0, then we must impose the additional requirement $\psi_{-1} = 0$ if $d_{-1} = +1$ and $\psi_{-1} = \pi$ if $d_{-1} = -1$, which also guarantees that ψ_0 is either 0 or π.

Now return to the description of DBPSK in terms of amplitude modulation. The information conveyed by the signal defined in (7.66) is the value of the symbol b_0. A sequence of data symbols can be sent by using similar waveforms in subsequent intervals. For example, analogous to (7.64), the waveform

$$\sqrt{2}\, A\, d_0\{p_T(t) + d_0\, d_1\, p_T(t - T)\} \cos(\omega_c t + \varphi)$$

can be used to represent the information symbol b_1 if d_1 is selected to satisfy $d_0\, d_1 = b_1$. If a 0 is to be sent, then $b_1 = +1$ and $d_1 = d_0$ (there is no polarity change between the two pulses involved), and the transmitted signal is

$$\sqrt{2}\, A\, d_0\{p_T(t) + p_T(t - T)\} \cos(\omega_c t + \varphi).$$

If a 1 is to be sent, then $b_1 = -1$ and $d_1 = -d_0$ (there is a polarity change between the two pulses involved), and the transmitted signal is

$$\sqrt{2}\, A\, d_0\{p_T(t) - p_T(t - T)\} \cos(\omega_c t + \varphi).$$

In either case, the signal

$$\sqrt{2}\, A\, d_0\, p_T(t) \cos(\omega_c t + \varphi)$$

is transmitted in the interval $[0, T]$, as required by the previous selection of d_0. Thus, analogous to (7.66), the signal for the interval $[0, 2T]$ is

$$\sqrt{2}\, A\, d_0\{p_T(t) + b_1\, p_T(t - T)\} \cos(\omega_c t + \varphi).$$

Now that d_1 is determined, 0 ($b_2 = +1$) or 1 ($b_2 = -1$) can be sent in the interval $[2T, 3T]$ by letting the amplitude d_2 satisfy $d_1 d_2 = b_2$, so the signal for the interval $[T, 3T]$ is given by

$$\sqrt{2} A d_1 \{ p_T (t - T) + d_1 d_2 \, p_T (t - 2T) \} \cos(\omega_c t + \varphi)$$
$$= \sqrt{2} A d_1 \{ p_T (t - T) + b_2 \, p_T (t - 2T) \} \cos(\omega_c t + \varphi).$$

This process can be continued to send as many binary digits as desired. In order to send N binary digits in this manner, $N + 1$ intervals of length T are needed. Thus, the data rate is $N/(N + 1)$ bits per interval, rather than one bit per interval as for standard BPSK. For large values of N, this slight reduction in data rate is not of any consequence in most applications.

The formal procedure for differentially encoding the data is as follows. Suppose that the sequence of symbols to be sent is denoted by $\beta_0, \beta_1, \beta_2, \ldots$. This is a sequence of 0s and 1s representing the information to be conveyed to the receiver. Map this sequence into a corresponding sequences of $+1$s and -1s by letting $b_n = +1$ if $\beta_n = 0$ and $b_n = -1$ if $\beta_n = 1$. If we were not using differential encoding, then the sequence $b_0, b_1, b_2, \ldots$ would be the sequence of polarities for the transmitted pulses. For differential encoding, the value of d_{-1} is chosen arbitrarily to be either $+1$ or -1, and the *differentially encoded data sequence* $d_0, d_1, d_2, \ldots$ is defined recursively by letting $d_n = d_{n-1} b_n$ for each nonnegative integer n. The sequence $d_0, d_1, d_2, \ldots$ is then the sequence of polarities for the transmitted pulses in the differentially encoded signal.

The corresponding representation in terms of phase modulation is

$$s(t) = \sqrt{2} A \cos(\omega_c t + \psi_n + \varphi), \quad nT \leq t < (n + 1)T,$$

where $\psi_n = \psi_{n-1} + \pi \beta_n$. To send 0 in the nth interval ($\beta_n = 0$), let $\psi_n = \psi_{n-1}$; and to send 1 in this interval ($\beta_n = 1$), let $\psi_n = \psi_{n-1} + \pi$. Thus, if two consecutive pulse amplitudes are of the same polarity, then the two corresponding phases are equal; that is, $d_{n-1} d_n = +1$ corresponds to $\psi_n = \psi_{n-1}$. Similarly, the condition that two consecutive pulse amplitudes differ in polarity (i.e., $d_{n-1} d_n = -1$) corresponds to $\psi_n = \psi_{n-1} + \pi$.

7.6.2   A Receiver for DBPSK and the Resulting Probability of Error

Return to the consideration of the two signals $s_0(t)$ and $s_1(t)$ defined in (7.65). From the preceding discussion, we see that these are of the same form as the signals that are sent in subsequent intervals, so it suffices to determine what happens when we attempt to recover the data symbol that is represented by the selection of $s_0(t)$ vs. $s_1(t)$. It is easy to show that the two signals $s_0(t)$ and $s_1(t)$ are orthogonal on the interval $[-T, T]$. Because $\omega_c T$ is a multiple of 2π, double-frequency components can be ignored. Orthogonality then follows from the fact that the baseband signals

$$a_0(t) = p_T(t + T) + p_T(t) \tag{7.67a}$$

and

$$a_1(t) = p_T(t + T) - p_T(t) \tag{7.67b}$$

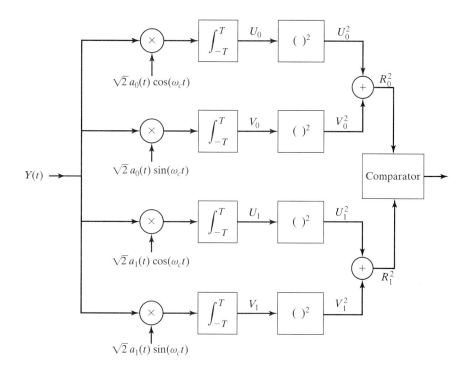

Figure 7-8: Differentially coherent receiver for DBPSK.

are orthogonal on $[-T, T]$. We know that it is possible to demodulate orthogonal carrier-modulated signals without having a phase reference, which is the principle that leads to one type of optimum demodulator for DBPSK. Notice that $d_{-1}a_0(t)$ and $d_{-1}a_1(t)$ are orthogonal regardless of the value of d_{-1}; moreover, $s_0(t)$ and $s_1(t)$ are orthogonal for all values of d_{-1} and φ. This formulation of the reception of DBPSK results in a problem of noncoherent reception of binary orthogonal signals on an interval of length $2T$.

This formulation suggests the use of the optimum noncoherent demodulator of Figure 7-2 with the interval $[0, T]$ replaced by $[-T, T]$ and with $\theta_0(t) = \theta_1(t) = 0$ for all t. The result of making these modifications to Figure 7-2 is shown in Figure 7-8. This is an optimum receiver for DBPSK if the processing interval is restricted to be of no greater duration than $2T$. The decision made by the receiver is as follows. If $R_0^2 > R_1^2$, then decide that 0 was sent; otherwise, decide that 1 was sent.

The receiver of Figure 7-8 is clearly not optimum if it is permissible to process the received signal over a longer time interval than $[-T, T]$ and if the phase angle φ is constant over this longer interval. In fact, given a long enough processing interval, the coherent receiver discussed in Section 6.2 can be employed. However, in some applications the phase variations are too rapid to permit coherent reception, but they are slow enough to permit use of differentially coherent reception over intervals of duration $2T$. Furthermore, the coherent receiver requires the use of a squaring loop (Figure 6-9) or equivalent, which adds to the cost and complexity of the receiver. Finally, as we are about to show, if the signal-to-noise ratio is large, then the performance of the differentially coherent receiver is nearly as good as that of the coherent receiver. Thus, for many applications, differentially coherent reception of DBPSK is an attractive alternative to coherent reception of BPSK.

The performance of the differentially coherent receiver of Figure 7-8 is determined easily from the general expression for the probability of error for noncoherent reception of binary orthogonal signals. From (7.8), we know that the probability of error in deciding between two equal-energy, orthogonal signals is

$$P_e = \tfrac{1}{2} \exp(-\mathcal{E}'/2N_0)$$

if the channel is an additive white Gaussian noise channel with spectral density $N_0/2$ and the energy in each signal is $\mathcal{E}'$. For the present application, $\mathcal{E}'$ is the energy in the signals $s_0(t)$ and $s_1(t)$, which is twice the energy of the basic BPSK pulse. That is, if $\mathcal{E} = A^2 T$ is the energy in the waveform

$$\sqrt{2}\, A\, p_T(t) \cos(\omega_c t + \varphi),$$

then the energy in $s_0(t)$ is $\mathcal{E}' = 2A^2 T = 2\mathcal{E}$. This doubling of the effective energy is the result of processing the signal over an interval of duration $2T$ rather than over an interval of duration T.

It follows from the discussion in the previous paragraph that the error probability for the receiver of Figure 7-8 is

$$P_e = \tfrac{1}{2} \exp(-\mathcal{E}/N_0). \tag{7.68}$$

The first observation to make about (7.68) is that differentially coherent reception of DBPSK achieves a given error probability with exactly one-half the energy per pulse required by noncoherent reception of BFSK. This represents an energy savings of 3 dB.

The error probability given by (7.68) for differentially coherent reception of DBPSK and the error probability for noncoherent reception of BFSK, which is $\exp(-\mathcal{E}/2N_0)/2$, are shown in Figure 7-9. These two curves are 3 dB apart for each value of the error probability. Also shown in Figure 7-9 are the error probabilities $Q(\sqrt{2\mathcal{E}/N_0})$ and $Q(\sqrt{\mathcal{E}/N_0})$ for ideal coherent reception of BPSK and BFSK, respectively. Notice that if $\mathcal{E}/N_0$ is greater than about 5 dB, differentially coherent reception of DBPSK provides a substantially smaller probability of error than either noncoherent or coherent reception of BFSK. Furthermore, for large values of $\mathcal{E}/N_0$, differentially coherent reception of DBPSK gives nearly as small an error probability as ideal coherent reception of BPSK, but without the need to obtain the accurate phase estimate required for coherent reception.

It should be observed from Figure 7-8, that differentially coherent reception does require an accurate estimate of the carrier *frequency*, however, so its usefulness is limited in certain channels, such as those that exhibit a large Doppler shift or similar kind of disturbance. But, for channels with a stable carrier frequency and receivers with a reliable frequency reference, differentially coherent reception of DBPSK provides good performance, and it can be demodulated by a simpler receiver than required for coherent reception.

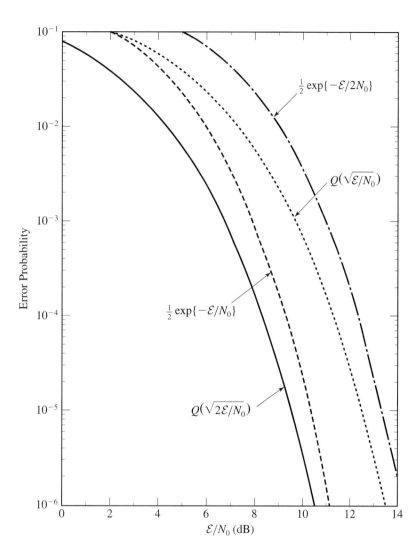

Figure 7-9: Error probabilities for four binary communication systems.

If the communication channel does result in a shift of the carrier frequency, it is still possible to use the idea of communication by means of phase differences. *Double-differential BPSK* (DDBPSK), also known as *second-order phase-difference modulation*, has been shown to be an effective modulation technique for channels that cause a fixed, but unknown, offset in the carrier frequency [7.3]. For both DDBPSK and standard DBPSK (also known as first-order phase-difference modulation), the transmitted signal is

$$s(t) = \sqrt{2}\,A\cos(\omega_c t + \psi_n + \varphi), \quad nT \le t < (n+1)T.$$

In order to represent the information sequence $\beta_0, \beta_1, \beta_2, \ldots$, which is a sequence of 0s and 1s, the phase angles ψ_n must satisfy $\psi_n = \psi_{n-1} + \pi\beta_n$ for standard DBPSK. If we define $\Delta_n = \pi\beta_n$, then the phase modulation in the nth interval is related to the phase modulation in the $(n-1)$st interval by $\psi_n = \psi_{n-1} + \Delta_n$. It is convenient to write this as $\Delta_n = \psi_n - \psi_{n-1}$ in order to display the fact that Δ_n is the phase difference for two successive pulses. The second-order phase difference, which is obtained by taking differences of successive values of Δ_n, is defined as $\Gamma_n = \Delta_n - \Delta_{n-1}$. Substituting for both Δ_n and Δ_{n-1} in terms of ψ_n, ψ_{n-1}, and ψ_{n-2}, we see that $\Gamma_n = \psi_n - 2\psi_{n-1} + \psi_{n-2}$, or, equivalently, the phase modulation for the signal in the nth interval is related to the phase modulation in the previous two intervals by

$$\psi_n = 2\psi_{n-1} - \psi_{n-2} + \Gamma_n.$$

Recall that ψ_n is either 0 or π, so $2\psi_{n-1} = 0$ (modulo 2π) in either case. It follows that for DDBPSK, $\Gamma_n = \psi_n + \psi_{n-2}$, or, equivalently, $\psi_n = -\psi_{n-2} + \Gamma_n$. Now ψ_{n-2} is either 0 or π, so $-\psi_{n-2} = +\psi_{n-2}$ since $-0 = 0$, of course, and $-\pi$ and $+\pi$ are equivalent phase angles (modulo 2π). Thus, for DDBPSK, we can write $\psi_n = \psi_{n-2} + \Gamma_n$.

In summary, for DBPSK and DDBPSK, the transmitted signal is of the form

$$s(t) = \sqrt{2}\,A\cos(\omega_c t + \psi_n + \varphi), \quad nT \le t < (n+1)T.$$

The phase of the waveform in the nth interval is $\psi_n = \psi_{n-1} + \Delta_n$ for DBPSK and $\psi_n = \psi_{n-2} + \Gamma_n$ for DDBPSK. The information is conveyed by setting $\Delta_n = \pi\beta_n$ in the DBPSK signal and $\Gamma_n = \Delta_n - \Delta_{n-1} = \pi(\beta_n - \beta_{n-1})$ in the DDBPSK signal. As shown in Problem 7.20, the DDBPSK signal conveys information properly in the presence of an offset in the carrier frequency, but the DBPSK signal does not. Additional information on DDBPSK is provided in [7.3].

7.6.3   Alternative Receivers for DBPSK

There are some implementation considerations that suggest alternative receivers to the one shown in Figure 7-8. Because this receiver must process waveforms of duration $2T$, it cannot be used as shown for the reception of a sequence of pulses. The problem is that this receiver requires $2T$ seconds to make a decision regarding a single symbol. As a result, two receivers of the type shown in Figure 7-8 would be required to demodulate a sequence of more than two pulses (representing more than one symbol). For instance, odd-numbered symbols could be demodulated by one receiver and even-numbered symbols by the other. However, a closer examination of Figure 7-8 reveals that two complete receivers are not really needed, and in fact we do not actually require the use of integrators that operate on intervals of length $2T$. This is demonstrated in the development that follows.

Consider the definition of $a_0(t)$ in (7.67a), and notice that U_0 in Figure 7-8 is given by

$$U_0 = \int_{-T}^{0} \sqrt{2}\,Y(t)\cos(\omega_c t)\,\mathrm{d}t + \int_{0}^{T} \sqrt{2}\,Y(t)\cos(\omega_c t)\,\mathrm{d}t. \tag{7.69}$$

If the first of the integrals in (7.69) is denoted by W_c and the second by Z_c, then we have $U_0 = W_c + Z_c$. An examination of Figure 7-8 and the definition of $a_1(t)$ shows that $U_1 = W_c - Z_c$.

By making the change of variable of integration $t = u - T$, the first integral of (7.69) can be written as

$$W_c = \int_0^T \sqrt{2} \, Y(u - T) \cos(\omega_c u - \omega_c T) \, du.$$

Because $\omega_c T$ is a multiple of 2π, $\cos(\omega_c u - \omega_c T) = \cos(\omega_c u)$; therefore,

$$W_c = \int_0^T \sqrt{2} \, Y(t - T) \cos(\omega_c t) \, dt.$$

Similarly, observe that V_0 is given by

$$V_0 = \int_{-T}^0 \sqrt{2} \, Y(t) \sin(\omega_c t) \, dt + \int_0^T \sqrt{2} \, Y(t) \sin(\omega_c t) \, dt. \qquad (7.70)$$

If the first of the integrals in (7.70) is denoted by W_s and the second by Z_s, then $V_0 = W_s + Z_s$ and $V_1 = W_s - Z_s$. Furthermore,

$$W_s = \int_0^T \sqrt{2} \, Y(t - T) \sin(\omega_c t) \, dt.$$

The decision for the receiver of Figure 7-8 is that 0 was sent if $R_0^2 > R_1^2$, and that 1 was sent otherwise. This is equivalent to deciding that 0 was sent if $R_0^2 - R_1^2 > 0$ and that 1 was sent if $R_0^2 - R_1^2 \le 0$. Next, notice that

$$R_0^2 = U_0^2 + V_0^2 = (W_c + Z_c)^2 + (W_s + Z_s)^2$$

and

$$R_1^2 = U_1^2 + V_1^2 = (W_c - Z_c)^2 + (W_s - Z_s)^2.$$

As a result,

$$R_0^2 - R_1^2 = 4(W_c Z_c + W_s Z_s). \qquad (7.71)$$

Thus, the same decision will be made if the receiver determines W_c, Z_c, W_s, and Z_s and then compares $W_c Z_c + W_s Z_s$ with 0 in order to make the decision. If

$$W_c Z_c + W_s Z_s > 0,$$

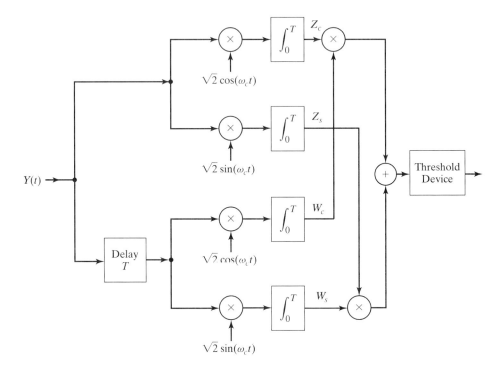

Figure 7-10: Alternative differentially coherent receiver for DBPSK.

then the decision is that 0 was sent, and if

$$W_c Z_c + W_s Z_s < 0,$$

then the decision is that 1 was sent. As usual, the event that equality occurs has probability zero, so we can ignore this event.

The advantage of the approach suggested by the preceding observations is that W_c, Z_c, W_s, and Z_s can all be obtained as the outputs of integrators that integrate over T seconds only. Integration over intervals of duration $2T$ is not really required in order to implement the differentially coherent receiver. A receiver that is equivalent to that shown in Figure 7-8 is illustrated in Figure 7-10.

Next we observe that if the delay is applied at the *output* of the integrators rather than at the input, then the number of integrators can be reduced from four to two. The resulting receiver is shown in Figure 7-11. This version of the receiver permits implementation of the delay element as a discrete-time device, rather than the continuous-time delay device required in Figure 7-10. For some applications, this is an additional advantage of the receiver shown in Figure 7-11.

Another receiver that is commonly suggested for demodulation of DBPSK involves delay and multiplication of the received continuous-time signal, as shown in Figure 7-12. Unfortunately, in much of the literature on the delay-and-multiply receiver, it is not explained that this receiver gives worse performance than the receivers illustrated in Figures 7-9 through 7-11.

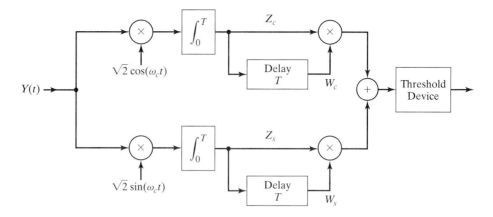

Figure 7-11: A differentially coherent receiver that employs only two integrators and permits use of discrete-time delay devices.

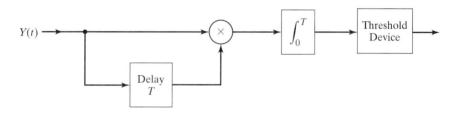

Figure 7-12: The delay-and-multiply receiver—a suboptimum receiver for DBPSK.

An examination of the output of the integrator in Figure 7-12 shows that it does not produce the same decision statistic as the receivers previously discussed. In particular, if the noise process is denoted by $X(t)$, observe that the output of this integrator contains a term of the form

$$\int_0^T X(t)\, X(t-T)\, dt,$$

which is not present in the decision statistic for the other receivers. It is true, however, that *at high signal-to-noise ratios*, the delay-and-multiply receiver gives an error probability that is nearly as small as for the other receivers discussed in this section. In fact, for very large signal-to-noise ratios, the term defined above is negligible compared with the other terms in the decision statistic.

Notice the similarity of the receiver of Figure 7-12 to the system of Figure 3-8 in Section 3.6.2. If $Y(t) = X(t)$ in Figure 3-8, then the system gives an estimate of the autocorrelation function. For this reason, the delay-and-multiply demodulator of Figure 7-12 is often referred to as an *autocorrelation demodulator* (ACD). An approximate analysis is given for this demodulator in [7.3], where its suboptimality is demonstrated quantitatively.

7.7   *M*-ary Noncoherent Correlation Receivers

The general form for the signals employed for M-ary noncoherent communications is the same as considered in Section 7.3. Thus, the signals are given by

$$s_i(t) = \sqrt{2}\, A\, a_i(t) \cos[\omega_c t + \theta_i(t) + \varphi_i] p_T(t)$$

for $0 \le i \le M - 1$. These can be written as

$$s_i(t) = \sqrt{2}\, m_i(t) \cos[\omega_c t + \theta_i(t) + \varphi_i], \qquad (7.72)$$

where

$$m_i(t) = A\, a_i(t)\, p_T(t).$$

Because the double-frequency terms are negligible, the energy in the ith signal is given by

$$\mathcal{E}_i = \int_0^T [m_i(t)]^2 \, dt.$$

One example is the signal set known as *M-ary frequency-shift keying* or *multiple frequency-shift keying* (MFSK). Both names are common, and each is abbreviated as MFSK. The MFSK signals are given by (7.72) with $a_i(t) = 1$ and $\theta_i(t) = (\omega_i - \omega_c)t$ for all t and for $0 \le i \le M - 1$. For these signals,

$$m_i(t) = A\, p_T(t),$$

so

$$s_i(t) = \sqrt{2}\, A \cos[\omega_i t + \varphi_i]\, p_T(t),$$

and $\mathcal{E}_i = A^2 T$. The individual MFSK signals are often referred to as tones. If attention is restricted to equally spaced tones, as in Example 6-6, then $\omega_i = \omega_0 + i\beta$ for $0 \le i \le M - 1$, where β is the frequency spacing between tones of the MFSK signal set.

The receiver under consideration is illustrated in Figure 7-13 for $M = 3$. In this illustration, there are three sections, one for each of the signals. In general, there are M such sections, one for each of the signals $s_i(t)$, $0 \le i \le M - 1$. Each section of the receiver is a noncoherent correlator in which the signals are correlated with

$$f_i(t) = \sqrt{2}\, w_i(t) \cos[\omega_c t + \theta_i(t)] \qquad (7.73)$$

and

$$g_i(t) = \sqrt{2}\, w_i(t) \sin[\omega_c t + \theta_i(t)], \qquad (7.74)$$

for each value of i. The integrator outputs, U_i and V_i, are squared and summed to produce the decision statistic R_i^2. The decision device then makes a decision based on the values of the statistics R_i^2, $0 \le i \le M - 1$.

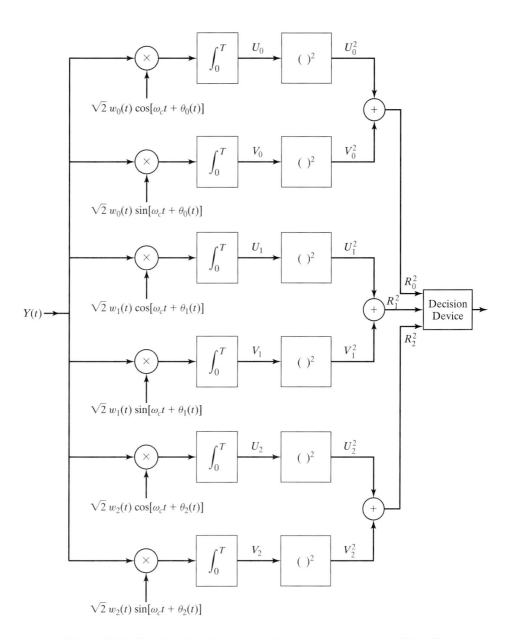

Figure 7-13: Receiver for *M*-ary noncoherent communications ($M = 3$).

The receiver of Figure 7-13 is a generalization of the optimum receiver. If $w_i(t)$ is not proportional to $m_i(t)$, then the correlators of Figure 7-13 are not optimum (i.e., they are not matched to the signal waveform). If $w_i(t)$ is proportional to $m_i(t)$ for each i, then the correlators are optimum. In fact, if the signals have equal energy and there is a constant c such that

$$w_i(t) = c\, m_i(t), \quad 0 \le i \le M - 1, \quad 0 \le t \le T,$$

then the receiver of Figure 7-13 is optimum.

The analysis presented in this section does not require that the waveforms $w_i(t)$ be proportional to $m_i(t)$, so that suboptimum receivers can be included in the development. However, the results of this section can also be used to determine the performance of optimum receivers by letting $w_i(t)$ be proportional to $m_i(t)$ in the final equations. In particular, the results for the noncoherent correlation receiver that gives the minimum probability of error for noncoherent reception of MFSK are obtained by setting

$$w_i(t) = a_i(t) = 1, \quad 0 \le t \le T$$

and

$$\theta_i(t) = (\omega_i - \omega_c)\,t, \quad 0 \le t \le T$$

for $0 \le i \le M - 1$.

The analysis that follows is valid only for signals $s_i(t)$, $f_i(t)$, and $g_i(t)$ that satisfy properties (1)–(4) given in Section 7.3. Examples of signals that satisfy these requirements can be obtained easily by generalizing the binary signals of Examples 7-3 and 7-4. We assume that $\|w_i\|^2 = \|w_0\|^2$ for each i, so the variance of the random variables X_i and Y_i is the same for each i.

The first part of the analysis follows very closely the analysis presented in Section 7.3, and so it is not repeated here. The results are summarized as follows: When conditioned on s_i being transmitted and $\varphi_i = \varphi$, the $2M$ random variables U_i, $0 \le i \le M - 1$, and V_i, $0 \le i \le M - 1$, are conditionally mutually independent and are given by

$$U_i = \alpha_i \cos(\varphi) + X_i,$$
$$V_i = -\alpha_i \sin(\varphi) + Y_i,$$
$$U_k = X_k, \ k \ne i,$$

and

$$V_k = Y_k, \ k \ne i,$$

where $\alpha_i = (m_i, w_i)$. The $2M$ random variables $X_k, 0 \le k \le M - 1$, and $Y_k, 0 \le k \le M - 1$, are mutually independent, zero-mean, Gaussian random variables, and X_k and Y_k each have variance $\sigma^2 = N_0 \|w_0\|^2/2$. If the correlators are optimum and the signals have equal energy $\mathcal{E}$, then for each i

$$\alpha_i = (m_i, w_i) = (m_i, c\, m_i) = c\|m_i\|^2 = \alpha,$$

where $\alpha = c\,\mathcal{E}$. In this case,

$$\sigma^2 = \tfrac{1}{2}N_0\|w_0\|^2 = \tfrac{1}{2}N_0\,c^2\mathcal{E}.$$

It follows that the signal-to-noise ratio is $\alpha^2/\sigma^2 = 2\,\mathcal{E}/N_0$.

The derivation of the optimum decision rule is an extension of the development in Section 7.5. Form the vector $\mathbf{R} = (R_0, R_1, \ldots, R_{M-1})$ of decision statistics, where R_i is the positive square root of R_i^2 for each i, and suppose that it is observed that $\mathbf{R} = \mathbf{r}$, where $\mathbf{r} = (r_0, r_1, \ldots, r_{M-1})$. The maximum-likelihood decision rule is to decide that s_k is the transmitted signal if

$$f_k(\mathbf{r}) = \max\{f_i(\mathbf{r}) : 0 \le i \le M - 1\}.$$

For equal-energy signals and optimum correlators, this reduces to deciding that s_k is transmitted if

$$r_k = \max\{r_i : 0 \le i \le M - 1\}.$$

Thus, as in Section 7.4, we find that the LS decision rule is the maximum-likelihood decision rule if equal-energy signals and optimum correlators are used.

The derivation of the probability of error for the LS decision rule parallels that given in Section 7.5. There are also some similarities to the analysis of Section 6.6 for coherent demodulation of *M*-ary orthogonal signals. To simplify the notation, let all probabilities be conditioned on the event that s_0 is transmitted. If 0 is sent, then the event that a correct decision is made is the event C_0 defined by

$$C_0 = \bigcap_{i=1}^{M-1} \{R_i < R_0\}. \tag{7.75}$$

Because we are not displaying the conditioning explicitly, the conditional probability of a correct decision given that s_0 is the transmitted signal is written as

$$P_{c,0} = P(C_0).$$

For convenience, let the intersection in (7.75) be denoted by $\bigcap_i$, so that C_0 can be written as

$$C_0 = \bigcap_i \{R_i < R_0\}.$$

Now replace R_0 with $\sqrt{U_0^2 + V_0^2}$ to give

$$C_0 = \bigcap_i \left\{ R_i < \sqrt{U_0^2 + V_0^2} \right\}.$$

If we condition on $U_0 = u$ and $V_0 = v$, we observe that

$$
\begin{aligned}
P(C_0 \mid U_0 = u, V_0 = v) &= P\!\left(\bigcap_i \left\{ R_i < \sqrt{U_0^2 + V_0^2} \right\} \,\middle|\, U_0 = u, V_0 = v \right) \\
&= P\!\left(\bigcap_i \left\{ R_i < \sqrt{u^2 + v^2} \right\} \,\middle|\, U_0 = u, V_0 = v \right) \\
&= P\!\left(\bigcap_i \left\{ R_i < \sqrt{u^2 + v^2} \right\} \right).
\end{aligned}
$$

Keep in mind that all of these probabilities are conditioned on the fact that s_0 is the transmitted signal.

Next observe that the $M - 1$ random variables R_i, $1 \leq i \leq M - 1$, are conditionally independent given that s_0 is the transmitted signal. Therefore,

$$P\left(\bigcap_i \left\{ R_i < \sqrt{u^2 + v^2} \right\} \right) = \prod_{i=1}^{M-1} P\left(R_i < \sqrt{u^2 + v^2} \right)$$

$$= \prod_{i=1}^{M-1} F_{R_1}\left(\sqrt{u^2 + v^2} \right). \tag{7.76}$$

In (7.76), we have used the fact that, when conditioned on the event that s_0 is the transmitted signal, each of the random variables R_i, $1 \leq i \leq M - 1$, has the same distribution as R_1.

From equation (7.50) in Section 7.5 we know that

$$F_{R_1}\left(\sqrt{u^2 + v^2} \right) = 1 - \exp\{-(u^2 + v^2)/2\sigma^2\},$$

so

$$P(C_0 \mid U_0 = u, V_0 = v) = [1 - \exp\{-(u^2 + v^2)/2\sigma^2\}]^{M-1}.$$

Because

$$P(C_0) = \int_{-\infty}^{\infty} \int_{-\infty}^{\infty} P(C_0 \mid U_0 = u, V_0 = v) \, f_{U_0}(u) \, f_{V_0}(v) \, du \, dv,$$

we have shown that the conditional probability of a correct decision given that s_0 is transmitted is

$$P_{c,0} = \int_{-\infty}^{\infty} \int_{-\infty}^{\infty} [1 - \exp\{-(u^2 + v^2)/2\sigma^2\}]^{M-1} f_{U_0}(u) \, f_{V_0}(v) \, du \, dv. \tag{7.77}$$

As in Section 6.6, the integral in (7.77) can be evaluated numerically. Unlike in Section 6.6, however, it can be expressed in terms of finite sum with no integrals. To obtain this closed-form expression, we use the fact that

$$(1 - x)^{M-1} = \sum_{i=0}^{M-1} \binom{M-1}{i} (-1)^i x^i$$

to write the integrand of (7.77) as

$$(1 - \exp\{-(u^2 + v^2)/2\sigma^2\})^{M-1} = \sum_{i=0}^{M-1} \binom{M-1}{i} (-1)^i \exp\{-i(u^2 + v^2)/2\sigma^2\}. \tag{7.78}$$

If the resulting expression is integrated term by term, the double integrals can be written as the product of two single integrals, each of the form

$$\mathcal{L}(u_0) = \int_{-\infty}^{\infty} \exp\{-iu^2/2\sigma^2\} \exp\{-(u - u_0)^2/2\sigma^2\} \, du/\sqrt{2\pi}\sigma. \tag{7.79}$$

In Problem 7.8, it is shown that this integral is equal to

$$\mathcal{L}(u_0) = \exp\{-iu_0^2/[2\sigma^2(i+1)]\}/\sqrt{i+1}. \tag{7.80}$$

As a consequence, (7.77) reduces to

$$P_{c,0} = \sum_{i=0}^{M-1} \binom{M-1}{i} (-1)^i (i+1)^{-1} \exp\{-i(u_0^2 + v_0^2)/2\sigma^2(i+1)\}. \tag{7.81}$$

Notice that the $i = 0$ term is equal to 1. Because the probability of error is

$$P_{e,0} = 1 - P_{c,0},$$

(7.81) implies that

$$P_{e,0} = -\sum_{i=1}^{M-1} \binom{M-1}{i} (-1)^i (i+1)^{-1} \exp\{-i(u_0^2 + v_0^2)/2\sigma^2(i+1)\}.$$

The term $(i+1)^{-1}$ can be absorbed into the binomial coefficient and the leading minus sign combined with the term $(-1)^i$ to give

$$\begin{aligned}
P_{e,0} &= M^{-1} \sum_{i=1}^{M-1} \binom{M}{i+1} (-1)^{i+1} \exp\{-i(u_0^2 + v_0^2)/2\sigma^2(i+1)\} \\
&= M^{-1} \sum_{n=2}^{M} \binom{M}{n} (-1)^n \exp\{-(n-1)(u_0^2 + v_0^2)/2n\sigma^2\}. \tag{7.82}
\end{aligned}$$

We can proceed further for a specific form of the parameters u_0 and v_0. If, when conditioned on 0 being sent, the random variables U_0 and V_0 are given by

$$U_0 = \alpha_0 \cos(\varphi) + X_0$$

and

$$V_0 = -\alpha_0 \sin(\varphi) + Y_0,$$

then

$$u_0^2 + v_0^2 = \alpha_0^2.$$

As a result, (7.82) becomes

$$P_{e,0} = M^{-1} \sum_{n=2}^{M} \binom{M}{n} (-1)^n \exp\{-(n-1)\alpha_0^2/2n\sigma^2\}. \tag{7.83}$$

Recall that the parameter α_0 is given by

$$\alpha_0 = (m_0, w_0)$$

and that

$$\sigma^2 = \tfrac{1}{2} N_0 \|w_0\|^2.$$

If the correlators are optimum and each signal has energy $\mathcal{E}$, then

$$\alpha_0 = (m_0, w_0) = (m_0, c\, m_0) = c\|m_0\|^2 = c\,\mathcal{E},$$

and

$$\sigma^2 = \tfrac{1}{2} N_0 \|w_0\|^2 = \tfrac{1}{2} N_0\, c^2 \mathcal{E}.$$

It follows that $\alpha_0^2/2\sigma^2 = \mathcal{E}/N_0$. Once we replace $\alpha_0^2/2\sigma^2$ with $\mathcal{E}/N_0$ in (7.83), there is no dependence on the fact that s_0 is the transmitted signal. Thus, for optimum correlators and equal-energy signals, the symbol error probability does not depend on which signal is transmitted, so we drop the subscript 0 and write

$$P_e = M^{-1} \sum_{n=2}^{M} \binom{M}{n} (-1)^n \exp\{-(1 - n^{-1})\mathcal{E}/N_0\}. \tag{7.84}$$

As discussed in Section 6.6.3, it is more useful for most applications to express the symbol error probability in terms of the energy per bit, rather than the energy per symbol. For M-ary signaling with $m = \log_2 M$ bits per symbol, the relationship is

$$\mathcal{E}_b = \mathcal{E}/m = \mathcal{E}/\log_2 M.$$

Replacement of $\mathcal{E}$ with $\mathcal{E}_b \log_2 M$ in (7.84) gives the desired expression for the symbol error probability as a function of $\mathcal{E}_b/N_0$ for equal-energy, M-ary orthogonal signaling and noncoherent demodulation.

This expression can be evaluated in a straightforward manner for values of M up to 32, although double-precision computations may be required for M larger than 8 or 16. For large values of M, the evaluation of (7.84) requires subtraction of pairs of very large numbers, and the accuracy of representing these numbers may be inadequate to give a reliable computation. For large values of M, it is not uncommon for a direct computation of the sum in (7.84) to give a negative value for P_e. If M is larger than 32, it is usually better to evaluate the error probability by applying numerical integration to (7.77), analogous to the evaluation in Section 6.6.3 of the error probabilities for M-ary orthogonal signaling with coherent demodulation.

Values of the symbol error probability for noncoherent demodulation of equal-energy, M-ary orthogonal signals are given in Table 7.2 for several values of $\mathcal{E}_b/N_0$ and three values of M. These numerical results were obtained by direct evaluation of (7.84) with double-precision computations. Notice that very low error probabilities can be obtained with 32-ary orthogonal signaling, even though noncoherent demodulation is employed.

The conversion from symbol error probability to bit error probability is exactly as derived in Section 6.6.3. If M is a power of 2 and a unique sequence of $m = \log_2 M$

Table 7.2: Symbol Error Probabilities for Equal-Energy, M-ary Orthogonal Signaling and Noncoherent Demodulation

$(\mathcal{E}_b/N_0)_{\mathrm{dB}}$	$M = 2$	$M = 8$	$M = 32$
3.0	1.84×10^{-1}	1.05×10^{-1}	5.03×10^{-2}
4.0	1.42×10^{-1}	5.38×10^{-2}	1.71×10^{-2}
5.0	1.03×10^{-1}	2.25×10^{-2}	4.03×10^{-3}
6.0	6.83×10^{-2}	7.25×10^{-3}	6.03×10^{-4}
7.0	4.08×10^{-2}	1.67×10^{-3}	5.08×10^{-5}
8.0	2.13×10^{-2}	2.53×10^{-4}	2.10×10^{-6}
9.0	9.42×10^{-3}	2.27×10^{-5}	3.64×10^{-8}
10.0	3.37×10^{-3}	1.06×10^{-6}	2.15×10^{-10}

binary digits is assigned to each of the M signals, then the bit error probability P_b is related to the symbol error probability P_e by

$$P_b = M P_e / [2(M - 1)].$$

The bit error probabilities are shown as a function of $\mathcal{E}_b/N_0$ for five different values of M in Figure 7-14. These curves are similar to those in Figure 6-44 for coherent demodulation. For large values of M, the performance at high signal-to-noise ratios is only a few tenths of a dB worse for noncoherent demodulation than for coherent demodulation. In the limit as $M \to \infty$, the bit and symbol error probabilities for noncoherent demodulation converge to zero, provided that $\mathcal{E}_b/N_0 > \ln(2)$, just as for coherent demodulation.

7.8   Envelope-Detector Implementation of the Noncoherent Receiver

In this section, an alternative implementation for the noncoherent receiver is presented. Recall that the signals are of the general form

$$s_i(t) = \sqrt{2}\, m_i(t) \cos(\omega_c t + \theta_i(t) + \varphi_i).$$

The signals $m_i(t)$ are time limited to the interval $[0, T]$, and they are often expressed as

$$m_i(t) = A\, a_i(t)\, p_T(t),$$

in which case the parameter A is used to control the power level of the signals and the waveforms $a_0(t)$ and $a_1(t)$ determine their shapes. Recall that under fairly general conditions the energy $\mathcal{E}_i$ in the signal $s_i(t)$ is just the energy in the waveform $m_i(t)$; that is, $\mathcal{E}_i = \|m_i\|^2$. Equivalently, $\mathcal{E}_i = A^2 \|a_i\|^2$.

In order to simplify the presentation in this section, we restrict attention to equal-energy binary signals for which the phase modulation is given by

$$\theta_i(t) = (\omega_i - \omega_c)\, t$$

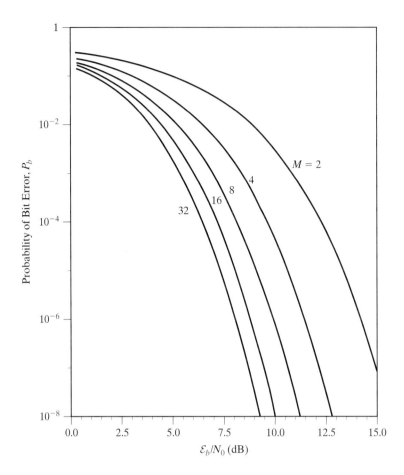

Figure 7-14: Bit error probabilities for M-ary orthogonal signals and optimum noncoherent demodulation.

for all t. As a result, the signals of interest in this section are of the form

$$s_i(t) = \sqrt{2}\, m_i(t) \cos(\omega_i t + \varphi_i) \tag{7.85}$$

for $i = 0$ and $i = 1$. The receiver structure that is described for binary communications in this section is generalized easily for M-ary communications.

The standard BFSK signals are obtained from (7.85) by letting

$$m_0(t) = m_1(t) = A\, p_T(t),$$

so that

$$s_i(t) = \sqrt{2}\, A \cos[\omega_i t + \varphi_i]\, p_T(t).$$

The form of the signals defined by (7.85) permits consideration of more general signal sets than BFSK, including pulse-shaped BFSK. The pulses may be shaped intentionally in order to reduce the spectral occupancy of the BFSK signals, or the filtering may be

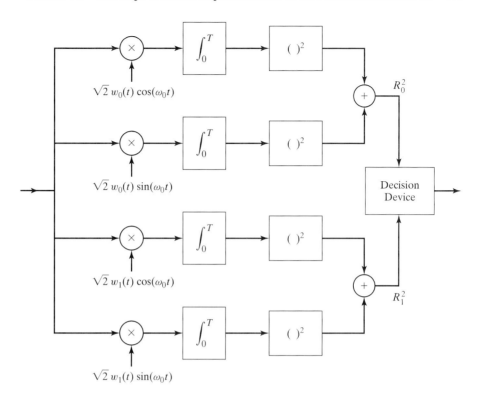

Figure 7-15: Noncoherent correlation receiver.

the result of bandwidth restrictions and other types of linear distortion that arise in the communication channel.

We assume that the bandwidth of $m_i(t)$ is small compared with the frequency ω_i. Such signals are referred to as *narrowband signals*. Suppose also that the noncoherent correlation receiver of Figure 7-15 is used to demodulate these signals. It is shown in the previous sections of this chapter that this receiver is optimum if the correlator waveforms $w_0(t)$ and $w_1(t)$ are matched to the signal waveforms $m_0(t)$ and $m_1(t)$; that is, the receiver waveforms must satisfy

$$w_0(t) = c\,m_0(t)$$

and

$$w_1(t) = c\,m_1(t)$$

for almost all t. The constant c is arbitrary. In particular, the optimum noncoherent correlation receiver for standard BFSK signals can be obtained by setting

$$w_i(t) = a_i(t) = 1, \quad 0 \le t \le T$$

for both $i = 0$ and $i = 1$.

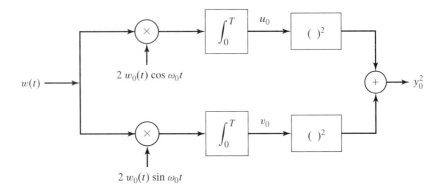

Figure 7-16: Noncoherent correlators.

More generally, the noncoherent correlation receiver may be used even if the wave-forms are not matched, and doing so may be desirable or even necessary for certain applications, as discussed in Section 7.3. Because the integration interval is $[0, T]$ for the correlators of Figure 7-15, there is no loss of generality in assuming that the correlator waveforms w_0 and w_1 are time limited to the interval $[0, T]$. We also assume that the bandwidths of these two waveforms are small compared with the frequencies ω_0 and ω_1. The goal of this section is to find an alternative implementation of the noncoherent correlation receiver for systems in which the transmitted signals are narrowband signals. This alternative implementation employs passive linear filters and envelope detectors rather than the correlators and squaring devices shown in Figure 7-15.

A portion of the noncoherent correlation receiver is illustrated in Figure 7-16. This system includes the two correlators that correspond to $s_0(t)$, except the constant multi-plier for the correlator waveforms has been changed from $\sqrt{2}$ to 2 in order to simplify some of the expressions that are obtained in this section. The system of Figure 7-16 is equivalent to the upper half of the noncoherent correlation receiver if each of the correlator waveforms in Figure 7-15 is multiplied by $\sqrt{2}$.

Suppose that the input to the system of Figure 7-16 is an ideal bandpass signal given by

$$w(t) = v_1(t) \cos(\omega_0 t) - v_2(t) \sin(\omega_0 t). \tag{7.86}$$

Recall from Section 4.6 that any ideal bandpass signal can be written in this form. If all double-frequency components are negligible, the outputs of the integrators of Figure 7-16 are given by

$$u_0 = \int_0^T v_1(t) \, w_0(t) \, dt \tag{7.87a}$$

and

$$v_0 = -\int_0^T v_2(t) \, w_0(t) \, dt. \tag{7.87b}$$

Figure 7-17: Filter and envelope detector.

It follows that the output of the noncoherent correlator structure is

$$y_0^2 = \left[\int_0^T v_1(t) \, w_0(t) \, dt \right]^2 + \left[\int_0^T v_2(t) \, w_0(t) \, dt \right]^2. \tag{7.88}$$

In (7.86)–(7.88), the waveform $w(t)$ can represent a BFSK waveform, a noise waveform, or the sum of the signal and the noise at the input to the demodulator.

Next, consider the system illustrated in Figure 7-17, which includes a linear, time-invariant bandpass filter with impulse response given by

$$h_0(t) = 2 \, g_0(t) \cos(\omega_0 t). \tag{7.89}$$

The bandwidth of $g_0(t)$ is assumed to be much smaller than ω_0. Recall from Section 4.6.2 that such an impulse response $h_0(t)$ is the impulse response of a locally symmetric bandpass frequency function (cf. (4.56)), and that $g_0(t)$ is the baseband equivalent of the bandpass filter. Let the input to this system be the same as the input to the non-coherent correlator structure of Figure 7-16. That is, the input is given by

$$w(t) = v_1(t) \cos(\omega_0 t) - v_2(t) \sin(\omega_0 t).$$

It follows from results in Section 4.6 that the output of the bandpass filter is

$$\hat{w}(t) = y_1(t) \cos(\omega_0 t) - y_2(t) \sin(\omega_0 t), \tag{7.90}$$

and the baseband signals $y_1(t)$ and $y_2(t)$ are the outputs of the lowpass equivalent filter if the inputs are $v_1(t)$ and $v_2(t)$, respectively. We can therefore write $y_1 = v_1 * g_0$ and $y_2 = v_2 * g_0$, where g_0 is the impulse response of the lowpass equivalent filter.

The fact that the input to the bandpass filter of Figure 7-17 is a narrowband signal implies that the output $\hat{w}(t)$ is also a narrowband signal. In particular, the bandwidths of $y_1(t)$ and $y_2(t)$ are much smaller than the frequency ω_0. Because the input to the envelope detector is the narrowband signal $\hat{w}(t)$ given by (7.90), the output is

$$e(t) = \sqrt{y_1^2(t) + y_2^2(t)}.$$

When sampled at time T, this gives

$$z_0 = e(T) = \sqrt{y_1^2(T) + y_2^2(T)}. \tag{7.91}$$

But $y_1 = v_1 * g_0$ and $y_2 = v_2 * g_0$; therefore,

$$y_1(T) = \int_{-\infty}^{\infty} v_1(t) \, g_0(T - t) \, dt \tag{7.92a}$$

and

$$y_2(T) = \int_{-\infty}^{\infty} v_2(t)\, g_0(T-t)\, dt. \tag{7.92b}$$

Hence, the output of the filter and envelope detector system shown in Figure 7-17 is

$$z_0^2 = \left[\int_{-\infty}^{\infty} v_1(t)\, g_0(T-t)\, dt \right]^2 + \left[\int_{-\infty}^{\infty} v_2(t)\, g_0(T-t)\, dt \right]^2. \tag{7.93}$$

Notice the similarity between (7.93) and

$$y_0^2 = \left[\int_0^T v_1(t)\, w_0(t)\, dt \right]^2 + \left[\int_0^T v_2(t)\, w_0(t)\, dt \right]^2,$$

which is the expression for the output of the noncoherent correlator structure.

The equivalence between the noncoherent correlator structure of Figure 7-16 and the filter and envelope detector system of Figure 7-17 is analogous to the equivalence demonstrated in Chapter 5 between the optimum correlator and the matched filter. If the correlator waveform w_0 and the bandpass filter's lowpass equivalent g_0 are related by

$$w_0(t) = g_0(T-t), \quad -\infty < t < \infty,$$

then g_0 is time limited to $[0, T]$ and (7.93) reduces to

$$\begin{aligned}
z_0^2 &= \left[\int_0^T v_1(t)\, g_0(T-t)\, dt \right]^2 + \left[\int_0^T v_2(t)\, g_0(T-t)\, dt \right]^2 \\
&= \left[\int_0^T v_1(t)\, w_0(t)\, dt \right]^2 + \left[\int_0^T v_2(t)\, w_0(t)\, dt \right]^2 \\
&= y_0^2.
\end{aligned}$$

Notice that $w_0(t) = g_0(T-t), 0 \le t \le T$, is equivalent to

$$g_0(t) = w_0(T-t), \quad 0 \le t \le T.$$

Thus, if $g_0(t) = w_0(T-t)$ for $0 \le t \le T$, the filter-and-envelope-detector structure of Figure 7-17 produces the same output as the noncoherent-correlator structure of Figure 7-16 provided that the input to each system is the same ideal bandpass waveform. This is true whether the bandpass waveform represents the communication signal, a sample function of the noise, or the sum of the signal and the noise.

The equivalence just demonstrated implies that the decision statistics produced at the outputs of the systems of Figures 7-16 and 7-17 are identical when the input is the sum of the communication signal and noise. By simply changing the subscripts in the equations, the preceding development shows that if

$$g_1(t) = w_1(T-t), \quad 0 \le t \le T,$$

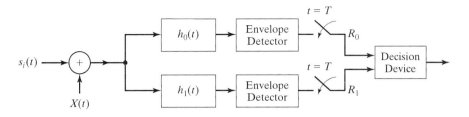

Figure 7-18: Noncoherent receiver with filters and envelope detectors.

the noncoherent correlators in the lower half of Figure 7-15 are equivalent to a system of the type illustrated in Figure 7-17, but having a bandpass filter with impulse response

$$h_1(t) = 2\, g_1(t) \cos(\omega_1 t).$$

A receiver based on a comparison of the outputs of two systems of the type shown in Figure 7-17 does not actually have to compute the square. Because the output of the envelope detector is a positive square root, the comparison between the two branches is unaffected by the squaring operation performed in each branch. Therefore, we might as well save hardware and eliminate the squaring operation. The comparison is then made between the two outputs of the two envelope detectors, as illustrated in Figure 7-18. Each decision statistic produced by the system of Figure 7-18 is just the square root of the corresponding decision statistic of the noncoherent correlation receiver of Figure 7-15.

Notice that the equivalence is very strong: For the same input waveform (representing signal plus noise), the two implementations produce exactly the same values for the decision statistics R_0 and R_1. This implies not only that the two different implementations of the noncoherent receiver give the same probability of error, but also the two systems make the same decision for the same input waveform. In particular, one system makes an error if and only if the other makes an error.

We can now provide the expressions for the error probabilities $P_{e,0}$ and $P_{e,1}$ in terms of the lowpass equivalent impulse responses for the filters shown in Figure 7-18. Recall that the signals are given by

$$s_i(t) = \sqrt{2}\, m_i(t) \cos(\omega_i t + \varphi_i)$$

and the impulse responses of the filters are of the form

$$h_i(t) = 2\, g_i(t) \cos(\omega_i t),$$

for $i = 0$ and $i = 1$. The expressions for the error probabilities $P_{e,0}$ and $P_{e,1}$ are given in terms of the lowpass equivalents $g_0(t)$ and $g_1(t)$. Define $\hat{m}_i = m_i * g_i$; that is,

$$\hat{m}_i(t) = \int_{-\infty}^{\infty} m_i(\tau)\, g_i(t - \tau)\, d\tau, \qquad (7.94)$$

which is the output of the ith lowpass equivalent filter if the input is $m_i(t)$.

The error probability expressions for the system of Figure 7-18 are obtained from (7.57) by substituting $\hat{m}_i(T)$ for the inner product (m_i, w_i) and the norm $\|g_i\|$ for the norm $\|w_i\|$. This gives

$$P_{e,0} = \tfrac{1}{2} \exp\{-[\hat{m}_0(T)]^2/2N_0\|g_0\|^2\} \tag{7.95}$$

and

$$P_{e,1} = \tfrac{1}{2} \exp\{-[\hat{m}_1(T)]^2/2N_0\|g_1\|^2\}. \tag{7.96}$$

Note that the condition $\|w_0\|^2 = \|w_1\|^2$, which is needed for the validity of (7.57) and (7.58), implies that the lowpass equivalent filters must satisfy $\|g_0\|^2 = \|g_1\|^2$ in order for (7.95) and (7.96) to be valid.

The conclusion is that if the communication signals are narrowband, an envelope detector and filter can replace a pair of correlators in the noncoherent correlation receiver. If the impulse responses of the filters are related to the correlator waveforms by $g_i(t) = w_i(T - t)$, $0 \le t \le T$, then the two different implementations of the noncoherent receiver give the same performance. In the next exercise, conditions for optimality of the envelope-detector implementation are explored.

Exercise 7-3
Suppose that the lowpass equivalent filters in the system of Figure 7-18 satisfy $\|g_0\|^2 = \|g_1\|^2$. Show that if the filter impulse response g_0 is matched to the waveform m_0, then

$$P_{e,0} = \tfrac{1}{2} \exp\{-\mathcal{E}_0/2N_0\}.$$

Similarly, if g_1 is matched to the waveform m_1, then

$$P_{e,1} = \tfrac{1}{2} \exp\{-\mathcal{E}_1/2N_0\}.$$

Solution.   The condition that the filter impulse response is matched to m_0 requires that, for some constant c_0,

$$g_0(t) = c_0 \, m_0(T - t)$$

for almost all t. For this filter,

$$\|g_0\|^2 = c_0^2 \|m_0\|^2.$$

It follows from the definition of $\hat{m}_0$ that

$$[\hat{m}_0(T)]^2 = c_0^2 \|m_0\|^4,$$

so

$$[\hat{m}_0(T)]^2/\|g_0\|^2 = \|m_0\|^2.$$

As pointed out at the beginning of this section, $\mathcal{E}_i = \|m_i\|^2$ for each i, so

$$[\hat{m}_0(T)]^2/\|g_0\|^2 = \mathcal{E}_0. \tag{7.97}$$

Replacing the subscript 0 by 1, we see also that if g_1 is matched to m_1, then

$$[\hat{m}_1(T)]^2/\|g_1\|^2 = \mathcal{E}_1. \tag{7.98}$$

Subject only to the condition that $\|g_0\|^2 = \|g_1\|^2$, equations (7.95) and (7.96) are valid with the substitutions from (7.97) and (7.98). We have shown that if the lowpass equivalent filters are matched to the signal waveforms, the resulting error probabilities are

$$P_{e,0} = \tfrac{1}{2} \exp\{-\mathcal{E}_0/2N_0\}$$

and

$$P_{e,1} = \tfrac{1}{2} \exp\{-\mathcal{E}_1/2N_0\},$$

the same as for the noncoherent correlation receiver.   ■

7.9   Noncoherent Communications with Nonselective Fading

In some communication channels, referred to as *multipath channels*, the transmission of a single signal produces two or more signals at the receiver. In such a situation, we refer to the individual signals as the multipath components and the combination of these components as the received signal. It may be that the multipath components differ only in their amplitudes and times of arrival; in fact, in a *specular multipath channel*, each component is a delayed, attenuated version of the transmitted signal, and the time offsets among the components are sufficiently large that the components can be distinguished. In a *diffuse multipath channel*, the time offsets are sufficiently small that the individual components cannot be distinguished, and the received signal is modeled as a continuum of multipath components.

From the point of view of the designer of the communication system, it is reasonable to model a multipath channel as consisting of multiple paths with a replica of the transmitted signal propagating along each path, as illustrated in Figure 7-19. Although this model may not be suitable for those wishing to apply electromagnetic wave theory to obtain a precise characterization of multipath propagation, it is an appropriate model for the design and analysis of modulation and demodulation systems [7.4].

The multipath components may combine constructively or destructively to form the received signal. A simple illustration of the destructive interference that can result from multipath propagation is obtained by considering the transmission of an unmodulated carrier at frequency f_0 over a channel that has two propagation paths. The signal at the output of the channel has two components, each of which is a sinusoidal signal at

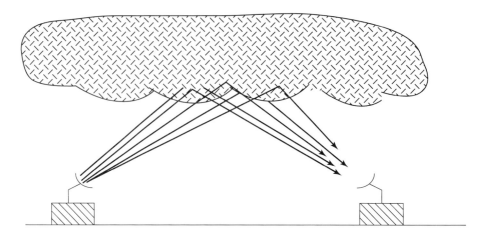

Figure 7-19: Illustration of the multipath model.

frequency f_0. Because the two paths may differ in length, the arrival times may differ for the two components, in which case the two multipath components are offset in time at the input to the receiver.

Suppose that the difference in the propagation times for the two paths, referred to as the *differential propagation delay*, is $\tau_0 = 1/(2f_0)$. Then one of the components is offset by an amount τ_0 relative to the other, and the received signal is of the form

$$r(t) = A_1 \cos[2\pi f_0 t + \varphi] + A_2 \cos[2\pi f_0(t - \tau_0) + \varphi].$$

Because $f_0\tau_0 = 1/2$, this signal is equivalent to

$$r(t) = A_1 \cos[2\pi f_0 t + \varphi] + A_2 \cos[2\pi f_0 t + \varphi - \pi].$$

Thus, for this example, the two components differ in phase by π radians. Combining the two terms, we see that the received signal is given by

$$r(t) = (A_1 - A_2) \cos[2\pi f_0 t + \varphi].$$

If A_1 and A_2 are positive, then the amplitude of the sum of the two signals is less than the larger of the two amplitudes of the individual signals; therefore, this is an example of destructive interference. In fact, the amplitude of the sum may be smaller than either of the component amplitudes. If the propagation loss is the same for the two paths, then $A_1 = A_2$ and the two signals cancel to give $r(t) = 0$. More generally, if $A_1 \approx A_2$, then the received signal is very weak and the multipath propagation is said to have caused a *deep fade* in the received signal. The difficulty in tracking a weak carrier signal makes coherent demodulation impractical for most channels in which deep fades occur frequently.

Notice that a differential propagation delay of $\tau_0 = 0$ or $\tau_0 = 1/f_0$ in the preceding example produces multipath components that are in phase. That is, for either of these two values of τ_0, the received signal is

$$
\begin{aligned}
r(t) &= A_1 \cos[2\pi f_0 t + \varphi] + A_2 \cos[2\pi f_0 t + \varphi] \\
&= (A_1 + A_2)\cos[2\pi f_0 t + \varphi].
\end{aligned}
$$

If A_1 and A_2 are positive, then the resulting amplitude is greater than the amplitude of each component. In this situation, the multipath components are said to add constructively.

It is convenient to model a multipath channel as a linear, time-invariant filter. The multipath channel can then be characterized by either the impulse response or the transfer function of this filter. Suppose the transfer function for the multipath channel has a magnitude that is flat (i.e., constant) and a phase that is linear over the frequency band of the signal. The only effect of passing a signal through such a multipath channel is to introduce a delay in the signal. The amount of delay is proportional to the slope of the phase as a function of frequency; that is, a phase function $\exp(-j2\pi f t_0)$ corresponds to a delay of t_0. A multipath channel that has a flat gain and linear phase does not cause any distortion of the signal. If the gain of the channel is not flat over the frequency band of the signal, the multipath channel distorts the signal by increasing the amplitudes of some frequency components while decreasing the amplitudes of others. This phenomenon is illustrated in Problem 7.17 for a two-component multipath channel.

The term *fading* is used in the communications literature to refer to the distortion and change in signal strength that can result from the constructive and destructive interference that occurs in a multipath channel. If the channel causes no distortion to the communication signal, and the only influence of the multipath channel is a possible change in signal strength, the channel is said to exhibit *nonselective fading*. A channel with nonselective fading can be modeled as a linear filter with a flat gain and linear phase over the frequency band occupied by the communication signal. In fact, an alternative name for nonselective fading is *flat fading*. A nonselective fading channel does not change the spectrum of the signal, but it does attenuate the signal if the gain is less than one.

If the simple two-component multipath channel has a differential propagation delay that is small compared with the inverse of the bandwidth of the signal, then the channel exhibits nonselective fading. (See part **(e)** of Problem 7.17.) However, if the differential propagation delays are sufficiently large, even this simple two-component multipath channel can produce fading that is frequency dependent. (See parts **(b)** and **(c)** of Problem 7.17.) In general, fading that affects signal components at one frequency in a significantly different way than signal components at another frequency is referred to as *frequency-selective fading*. More precisely, frequency-selective fading occurs if the channel gain is not flat or its phase response is not linear.

A multipath channel may have several paths, each with a different propagation delay. The range or spread of the propagation delays for the different paths is referred to as the *multipath spread* for the channel. The multipath channel exhibits nonselective fading if the multipath spread is small compared with the inverse of the bandwidth of the signal, but it exhibits frequency-selective fading if the multipath spread is large compared to the inverse bandwidth. In this section, we investigate the effects of nonselective fading on noncoherent communications. For more information on selective fading, consult [7.2] or [7.4].

In nonselective fading, the different paths have propagation delays that are approximately the same. As a result, the channel does not distort the signal, but it can alter the amplitude of the signal as demonstrated earlier. If the multipath channel exhibits nonselective fading, the received waveform has the same shape as the transmitted waveform, but the amplitudes of these two waveforms may differ.

A very important model in the analysis of the effects of nonselective fading is the *nonselective Rayleigh fading channel*. The Rayleigh fading model has a great deal of intuitive appeal, and it has been found to provide an accurate model for many physical channels. The following sketch of the development of the Rayleigh model is intended to provide an intuitive understanding of the origin of the model and to motivate its use in the analysis of the effects of nonselective fading. However, it is by no means intended as a rigorous derivation.

Suppose that the signal

$$s(t) = \cos(2\pi f_c t)$$

is transmitted over a nonselective fading channel. The signal that arrives at the receiver via the ith path is

$$r_i(t) = V_i \cos(2\pi f_c t + \Theta_i).$$

If we model the propagation paths as resulting from random phenomena, such as scattering from randomly located objects, then V_i and Θ_i are random variables. For most applications, V_i and Θ_i are independent and the phase angle Θ_i is uniformly distributed on the interval $[0, 2\pi]$. The signal received via the ith path can be expressed in inphase-quadrature form as

$$r_i(t) = X_i \cos(2\pi f_c t) + Y_i \sin(2\pi f_c t),$$

where $X_i = V_i \cos(\Theta_i)$ and $Y_i = -V_i \sin(\Theta_i)$. Notice that X_i and Y_i are uncorrelated, the mean of each is zero, and they have the same variance.

In many common models, such as the independent-scattering model, the parameters for different paths are independent. In the independent-scattering model, the locations of the randomly placed objects are assumed to be independent. Independence of the parameters for different paths implies that (V_i, Θ_i) is independent of (V_j, Θ_j) for $i \neq j$. This in turn implies independence of (X_i, Y_i) and (X_j, Y_j) for $i \neq j$.

Because the received signal is the sum of the individual components, it can be written as

$$r(t) = X \cos(2\pi f_c t) + Y \sin(2\pi f_c t),$$

where X is the sum $X_1 + X_2 + \cdots$ and Y is the sum $Y_1 + Y_2 + \cdots$. The expression of X and Y as sums of independent random variables suggests that they are approximately jointly Gaussian if the number of terms is sufficiently large (central-limit theorem, Section 1.4.2). Recall that X_i and Y_j are independent for $i \neq j$ and uncorrelated for $i = j$, each has mean equal to zero, and they have the same variance. Thus, the resulting Gaussian random variables X and Y are uncorrelated (hence independent), each has mean equal to zero, and they have the same variance. From the results of Section 7.5.1, it follows that the distribution of $V = \sqrt{X^2 + Y^2}$ is Rayleigh. Because $r(t)$ can be written as

$$r(t) = V \cos(2\pi f_c t + \Theta),$$

it follows that the amplitude of the received signal has a Rayleigh distribution. (The density and distribution functions are given in expressions (7.49) and (7.50).) It can be shown that Θ is uniformly distributed on $[0, 2\pi]$, but this is of less significance in the analysis of noncoherent communications than the distribution of the amplitude. Because the transmitted signal is $s(t) = \cos(2\pi f_c t)$ and the received signal is $r(t) = V \cos(2\pi f_c t + \Theta)$, V represents the channel gain.

Now we consider a binary noncoherent communication system with a nonselective fading channel and a binary orthogonal signal set. The general form for the communication signal that represents symbol i is

$$s_i(t) = \sqrt{2}\, m_i(t) \cos(\omega_c t + \theta_i(t) + \varphi_i).$$

Because the fading is nonselective, the signal that arrives at the receiver can be modeled as

$$r_i(t) = \sqrt{2} V\, m_i(t) \cos(\omega_c t + \theta_i(t) + \varphi_i + \Theta). \tag{7.99}$$

The amplitude parameter V is a nonnegative random variable with density function denoted by f_V. The phase shift Θ does not play an important role, because the noncoherent demodulation process does not depend on the phase angle of the signal. Also, as outlined in Problem 7.1, the addition modulo 2π of two independent phase angles, at least one of which is uniformly distributed on $[0, 2\pi]$, produces a phase angle that is uniformly distributed on $[0, 2\pi]$. As a result of this fact, we might as well replace $\varphi_i + \Theta$ with φ_i in (7.99) and then model φ_i as a random variable with uniform distribution on $[0, 2\pi]$.

For a nonselective Rayleigh fading channel, the density function for V is given by

$$f_V(v) = 2(v/\beta^2) \exp(-v^2/\beta^2), \quad 0 \le v < \infty, \tag{7.100}$$

where $\beta^2 = E\{V^2\}$ is the second moment of the gain of the channel. In order to obtain the expression for the Rayleigh density that is given in (7.49) of Section 7.5.1, replace β^2 with $2\sigma^2$ in (7.100).

Because the amplitude of the received signal is now a random variable, the energy in the received signal is also a random variable. Recall that under fairly general conditions the energy $\mathcal{E}_i$ in the transmitted signal $s_i(t)$ is just the energy in the waveform $m_i(t)$; that is, $\mathcal{E}_i = \|m_i\|^2$. Similarly, the energy in the received waveform is the random variable $V^2 \mathcal{E}_i = V^2 \|m_i\|^2$. The expected value of this random variable is the average energy in the received signal, which is given by $\bar{\mathcal{E}}_i = E\{V^2\}\mathcal{E}_i = \beta^2 \mathcal{E}_i$.

If we condition on $V = \alpha$, the received energy is $\alpha^2 \mathcal{E}_i$. The receiver does not know the value of this energy, so we assume that the correlator waveforms are matched to the shapes of the modulation waveforms, but not necessarily to their amplitudes. According to results given in Section 7.5.2, the probability of error for the LS decision rule in a binary noncoherent communication system is given by

$$P_{e,i}(\alpha) = \tfrac{1}{2} \exp\{-\alpha^2 \mathcal{E}_i / 2N_0\} \tag{7.101}$$

if the received energy is $\alpha^2 \mathcal{E}_i$. This probability depends on the amplitude α; in fact, it is the conditional probability of error given that symbol i was sent and that the received amplitude is α.

The most common performance measure for a nonselective fading channel is the average probability of error. This probability is defined as

$$\overline{P}_{e,i} = \int_0^\infty P_{e,i}(\alpha)\, f_V(\alpha)\, d\alpha. \tag{7.102}$$

Substituting from (7.100) and (7.101) into (7.102), we obtain

$$\overline{P}_{e,i} = \int_0^\infty \alpha \exp\{-\alpha^2 \mathcal{E}_i / 2N_0\} \exp(-\alpha^2/\beta^2)\, d\alpha/\beta^2.$$

Combining the exponential terms, we get

$$\overline{P}_{e,i} = \int_0^\infty \alpha \exp\{-\alpha^2 \lambda\}\, d\alpha/\beta^2 = (2\lambda\beta^2)^{-1}, \tag{7.103}$$

where

$$\lambda = (\mathcal{E}_i/2N_0) + \beta^{-2}.$$

Substituting for λ in (7.103) and simplifying, we find that

$$\overline{P}_{e,i} = \{2 + (\beta^2 \mathcal{E}_i/N_0)\}^{-1}.$$

The average energy is given by $\bar{\mathcal{E}}_i = \beta^2 \mathcal{E}_i$, so the average probability of error can be written as

$$\overline{P}_{e,i} = \{2 + (\bar{\mathcal{E}}_i/N_0)\}^{-1}. \tag{7.104a}$$

If $\mathcal{E}_0 = \mathcal{E}_1$, the average energy is the same for the two signals, and the average probability of error is given by

$$\overline{P}_e = \{2 + (\bar{\mathcal{E}}/N_0)\}^{-1}, \tag{7.104b}$$

where $\bar{\mathcal{E}}$ denotes the common value of the average energy and $\overline{P}_e$ denotes the common value of the average probability of error for the two signals.

Recall that the probability of error for noncoherent demodulation of binary orthogonal signaling over a channel with no fading is

$$P_e = \tfrac{1}{2} \exp\{-\mathcal{E}/2N_0\}.$$

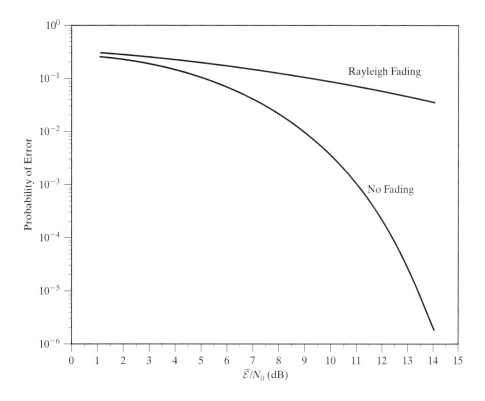

Figure 7-20: Error probabilities for binary orthogonal signaling, noncoherent demodulation, and Rayleigh fading.

For a channel with no fading, the average energy $\bar{\mathcal{E}}$ is just the energy $\mathcal{E}$ in the signals s_0 and s_1 (i.e., there is no variation in the amount of energy in the received signal). A comparison of the error probabilities for equal-energy, orthogonal signaling and noncoherent demodulation is presented in Figure 7-20. It is clear that for values of $\bar{\mathcal{E}}/N_0$ greater than about 12 dB, the probability of error for a channel with nonselective Rayleigh fading is several orders of magnitude larger than the probability of error for a channel that has no fading. Even at 10 dB, the probability of error for a channel with no fading is approximately 0.337×10^{-2}, and the average probability of error for a channel with Rayleigh fading is approximately 8.33×10^{-2} (more than a factor of 20 larger).

It may appear to the reader that the curve for Rayleigh fading is approaching a straight line as $\bar{\mathcal{E}}/N_0$ increases. This is true, and it is even more noticeable if the graph is extended to include larger values of $\bar{\mathcal{E}}/N_0$ than shown in Figure 7-20. To see that the curve is linear for large values of $\bar{\mathcal{E}}/N_0$, first notice that the probability of error is shown on a logarithmic scale. So, we should write (7.104b) as

$$\log_{10}(\overline{P_e}) = -\log_{10}\{2 + (\bar{\mathcal{E}}/N_0)\}$$

to match the graph. If $\bar{\mathcal{E}}/N_0$ is large, $2 + (\bar{\mathcal{E}}/N_0) \approx \bar{\mathcal{E}}/N_0$; therefore,

$$\log_{10}(\overline{P}_e) \approx -\log_{10}(\bar{\mathcal{E}}/N_0) = -0.1(\bar{\mathcal{E}}/N_0)_{\text{dB}}. \tag{7.105}$$

So, for large values of $(\bar{\mathcal{E}}/N_0)$, the plot of $\log_{10}(\overline{P}_e)$ vs. $(\bar{\mathcal{E}}/N_0)_{\text{dB}}$ is linear with a slope of $-1/10$.

Suppose that we consider error probabilities of the form 10^{-n} and examine the change that must occur in $(\bar{\mathcal{E}}/N_0)_{\text{dB}}$ in order to increase n. Note that increasing n corresponds to moving down the probability axis of Figure 7-20. First, observe that if $\overline{P}_e = 10^{-n}$, then $n = -\log_{10}(\overline{P}_e)$. Therefore, (7.105) gives

$$n \approx 0.1(\bar{\mathcal{E}}/N_0)_{\text{dB}}. \tag{7.106}$$

Our conclusion is that for binary, orthogonal, equal-energy signaling with noncoherent demodulation and Rayleigh fading, n increases only linearly in $(\bar{\mathcal{E}}/N_0)_{\text{dB}}$ for large values of $(\bar{\mathcal{E}}/N_0)_{\text{dB}}$. A comparison of the relationships between $-\log_{10}(\overline{P}_e)$ and $(\bar{\mathcal{E}}/N_0)_{\text{dB}}$ for a channel with Rayleigh fading and a channel with no fading shows why there is such a dramatic difference between the two curves in Figure 7-20. (See Problem 7.18.)

Expression (7.106) implies that, for large values of n, $\overline{P}_e = 10^{-n}$ is achieved on a channel with Rayleigh fading for

$$(\bar{\mathcal{E}}/N_0)_{\text{dB}} \approx 10\,n \text{ dB}. \tag{7.107}$$

For example, $\overline{P}_e = 10^{-5}$ is achieved for $(\bar{\mathcal{E}}/N_0)_{\text{dB}} \approx 50$ dB. The precise value of $(\bar{\mathcal{E}}/N_0)_{\text{dB}}$ required to give $\overline{P}_e = 10^{-5}$ is

$$(\bar{\mathcal{E}}/N_0)_{\text{dB}} = 10\log_{10}(10^5 - 2),$$

which is approximately 49.999913 dB. Even for $\overline{P}_e = 10^{-2}$, $(\bar{\mathcal{E}}/N_0)_{\text{dB}}$ is slightly greater than 19.91 dB, which is close to the approximation given in (7.107) for $n = 2$.

In the model for fading considered thus far in this section, the entire received signal experiences fading. In some applications, there may be a strong signal component at the receiver that is not faded. Such a component is referred to as a *specular component* of the received signal. A specular component is present, for example, if there is a line-of-sight path between the transmitter and receiver, possibly in addition to a number of reflected paths. This specular component does not fluctuate, and so it is modeled as a signal with a deterministic amplitude.

For Rayleigh fading, the amplitudes of the inphase and quadrature parts of each component of the received signal are zero-mean random variables. Thus, the amplitudes X and Y of the inphase and quadrature parts of the received signal are also zero-mean random variables, and this leads to the conclusion that the amplitude $V = \sqrt{X^2 + Y^2}$ of the received signal has a Rayleigh distribution (based on the results of Exercise 7-2 of Section 7.5.1). If a specular component is present, then the random variables X and Y have mean values that are equal to the amplitudes of the specular signal's inphase and quadrature components, respectively. As shown in Problem 7.3, this means that the random variable V has a Rician density. We conclude that if the received signal has both a specular component and a Rayleigh-faded component, then the amplitude of the received signal has a Rician distribution. The channel that produces such components is referred to as a *Rician fading channel*. It is convenient and conventional to refer to the Rayleigh-faded component as the *diffuse component*, and we shall do so in what follows.

The density function for V in the Rician fading channel is given by

$$f_V(\alpha) = 2(\alpha/\beta^2) \exp\{-(\alpha^2 + c^2)/\beta^2\} I_0(2c\alpha/\beta^2), \quad 0 \le \alpha < \infty, \qquad (7.108)$$

where I_0 is the zero-order, modified Bessel function of the first kind, defined in Section 7.4. The parameter c is related to the strength of the specular component of the received signal. Just as $\beta^2 \mathcal{E}_i$ is the average energy in the diffuse component, $c^2 \mathcal{E}_i$ is the energy in the specular component of the received signal. The average energy in the received signal is $\bar{\mathcal{E}}_i = (\beta^2 + c^2)\mathcal{E}_i$. Notice that for $c = 0$, the density of (7.108) reduces to the Rayleigh density, as it should, because $c = 0$ implies that there is no specular component and all of the energy is in the diffuse component.

By substituting the density of (7.108) into (7.102), we obtain the expression for the average probability of error for binary orthogonal signaling with Rician fading and noncoherent demodulation. For equal-energy binary signals, the expression is

$$\overline{P}_e = \{2 + (\beta^2 \mathcal{E}/N_0)\}^{-1} \exp\{-(c^2 \mathcal{E}/N_0)/[2 + (\beta^2 \mathcal{E}/N_0)]\}. \qquad (7.109)$$

Let $\mathcal{E}_{sp}$ denote the energy in the specular component of the received signal and $\mathcal{E}_d$ denote the average energy in the diffuse component. Making these substitutions in (7.109), we see that the average probability of error is

$$\overline{P}_e = \{2 + (\mathcal{E}_d/N_0)\}^{-1} \exp\{-(\mathcal{E}_{sp}/N_0)/[2 + (\mathcal{E}_d/N_0)]\}. \qquad (7.110)$$

This expression involves the individual energies for the diffuse and specular components, but it can also be written in terms of the average total energy

$$\bar{\mathcal{E}} = (\beta^2 + c^2)\mathcal{E}.$$

Let $\Lambda = \bar{\mathcal{E}}/N_0$ and $\gamma^2 = \mathcal{E}_d/\mathcal{E}_{sp} = \beta^2/c^2$. The parameter γ^2 is the ratio of the average energy in the diffuse component to the energy in the specular component. Substitution into (7.110) for $\mathcal{E}_d/N_0$ and $\mathcal{E}_{sp}/N_0$ in terms of Λ and γ^2 gives

$$\overline{P}_e = (\gamma^2 + 1)\{2(\gamma^2 + 1) + \gamma^2\Lambda\}^{-1} \exp\{-\Lambda/[2(\gamma^2 + 1) + \gamma^2\Lambda]\}. \qquad (7.111)$$

If c^2 is held constant and $\beta^2 \to 0$, then $\gamma^2 \to 0$. Examination of (7.111) for small values of γ^2 shows that

$$\overline{P}_e \to \tfrac{1}{2} \exp\{-\Lambda/2\}$$

as $\gamma^2 \to 0$. Note also that $\Lambda = (\beta^2 + c^2)\mathcal{E}/N_0$, so $\Lambda \to c^2\mathcal{E}/N_0$ as $\beta^2 \to 0$. Because $c^2\mathcal{E}/N_0 = \mathcal{E}_{sp}/N_0$, it follows that

$$\overline{P}_e \to \tfrac{1}{2} \exp\{-\mathcal{E}_{sp}/2N_0\}$$

as $\beta^2 \to 0$. But this limiting expression is the probability of error for a channel that has no fading. This makes sense, because in the limit as $\beta^2 \to 0$ (and hence $\gamma^2 \to 0$), the diffuse component disappears, the channel does not exhibit fading, and the total energy in the received signal is $\mathcal{E}_{sp}$.

If β^2 is held constant and $c^2 \to 0$, then $\gamma^2 \to \infty$. Notice that

$$(\gamma^2 + 1)\{2(\gamma^2 + 1) + \gamma^2\Lambda\}^{-1} = \{2 + (1 + \gamma^{-2})^{-1}\Lambda\}^{-1},$$

which converges to $\{2 + \Lambda\}^{-1}$ as $\gamma^2 \to \infty$. Notice also that as $\gamma^2 \to \infty$,

$$\exp\{-\Lambda/[2(\gamma^2 + 1) + \gamma^2\Lambda]\} \to \exp\{0\} = 1.$$

It follows from these two observations that as $\gamma^2 \to \infty$,

$$\overline{P}_e \to \{2 + \Lambda\}^{-1}.$$

But $\Lambda = (\beta^2 + c^2)\mathcal{E}/N_0$, so $\Lambda \to \beta^2\mathcal{E}/N_0$ as $c^2 \to 0$; therefore,

$$\overline{P}_e \to \{2 + (\mathcal{E}_d/N_0)\}^{-1},$$

which is consistent with (7.104b), because $\bar{\mathcal{E}} = \beta^2\mathcal{E} = \mathcal{E}_d$ if $c^2 = 0$. In the limit as $c^2 \to 0$ (and hence $\gamma^2 \to \infty$), the received signal has no specular component, the channel exhibits pure Rayleigh fading, and the total average energy in the received signal is $\mathcal{E}_d$.

Error probabilities for equal-energy, orthogonal signaling in a system with Rician fading and noncoherent demodulation are illustrated in Figure 7-21. The limiting error probabilities for $\gamma^2 \to \infty$ (Rayleigh fading) and $\gamma^2 \to 0$ (no fading) are included. The error probabilities converge to their limiting values rather quickly as γ^2 increases or decreases. In particular, the parameter γ^2 does not have to be very large in order for the error probability for Rician fading with parameter γ to be approximately the same as the error probability for Rayleigh fading. Notice in Figure 7-21 that the curve that corresponds to $\gamma^2 = 1.0$ is approximately the same as the curve for Rayleigh fading. Similarly, γ^2 does not have to be very small in order to have the error probability for Rician fading approximate the error probability for a channel with no fading. As γ^2 decreases from 1.0 to 0.01, the error probabilities decrease from approximately the same as for Rayleigh fading to approximately the same as for a channel with no fading.

All of these results for binary orthogonal signaling (e.g., binary FSK) can be applied to DBPSK signaling as well. All that is needed is to replace the energy by twice the energy in each of the expressions for the probability of error. For example, for DBPSK signaling with Rayleigh fading and optimum noncoherent demodulation the average probability of error is

$$\overline{P}_e = \{2 + (2\,\bar{\mathcal{E}}/N_0)\}^{-1}. \tag{7.112}$$

For Rician fading with DBPSK signaling, the error probability is

$$\overline{P}_e = \{2 + (2\,\mathcal{E}_d/N_0)\}^{-1} \exp\{-(\mathcal{E}_{sp}/N_0)/[1 + (\mathcal{E}_d/N_0)]\}. \tag{7.113}$$

The symbol error probability for M-ary orthogonal signaling, noncoherent demodulation, and a Rician fading channel can be derived by first conditioning on the received

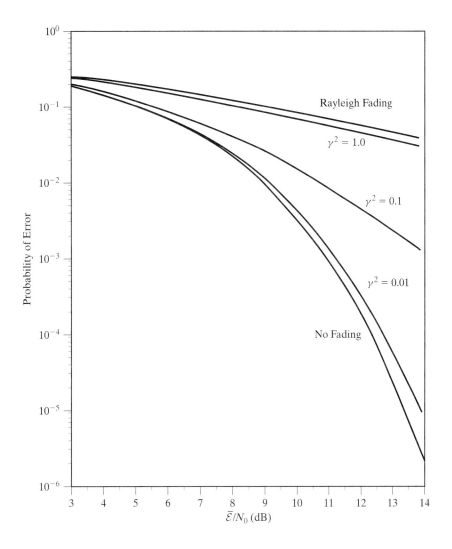

Figure 7-21: Error probabilities for binary orthogonal signaling, noncoherent demodulation, and Rician fading.

signal energy and then averaging the resulting expression. The result is a combination of (7.84), (7.102), and (7.108). We begin by recalling from (7.84) that the conditional probability of symbol error given that the received energy per symbol is $\alpha^2 \mathcal{E}$ is

$$P_e(\alpha) = M^{-1} \sum_{n=2}^{M} \binom{M}{n} (-1)^n \exp\{-(1 - n^{-1})\alpha^2 \mathcal{E}/N_0\}. \qquad (7.114)$$

The average probability of symbol error is then given by

$$\overline{P}_e = \int_0^\infty P_e(\alpha)\, f_V(\alpha)\, d\alpha, \tag{7.115}$$

where the density function f_V is the Rician density of (7.108). The next step is to substitute from (7.114) and (7.108) into (7.115) and then interchange the order of the summation and integration (i.e., integrate term by term). The result is

$$\overline{P}_e = M^{-1} \sum_{n=2}^{M} \binom{M}{n} (-1)^n\, G(1 - n^{-1}, \mathcal{E}_{sp}/N_0, \mathcal{E}_d/N_0), \tag{7.116}$$

where the function G is defined by

$$G(\lambda, x, y) = (1 + \lambda y)^{-1} \exp\{-\lambda x/(1 + \lambda y)\}$$

for positive values of the variables λ, x, and y. Notice that setting $M = 2$ in (7.116) gives

$$\overline{P}_e = \tfrac{1}{2} G(\tfrac{1}{2}, \mathcal{E}_{sp}/N_0, \mathcal{E}_d/N_0),$$

which agrees with (7.110).

The probability of symbol error can be expressed in terms of the energy per bit by dividing the energies $\mathcal{E}_{sp}$ and $\mathcal{E}_d$ by $\log_2 M$, as discussed in Section 7.7 for channels with no fading. Expressions for the bit error probability are also obtained in the manner described in Section 7.7. If M is a power of 2 and a unique sequence of $m = \log_2 M$ binary digits is assigned to each of the M signals, then the average probability of bit error is given by

$$\overline{P}_b = M\, \overline{P}_e/[2\,(M - 1)].$$

References and Suggestions for Further Reading

[7.1] M. Abramowitz and I. A. Stegun, *Handbook of Mathematical Functions*, National Bureau of Standards, Applied Mathematics Series, vol. 55, 1964; reprinted by Dover, 1965.

[7.2] J. G. Proakis, *Digital Communications* (4th ed.). New York: McGraw-Hill, 2001.

[7.3] M. K. Simon, S. M. Hinedi, and W. C. Lindsey, *Digital Communication Techniques*. Englewood Cliffs, NJ: Prentice-Hall, 1995.

[7.4] S. Stein, Part III of *Communication Systems and Techniques*. New York: McGraw-Hill, 1966.

[7.5] G. L. Turin, "Error probabilities for binary symmetric ideal reception through nonselective slow fading and noise," *Proceedings of the IRE*, vol. 46, September 1958, pp. 1603–1619.

Problems

7.1 Suppose that Θ_0 is a random variable that is uniformly distributed on the interval $[0, 2\pi]$, and let Θ_1 be a random variable that is independent of Θ_0. Let Θ be the modulo-2π sum of Θ_0 and Θ_1. Then show that Θ is also uniformly distributed on the interval $[0, 2\pi]$. The random variable Θ_1 could represent the sum of several components of the phase shift incurred by a signal during transmission from the source to the destination, as discussed in Section 7.1.

One approach to this problem is to consider first the conditional density of Θ given that $\Theta_1 = \xi$. If you can show that this conditional density is the uniform density on $[0, 2\pi]$ for all values of ξ, then the desired result follows from averaging the conditional density with respect to the density for Θ_1.

7.2 Show that for any real numbers α and β,

$$\int_0^{2\pi} \exp\{\alpha \cos(\varphi) - \beta \sin(\varphi)\}\, d\varphi/2\pi = I_0\left(\sqrt{\alpha^2 + \beta^2}\,\right).$$

Hint: Write $\alpha \cos(\varphi) - \beta \sin(\varphi)$ as $\sqrt{\alpha^2 + \beta^2} \cos(\varphi + \psi)$, where ψ is defined by

$$\sin(\psi) = \beta / \sqrt{\alpha^2 + \beta^2}$$

and

$$\cos(\psi) = \alpha / \sqrt{\alpha^2 + \beta^2}.$$

In other words, $\psi = \tan^{-1}(\beta/\alpha)$. This shows that the original integral is equal to

$$\int_0^{2\pi} \exp\{\sqrt{\alpha^2 + \beta^2} \cos(\varphi + \psi)\}\, d\varphi/2\pi.$$

Now use the fact that $\cos(\theta)$ is periodic in θ to obtain the desired result.

7.3 In Section 7.5, the random variable R_0 is defined by

$$R_0 = \sqrt{U_0^2 + V_0^2},$$

where U_0 and V_0 are independent Gaussian random variables, $\text{Var}\{U_0\} = \text{Var}\{V_0\} = \sigma^2$, $E\{U_0\} = u_0$, and $E\{V_0\} = v_0$.

(a) For the random variable R_0 in place of R_1, repeat the steps used in going from (7.42) to (7.49) and apply the results of Problem 7.2 to conclude that the density function for R_0 is given by

$$f_{R_0}(r) = \begin{cases} (r/\sigma^2) \exp\{-(r^2 + c^2)/2\sigma^2\} I_0(c\,r/\sigma^2), & r \geq 0, \\ 0, & r < 0, \end{cases}$$

where $c = \sqrt{u_0^2 + v_0^2}$. This density function is known as the *Rician density*.

(b) Show that the Rician density reduces to the Rayleigh density if $u_0 = v_0 = 0$.

7.4 Consider the receiver shown in Figure 7-6 with $w_0(t) = w_1(t) = p_T(t)$ and $\theta_i(t) = (\omega_i - \omega_c) t$. Assume that the LS decision rule is used. The transmitted signals are given by $s_i(t) = \sqrt{2} \, A \, a_i(t) \, p_T(t) \cos(\omega_i t + \varphi_i)$ for $i = 1$ and $i = 0$. The amplitude shaping is not the same for the two signals. The signal $a_0(t)$ is the sine pulse of duration T and $a_1(t)$ is the unit-amplitude triangular pulse of duration T; that is, $a_0(t) = \sin(\pi t/T) \, p_T(t)$ and $a_1(t) = 2t/T$ for $0 \le t \le \frac{1}{2}T$, and $a_1(t) = 2 - (2t/T)$ for $\frac{1}{2}T < t \le T$. Assume that $\omega_1 > \omega_0 \gg 1/T$, $\omega_i = 2\pi n_i/T$ for some integer n_i, and $\omega_1 - \omega_0$ is sufficiently large that, for any phase angle φ,

$$\int_0^T s_i(t) \cos(\omega_k t + \varphi) \, dt = 0$$

if $i \ne k$. The channel is an additive white Gaussian noise channel with spectral density $N_0/2$.

(a) Give a simple closed-form expression (no integrals) for the conditional error probability $P_{e,0}$ in terms of A, T, and N_0.

(b) Give a simple closed-form expression for $P_{e,1}$ in terms of A, T, and N_0.

(c) Are these two error probabilities the same? Should they be the same for the given system and signal set? Explain why or why not.

(d) Does the LS decision rule give a minimax decision based on the statistics R_1^2 and R_0^2? Explain your answer.

7.5 Consider the same receiver, transmitted signals, and system assumptions as in Problem 7.4, except that $A = 100$ and $a_i(t) = \sin(\pi t/T)$ for both values of i. The channel is an additive white Gaussian noise channel with spectral density $N_0/2 = 4$. The data rate is 100 bits per second.

(a) Give a simple closed-form expression for $P_{e,0}$.

(b) Is this receiver the optimum noncoherent receiver for the given signals and channel? Explain why or why not.

7.6 The noncoherent communication receiver of Figure 7-4 is to be used for a packet radio communication system. Each packet consists of N binary digits that are transmitted using binary FSK modulation. The transmission rate is $1/T$ bits/s. In the absence of fading, the received signal for the nth binary digit is given by

$$s(t) = \sqrt{2} \, A \cos(\omega_i t + \varphi_i), \quad nT \le t < (n+1)T,$$

for $i = 0$ (if 0 is sent) or $i = 1$ (if 1 is sent). Assume that $\omega_0 \ne \omega_1$ and that ω_0 and ω_1 are multiples of $2\pi/T$. The front end of the receiving radio adds white Gaussian noise $X(t)$ of spectral density $N_0/2$ to the received signal, so the input to the receiver is $Y(t) = s(t) + X(t)$.

(a) In the absence of fading, A is just a deterministic constant. What is the probability of bit error for this receiver? What is the probability of packet error for this receiver? Answer in terms of A and the other parameters given.

(b) Give a high signal-to-noise ratio approximation to the packet error probability of part (a).

(c) For a particular fading channel, the parameter A is constant over the packet, but it is a random variable that has a Gaussian distribution with mean 0 and variance β^2. Find the average bit error probability for this channel and receiver (average over the amplitudes of the received signal).

(d) In this part, the packet is deemed unacceptable if its bit error probability exceeds a specified probability p. Find an expression for the outage probability, which is the probability that the amplitude A is such that the given packet is unacceptable. Simplify as much as possible.

(e) Consider the same receiver, but a different fading channel that also distorts the pulse shape. The received signal is now of the form

$$s(t) = \sqrt{2} A \sin(\pi t/T) \cos(\omega_i t + \varphi_i), \quad nT \le t < (n+1)T.$$

Assume that $T^{-1} \ll |\omega_1 - \omega_0| < \omega_0 < \omega_1$, so that, among other implications, double-frequency terms can be ignored and the signals are orthogonal. Find the average probability of bit error if the amplitude A is as described in part (c).

7.7 Compare the decision statistics for the delay-and-multiply receiver of Figure 7-12 and the other receivers of Section 7.6.

7.8 Generalize the method used to evaluate (7.52) in order to evaluate

$$\mathcal{L}(\beta) = \int_{-\infty}^{\infty} \exp\{-\rho x^2/2\sigma^2\} \exp\{-(x-\beta)^2/2\sigma^2\} \, dx/\sqrt{2\pi}\sigma,$$

where ρ and β are arbitrary real numbers. Show that

$$\mathcal{L}(\beta) = \exp\{-\rho\beta^2/[2\sigma^2(\rho+1)]\}/\sqrt{\rho+1},$$

and check this result for $\rho = 1$. (For $\rho = 1$, you should get $\mathcal{L}(\beta) = \mathcal{I}(\beta)$.) One approach for the evaluation of $\mathcal{L}(\beta)$ is to write the integrand in a more convenient form by collecting terms inside a single exponential and making the substitution $\lambda = \sigma/\sqrt{\rho+1}$. Then complete the square and proceed as in Section 7.5.1. Use the fact that

$$\int_{-\infty}^{\infty} \exp\{-[u - (\rho+1)^{-1}\beta]^2/2\lambda^2\} \, du/\sqrt{2\pi}\lambda = 1.$$

7.9 Compare the required $\mathcal{E}/N_0$ to the nearest tenth of a dB for coherent reception of BPSK, differentially coherent detection of DBPSK, and noncoherent detection of BFSK if the bit error probability for each is to be 10^{-6}. Use $Q(\sqrt{2\mathcal{E}/N_0}) = 10^{-6}$ for $(\mathcal{E}/N_0)_{dB} = 10.5$ dB. You get to use your calculators for this one!

7.10 The signals for the noncoherent filter/envelope-detector BFSK system shown in Figure 7-18 are given by $s_i(t) = v(t) \cos(\omega_i t + \varphi_i)$, for $i = 0$ or 1 and $0 \le t \le T$. The system satisfies the usual conditions:

(1) $\omega_0 < \omega_1$.

(2) $H_i(\omega) = G(\omega - \omega_i) + G(\omega + \omega_i)$ where $G(\omega) = G^*(-\omega)$ and the bandwidth of $G(\omega)$ is much smaller than both ω_0 and $\omega_1 - \omega_0$.

(3) $H_0(\omega) H_1(\omega) \approx 0$ for all ω.

The waveform $v(t)$ is a unit-amplitude rectangular pulse of duration T. The channel noise is a white Gaussian noise process with spectral density $N_0/2 = 32$, and the transfer function $G(\omega)$ is given in terms of its inverse Fourier transform $g(t)$ by

$$g(t) = \begin{cases} 3\exp(-3t), & t \geq 0, \\ 0, & t < 0. \end{cases}$$

(a) Find the probability of error for this system (in terms of T).

(b) Repeat part (a) for a system that is identical to the one just described except that the sampling in the receiver is at time $T_0 \neq T$ rather than at time T. Consider both $T_0 > T$ and $T_0 < T$.

7.11 Consider narrowband signals $s_0(t)$ and $s_1(t)$ of the form

$$s_i(t) = v(t)\cos(2\pi f_i t + \varphi_i), \quad 0 \leq t \leq T.$$

Assume that $f_i T$ is an integer for each i. Suppose the signals are used in the non-coherent communications system shown in Figure 7-18. The filters $h_0(t)$ and $h_1(t)$ are narrowband filters of the form

$$h_i(t) = 2\,g(t)\cos(2\pi f_i t).$$

The channel is an AWGN channel with spectral density $N_0/2$.

(a) Suppose that $g(t)$ has Fourier Transform

$$G(f) = \begin{cases} 1, & |f| \leq 1/T, \\ 2 - |f|T, & 1/T < |f| \leq 2/T, \\ 0, & |f| > 2/T. \end{cases}$$

It is known that $(g*v)(T_0) = 2$. Find the error probabilities $P_{e,0}$ and $P_{e,1}$ for this system.

(b) Suppose that $g(t) = \sin(\pi t/T), 0 \leq t \leq T$, and $v(t) = A, 0 \leq t \leq T$. Assume that the sampling time is chosen to maximize $(v*g)(T_0)$. That is,

$$\hat{v}(T_0) = \max\{\hat{v}(t) : 0 < t < T\},$$

where $\hat{v} = v*g$. Find the error probabilities.

Hint: It is not necessary to evaluate the convolution $(v*g)(t)$ for all t in order to determine $\hat{v}(T_0)$.

7.12 A filter/envelope-detector receiver (Figure 7-18) for noncoherent FSK has bandpass filters with transfer functions given by

$$H_i(f) = \begin{cases} 1, & |f - f_i| \leq W \text{ or } |f + f_i| \leq W, \\ 0, & \text{otherwise,} \end{cases}$$

for $i = 0$ and $i = 1$. The signals are

$$s_i(t) = v(t)\cos(2\pi f_i t + \varphi_i),$$

where $v(t) = A\,p_T(t)$. The AWGN channel has spectral density $N_0/2$. Assume that $f_1 > f_0 \gg 1/T$, $f_1 - f_0 > 2W$, and $f_0 \gg W$.

(a) What is the transfer function $G(f)$ of the lowpass equivalent for these bandpass filters?

(b) Find the impulse response $g(t)$ for the lowpass equivalent filter. Express your answer in terms of the function sinc.

(c) Find the impulse response for each of the bandpass filters.

(d) Find the variance σ^2 of the noise at the outputs of the bandpass filters.

(e) Evaluate $\hat{v}(T)$ where $\hat{v} = v * g$. Express your answer in terms of the tabulated sine integral

$$\text{Si}(x) = \int_0^x y^{-1} \sin(y)\, dy.$$

(f) Find the error probability for this receiver.

7.13 Consider a noncoherent BFSK communication system with the receiver of Figure 7-18. Let $\omega_i = 2\pi f_i$ and $g_i(t) = p_T(t)$ for $i = 0$ and $i = 1$. The signals are of the form

$$s_i(t) = \sqrt{2}\, A \cos(2\pi f_i t + \varphi_i)\, p_T(t),$$

for each value of i. The signal $s_0(t)$ is referred to as the *mark* signal, and $s_1(t)$ is called the *space* signal. Assume that the two signals are orthogonal.

Suppose that thermal noise is negligible in this system, and the only noise that affects the performance of the noncoherent receiver is *bandlimited* white Gaussian noise with two-sided spectral density $N_1/2$. This noise may be present at none, one, or both of the frequencies used for the two signals. The bandwidth of this noise is sufficiently large that, when it is present at a given frequency, it produces the same effect on the corresponding branch of the receiver as white Gaussian noise would produce; however, the frequencies f_0 and f_1 are sufficiently far apart that the presence of noise at frequency f_0 does not affect the space filter, and the presence of noise at frequency f_1 does not affect the mark filter. In addition, the noise processes at the two frequencies are statistically independent.

The presence or absence of noise at a given frequency is a random phenomenon. The probability that noise is present at frequency f_0 is β_0, and the probability that noise is present at frequency f_1 is β_1. The event that noise is present at f_0 is statistically independent of the event that noise is present at f_1.

Give an expression for $P_{e,0}$, the probability that the receiver makes an error given that the mark signal is transmitted. Also, give an expression for $P_{e,1}$, the probability that the receiver makes an error given that the space signal is transmitted. These expressions should be in terms of A, T, N_1, β_0, and β_1.

Explain how to solve the problem if the thermal noise is *not* negligible. A detailed solution is acceptable, but not required. It is sufficient to describe the steps needed to obtain a solution.

7.14 The signals for a noncoherent filter/envelope-detector BFSK system shown in Figure 7-18 are denoted by $s_0(t)$ and $s_1(t)$. It is known that if $s_i(t)$ is the transmitted signal (for $i = 0$ or $i = 1$), then the decision statistic R_i can be written as

$$R_i = \sqrt{U_i^2 + V_i^2}$$

and the other decision statistic R_j $(j \neq i)$ can be written as

$$R_j = \sqrt{U_j^2 + V_j^2},$$

where U_i, V_i, U_j, and V_j are jointly Gaussian and each has variance σ^2; U_i has mean u and V_i has mean v, where u and v always satisfy $u^2 + v^2 = c^2$ for a known constant c; and U_j and V_j have mean 0. Notice that σ and c do not depend on which signal is sent. The decision is made according to the following rule: If $R_0 > \lambda R_1$ decide that s_0 was sent; if $R_1 > \lambda R_0$ decide that s_1 was sent; otherwise, declare an erasure. The parameter λ is a real number that satisfies $\lambda > 1$. Find an expression in terms of the parameters λ, c, and σ for the conditional probability given that s_0 was sent that either an erasure is made or the decision is that s_1 was sent (i.e., the probability of not deciding s_0 was sent when in fact s_0 was sent).

7.15 Consider a noncoherent correlation receiver for FSK signals. The signals are of the form

$$s_i(t) = A\, p_T(t)\cos(\omega_i t + \varphi_i)$$

for $i = 0$ and $i = 1$, and the channel is an AWGN channel (spectral density $N_0/2$). The decision statistics are R_0 and R_1, where $R_k = +\sqrt{U_k^2 + V_k^2}$ for $k = 0$ and $k = 1$. The random variables U_k and V_k are given in terms of the receiver input $Y(t)$ by

$$U_k = \beta_k \int_0^T Y(t)\cos(\omega_k t)\,dt$$

and

$$V_k = \beta_k \int_0^T Y(t)\sin(\omega_k t)\,dt,$$

for $k = 0$ and $k = 1$. The gains β_0 and β_1 are not necessarily equal. Assume throughout the problem that signal s_0 was sent.

(a) Give an expression for $F_{R_1}(r)$, the *distribution* function for R_1. Relate any parameters in this distribution function to A, β_0, β_1, N_0, and T.

(b) Give an expression for $f_{R_0}(r)$, the *density* function for R_0. Relate any parameters in this density function to A, β_0, β_1, N_0, and T.

(c) As usual, the receiver decides that s_1 was sent if $R_1 > R_0$; otherwise, it decides that s_0 was sent. Give an expression for $P_{e,0}$ in terms of $F_{R_1}(r)$ and $f_{R_0}(r)$, the functions defined in parts (a) and (b), respectively.

(d) Evaluate the integral in part (c) to obtain a closed form expression for $P_{e,0}$ in terms of the parameters A, β_0, β_1, N_0, and T.

(e) Does your work in part (d) show how to generalize (7.61) to give (7.62)?

7.16 (a) Compute the symbol error probability P_e for an MFSK system with an optimum noncoherent receiver for $M = 4$ and $(\mathcal{E}_b/N_0)_{\mathrm{dB}} = 9.5$ dB.

(b) Evaluate the union bound on P_e for the parameters in part (a), and compare the value of the bound with the value you computed for P_e in part (a).

7.17 Define $s(t)$ by

$$s(t) = s_1(t, \varphi_1) + s_2(t, \varphi_2),$$

where

$$s_i(t, \theta) = \cos(2\pi f_i t + \theta), \quad 0 \le \theta < 2\pi,$$

for $i = 1$ and $i = 2$. Suppose $s(t)$ is transmitted over a multipath channel that has two paths. The differential propagation delay for these two paths is τ_0, so the received signal is

$$
\begin{aligned}
r(t) = \ & \beta_1 \{\cos(2\pi f_1 t + \varphi_1) + \cos(2\pi f_2 t + \varphi_2)\} \\
& + \beta_2 \{\cos[2\pi f_1 (t - \tau_0) + \varphi_1] + \cos[2\pi f_2 (t - \tau_0) + \varphi_2]\}.
\end{aligned}
$$

The parameters β_1 and β_2 account for the propagation losses for the two paths.

(a) Show that this signal can be written as

$$
\begin{aligned}
r(t) = \ & I(f_1) \cos(2\pi f_1 t + \varphi_1) + Q(f_1) \sin(2\pi f_1 t + \varphi_1) \\
& + I(f_2) \cos(2\pi f_2 t + \varphi_2) + Q(f_2) \sin(2\pi f_2 t + \varphi_2),
\end{aligned}
$$

where $I(f) = \beta_1 + \beta_2 \cos(2\pi f \tau_0)$ and $Q(f) = \beta_2 \sin(2\pi f \tau_0)$ for $-\infty < f < \infty$.

(b) Suppose that $\beta_1 = \beta_2 = 1/\sqrt{2}$. Find values for τ_0, f_1, and f_2 for which $I(f_1) = \sqrt{2}$, $I(f_2) = 1/\sqrt{2}$, $Q(f_1) = 0$, and $Q(f_2) = 1/\sqrt{2}$.

(c) For the values of β_1, β_2, τ_0, f_1, and f_2 from part (b), show that the multipath channel increases the amplitude of $s_1(t, \varphi_1)$ by a factor of $\sqrt{2}$ (a factor of 2 increase in power). Show that it does not change the amplitude of $s_2(t, \varphi_2)$, but it shifts its phase by $\pi/4$ radians. Thus, the effects of the channel are frequency dependent for the parameter values in part (b).

(d) Suppose that $\beta_1 = \beta_2$, $f_2 = 15$ Mhz, and $\tau_0 = 100$ ns. What is the output of the channel due to the transmitted signal $s_2(t, \varphi_2)$?

(e) Suppose $0 < \beta_2 \le \beta_1$ and $\tau_0 \ll |f_2 - f_1|^{-1}$. Show that the effects of this multipath channel are approximately independent of frequency by proving that $I(f_2) \approx I(f_1)$ and $Q(f_2) \approx Q(f_1)$ in the sense that $|I(f_2) - I(f_1)|$ and $|Q(f_2) - Q(f_1)|$ are each much smaller than β_1 and β_2.
Hint: First show that

$$|I(f_2) - I(f_1)| \le 2|\beta_2 \sin[\pi(f_2 - f_1)\tau_0]|$$

and

$$|Q(f_2) - Q(f_1)| \le 2|\beta_2 \sin[\pi(f_2 - f_1)\tau_0]|.$$

(f) Suppose again that $\beta_1 = \beta_2 = 1/\sqrt{2}$. As an example for which $\tau_0 \ll |f_2 - f_1|^{-1}$, suppose $\tau_0 = 1$ μs and $f_2 - f_1 = 10$ kHz. Evaluate the bound on $|I(f_2) - I(f_1)|$ given in part (e).

7.18 Consider the value of $\mathcal{E}/N_0$ required to achieve a bit error probability of $P_e = 10^{-n}$ for equal-energy, binary, orthogonal signaling with optimum noncoherent demodulation and a channel with no fading. Find an equation for n in terms of $\mathcal{E}/N_0$, and show that n increases linearly in $\mathcal{E}/N_0$. Find an equation for n in terms of $(\mathcal{E}/N_0)_{dB}$, and show that n is an exponentially increasing function of $(\mathcal{E}/N_0)_{dB}$.

7.19 Suppose that the random variables X and Y are independent, Gaussian random variables, each having variance σ^2 and mean zero. The random variable R is defined by $R = \sqrt{X^2 + Y^2}$. In the following, evaluate all integrals without the use of tables, and express all answers in terms of the parameter σ:

(a) Give an expression for $f_R(r)$, the density function for R.

(b) From this expression, use integration by parts to integrate $r f_R(r)$, and find the mean value of R.

(c) Use the answer to part **(a)** to find the second moment of R directly from the density.

(d) Express the integral in part **(b)** in terms of an integral for the second moment of a Gaussian density with zero mean and variance σ^2. Use this integral expression to find the mean of R without evaluating an integral. *Hint:* The second moment of a zero-mean Gaussian density is its variance.

(e) Now use the definition of the random variable R to find its second moment. Which method is easier, part **(c)** or part **(e)**?

(f) Find the variance of R.

(g) In Section 7.9, it is stated that β^2 is the second moment of the gain of the channel. Use your answers to parts **(c)** and **(e)** to verify this statement.

7.20 Consider the DBPSK signal given by

$$\sqrt{2}\, A \cos[(\omega_c + \omega_0)\, t + \psi_n + \varphi]$$

for $nT \leq t < (n+1)T$, which we refer to as pulse n. For DBPSK, the phase angles ψ_n and ψ_{n-1} for pulse n and pulse $n-1$ satisfy $\psi_n = \psi_{n-1} + \pi\beta_n$. The radian frequency ω_c is the nominal carrier frequency, and the radian frequency ω_0 is an unknown frequency offset. If pulse $n - 1$ is delayed by T units of time, it is given by

$$\sqrt{2}\, A \cos[(\omega_c + \omega_0)(t - T) + \psi_{n-1} + \varphi].$$

The instantaneous phase difference (modulo 2π) for pulse n and the delayed version of pulse $n - 1$ is

$$\begin{aligned}
\theta_n &= [(\omega_c + \omega_0)\, t + \psi_n + \varphi] - [(\omega_c + \omega_0)(t - T) + \psi_{n-1} + \varphi]\\
&= \omega_c T + \omega_0 T + \psi_n - \psi_{n-1} = \omega_0 T + \psi_n - \psi_{n-1}.
\end{aligned}$$

The last step follows from $\omega_c T = 0$ (modulo 2π). Now use $\Delta_n = \psi_n - \psi_{n-1} = \pi\beta_n$ to conclude the instantaneous phase difference is $\theta_n = \omega_0 T + \Delta_n = \omega_0 T + \pi\beta_n$, from which we cannot recover the data β_n, because ω_0 is unknown. Without knowing ω_0, we cannot determine whether $\beta_n = 1$ or $\beta_n = 0$. For example, if $\theta_n = \pi$, the data symbol could be $\beta_n = 1$ (if $\omega_0 = 0$) or $\beta_n = 0$ (if $\omega_0 = \pi/T$). Since β_n represents the information that the transmitter is attempting to convey to the receiver, the conclusion is that this information is not conveyed by DBPSK if there is an unknown frequency offset.

Now consider DDBPSK. The signal format is the same as for DBPSK, but the phase modulation is given by $\psi_n = 2\psi_{n-1} - \psi_{n-2} + \Gamma_n$, where $\Gamma_n = \Delta_n - \Delta_{n-1} = \pi(\beta_n - \beta_{n-1})$. Assume that an initial data variable is known to both the transmitter and receiver (i.e., always set $\beta_{-1} = 1$ at the beginning of the message that is used to convey $\beta_0, \beta_1, \beta_2, \ldots$). Examine the signal over three consecutive intervals, and show that the difference of two consecutive values of the instantaneous phase difference (i.e., $\theta_n - \theta_{n-1}$) does convey the desired information, even if there is a unknown frequency offset. Explain how this information is conveyed.

7.21 It is desired to develop an expression for the probability of error that is valid for the noncoherent receiver of Figure 7-18 with the LS decision rule, even if the bandpass filters are not locally symmetric (e.g., the impulse response $h_0(t)$ does not satisfy (7.89)). The two filters are ideal bandpass filters, however, so their impulse responses can be written in a form analogous to (4.48):

$$h_i(t) = 2\, g_i(t) \cos(\omega_i t) - 2\, \hat{g}_i(t) \sin(\omega_i t)$$

for $i = 0$ and $i = 1$. Note that the factor of 2 is included, as in (4.56) and (7.89). Assume that the bandwidths of all signals and filters are small compared to ω_0 and ω_1, and ignore double-frequency terms in all that follows:

    (a) Suppose the input to the filter with impulse response $h_0(t)$ is the ideal bandpass signal given by

$$w(t) = v_1(t) \cos(\omega_0 t) - v_2(t) \sin(\omega_0 t).$$

The output signal is still given by (7.90). After all, the output is still an ideal bandpass signal, so it must have this form. Find expressions for the functions y_1 and y_2 of (7.90) in terms of the functions v_1, v_2, g_0, and $\hat{g}_0$ only.

    (b) Give a block diagram of a noncoherent correlator that performs the same operation as the filter and envelope detector of Figure 7-17 if

$$h_0(t) = 2\, g_0(t) \cos(\omega_0 t) - 2\, \hat{g}_0(t) \sin(\omega_0 t).$$

The block diagram should be similar to Figure 7-16, but there are two important differences. The most important difference is that $2\, w_0(t) \cos(\omega_0 t)$ and $2\, w_0(t) \sin(\omega_0 t)$ must be replaced by signals that depend on the functions g_0 and $\hat{g}_0$ and that may have both inphase and quadrature components. In addition, the limits on the integrators may differ from those in Figure 7-16, because $g_0(t)$ and $\hat{g}_0(t)$ are not required to be time limited.

    (c) Now assume that $g_i(t)$ and $\hat{g}_i(t)$ are time limited to the interval $[0, T]$. Analogous to the comparison of the systems of Figures 7-17 and 7-16, show that your noncoherent correlator is equivalent to the filter and envelope detector system of Figure 7-17 for the filter impulse response $h_0(t)$ as given in part **(b)**. That is, show that the two systems produce the same output for the same ideal bandpass signal at the input.

    (d) Assume that the transmitted signal is given by (7.85) for $i = 0$, and obtain the desired expression for the probability of error $P_{e,0}$. Express your answer in terms of $\|\hat{g}_0\|$, $\|g_0\|$, and the parameters $\hat{\beta}_0$ and β_0, which are defined by $\hat{\beta}_0 = (m_0 * \hat{g}_0)(T)$ and $\beta_0 = (m_0 * g_0)(T)$.

    (e) Show that if the filter $h_0(t)$ is locally symmetric around frequency ω_0, your expression in part **(d)** reduces to (7.95).

7.22 Suppose that $\omega_0 = 2\pi n/T$ for some integer n and $\omega_\delta = 2\pi/T$. The 16-ary FSK signal set defined by $s_i(t) = \sqrt{2}\, A\, p_T(t) \cos[(\omega_0 + i\omega_\delta)t + \varphi]$, $0 \le i \le 15$, is used for communication over an AWGN channel with spectral density $N_0/2$.

(a) Suppose $10 \log_{10}(A^2 T / N_0) = 9.0$, and the receiver is the maximum-likelihood coherent receiver. Use Table 7.1 to evaluate the union bound on the probability of symbol error.

(b) Suppose $A^2 T / N_0 = 10$, and the receiver is the maximum-likelihood noncoherent receiver. Use Table 7.1 to evaluate the union bound on the probability of symbol error.

7.23 A standard binary, equal-energy, orthogonal FSK signal set $\{s_0, s_1\}$ is employed with the receiver shown in Figure 7-4. In the absence of fading, the received signal is given by $s_i(t) = \sqrt{2}\beta \cos(\omega_i t + \varphi_i)$, $0 \le t < T$, for $i = 0$ if s_0 is sent or $i = 1$ if s_1 is sent. Assume that $\omega_0 \ne \omega_1$ and that ω_0 and ω_1 are multiples of $2\pi/T$. The channel is an AWGN channel with spectral density $N_0/2$.

(a) If the channel exhibits no fading, what is the probability of bit error? Answer in terms of the appropriate function and the parameters β, N_0, and T.

For parts **(b)–(f)**, the channel is a nonselective fading channel for which the received signal is modeled as $V s_i(t)$. The random variable V is Gaussian with mean 0 and variance λ^2, it is independent of $X(t)$, and it is independent of which signal is transmitted.

(b) Give an expression for $\bar{\mathcal{E}}_b$, the *average* energy per bit in the received signal. Your answer must be in terms of the parameters β, λ, and T.

(c) Derive an expression for $\overline{P}_{e,0}$, the average probability of error when s_0 is sent. Express your answer in terms of the appropriate function and the parameters β, λ, N_0, and T.

(d) Define the parameter ζ by $\zeta = \bar{\mathcal{E}}_b/N_0$, and give an expression for $\overline{P}_{e,0}$ in terms of the parameter ζ only.

For parts **(e)** and **(f)**, suppose that β and N_0 are unknown. However, it is determined that if s_0 is sent, then $E\{R_0^2\} = \rho_0$ and $E\{R_1^2\} = \rho_1$. For parts **(e)** and **(f)**, you must express your answer in terms of the appropriate function and the parameters ρ_0, ρ_1, and λ only. You may not use the parameters β, N_0, or T in your answer.

(e) Give an expression for $\overline{P}_{e,0}$, the average probability of error when s_0 is sent.

(f) Give an expression for $\overline{P}_{e,1}$, the average probability of error when s_1 is sent.

Chapter 8

Intersymbol Interference

In practical communication systems, the communication signal is filtered by a number of linear systems. These linear systems may include filters and linear amplifiers in the transmitter, the propagation medium, filters and amplifiers in the front end of the receiver, and filters that are used in the demodulation process. The combined effects of all such linear systems can be represented by a single linear filter, and we can consider the impulse response and the transfer function of this filter. In most practical communication systems, this filter is not the optimum filter for processing the communication signal. This fact leads us to consider the effects of suboptimum filtering on the performance of digital communication systems.

There are two ways in which suboptimum filtering increases the probability of error. First, because the filter is not matched to the communication waveform, there is a degradation in performance even if only a single pulse is sent over the channel. This is the result of a decrease in the signal component of the sampled waveform at the filter output, an increase in the variance of the noise component of the sampled waveform, or a combination of the two. Second, if the transmitted signal consists of a sequence of pulses, as is always the case in practice of course, the suboptimum filter may produce interference among the pulses in the sequence. This latter phenomenon is known as *intersymbol interference*.

We begin this chapter by describing intersymbol interference and its effects on communication system performance. Methods are presented for determining the probability of error in systems that have intersymbol interference. We then describe conditions under which intersymbol interference does not occur in a communication system, and we conclude the chapter by presenting an introduction to equalization techniques that reduce the effects of intersymbol interference.

8.1   Intersymbol Interference Components in Digital Communications

Let $\beta(t)$ be a baseband pulse waveform, and define the baseband signal

$$d(t) = \sum_{n=n_1}^{n_2} b_n\, \beta(t - nT),\tag{8.1}$$

where each b_n is either -1 or $+1$. At the transmitter, we let b_n be -1 if the nth binary digit is 1, and we set $b_n = +1$ if the nth binary digit is 0. Thus, the continuous-time waveform $d(t)$ that is defined in (8.1) can convey as many as $n_2 + 1 - n_1$ bits of information by a proper choice of the sequence $(b_n) = b_{n_1}, \ldots, b_{n_2}$ of pulse amplitudes. The pulse $b_n\, \beta(t - nT)$ is referred to as the nth pulse, and b_n is called the amplitude of the nth pulse. In particular, the 0th pulse is $b_0\, \beta(t)$. It is common that we can fix the value of b_0 to be $+1$ in a particular discussion, in which case $\beta(t)$ is often referred to as the 0th pulse.

For some applications, it is desirable to consider an infinite sequence of pulse amplitudes by letting $n_1 = -\infty$, $n_2 = +\infty$, or both. If $n_1 = -\infty$ and $n_2 = +\infty$, the sequence (b_n) is referred to as a *doubly infinite* sequence, and it is often denoted by $\ldots, b_{-1}, b_0, b_1, b_2, \ldots$. For a doubly infinite sequence of pulse amplitudes, (8.1) is the infinite series

$$d(t) = \sum_{n=-\infty}^{\infty} b_n\, \beta(t - nT).\tag{8.2}$$

An example of the pulse waveform $\beta(t)$ is the rectangular pulse $p_T(t)$, which is often used to model the communication signal before any filtering has been applied. If the pulse waveform is rectangular and the pulse amplitude sequence is doubly infinite, then the signal $d(t)$ is given by

$$d(t) = \sum_{n=-\infty}^{\infty} b_n\, p_T(t - nT).\tag{8.3}$$

If the communication system employs a carrier-modulated signal, the received signal in the communication system is

$$s(t) = \sqrt{2}\, A\, d(t) \cos(\omega_c t + \theta).\tag{8.4}$$

The communication channel is an additive Gaussian noise channel, so the input to the receiver is $s(t) + X(t)$, where $X(t)$ represents Gaussian noise. A model for a coherent receiver is illustrated in Figure 8-1. It consists of a multiplier, a time-invariant linear filter with impulse response $h(t)$, a sampler that samples the signal every T time units, and a threshold device. The input to the receiver is $s(t) + X(t)$, the sum of the signal and the noise. The linear filter is a baseband filter; for example, it could be matched to the waveform $\beta(t)$.

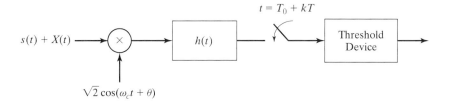

Figure 8-1: Model for a coherent communication system.

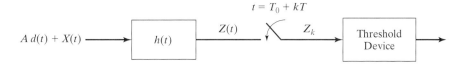

Figure 8-2: Model for a baseband communication system.

For a baseband communication system, the signal component of the received waveform is

$$s(t) = A\,d(t), \tag{8.5}$$

and the input to the receiver is $s(t) + X(t)$. The baseband receiver is illustrated in Figure 8-2. It consists of a time-invariant linear filter with impulse response $h(t)$, a sampler that samples the signal every T time units, and a threshold device, all of which are also included in the receiver of Figure 8-1.

The constant A that appears in (8.4) and (8.5) accounts for the transmitter power level, antenna gain, propagation loss, etc., but it does not play a significant role in our discussions of intersymbol interference and related matters. As a result, we focus on the baseband signal $d(t)$ in most of what follows, and we ignore the amplitude factor A. Only when we wish to evaluate the error probability do we need to account for this amplitude factor.

The output of the multiplier of Figure 8-1 has one component due to the signal at the multiplier input and one component due to the noise. The signal portion of the output of the multiplier has two components: a baseband component, which is proportional to $d(t)$, and a double-frequency component. We ignore the double-frequency component, because it is eliminated by the baseband filter. Thus, for our purposes, the signal portion of the input to the filter in the system of Figure 8-1 is the same as the signal portion of the input to the filter in the system of Figure 8-2. In this sense, the receiver of Figure 8-2 is the baseband equivalent of the receiver of Figure 8-1.

One could argue that there is an abuse of notation with respect to the noise in these two systems, because $X(t)$ is used to denote the noise at the input to each, even though one is a carrier-modulated system, and the other is a baseband system. The intent is that, if the original system is a carrier-modulated communication system and Figure 8-2 is its baseband equivalent, then the noise shown in Figure 8-2 is the baseband equivalent of the original noise process. Note that if the noise at the input to the multiplier in Figure 8-1 is white Gaussian noise with spectral density $N_0/2$, and if the phase angle

θ is a random variable that is independent of the noise and uniformly distributed on $[0, 2\pi]$, then the noise at the output of the multiplier is also white Gaussian noise with spectral density $N_0/2$.

The key issues that are of interest in this section are not dependent on the noise, but instead they relate to the effects of the filter on the baseband signal $d(t)$. Thus, for most of our development, we think of $d(t)$ as being the transmitted signal as well as the input to the filter. When it is necessary to determine the probability of error, we account for the noise and the constant A.

The baseband signal $d(t)$ is a sequence of pulses, as described mathematically by (8.1). As a result, the effects of the filter on $d(t)$ can be determined by evaluating the response of the filter to one of these pulses. From this response, we can determine anything we need to know about the output of the filter when $d(t)$ is the input. As a result, the response of the filter to the waveform $\beta(t)$ is of primary importance to our development.

Let $v(t)$ denote the output of the time-invariant linear filter with impulse response $h(t)$ when $\beta(t)$ is the input. That is, $v(t)$ is the convolution of $\beta(t)$ and $h(t)$. We refer to $v(t)$ as the *pulse-response waveform* or simply the *pulse response*.

Corresponding to each input pulse $b_k \beta(t - kT)$ there is an output pulse $b_k v(t - kT)$; that is, $v(t - kT)$ is the response of the filter to the input $\beta(t - kT)$, and b_k is the amplitude associated with the kth pulse. The output is sampled at time $T_0 = kT$ in order to provide the decision statistic Z_k, which is used by the receiver in deciding whether $b_k = +1$ or $b_k = -1$.

The decision statistic for the kth pulse, which we have denoted by Z_k, has a signal component and a noise component. We denote the signal component by u_k and the noise component by Y_k. From Chapters 5 and 6, we know that if the noise is white then Y_k is Gaussian, its mean value is zero, and its variance is $\sigma^2 = N_0\|h\|^2/2$. There is nothing new about the noise component in our present discussion. Our interest is in the signal component and the influence the filtering has on the signal component. Thus, our discussion focuses on u_k, the signal component of the decision statistic.

If the different pulses of (8.1) do not interfere with each other, then u_k depends on $d(t)$ through the kth term only. In practice, however, the filtering in the system often results in u_k depending on other terms in (8.1) in addition to the kth. This leads us to the central question for this chapter: To what extent do the pulses in the data signal $d(t)$ interfere with each other as they are passed through the linear filter? More precisely, the issue is the extent to which u_k, the signal component of the decision statistic for the kth pulse, depends on elements of the sequence (b_n) other than b_k.

If the pulses do interfere with each other (i.e., if u_k depends on b_n for one or more values of n other than k), then the communication system is said to produce *intersymbol interference*. Before considering a system that produces intersymbol interference, we first examine a simple example in which no intersymbol interference occurs.

Example 8-1 The Rectangular Data Pulse and Its Matched Filter

Suppose the filter impulse response for either of the receivers shown in Figures 8-1 and 8-2 is $h(t) = p_T(t)$, the rectangular function of duration T that begins at $t = 0$. Suppose that the baseband pulse waveform $\beta(t)$ is also a rectangular pulse of duration T. The

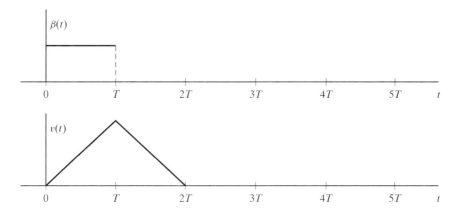

Figure 8-3: Filter input and output for a single pulse.

pulse-response waveform $v(t)$ is an isosceles triangle that has a base of length $2T$. In particular, if the input rectangular pulse also begins at $t = 0$, the pulse response is as illustrated in Figure 8-3. An important feature of this example is that the pulse-response waveform has its peak value at time T and it satisfies $v(nT) = 0$ for all values of n except for $n = 1$.

If the input to this filter is a sequence of rectangular pulses, the output consists of the superposition of the corresponding triangular pulses, as illustrated in Figure 8-4. The optimum sampling times are T for the first pulse, $2T$ for the second pulse, and $3T$ for the third pulse. At these sampling times, there is no intersymbol interference. That is, at time T only the first pulse has a nonzero response, at time $2T$ only the second pulse has a nonzero response, and at time $3T$ only the third pulse has a nonzero response. For any number of input pulses in the sequence, the output is free of intersymbol interference if the sampling times are at multiples of T.
■

Returning to the general situation, the pulse-response waveform $v(t)$ is the output of the filter that represents the channel when the input is $\beta(t)$. Examples of the waveforms $\beta(t)$ and $v(t)$ are shown in Figure 8-5. These examples are used in the following paragraphs to illustrate the definitions and concepts as they are introduced.

From (8.1), we see that the baseband data signal is the sum of a series of positive or negative, time-shifted versions of $\beta(t)$. For the present discussion, let $n_1 = -\infty$ and $n_2 = +\infty$, so there are infinitely many pulses in (8.1). The 0th input pulse in the sequence is $b_0 \beta(t)$, and the response of the filter to the 0th input pulse is $b_0 v(t)$, as illustrated in Figure 8-6.

Suppose the receiver samples the filter output at time T_0 to obtain the decision statistic for making a decision regarding b_0. The value of T_0 may be selected to maximize the difference between $+v(T_0)$ and $-v(T_0)$, for example. No matter how the sampling time for the 0th pulse is selected, once it is decided that the response to the 0th pulse is to be sampled at time T_0, the sampling time that provides the decision statistic for the nth pulse is $T_0 + nT$. This is because the output due to the nth pulse is just the output due to the 0th pulse delayed by nT. Thus, if the input to the filter is $d(t)$, the *sampling times* for the signal at the output of the filter are $T_0 + nT$ for all integers n.

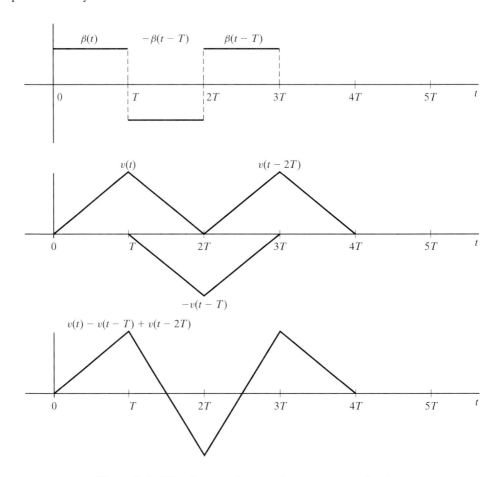

Figure 8-4: Filter input and output for a sequence of pulses.

Now return to Example 8-1 in which the pulse $\beta(t)$ and the impulse response $h(t)$ are rectangular pulses of duration T. The feature of Example 8-1 that guarantees there is no intersymbol interference is that the response of the filter to a single pulse is zero at all of the sampling times except for the sampling time used for that pulse. In Example 8-1, the obvious choice for the sampling time is $T_0 = T$. (Fortunately, this obvious choice is also optimum.) The sampling times for the other pulses are $T + nT = (n+1)T$ for all $n \neq 0$. It is clear from Figure 8-3 that $v((n+1)T) = 0$ for all $n \neq 0$. For an arbitrary pulse waveform $\beta(t)$ and an arbitrary time-invariant linear filter, there is no intersymbol interference if and only if the pulse-response waveform $v(t)$ satisfies $v(T_0 + nT) = 0$ for all $n \neq 0$.

For the illustration in Figure 8-7, the response of the filter is not zero at some of the sampling times for the other pulses. As a result, the corresponding communication system exhibits intersymbol interference. The sample $v(T_0)$ is referred to as the *desired signal component*, and the nonzero values of $v(T_0 + nT)$ for $n \neq 0$ are referred to as *intersymbol interference components*. The intersymbol interference components

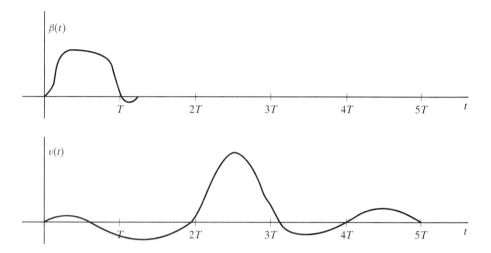

Figure 8-5: Filter input $\beta(t)$ and output $v(t)$ for a single pulse.

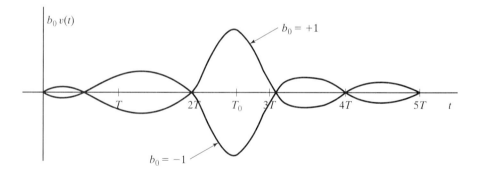

Figure 8-6: Filter output for input $b_0\,\beta(t)$.

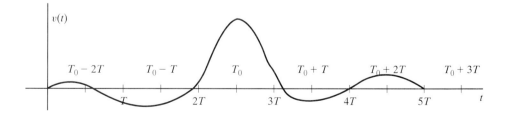

Figure 8-7: Sampling times for the pulse-response waveform.

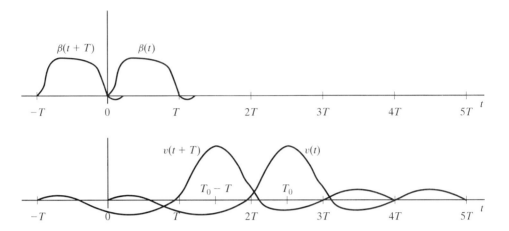

Figure 8-8: The response to two consecutive pulses.

illustrated in Figure 8-7 are $v(T_0 - 2T)$, $v(T_0 - T)$, $v(T_0 + T)$, and $v(T_0 + 2T)$. For this pulse-response waveform, $v(T_0 + nT) = 0$ for $n \geq +3$ or $n \leq -3$. It simplifies the discussion and introduces no loss of generality to assume that the desired signal component is positive; that is, $v(T_0) > 0$. This is essentially the same as the assumption in Chapter 5 that the signal $s_0(t)$ produces a larger output at the sampling instant than the signal $s_1(t)$ produces.

There is a feature of Figure 8-7 that deserves some special attention, because it often causes confusion among those who are encountering it for the first time. For the pulse response illustrated in Figure 8-7, it is shown in what follows that a pulse that is currently being transmitted is interfered with by two pulses that are transmitted in the future. Upon noticing this fact, one might be led to the conclusion that a system with this pulse response is not causal. In reality, however, it is not uncommon in causal systems for one pulse to interfere with another pulse that was transmitted earlier.

In order to explain the phenomenon just described, we first direct the reader's attention to Figure 8-5 and point out that the pulse response $v(t)$ is zero for $t < 0$. Thus, the requirements for a causal system are not violated. Now consider the intersymbol interference components shown in Figure 8-7. The sampling time for the desired signal component is T_0, and there are intersymbol interference components at $T_0 - 2T$, $T_0 - T$, $T_0 + T$, and $T_0 + 2T$. As explained in what follows, the fact that there are intersymbol interference components at times $T_0 - 2T$ and $T_0 - T$ implies that the pulse $\beta(t)$ causes intersymbol interference to two pulses that are transmitted prior to the transmission of $\beta(t)$. In other words, $\beta(t)$ interferes with two earlier pulses.

The pulse that immediately precedes $\beta(t)$ is $\beta(t + T)$, as illustrated in Figure 8-8, and the sampling time for this preceding pulse is $T_0 - T$. At this sampling time, $\beta(t)$ contributes an intersymbol interference component (which is negative for the pulse-response waveform in Figures 8-5 through 8-8). Similarly, the sampling time is $T_0 - 2T$ for the pulse $\beta(t + 2T)$ that is transmitted two time slots before $\beta(t)$, and $\beta(t)$ also contributes an intersymbol interference component at this time, as shown in Figure 8-7. So, for the pulse response of Figure 8-7, $\beta(t + 2T)$ and $\beta(t + T)$ are interfered with by a pulse that is transmitted at a later time, even though the system is causal.

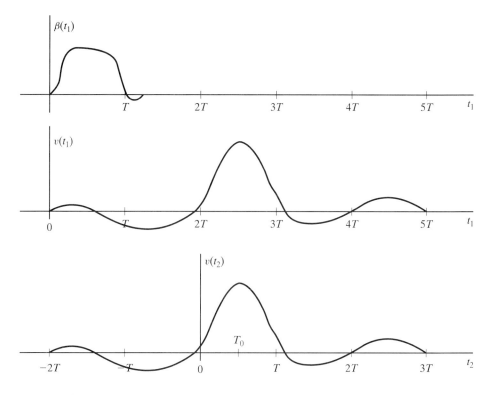

Figure 8-9: The pulse response as seen by the transmitter and receiver.

The illustration in Figure 8-9 may be helpful in understanding how pulses transmitted in the future can interfere with pulses transmitted in the past. Let t_1 represent time as measured by the transmitter's clock, and suppose that the time reference $t_1 = 0$ corresponds to the start of the transmission of the pulse $\beta(t)$. The pulse response according to the transmitter's clock is $v(t_1)$. The receiver must derive a time reference from the received signal and determine the sampling time T_0. In doing so, it establishes a different time reference than the one used by the transmitter. In fact, the time at which the pulse transmission began is of no concern to the receiver; the receiver is concerned only with the arrival time of the pulse. In practice, this arrival time is later than the time at which the pulse is transmitted because of delays introduced in the filters and propagation delays in the communication channel. Since the pulse shape is changed by the filtering operation, there is no precise definition of the arrival time. The receiver may use the peak of the received pulse to establish a time reference, or it may rely on zero crossings or other features of the pulse. All we are concerned about in this chapter is the fact that a time reference is established at the receiver, and this time reference is usually later than the time reference used by the transmitter. In Figure 8-9, time is measured by the parameter t_2 in the receiver, and the sampling time for the received pulse is $t_2 = T_0$.

In most publications, the pulse response is plotted as a function of time as measured by the receiver's clock. In Figure 8-9 this corresponds to the graph of $v(t_2)$. The desired signal component is the sample at $t_2 = T_0$, which occurs in the interval $[0, T]$ as measured by the receiver's clock. The intersymbol interference components are the samples at $t_2 = T_0 - 2T$, $t_2 = T_0 - T$, $t_2 = T_0 + T$, and $t_2 = T_0 + 2T$, which are in the two adjacent intervals on either side of the interval containing the desired signal component.

Thus, we see that even in a causal system the intersymbol interference for a given pulse can result from pulses that are transmitted after the pulse in question as well as from pulses that are transmitted before that pulse. In particular, if time is measured by the receiver's clock, the intersymbol interference caused by the 0th input pulse can occur at both positive and negative sampling times, as illustrated in Figure 8-9.

8.2   Intersymbol Interference Effects in Binary Digital Communications

In this section, we investigate the consequences of transmitting multiple pulses in the baseband communication system of Figure 8-2. The baseband signal is

$$d(t) = \sum_{n=n_1}^{n_2} b_n\, \beta(t - nT).\tag{8.6}$$

As discussed in Section 8.1, we might as well consider this to be the transmitted signal and also the input to the filter, provided that we account for the amplitude factor A and the noise variance whenever we evaluate the error probability.

The limits on the summation of (8.6) can be finite or infinite. If an infinite sequence of pulses is transmitted, $n_1 = -\infty$ and $n_2 = +\infty$. Although the basic concepts are all valid for any set of values for the data variables b_n, we restrict attention to binary antipodal communications in this section. For binary antipodal communications, there is no loss in generality if we let the data alphabet be the set $\{+1, -1\}$.

8.2.1   One-Shot Analysis

In order to illustrate the effects of intersymbol interference, it is better to employ a simpler pulse-response waveform than the one used for Figures 8-5 through 8-9. Suppose that the response to $\beta(t)$ is the waveform $v(t)$ shown in Figure 8-10. If $\beta(t)$ is the only pulse that is transmitted, then the signal component of the filter output is proportional to $v(T_0)$, the desired signal component.

The analysis of the communication system for a transmitted signal consisting of one pulse is often referred to as a *one-shot analysis*. Such an analysis does not account for the intersymbol interference that may result when the transmitted signal consists of multiple pulses; however, the one-shot analysis is useful in bounding certain probabilities and approximating others. If the transmitted signal is $d(t) = b_0\, \beta(t)$, the pulse-response waveform is $v(t)$ for the input pulse $\beta(t)$, and the sample time is T_0, then the signal component of the filter output at time T_0 is $b_0\, v(T_0)$.

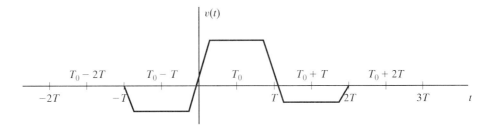

Figure 8-10: A pulse-response waveform.

The signal component of the decision statistic is $b_0 \, v(T_0)$. Recall that in order to evaluate the probability of error, we must account for the amplitude factor A in (8.4) and (8.5) and the variance σ^2 of the noise at the output of the filter. If the threshold in the threshold device is zero, the probability of error for a transmitted signal consisting of a single pulse is $Q(A \, v(T_0)/\sigma)$. This is referred to the *one-shot probability of error*, and it is denoted by $P_{e,\text{os}}$.

8.2.2   The Probability of Error for Systems with Intersymbol Interference

If the transmitted baseband signal is $b_0 \, \beta(t) + b_1 \, \beta(t - T)$, then the corresponding response consists of a linear combination of the two waveforms shown in Figure 8-11. Because the linear filter is time invariant, the response due to $\beta(t - T)$ is just a delayed version of the response due to $\beta(t)$. Hence, if the response to $\beta(t)$ is $v(t)$, then the response due to $\beta(t - T)$ is $v(t - T)$.

Linearity and time invariance of the filter guarantee the response of the filter to input $b_0 \, \beta(t) + b_1 \, \beta(t - T)$ is $b_0 \, v(t) + b_1 \, v(t - T)$. Because of the symmetry of the problem, we can assume without loss of generality that $b_0 = +1$. The issue is actually whether b_0 and b_1 have the same polarity or have different polarities. In one case, the outputs due to the two pulses add, and in the other case they subtract. If $b_1 = +1$, the filter response is the sum of the two waveforms shown in Figure 8-12(a); if $b_1 = -1$, the response is the difference of these two waveforms. The difference of the two waveforms of Figure 8-12(a) is the sum of the two waveforms shown in Figure 8-12(b).

The signal obtained by adding the two waveforms of Figure 8-12(a) is illustrated in Figure 8-13(a), and the result of the addition of the two waveforms of Figure 8-12(b) is shown in Figure 8-13(b). The most important features of the two waveforms shown in Figure 8-13 are the samples of each waveform at time T_0, because these samples represent the two possible signal components of the decision statistic that is used for the 0th pulse. If $b_1 = +1$, the signal component of the decision statistic is $v(T_0)+v(T_0-T)$; if $b_1 = -1$, the signal component of the decision statistic is $v(T_0)-v(T_0-T)$. Thus, the signal component of the decision statistic for the 0th pulse depends on b_1, the polarity of the first pulse. Such dependence on one or more other pulses is the key feature of intersymbol interference.

The intersymbol interference due to the presence of the pulse $b_1 \, \beta(t - T)$ can be either beneficial or detrimental to the signal component of the decision statistic for the

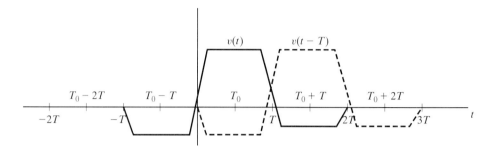

Figure 8-11: The individual pulse-response waveforms for $\beta(t)$ and $\beta(t - T)$.

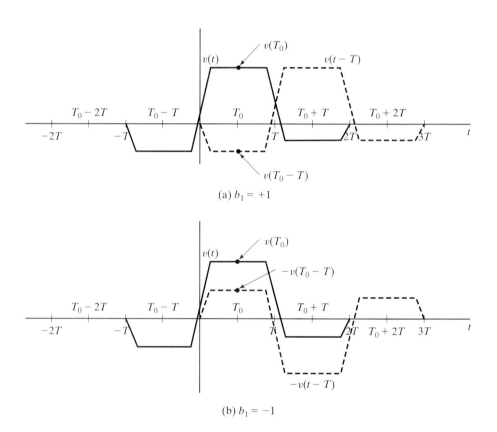

(a) $b_1 = +1$

(b) $b_1 = -1$

Figure 8-12: The individual pulse-response waveforms for $\beta(t)$ and $b_1 \beta(t - T)$.

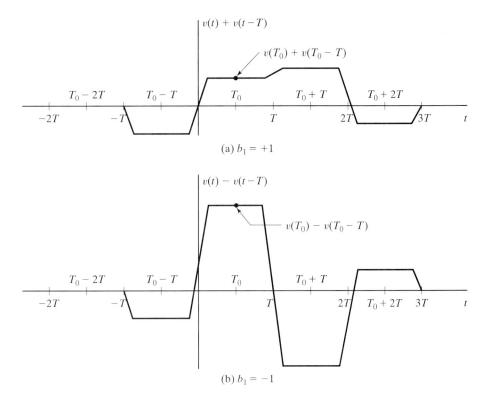

Figure 8-13: The composite response waveforms for $\beta(t) + b_1\beta(t - T)$.

0th pulse. If $b_1 = +1$, the response is as shown in Figure 8-13(a), and it is clear that the signal component at time T_0 is less than $v(T_0)$; in fact, it is $v(T_0) + v(T_0 - T)$, and $v(T_0 - T)$ is negative. The noise is unaffected by the transmission of the additional pulse, so the variance of the noise component of the decision statistic is σ^2. Hence, if $b_1 = +1$, the error probability for detecting the 0th pulse is

$$Q(A[v(T_0) + v(T_0 - T)]/\sigma),$$

which is greater than $Q(A\,v(T_0)/\sigma)$, the one-shot probability of error. On the other hand, if $b_1 = -1$, the response is as shown in Figure 8-13(b), the signal component is $v(T_0) - v(T_0 - T)$, and the error probability is

$$Q(A[v(T_0) - v(T_0 - T)]/\sigma).$$

Since $v(T_0 - T)$ is negative, this error probability is smaller than $Q(A\,v(T_0)/\sigma)$.

The restriction of the transmitted sequence to consist of only two pulses is artificial, and it is done solely to simplify the illustrations. The pulse-response waveform of Figure 8-10 spans three consecutive pulse intervals, each of duration T. In order to evaluate the effects of intersymbol interference for such a pulse-response waveform, we must consider pulse sequences of length three. In order to evaluate the effects of

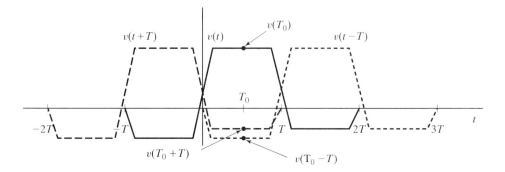

Figure 8-14: The individual pulse-response waveforms for $\beta(t+T)$, $\beta(t)$, and $\beta(t-T)$.

intersymbol interference, we can restrict attention to the transmitted baseband signal

$$d(t) = b_{-1}\,\beta(t+T) + b_0\,\beta(t) + b_1\,\beta(t-T) \qquad (8.7)$$

and examine the different samples that can be obtained at the filter output at time T_0 for different choices for the data variables b_{-1}, b_0, and b_1. If the transmitted signal is given by (8.7), the output of the filter is

$$\hat{d}(t) = b_{-1}\,v(t+T) + b_0\,v(t) + b_1\,v(t-T). \qquad (8.8)$$

The response waveforms $v(t+T)$, $v(t)$, and $v(t-T)$ are shown in Figure 8-14. The quantities $v(T_0)$, $v(T_0-T)$, and $v(T_0+T)$ combine to give the signal component of the decision statistic for the 0th pulse. The term $v(T_0)$ is the desired signal, and the terms $v(T_0-T)$ and $v(T_0+T)$ are intersymbol interference components.

It is clear from Figure 8-14 that if b_{-1}, b_0, and b_1 are each $+1$, the signal component of the decision statistic for the 0th pulse is $v(T_0) + v(T_0-T) + v(T_0+T)$, which is smaller than $v(T_0)$. The error probability for this same sequence of data variables is

$$Q(A[v(T_0) + v(T_0-T) + v(T_0+T)]/\sigma),$$

which is larger than the error probability that results if a single pulse is transmitted. On the other hand, if $b_{-1} = b_1 = -1$ and $b_0 = +1$, the signal component of the decision statistic for the 0th pulse is $v(T_0) - v(T_0-T) - v(T_0+T)$, which is larger than $v(T_0)$, and the error probability is

$$Q(A[v(T_0) - v(T_0-T) - v(T_0+T)]/\sigma),$$

which is smaller than the error probability for a single transmitted pulse.

Because the error probability depends on the sequence of pulses that is transmitted, it is often convenient to display this dependence in the notation. The sequence of pulse amplitudes, which we denote by $\mathbf{b}$, can be a finite sequence, denoted by $\mathbf{b} = (b_{n_1}, \ldots, b_{n_2})$, or an infinite sequence, denoted by $\mathbf{b} = (b_n) = \ldots, b_{-1}, b_0, b_1, b_2, \ldots$. The illustrations in Figures 8-11 through 8-13 are for the finite sequence $\mathbf{b} = (b_0, b_1)$, while the discussion concerning Figure 8-14 is for the finite sequence $\mathbf{b} = (b_{-1}, b_0, b_1)$. The sequence $\mathbf{b}$ is referred to as the *pulse pattern*.

The error probability for a pulse pattern $\mathbf{b}$ is denoted by $P_e(\mathbf{b})$. In this notation, the results for the three-pulse sequence are written as

$$P_e(+1, +1, +1) = Q(A[v(T_0) + v(T_0 - T) + v(T_0 + T)]/\sigma)$$

and

$$P_e(-1, +1, -1) = Q(A[v(T_0) - v(T_0 - T) - v(T_0 + T)]/\sigma).$$

Because of symmetry, $P_e(-1, -1, -1) = P_e(+1, +1, +1)$ and $P_e(+1, -1, +1) = P_e(-1, +1, -1)$. The remaining error probabilities are

$$P_e(+1, +1, -1) = Q(A[v(T_0) - v(T_0 - T) + v(T_0 + T)]/\sigma)$$

and

$$P_e(-1, +1, +1) = Q(A[v(T_0) + v(T_0 - T) - v(T_0 + T)]/\sigma).$$

Symmetry implies that $P_e(-1, -1, +1) = P_e(+1, +1, -1)$ and $P_e(+1, -1, -1) = P_e(-1, +1, +1)$. Because of this symmetry, there is no loss of generality in assuming that $b_0 = +1$ in our evaluation of the error probability, and we often take advantage of this fact in order to simplify the presentation.

The one-shot probability of error is less than $P_e(+1, +1, +1)$ and greater than $P_e(-1, +1, -1)$. As illustrated in this example, the intersymbol interference is sometimes beneficial in the sense that the signal component of the decision statistic is enhanced for certain pulse patterns. In practical communication systems, however, intersymbol interference effects always produce an increase in both the maximum probability of error and the average probability of error. Thus, according to either of these two performance measures, the intersymbol interference is always detrimental in communication systems of practical interest.

The *maximum probability of error* is defined as the maximum value of $P_e(\mathbf{b})$ over all sequences $\mathbf{b}$, and it is denoted by $P_{e,\max}$. If a communication system has intersymbol interference, the maximum probability of error is always greater than the one-shot probability of error. The *worst pulse pattern* is defined as the sequence $\mathbf{b}$ of pulse polarities that produces the maximum probability of error.

Consider a sequence $\mathbf{b}$ for which $b_0 = +1$, and recall that $v(T_0) > 0$. For this value of b_0, the worst pulse pattern is the set of pulse polarities b_n for $n \neq 0$ for which the signal component of the decision statistic for the 0th pulse is minimized. It is possible for this signal component to be negative, in which case the maximum probability of error is greater than $1/2$. For example, if the pulse-response waveform illustrated in Figure 8-14 satisfies

$$0 < v(T_0) < -v(T_0 - T) - v(T_0 + T), \tag{8.9}$$

then $P_e(+1, +1, +1) > 1/2$. This is because the signal component of the decision statistic for the 0th pulse is $v(T_0) + v(T_0 - T) + v(T_0 + T)$ for the pulse pattern $(+1, +1, +1)$, and (8.9) implies that

$$v(T_0) + v(T_0 - T) + v(T_0 + T) < 0.$$

Because $v(T_0) > 0$, the decision device decides that $b_0 = +1$ if the decision statistic is positive or $b_0 = -1$ if the decision statistic is negative. Thus, (8.9) implies that

$$P_e(+1, +1, +1) = Q(A[v(T_0) + v(T_0 - T) + v(T_0 + T)]/\sigma) > Q(0) = \tfrac{1}{2}.$$

Notice also that (8.9) implies that if there is no noise in the system, the receiver always makes an error when the pulse pattern is $(+1, +1, +1)$. Regardless of whether there is noise in the system, the intersymbol interference changes the polarity of the signal component of the decision statistic.

For the pulse-response waveform of Figure 8-10, $P_{e,\max}$ is the maximum of the four error probabilities, each of which corresponds to one of the four possible values for the pair (b_{-1}, b_1) with $b_0 = +1$. If there are two negative intersymbol interference components, one on either side of the desired signal (as in Figure 8-10), then

$$P_{e,\max} = Q(A[v(T_0) + v(T_0 - T) + v(T_0 + T)]/\sigma).$$

This probability is greater than $Q(A\,v(T_0)/\sigma)$, because $v(T_0 - T)$ and $v(T_0 + T)$ are negative. The worst pulse pattern for this situation is $\mathbf{b} = (+1, +1, +1)$.

Although it is much less important than the maximum probability of error, we define *the minimum probability of error* to be the minimum value of $P_e(\mathbf{b})$ over all sequences $\mathbf{b}$, and we denote it by $P_{e,\min}$. The minimum probability of error is always less than the one-shot probability of error. The best pulse pattern is the sequence $\mathbf{b}$ of pulse polarities that produces the minimum probability of error.

We are also interested in the *average probability of error*, the performance measure that is obtained by averaging $P_e(\mathbf{b})$ over the different pulse sequences. This performance measure is normally defined only for systems that have a finite number of nonzero intersymbol interference components. For the pulse-response waveform of Figure 8-10, there are two nonzero intersymbol interference components, and the average probability of error is

$$\begin{aligned}
P_{e,\text{ave}} = &\tfrac{1}{4}Q(A[v(T_0) + v(T_0 - T) + v(T_0 + T)]/\sigma) \\
&+ \tfrac{1}{4}Q(A[v(T_0) - v(T_0 - T) - v(T_0 + T)]/\sigma) \\
&+ \tfrac{1}{4}Q(A[v(T_0) - v(T_0 - T) + v(T_0 + T)]/\sigma) \\
&+ \tfrac{1}{4}Q(A[v(T_0) + v(T_0 - T) - v(T_0 + T)]/\sigma),
\end{aligned}$$

which is greater than $P_{e,\text{os}}$ for pulse responses that arise in practical communication systems. This average can be thought of as the numerical average of the four different error probabilities, or it can be viewed as the expected value of the error probability if the data variables are random, independent, and satisfy $P(b_n = +1) = P(b_n = -1) = 1/2$ for each n.

Exercise 8-1

Consider the baseband communication system of Figure 8-2 in which the communication signal is

$$d(t) = \sum_{n=-\infty}^{\infty} b_n \, \beta(t - nT).$$

Suppose the response $v(t)$ of the filter in Figure 8-2 to the input pulse $\beta(t)$ has sample values $v(T_0) = 7$, $v(T_0 - T) = 2$, $v(T_0 + T) = -1$, and $v(T_0 + nT) = 0$ for all integers n for which $|n| > 1$. It is determined that the variance of the noise at the output of the filter is $\sigma^2 = 4$. Find the maximum probability of error, the minimum probability of error, and the average probability of error if an infinite sequence of pulses is transmitted over the channel. Determine the sequence of pulses that produces the maximum probability of error and the sequence of pulses that produces the minimum probability of error.

Solution.   The input to the filter is $d(t)$, so the output is

$$\hat{d}(t) = \sum_{n=-\infty}^{\infty} b_n \, v(t - nT).$$

At sampling time T_0, the output is

$$\hat{d}(T_0) = b_{-1} \, v(T_0 + T) + b_0 \, v(T_0) + b_1 \, v(T_0 - T),$$

because $v(T_0 + nT) = 0$ for $|n| > 1$. Thus, the intersymbol interference that affects the decision on the 0th pulse is caused by two pulses, the one before the 0th pulse and the one that follows the 0th pulse. This situation is illustrated in Figure 8-14 for a specific pulse-response waveform. Because of the symmetry of the problem, we can assume, without loss of generality, that $b_0 = +1$ in our evaluation of the maximum, minimum, and average probabilities of error for the 0th pulse. If $b_0 = +1$, the signal component at sampling time T_0 is

$$b_{-1} \, v(T_0 + T) + v(T_0) + b_1 \, v(T_0 - T) = -b_{-1} + 7 + 2 \, b_1$$

for $b_{-1} = \pm 1$ and $b_1 = \pm 1$, so the possible values are $10, 8, 6,$ and 4. Since the minimum of these is 4, the maximum probability of error is $P_{e.\max} = Q(4A/\sigma) = Q(2A)$. The pulse sequence that gives rise to this minimum signal value is defined by $b_{-1} = +1$, $b_0 = +1$, and $b_1 = -1$. Because of the symmetry of the problem, the error probability that results if the sequence is $b_{-1} = -1$, $b_0 = -1$, and $b_1 = +1$ is also $Q(2A)$. The minimum probability of error is $P_{e.\min} = Q(10A/\sigma) = Q(5A)$, and this results if $b_{-1} = -1$, $b_0 = +1$, and $b_1 = +1$. The average probability of error is

$$P_{e.\text{ave}} = [Q(10A/\sigma) + Q(8A/\sigma) + Q(6A/\sigma) + Q(4A/\sigma)]/4$$
$$= [Q(5A) + Q(4A) + Q(3A) + Q(2A)]/4. \qquad \blacksquare$$

Exercise 8-2

In the communication system of Exercise 8-1, suppose that $A/\sigma = 0.25$. Compare the maximum probability of error, the minimum probability of error, the average probability of error, and the one-shot probability of error.

Solution.   From the solution to Exercise 8-1, the maximum probability of error is

$$P_{e,\max} = Q(4A/\sigma) = Q(1) \approx 0.15865,$$

the minimum probability of error is

$$P_{e,\min} = Q(10A/\sigma) = Q(2.5) \approx 0.00621,$$

and the average probability of error is

$$P_{e,\text{ave}} = [Q(10A/\sigma) + Q(8A/\sigma) + Q(6A/\sigma) + Q(4A/\sigma)]/4$$
$$= [Q(2.5) + Q(2) + Q(1.5) + Q(1)]/4 \approx 0.06361.$$

Because $v(T_0) = 7$, the one-shot probability of error is

$$P_{e,\text{os}} = Q(7A/\sigma) = Q(1.75) \approx 0.04006.$$

We conclude that $P_{e,\min} < P_{e,\text{os}} < P_{e,\text{ave}} < P_{e,\max}$.   ∎

As mentioned previously, and as illustrated by Exercise 8-2, the intersymbol interference due to one or more pulses can be either beneficial or detrimental to the decision statistic for the 0th pulse. However, when averaged over the polarities for the pulses involved, the intersymbol interference is always detrimental.

8.2.3   Bounds for Intersymbol Interference

The magnitude of the intersymbol interference that corresponds to the maximum probability of error, and therefore to the worst pulse pattern, is referred to as the *maximum intersymbol interference*. If $b_0 = +1$, it follows from $v(T_0) > 0$ that the worst pulse pattern is the one that makes each intersymbol interference contribution subtract from the desired signal. If the baseband signal to be transmitted is

$$d(t) = \sum_{n=n_1}^{n_2} b_n \, \beta(t - nT),$$

the signal component of the output of the filter is given by

$$\hat{d}(t) = \sum_{n=n_1}^{n_2} b_n \, v(t - nT).$$

This is sampled at time T_0 to detect the 0th pulse.

Assume for the moment that $b_0 = +1$ (i.e., the 0th pulse is positive). The intersymbol interference contribution of the nth pulse is $b_n \, v(T_0 - nT)$. To make this contribution negative, and therefore subtract from the desired signal, simply let the polarity of b_n be the opposite of the polarity of $v(T_0 - nT)$. Thus, the worst pulse pattern is obtained by letting

$$b_n = -\,\text{sgn}\{v(T_0 - nT)\},$$

where $\text{sgn}(x) = +1$ for $x \geq 0$ and $\text{sgn}(x) = -1$ for $x < 0$. For this choice of b_n,

$$b_n\, v(T_0 - nT) = -|v(T_0 - nT)|$$

for each $n \neq 0$.

The output of the filter at sampling time T_0 is

$$\hat{d}(T_0) = \sum_{n=n_1}^{n_2} b_n\, v(T_0 - nT), \tag{8.10}$$

which consists of the desired signal $b_0\, v(T_0) = +v(T_0)$ plus the intersymbol interference. The intersymbol interference is

$$I(\mathbf{b}, T_0) = \sum_{n=n_1}^{n_2}{}' b_n\, v(T_0 - nT), \tag{8.11}$$

where $\sum'$ denotes the sum over the indicated range ($n_1 \leq n \leq n_2$) except the $n = 0$ term is omitted. So the signal component of the decision statistic for the 0th pulse is

$$\hat{d}(T_0) = v(T_0) + I(\mathbf{b}, T_0).$$

We conclude that the worst value for intersymbol interference is the minimum value of $I(\mathbf{b}, T_0)$, minimized over all sequences $\mathbf{b}$. From (8.11) and the preceding discussion, we see that this minimum value is

$$I_{\min} = -\sum_{n=n_1}^{n_2}{}' |v(T_0 - nT)|.$$

Equivalently, the minimum value is $-I_m(T_0)$, where $I_m(T_0)$ is the maximum value of $I(\mathbf{b}, T_0)$ over all sequences $\mathbf{b}$. In other words,

$$I_m(T_0) = \sum_{n=n_1}^{n_2}{}' |v(T_0 - nT)|.$$

Thus, if $b_0 = +1$, the minimum value of $\hat{d}(T_0)$ is

$$\min\{\hat{d}(T_0) : b_0 = +1\} = v(T_0) - I_m(T_0).$$

It is possible for this minimum to be negative. If $I_m(T_0) > v(T_0)$, we say that *the maximum intersymbol interference exceeds the desired signal*.

If, instead of $b_0 = +1$, we have $b_0 = -1$, then the desired signal is $b_0 v(T_0) = -v(T_0)$, which is negative. We then invert the signs in the appropriate expressions to conclude that the maximum value of $\hat{d}(T_0)$ subject to $b_0 = -1$ is

$$\max\{\hat{d}(T_0) : b_0 = -1\} = -v(T_0) + I_m(T_0).$$

If the gain of the filter in the receiver increases, then the desired signal and each of the intersymbol interference components increase. In particular, both $v(T_0)$ and $I_m(T_0)$ increase. One disadvantage of $I_m(T_0)$ as a measure of intersymbol interference is its dependence on gains and losses in the receiver (e.g., the gain of the filter in Figure 8-1). This dependence can be eliminated by proper normalization. The *peak distortion* is defined as

$$d_0(T_0) = I_m(T_0)/v(T_0),$$

which is the maximum intersymbol interference divided by the desired signal. The peak distortion is invariant with respect to changes in gains and losses in the receiver; consequently, it is a better measure to use in the comparison of different systems.

In summary, the *maximum intersymbol interference* is

$$I_m(T_0) = \sum_{n=n_1}^{n_2}{}' |v(T_0 - nT)|. \tag{8.12}$$

The minimum value of $\hat{d}(T_0)$ subject to $b_0 = +1$ is

$$\min\{\hat{d}(T_0) : b_0 = +1\} = v(T_0) - I_m(T_0), \tag{8.13}$$

and the maximum value of $\hat{d}(T_0)$ subject to $b_0 = -1$ is

$$\max\{\hat{d}(T_0) : b_0 = -1\} = -v(T_0) + I_m(T_0). \tag{8.14}$$

The maximum probability of error for $b_0 = +1$ corresponds to the minimum value of the signal component $\hat{d}(T_0)$ subject to $b_0 = +1$, so

$$P_{e,\max} = Q(A\,[v(T_0) - I_m(T_0)]/\sigma). \tag{8.15}$$

The peak distortion is $d_0(T_0) = I_m(T_0)/v(T_0)$, so (8.15) can also be written as

$$P_{e,\max} = Q(A\,v(T_0)\,[1 - d_0(T_0)]/\sigma). \tag{8.16}$$

Exercise 8-3

If the pulse-response waveform for a communication system is $v(t)$, find the maximum probability of error. If $v(t)$ has the sample values given in Exercise 8-1, find the maximum intersymbol interference and the maximum probability of error. Assume that the value of σ is the same as in Exercise 8-1.

Solution. Assume that $b_0 = +1$. The maximum probability of error is given by (8.15). According to Exercise 8-1, $\sigma = 2$ and $v(T_0) = 7$, $v(T_0 - T) = 2$, $v(T_0 + T) = -1$, and $v(T_0 + nT) = 0$ if $|n| \geq 2$. Therefore $I_m(T_0) = 3$ and $v(T_0) - I_m(T_0) = 4$. Because $\sigma = 2$, the maximum probability of error is $P_{e,\max} = Q(2A)$, as determined in Exercise 8-1. ∎

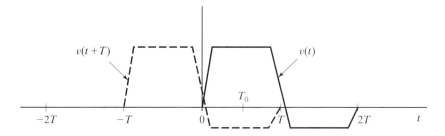

Figure 8-15: A simple pulse-response waveform.

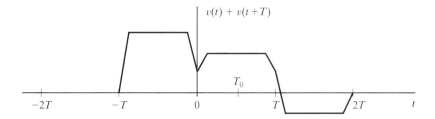

Figure 8-16: The response to two consecutive positive pulses.

8.2.4   The Eye Pattern

A very useful tool in the evaluation of communication system performance is the *eye pattern*. In this section, the eye pattern for intersymbol interference is described and illustrated. The eye pattern can be used to determine the peak distortion and other measures of performance in a system with intersymbol interference.

Consider the very simple pulse response $v(t)$ illustrated in Figure 8-15. The sampling time T_0 is used to obtain the decision statistic for the 0th pulse, and for the pulse response of Figure 8-15 the only intersymbol interference component at time T_0 is due to the preceding pulse. The pulse response for the preceding pulse is $v(t + T)$, which is also illustrated in Figure 8-15. In the interval from 0 to T, $v(t)$ represents the desired signal and $v(t + T)$ represents the intersymbol interference. If the 0th pulse and the preceding pulse are both positive, then the response waveform for the sum of these two pulses is $v(t) + v(t + T)$, which is illustrated in Figure 8-16. Consider the intersymbol interference component at time T_0 due to the preceding positive pulse. We see from Figure 8-15 that this intersymbol interference component is negative, which results in a reduction in the sample at time T_0, as illustrated in Figure 8-16. That is, $v(T_0) + v(T_0 + T)$ is less than $v(T_0)$.

Next suppose the preceding pulse is negative. The response to a negative pulse followed by a positive pulse is the difference $v(t) - v(t + T)$, which is illustrated in Figure 8-17. In this situation, the intersymbol interference results in an increase in the sample at time T_0. We conclude that if the 0th pulse is positive, the response can be either $v(t) - v(t + T)$ or $v(t) + v(t + T)$, depending on the polarity of the preceding pulse. Similarly, if the 0th pulse is negative, the response to the two consecutive pulses is $-v(t) - v(t + T)$ if the preceding pulse is negative and $-v(t) + v(t + T)$ if the

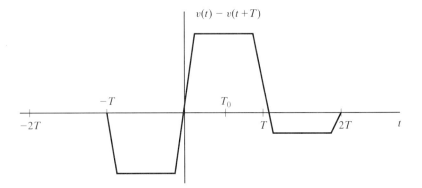

Figure 8-17: The response to a negative pulse followed by a positive pulse.

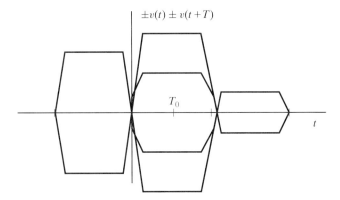

Figure 8-18: The response waveforms $\pm v(t) \pm v(t + T)$.

preceding pulse is positive. There are four possible combinations of polarities for the two consecutive pulses, and the four possible response functions are $\pm v(t) \pm v(t + T)$, as shown in Figure 8-18. In order to focus on the effects of intersymbol interference on the decision statistic for the 0th pulse, we restrict attention to the time interval around T_0, the sampling time for the 0th pulse. This leads us to the consideration of the eye pattern, which is illustrated in Figure 8-19 for the pulse-response waveform $v(t)$ of Figure 8-15.

The eye pattern is a set of waveforms on an interval that approximates $[0, T]$. Each of the waveforms corresponds to a different pulse pattern. If the pulse-response function $v(t)$ spans k pulse intervals, then the eye pattern has 2^k different waveforms. The response function shown in Figure 8-15 spans two pulse intervals, so there are four waveforms in its eye diagram, as shown in Figure 8-19. In this eye pattern, the *upper* waveforms are given by $v(t) \pm v(t + T)$ and the *lower* waveforms are given by $-v(t) \pm v(t + T)$. In other words, the upper waveforms in the eye pattern correspond to the 0th pulse being positive, and the lower waveforms correspond to the 0th pulse being negative.

$$\pm v(t) \pm v(t + T)$$

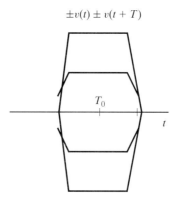

Figure 8-19: The eye pattern for the pulse response of Figure 8-15.

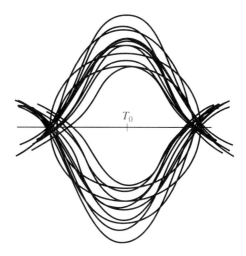

Figure 8-20: A typical eye pattern.

The sample of a given waveform at time T_0 is $\hat{d}(T_0)$ for the pulse pattern that generates that waveform, and the collection of the samples of each of the waveforms at time T_0 is the set of all possible values of $\hat{d}(T_0)$. In other words, the collection of samples obtained from the eye pattern is the set

$$\{b_{-1}\, v(T_0 + T) + b_0\, v(T_0) : b_{-1} = \pm 1, b_0 = \pm 1\}.$$

Notice that this is consistent with (8.10) if we let $n_1 = -1$ and $n_2 = 0$. If the pulse response $v(t)$ spans a larger number of pulse intervals, there are more waveforms in the eye pattern, as illustrated in Figure 8-20.

The eye pattern can be generated experimentally. Suppose the pulse-response waveform $v(t)$ spans k pulse intervals. We wish to apply to the filter input a binary data stream that has the property that each possible sequence of k consecutive binary symbols

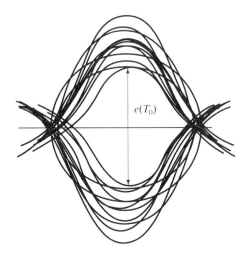

Figure 8-21: Illustration of the eye opening.

appears once and only once. As illustrated in Section 6.6.1 for $k = 3$, an m-sequence is a sequence of zeros and ones that has the desired property except for the lack of a sequence of k zeros. That is, each of the $2^k - 1$ nonzero binary words of length k appears exactly once in each period of an m-sequence of period $2^k - 1$. If a single zero is added in the proper place within each period of an m-sequence, an *extended m-sequence* of period 2^k is obtained. The extended m-sequence is derived from the original m-sequence by adding a zero after each string of $k - 1$ consecutive zeros in m-sequence. Note that a string of $k - 1$ consecutive zeros occurs once in each period of the original m-sequence, and the addition of a zero at the end of this string converts it to a string of k consecutive zeros. Thus, the extended m-sequence has the property that each of the 2^k binary words of length k appears exactly once in each period of the sequence. One of the consequences of this fact is that an extended m-sequence cannot be an m-sequence. The extended m-sequence is converted to a pulse pattern (b_n) by the usual mapping from $\{0, 1\}$ to $\{+1, -1\}$, as described in Section 6.6.1.

When converted to a pulse pattern and applied to the input of the filter, each of the k-bit binary words corresponds to a unique waveform at the output of the filter, and each such waveform is one of the 2^k waveforms that make up the eye pattern. The eye pattern can be displayed on an oscilloscope by using the filter output as the vertical input to the oscilloscope and letting the clock for the filter input waveform control the oscilloscope's horizontal sweep.

The eye pattern has all of the information about the filter output that is needed to evaluate the performance measures we have considered for intersymbol interference (e.g., the maximum intersymbol interference and the average probability of error). The use of the eye pattern to evaluate the maximum intersymbol interference (and hence the maximum probability of error) is particularly simple. Let $e(T_0)$ denote the *eye opening* at time T_0, as illustrated in Figure 8-21. Notice that $e(T_0)/2$ is the minimum of $\hat{d}(T_0)$ over all pulse patterns; that is, one-half of the eye opening at time T_0 is the minimum value of the signal, minimized over all pulse patterns. It follows that the eye opening is related to the maximum intersymbol interference by

$$e(T_0) = 2[v(T_0) - I_m(T_0)].$$

The eye opening can also be related to the peak distortion $d_0(T_0)$. If we normalize the eye opening by dividing by the desired signal, we obtain

$$e(T_0)/v(T_0) = 2[1 - d_0(T_0)].$$

For a fixed value of $v(T_0)$, an increase in the eye opening corresponds to a decrease in peak distortion. One important application of the eye pattern is in the selection of the sampling time that minimizes the maximum probability of error. The maximum probability of error is decreased by decreasing $d_0(T_0)$, and decreasing $d_0(T_0)$ corresponds to increasing $e(T_0)$. Therefore, if the goal is to minimize the maximum probability of error, the sampling time should be selected to give the largest possible eye opening.

Consider a system in which the impulse response is varied in a way that increases the maximum intersymbol interference. In particular, suppose that over a period of time we make a sequence of modifications to the impulse response, and each of these modifications decreases each of the upper waveforms in the eye pattern and increases each of the lower waveforms. Throughout this sequence of modifications the eye opening is becoming smaller. The modifications can be designed in such a way that eventually the minimum of the upper waveforms at time T_0 becomes zero. Because of symmetry, the maximum of the lower waveforms at time T_0 is also zero, so $e(T_0) = 0$. At this point, we say that the *eye pattern is closed at T_0*. If the intersymbol interference increases beyond this point, the eye pattern remains closed at T_0; in fact, some of the "upper" waveforms drop below the axis at T_0, and an equal number of the "lower" waveforms rise above the axis at T_0. In other words, there is a crossover between the upper waveforms and the lower waveforms in the eye pattern. When this occurs, the maximum intersymbol interference exceeds the desired signal; that is, $I_m(T_0) > v(T_0)$.

If $I_m(T_0) \geq v(T_0)$, the eye pattern is closed at time T_0. If the eye pattern is closed at T_0 and the sampling time in the receiver is T_0, the maximum probability of error is $1/2$ or larger. If $I_m(t) \geq v(t)$ for *all* values of t in the range $0 \leq t \leq T$, we say *the eye pattern is closed*. If the eye pattern is closed, the maximum intersymbol interference is at least as large as the desired signal regardless of the sampling time.

8.2.5   Output Signal-to-Noise Ratio

In order to evaluate the probability of error, it is often convenient to begin by determining the signal-to-noise ratio at the output of the filter (e.g., the filter in Figure 8-2). It is straightforward to evaluate the error probability from this signal-to-noise ratio. Recall that the output of the filter is a function of the pulse pattern $\mathbf{b}$, and the pulse pattern may be a finite sequence $\mathbf{b} = (b_{n_1}, \ldots, b_{n_2})$ or an infinite sequence $\mathbf{b} = (b_n) = \ldots, b_{-1}, b_0, b_1, b_2, \ldots$. In the absence of noise, the output of the filter at sampling time T_0 is $\hat{d}(T_0)$, which is given by (8.10). The lower limit in (8.10) can be finite or it can be $-\infty$, and the upper limit can be finite or $+\infty$. It is clear from (8.10) that $\hat{d}(T_0)$ depends on the pulse pattern $\mathbf{b}$. As a result, the signal-to-noise ratio also depends on $\mathbf{b}$.

For a given pulse pattern **b**, we define the output signal-to-noise ratio by

$$S_{\mathbf{b}}(T_0) = A\, b_0\, \hat{d}(T_0)/\sigma, \tag{8.17}$$

where σ^2 is the variance of the noise at the output of the filter. The probability of error for pulse pattern **b** is then given by

$$P_e(\mathbf{b}) = Q(S_{\mathbf{b}}(T_0)). \tag{8.18}$$

The purpose of the term b_0 in the expression for $S_{\mathbf{b}}(T_0)$ is to ensure that the signal-to-noise ratio has the correct sign.

For example, if $b_0 = -1$ and $\hat{d}(T_0) < 0$, we want $S_{\mathbf{b}}(T_0) > 0$. Recall that $v(T_0) > 0$, so $b_0 = -1$ implies that $b_0\, v(T_0) < 0$, and thus the desired signal is negative. Because $\hat{d}(T_0) < 0$ the output signal at the sampling instant is also negative. This implies that $P_e(\mathbf{b})$ should be less than $1/2$, as indeed it is, because $S_{\mathbf{b}}(T_0) > 0$ ensures $P_e(\mathbf{b}) < 1/2$.

On the other hand, if $b_0 = +1$ and $\hat{d}(T_0) < 0$, it must be that $S_{\mathbf{b}}(T_0) < 0$ and $P_e(\mathbf{b}) > 1/2$. In this situation, the signal sample is negative even though the transmitted pulse is positive. One consequence is that, if there is no noise in the system, the transmission of the pulse pattern **b** results in the receiver deciding that $b_0 = -1$ when in fact $b_0 = +1$ was transmitted. This occurs because $b_0\, v(T_0) > 0$, but $\hat{d}(T_0) < 0$. In other words, for this pulse pattern the sum of the desired signal and the intersymbol interference is negative even though the desired signal is positive. In order to have $v(T_0) + I(\mathbf{b}, T_0) < 0$, it must be that $I(\mathbf{b}, T_0) < -v(T_0)$. It follows that $|I(\mathbf{b}, T_0)| > v(T_0)$, and this implies that $I_m(T_0) > v(T_0)$; that is, the maximum intersymbol interference exceeds the desired signal. Notice that $I_m(T_0) > v(T_0)$ is equivalent to $d_0(T_0) > 1$, so the maximum intersymbol interference exceeds the desired signal if and only if the peak distortion exceeds one. Notice also that (8.15) and (8.16) show that each of the equivalent conditions $I_m(T_0) > v(T_0)$ and $d_0(T_0) > 1$ is a necessary and sufficient condition for $P_{e,\max} > 1/2$.

Exercise 8-4

Consider a system for which the pulse response waveform $v(t)$ has sample values $v(T_0) = 5$, $v(T_0 - T) = 4$, $v(T_0 + T) = -2$, and $v(T_0 + nT) = 0$ for all integers n for which $|n| > 1$. Find the probability of error for a data sequence **b** for which $b_{-1} = +1$, $b_0 = +1$, and $b_1 = -1$. Also, find the probability of error for $b_{-1} = -1$, $b_0 = -1$, and $b_1 = +1$

Solution.   First, we use the method that was employed in the solution to Exercise 8-1. The signal component at sampling time T_0 is

$$\hat{d}(T_0) = b_{-1}\, v(T_0 + T) + b_0\, v(T_0) + b_1\, v(T_0 - T) = -2\, b_{-1} + 5\, b_0 + 4\, b_1 = -1.$$

Thus, the probability of error is $Q(-A/\sigma)$, which is greater than $1/2$. Because of the symmetry of the problem, $Q(-A/\sigma)$ is also the error probability that results if $b_{-1} = -1$, $b_0 = -1$, and $b_1 = +1$. Now let us use the method in which we first calculate the signal-to-noise ratio. If the sequence $\mathbf{b}$ has $b_{-1} = +1$, $b_0 = +1$, and $b_1 = -1$, then $\hat{d}(T_0) = -1$ and (8.17) implies that

$$S_{\mathbf{b}}(T_0) = A\, b_0\, \hat{d}(T_0)/\sigma = -A/\sigma.$$

Thus, the signal-to-noise ratio is negative, as it should be. The resulting probability of error, obtained from (8.18), is $P_e(b) = Q(-A/\sigma)$. Notice that if $b_{-1} = -1$, $b_0 = -1$, and $b_1 = +1$, then $\hat{d}(T_0) = +1$, so $b_0\,\hat{d}(T_0) = -\hat{d}(T_0)$ and $S_{\mathbf{b}}(T_0) = -A/\sigma$, the same as for $b_{-1} = +1$, $b_0 = +1$, and $b_1 = -1$.   ∎

As illustrated in Exercise 8-4, the sum of the desired signal and the intersymbol interference may have a different sign than the desired signal. This happens for each pulse pattern $\mathbf{b}$ for which $v(T_0) < |I(\mathbf{b}, T_0)|$ and the sign of $b_0\, v(T_0)$ differs from the sign of $I(\mathbf{b}, T_0)$. In this situation, the receiver makes an error due to intersymbol interference alone; that is, if there is no noise in the system, the decision that is made in the threshold device is wrong each time the pulse pattern $\mathbf{b}$ is transmitted. For a system with zero-mean Gaussian noise, the error probability exceeds $1/2$ each time this pulse pattern is transmitted. This is demonstrated in Exercise 8-4 for any pulse pattern $\mathbf{b}$ for which $b_{-1} = +1$, $b_0 = +1$, and $b_1 = -1$. Also illustrated in Exercise 8-4 is the fact that $P_e(\mathbf{b}) = P_e(-\mathbf{b})$; that is, the pulse pattern

$$\mathbf{b} = (b_n) = \ldots, b_{-1}, b_0, b_1, b_2, \ldots$$

gives the same probability of error as the pulse pattern

$$-\mathbf{b} = (-b_n) = \ldots, -b_{-1}, -b_0, -b_1, -b_2, \ldots \ .$$

It is desirable to have performance measures that can be used to compare different systems or to evaluate alternative filters for a given system. The first thought might be to determine the probability of error for each pulse pattern, but this is not feasible in general. If there are an infinite number of nonzero intersymbol interference components (i.e., if $v(T_0 - nT) \neq 0$ for infinitely many values of n), then an infinite number of different pulse patterns must be considered. This means that we must evaluate an infinite number of probabilities in order to determine the probability of error for each of the pulse patterns.

Even if the number of nonzero intersymbol interference components is finite, the number of different pulse patterns may be so large that it is impractical to use the set of all error probabilities as a performance measure. Suppose, for example, that $v(T_0 - nT) \neq 0$ for $-3 \leq n \leq +3$, and $v(T_0 - nT) = 0$ for all other values of n. Then there are six nonzero intersymbol interference components, and we must consider $2^6 = 64$ different pulse patterns and evaluate the error probability for each. In most applications, it is impractical to compare different communication systems on the basis of 64 error probabilities for each system. For some pulse patterns, the error probabilities may be larger in one system than in the other, but for other pulse patterns it may be the other way around, which makes it difficult to determine which system is preferred. Two performance measures that we have already considered overcome this difficulty. They are the maximum probability of error and the average probability of error.

The maximum probability of error can be determined from (8.15). Alternatively, we can first evaluate the minimum signal-to-noise ratio, and from this we can determine the maximum probability of error. If $S_{\min}$ is the minimum of $S_{\mathbf{b}}(T_0)$ over all pulse patterns $\mathbf{b}$, then $P_{e,\max} = Q(S_{\min})$. If there are a finite number of nonzero intersymbol interference components, only a finite number of pulse patterns need be considered, and the signal-to-noise ratio can be computed for each pattern. It is then a simple matter to determine the smallest of these signal-to-noise ratios. If there are an infinite number of nonzero intersymbol interference components, we may be able to use an exact analytical method (e.g., see Problems 8.1 and 8.2), or we might have to employ an approximation or bound. A typical approximation is to ignore intersymbol interference components that are smaller than some given level, thereby reducing the number of nonzero components to a finite number.

If there are only a finite number of intersymbol interference components, we can assign a probability to each of the finite number of pulse patterns that must be considered. In this situation, it is possible to evaluate the average probability of error. Suppose that the only nonzero values of $v(T_0 - nT)$ are for $n_1 \leq n \leq n_2$, so $\mathbf{b}$ is the finite sequence $(b_{n_1}, \ldots, b_{n_2})$, and the output of the filter at the sampling time is given by

$$\hat{d}(T_0) = \sum_{n=n_1}^{n_2} b_n \, v(T_0 - nT).$$

Recall that $S_{\mathbf{b}}(T_0) = A \, b_0 \, \hat{d}(T_0)/\sigma$ and $P_e(\mathbf{b}) = Q(S_{\mathbf{b}}(T_0))$. If $p(\mathbf{b})$ denotes the probability that the pulse pattern is $\mathbf{b}$, the average probability of error is given by

$$\overline{P}_{e,\text{ave}} = \sum_{\mathbf{b}} P_e(\mathbf{b}) \, p(\mathbf{b}), \tag{8.19}$$

where the sum is over all binary sequences of the form $(b_{n_1}, \ldots, b_{n_2})$. Since $\mathbf{b}$ has $n_2 - n_1 + 1$ components, there are $2^{n_2 - n_1 + 1}$ terms in the sum of (8.19). In practice, there may be additional nonzero intersymbol interference components, but in order to approximate the average probability of error we may choose n_1 and n_2 in such a way that $v(T_0 - nT)$ is negligible if n is less than n_1 or greater than n_2.

8.3   Digital Communication Systems with No Intersymbol Interference

Suppose that the pulse-response waveform for a communication system is $v(t)$ and the sampling times for the data pulses are of the form $T_0 + nT$ for all integers n. It is shown in the previous section that the intersymbol interference components for this system are the samples $v(T_0 + nT)$ for all $n \neq 0$. Suppose we insist on having no intersymbol interference; that is, suppose we require that $v(T_0 + nT) = 0$ for all $n \neq 0$. Is it possible for a system to satisfy this requirement? If so, what are the necessary and sufficient conditions for the system transfer function (or impulse response) to guarantee there is no intersymbol interference? Such questions are answered in this section.

For our discussion of systems with no intersymbol interference, the measure of bandwidth we use for a baseband filter or a baseband signal is its cutoff bandwidth. The cutoff bandwidth W for a bandpass filter is defined in Section 4.6.1, and the definition for a baseband filter is analogous: The *cutoff bandwidth for a baseband filter* with transfer function $H(f)$ is the maximum frequency W for which $H(W) \neq 0$. Because we are considering only real impulse responses, $H(-f) = H^*(f)$ for all f. So it is also true that $-W$ is the minimum frequency for which the transfer function is nonzero. Similarly, if $V(f)$ is the Fourier transform of the baseband signal $v(t)$, the *cutoff bandwidth for the baseband signal* is the maximum frequency W for which $V(W) \neq 0$.

As described in Section 8.1, the transmitted baseband signal is $s(t) = A x(t)$, where $x(t)$ represents a sequence of pulses whose amplitudes are given by the sequence

$$\mathbf{x} = (x_n) = \ldots, x_{-1}, x_0, x_1, x_2, \ldots \ .$$

We employ different notation in this section for the baseband signal and the sequence of pulse amplitudes: $d(t)$ is replaced by $x(t)$ and b_n is replaced by x_n. This is to highlight the fact that in this section the amplitudes x_n are allowed to be any real numbers, whereas b_n is either -1 or $+1$ in Sections 8.1 and 8.2.

The baseband pulse waveform of primary interest is the received waveform that corresponds to the transmission of a single pulse of unit amplitude. As in the previous sections, we denote this received waveform by $v(t)$. The shape of the transmitted pulse is unimportant for the present discussion, because the intersymbol interference in the receiver depends only on the waveform $v(t)$. With no loss in generality (since we can always shift the time origin), we can let the sampling time for $v(t)$ be $T_0 = 0$. This sampling time may correspond to the maximum value of $v(t)$, or it may be selected according to some other criterion.

Suppose a sequence of pulses is transmitted at the rate of $R = 1/T$ symbols per second. In the absence of noise, the received signal is $A \hat{x}(t)$, where $\hat{x}(t)$ is given by

$$\hat{x}(t) = \sum_{n=-\infty}^{\infty} x_n \, v(t - nT). \tag{8.20}$$

In an actual communication system, the signal is received in the presence of noise, of course, but the noise can be ignored for the present discussion. Consequently, the amplitude factor A can also be ignored for most of the issues covered in this section.

The 0th pulse of (8.20) is $x_0 \, v(t)$. In order to determine the amplitude of the 0th pulse, the receiver samples $\hat{x}(t)$ at time 0. This gives

$$\hat{x}(0) = \sum_{n=-\infty}^{\infty} x_n \, v(-nT) = \sum_{n=-\infty}^{\infty} x_{-n} \, v(nT).$$

If there is no intersymbol interference, the sampled value of the received signal is $x_0 \, v(0)$. The remaining components of $\hat{x}(0)$ are intersymbol interference components. Thus, there is no intersymbol interference if and only if $v(nT) = 0$ for each $n \neq 0$.

Let us examine the requirement that $v(nT) = 0$ for each $n \neq 0$, the necessary and sufficient condition for intersymbol-interference-free reception. A convenient way to express this condition is

$$v(0)\,\delta(t) = \sum_{n=-\infty}^{\infty} v(t)\,\delta(t - nT). \tag{8.21}$$

To see that (8.21) holds if and only if there is no intersymbol interference, first notice that

$$v(t)\,\delta(t - nT) = v(nT)\,\delta(t - nT)$$

for each value of t and each integer n. If each side of this equation is summed over all n, we have

$$\sum_{n=-\infty}^{\infty} v(t)\,\delta(t - nT) = \sum_{n=-\infty}^{\infty} v(nT)\,\delta(t - nT). \tag{8.22}$$

Because of (8.22), we see that (8.21) is equivalent to

$$v(0)\,\delta(t) = \sum_{n=-\infty}^{\infty} v(nT)\,\delta(t - nT). \tag{8.23}$$

Clearly, the two sides of (8.23) are equal if and only if $v(nT) = 0$ for each nonzero integer n, but this is just the condition that there is no intersymbol interference. Therefore, (8.21) is a necessary and sufficient condition for there to be no intersymbol interference.
    Next, we write (8.21) as

$$v(0)\,\delta(t) = v(t)\sum_{n=-\infty}^{\infty} \delta(t - nT). \tag{8.24}$$

Equation (8.24) is equivalent to

$$v(0)\,\delta(t) = v(t)\,d_I(t), \tag{8.25}$$

where $d_I(t)$ is the *Dirac comb*, which is defined by

$$d_I(t) = \sum_{n=-\infty}^{\infty} \delta(t - nT). \tag{8.26}$$

The Dirac comb is just a sequence of equally spaced, unit-amplitude, Dirac delta functions. Because of its role in the analysis of sampling, it is often referred to as the *ideal sampling function*. As shown in most textbooks on linear systems, the Fourier transform of the Dirac comb is given by

$$D_I(f) = T^{-1}\sum_{n=-\infty}^{\infty} \delta(f - nT^{-1}). \tag{8.27}$$

Take the Fourier transform of both sides of (8.25), and use the fact that multiplication in the time domain corresponds to convolution in the frequency domain. This shows that the condition for no intersymbol interference is equivalent to

$$v(0) = (V * D_I)(f), \quad -\infty < f < \infty. \tag{8.28}$$

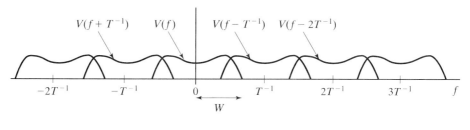

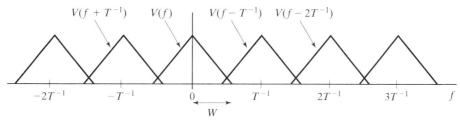

Figure 8-22: Two examples that do not satisfy Nyquist's criterion.

Next use the fact that the convolution of V with the delta function $\delta(f - nT^{-1})$ is $V(f - nT^{-1})$, so that (8.27) and (8.28) imply that

$$v(0) = T^{-1} \sum_{n=-\infty}^{\infty} V(f - nT^{-1}), \quad -\infty < f < \infty, \tag{8.29}$$

is a necessary and sufficient condition for the pulse response waveform $v(t)$ to have no intersymbol interference. It is helpful to define the frequency function $V_s(f)$ by

$$V_s(f) = \sum_{n=-\infty}^{\infty} V(f - nT^{-1}), \quad -\infty < f < \infty. \tag{8.30a}$$

The important property obtained from (8.29) is the necessary and sufficient condition for reception without intersymbol interference; namely, $v(t)$ must satisfy

$$V_s(f) = c, \quad -\infty < f < \infty, \tag{8.30b}$$

for some constant c. Although the value of the constant is not important, (8.29) and (8.30) imply that it must be $c = v(0)T$. The key feature of $v(t)$ that guarantees intersymbol-interference-free reception is that the sum on the right-hand side of (8.29) does not fluctuate over the range $-\infty < f < \infty$, which we refer to as *Nyquist's criterion*. (The criterion is actually *Nyquist's first criterion*, but since Nyquist's other criteria are not considered in this book, the more concise designation is preferred.) Two examples that do not satisfy Nyquist's criterion are illustrated in Figure 8-22. See also Figure 8-23 for an example of a graph of the function $V_s(f)$.

For maximum spectral efficiency, we wish to transmit as much information as possible using the smallest possible bandwidth. If we insist that there be no intersymbol interference, the goal is to make the bandwidth of $v(t)$ as small as possible subject to

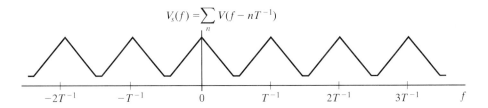

$$V_s(f) = \sum_n V(f - nT^{-1})$$

Figure 8-23: The function $V_s(f)$ for the second example of Figure 8-22.

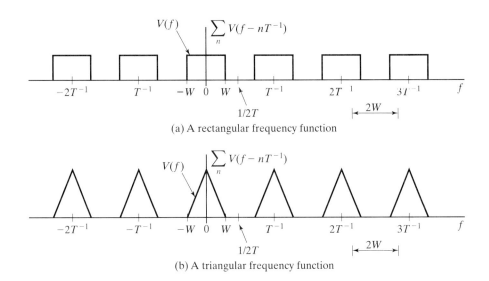

(a) A rectangular frequency function

(b) A triangular frequency function

Figure 8-24: Two examples that satisfy $W < 1/2T$.

the constraint that Nyquist's criterion must be satisfied. Suppose the function V has cutoff bandwidth W, and suppose we want $W \le 1/2T$. Because the transmission rate is $R = 1/T$, the condition $W \le 1/2T$ is equivalent to $W \le R/2$.

Notice that each of the frequency functions illustrated in Figure 8-22 has a cutoff bandwidth W that does not satisfy $W \le 1/2T$. Functions that satisfy $W \le 1/2T$ (and in fact satisfy $W < 1/2T$) are illustrated in Figure 8-24 for rectangular and triangular frequency functions. In the investigation of Nyquist's criterion, there is a fundamental difference between rectangular frequency functions and any other frequency functions. A rectangular frequency function with cutoff bandwidth W satisfies $V(f) = c$ for $-W \le f \le W$, whereas $V(f)$ is not constant in the range $-W \le f \le W$ for any other frequency function that has cutoff bandwidth W.

In order to satisfy Nyquist's criterion with a rectangular frequency function, it is necessary and sufficient that there be no gaps between adjacent rectangles. This is accomplished by letting $W = 1/2T$, as illustrated in Figure 8-25(a). For frequency functions of other shapes, however, it is not possible to satisfy Nyquist's criterion for $W = 1/2T$. If the criterion can be satisfied at all for frequency functions that are

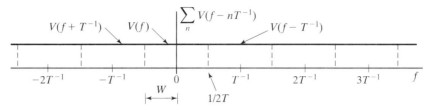

(a) A rectangular frequency function with $W = 1/2T$

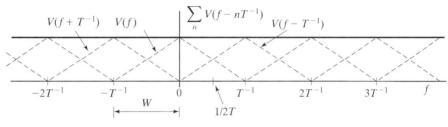

(b) A triangular frequency function with $W = 1/T$

Figure 8-25: Two examples that satisfy Nyquist's criterion.

not rectangular, a larger bandwidth than $1/2T$ is required. Thus, the spectral efficiency is necessarily worse if $V(f)$ is not rectangular. For example, as illustrated in
Figure 8-25(b), in order to satisfy Nyquist's criterion using a triangular frequency function, the bandwidth must be $1/T$, which is twice the bandwidth required by a rectangular
frequency function. Notice from Figure 8-25(b) that the sums of the overlapping spectra
are all the same, so that

$$\sum_{n=-\infty}^{\infty} V(f - nT^{-1}) = c, \quad -\infty < f < \infty,$$

as required. For arbitrary shapes, this is not necessarily true, and there may be no choice
of signal bandwidth for which Nyquist's criterion is satisfied.

We next examine the implications of the preceding findings on the minimum bandwidth requirement for the communication channel. In order for $V(f)$ to have bandwidth
W, the communication channel must have bandwidth at least as large as W. The channel bandwidth can be larger, but not smaller, than the bandwidth of the output signal.
Thus, the minimum channel bandwidth for intersymbol-interference-free reception is
$W = 1/2T$, and it can be this small if and only if the channel produces an output
waveform whose Fourier transform is a rectangle on the interval $-W \le f \le W$.

In conclusion, we see that it is not possible to satisfy Nyquist's criterion if the bandwidth is less than $1/2T$. Thus, the minimum bandwidth for intersymbol-interference-
free communication is $W = 1/2T$, which is referred to as the *Nyquist bandwidth*. The
only frequency function that satisfies Nyquist's criterion for $W = 1/2T$ is the rectangu-
lar frequency function: $V(f) = c$ for $|f| \le 1/2T$ for some constant c and $V(f) = 0$
for $|f| > 1/2T$. A frequency function with any other shape requires greater bandwidth
in order to avoid intersymbol interference, and frequency functions of some shapes have

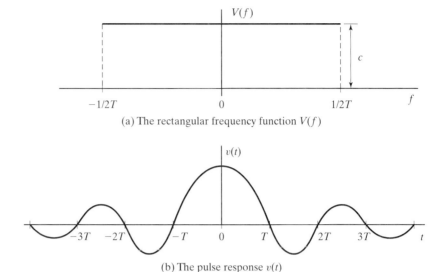

(a) The rectangular frequency function $V(f)$

(b) The pulse response $v(t)$

Figure 8-26: The rectangular frequency function and the corresponding pulse response (the *ideal* pulse-response waveform).

intersymbol interference no matter what bandwidth is used. The rectangular frequency function that satisfies Nyquist's criterion is illustrated in Figure 8-26 along with its inverse Fourier transform, the pulse response waveform $v(t) = c\,T^{-1} \mathrm{sinc}(t/T)$, where $\mathrm{sinc}(x) = \sin(\pi x)/(\pi x)$. A channel that produces this pulse-response waveform is referred to as a *Nyquist channel*.

Our conclusions can also be described in terms of the transmission rate $R = 1/T$. The smallest bandwidth required to transmit at rate R without intersymbol interference is $W = R/2$. Equivalently, if the bandwidth is W, the maximum rate at which data can be transmitted without intersymbol interference is $R = 2W$. This maximum rate is called the *Nyquist rate*.

If the bandwidth is measured in Hz, and the transmission rate is in symbols per second, then transmission at the Nyquist rate corresponds to $2W$ symbols per second being sent in a bandwidth of W Hz. The spectral efficiency of a communication system can be defined as R/W, the transmission rate per unit bandwidth. If the output of the communication system is characterized by the rectangular frequency function illustrated in Figure 8-25(a), then $R = 2W$, and the spectral efficiency of the communication system is two symbols per second per hertz of bandwidth. This efficiency is abbreviated as 2 symbols/s/Hz, and it represents the best possible efficiency that can be achieved without intersymbol interference. By comparison, a communication system for which the output pulses have the triangular frequency function illustrated in Figure 8-25(b) has a spectral efficiency of 1 symbol/s/Hz. The spectral efficiency is often expressed in terms of bits rather than symbols. If M-ary symbols are transmitted at the Nyquist rate, then the spectral efficiency is $2 \log_2(M)$ bits per second per hertz, which is abbreviated as $2 \log_2(M)$ b/s/Hz. Thus, binary transmission at the Nyquist rate has a spectral efficiency of 2 b/s/Hz.

For the rectangular frequency function shown in Figure 8-26(a), recall that the pulse-response function is

$$v(t) = c\,T^{-1}\,\text{sinc}(t/T). \tag{8.31}$$

It is easy to see from Figure 8-26(b) that the optimum sampling time for this pulse-response function is $T_0 = 0$. The necessary and sufficient condition for intersymbol-interference-free reception is $v(T_0 - nT) = 0$ for each nonzero integer n, and for $T_0 = 0$ this is equivalent to $v(-nT) = 0$ for each $n \neq 0$.

The waveform of Figure 8-26(b) clearly satisfies the requirement $v(-nT) = 0$ for each $n \neq 0$, since it has a zero-crossing at each nonzero multiple of T. We refer to this pulse-response waveform as the *ideal* pulse-response waveform. If the pulse response for a communication system is ideal, and if the sampling times for the system are multiples of T, there is no intersymbol interference in the system. Mathematically, this is just the observation that if n is any nonzero integer and $t = -nT$, then

$$v(t) = v(-nT) = c\,T^{-1}\,\text{sinc}(-n) = c\,T^{-1}\,\sin(n\pi)/(n\pi) = 0.$$

We conclude that if the sampling is perfect (i.e., if the sampling times are precisely at integer multiples of T), then the ideal response waveform of Figure 8-26 provides intersymbol-interference-free reception.

Next we turn our attention to the situation in which the pulse response is ideal, but the sampling is not perfect. Suppose the intent is to sample at time T_0 for the 0th pulse, but the actual sampling time is $T_0 + \delta$ for δ in the range $0 < |\delta| < T$. If the pulse-response waveform is $v(t)$, the intersymbol interference components in a system with sampling error δ are $v(T_0 + \delta - nT)$ for $n \neq 0$. If $T_0 = 0$, the sampling time for the 0th pulse is δ, and the intersymbol interference components are $v(\delta - nT)$ for $n \neq 0$. It is clear from Figure 8-26(b) that $0 < |\delta| < T$ implies that $v(\delta - nT) \neq 0$ for each n, so imperfect sampling produces intersymbol interference in a system with the ideal pulse-response waveform. In fact, it is shown in the exercise that follows that if an infinite sequence of pulses is transmitted, then the maximum intersymbol interference is infinite for the ideal pulse response. Any nonzero sampling error, no matter how small, gives an infinite value for the maximum intersymbol interference. This is one of the reasons why the pulse response of Figure 8-26 is not a good candidate for a practical communication system.

Exercise 8-5

Suppose that an infinite sequence of pulses is transmitted in a system that has the ideal pulse-response waveform. If the sampling time for the 0th pulse is δ, and δ is in the range $0 < \delta < T$, show that the maximum intersymbol interference is infinite.

Solution.   If the sampling time is $T_0 + \delta$, the expression for the maximum intersymbol interference becomes

$$I_{\max} = \sideset{}{'}\sum_{n=n_1}^{n_2} |v(T_0 + \delta - nT)|.$$

Since an infinite sequence of pulses is transmitted, $n_1 = -\infty$ and $n_2 = +\infty$. Also, $T_0 = 0$ for the ideal pulse-response waveform, so

$$I_{\max} = \sideset{}{'}\sum_{n=-\infty}^{\infty} |v(\delta - nT)| \geq \sum_{n=1}^{\infty} |v(\delta - nT)|. \tag{8.32}$$

The ideal pulse response is $v(t) = c\, T^{-1} \operatorname{sinc}(t/T)$ for $-\infty < t < +\infty$, where c is an arbitrary positive constant. The intersymbol interference components are

$$v(\delta - nT) = c\, T^{-1}[\operatorname{sinc}(\delta T^{-1} - n)] = c\,[\sin(\pi \delta T^{-1} - \pi n)]/[\pi(\delta - nT)],$$

for each $n \neq 0$. Next, use the fact that

$$\sin(\pi \delta T^{-1} - \pi n) = \sin(\pi \delta T^{-1})\cos(\pi n) - \cos(\pi \delta T^{-1})\sin(\pi n),$$

and observe that $\sin(\pi n) = 0$ and $\cos(\pi n) = (-1)^n$. This gives

$$v(\delta - nT) = c\,(-1)^n \sin(\pi \delta T^{-1})]/[\pi(\delta - nT)] = d(-1)^n/(\delta - nT),$$

where $d = c \sin(\pi \delta T^{-1})/\pi$. We know that $\sin(\pi \delta T^{-1}) > 0$, because $0 < \delta < T$ implies that $0 < \pi \delta T^{-1} < \pi$. Therefore, $d > 0$ and

$$|v(\delta - nT)| = d/|\delta - nT|.$$

For $n \geq 1$, $|\delta - nT| = nT - \delta \leq nT$; therefore,

$$|v(\delta - nT)| \geq d/nT.$$

Using this inequality in (8.32), we obtain

$$I_{\max} \geq d\, T^{-1} \sum_{n=1}^{\infty}(1/n) = \infty. \tag{8.33}$$

∎

Even if the transmitted sequence of pulses is finite in duration, the maximum intersymbol interference in a system that has an ideal pulse response can be quite large if there is a sampling error. There are other reasons why the pulse response depicted in Figure 8-26 has limited practical value. The ideal frequency function illustrated in Figure 8-26(a) has discontinuities at two frequencies, which makes it difficult to approximate this function with practical waveforms and filters. Furthermore, any system that produces the pulse response shown in Figure 8-26(b) cannot be causal. Recall that $v(t)$ is the response of the system to the 0th pulse, and notice that the pulse response in Figure 8-26(b) is nonzero for $t < 0$. A causal approximation to the ideal system could be obtained by introducing a large delay τ, thereby shifting $v(t)$ to the right to give $v(t - \tau)$, and then truncating the pulse response at time 0 to make it causal (i.e., setting $v(t - \tau) = 0$ for $t < 0$). In order to provide a good approximation, however, the delay would have to be very large, because $\operatorname{sinc}(t/T)$ converges to zero very slowly

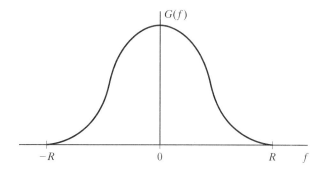

Figure 8-27: The raised-cosine frequency function.

as $t \rightarrow -\infty$. A large delay in the delivery of data at the output of the receiver is unacceptable for most applications, including the digital transmission of speech.

Some of these problems can be overcome by using frequency functions that are easier to implement or approximate. For example, consider the frequency function defined by

$$G(f) = \begin{cases} \cos^2(\pi f/2R), & |f| \leq R, \\ 0, & |f| > R. \end{cases} \qquad (8.34a)$$

This frequency function does not have a discontinuity, as illustrated in Figure 8-27. For transmission rate R, let $T = 1/R$. The inverse Fourier transform of $G(f)$ satisfies $g(nT) = 0$ for $n \neq 0$, as shown in Problem 8.7, so the pulse response $g(t)$ satisfies Nyquist's criterion for intersymbol-interference-free reception at a rate of $R = 1/T$ symbols/s. Since the bandwidth is R Hz, the spectral efficiency is 1 symbol/s/Hz, which is only half the efficiency of the ideal pulse response.

The frequency function of (8.34a) can also be written as

$$G(f) = \begin{cases} [1 + \cos(\pi f/R)]/2, & |f| \leq R, \\ 0, & |f| > R, \end{cases}$$

so the graph of $G(f)$ is one period of a cosine wave that is raised just enough to make it nonnegative for $-R \leq f \leq R$, as can be seen from Figure 8-27. Thus, the frequency function defined by (8.34a) is referred to as the *raised-cosine frequency function*.

The raised-cosine frequency function can be viewed as one member of a family of frequency functions. Let W denote the cutoff bandwidth for the baseband frequency functions, and consider values of W in the range $R/2 \leq W \leq R$. The *raised-cosine family* of frequency functions for transmission rate R is defined by

$$G(f) = \begin{cases} 1, & 0 \leq |f| < R - W, \\ \cos^2\{\pi(|f| + W - R)/(4W - 2R)\}, & R - W < |f| < W, \\ 0, & |f| > W. \end{cases} \qquad (8.34b)$$

At one extreme we have $W = R/2$; that is, the cutoff bandwidth is equal to the Nyquist bandwidth. For $W = R/2$, $G(f) = 1$ for $0 \leq |f| < R/2$ and $G(f) = 0$ for $|f| > R/2$,

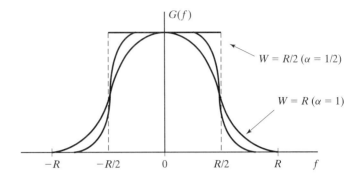

Figure 8-28: The raised-cosine family of frequency functions.

so $G(f)$ is the ideal rectangular frequency function of Figure 8-26(a). Notice that if $W - R/2$, there are no values of f for which $R - W < |f| < W$. At the other extreme, the cutoff bandwidth is twice the Nyquist bandwidth (i.e., $W = R$), there are no values of f for which $0 < |f| < R - W$, and (8.34b) reduces to (8.34a). The two extremes are shown in Figure 8-28 along with a typical frequency function in the raised-cosine family for a value of W approximately halfway between the two extremes. In Problem 8.7, you are asked to prove that each member of the raised-cosine family satisfies Nyquist's criterion for transmission at rate R.

If we denote the Nyquist bandwidth by B, then $B = R/2$ if the transmission rate is R. The amount by which the bandwidth of a frequency function exceeds the Nyquist bandwidth (i.e., $W - B$), is referred to as the *excess bandwidth*. For example, the excess bandwidth for the raised-cosine frequency function is equal to $R/2$. For the raised-cosine family, $R/2 \leq W \leq R$, which is equivalent to $B \leq W \leq 2B$.

The parameter $\alpha = (W - B)/B$ represents the fractional excess bandwidth, and it is typically referred to as the *roll-off factor* or *excess-bandwidth factor*. Note that $0 \leq \alpha \leq 1$ for the raised-cosine family, and $\alpha \geq 0$ for any frequency function that satisfies Nyquist's criterion. Solving $\alpha = (W - B)/B$ for W, we obtain $W = (1 + \alpha)B$, so α can be interpreted as the fractional increase in bandwidth above the minimum possible bandwidth needed for intersymbol-interference-free reception. As the value of α is decreased, the required bandwidth decreases and the spectral efficiency improves. However, for very small values of α, the implementation of a system to provide the required frequency function is more difficult and the tails of the corresponding time function are larger. Indeed, in the limit as $\alpha \to 0$, the frequency function approaches the ideal rectangular frequency function, and the corresponding time function is the sinc function. A value of $\alpha = 0.3$ is often used in practice as a compromise between spectral efficiency and implementation complexity. It is common to refer to a system with $\alpha = 0.3$ as having 30% excess bandwidth.

Intersymbol-interference-free reception is achieved if the Fourier transform of the pulse at the input to the receiver's sampler satisfies Nyquist's criterion. Thus, if the pulse at the input to the receiver has Fourier transform $V(f)$ and the receiver filters this pulse with a filter that has transfer function $H(f)$, we can achieve intersymbol-interference-free reception if the product $V(f) H(f)$ satisfies Nyquist's criterion. Thus,

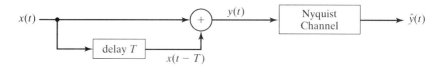

Figure 8-29: Duobinary transmission.

if $V(f)\,H(f) = G(f)$ and $G(f)$ is a member of the raised-cosine family, pulses can be transmitted at rate R with intersymbol-interference-free reception. One such system is obtained if we let $V(f) = H(f) = \sqrt{G(f)}$. If $G(f)$ is a member of the raised-cosine family, $\sqrt{G(f)}$ is referred to as a *square-root raised-cosine frequency function*, or simply a *root-raised-cosine frequency function*. The inverse transform of $\sqrt{G(f)}$ is given in [8.2].

In practice, there may be some delay associated with the pulse or the receiver's filter, in which case

$$V(f) = \sqrt{G(f)}\exp\{-j\,2\pi f\,T_s\}$$

and

$$H(f) = \sqrt{G(f)}\exp\{-j\,2\pi f\,T_r\}.$$

For example, the delay T_r may be required to provide a causal filter in the receiver. If such delays are present, the sampling times must be shifted accordingly, of course.

The use of shaping in the frequency domain can provide a continuous frequency function that is easier to implement than the ideal rectangular frequency function. If we obtain a continuous frequency function from the raised-cosine family, it must be that $W > B$ and the spectral efficiency is not as good as for the ideal rectangular frequency function. For example, the spectral efficiency of a system that has the ideal rectangular frequency function is twice that of a system with the continuous frequency function shown in Figure 8-27. Partial-response signaling is a general method that can provide a continuous spectrum without reducing spectral efficiency.

An example of partial-response signaling, known as *duobinary transmission*, is illustrated in Figure 8-29. In the system of Figure 8-29, the information signal $x(t)$ consists of a sequence of binary pulses, each pulse having amplitude $+1$ or -1; that is,

$$x(t) = \sum_{n=n_1}^{n_2} x_n\,\beta(t - nT), \tag{8.35}$$

where, for each n, $x_n = +1$ or $x_n = -1$. The pulse rate is $1/T$ pulses per second so the data rate is $1/T$ b/s. The channel transfer function is such that its output is the ideal pulse-response waveform $v(t) = T^{-1}\operatorname{sinc}(t/T)$ when the input is $\beta(t)$. Thus, the Fourier transform of the pulse response is the ideal rectangular frequency function $V(f)$ illustrated in Figure 8-26(a) with $c = 1$, and its bandwidth is the Nyquist bandwidth $B = 1/2T$.

The binary signal $x(t)$ is converted to a ternary signal $y(t)$, which is also a sequence of pulses at a rate of $1/T$ pulses per second. The possible pulse amplitudes for $y(t)$ are $+2$, 0, and -2, because $y(t) = x(t) + x(t - T)$. Thus,

$$y(t) = \sum_{n=n_1}^{n_2} y_n \, \beta(t - nT), \tag{8.36}$$

where the sequence (y_n) is produced from (x_n) by *duobinary encoding*: $y_n = x_n + x_{n-1}$ for $n_1 \le n \le n_2$. The transmitter and receiver are each given the value of x_{n_1-1} as an initial condition to be used in the encoding and decoding. The elements y_n and y_{n+1} each depend on x_n, so the elements of the sequence (y_n) are correlated, even if the elements of (x_n) are independent of each other. For example, if y_n is $+2$, then x_n must be $+1$, so y_{n+1} has to be either 0 or $+2$. Thus, $y_n = +2$ implies that $y_{n+1} \ne -2$. Similarly, $y_n = -2$ implies that $y_{n+1} \ne +2$. Partial-response signaling is also known as *correlative coding*, because of the correlation that is introduced into the transmitted sequence.

We know that a sequence (y_n) can be transmitted over a channel at a rate of $2/T$ symbols/s using a bandwidth of $1/T$ Hz. To accomplish this, the pulse response must be the ideal pulse response. Let the channel response to $\beta(t)$ be the ideal pulse-response waveform $v(t)$ shown in Figure 8-26(b), for which the corresponding frequency function is the rectangular frequency function $V(f)$ shown in Figure 8-26(a). The time-domain operation $y(t) = x(t) + x(t - T)$ is equivalent to passing $x(t)$ through a linear filter with impulse response $w(t) = \delta(t) + \delta(t - T)$, which corresponds to the transfer function

$$W(f) = 1 + e^{-j2\pi f T}.$$

Thus, the overall frequency function for the system shown in Figure 8-29 is given by

$$\begin{aligned} G(f) &= V(f)\,W(f) = V(f)\,[1 + e^{-j2\pi f T}] \\ &= V(f)\,[e^{+j\pi f T} + e^{-j\pi f T}]\,e^{-j\pi f T} \\ &= 2\cos(\pi T f)\,V(f)\,e^{-j\pi f T}. \end{aligned}$$

Recall that $V(f) = 1$ for $-B \le f \le B$ and $V(f) = 0$ for $|f| > B$, so the bandwidth of $G(f)$ cannot exceed the Nyquist bandwidth. The magnitude of the frequency function is

$$|G(f)| = \begin{cases} 2\cos(\pi T f), & |f| \le B, \\ 0, & |f| > B, \end{cases}$$

which is continuous, as illustrated in Figure 8-30.

At this stage it may seem as though nothing is gained, because the duobinary system appears to require the rectangular frequency function. However, we do not implement the system as it is drawn in Figure 8-29, which shows the system as the cascade of two linear systems. To do so requires implementing a system with frequency function $V(f)$, which is not continuous. Instead, the term $1 + e^{-j2\pi f T}$ is incorporated into the overall frequency function to give the system illustrated in Figure 8-31. The frequency function for this system is $G(f)$, which is continuous. Although the illustration in Figure 8-29

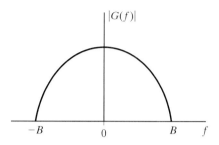

Figure 8-30: The frequency function $2\cos(\pi T f)$ for $-B \le f \le B$.

Figure 8-31: Duobinary transmission with a continuous frequency function.

is very useful as an aid in understanding how the system accomplishes its objectives, a transmission system based on Figures 8-30 and 8-31 is preferable for implementation.

One difficulty that arises in duobinary transmission and related signaling methods is that even if the original signal $x(t)$ is binary, the signal $\hat{y}(t)$ is not. If each x_n is either $+1$ or -1 in (8.35), then the sequence (y_n) of pulse amplitudes for the signal $y(t)$ of (8.36) takes values in the set $\{-2, 0, +2\}$. Thus, the binary sequence (x_n) produces a ternary sequence (y_n).

If $v(t)$ denotes the response of the Nyquist channel to the input pulse $\beta(t)$, then the output of the Nyquist channel in the system of Figure 8-29 is

$$\hat{y}(t) = \sum_{n=n_1}^{n_2} y_n\, v(t - nT) \tag{8.37}$$

if $x(t)$ is the input. Noise is added in the channel and the front end of the receiver, of course, so the input to the demodulator is actually the sum of the signal $\hat{y}(t)$ and noise. The fact that y_n is ternary rather than binary implies that the decision device must be more complex than the simple threshold device that is used for binary antipodal signals. The decision device must distinguish among the pulse amplitudes -2, 0, and $+2$. In addition to other complications, this requires knowledge of the signal amplitude A in order to determine the decision boundaries, and any fluctuations in the propagation loss or system gain must be accounted for in setting these decision boundaries. (This issue is discussed in Chapter 5.)

Based on receiving $\hat{y}(t)$ plus noise, the demodulator in the receiver makes a decision $\hat{y}_n$ regarding the value of y_n. This ternary decision must be converted to a binary decision in order to decide which pulse amplitude x_n was sent. Recall that

$$y_n = x_n + x_{n-1}, \quad n_1 \le n \le n_2, \tag{8.38}$$

and that the transmitter and receiver are given the value of x_{n_1-1} as an initial condition. It follows that the sequence (x_n) can be recovered from the sequence (y_n) by reversing this operation; that is,

$$x_n = y_n - x_{n-1}, \quad n_1 \leq n \leq n_2. \tag{8.39}$$

Example 8-2   Duobinary Encoding and Decoding

Let $n_1 = 1$, and assume that $x_0 = +1$ is known to the transmitter and receiver. Suppose it is desired to send $x_1 = -1$, $x_2 = -1$, $x_3 = +1$, $x_4 = +1$, and $x_5 = +1$. From (8.38), we see that $y_1 = 0$, $y_2 = -2$, $y_3 = 0$, $y_4 = +2$, and $y_5 = +2$. We can use (8.39) and the fact that $x_0 = +1$ to determine x_1, x_2, x_3, x_4, and x_5 from y_1, y_2, y_3, y_4, and y_5. First note that $x_1 = y_1 - x_0 = -1$, and then use this value for x_1 to determine x_2 according to $x_2 = y_2 - x_1 = -1$, etc. ∎

The receiver does not know the sequence (y_n) of course, but it does have an estimate $\hat{y}_n$ for each y_n in the sequence. This estimate is the decision produced by the decision device in the receiver. From the sequence of estimates $(\hat{y}_n)$, the receiver can determine a corresponding sequence of estimates of the transmitted sequence. To begin, the transmitter and receiver are each given the value of x_{n-1} as an initial condition, so the receiver sets $\hat{x}_{n_1-1} = x_{n_1-1}$. The receiver then estimates the value of y_{n_1} by $\hat{y}_{n_1}$, and it uses $\hat{y}_{n_1}$ in the expression

$$\hat{x}_{n_1} = \hat{y}_{n_1} - \hat{x}_{n_1-1}$$

to obtain an estimate of $\hat{x}_{n_1}$. If the receiver then decides the value of y_{n_1+1} is $\hat{y}_{n_1+1}$, it uses this to estimate $\hat{x}_{n_1+1}$, etc. In general, for the nth element of the data sequence, suppose the receiver has estimated the value of previous binary symbol x_{n-1} by $\hat{x}_{n-1}$, and it makes decision $\hat{y}_n$ for the present ternary symbol. It can convert this decision to a decision on the present binary symbol by computing

$$\hat{x}_n = \hat{y}_n - \hat{x}_{n-1}.$$

If all of the decisions $\hat{y}_n$ are correct, the operation

$$\hat{x}_n = \hat{y}_n - \hat{x}_{n-1}, \quad n_1 \leq n \leq n_2, \tag{8.40}$$

reverses the duobinary encoding that is carried out at the transmitter. That is, if there is no noise in the system, $\hat{y}_n = y_n$ and $\hat{x}_n = x_n$ for each n, so (8.39) and (8.40) are identical. In a system with noise, it may be that $\hat{y}_n \neq y_n$ for one or more values of n. Suppose that $\hat{y}_n = y_n$ for $n_1 \leq n \leq j-1$ but $\hat{y}_j \neq y_j$. Because $\hat{y}_n = y_n$ for $n_1 \leq n \leq j-1$, then $\hat{x}_n = x_n$ for $n_1 \leq n \leq j-1$. In particular, $\hat{x}_{j-1} = x_{j-1}$. Letting $n = j$, we see that (8.40) implies that

$$\hat{x}_j = \hat{y}_j - \hat{x}_{j-1} = \hat{y}_j - x_{j-1}$$

and (8.39) implies that

$$x_j = y_j - x_{j-1}.$$

It follows that if $\hat{y}_j \neq y_j$ then $\hat{x}_j$ cannot be equal to x_j. It is not surprising that an error in the estimate of y_j produces an error in the estimate of x_j, and this by itself does not cause any unusual difficulty. The problem is that once an error occurs, it influences future decisions as well, and so it typically causes additional errors in the estimates of the binary information sequence (x_n).

In our example, suppose that $\hat{y}_n = y_n$ for $n_1 \leq n \leq j - 1$ and $j + 1 \leq n \leq n_2$, but $\hat{y}_j \neq y_j$. We have already seen that $\hat{x}_j \neq x_j$. Let $n = j + 1$ in (8.40), and use the fact that $\hat{y}_{j+1} = y_{j+1}$ to show that

$$\hat{x}_{j+1} = \hat{y}_{j+1} - \hat{x}_j = y_{j+1} - \hat{x}_j.$$

But (8.39) implies that

$$x_{j+1} = y_{j+1} - x_j,$$

and $\hat{x}_j \neq x_j$. So it must be that $\hat{x}_{j+1} \neq x_{j+1}$. Thus, $\hat{x}_j \neq x_j$ and $\hat{y}_{j+1} = y_{j+1}$ implies that $\hat{x}_{j+1}$ is incorrect. Similarly, $\hat{x}_{j+1} \neq x_{j+1}$ and $\hat{y}_{j+2} = y_{j+2}$ implies that $\hat{x}_{j+2}$ is incorrect. A single error in estimating y_j leads to errors in $\hat{x}_j$, $\hat{x}_{j+1}$, $\hat{x}_{j+2}$, etc. In summary, if $\hat{y}_n = y_n$ for $n_1 \leq n \leq j - 1$ and $j + 1 \leq n \leq n_2$, but $\hat{y}_j \neq y_j$, then $\hat{x}_n = x_n$ for $n_1 \leq n \leq j - 1$, but $\hat{x}_n \neq x_n$ for $j \leq n \leq n_2$. Note that if $n_2 = +\infty$, there are an infinite number of errors in the estimated data stream $(\hat{x}_n)$. This type of phenomenon, in which a single decision error produces a large number of errors in the data stream, is referred to as *error propagation*.

Fortunately, there is a remedy for the error propagation that can arise in duobinary transmission systems. Error propagation can be avoided by *precoding* the data before duobinary encoding is applied. Suppose the data sequence is a binary sequence $(x_n) = x_{n_1}, x_{n_1+1}, \ldots, x_{n_2}$ of elements from the set $\{-1, +1\}$. The precoding operation maps this data sequence into a new sequence (α_n) to which duobinary encoding is applied. To start the process, we fix a value for α_{n_1-1} from $\{-1, +1\}$ to serve as the initial value for the sequence (α_n). The precoding operation is defined by

$$\alpha_n = x_n \alpha_{n-1}, \quad n_1 \leq n \leq n_2.$$

Notice that the sequence (α_n) is also a sequence of elements from the set $\{-1, +1\}$. Duobinary encoding is applied to (α_n) to give the sequence (y_n); that is,

$$y_n = \alpha_n + \alpha_{n-1}, \quad n_1 \leq n \leq n_2.$$

As before, (y_n) is a sequence of elements from the set $\{-2, 0, +2\}$. At the receiver, an estimate $\hat{y}_n$ is obtained for each y_n, and the sequence $(\hat{y}_n)$ of estimates is decoded into the sequence $(\hat{x}_n)$ by the following decoding rule: If $\hat{y}_n = 0$, then $\hat{x}_n = -1$; otherwise, $\hat{x}_n = +1$. That is, $\hat{x}_n = +1$ if $\hat{y}_n = -2$ or $\hat{y}_n = +2$. Notice that $\hat{x}_n$ depends on $\hat{y}_n$ only. In particular, $\hat{x}_n$ does not depend on $\hat{x}_{n-1}$ or any other previous estimates. As a result, there is no error propagation. A single error in the sequence $(\hat{y}_n)$ produces only a single error in the sequence $(\hat{x}_n)$. More generally, the number of errors in the estimated data sequence $(\hat{x}_n)$ is the same as the number of errors in $(\hat{y}_n)$. The precoding procedure is illustrated in the following example:

Example 8-3   The Precoding Procedure

Suppose that $n_1 = 0$ and $n_2 = 7$, and the data sequence is

$$(x_0, x_1, x_2, x_3, x_4, x_5, x_6, x_7) = (+1, -1, +1, +1, -1, +1, -1, -1).$$

If $\alpha_{-1} = +1$, the sequence (α_k) obtained by precoding is

$$(\alpha_0, \alpha_1, \alpha_2, \alpha_3, \alpha_4, \alpha_5, \alpha_6, \alpha_7) = (+1, -1, -1, -1, +1, +1, -1, +1),$$

and the sequence (y_n) obtained from the duobinary encoding of (α_k) is

$$(y_0, y_1, y_2, y_3, y_4, y_5, y_6, y_7) = (+2, 0, -2, -2, 0, +2, 0, 0).$$

By comparing y_n with x_n for each n, we see that $x_n = -1$ corresponds to $y_n = 0$ and $x_n = +1$ corresponds to $|y_n| = 2$. ∎

8.4   Introduction to Linear Equalization

The goal in Section 8.3 is to employ a communication system that does not generate any intersymbol interference. This is accomplished if the pulse response waveform $v(t)$ and the sampling times $T_0 + nT$ satisfy $v(T_0 + nT) = 0$ for all $n \neq 0$. Based on the necessary and sufficient conditions for the system to have this property, we conclude in Section 8.3 that for many applications the goal of intersymbol-interference-free transmission is not practical. The required filters are difficult (or impossible) to implement, and the resulting performance is very sensitive to timing errors and other imperfections.

In the present section, we do not require intersymbol-interference-free transmission, but instead we seek a method to reduce the intersymbol interference that arises in a communication system. One approach to intersymbol interference reduction is the inclusion in the communication receiver of a subsystem known as a *linear equalizer*. Such an equalizer can improve the performance of the system in which the intersymbol interference is due to linear distortion (e.g., a bandwidth limitation in the linear filters of the receiver).

This section is intended to be an introduction to linear equalization rather than a comprehensive exposition of the wide range of equalization methods that have been developed to combat intersymbol interference (e.g., [8.5], [8.7], and [8.9]). Accordingly, attention is restricted to systems with known transfer functions and to the type of equalizer known as a *tapped-delay-line equalizer* or *linear transversal equalizer*. An example of such an equalizer is applied at the output of a filter with impulse response $h(t)$ in the illustration of Figure 8-32. Although the equalizer shown in Figure 8-32 is a linear time-invariant filter, confusion is avoided in our discussions by reserving the term *filter* for the filter that has impulse response $h(t)$. Hence, we describe the receiver of Figure 8-32 as a linear filter followed by a linear equalizer.

The equalizer of Figure 8-32 has three taps, and the tap gains are denoted by λ_1, λ_2, and λ_3. There are two delay elements, each of which provides a delay of T seconds, where $1/T$ is symbol rate for the communication signal. Equalizers that use delays

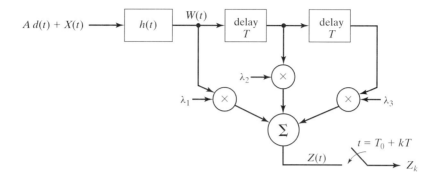

Figure 8-32: A three-tap linear equalizer.

that are less than T, known as fractionally spaced equalizers, are not considered here. Because the impulse response $h(t)$ is known, the intersymbol interference components at the output of this filter can be determined. If the impulse response of the communication system is not known, standard procedures may be employed to estimate the intersymbol interference components. The intersymbol interference components are then used to determine the tap gains.

The tap gains λ_1, λ_2, and λ_3 are chosen to optimize the system according to some performance measure. Although it might be desirable to select the tap gains to minimize the average probability of error, it is typically very difficult to do so in practice. As a result, various other criteria have been employed to choose the tap gains. The method described here for the selection of the tap gains is known as the *zero-forcing method*, and the resulting equalizer is known as a *zero-forcing linear equalizer*. In this method, the tap gains are chosen in a way that forces the response waveform at the equalizer output to be zero at certain sampling times. Under certain conditions, this choice of tap gains minimizes the peak distortion.

Once again, the issue of the selection of a convenient time reference arises. Just as the channel and the filter typically introduce a time delay, the equalizer may also introduce a time delay. As a simple illustration, consider a system in which the response of the filter to the signal pulse $\beta(t)$ has filter-output samples that are all equal to zero except for the sample at time T_0 (i.e., the desired signal component). Let this sample serve as a reference signal for the moment. Suppose the tap gains for the equalizer in Figure 8-32 are $\lambda_1 = \lambda_3 = 0$ and $\lambda_2 = 1$. The signal samples at the output of the equalizer are all equal to zero except at time $T_0 + T$; that is, a delay of T has been introduced by the equalizer. The most convenient time reference to use in the analysis of the effects of the equalizer corresponds to the time of arrival of this reference signal at the output of the equalizer. Using such a time reference amounts to shifting the time axis by an amount T for the system considered in this illustration.

In this simple illustration, the time of the reference signal's arrival at the equalizer output is $T_0 + T$ according to the natural choice for the filter-output clock. If the timing clock used for the equalizer is T seconds behind the filter-output clock, then the reference signal arrives at the sampler at time T_0 according to the equalizer's timing clock. Based on the equalizer's clock, the correct sampling time for the reference signal in this illustration is T_0 if the input is the single pulse $\beta(t)$. This sampling time corresponds to $k = 0$ in Figure 8-32.

Let $v(t)$ denote the pulse response at the output of the equalizer if time is measured according the equalizer's clock. In all that follows, we set the equalizer's clock in such a way that $k = 0$ gives the correct sampling time for the desired signal component. In other words, $v(t)$ is the output of the equalizer when $\beta(t)$ is the input to the filter, and a time offset is employed as needed to make the sample at time T_0 correspond to the desired signal component. Usually this is equivalent to choosing the time reference in such a way that the maximum value of $|v(T_0+nT)|$ over all integers n occurs for $n = 0$. Analogous to the terminology in Section 8.1, the sample $v(T_0)$ at the equalizer output is referred to as the *desired signal component of the equalizer output*. The remaining samples, $v(T_0 + nT)$ for $n \neq 0$, are called the *intersymbol interference components of the equalizer output*.

Instead of requiring that $v(T_0 + nT) = 0$ for all $n \neq 0$, as in Section 8.3, the objective of the zero-forcing method for a linear equalizer with a finite number of taps is to achieve $v(T_0+nT) = 0$ for specific nonzero values of n. We define the zero-forcing equalizer as a tapped-delay-line equalizer for which $v(T_0 + nT) = 0$ for $-i_1 \leq n < 0$ and $0 < n \leq i_2$ for some choice of the integers i_1 and i_2. Such an equalizer forces at least $i_1 + i_2$ of the samples of the pulse response to be zero at the equalizer output. The maximum value of $i_1 + i_2$ is limited by the number of taps in the equalizer. For example, $i_1 + i_2 = 2$ can be achieved with the three-tap linear equalizer illustrated in Figure 8-32. If a communication system satisfies $v(T_0+nT) = 0$ for $-i_1 \leq n < 0$ and $0 < n \leq i_2$, then a given data pulse experiences no intersymbol interference from the i_2 pulses that precede that pulse or the i_1 pulses that follow it. In the following exercise, $i_1 = i_2 = 1$, so a zero is forced on each side of the desired signal component.

Exercise 8-6

In the system of Figure 8-32, let $W(t) = w(t)+\hat{X}(t)$, where $w(t)$ is the signal component at the output of the filter and $\hat{X}(t)$ is the filtered noise. Suppose it is determined that if the input to the filter is $\beta(t)$ only (in particular, there is no noise), then the signal at the output of the filter satisfies $w(T_0 - T) = -1$, $w(T_0) = +3$, $w(T_0 + T) = -1$, and $w(T_0 + nT) = 0$ for $|n| > 1$. Specify the tap gains that force a zero on each side of the desired signal component at the output of the equalizer.

Solution. Observe that if the timing clock for the equalizer output waveform $v(t)$ is T seconds behind the timing clock used to describe $w(t)$, then

$$v(t) = \lambda_1 \, w(t + T) + \lambda_2 \, w(t) + \lambda_3 \, w(t - T)$$

for all t. In particular, at sampling time $T_0 + kT$, the output is given by

$$v(T_0 + kT) = \lambda_1 \, w(T_0 + (k + 1)T) + \lambda_2 \, w(T_0 + kT) + \lambda_3 \, w(T_0 + (k - 1)T).$$

For the given values of $w(T_0 + nT)$, we find that

$$v(T_0 - 2T) = -\lambda_1,$$
$$v(T_0 - T) = 3\lambda_1 - \lambda_2,$$
$$v(T_0) = -\lambda_1 + 3\lambda_2 - \lambda_3,$$
$$v(T_0 + T) = -\lambda_2 + 3\lambda_3,$$
$$v(T_0 + 2T) = -\lambda_3,$$

and $v(T_0 + kT) = 0$ for $|k| \geq 3$. From the symmetry of this set of equations, it is natural to focus on $v(T_0)$ as the desired signal component and treat the other components as intersymbol interference components. In particular, if λ_1 and λ_3 are small compared to λ_2, the largest sample is $v(T_0)$. Based on this observation, we can rule out $\lambda_2 = 0$ in our selection of the tap gains. To force zeros at $T_0 - T$ and $T_0 + T$, we require that $3\lambda_1 - \lambda_2 = 0$ and $-\lambda_2 + 3\lambda_3 = 0$. In other words, $\lambda_1 = \lambda_2/3$ and $\lambda_3 = \lambda_2/3$. The received waveform $W(t)$ can be multiplied by any positive constant without affecting the performance of the system. This simply multiplies the signal and noise by the same constant, so the signal-to-noise ratio is unchanged (as discussed in Chapter 5). In addition, the desired signal component and the intersymbol interference components are each multiplied by the same constant, so the effects of intersymbol interference are also unchanged. Therefore, there is no loss in generality in letting the center tap have gain $\lambda_2 = 1$. This normalization is equivalent to multiplying the received waveform by $1/\lambda_2$. If $\lambda_2 = 1$, then the zero-forcing solution for the remaining tap gains is $\lambda_1 = \lambda_3 = 1/3$. ∎

Recall from Section 8.2.3 that the maximum intersymbol interference is

$$I_m(T_0) = \sum_{n=n_1}^{n_2}{}' |v(T_0 - nT)| \qquad (8.41)$$

and the *peak distortion* is

$$d_0(T_0) = I_m(T_0)/v(T_0). \qquad (8.42)$$

One of the goals of incorporating an equalizer into the system is to give a smaller peak distortion than provided by the original system. As long as the maximum intersymbol interference is less than the desired signal, the zero-forcing equalizer is effective in reducing the peak distortion.

Exercise 8-7
Consider the system of Figure 8-32 with the features described in Exercise 8-6. Determine the peak distortion for this system with and without the equalizer.

Solution.   First consider the system without the equalizer. For the pulse response waveform given in Exercise 8-6, the maximum intersymbol interference is $I_m(T_0) = 2$, and desired signal component at the sampling time is $w(T_0) = 3$. The peak distortion for the system without an equalizer is $d_0(T_0) = 2/3$. Now consider the addition of the

equalizer with the tap gains specified in the solution to Exercise 8-6. The samples at
the output of the equalizer are $v(T_0 - 2T) = -1/3$, $v(T_0 - T) = 0$, $v(T_0) = 7/3$,
$v(T_0 + T) = 0$, $v(T_0 + 2T) = -1/3$, and $v(T_0 + kT) = 0$ for $|k| > 2$. The resulting
maximum intersymbol interference is $\hat{I}_m(T_0) = 2/3$, and the desired signal component
is $v(T_0) = 7/3$. The peak distortion for the system with the equalizer is $\hat{d}_0(T_0) = 2/7$,
which is much smaller than the peak distortion for the system without an equalizer.

∎

One byproduct of incorporating an equalizer into the system is that doing so changes
the autocorrelation function of the noise. That is, the noise at the output of the equalizer
in Figure 8-32 has a different autocorrelation function than the noise at the output of
the filter. In particular, the variance of the noise at the equalizer output is typically
different than the variance of the noise at the output of the filter. In order to evaluate the
probability of error for a system with an equalizer, it is necessary to find the variance
of the noise at the equalizer output. To determine the variance of noise at the equalizer
output, the autocorrelation of the noise at the filter output must be known. Knowledge
of the variance of the noise at the filter output is not sufficient for determination of the
variance at the output of the equalizer, as can be seen in the following exercise:

Exercise 8-8

Consider again the system of Figure 8-32 with the features described in Exercise 8-6.
The decision is made by a threshold device in which the threshold is zero. Suppose it
is determined that $\hat{X}(t)$ is a zero-mean Gaussian random process with autocorrelation
function $R(\tau) = \rho^2 \exp\{-(2\tau/T)^2\}$ for $-\infty < \tau < \infty$. The parameter ρ is positive.
Determine both $P_{e,\max}$ and $P_{e,\text{ave}}$ for the system with the equalizer and the system without
the equalizer.

Solution.   The maximum and average values of the probability of error for the system
without the equalizer are

$$P_{e,\max} = Q\left(A[w(T_0) - |w(t_0 - T)| - |w(t_0 + T)|]/\sqrt{R(0)} \right) = Q(A/\rho)$$

and

$$P_{e,\text{ave}} = [Q(5A/\rho) + 2\,Q(3A/\rho) + Q(A/\rho)]/4,$$

respectively. For the system with the equalizer, the noise component of the equalizer
output is given by $[\hat{X}(t + T) + 3\hat{X}(t) + \hat{X}(t - T)]/3$. Because the mean of this random
process is zero, its variance is equal to its second moment. The variance is given by

$$\begin{aligned}
\sigma^2 &= [11R(0) + 12R(T) + 2R(2T)]/9 \\
&= [11 + 12\,e^{-4} + 2\,e^{-16}]\rho^2/9 \approx 1.247\rho^2.
\end{aligned}$$

Notice that the variance of the noise at the equalizer output is almost 25% larger than the
variance of the noise at the filter output. Since the maximum intersymbol interference
is $\hat{I}_m(T_0) = 2/3$ and the desired signal component is $v(T_0) = 7/3$, the smallest value
of the signal at the sampling instant is $v(T_0) - \hat{I}_m(T_0) = 5/3$. Hence, the maximum

probability of error for the system with the equalizer is $P_{e,\max} = Q(5A/3\sigma)$. Each of the two intersymbol interference components at the equalizer output has magnitude $1/3$, so the possible signal components at the sampling time are of the form $(7/3)\pm(1/3)\pm(1/3)$. The resulting average probability of error is

$$P_{e,\mathrm{ave}} = [Q(3A/\sigma) + 2\,Q(7A/3\sigma) + Q(5A/3\sigma)]/4. \qquad \blacksquare$$

In the solution to Exercise 8-8, we see that the variance of the noise is increased by the equalizer. This phenomenon is referred to as *noise enhancement*, and it is a well-known property of zero-forcing equalizers. We can use this example to illustrate that the equalizer can decrease both the maximum probability of error and the average probability of error in many situations, even though the noise variance is increased.

Exercise 8-9
Suppose that $A/\rho = 2$ in Exercise 8-8. Find the maximum probability of error and the average probability of error for the original system and the system with the equalizer.

Solution.   For the original system, $P_{e,\max} = Q(A/\rho) \approx 2.3 \times 10^{-2}$ and

$$P_{e,\mathrm{ave}} = [Q(10) + 2\,Q(6) + Q(2)]/4 \approx 5.7 \times 10^{-3}.$$

For the system with the equalizer, $P_{e,\max} = Q(5A/3\sigma) \approx 1.4 \times 10^{-3}$ and $P_{e,\mathrm{ave}} \approx 3.6 \times 10^{-4}$. The maximum probability of error for the system with the equalizer is more than 16 times smaller than the maximum probability of error for the original system. The average probability of error is reduced by more than a factor of 15 by the equalizer.
$\blacksquare$

The fact that the noise variance is increased by the equalizer in the preceding exercise has implications for the relative values of the one-shot probability of error for the system with and without the equalizer. For the system of Exercise 8-9 without the equalizer, the one-shot probability of error is $P_{e,\mathrm{os}} = Q(3A/\rho)$, which is less than 10^{-9}. If the equalizer is used with the tap gains specified in the solution of Exercise 8-6, the one-shot probability of error is $P_{e,\mathrm{os}} = Q(7A/3\sigma) \approx 1.5 \times 10^{-5}$. This increase is due to a combination of noise enhancement (ρ^2 is replaced by σ^2) and reduction in the desired signal ($3A$ is replaced by $7A/3$). Although this observation regarding the one-shot probability of error may be of some interest and educational value, the one-shot probability of error is not a useful performance measure for an equalizer. The maximum and average values of the probability of error are the important performance measures, and we see that in the preceding exercise they are decreased significantly by the use of the zero-forcing equalizer. The amount by which they are decreased depends in part on the signal-to-noise ratio (e.g., A/ρ in Exercise 8-9), as illustrated in Problem 8.15.

We next return to the three-tap equalizer of Figure 8-32 and derive expressions for the zero-forcing tap gains for a general response waveform $w(t)$. Recall that the response waveform is the output of the filter when the input is the single pulse $\beta(t)$. Suppose that the samples $w(T_0 + kT)$ are known for all values of k, and it is desired to force zeros at $T_0 - T$ and $T_0 + T$. Before proceeding, we should mention that it is not always best to force the two zeros at $T_0 - T$ and $T_0 + T$, as is illustrated in Problem 8.10.

Without loss of generality, we can set $\lambda_2 = 1$, as discussed in the solution to Exercise 8-6. If for implementation purposes a different value of λ_2 is preferred, then we can simply interpret the gains λ_1 and λ_3 as relative gains, relative to the gain of the center tap. Let the output of the equalizer be denoted by $v(t)$ when the input to the equalizer is $w(t)$, and let the timing clock for the equalizer output waveform be T seconds behind the timing clock used to describe $w(t)$. Using these conventions for the center-tap gain and the timing clock, we can write the equalizer output samples as

$$v(t) = \lambda_1 \, w(t + T) + w(t) + \lambda_3 \, w(t - T)$$

for all t. In particular, at sampling time $T_0 + kT$ the equalizer output is given by

$$v(T_0 + kT) = \lambda_1 \, w(T_0 + (k + 1)T) + w(T_0 + kT) + \lambda_3 \, w(T_0 + (k - 1)T).$$

The tap gains λ_1 and λ_3 are the solutions to the equations

$$v(T_0 + T) = \lambda_1 \, w(T_0 + 2T) + w(T_0 + T) + \lambda_3 \, w(T_0) = 0 \qquad (8.43)$$

and

$$v(T_0 - T) = \lambda_1 \, w(T_0) + w(T_0 - T) + \lambda_3 \, w(T_0 - 2T) = 0. \qquad (8.44)$$

From (8.43), we see that

$$\lambda_3 = -[\lambda_1 \, w(T_0 + 2T) + w(T_0 + T)]/w(T_0). \qquad (8.45)$$

If we substitute the right-hand side of (8.45) for λ_3 in (8.44), we find that

$$\lambda_1 = \frac{[w(T_0 + T) \, w(T_0 - 2T) - w(T_0 - T) \, w(T_0)]}{[w^2(T_0) - w(T_0 + 2T) \, w(T_0 - 2T)]}. \qquad (8.46)$$

Substituting for λ_1 into (8.45) from (8.46) gives

$$\lambda_3 = \frac{[w(T_0 - T) \, w(T_0 + 2T) - w(T_0 + T) \, w(T_0)]}{[w^2(T_0) - w(T_0 + 2T) \, w(T_0 - 2T)]}. \qquad (8.47)$$

Equations (8.46) and (8.47) give the tap gains that force zeros at $T_0 - T$ and $T_0 + T$.

As a special case, suppose that the response waveform satisfies $w(T_0 + kT) = 0$ for $|k| \geq 2$. That is, the only samples that we need to consider are $w(T_0 + T)$, $w(T_0)$, and $w(T_0 - T)$. For this special case, (8.46) and (8.47) reduce to

$$\lambda_1 = -w(T_0 - T)/w(T_0) \qquad (8.48)$$

and

$$\lambda_3 = -w(T_0 + T)/w(T_0), \qquad (8.49)$$

respectively. The tap gains given by equations (8.48) and (8.49) force zeros in the equalizer output at times $T_0 + T$ and $T_0 - T$ if the response function $w(t)$ has only two nonnegligible intersymbol-interference components and these two components are $w(T_0 + T)$ and $w(T_0 - T)$.

References and Suggestions for Further Reading

[8.1] J. B. Anderson, *Digital Transmission Engineering*. New York: IEEE Press, 1999.

[8.2] S. Chennakeshu and G. J. Saulnier, "Differential detection of $\pi/4$-shifted-DQPSK for digital cellular radio," *IEEE Transactions on Vehicular Technology*, vol. 42, February 1993, pp. 46–57.

[8.3] J. J. Jones, "Filter distortion and intersymbol interference effects on PSK signals," *IEEE Transactions on Communications Technology*, vol. COM-19, April 1971, pp. 120–132.

[8.4] A. Lender, "The duobinary technique for high speed data transmission," *IEEE Transactions on Communications Electronics*, vol. 82, May 1963, pp. 214–218.

[8.5] R. W. Lucky, J. Salz, and E. J. Weldon, Jr., *Principles of Data Communication*. New York: McGraw-Hill, 1968.

[8.6] S. Pasupathy, "Correlative coding—A bandwidth efficient signaling scheme," *IEEE Communications Magazine*, July 1977, pp. 4–11.

[8.7] J. G. Proakis, *Digital Communications* (4th ed.). New York: McGraw-Hill, 2001.

[8.8] S. Qureshi, "Adaptive equalization," *IEEE Communications Magazine*, vol. 20, March 1982, pp. 9–16.

[8.9] S. Qureshi, "Adaptive equalization," *Proceedings of IEEE*, vol. 53, September 1985, pp. 1349–1387.

[8.10] M. K. Simon, S. M. Hinedi, and W. C. Lindsey, *Digital Communication Techniques*. Englewood Cliffs, NJ: Prentice-Hall, 1995.

Problems

8.1 Consider the communication system of Figure 8-1 with the input signal given by

$$s(t) = \sqrt{2}\, A\, d(t) \cos(\omega_c t + \theta).$$

$X(t)$ is white Gaussian noise with spectral density $N_0/2$. The filter is a single-pole RC filter with impulse response

$$h(t) = (RC)^{-1} \exp(-t/RC)\, u(t),$$

where $u(t)$ is the unit step function: $u(t) = 1, t \geq 0$; $u(t) = 0, t < 0$.

  (a) Show that the transfer function for this filter is

$$H(f) = [1 + j\,2\pi f RC]^{-1}, \quad -\infty < f < \infty.$$

  (b) The 3-dB bandwidth, also known as the half-power bandwidth, is defined by $|H(f_{3\text{dB}})|^2 = |H(0)|^2/2$. Show that the 3-dB bandwidth for this filter is given by $f_{3\text{dB}} = (2\pi RC)^{-1}$.

  (c) Suppose $d(t) = p_T(t)$ is the input to this RC filter, and consider the signal at the output of the filter. Show that $T_0 = T$ is the optimum sampling time if the goal is to maximize the signal at the filter output.

(d) Let $d(t)$ be the single pulse $b_0 \, p_T(t)$, where b_0 is either $+1$ or -1. Then the transmitted signal $s(t)$ is time limited to $[0, T]$. Let the signal-to-noise ratio be defined as in Chapter 5, and show that if the sampling time is $T_0 = T$, the signal-to-noise ratio satisfies

$$(\text{SNR})^2 = (4A^2/N_0) \, RC\{1 - \exp(-T/RC)\}^2.$$

(e) Let $d(t)$ be the baseband signal given by (8.1) with $n_1 = -\infty$ and $n_2 = +\infty$. The baseband pulse waveform is $\beta(t) = p_T(t)$. If $d(t)$ is the input to the RC filter, there will be intersymbol interference. Find the tightest possible upper and lower bounds on $\hat{d}(T)$, the output signal at time $T_0 = T$, if $b_0 = +1$ and b_n is arbitrary for $n \neq 0$ (subject only to $b_n = \pm 1$). Your answer should be in terms of T and the product of R and C.

(f) From your bounds in part (e), determine the corresponding bounds on the output signal-to-noise ratio, which is defined as the ratio of the output signal at the sampling time to the standard deviation of the noise. First, express the bounds in terms of RC, T, A, and N_0, and then give an equivalent expression in terms of the time-bandwidth product $T f_{3\text{dB}}$ and the energy-to-noise-density ratio $\mathcal{E}/N_0$, where $\mathcal{E}$ is the energy per bit in the signal $s(t)$.

8.2 Repeat Problem 8.1 for a Gaussian filter with impulse response

$$h(t) = \sqrt{\pi/\alpha} \, \exp(-\pi^2 t^2/\alpha), \quad -\infty < t < \infty.$$

(a) Show that the transfer function for this filter is

$$H(f) = \exp\{-\alpha f^2\}, \quad -\infty < f < \infty.$$

(b) Show that the 3-dB bandwidth is $f_{3\text{dB}} = \{(\ln 2)/2\alpha\}^{1/2}$.

(c) Suppose $d(t) = p_T(t)$ is the input to this filter. Show that $T_0 = T/2$ is the optimum sampling time if the goal is to maximize the signal at the filter output.

(d) Let $d(t)$ be the single pulse $b_0 \, p_T(t)$, where b_0 is either $+1$ or -1. Show that if the sampling time is $T_0 = T/2$, the signal-to-noise ratio satisfies

$$(\text{SNR})^2 = (2A^2/N_0) \, \sqrt{2\alpha/\pi} \, \left\{2 \, \Phi(\pi T/\sqrt{2\alpha}) - 1\right\}^2.$$

(e) Let $d(t)$ be the baseband signal given by (8.1) with $n_1 = -\infty$ and $n_2 = +\infty$. The baseband pulse waveform is $\beta(t) = p_T(t)$. If $d(t)$ is the input to the Gaussian filter, there will be intersymbol interference. Find the tightest possible upper and lower bounds on $\hat{d}(T/2)$, the output signal at time $T_0 = T/2$, if $b_0 = +1$ and b_n is arbitrary for $n \neq 0$ (subject only to $b_n = \pm 1$). Give your answer in terms of α and T.

(f) Now consider the communication system of Figure 8-1 with the Gaussian filter. The sampling time is $T_0 = T/2$. From your bounds in part (e), determine the corresponding bounds on the output signal-to-noise ratio, which is defined as the ratio of the output signal at the sampling time to the standard deviation of the noise. First, express the bounds in terms of α, T, A, and N_0, and then give the bounds in terms of $f_{3\text{dB}}$ and the ratio $\mathcal{E}/N_0$, where $\mathcal{E}$ is the energy per bit in the signal $s(t)$.

8.3 The baseband equivalent model for a coherent communication system is as shown Figure 8-2. The noise $X(t)$ is white Gaussian noise with unknown spectral density, and the threshold is zero. It is known that the noise $\hat{X}(t)$ at the output of the linear filter has autocorrelation function $R_{\hat{X}}(\tau) = 4\exp(-|\tau|)$. When $d(t) = p_T(t)$ and there is no noise in the system, it is observed that $Z(T_0 - T) = -1$, $Z(T_0) = +6$, $Z(T_0 + T) = -1$, and $Z(T_0 + kT) = 0$ for $|k| > 1$.

Give an expression for the average probability of error for the system if $d(t)$ is given by (8.1) with $\beta(t) = p_T(t)$, $n_1 = -\infty$, and $n_2 = +\infty$. Assume that (b_n) is a sequence of independent random variables with $P(b_n = +1) = P(b_n = -1) = 1/2$ for each k. Express your answer in terms of the function Q.

8.4 The linear filter in the coherent communication system of Figure 8-1 has transfer function defined by $H(\omega) = T^2/(1 + j\omega T)(2 + j\omega T)$ for $-\infty < \omega < \infty$. The transmitted signal is

$$s(t) = \sqrt{2}\, A\, d(t)\cos(\omega_c t + \theta),$$

where $d(t)$ is given by (8.1) with $n_1 = -\infty$ and $n_2 = +\infty$. The noise process $X(t)$ is white Gaussian noise with spectral density $N_0/2$, $\beta(t)$ is the rectangular pulse of duration T, the sampling time is $T_0 = T$, and the threshold is zero.

(a) Suppose that $b_0 = +1$. Which values of b_n for $n \neq 0$ give the *largest* probability of error?

(b) Find the filter output at time T when the input to the filter is the rectangular pulse $p_T(t)$.

(c) Give a simple closed-form expression (no integrals or infinite series) in terms of A, T, and N_0 for the maximum probability of error for this system.

(d) If $\mathcal{E}$ is the energy per bit, the answer to part **(c)** can be written in the form $P_{e,\max} = Q\big(\sqrt{2\mathcal{E}/N_0}\,\delta\big)$. Give an expression for δ. (*Hint:* δ does not depend on A, T, or N_0.)

8.5 A binary baseband data transmission system of Figure 8-2 uses rectangular pulses of duration T, and $X(t)$ is white Gaussian noise with spectral density $N_0/2$. Assume that $A = 1$. The filter impulse response is

$$h(t) = p_{2T}(t).$$

The output $Z(t)$ of the filter consists of a signal component plus a noise component. The noise component $W(t)$ is the output of the filter when the input is the noise process $X(t)$ alone.

(a) Find the autocorrelation function $R_W(\tau)$ for the noise process at the filter output. In addition, give the value of the output noise variance $\sigma^2 = R_W(0)$.

(b) Apply an input that consists of a single rectangular pulse $p_T(t)$. Find the output $v(t)$ due to this pulse.

(c) For $v(t)$ defined as in part **(b)**, give the values of $v_k = v(T_0 - kT)$ for all values of k if $T_0 = 3T/2$.

(d) Now assume the input data signal $d(t)$ is given by

$$d(t) = \sum_{j=-\infty}^{\infty} b_j \, p_T(t - jT).$$

The data variables b_j take values $+1$ and -1 as usual. Give an expression for the average probability of error for this system if the variables b_j are independent random variables with $P(b_j = +1) = P(b_j = -1) = 1/2$. First give your answer in terms of T, σ, and, if needed, the correlation function $R_W(\tau)$. Next, convert your answer to an expression in terms of N_0 and T.

(e) Find the limiting value of the average probability of error from part **(d)** as $N_0 \rightarrow 0$ (i.e., a noiseless channel).

8.6 A binary baseband communication uses antipodal signaling with rectangular pulses of duration T. The receiver is as shown in Figure 8-2. The receiver input is $Y(t) = s(t) + X(t)$, where $X(t)$ is a white Gaussian noise process with spectral density $N_0/2$ and $s(t) = A\, d(t)$ is the transmitted signal. The output $Z(t)$ of the filter consists of a signal component $\hat{s}(t)$ plus a noise component $W(t)$. $W(t)$ is the filter output when the input is the noise process $X(t)$ alone. The filter impulse response is given by $h(t) = 1$ for $-T/8 \leq t \leq 3T/4$ and $h(t) = (5T - 4t)/2T$ for $3T/4 \leq t \leq 5T/4$, as illustrated in Figure 8-33.

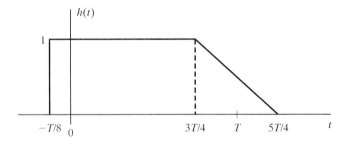

Figure 8-33: The impulse response $h(t)$.

(a) If the transmitted signal is $s(t) = A\, b_0 \, p_T(t)$, where b_0 is $+1$ or -1, what is the optimum sampling time? Give a *brief* justification. Note that you are being asked for the optimum sampling time based only on the one-shot analysis.

(b) Find an integral expression for the variance σ^2 for the noise process $W(t)$ at the output of the linear filter. Integrands and limits on the integrals can depend on the parameters N_0 and T only. You need not evaluate the integrals.

(c) Evaluate the integral expression obtained in part **(b)** for the variance σ^2. Express your answer in terms of the parameters N_0 and T only.

Note: The solutions for parts **(a)**–**(c)** are *not* required in the remaining parts of this problem. In parts **(d)**–**(h)**, express your answer in terms of the appropriate functions and the parameters A, T, and σ. Do *not* substitute for σ from part **(b)** or part **(c)**.

(d) Apply the input $A\,p_T(t)$ to the filter, and find the output $v(t)$ at time $t = T$ that is due to this pulse being applied at the input.

(e) Find the probability of error if the transmitted signal is as given in part (a) and the sampling time is T (i.e., $T_0 = T$). This is still a one-shot analysis.

(f) For $v(t)$ defined as in part (d), give the values of $v_k = v(T_0 - kT)$ for all integers k if $T_0 = T$.

(g) Now the transmitted signal is a sequence of pulses of the form

$$s(t) = A \sum_{n=-\infty}^{\infty} b_n\, p_T(t - nT),$$

where, for each n, b_n is $+1$ or -1. What is the maximum probability of error (maximized over all possible pulse patterns) if the sampling time is T?

(h) Suppose that the transmitted signal is as in part (g), and the sequence (b_n) is a sequence of independent random variables with $P(b_n = +1) = P(b_n = -1) = 1/2$ for each n. What is the average probability of error (averaged over all pulse patterns) if the sampling time is T?

8.7 (a) Show that the raised-cosine frequency function

$$G(f) = \begin{cases} \cos^2(\pi f/2R), & 0 \le |f| < R, \\ 0, & \text{otherwise,} \end{cases}$$

satisfies Nyquist's criterion for communication at a rate of R symbols per second.

(b) For each value of W in the range $R/2 \le W \le R$, prove that the frequency function

$$G(f) = \begin{cases} 1, & 0 \le |f| < R - W, \\ \cos^2\{\pi(|f| + W - R)/(4W - 2R)\}, & R - W < |f| < W, \\ 0, & |f| > W, \end{cases}$$

satisfies Nyquist's criterion for communication at a rate of R symbols per second.

8.8 A binary baseband communication system uses antipodal signaling with rectangular pulses of duration T. The noise is Gaussian with spectral density $N_0/2$. The receiver consists of a time-invariant linear filter followed by a sampler and a threshold decision device with threshold $\gamma = 0$. The impulse response of the linear filter is given by $h(t) = a$ for $0 \le t < T$, $h(t) = b$ for $T \le t < 2T$, and $h(t) = 0$, otherwise. Assume that $a > b > 0$.

(a) The filter output is sampled at time T. What is the variance σ^2 for the noise component of the sample? Express your answer in terms of a, b, T, and N_0.

Note: The solution for part (a) is not required for the remaining parts of this problem, because all subsequent answers are to be given in terms of σ.

(b) Apply the input $p_T(t)$ to the filter, and find the output $v(t)$ due to this pulse alone. A mathematical expression is not required. Give your answer in terms of a *sketch* of the function $v(t)$ for all t. Label each point in the sketch that is necessary to define the function $v(t)$.

(c) Sketch the function $v(t)$ from part (b) for $a = 4$ and $b = 3$. Label key points in the sketch in order to define the function $v(t)$. Repeat for $a = 4$ and $b = 2$.

(d) Give an expression for the sample value $v(nT)$ in terms of a, b, and T for each choice of the integer n.

Now suppose that the transmitted signal is a sequence of rectangular pulses of the form

$$s(t) = A \sum_{k=-\infty}^{\infty} b_k \, p_T (t - kT),$$

where, for each integer k, b_k is $+1$ or -1. The filter output is sampled at time nT for each integer n. Express your answers to parts (e)–(j) of this problem in terms of the appropriate function and parameters a, b, T, σ, and A.

(e) What is the *maximum* probability of error (maximized over all pulse patterns)?

(f) Give a block diagram of a tapped-delay-line equalizer with *four* taps for this communication system. The tap gains are λ_1, λ_2, λ_3, and λ_4, in that order, where λ_1 corresponds to the tap with the least delay and λ_4 corresponds to the tap with the greatest delay.

(g) Specify the tap gains for the four-tap zero-forcing equalizer that minimizes the maximum intersymbol interference for the given filter and signal. Label your diagram with the proper values for the gains and delays.

(h) If the equalizer described in part (g) is inserted after the filter and before the sampler, what are the possible values for the sampled signal at the output of the equalizer?

(i) What is the impulse response of the combination of the filter followed by the equalizer if the tap gains are as described in part (g)? Give a sketch of this impulse response. Label all points on your sketch that are necessary to define the impulse response.

(j) Find the variance β^2 of the noise at the output of the equalizer. Again assume that the tap gains are as described in part (g).

(k) What is the *maximum* probability of error (maximized over all pulse patterns) for the system with the equalizer described in part (g)? Express your answer in terms of the appropriate function and parameters a, b, T, β, and A.

8.9 The model for a baseband communication system is as shown in Figure 8-2. The noise $X(t)$ is white Gaussian noise with unknown spectral density. It is known that the noise $\tilde{X}(t)$ at the filter output has autocorrelation function $R_{\tilde{X}}(\tau) = 6 \exp(-3|\tau|)$. If $d(t) = p_T(t)$ and there is no noise in the system, it is observed that $Z(T_0 - T) = 0$, $Z(T_0) = +1$, $Z(T_0 + T) = +6$, and $Z(T_0 + kT) = 0$ for $|k| > 1$.

(a) Suppose that $d(t)$ is $\pm p_T(t)$ and the objective is to make the minimax decision as to whether the pulse is positive or negative. What is the optimum sampling time T_s among the times of the form $T_s = T_0 + kT$ for some integer k?

(b) What is the probability of error for the reception of a single pulse using the sampling time you chose in part (a)? Express your answer in terms of the function Q.

(c) Give an expression for the average probability of error for the system if

$$d(t) = \sum_{n=-\infty}^{\infty} b_n \, p_T(t - nT)$$

and (b_n) is a sequence of independent random variables with $P(b_n = +1) = P(b_n = -1) = 1/2$ for each n. Express your answer in terms of the function Q and the parameter T.

(d) Give a block diagram for a three-tap linear transversal filter to insert after the filter and before the sampler. Specify the tap gains that minimize the maximum intersymbol interference. Label the diagram with the values for the gains and delays, and specify the way that the decision is made in the threshold device.

(e) Find the variance β^2 of the noise at the output of the equalizer of part **(d)**.

(f) What is the average probability of error (averaged over all pulse patterns) for the system with the equalizer described in part **(g)**? Express your answer in terms of the function Q and the parameter β only.

8.10 The model for a certain multipath channel is illustrated in Figure 8-34. The reflected path has a propagation time that is τ seconds longer than the direct path, and the gain of the reflected path is α. (Assume that $0 < \alpha < 1$.)

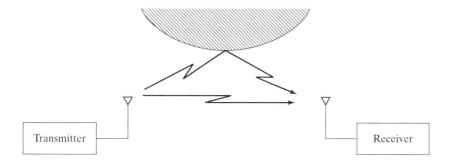

Figure 8-34: The model for the multipath channel.

Thus, if the transmitted signal is $s(t)$, the signal at the input to the receiver is

$$Y(t) = s(t) + \alpha \, s(t - \tau) + X(t),$$

where $X(t)$ is a white Gaussian noise process with spectral density $N_0/2$. The receiver is as illustrated in Figure 8-35. For receiving the kth symbol, the receiver integrates the received waveform during the interval from kT to $(k + 1)T$, and the decision is based on the comparison of the output of the integrator with zero (the usual minimax decision rule).

(a) The transmitted signal is a sequence of pulses of the form

$$s(t) = A \sum_{n=-\infty}^{\infty} b_n \, p_T(t - nT),$$

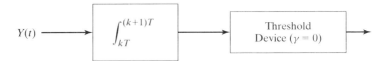

Figure 8-35: The receiver.

where, for each n, b_n is $+1$ or -1. What is the maximum probability of error (maximized over all possible pulse patterns) if $\tau = 3T/4$? Express your answer in terms of the parameters A, T, α, and N_0.

(b) Repeat part **(a)** for $\tau = T$.

(c) Suppose that the transmitted signal is as in part **(a)**, and (b_n) is a sequence of independent random variables with $P(b_n = +1) = P(b_n = -1) = 1/2$ for each n. What is the average probability of error (averaged over all pulse patterns) if $\tau = 3T/4$?

(d) Repeat part **(c)** for $\tau = T$.

(e) Give a block diagram of a tapped-delay-line equalizer with three taps for this communication system if the path delay on the multipath channel satisfies $\tau = T$. Specify the tap gains for a zero-forcing equalizer if the sampling time is T, and *label your diagram* with the values for the gains.

(f) What is the average probability of error for the receiver with the zero-forcing equalizer if $\tau = T$ (i.e., repeat part **(d)** for the receiver with the equalizer). Express your answer in terms of the parameters A, T, α, and N_0.

8.11 For the system described in Problem 8.3, we wish to use a three-tap transversal filter to provide equalization. Suppose that the center tap has unit gain and the other taps have gain $1/6$. Find the average probability of error for the system in Problem 8.3 if this transversal filter is employed.

8.12 Apply equalization to the system described in Problem 8.5.

(a) Specify the tap gain coefficients for a three-tap zero-forcing equalizer.

(b) What is the average probability of error for the receiver with this equalizer? Use the tap gain coefficients you obtained in part **(a)**.

8.13 Consider a tapped-delay-line equalizer with three taps for the communication system of Problem 8.6.

(a) Give a block diagram for the equalizer, specify the tap gains for a zero-forcing equalizer if the sampling time is T, and label your diagram with the proper numerical values for the gains.

(b) Repeat part **(g)** of Problem 8.6 for the receiver with the zero-forcing equalizer. Express your answer in terms of the parameters A and T and the appropriate functions, including the function R_W, the autocorrelation function for the noise $W(t)$ at the filter output.

8.14 Prove that the zero-forcing tap gains derived in the solution to Exercise 8-6 give the minimum peak distortion for any tap gains. *Hint*: Use symmetry to conclude that $\lambda_1 = \lambda_3$, so the peak distortion can be expressed as a function of a single parameter. Show that the zero-forcing method gives the optimum choice for this parameter.

8.15 Consider Exercise 8-9, but let $A/\rho = 1$. Find the maximum probability of error and the average probability of error for the original system and the system with the equalizer.

Chapter 9

Spread-Spectrum Communications

Spread-spectrum modulation is one of many signaling methods for which there is a tradeoff between bandwidth and performance. For most other such signaling methods, the tradeoff is between the bandwidth and the probability of error for a channel in which the major disturbance is thermal noise. For example, M-ary orthogonal signaling provides such a tradeoff (Sections 6.6 and 7.7). By increasing M, the number of signals in the orthogonal signal set, the bandwidth requirement increases. If the only interference is white Gaussian noise of spectral density $N_0/2$ and if $\mathcal{E}_b/N_0 > \ln(2)$, then the error probability decreases to zero as M increases (e.g., See Section 6.6.3).

For spread-spectrum modulation, the performance gains are for interference sources other than thermal noise. In fact, classical spread-spectrum modulation provides no performance improvement against thermal noise. The advantages of spread spectrum result from its ability to discriminate against narrowband interference, multipath interference, multiple-access interference, and other types of structured interference that arise in radio-frequency (RF) communication channels. Spread spectrum also provides some advantages for wireline, optical, and other communication media, particularly if several transmitters have simultaneous access to the same frequency band.

In addition to protection against various types of interference, spread-spectrum signals are difficult for unauthorized receivers to demodulate or even to detect. This feature makes spread spectrum desirable for military systems and for civilian communication services that wish to offer a certain degree of privacy. Spread-spectrum modulation also distributes the transmitted power across a wide range of frequencies in a way that permits coexistence with other forms of communications that share the frequency band with the spread-spectrum system. Some of the advantages can be illustrated conveniently if the spread-spectrum signal is modeled by a random process, and that is the approach taken in the next section.

9.1   Random Signal Model for Spread Spectrum

The modeling of a spread-spectrum signal as a random signal provides a simple frame-work in which to illustrate the basic concepts of spread-spectrum communications, especially the form of spread spectrum known as direct-sequence (DS) spread spectrum. In the random signal model, the spectral spreading signals are random processes. In addition to providing an illustration of the benefits of spread spectrum, the modeling of spread-spectrum signals by random processes has an historical connection to the early development of spread spectrum [9.44]. Much of the early work on spread spectrum relates to the noise-like properties of DS spread-spectrum signals, and the sequences employed in DS spread-spectrum systems are sometimes referred to as pseudonoise sequences, even in current literature.

In the analysis of the random signal model, no attempt is made to be mathematically precise, because random processes of the type used in the illustrations are not of interest in actual implementations of spread spectrum. The analysis presented in this section is only for the purpose of developing a physical insight into the basic concepts of spread-spectrum communications [9.30]. In later sections, more precise analyses are presented, and these analyses are for models that are suitable for implementation. In a DS spread-spectrum signal, each data symbol is represented by a sequence of N elemental pulses, which are referred to as *chips*. If T is the symbol duration, we let $T_c = T/N$. In the most common example, the chip waveform is a rectangular pulse of duration T_c; consequently, T_c is often referred to as the *chip duration*. In general, however, the chip waveform can have any shape, and it need not be time limited.

The randomness in a spread-spectrum signal is primarily in its baseband spectral spreading signal. The spectral spreading signal is independent of the data to be transmitted, and its function is to spread the data signal over a wide range of frequencies. In this section, the spectral spreading signal is modeled as a random process, denoted by $V(t)$. It is convenient for our illustration to let $V(t)$ be a zero-mean, stationary random process with autocorrelation function $R_V(\tau)$ given by

$$R_V(\tau) = \begin{cases} (T_c - |\tau|)/T_c, & |\tau| \le T_c, \\ 0, & |\tau| > T_c. \end{cases} \tag{9.1}$$

The actual shape of the autocorrelation function is not particularly important, however. All that is required is that $R_V(\tau) \approx 0$ for $\tau > T_c$ and $R_V(0) = 1$.

There are a number of different random signal models that can be used to illustrate the features of DS spread-spectrum communications. The essential requirement is that certain time averages can be accurately approximated by the corresponding probabilistic averages (i.e., the random process must be ergodic in the appropriate sense). If $V(t)$ is a Gaussian random process, for example, (9.1) is enough to guarantee that $V(t)$ is ergodic. But certain other random processes are of even greater interest for modeling DS spread-spectrum signals. For example, a more accurate random signal model for DS spread spectrum is a sequence of pulses with random amplitudes and a suitable random time offset, such as

$$V(t) = \sum_{n=-\infty}^{\infty} A_n \, \psi(t - nT_c - U), \qquad (9.2)$$

where $\psi(t)$ is the chip waveform, $\{A_n : -\infty < n < \infty\}$ is a sequence of independent random variables with $P(A_n = +1) = P(A_n = -1) = 1/2$ for each n, and U is a random variable that is uniformly distributed on the interval $[0, T_c]$ and independent of $\{A_n : -\infty < n < \infty\}$. If $\psi(t)$ is the rectangular pulse of duration T_c, the resulting autocorrelation function is given by (9.1).

The power spectral density for a random process $V(t)$ that has the autocorrelation function of (9.1) is

$$S_V(f) = T_c \, [\, \text{sinc}(f T_c) \,]^2. \qquad (9.3)$$

where $\text{sinc}(x) = \sin(\pi x)/(\pi x)$ for $x \neq 0$ and $\text{sinc}(0) = 1$. For purposes of this illustrative discussion, the *bandwidth* of a random process is defined as the first zero of its spectral density. More precisely, the bandwidth B of a baseband random process $V(t)$ is the smallest positive value of f for which $S_V(f) = 0$. For a random process that has an autocorrelation function given by (9.1) and a spectral density given by (9.3), the bandwidth is $B = 1/T_c$.

Let $d(t)$ be a binary data signal that consists of a sequence of rectangular pulses, each of which is of duration T and amplitude $\pm A$ ($A > 0$). Let $p_T(t)$ denote the unit-amplitude, rectangular pulse of duration T that begins at $t = 0$. Define

$$u_i(t) = (-1)^i \, A \, p_T(t)$$

for $i = 0$ and $i = 1$; that is, $u_0(t) = +A \, p_T(t)$ and $u_1(t) = -A \, p_T(t)$. Either

$$d(t) = u_0(t - nT)$$

or

$$d(t) = u_1(t - nT),$$

in the nth signaling interval, $nT \leq t < (n+1)T$, depending on which binary digit, 0 or 1, is being sent over the channel in that interval. Analogous to the definition of the bandwidth of $V(t)$, the bandwidth of the data signal is defined as the first zero of the Fourier transform of $p_T(t)$, which is $1/T$.

The value of N is typically in the range of tens or hundreds for civilian systems, and it may be much larger for military systems. It follows from $N \gg 1$ that $B \gg 1/T$: The bandwidth of $V(t)$ is much greater than the bandwidth of the data signal $d(t)$. The transmitted signal is $V(t) \, d(t)$, which for most applications has bandwidth approximately equal to the bandwidth of $V(t)$.

The channel model that is employed in this section is a baseband channel that has no thermal noise, but various other types of interference are present. This channel model is sufficient to illustrate several of the most important benefits of spread spectrum. The

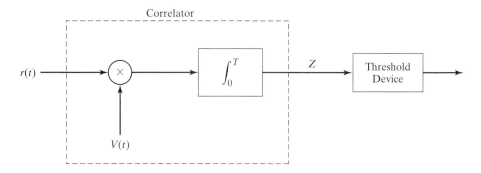

Figure 9-1: Correlation receiver for $V(t) d(t)$.

baseband receiver is a correlator that multiplies the received signal $r(t)$ by $V(t)$ and then integrates the product, as illustrated in Figure 9-1.

For the data pulse transmitted during the 0th signaling interval, the decision statistic is

$$Z = \int_0^T V(t) r(t) \, dt.$$

The decision to be made is whether $d(t) = A$ or $d(t) = -A$ during a given signaling interval. If $Z > 0$, the receiver decides that a positive pulse was transmitted in the interval $[0, T]$; if $Z \leq 0$, the receiver decides that a negative pulse was transmitted. In the absence of interference, the received signal is $r(t) = V(t) d(t)$, and so the decision statistic is

$$Z = \int_0^T V^2(t) d(t) \, dt.$$

Denote by m the binary digit that is transmitted in the 0th signaling interval. The decision statistic for the 0th interval is given by

$$Z = (-1)^m A \int_0^T V^2(t) \, dt$$

for $m = 0$ or $m = 1$.

If $V(t)$ is ergodic in the appropriate sense, then, with high probability,

$$T^{-1} \int_0^T V^2(t) \, dt \approx E\{V^2(t)\} = R_V(0) = 1,$$

because the integration time is long compared with the inverse of the bandwidth of the random process $V(t)$ (i.e., $T \gg T_c$). As a result, if the binary digit m is sent over the channel, the decision statistic satisfies

$$Z \approx (-1)^m A T. \tag{9.4}$$

It follows that if $m = 0$, then $Z > 0$, and the receiver correctly decides that a positive pulse was sent. If $m = 1$, then $Z \leq 0$, and the receiver correctly decides that a negative pulse was sent. Thus, in the absence of any interference in the channel, the transmitted information is recovered from $V(t) d(t)$.

Notice that the receiver must know the random process $V(t)$ exactly, which is the reason that this conceptual model is not of interest for practical implementations. Many of the early attempts to develop spread-spectrum systems involved the use of a noise signal as a spectral spreading signal [9.44]. In such an approach, it is necessary to have the same noise signal at the transmitter and receiver, and several mechanisms were invented to accomplish this. In practical implementations, the characteristics of $V(t)$ are approximated by a noise-like or pseudonoise waveform. For instance, many modern spread-spectrum systems use noise-like signals that are derived from shift-register sequences (e.g., see [9.14], [9.15], and [9.43]). Such signals give the necessary randomness, yet they permit duplication of the spectral spreading signal at the receiver.

The analysis leading to (9.4) does not account for the possibility of interference in the channel. Such interference can result from authorized transmitters that operate in the same frequency band (e.g., in a multiple-access spread-spectrum system) or other systems that emit energy in that frequency band (e.g., intentional jamming). The effects of these and other types of interfering signals can be illustrated by use of the random signal model.

9.1.1   Multiple-Access Capability

Suppose that there are K transmitted signals, and each is of the form described in the previous section. For simplicity, assume that the K signals are synchronous in the sense that the data pulses for the different signals are aligned in time. The data signals are $d_k(t), 1 \leq k \leq K$, and each has the same form as the data signal $d(t)$ defined previously. Each transmitter is assigned a spectral spreading signal $V_k(t), 1 \leq k \leq K$, and each spectral spreading signal is modeled as a zero-mean, stationary, random process with autocorrelation function $R_V(\tau)$. The processes are mutually independent. The received signal is now

$$r(t) = \sum_{k=1}^{K} V_k(t) d_k(t).$$

Under the assumption that the binary digit m is sent by the ith transmitter during the 0th signaling interval, the output of the ith correlation receiver is given by

$$Z_i \approx (-1)^m A T + \sum_{k \neq i} b_k A \int_0^1 V_k(t) V_i(t) dt,$$

where b_k is $+1$ or -1 depending on whether a 0 or 1 is sent by the kth transmitter in the 0th signaling interval.

Because $V_k(t)$ and $V_i(t)$ are independent, zero-mean, random processes for $k \neq i$, then

$$T^{-1} \int_0^T V_k(t)\, V_i(t)\, dt \approx E\{V_k(t)\, V_i(t)\} = 0, \tag{9.5}$$

provided that the random processes $V_k(t)$ and $V_i(t)$ are jointly ergodic in the appropriate sense. It follows that the decision statistic satisfies

$$Z \approx (-1)^m\, A\, T,$$

which is the same as (9.4), even though other signals are now sharing the frequency band. To within the accuracy of the approximation given in (9.5), the other signals in the spread-spectrum system do not interfere with the transmission of the ith signal. This demonstrates the multiple-access capability of spread spectrum.

9.1.2   Interference Rejection Capability

Suppose that the received signal is

$$r(t) = V(t)\, d(t) + A';$$

that is, a DC signal has been added to the desired signal. The DC signal level A' can be either positive or negative. A DC signal in a baseband model corresponds to a sinusoidal signal with frequency equal to the carrier frequency in an RF system, and so this is an appropriate baseband model for an interference tone at the carrier frequency. If the binary digit m is sent during the 0th interval, the output of the correlation receiver is

$$Z \approx (-1)^m\, A\, T + A' \int_0^T V(t)\, dt. \tag{9.6}$$

The second term of the right-hand side of (9.6) represents the interference component at the output of the correlator. Assuming that it is valid to equate the time average and the expectation, we see that

$$T^{-1} \int_0^T V(t)\, dt \approx 0$$

because $V(t)$ has zero mean and T is much larger than the inverse of the bandwidth of $V(t)$. The interference component in (9.6) is negligible if

$$(A'/A) \left\{ T^{-1} \int_0^T V(t)\, dt \right\} \approx 0,$$

which is true provided that A' is not much larger than A. This demonstrates the DC interference rejection capability of a baseband spread-spectrum system. Sinusoidal interference rejection can be demonstrated via a similar argument that uses

$$T^{-1} \int_0^T V(t) \cos(2\pi f_0 t)\, dt \approx 0.$$

Sinusoidal interference at frequency f_0 in the baseband model corresponds to sinusoidal interference at frequency $f_c + f_0$ or $f_c - f_0$ in an RF system with carrier frequency f_c.

9.1.3   Anti-Multipath Capability

In a simplified baseband model of a multipath channel, the received signal might be

$$r(t) = V(t)\,d(t) - \beta\,V(t - \tau)\,d(t - \tau), \tag{9.7}$$

where β is in the range $0 \le \beta \le 1$. The second term on the right-hand side of (9.7) represents a multipath component that arrives with a delay of τ relative to the primary component. If $\tau \ll T$ and $d(t) = +A$ for $0 \le t < T$, the decision statistic is

$$Z \approx A\,T - A\,\beta \int_0^T V(t - \tau)\,V(t)\,dt \approx A\,T\{1 - \beta\,R_V(\tau)\}.$$

Recall that $R_V(\tau) = 0$ if $\tau \ge T_c$ so the multipath component does not interfere with the primary component if the delay in the multipath component is T_c or greater.

9.1.4   Time- and Frequency-Domain Interpretations of Despreading

Just as the modulation process in spread-spectrum communications is often referred to as *spreading* the data signal, the demodulation process is often referred to as *despreading*. For each of the types of interference considered earlier, the key feature is that the interfering signal has small correlation with $V(t)$. As a result, the interfering signal produces only a negligible change in the output of the correlator of Figure 9-1. The integrator in the correlation receiver tends to smooth and "average out" the interfering signals.

The effect of the correlator on the desired signal is quite different. It produces a large peak in response to the desired signal, because it is matched to the shape of the rectangular data pulse. The output of the integrator is equivalent to an appropriate sample of the output of a time-invariant linear filter with a rectangular impulse response (Chapter 5). Thus, from a time-domain point of view, the correlation receiver (or its equivalent matched-filter receiver) averages out the interference while enhancing the desired signal, thereby increasing the signal-to-interference ratio.

As viewed in the frequency domain, the multiplication by the local replica of the spectral spreading signal despreads (i.e., reduces the bandwidth of) the desired signal at the same time that it spreads the interference. The receiver "strips off" the spectral spreading signal from the data signal by multiplying the received signal by $V(t)$. This is particularly easy to see if the spectral spreading signal is given by (9.2) and the chip waveform is rectangular, because for this model $V^2(t) = 1$ for all t. Thus, the multiplication of the spread spectrum signal $V(t)\,d(t)$ by $V(t)$ produces

$$V^2(t)\,d(t) = d(t),$$

which is the data signal, and it has a much narrower bandwidth than the original spread-spectrum signal.

After the multiplication process, the receiver passes the data signal through the integrator of Figure 9-1, and this integration process has a bandwidth that is much smaller than B, the bandwidth of the spread-spectrum signal, but approximately the same as $d(t)$, the data signal. On the other hand, the result of the multiplication process on the interference signal is to increase the bandwidth of the interference. The product of the interfering signal and $V(t)$ has a bandwidth that is at least as large as B. The higher frequency components of this product signal are then filtered out by the integrator. The despreading process for the communication signal actually spreads the spectrum of the interference, and the subsequent narrowband filtering effect of the integrator therefore reduces the power in the interference.

9.1.5   Identification of Linear Systems

For many of the beneficial features of spread spectrum, the ideal choice for the transmitter's spectral spreading signal $V(t)$ would be white noise if it were not for the extreme difficulty (impossibility in most situations) of reproducing the noise waveform at the receiver. However, implementation issues are not of concern in this section, since the objective is to develop an intuitive understanding of the benefits of spread spectrum. A white-noise spreading signal is a good test signal for linear system identification (i.e., estimation of the unknown impulse response of a linear system [9.3]). The spreading signal serves as the input signal, and standard methods for estimation of the cross-correlation function for the input and output signals give an estimate of the impulse response. (See Section 3.6 and Example 3-6 in Section 3.7.)

Consider a white-noise random process $X(t)$ with autocorrelation function $R_X(\tau) = \delta(\tau)$, a Dirac delta function centered at the origin. Notice that if $R_V(\tau)$ is given by (9.1), the autocorrelation function $T_c^{-1} R_V(\tau)$ approximates $R_X(\tau)$ if T_c is very small.

Suppose that $X(t)$ is the input to a time-invariant linear system with unknown impulse response $h(t)$. The goal is to estimate this impulse response. Let $Y(t)$ denote the output of the system. The crosscorrelation function for the two random processes $X(t)$ and $Y(t)$ is

$$R_{Y.X}(\tau) = E\{Y(t+\tau)\,X(t)\}, \quad -\infty < \tau < \infty.$$

Because $Y(t)$ is the output of the linear system when $X(t)$ is the input, the cross-correlation function can be written as

$$R_{Y.X}(\tau) = \int_{-\infty}^{\infty} h(\tau - \alpha)\, R_X(\alpha)\, d\alpha.$$

If $R_X(\alpha) = \delta(\alpha)$, this result implies that

$$R_{Y.X}(\tau) = h(\tau), \quad -\infty < \tau < \infty.$$

Because the crosscorrelation function is equal to the unknown impulse response, the problem of estimating the impulse response can be solved by estimating the cross-correlation function for the random processes $X(t)$ and $Y(t)$. One system that provides an estimate for this crosscorrelation function is illustrated in Figure 9-2. The output of the integrator shown in Figure 9-2 is given by

$$Z(\tau) = T_0^{-1} \int_{\tau}^{T_0+\tau} Y(t)\, X(t-\tau)\, dt = T_0^{-1} \int_{0}^{T_0} Y(t+\tau)\, X(t)\, dt,$$

where τ is arbitrary, but held constant for the moment.

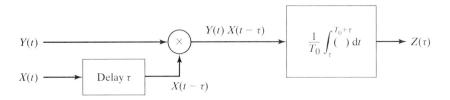

Figure 9-2: Estimation of the crosscorrelation function for $X(t)$ and $Y(t)$.

For sufficiently large values of T_0, $Z(\tau)$ is a good approximation to $R_{YX}(\tau)$. Notice that

$$E\{Z(\tau)\} = T_0^{-1} \int_0^{T_0} E\{Y(t+\tau)\,X(t)\}\,dt$$
$$= R_{YX}(\tau)$$

for any $T_0 > 0$. A commonly used measure of accuracy for estimation is the mean-squared error $E\{[Z(\tau) - R_{YX}(\tau)]^2\}$. For ergodic processes, the mean-squared error that results if $Z(\tau)$ is used to estimate $R_{YX}(\tau)$ converges to zero as $T_0 \to \infty$. See Section 3.6 for further discussion and analysis.

The value of τ is varied in order to estimate the impulse response $h(\tau)$ for the range of values of τ of interest for the particular system. One application of this approach to the estimation of an unknown impulse response is in mobile communications with fading channels [9.51], where it may be necessary to estimate the impulse response of the channel in order to perform multipath combining or equalization. In such an application, a wideband direct-sequence spread-spectrum signal is a good probe signal to transmit over the channel, because such a signal can be designed to approximate a white-noise random process.

9.2   Direct-Sequence Spread-Spectrum Signals

As illustrated in the previous section, spread spectrum can provide a means of reliable communication in the presence of various types of interference. If a correlation receiver is employed, the goal in the design of direct-sequence (DS) spread-spectrum signals is to make the interference orthogonal to the communication signal. In many systems, true orthogonality is impossible to achieve, and the best that can be done is to make the interference approximately orthogonal to the communication signal. The greater the departure from orthogonality, the more that can be gained by using a receiver with more capability than a correlation receiver.

One segment $\xi(t)$ of a typical baseband spectral spreading signal is illustrated in Figure 9-3. In this example, the segment is *time limited* to the interval $[0, T]$; that is, $\xi(t) = 0$ if $t < 0$ or $t \geq T$. Each of the seven shorter pulses in the segment is referred to as a *chip*. For the simple example shown in Figure 9-3, the number of chips per segment is seven, but in practice this number is typically much larger.

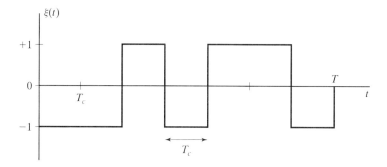

Figure 9-3: Baseband spectral spreading signal with seven chips.

If the waveform $\xi(t)$ is repeated every T seconds, a periodic signal $v(t)$ is generated. This signal, referred to as the *spectral spreading signal*, satisfies $v(t) = \xi(t - nT)$ for the nth signaling interval, $nT \leq t < (n+1)T$, and it can be represented as

$$v(t) = \sum_{n=-\infty}^{\infty} \xi(t - nT).$$

This last expression is a valid representation for a spectral spreading signal even if the original waveform $\xi(t)$ is not time limited to the interval $[0, T]$.

Notice that once the chip waveform is selected (in the present example it is rectangular and has duration T_c), the spectral spreading signal $v(t)$ is defined completely by specifying the sequence of pulse amplitudes for the chips in the time interval $[0, T]$. This sequence of pulse amplitudes is known as the *signature sequence* or *spreading sequence*. The former term is especially appropriate for multiple-access systems, because the sequence is used to identify the signal [9.29].

The signature sequence for the waveform of Figure 9-3 is $-1, -1, +1, -1, +1, +1, -1$. As an alternative to listing the sequence of polarities for the chips, it is common to use a binary notation in which the positive polarity is denoted by the binary 0 and the negative polarity denoted by the binary 1. In binary form, the signature sequence is written as 1101001.

If the sequence of chip amplitudes is denoted by $\mathbf{x} = (x_0, x_1, \ldots, x_{N-1})$, one segment of the spectral spreading signal is given by

$$\xi(t) = \sum_{i=0}^{N-1} x_i \, \psi(t - iT_c),$$

where $1/T_c$ is the chip rate and $\psi(t)$ is the chip waveform. Typical examples for the chip waveform are a *rectangular pulse of duration T_c*, such as

$$\psi(t) = 1, \quad 0 \leq t < T_c,$$

and a *sine pulse of duration T_c*, which could be given by

$$\psi(t) = \sin(\pi t / T_c), \quad 0 \leq t < T_c.$$

In order to have unit energy, $\psi(t)$ may be multiplied by an appropriate constant (e.g., $\sqrt{1/T_c}$ for a rectangular pulse or $\sqrt{2/T_c}$ for a sine pulse). For the example illustrated in Figure 9-3, the number of chips is $N = 7$, $\psi(t)$ is a unit-amplitude rectangular pulse of duration T_c, and the signature sequence is

$$\mathbf{x} = (-1, -1, +1, -1, +1, +1, -1).$$

A finite sequence $\mathbf{x} = (x_0, x_1, \ldots, x_{N-1})$ can be extended to an infinite periodic sequence by letting $x_{j+Nk} = x_j$ for each integer k and each integer j that is in the range $0 \le j \le N - 1$. We denote the periodic extension of $\mathbf{x}$ by (x_i), or, if explicit display of the index i is not necessary, we use the simpler notation x. The introduction of the infinite sequence permits us to write the spectral spreading signal as

$$v(t) = \sum_{i=-\infty}^{\infty} x_i \, \psi(t - iT_c). \tag{9.8}$$

This formulation is sufficiently general to include signature sequences that are not periodic; in fact, (x_i) can be any infinite sequence of real numbers. If (x_i) is periodic with period N, however, then $v(t)$ is periodic with period $T = NT_c$.

Let S be the *shift operator* on infinite sequences; specifically, for each sequence $x = (x_i)$, the infinite sequence Sx is the sequence $y = (y_i)$ defined by $y_i = x_{i+1}$ for $-\infty < i < \infty$. More generally, for each integer j, define $S^j x$ to be the sequence y for which $y_i = x_{i+j}$ for $-\infty < i < \infty$. The *period* of a sequence x is the smallest positive integer M for which $S^M x = x$. If the sequence x is the periodic extension of $\mathbf{x} = (x_0, x_1, \ldots, x_{N-1})$, then $S^N x = x$. Therefore, the period of the sequence x must be a divisor of N. For example, if N is even, the period of x could be $N/2$, as illustrated by the periodic extension of the vector

$$\mathbf{x} = (-1, +1, +1, -1, -1, +1, +1, -1),$$

which has length 8, but produces an infinite sequence of period 4. If

$$\mathbf{x} = (-1, +1, -1, +1, -1, +1, -1, +1,),$$

the period of x is 2, and if

$$\mathbf{x} = (+1, +1, -1, +1, -1, -1, -1, +1),$$

the period is 8. All three vectors have length 8.

If the sequence x is the periodic extension of $\mathbf{x} = (x_0, x_1, \ldots, x_{N-1})$, we refer to the vector $(x_0, x_1, \ldots, x_{N-1})$ as the *segment* of the infinite sequence x. The period of x cannot exceed the length of the segment. If the signature sequence is periodic, the spectral spreading waveform $\xi(t)$ represents one segment of the baseband spectral spreading signal $v(t)$, and the baseband spectral spreading signal is a periodic continuous-time signal that has period MT_c for some positive integer M that divides N.

Data can be communicated on the spectral spreading signal by superimposing amplitude modulation on $v(t)$. For purposes of illustration, consider a system with a data transmission rate of $1/T$ symbols per second, and suppose one data symbol is to be transmitted during each segment of the spectral spreading signal. If

$$\mathbf{b} = \ldots, b_{-1}, b_0, b_1, \ldots$$

is the sequence of data symbols to be sent to the receiver, the baseband spread-spectrum signal is

$$s(t) = A \sum_{n=-\infty}^{\infty} b_n \xi(t - nT),$$

where A is a fixed amplitude. Notice that if $\xi(t)$ is time limited to the interval $[0, T]$, the nth term of the sum is nonzero only if $nT \leq t < (n + 1)T$. This formulation corresponds to a spectral spreading signal for which the duration of a segment is equal to the data symbol duration, and the period of the spectral spreading signal cannot exceed the duration of the segment. For some applications, it is necessary to use a spectral spreading signal with a longer period, in which case an alternative description of the spread-spectrum signal is preferred.

To obtain a more general formulation, and to develop an illustration of the bandwidth spreading accomplished by DS spread-spectrum modulation, define the data signal $d(t)$ as

$$d(t) = \sum_{n=-\infty}^{\infty} b_n \, p_T(t - nT), \tag{9.9}$$

where p_T is the rectangular pulse function defined by

$$p_T(u) = \begin{cases} 1, & 0 \leq u < T, \\ 0, & \text{otherwise.} \end{cases}$$

The spread-spectrum signal can then be written as

$$s(t) = A \, v(t) \, d(t). \tag{9.10}$$

This formulation is valid even if the period of the spectral spreading signal is not the same as the duration T of the data pulse; in fact, the spectral spreading signal is not required to be periodic in this formulation.

The time- and frequency-domain relationships between the data signal and the resulting spread-spectrum signal are illustrated in Figure 9-4. In this figure, $D(f)$ is the Fourier transform of the data pulse $d(t) = p_T(t)$, and $S(f)$ is the Fourier transform of the chip waveform, which for this illustration is a rectangular pulse of duration T_c. Only the main lobe and the two adjacent side lobes of $D(f)$ and $S(f)$ are shown in Figure 9-4. For some purposes, it is convenient to model the signature sequence and data sequence as independent sequences of independent random variables, each of which takes the value $+1$ with probability $1/2$ and the value -1 with probability $1/2$. If this random sequence model is adopted, and if an appropriate random time offset is included

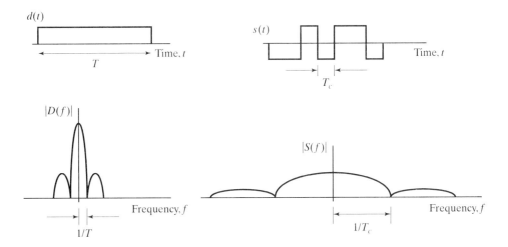

Figure 9-4: Illustration of bandwidth expansion in DS spread spectrum.

in the signals, the power spectral densities for the data and spread-spectrum signals are proportional to $|D(f)|^2$ and $|S(f)|^2$, respectively.

Notice that the first null in $D(f)$ is at $1/T$, and the spread-spectrum signal has its first null at $1/T_c$. The bandwidth of a DS spread-spectrum signal with N chips per data pulse is therefore N times the bandwidth of the data signal. Hence, the spectrum of the data signal is spread by a factor of N when it is multiplied by a spectral spreading signal.

For transmission on RF or optical channels, it is necessary that the baseband spread-spectrum signal be modulated onto an appropriate sinusoidal carrier. Because both inphase and quadrature modulation can be employed, two baseband DS spread-spectrum signals can be incorporated into a single DS spread-spectrum carrier modulated signal. One form of DS spread-spectrum signal is given by

$$s(t) = A\, v_1(t)\, d_1(t) \cos(\omega_c t + \varphi) - A\, v_2(t)\, d_2(t) \sin(\omega_c t + \varphi), \qquad (9.11)$$

where $v_1(t)$ and $v_2(t)$ are baseband spectral spreading signals of the type defined in (9.8), and $d_1(t)$ and $d_2(t)$ are data signals of the type defined in (9.9). The baseband spectral spreading signals are obtained from two possibly different signature sequences, but we assume that they employ the same chip waveform $\psi(t)$ and therefore have the same chip duration. If the chip waveform is a rectangular pulse of duration T_c, the signal defined in (9.11) is a QPSK DS spread-spectrum signal. A time offset of $t_0 = T_c/2$ can be introduced in the inphase component to give an offset QPSK DS spread-spectrum signal. If this time offset is employed and the chip waveform is the sine pulse, the resulting signal represents MSK DS spread spectrum.

One example of DS spread spectrum is the low-cost packet radio (LPR) developed with the support of the Defense Advanced Research Projects Agency (DARPA) and the U.S. Army Communications and Electronics Command (CECOM). A description of this radio and a block diagram of the demodulator are given in [9.7]. Some of the most important features are as follows. The chip rate is 12.8×10^6 chips per second,

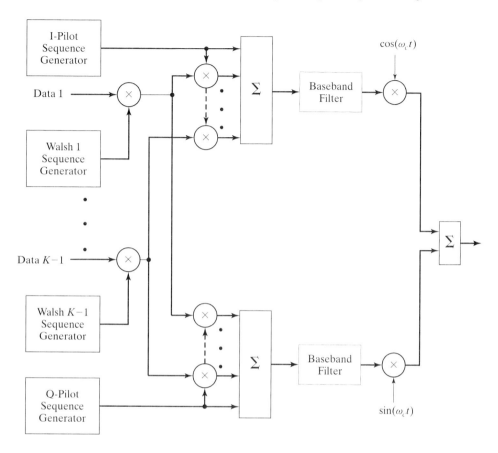

Figure 9-5: Base station transmitted signal generation for DS spread-spectrum mobile cellular communications.

and the spread bandwidth is 20 MHz. The DS spread-spectrum modulation is MSK modulation, so the chip waveform $\psi(t)$ is a sine pulse. The RF signal is transmitted in one of twenty channels that span a frequency band from approximately 1.718 GHz to approximately 1.840 GHz. The LPR can transmit the encoded data at one of two transmission rates: 100 kbps and 400 kbps. The number of chips per bit is 128 at a transmission rate of 100 kbps and 32 at a transmission rate of 400 kbps.

A second example of DS spread spectrum is the wideband spread-spectrum cellular system known as the mobile cellular code-division multiple-access (CDMA) system. The technical requirements for the mobile cellular CDMA system are given in [9.52], which is the documentation for the standard known as TIA-95. Additional information on this system can be found in [9.40] and [9.58].

The generation of the composite spread-spectrum signal for the *forward link* of the TIA-95 system, which is the communication link from the base station to the mobile stations, is illustrated in Figure 9-5. The maximum value of K is 64 in TIA-95, which corresponds to one pilot channel and 63 channels for data, paging, and synchronization.

574   Chapter 9   Spread-Spectrum Communications

For simplicity, we refer to all forward-link channels other than the pilot channel as data channels. Also, no scrambling operation is shown in Figure 9-5. If scrambling is employed in a particular data channel, the corresponding data signal in Figure 9-5 should be interpreted as a scrambled data signal.

The pilot channel consists of a spectral spreading signal with no other modulation. The spreading sequence for the inphase component of the pilot signal is the *inphase pilot sequence* (I-pilot sequence), and the spreading sequence for the quadrature component of the pilot signal is the *quadrature pilot sequence* (Q-pilot sequence). The pilot signal provides the mobile stations with a convenient reference for use in acquisition, tracking, and demodulation of the other forward-link signals.

If the baseband spectral spreading signal derived from the I-pilot sequence is denoted by $x_1(t)$ and the baseband spectral spreading signal derived from the Q-pilot sequence is denoted by $x_2(t)$, then the RF pilot signal can be expressed as

$$s_0(t) = A_0\{x_1(t)\cos(\omega_c t) + x_2(t)\sin(\omega_c t)\}. \tag{9.12}$$

(The plus sign in (9.12) is consistent with [9.52].) If $b_k(t)$ represents the kth baseband data signal ($1 \le k \le K - 1$), then the kth RF data signal is given by

$$s_k(t) = A_k\, w_k(t)\, b_k(t)\,\{x_1(t)\cos(\omega_c t) + x_2(t)\sin(\omega_c t)\}, \tag{9.13}$$

where $w_k(t)$ is a baseband spectral spreading signal that is obtained from a special signature sequence that is referred to as a Walsh function in TIA-95 [9.52]. The Walsh functions of TIA-95 are just the rows of the appropriate 64×64 Hadamard matrix (described in Chapter 6), and so any two of them are orthogonal. The amplitudes $A_0, A_1, \ldots, A_{K-1}$ can be adjusted to give the desired amount of power in the data signals relative to the pilot signal.

The description of the kth data signal in (9.13) can be cast in the general form of (9.11) by letting the baseband spectral spreading signals be defined by

$$v_1(t) = w_k(t)\, x_1(t)$$

and

$$v_2(t) = -w_k(t)\, x_2(t).$$

The baseband data signals of (9.11) are defined by $d_1(t) = d_2(t) = b_k(t)$. There is only one baseband data signal for each RF data signal in the TIA-95 forward link.

The signals $s_0(t)$ and $s_k(t)$ of (9.12) and (9.13) can be written in a common form by defining the spectral spreading signals $g_{0,1}(t) = x_1(t)$ and $g_{0,2}(t) = x_2(t)$ for the pilot signal and $g_{k,1}(t) = w_k(t)\, x_1(t)$ and $g_{k,2}(t) = w_k(t)\, x_2(t)$ for the data signals. The pilot signal can then be written as

$$s_0(t) = A_0\{g_{0,1}(t)\cos(\omega_c t) + g_{0,2}(t)\sin(\omega_c t)\},$$

and the kth RF data signal is given by

$$s_k(t) = A_k\, b_k(t)\,\{g_{k,1}(t)\cos(\omega_c t) + g_{k,2}(t)\sin(\omega_c t)\}.$$

The transmitted RF signal on the forward link is

$$s(t) = \sum_{k=0}^{K-1} s_k(t).$$

If we define $b_0(t) = 1$ for all t, the forward-link transmitted signal can be expressed as

$$s(t) = \sum_{k=0}^{K-1} A_k \, b_k(t)\{g_{k,1}(t) \cos(\omega_0 t) + g_{k,2}(t) \sin(\omega_0 t)\}.$$

We let $b_0(t) = 1$ for all t to reflect the absence of data in the pilot signal.

The signals for the reverse link (mobile to base station) of the TIA-95 cellular system are derived from a 64×64 Hadamard matrix in the manner described in Section 6.6.1, except the chip waveforms are not rectangular. Thus, the data waveforms form a set of 64 orthogonal signals that correspond to the rows of the Hadamard matrix. The 64 data waveforms, each of which is a different sequence of 64 chips, are referred to as 64-ary Walsh signals in the TIA-95 documentation [9.52]. A signature sequence is applied to the data waveform at a rate of four chips per waveform chip.

A somewhat similar scheme was employed much earlier [9.42] in the Joint Tactical Information Distribution System (JTIDS). In JTIDS, the data waveforms are derived from the cyclic shifts of a sequence of length 32 in the manner described in Section 6.6.1 (e.g., see [9.4] or [9.20]). In the literature on JTIDS, the modulation used to obtain data waveforms is referred to as cyclic code-shift key (CCSK). The 32-chip sequence is not an m-sequence, of course, since 32 is not one less than a power of two. In fact, the set of data waveforms for JTIDS is neither an orthogonal set nor a simplex set, but the signal design for JTIDS permits a simpler demodulator than a system with 32 parallel correlators. (JTIDS can use a demodulation scheme similar to pulse-position demodulation.)

Another example of DS spread spectrum is the Global Positioning System (GPS), which consists of 24 satellites in six orbital planes and a ground control station [9.6, 9.12]. Each satellite transmits two DS spread-spectrum navigation signals. The first is an I–Q modulated DS spread-spectrum signal with carrier frequency $f_1 = 1575.42$ MHz. Two sequences are employed to spread the first navigation signal. One of the sequences is referred to as the *clear/acquisition* sequence by some authors and the *coarse acquisition* sequence by others. Fortunately, either can be abbreviated as *C/A sequence*, which is the most common designation in the literature. The second sequence also has two names: *precision* sequence and *protected* sequence. Each is abbreviated as *P sequence*.

Each satellite transmits a unique C/A sequence and a unique P sequence. The spreading sequences for the C/A components that are transmitted by different satellites are different sequences from a set of Gold sequences of period 1023 [9.43]. The P sequences are very long segments of a single sequence that has a period in excess of 38 weeks [9.49]. Each P sequence runs for one week without repeating, and it is reset at the end of each week.

If the spreading signal corresponding to the P sequence is denoted by $p(t)$ and the spreading signal corresponding to the C/A sequence is denoted by $x(t)$, then the first navigation signal transmitted by a particular satellite can be expressed as

$$s_1(t) = \sqrt{2} \, A_0 \, d(t) \, p(t) \cos(2\pi f_1 t + \theta_1) + \sqrt{2} \, A_1 \, d(t) \, x(t) \sin(2\pi f_1 t + \theta_1).$$

The baseband signal $d(t)$ represents a sequence of rectangular pulses that convey the navigation data from that satellite. The data rate is 50 bits per second.

The P component of the first navigation signal is

$$\sqrt{2}\, A_0\, d(t)\, p(t) \cos(2\pi f_1 t + \theta_1),$$

and the C/A component is

$$\sqrt{2}\, A_1\, d(t)\, x(t) \sin(2\pi f_1 t + \theta_1).$$

The power in the P component is half that in C/A component; that is, $A_0^2 = A_1^2/2$. As a result, the first navigation signal can be viewed as an imbalanced QASK spread-spectrum signal or, equivalently, two unequal-power BPSK spread-spectrum signals on quadrature carriers. The chip rate for the C/A spreading signal is 1.023 Mc/s, and the chip rate for the P spreading signal is 10.23 Mc/s. Thus, the bandwidth of the P component is ten times the bandwidth of the C/A component. Because the C/A spreading signal has chip rate 1.023 Mc/s, and the period of the C/A sequence is 1023, the waveform $x(t)$ repeats every 1 ms. The data rate of 50 bps corresponds to a data symbol duration of 20 ms, so there are 20 periods of the C/A spreading sequence per data pulse.

In order to reduce the range measurement errors that are caused by ionospheric refraction, a second navigation signal is transmitted at a different frequency than the first. The second navigation signal has carrier frequency $f_2 = 1227.60$ MHz, and it is a standard BPSK direct-sequence spread-spectrum signal that carries the P sequence only. It can be expressed as

$$s_2(t) = \sqrt{2}\, A_2\, d(t)\, p(t) \sin(2\pi f_2 t + \theta_2).$$

The power in the second navigation signal is 7.9 dB below the power in the C/A component of the first navigation signal [9.6].

All GPS receivers have access to the C/A component of the first signal, and even those receivers that have access to the P component typically use the C/A component to acquire an initial timing reference (i.e., coarse acquisition). The P sequence may not be available to all receivers at all times, since it may be switched to a more secure sequence by the U.S. Department of Defense. For this and other reasons, most commercial GPS receivers rely on the C/A component only. The P component is present in the navigation signals primarily for military receivers that require greater accuracy, more security, and better protection against jamming and incidental interference.

9.3 Sequences and Their Periodic Correlation Functions

A baseband spectral spreading signal $v(t)$ is a continuous-time waveform that is defined completely by its chip waveform $\psi(t)$ and its signature sequence (x_i). The mathematical relationship is given by

$$v(t) = \sum_{i=-\infty}^{\infty} x_i \, \psi(t - iT_c). \tag{9.14}$$

If $v(t)$ has period T, its continuous-time periodic autocorrelation function is

$$r_v(\tau) = \int_0^T v(t) \, v(t + \tau) \, dt. \tag{9.15}$$

Because $v(t)$ has period T, the only requirement on the limits for this integral is that the difference between the upper and lower limits must be T. It is often convenient to normalize this autocorrelation function, so we define $R_v(\tau) = r_v(\tau)/T$ for all τ.

The integral in the above definition of the continuous-time periodic autocorrelation function is referred to as a *correlation*. It is very similar to a convolution in many respects. In fact, the only difference between the two is that the integrand in the correlation is $v(t)\,v(t + \tau)$, whereas the integrand is $v(t)\,v(\tau - t)$ for a convolution. We make use of this similarity in determining the graph of the autocorrelation function. The first step is to examine $r_v(\tau)$ for values of τ of the form jT_c. Toward this end, we introduce the discrete autocorrelation function.

Recall that the *inner product* of two vectors $\mathbf{x} = (x_0, x_1, \ldots, x_{N-1})$ and $\mathbf{y} = (y_0, y_1, \ldots, y_{N-1})$ of length N is defined by

$$(\mathbf{x}, \mathbf{y}) = \sum_{n=0}^{N-1} x_n \, y_n.$$

The *discrete periodic autocorrelation function* for a sequence $x = (x_i)$ of period N is defined by

$$\theta_x(j) = \sum_{n=0}^{N-1} x_n \, x_{n+j}, \quad 0 \le j \le N - 1. \tag{9.16}$$

Notice that if the sequence y is defined by $y = S^j x$, and if $\mathbf{y}$ is the segment $(y_0, y_1, \ldots, y_{N-1})$, then $\theta_x(j) = (\mathbf{x}, \mathbf{y})$.

The continuous-time periodic autocorrelation function can be determined from knowledge of the chip waveform and the discrete periodic autocorrelation function. For example, suppose that $\psi(t) = 0$ if $t < 0$ or $t > T_c$, and let

$$\lambda = \int_0^{T_c} \psi^2(t) \, dt.$$

We first show that for values of τ of the form jT_c, the value of $r_v(\tau)$ can be determined from the relationship $r_v(jT_c) = \lambda \, \theta_x(j)$.

Let $\tau = jT_c$, and substitute from (9.14) into (9.15) to obtain

$$r_v(jT_c) = \int_0^T \left\{ \sum_{i=-\infty}^{\infty} x_i \, \psi(t - iT_c) \sum_{k=-\infty}^{\infty} x_k \, \psi(t + jT_c - kT_c) \right\} dt$$

$$= \sum_{i=-\infty}^{\infty} x_i \sum_{k=-\infty}^{\infty} x_k \int_0^T \psi(t - iT_c) \, \psi(t + jT_c - kT_c) \, dt.$$

Next, we use the observation that if $k \neq i + j$, then

$$\psi (t - iT_c) \psi (t + jT_c - kT_c) = 0$$

for all values of t. This follows from the fact that $\psi (u) = 0$ except for $0 \leq u \leq T_c$. Consequently, the expression for $r_v(jT_c)$ reduces to

$$r_v(jT_c) = \sum_{i=-\infty}^{\infty} x_i \, x_{i+j} \int_0^T \psi^2(t - iT_c) \, dt.$$

Again we use the fact that $\psi (u) = 0$ except for $0 \leq u \leq T_c$ to conclude that the integrand is zero for $0 \leq t \leq T$ unless $0 \leq i \leq N - 1$. For i in the range $0 \leq i \leq N - 1$, the integral is equal to λ, so

$$r_v(jT_c) = \lambda \sum_{i=0}^{N-1} x_i \, x_{i+j} = \lambda \, \theta_x(j).$$

The mathematical form for $r_v(\tau)$ in the intervals between multiples of T_c depends on the chip waveform. As an example, let the chip waveform be a rectangular pulse of duration T_c. To determine the graph of $r_v(\tau)$, we can examine the contribution to $r_v(\tau)$ of each chip in the waveform $v(t)$. The reader should recall that the convolution of two rectangular pulses of the same duration is a triangular pulse. Similarly, the correlation of two rectangular pulses of the same duration is a triangular pulse. For the unit-amplitude rectangular pulse of duration T_c, the resulting triangular pulse is an isosceles triangle that has base $2T_c$ and height T_c. Since each chip is a rectangular pulse of duration T_c, its contribution to $r_v(\tau)$ is such an isosceles triangle. Because $v(t)$ is the sum of rectangular pulses of duration T_c, $r_v(\tau)$ must be the sum of these triangles. Thus, in each interval of the form $jT_c \leq \tau \leq (j + 1)T_c$, the graph of $r_v(\tau)$ is the sum of straight lines on this interval, each of which represents one half of one of these isosceles triangles. The sum of a number of straight lines is a straight line, so the graph of $r_v(\tau)$ for the interval from jT_c to $[j + 1]T_c$ is the straight line that connects the point with coordinates $(jT_c, r_v(jT_c))$ to the point with coordinates $([j + 1]T_c, r_v([j + 1]T_c))$. Since this is true for each integer j, the graph of $r_v(\tau)$ is a sequence of straight lines connecting consecutive points with coordinates $(jT_c, r_v(jT_c))$ for $j = 0, 1, \ldots, N$. A mathematical derivation of this result is given in [9.30], which also provides results for arbitrary chip waveforms of duration T_c.

For a *binary* signature sequence, each element of the sequence (x_i) is either $+1$ or -1, and so $x_i^2 = 1$ for each i. Thus, $\theta_x(0) = N$ for each binary signature sequence. For most applications, it is desirable for the discrete periodic autocorrelation function to be small for $j \neq 0$. The values of $\theta_x(j)$ for $j \neq 0$ are often referred to as the side lobes of the discrete periodic autocorrelation function. Similarly, the values of $r_v(\tau)$ and $R_v(\tau)$ for $\tau \neq 0$ are referred to as the *side lobes* of the continuous-time periodic autocorrelation function.

Binary sequences are usually generated as sequences of 0s and 1s, and they can be converted to sequences of $+1$s and -1s by the mapping $0 \to +1$ and $1 \to -1$. The mathematical operation used in the generation of the binary sequences is addition

$\oplus$	0	1
0	0	1
1	1	0

$\times$	$+1$	-1
$+1$	$+1$	-1
-1	-1	$+1$

Figure 9-6: Relationship between $\{0, 1\}$ with mod-2 addition and $\{+1, -1\}$ with multiplication.

modulo 2, for which the symbol is $\oplus$. Addition modulo 2, often referred to as mod-2 addition, is equivalent to exclusive OR (XOR). Modulo-2 addition is defined by $0 \oplus 0 = 1 \oplus 1 = 0$ and $0 \oplus 1 = 1 \oplus 0 = 1$, as shown in Figure 9-6. The equivalent operation between elements of $\{+1, -1\}$ is multiplication, and the correspondence between mod-2 addition for $\{0, 1\}$ and multiplication for $\{+1, -1\}$ is illustrated in Figure 9-6. From Figure 9-6, it is clear that 0 is the identity element under mod-2 addition, just as $+1$ is the identity element for multiplication, which is the reason for the association $0 \leftrightarrow +1$ and $1 \leftrightarrow -1$ between the elements of the two sets. It is common that mod-2 addition and multiplication are used interchangeably in block diagrams and verbal descriptions of various subsystems for DS spread-spectrum communications. For example, a correlator may use term-by-term mod-2 addition (perhaps implemented with an XOR gate) for two binary sequences if they are in $\{0, 1\}$ format or term-by-term multiplication if they are in $\{+1, -1\}$ format.

One family of binary sequences with a particularly good periodic autocorrelation function is the family of *maximal-length linear feedback shift-register sequences*, which are commonly referred to as *m-sequences* in the literature. The important properties of such sequences are given in [9.15] and [9.43]. The period N of an m-sequence must be of the form $N = 2^n - 1$ for some integer n. As discussed in what follows, this is the largest possible period for any sequence generated by a linear feedback shift register with n storage elements (hence, the terminology *maximal length*).

Two examples of linear feedback shift registers that can generate m-sequences are shown in Figures 9-7 and 9-8. (The first of these is from Figure 6-40 of Section 6.6.1.) Each shift register has three storage elements and a single mod-2 adder. If the initial loading for the shift register of Figure 9-8 is 101, as shown, then the sequence produced by this shift register is

$$1010011 1010011 101 \ldots,$$

which is a periodic sequence with period seven. As discussed in Section 6.6.1, the shift register of Figure 9-7 with the indicated initial loading also generates a periodic sequence of period seven, and this is largest period that can be obtained from a linear feedback shift register with three storage elements. If the register contents are viewed in sequence, then it is clear that the output begins to repeat once the sequence of register contents repeats. Since there are only seven different nonzero patterns of three binary digits, the longest span between consecutive occurrences of a given nonzero pattern in

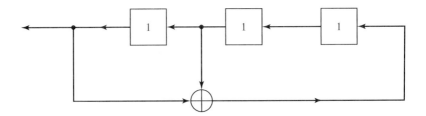

Figure 9-7: A linear feedback shift register with initial loading 111.

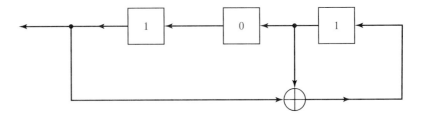

Figure 9-8: A linear feedback shift register with initial loading 101.

the register is seven. If the all-zeros pattern appears in the register, the output sequence consists of all zeros; thus, the generated sequence has period one rather than seven. Since this last observation is independent of the number of storage elements, we conclude that the largest possible period for the sequence generated by a binary linear feedback shift register with n storage elements is $2^n - 1$.

If x is an m-sequence of period $N = 2^n - 1$, its discrete periodic autocorrelation function is given by $\theta_x(0) = N$ and $\theta_x(j) = -1$ for $1 \leq j \leq N - 1$. The magnitudes of the periodic autocorrelation side lobes for an m-sequence are the smallest possible for any binary sequence of the same period. (Note that the period of an m-sequence is necessarily an odd integer.) If the chip waveform is a unit-amplitude rectangular pulse of duration T_c, the corresponding continuous-time periodic autocorrelation function for a spectral spreading signal derived from an m-sequence of period N is given by

$$r_v(\tau) = \begin{cases} NT_c - (N+1)|\tau|, & |\tau| \leq T_c, \\ -T_c, & T_c < |\tau| \leq (N-1)T_c. \end{cases} \tag{9.17}$$

This autocorrelation function is illustrated in Figure 9-9.

Because $v(t)$ is periodic with period $T = NT_c$, the continuous-time periodic auto-correlation function satisfies $r_v(\tau) = r_v(\tau + iNT_c)$ for each integer i. In particular, $r_v(0) = r_v(iNT_c) = NT_c$ for each integer i. Notice also that $r_v(jT_c) = -T_c$ for $1 \leq j \leq N - 1$. The normalized continuous-time periodic autocorrelation function for a spectral spreading signal derived from an m-sequence of period N satisfies $R_v(0) = 1$ and $R_v(jT_c) = -1/N$ for $1 \leq j \leq N - 1$. Note the similarity between this normalized continuous-time periodic autocorrelation function and the autocorrelation function for the random process $V(t)$ considered in Section 9.1 (equation (9.1) in particular).

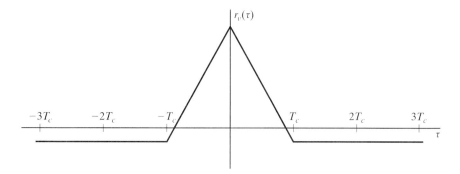

Figure 9-9: Continuous-time periodic autocorrelation function for an m-sequence.

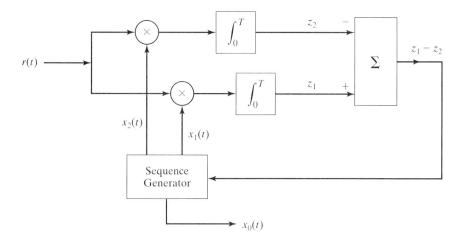

Figure 9-10: Tracking loop for a DS spread-spectrum receiver.

The features of the autocorrelation function can be exploited in order to provide an appropriate error signal for a feedback system to track the spread-spectrum waveform. The receiver must know the timing of the incoming spectral spreading signal in order to demodulate the signal and recover the data. This timing changes because of timing drift in the clock at the transmitter, motion of the transmitter and receiver, and changes in the propagation conditions on the communication link. A system that can track these changes is shown in Figure 9-10. This is a version of a delay-lock loop [9.48].

Suppose that the spectral spreading signal has a time offset τ as it arrives at the receiver; that is, the received spreading signal is $v(t-\tau)$. The input $r(t)$ to the delay-lock loop normally consists of the sum of the desired signal $v(t-\tau)$, other communication signals, and noise. To illustrate the operation of the delay-lock loop, consider a situation in which the received signal is the desired signal only; that is, $r(t) = v(t-\tau)$. The signal $x_0(t)$ shown in Figure 9-10 is known as the punctual signal. It is generated by the same sequence generator that generates $v(t)$, so $x_0(t) = v(t-\hat{\tau})$. The only possible difference between $x_0(t)$ and $v(t-\tau)$ is an offset in timing. If the loop is locked and

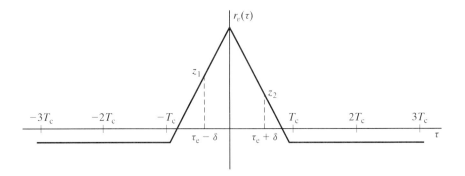

Figure 9-11: Early and late samples for a delay-lock loop.

tracking well, any timing offset is small: $\hat{\tau} \approx \tau$. In particular, the *error* $\tau_e = \tau - \hat{\tau}$ in the receiver's estimate of the timing offset should be small compared to the chip duration T_c.

The signal $x_1(t)$ is known as the *late signal*, and it is a delayed version of the punctual signal: $x_1(t) = x_0(t-\delta)$. The signal $x_2(t)$ is known as the *early signal*. It is an advanced version of the punctual signal: $x_2(t) = x_0(t + \delta)$. Thus, $x_1(t) = v(t - \hat{\tau} - \delta)$ and $x_2(t) = v(t - \hat{\tau} + \delta)$. The parameter δ is selected by the system designer to meet the requirements of the tracking system. A common choice is $\delta = T_c/2$. The outputs of the integrators are the correlations between the received signal and the early and late signals. Since all of these signals are just time offsets of the same spreading signal, these correlations are autocorrelations. In particular, the late sample is

$$z_1 = \int_0^T v(t - \tau)\, x_1(t)\, dt = \int_0^T v(t - \tau)\, v(t - \hat{\tau} - \delta)\, dt,$$

and the early sample is

$$z_2 = \int_0^T v(t - \tau)\, x_2(t)\, dt = \int_0^T v(t - \tau)\, v(t - \hat{\tau} + \delta)\, dt.$$

Because $v(t)$ has period T, it follows that the late sample is given by

$$z_1 = r_v(\tau - \hat{\tau} - \delta) = r_v(\tau_e - \delta),$$

and the early sample is given by

$$z_2 = r_v(\tau - \hat{\tau} + \delta) = r_v(\tau_e + \delta).$$

The resulting error signal is $z_1 - z_2 = r_v(\tau_e - \delta) - r_v(\tau_e + \delta)$. The early and late samples are illustrated in Figure 9-11 for an m-sequence and a correlator spacing of $\delta = T_c/2$.

Notice that if the error in the timing estimate $\hat{\tau}$ is positive (i.e., $\tau_e > 0$), the error signal is also positive. This is the situation illustrated in Figure 9-11. If the error is positive, the estimate $\hat{\tau}$ is smaller than the actual time offset τ, and so the positive feedback results in an increase of the estimate, thereby reducing the error. On the other hand, if the error in the timing estimate $\hat{\tau}$ is negative ($\tau_e < 0$), the error signal is negative, and the estimate is reduced by the feedback mechanism. This in turn reduces the error.

For multiple-access communications and certain other applications, the crosscorrelation functions for the sequences are at least as important as the autocorrelation functions. The *discrete periodic crosscorrelation function* for two sequences $x = (x_i)$ and $y = (y_i)$, each of period N, is defined by

$$\theta_{x,y}(j) = \sum_{n=0}^{N-1} x_n \, y_{n+j}, \quad 0 \le j \le N-1.$$

For $j = 0$, the periodic crosscorrelation is the inner product

$$\theta_{x,y}(0) = (\mathbf{x}, \mathbf{y}) = \sum_{n=0}^{N-1} x_n \, y_n. \tag{9.18}$$

If $\theta_{x,y}(0) = 0$, the sequences (x_i) and (y_i) are said to be *orthogonal*. Orthogonal sequences are of particular interest for code-division multiplexing (e.g., in the base station transmitter of the TIA-95 mobile cellular communication system). If the sequence z is defined by $z = S^j y$ then

$$\theta_{x,y}(j) = \theta_{x,z}(0) = (\mathbf{x}, \mathbf{z}),$$

so the discrete periodic crosscorrelation for two sequences is the inner product of one of the sequences and a shifted version of the other.

Sets of orthogonal sequences can be obtained from the rows of Hadamard matrices, as discussed in Section 6.6.1. The 4×4 Hadamard matrix is

$$H_2 = \begin{bmatrix} +1 & +1 & +1 & +1 \\ +1 & -1 & +1 & -1 \\ +1 & +1 & -1 & -1 \\ +1 & -1 & -1 & +1 \end{bmatrix},$$

which has orthogonal row vectors $(+1, +1, +1, +1)$, $(+1, -1, +1, -1)$, $(+1, +1, -1, -1)$, and $(+1, -1, -1, +1)$. Each vector is repeated periodically to form an infinite-length periodic sequence. For example, the vector $(+1, -1, -1, +1)$ produces the sequence

$$\ldots, +1, -1, -1, +1, +1, -1, -1, +1, +1, -1, -1, +1, \ldots$$

and the vector $(+1, -1, +1, -1)$ produces the sequence

$$\ldots, +1, -1, +1, -1, +1, -1, +1, -1, +1, -1, +1, -1, \ldots \, .$$

We refer to sequences generated in this manner as *Hadamard sequences*. Alternative names are Walsh functions [9.52] and Hadamard–Walsh sequences [9.58]. A different set of orthogonal sequences can be obtained from m-sequences in the manner described in Section 6.6.1.

Such sets of sequences have the property that if (x_i) and (y_i) are sequences from a given set, then $\theta_{x,y}(0) = (\mathbf{x}, \mathbf{y}) = 0$, which is just the condition for orthogonality. However, it is not true in general that $\theta_{x,y}(j) = 0$ for $j \neq 0$. In fact, $\theta_{x,y}(j)$ can be very large if $j \neq 0$. For binary sequences of period 4, the largest possible value for $\theta_{x,y}(j)$ is 4, and this value is achieved if and only if $x = S^j y$ (i.e., x is a shifted version of y, shifted by j units). But this is exactly the situation for the sequences we have considered thus far.

For example, suppose $x = (x_i)$ is the Hadamard sequence

$$\ldots, +1, -1, -1, +1, +1, -1, -1, +1, +1, -1, -1, +1, \ldots$$

which is generated by the vector $(+1, -1, -1, +1)$, and $y = (y_i)$ is the Hadamard sequence

$$\ldots, +1, +1, -1, -1, +1, +1, -1, -1, +1, +1, -1, -1, \ldots,$$

which is generated by the vector $(+1, +1, -1, -1)$, then it is easy to see that $x = Sy$. The Hadamard sequence (x_i) is just a shifted version of the Hadamard sequence (y_i), shifted by one time unit. Consequently, for this example, $\theta_{x,y}(1) = 4$. Alternatively, since the segments are of length 4, the definition of the periodic crosscorrelation function implies that

$$\theta_{x,y}(1) = \sum_{n=0}^{3} x_n\, y_{n+1},$$

which gives $\theta_{x,y}(1) = 4$ for these two sequences.

The orthogonal sequences generated from an m-sequence in the manner described in Section 6.6.1 also have poor periodic crosscorrelation functions. Consider the two vectors

$$0\,0\,1\,0\,1\,1\,1\,0$$

and

$$0\,1\,0\,1\,1\,1\,0\,0.$$

Because these are cyclic shifts of each other, the two infinite sequences generated from these vectors are shifts of each other. Consequently, when converted to sequences of $+1$s and -1s, their periodic crosscorrelation function must take on the value 8 for some value of its argument (which is the largest possible value for sequences generated by vectors of length 8). In particular, if x denotes the infinite sequence obtained from the first vector and y denotes the sequence obtained from the second, then $\theta_{y,x}(1) = \theta_{x,y}(7) = 8$.

Fortunately, there are sequences for which the periodic crosscorrelation function $\theta_{x,y}(j)$ is relatively small for all values of j. Since the sequences described in this section are generated by linear feedback shift registers, each sequence is identified by specifying the feedback connections and initial conditions for the shift register that generates the sequence. A linear feedback shift register with n storage elements is illustrated in

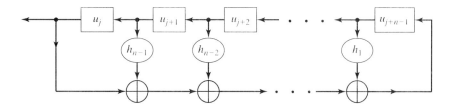

Figure 9-12: General n-stage linear feedback shift register.

Figure 9-12. The sequence generated by this register is denoted by $u = (u_i)$. Each of the tap coefficients $h_{n-1}, h_{n-2}, \ldots, h_1$ has value 0 or 1. If $h_k = 0$, there is no feedback through the kth tap (numbered from right to left); however, if $h_k = 1$, there is feedback through the kth tap. The output sequence is described by the equation

$$u_{j+n} = u_j \oplus h_{n-1} u_{j+1} \oplus h_{n-2} u_{j+2} \oplus \cdots \oplus h_1 u_{j+n-1}.$$

The two most common ranges for the integer j in this equation are $-\infty < j < \infty$, which gives a doubly infinite sequence, and $0 \leq j < \infty$, which gives a one-sided sequence that begins at time 0. Notice that u_{j+n} is the element that is being fed back into the rightmost storage element as u_j is shifting out the left end of the register.

For convenience, we define $h_n = 1$, which corresponds to having feedback from the leftmost storage element, and $h_0 = 1$, which corresponds to having feedback into the rightmost storage element, as shown in Figure 9-12. Notice that if $h_n = 0$, the leftmost storage element has no role in determining which sequence is generated, and so this storage element might as well be discarded. If $h_0 = 0$, there is no feedback at all. Consequently, the coefficients h_n and h_0 always take the value 1 in this section, and so they are not shown as variables in Figure 9-12.

A linear feedback shift register with n storage elements can be represented by the binary vector $h = (h_0, h_1, \ldots, h_n)$, where $h_0 = h_n = 1$ for all shift registers with n storage elements. This vector is referred to as the *feedback tap coefficient vector*. Because the only values for the coefficients in question are 0 or 1, the tap coefficients merely indicate which storage elements are connected to mod-2 adders in the feedback circuit and which are not. As a result, a simpler name for the feedback tap coefficient vector is *connection vector*. If $h_i = 1$, the output of ith storage element (again, numbered from right to left) has a connection to the feedback circuit; and if $h_i = 0$, it does not.

In order to provide a compact designation for these shift registers, the connection vectors are often represented in octal form [9.43]. For example, if $n = 5$, the vector is $(h_0, h_1, h_2, h_3, h_4, h_5)$, which is grouped as $h_0 h_1 h_2$ and $h_3 h_4 h_5$. These two triples of binary numbers are then converted to octal. For example, the linear feedback shift register that has five storage elements and feedback taps given by the binary vector $(1, 0, 0, 1, 0, 1)$ is represented in octal form as 45. The grouping process begins with the rightmost three components of $(h_0, h_1, \ldots, h_n)$ and proceeds to the left. If the number of components is not a multiple of 3, then the leftmost group does not have the full three elements. For example, if $n = 4$, the vector is $(h_0, h_1, h_2, h_3, h_4)$, and the grouping is $h_0 h_1$ and $h_2 h_3 h_4$. Thus, the linear feedback shift register that has four storage elements and feedback taps given by the binary vector $(1, 0, 1, 0, 1)$ is represented in octal form as 25. As a final example, consider the linear feedback shift registers illustrated in Figures 9-7 and 9-8, each of which has three storage elements $(n = 3)$. The connection vector for the shift register of Figure 9-7 is $(1, 0, 1, 1)$, which is expressed in octal form as 13, while the connection vector for the shift register of Figure 9-8 is $(1, 1, 0, 1)$, which is 15 in octal.

It is convenient to represent a linear feedback shift register with n storage elements by a polynomial of degree n. The shift register illustrated in Figure 9-12 corresponds to the polynomial

$$h(x) = h_0 x^n + h_1 x^{n-1} + h_2 x^{n-2} + \cdots + h_{n-1} x + h_n.$$

For example, the polynomial for the linear feedback shift register of Figure 9-7 is $x^3 + x + 1$ and the polynomial for the shift register of Figure 9-8 is $x^3 + x^2 + 1$. These two linear feedback shift registers each generate an m-sequence; moreover, one sequence is just the reverse of the other. To see this, consider the initial loading 111 in the shift register of Figure 9-7 and the initial loading of 010 in the shift register of Figure 9-8. (Note that this is a different loading than shown in Figure 9-8.) A comparison of the first seven bits of each of the two resulting sequences shows that one is the reverse of the other. Of all linear feedback shift registers with three storage elements, these are the only two that generate sequences of period seven.

From the properties of the polynomial that represents a particular linear feedback shift register, certain properties can be determined for the sequence that is generated by that shift register. A simple example is the fact that if the polynomial has degree n, the period of the sequence generated by the corresponding shift register cannot exceed $2^n - 1$. Those sequences of period $2^n - 1$ that are generated by a shift register whose polynomial has degree n are precisely the m-sequences, and the corresponding polynomial is a *primitive polynomial* [9.27]. In multiplying two polynomials or factoring a polynomial, the coefficients of the polynomials are binary and their addition is defined using mod-2 arithmetic. For example, in this arithmetic $x + x = (1 \oplus 1)x = 0$. Thus, the polynomial $x^2 + 1$ factors as $(x + 1)(x + 1)$. Similarly, $x^4 + 1 = (x^2 + 1)(x^2 + 1) = (x + 1)^4$ and $x^3 + x^2 + x + 1 = (x + 1)(x^2 + 1) = (x + 1)^3$.

If a polynomial cannot be factored, it is referred to as an *irreducible polynomial*. Primitive polynomials are irreducible, but there are many irreducible polynomials that are not primitive. Tables of irreducible polynomials, expressed in the octal notation defined earlier, are given in [9.27]. The irreducible polynomials that are also primitive are identified in these tables.

The shift registers of Figures 9-7 and 9-8 correspond to the polynomials $x^3 + x + 1$ and $x^3 + x^2 + 1$, respectively. The only other polynomials of degree three with binary coefficients are $x^3 + 1$ and $x^3 + x^2 + x + 1$, and neither of these corresponds to an m-sequence generator for a sequence of period seven. Observe that each of these polynomials factors:

$$x^3 + 1 = (x + 1)(x^2 + x + 1)$$

and

$$x^3 + x^2 + x + 1 = (x + 1)^3.$$

We conclude that the two polynomials x^3+1 and x^3+x^2+x+1 are not even irreducible, let alone primitive.

Although each individual m-sequence has good periodic autocorrelation properties, not all pairs of m-sequences have good periodic crosscorrelation properties. Among the set of all m-sequences of a given period, however, certain pairs do have relatively small periodic crosscorrelation values. These pairs are referred to as *preferred pairs* of m-sequences and the associated primitive polynomials are referred to as *preferred pairs* of polynomials [9.43]. For m-sequences of period 31, for example, the maximum periodic crosscorrelation for an arbitrary pair of m-sequences is 11, but the maximum periodic crosscorrelation for a preferred pair of m-sequences is only 9. The situation for m-sequences of period 127 is much more dramatic: For an arbitrary pair of m-sequences, the periodic crosscorrelation can be as large as 41, but for a preferred pair of m-sequences, the maximum periodic crosscorrelation is 17. For m-sequences of period 4095, the maximum periodic crosscorrelation for an arbitrary pair of m-sequences is more than an order of magnitude greater than for a preferred pair of m-sequences.

The periodic crosscorrelation function for a preferred pair of m-sequences is known to take on three values only. For example, if x and y are a preferred pair of m-sequences of period 31, $\theta_{x,y}(j)$ is equal to -1 for 15 values of j, equal to $+7$ for 10 values of j, and equal to -9 for 6 values of j. If the preferred pair of m-sequences has period 127, the three values taken on by the periodic crosscorrelation function are -1, $+15$, and -17. Preferred pairs of m-sequences are of interest for applications, not so much because the number of different crosscorrelation values is small, but primarily because the magnitudes of these values are relatively small compared with the sequence period and compared with the crosscorrelation values for pairs of m-sequences that are not preferred pairs. For each positive integer n, let $t(n)$ be defined by

$$t(n) = 1 + 2^{\lfloor (n+2)/2 \rfloor},$$

where $\lfloor (n + 2)/2 \rfloor$ denotes the integer part of $(n + 2)/2$. Notice that $t(5) = 9$ and $t(7) = 17$, and also notice that these are the maximum values of the magnitudes of the crosscorrelation functions for preferred pairs of m-sequences of period 31 and 127, respectively. If the discrete periodic crosscorrelation function for two m-sequences of period $2^n - 1$ takes on the three values -1, $t(n) - 2$, and $-t(n)$ only, then it is said to be a *preferred three-valued crosscorrelation function* [9.43]. A preferred pair of m-sequences of period $2^n - 1$ has a preferred three-valued crosscorrelation function.

A *maximal connected set* of m-sequences is a collection of m-sequences with the property that each pair in the set is a preferred pair. By requiring that all of the sequences for a given application (e.g., a spread-spectrum multiple-access communication system) are m-sequences from the same maximal connected set, one can be sure that no matter which two signals are considered, the discrete periodic crosscorrelation function for the two signals is a preferred three-valued crosscorrelation function. The reason that we cannot restrict attention to maximal connected sets for all applications is that the sizes of the sets are too small. For example, a maximal connected set of m-sequences of period 31 has only three sequences, and a maximal connected set of m-sequences of period 127 has only six sequences. For periods other than 127 that are of interest for most applications (e.g., from 7 to 65,535), none of the maximal connected sets has more than four sequences. Even though the total number of m-sequences of period N is quite large for large values of N, the size of a maximal connected set of such sequences remains small (e.g., for period 8191 there are a total of 630 m-sequences, but there are only four sequences in a maximal connected set).

Sets of Gold sequences [9.13] are much larger than maximal connected sets, and their crosscorrelation functions take on the same three values. However, the periodic autocorrelation functions for most of the Gold sequences are not as good as the periodic autocorrelation functions for m-sequences. Also, Gold sequences of period $2^n - 1$ can be generated by linear feedback shift registers with $2n$ storage elements, and so their periods are approximately the square root of the maximum periods for linear sequences generated with the same number of storage elements. However, if a large number of sequences is required for a given application, and if the crosscorrelation function is more important than the autocorrelation function, then the Gold sequences are much better than m-sequences for that application.

The simplest way to describe how to generate Gold sequences is to use the polynomial representation. If $f_1(x)$ and $f_2(x)$ are a preferred pair of primitive polynomials of degree n, the set of sequences generated by the product polynomial $f(x) = f_1(x) f_2(x)$ is a set of Gold sequences of period $2^n - 1$. Because the product polynomial has degree $2n$, the corresponding shift register has $2n$ storage elements. For example, for $n = 5$ the m-sequences of period 31 that have octal representations 45 and 75 are a preferred pair of m-sequences. The connection vector corresponding to 45 is $(1, 0, 0, 1, 0, 1)$, as previously mentioned, and the connection vector corresponding to 75 is $(1, 1, 1, 1, 0, 1)$. If $f_1(x)$ is the polynomial corresponding to 45 then $f_1(x) = x^5 + x^2 + 1$, and if $f_2(x)$ is the polynomial corresponding to 75, then $f_2(x) = x^5 + x^4 + x^3 + x^2 + 1$. The product polynomial is

$$f(x) = f_1(x) f_2(x) = x^{10} + x^9 + x^8 + x^6 + x^5 + x^3 + 1.$$

This polynomial corresponds to a linear feedback shift register with ten storage elements and connection vector $(1, 1, 1, 0, 1, 1, 0, 1, 0, 0, 1)$. The octal representation for this shift register is 3551.

When loaded with appropriate initial conditions, this linear feedback shift register generates a sequence of period 31. Note that such a sequence is not a maximal-length sequence for this number of storage elements. The maximum period of a sequence generated by a linear feedback shift register with ten storage elements is $2^{10} - 1 = 1023$.

An alternative method for generating Gold sequences is derived from the fact that if two polynomials $f_1(x)$ and $f_2(x)$ have no common factors (e.g., if at least one of them is irreducible), a sequence (w_i) can be generated by the shift register with polynomial $f_1(x) f_2(x)$ if and only if $w_i = u_i \oplus v_i$ for each integer i, where the sequence (u_i) is generated by the shift register with polynomial $f_1(x)$ and (v_i) is a sequence generated by the shift register with polynomial $f_2(x)$. In this special situation, one of the sequences (u_i) or (v_i) can be the all-zeros sequence.

Now suppose that $f_1(x)$ and $f_2(x)$ are a preferred pair of primitive polynomials, and let (u_i) be the sequence generated by the shift register corresponding to $f_1(x)$ and (v_i)

be the sequence generated by the shift register corresponding to $f_2(x)$. The set of Gold sequences generated by the product polynomial $f_1(x) \, f_2(x)$ is the set of all sequences (w_i) that satisfy

$$w_i \ = \ u_i \text{ for each integer } i,$$
$$w_i \ = \ v_i \text{ for each integer } i,$$

or, for some integer k,

$$w_i = u_i \oplus v_{i+k} \text{ for each integer } i.$$

That is, a set of Gold sequences consists of each of the original two m-sequences (u_i) and (v_i) together with term-by-term mod-2 sums of (u_i) and shifted versions of (v_i). A more thorough discussion of Gold sequences and this method for generating them is given in [9.43].

There are many other classes of linear feedback shift register sequences that provide different set sizes and different correlation properties. Most notable among these are the small and large sets of Kasami sequences [9.43]. For sequence period 63, for example, the small set of Kasami sequences has a smaller maximum periodic cross-correlation function than the Gold sequences. For this same period, the large set of Kasami sequences has a larger number of sequences than a set of Gold sequences of period 63, yet the maximum value of the periodic crosscorrelation function is the same for the two sets. These and other sets of sequences are described in [9.43].

9.4   DS Spread-Spectrum Multiple-Access Communications

Suppose the signal $s(t)$ is transmitted over a linear channel in which interference and noise are added, and let $r(t)$ denote the input to the communication receiver. For this channel model,

$$r(t) = s(t) + \eta(t) + X(t), \tag{9.19}$$

where $\eta(t)$ represents interference present on the channel and $X(t)$ represents thermal noise, which is modeled as a white Gaussian noise process. The signal $s(t)$ contains the information that is being conveyed to the receiving terminal, and so it is referred to as the *desired signal*. It is a DS spread-spectrum signal of the type described in the previous sections. For example, the communication signal may be of the form

$$s(t) = \sqrt{2} \, A \, x(t) \, d(t) \cos(\omega_c t + \varphi),$$

where $d(t)$ is a data signal defined by (9.9) and $x(t)$ is a baseband spectral spreading signal given by

$$x(t) = \sum_{i=-\infty}^{\infty} x_i \, \psi(t - iT_c).$$

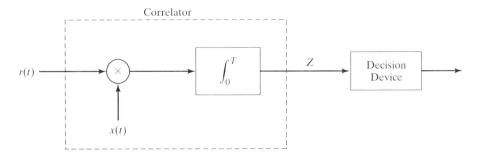

Figure 9-13: Correlation receiver for the baseband signal $s(t)$.

In order to simplify the following presentation, we assume that the chip waveform $\psi(t)$ is time limited to the interval $[0, T_c]$.

The receiver is a correlation receiver that is matched to the desired signal. The basic concepts can be illustrated with greater simplicity by considering baseband models for the communication signals and the receiver. The baseband model for the desired signal is $s(t) = A x(t) d(t)$. The noise and interference are also described in terms of their baseband models. The baseband model for the correlation receiver is shown in Figure 9-13. The correlation receiver has a replica of the spectral spreading signal $x(t)$, and it processes the received signal by multiplying it by the spectral spreading signal and integrating the product of the two signals over the duration of the data pulse. The output of the integrator is Z, the statistic upon which the receiver bases its decision as to which data symbol was transmitted (e.g., the decision may depend on whether Z exceeds some threshold).

The statistic Z has three components: a desired signal component μ, which is the part of the output that is due to $s(t)$; an interference component I, which is the output of the integrator if the input is $\eta(t)$ only; and the noise component, which is the result of thermal noise at the input to the receiver. The desired signal component for the interval $[0, T]$ is given by $\mu = b_0 A T$.

The influence of thermal noise is no different than in a baseband digital communication system that does not employ spread spectrum. This can be seen from an examination of the noise component at the output of the correlator. The noise component is the zero-mean Gaussian random variable defined by

$$W = \int_0^T X(t) x(t) \, dt.$$

If α^2 is T^{-1} times the energy in $x(t)$, $0 \le t \le T$, then W has the same probability distribution as

$$W' = \alpha \int_0^T X(t) \, dt,$$

which is the noise component of the output of the correlator or matched filter for a system with $d(t)$ as the transmitted signal (i.e., a system that does not use spread spectrum).

As a result, the novel features of spread spectrum are demonstrated by focusing on the interference component of the decision statistic. The interference component is defined by

$$I = \int_0^T \eta(t)\, x(t)\, dt. \tag{9.20}$$

If the interference is due exclusively to the presence of one or more additional DS spread-spectrum signals in the same frequency band as the desired signal, then the interference component I is referred to as the *multiple-access interference*.

Suppose that the only interference on the channel is another DS spread-spectrum signal $s'(t)$ with the same general characteristics as the desired signal (e.g., the same chip waveform, chip rate, and data rate). The multiple-access interference signal differs from the desired signal in that its amplitude is A' rather than A, its signature sequence is (y_i) rather than (x_i), and it arrives at the receiver with a time delay of τ units relative to the desired signal. The time delay is used to account for any offsets between the timing references for the two signals and any differences in the propagation times from the two transmitters to the common receiver. The amplitudes A' and A are used to account for the power levels in the transmitted signals and the propagation losses in the two communication paths. If A' is larger than A, the interference signal has more power than the desired signal. Such a situation can result if the source of the interfering signal is closer to the receiver than the source of the desired signal, and so it is often referred to as the *near–far condition*. An unfavorable near–far condition, whether it is a result of greater power in the transmitted interfering signal or shorter range from the interfering transmitter to the receiver, can severely limit the multiple-access capability of a DS spread-spectrum system in which the demodulation is accomplished by correlation receivers.

Suppose that the signature sequences for the desired signal and the interference signal each have period N. Analogous to (9.8) and (9.9), the spectral spreading signal for the interference is expressed as

$$y(t) = \sum_{i=-\infty}^{\infty} y_i\, \psi(t - iT_c),$$

and its data signal is

$$d'(t) = \sum_{n=-\infty}^{\infty} d_n\, p_I(t - nT).$$

The interference signal is given by

$$s'(t) = A'\, y(t)\, d'(t).$$

The spectral spreading waveform for the interference is

$$\zeta(t) = \sum_{i=0}^{N-1} y_i\, \psi(t - iT_c),$$

and so the interference signal can also be written as

$$s'(t) = A' \sum_{n=-\infty}^{\infty} d_n \zeta(t - nT).$$

In this section, the chip waveform $\psi(t)$ is required to be time limited to the interval $[0, T_c]$ and there is one period of the signature sequence per data pulse (i.e., $T = NT_c$). It follows that $\zeta(t)$ is time limited to the interval $[0, T]$, and so the nth term of the preceding expression for $s'(t)$ is nonzero only if $nT \leq t < (n+1)T$.

Because $\eta(t) = s'(t - \tau)$, it follows from (9.19) that the received signal is given by

$$r(t) = s(t) + s'(t - \tau) + X(t), \tag{9.21}$$

and the interference component is

$$I = \int_0^T s'(t - \tau) x(t) \, dt. \tag{9.22}$$

The signals are of infinite duration in the model, so it suffices for most purposes to restrict the time delay to the range $0 \leq \tau < T$. This means that during the interval $[0, T]$, the only nonzero terms of the interference signal $s'(t - \tau)$ are $d_{-1} \zeta(t - \tau + T)$ and $d_0 \zeta(t - \tau)$. If τ is in the interval $[0, T]$, the only values of t in $[0, T]$ for which $\zeta(t - \tau + T)$ can be nonzero are those that satisfy $0 \leq t < \tau$. Similarly, $\zeta(t - \tau)$ can be nonzero only for $\tau \leq t \leq T$. By substituting for $s'(t - \tau)$ in (9.22), it is easy to see that

$$I = A'\{d_{-1} R_{y,x}(\tau) + d_0 \hat{R}_{y,x}(\tau)\}, \tag{9.23}$$

where the *continuous-time partial crosscorrelation functions* $R_{y,x}$ and $\hat{R}_{y,x}$ are defined by

$$R_{y,x}(\tau) = \int_0^\tau y(t - \tau) x(t) \, dt \tag{9.24}$$

and

$$\hat{R}_{y,x}(\tau) = \int_\tau^T y(t - \tau) x(t) \, dt. \tag{9.25}$$

Notice that if $\tau \neq 0$, the amplitudes of two consecutive data symbols of the interference signal influence the output of the correlator. These amplitudes are the terms d_{-1} and d_0 in (9.23). For example, if the sequence (d_n) is a binary sequence from the set $\{-1, +1\}$, it is clear that the polarity of the interference depends on the values of d_{-1} and d_0. In most situations, even the magnitude of the interference depends on the relative polarities of these two consecutive data pulses. That is, in most situations, the interference magnitudes are different for the two cases $d_{-1} = d_0$ and $d_{-1} = -d_0$.

If the two signals are aligned perfectly in time (i.e., the data symbols of the two signals are synchronized), then $\tau = 0$ and (9.23) simplifies greatly. Notice that $R_{y,x}(0) = 0$ and

$$\hat{R}_{y,x}(0) = \int_0^T y(t) x(t) \, dt,$$

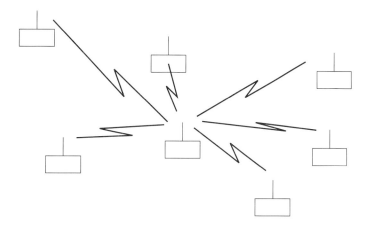

Figure 9-14: A radio network with a star topology.

which is just the inner product $(\mathbf{y}, \mathbf{x})$ for the two baseband spectral spreading signals. Thus, for $\tau = 0$, (9.23) becomes

$$I = A'\{d_{-1}\, R_{y,x}(0) + d_0\, \hat{R}_{y,x}(0)\} = A' d_0\, \hat{R}_{y,x}(0) = A' d_0\, (\mathbf{y}, \mathbf{x}).$$

It follows that if the two spectral spreading signals $x(t)$ and $y(t)$ are orthogonal, and if the desired signal $s(t)$ and the interference $s'(t)$ are synchronous, then there is no multiple-access interference.

As described in Section 9.2, the mobile cellular CDMA system (i.e., the TIA-95 system) has the feature that the signature sequences (and hence, the spectral spreading signals) are orthogonal in the composite signal that is transmitted by the base station. This orthogonality implies there is no multiple-access interference at the mobile station if channel conditions are perfect. The cellular system is an example of a *star topology*, which is the term used for a network in which all communication takes place between a single terminal and each of two or more other terminals. That is, one designated terminal is transmitting to multiple receivers or it is receiving from multiple transmitters. A radio network with star topology is illustrated in Figure 9-14. A single cell of a cellular telephone system is an example of a star topology: The base station transmits to the mobile phones within the cell on the forward link and receives from the mobile phones on the reverse link. The mobile phones do not transmit directly to each other.

The data signals to be transmitted from the base station to the mobile phones of the cellular system are modulated and combined into one composite signal in a single system within the base station, as illustrated in Figure 9-5. Since all of the sequence generators are also within the base station, it is easy to align the chips of the signature sequences for the different signals that are combined into the composite signal. Consequently, the term *multiplexing* may be more precise than the term *multiple access* for the process of combining these spread-spectrum signals into a single composite signal within the base station. It is helpful to make this distinction, because it is relatively easy to have precise timing synchronism among the individual signals when they are all generated within the same system. Using this terminology, the forward link of a mobile cellular system that uses spread spectrum may be more accurately referred to as *spread-spectrum multiplexing* or *code-division multiplexing*. If the channel is ideal (e.g., no multipath propagation), the synchronism that is present in the transmitter is preserved as the signals propagate over the channel, and so the mobile receiver sees a composite signal in which the individual spread-spectrum signals are synchronized with each other (i.e., their data pulses and chip waveforms are aligned). In particular, if the spreading signals are orthogonal at the transmitter, they remain so during propagation over an ideal channel, and so they are orthogonal at the receiver as well.

The timing on the reverse links (from the mobile stations to the base station) is more difficult to control. The mobile terminals are geographically distributed, and they do not have a precise common timing reference. If they wish to coordinate their transmissions in a way that guarantees the corresponding signals at the base station receiver are in synchronism, the mobile transmitters must not only coordinate their timing references, they must also account for the different propagation times from the mobile terminals to the base station. In addition, each mobile terminal would have to adjust its timing to compensate for changes in the propagation time as it moves relative to the base station. The various signals transmitted on the reverse links by the mobile terminals are combined in the propagation medium; thus, the combining process is not under the control of the communication system as it is on the forward link.

The mechanism by which the different terminals have access to the communication channel is therefore quite different on the reverse links than on the forward links. The combining process on the reverse links is what most communication engineers think of as *multiple access* rather than *multiplexing*. In particular, because the timing among the reverse links cannot be controlled precisely in most applications, the use of orthogonal signature sequences for the reverse links in the mobile cellular system does not eliminate interference.

Up to this point, the discussion has been focused on an ideal communication channel. Mobile cellular CDMA systems typically must contend with multipath propagation on the channel. As a result of multipath propagation and other sources of channel distortion, the received versions of the different forward-link signals are not orthogonal, even though they may be orthogonal at the transmitter. This is because different propagation paths have different propagation times. A signal that arrives at the receiver via one path may not be orthogonal to a signal that arrives at the receiver via another path, even if the two signals are orthogonal at the transmitter. Thus, even on the forward link, the signals for different mobile terminals may not be orthogonal at any one of the mobile terminals, in spite of the star topology of the network, the orthogonality of the signature sequences, and the precise timing control within the base station transmitter. As a result, in practice there is some interference among signals in the forward link of the mobile cellular CDMA system, even if orthogonal signature sequences are employed.

The ability of a DS spread-spectrum system to benefit from the use of orthogonal signature sequences depends critically on being able to maintain time synchronism among the different signature sequences during the modulation, transmission, propagation, and reception of the signals. As discussed earlier, in the absence of multipath and other channel disturbances, this time synchronism can be accomplished in the forward link of the mobile cellular CDMA system. For mobile radio systems and networks in which different data signals are modulated onto different RF carriers at different loca-

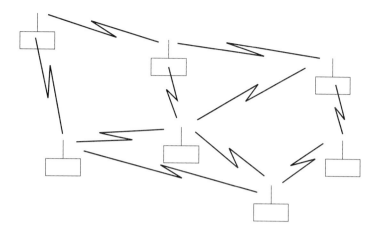

Figure 9-15: A radio network with a distributed topology.

tions throughout the network, it is rarely feasible to provide the synchronism needed to guarantee orthogonality among the transmitted spread-spectrum signals. Even if there is synchronism among the mobile transmitters, it is still impossible to provide orthogonality among the *received* spread-spectrum signals at each of the terminals in a network that does not have a star topology, such as the network illustrated in Figure 9-15. The fact that the distances between the different pairs of transmitters and receivers can vary greatly throughout the network is enough to make it impossible to have all of the signature sequences aligned at each of the receivers, regardless of the synchronism that may exist among the transmitters.

A mobile, *distributed*, spread-spectrum multiple-access radio network is an example of a network in which precise synchronism among the signals arriving at each receiver is not possible. In such a network, there is not a single base station that serves as the recipient of all of the signals from the mobile terminals; in fact, there may be no base stations at all. Because the distributed network does not have a star topology, different communication links may have significantly different propagation delays. Furthermore, the delays for different links are changing independently of each other as a result of the movement of the terminals in the network.

In systems and networks for which precise synchronism among the received signals is impossible or impractical, the *aperiodic* correlation properties of the sequences determine the multiple-access performance of the system. Unfortunately, the standard sets of orthogonal sequences, such as the Hadamard sequences employed in forward link for the TIA-95 mobile cellular CDMA system, have very poor aperiodic correlation properties. As a result, they are not good choices for signature sequences for other types of spread-spectrum systems, such as distributed radio networks.

Orthogonal sequences are also not good choices for use in the GPS system, which is described in Section 9.2. For the GPS system, both the autocorrelation and cross-correlation properties of the signature sequences are important. The autocorrelation functions for the sequences play a role in the acquisition performance of the receiver and in the ability of the receiver to discriminate against multipath interference. The crosscorrelation functions among the sequences employed by the different satellites determine the multiple-access capability of the GPS system.

Recall that if two DS spread-spectrum signals are transmitted, the multiple-access interference in a correlation receiver is given by the expression in (9.23), which is reprinted as follows for convenience:

$$I = A'\{d_{-1}\, R_{y,x}(\tau) + d_0\, \hat{R}_{y,x}(\tau)\}.$$

If there is a timing offset between the two spread-spectrum signals, then $0 < \tau < T$ and both continuous-time partial crosscorrelation functions are needed in order to determine the multiple-access interference. If $d_{-1} = d_0$, the multiple-access interference is proportional to $R_{y,x}(\tau) + \hat{R}_{y,x}(\tau)$. It is easy to see from (9.24) and (9.25) that

$$R_{y,x}(\tau) + \hat{R}_{y,x}(\tau) = \int_0^T y(t - \tau)\, x(t)\, dt, \tag{9.26}$$

which is just the continuous-time *periodic* crosscorrelation function. This continuous-time crosscorrelation function can be written in terms of the discrete periodic cross-correlation function of (9.18) in the same way as presented in Section 9.2 for writing the continuous-time autocorrelation function in terms of the discrete periodic auto-correlation function.

An important observation is that if $d_{-1} \neq d_0$, then the multiple-access interference cannot be expressed in terms of the periodic crosscorrelation function only. The appropriate function for characterizing the interference in an asynchronous DS spread-spectrum multiple-access system is the aperiodic crosscorrelation function [9.29, 9.43], since it can handle both $d_{-1} = d_0$ and $d_{-1} \neq d_0$.

The *aperiodic crosscorrelation function* for two vectors $\mathbf{u} = (u_0, u_1, \ldots, u_{N-1})$ and $\mathbf{v} = (v_0, v_1, \ldots, v_{N-1})$ of length N is defined by

$$C_{u,v}(i) = \sum_{j=0}^{N-1-i} u_j\, v_{j+i}, \qquad 0 \le i \le N - 1, \tag{9.27a}$$

$$C_{u,v}(i) = \sum_{j=0}^{N-1+i} u_{j-i}\, v_j, \qquad -(N-1) \le i < 0, \tag{9.27b}$$

and $C_{u,v}(i) = 0$ if $i \ge N$ or $i \le -N$. This is also the definition of the aperiodic cross-correlation function for the infinite sequences u and v that are the periodic extensions of the vectors $\mathbf{u} = (u_0, u_1, \ldots, u_{N-1})$ and $\mathbf{v} = (v_0, v_1, \ldots, v_{N-1})$, respectively. The sequence elements that are involved in the expressions for the crosscorrelation function $C_{u,v}(i)$ for $i = m$ and for $i = m - N$ are illustrated in Figure 9-16 for an arbitrary integer m in the range $0 < m < N - 1$. As shown in Figure 9-16, the integer m represents the offset of the sequence u relative to the sequence v. Notice from the definition or from the illustration in Figure 9-16 that $C_{u,v}(0) = \theta_{u,v}(0) = (\mathbf{u},\mathbf{v})$ and that $C_{u,v}(i) = C_{v,u}(-i)$ for each integer i.

The crosscorrelation function is somewhat simpler to illustrate for a specific value of m. The elements of $\mathbf{u} = (u_0, u_1, \ldots, u_{N-1})$ and $\mathbf{v} = (v_0, v_1, \ldots, v_{N-1})$ that are

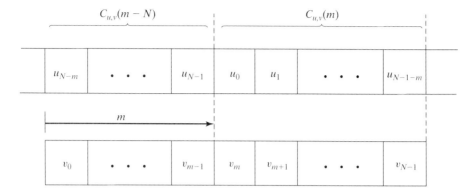

Figure 9-16: Illustration of the crosscorrelations $C_{u,v}(m - N)$ and $C_{u,v}(m)$.

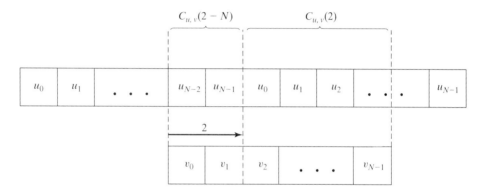

Figure 9-17: Illustration of the crosscorrelations $C_{u,v}(2 - N)$ and $C_{u,v}(2)$.

involved in the expressions for $C_{u,v}(m - N)$ and $C_{u,v}(m)$ are shown in Figure 9-17 for $m = 2$. Notice from Figures 9-16 and 9-17 that the aperiodic crosscorrelation function $C_{u,v}(i)$ is always a correlation of a prefix (i.e., first part) of one vector with a suffix (i.e., last part) of the other vector. The vector from which the prefix is used is **u** if $0 < i \leq N - 1$, but the prefix is from **v** if $1 - N \leq i < 0$.

The aperiodic autocorrelation function is a special case of the aperiodic cross-correlation function. If the vector **v** is replaced by the vector **u** in the preceding definition of the aperiodic crosscorrelation function, the result is the *aperiodic autocorrelation function $C_u(i)$* for the vector **u**. In other words, the aperiodic autocorrelation function C_u is defined by letting $C_u(i) = C_{u,u}(i)$ for each integer i.

The next step is to derive an expression for the multiple-access interference in terms of the aperiodic crosscorrelation function for the signature sequences involved. Toward this end, we return to the expression

$$I = A'\{d_{-1} R_{y,x}(\tau) + d_0 \hat{R}_{y,x}(\tau)\},$$

which gives the multiple-access interference in terms of the continuous-time partial crosscorrelation functions.

It is shown in [9.30] that the continuous-time crosscorrelation functions can be determined from the aperiodic crosscorrelation function for the signature sequences x and y. In particular, the continuous-time crosscorrelation functions can be expressed in terms of functions that depend on the signature sequences only and functions that depend on the chip waveform only. The waveform-dependent parts can be expressed in terms of the partial autocorrelation functions for the chip waveform, which are

$$R_\psi(s) = \int_0^s \psi(t)\,\psi(t + T_c - s)\,dt, \quad 0 \le s \le T_c, \tag{9.28}$$

and

$$\hat{R}_\psi(s) = \int_s^{T_c} \psi(t)\,\psi(t - s)\,dt, \quad 0 \le s \le T_c. \tag{9.29}$$

These two functions are related by $R_\psi(s) = \hat{R}_\psi(T_c - s)$ for $0 \le s \le T_c$, and we see that $R_\psi(0) = 0$ and $\hat{R}_\psi(0) = \mathcal{E}_\psi$, which is the energy in the chip waveform $\psi(t)$.

If the chip waveform is a unit-amplitude rectangular pulse of duration T_c, the partial autocorrelation functions are given by $R_\psi(s) = s$ and $\hat{R}_\psi(s) = T_c - s$ for $0 \le s \le T_c$. Note that the graph of each of these functions is a straight line on the interval $[0, T_c]$. Also observe that $R_\psi(0) = 0$ and $\hat{R}_\psi(0) = T_c$.

If τ is in the interval $iT_c \le \tau < (i + 1)T_c$ for $0 \le i \le N - 1$, the continuous-time partial crosscorrelation functions of (9.24) and (9.25) can be written as [9.30]

$$R_{y,x}(\tau) = C_{y,x}(i - N)\,\hat{R}_\psi(\tau - iT_c) + C_{y,x}(i + 1 - N)\,R_\psi(\tau - iT_c) \tag{9.30}$$

and

$$\hat{R}_{y,x}(\tau) = C_{y,x}(i)\,\hat{R}_\psi(\tau - iT_c) + C_{y,x}(i + 1)\,R_\psi(\tau - iT_c). \tag{9.31}$$

As illustrated by these expressions, it is the aperiodic (rather than the periodic) crosscorrelation that determines the multiple-access interference between two DS spread-spectrum signals whose symbols are not aligned perfectly at the receiver. As discussed previously in this section, it is not possible to synchronize the symbol timing for all of the received signals in a distributed wireless communication network with multiple transmitters and receivers. If the chip waveform is a rectangular pulse of duration T_c, each of the terms in (9.30) and (9.31) that depends on τ has a graph that is a straight line for each interval of the form $iT_c \le \tau < (i + 1)T_c$. It follows from $R_\psi(0) = 0$ and $\hat{R}_\psi(0) = \mathcal{E}_\psi$ that for any chip waveform that is time limited to $[0, T_c]$, the continuous-time partial crosscorrelation functions satisfy

$$R_{y,x}(iT_c) = C_{y,x}(i - N)\,\mathcal{E}_\psi$$

and

$$\hat{R}_{y,x}(iT_c) = C_{y,x}(i)\,\mathcal{E}_\psi.$$

Thus, if the time offset τ is a multiple of the chip duration, then the continuous-time partial crosscorrelation functions are equal to the chip energy times the discrete aperiodic crosscorrelation functions for the signature sequences x and y. Hence, for the rectangular chip waveform, each of the continuous-time partial crosscorrelation functions can be determined by drawing straight lines between consecutive values of the discrete aperiodic crosscorrelation function. This is a generalization of the observation made in Section 9.3 for the continuous-time periodic autocorrelation function. There are several important implications of this functional form for the continuous-time partial crosscorrelation functions, including the fact that the maximum magnitude of a continuous-time partial crosscorrelation function occurs at a value of τ that is a multiple of the chip duration. If the performance measure of interest is the maximum interference due to another DS spread-spectrum signal, we conclude that the poorest performance occurs when the chips of the interfering signal are aligned with those of the desired signal (i.e., *chip-synchronous interference*).

In DS spread-spectrum multiple access, which is sometimes referred to as code-division multiple access or CDMA (see Section 9.2), there are a number of different spread spectrum signals that occupy the same frequency band simultaneously [9.33]. Rather than having just one interfering DS spread-spectrum signal, as we have discussed so far in this section, there are several interfering signals. However, the interference caused by each signal is precisely of the type just described. If the system has K transmitted signals, the receiver that is matched to one of them experiences interference from the other $K-1$ signals, and $K-1$ aperiodic crosscorrelation functions are required to characterize this interference.

Not only is the multiple-access capability enhanced by careful selection of the signature sequences, but the ability of the DS spread-spectrum system to provide protection against tone jamming and multipath is also improved by employing signature sequences that have good correlation properties. The correlation properties that influence the performance of DS spread-spectrum communication systems are derived from the aperiodic autocorrelation and crosscorrelation functions for the signature sequences employed in the system.

In order to examine the influence of sequence selection on the performance of the system, we consider the processing that takes place in a correlation receiver for a spread-spectrum communication system. Suppose that a DS spread-spectrum system has N chips per data pulse. If binary data modulation is employed (e.g., binary PSK), each data pulse represents a binary channel symbol. Alternatively, each data pulse can represent a binary symbol that is sent on one component of a spread-spectrum signal that has both inphase and quadrature modulation (e.g., QPSK or MSK).

Consider a communication receiver that is matched to the desired signal, and suppose that the desired signal uses the signature sequence $\mathbf{x} = x_0, x_1, \ldots, x_{N-1}$ to spread the pulse that is being demodulated. Because of the asynchronism between the desired signal and the interference, two consecutive pulses of the interference signal affect the outcome of the demodulation of a single symbol of the desired signal. Suppose that the interference signal uses sequences $\mathbf{y} = y_0, y_1, \ldots, y_{N-1}$ and $\mathbf{z} = z_0, z_1, \ldots, z_{N-1}$ for spreading these two consecutive pulses, as illustrated in Figure 9-18. This figure portrays the time alignment between the interference and the receiver's reproduction of the spreading signal $x(t)$. It is a generalization of Figure 9-16, because the sequences are not required to be periodic in Figure 9-18. The two figures are equivalent if $\mathbf{x} = \mathbf{v}$, $\mathbf{y} = \mathbf{u}$, and $\mathbf{z} = \mathbf{u}$.

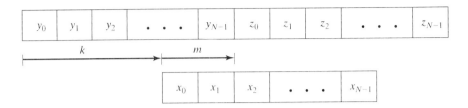

Figure 9-18: Sequence alignment for an offset of m chips.

In Figure 9-18, the offset from the beginning of the sequence **y** to the beginning of the sequence **x** is k times the chip duration. Similarly, the offset from the beginning of the sequence **x** to the beginning of the sequence **z** is m times the chip duration. (The offset index m is the same in Figure 9-16 and Figure 9-18.) Notice that $k + m = N$. We have already pointed out that for any sequences **u** and **v**, $C_{u,v}(i) = C_{v,u}(-i)$ for each integer i. Thus, it follows that

$$C_{z,x}(m) = C_{x,z}(-m) = C_{x,z}(k - N),$$

where we have used the fact that $-m = k - N$. Also,

$$C_{y,x}(m - N) = C_{x,y}(N - m) = C_{x,y}(k),$$

because $k = N - m$.

We focus attention on offsets that are multiples of the chip duration, because the complete correlation function can be determined from the correlations that correspond to such offsets. As noted previously, if the chip waveform is a rectangular pulse of duration T_c, the continuous-time correlation functions are straight lines on the interval $[jT_c, (j+1)T_c]$ for each integer j. If the chip waveform is not rectangular, the correlation functions are not straight lines in general, but their maximum values are still attained at multiples of the chip duration [9.30].

The correlation process that takes place in the receiver produces two correlations between segments of the sequences involved: A suffix of **y** is correlated against a prefix of **x**, and a prefix of **z** is correlated against a suffix of **x**. These two correlations are the two aperiodic crosscorrelations $C_{x,y}(k)$ and $C_{x,z}(k - N)$, respectively. Because the polarity of each pulse involved can be either positive or negative, the two aperiodic correlations may either add or subtract to produce the correlator output for the pulse being demodulated. That is, the polarity of one data pulse for the interference signal may result in $\mathbf{y} = y_0, y_1, \ldots, y_{N-1}$ being replaced by $-\mathbf{y} = -y_0, -y_1, \ldots, -y_{N-1}$, and the polarity of the other data pulse for the interference signal may result in $\mathbf{z} = z_0, z_1, \ldots, z_{N-1}$ being replaced by $-\mathbf{z} = -z_0, -z_1, \ldots, -z_{N-1}$. The correlator output that results from the presence of interference is therefore a constant times one of the quantities of the form

$$\pm C_{x,y}(k) \pm C_{x,z}(k - N),$$

and the choices for the two signs depend on the polarities of the pulses involved. Because we have no control over these polarities (they are determined by the data symbols being

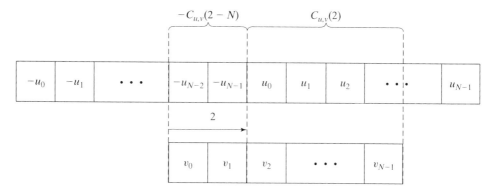

Figure 9-19: Illustration of the odd crosscorrelation $\hat{\theta}_{u,v}(2)$.

transmitted), it is desirable to choose the sequences in a way that guarantees both $|C_{x,y}(k) + C_{x,z}(k - N)|$ and $|C_{x,y}(k) - C_{x,z}(k - N)|$ are small.

Nearly all of the investigations of sequences with small correlation have been carried out for systems in which $\mathbf{z} = \mathbf{y}$; that is, the interference signal uses a single periodic sequence of period N, so each pulse of the interfering signal has the same signature sequence. For investigations of DS spread-spectrum multiple-access capability in systems for which each signal has a signature sequence of period N, then $\mathbf{z} = \mathbf{y}$ and the correlations involved can be written in terms of the periodic crosscorrelation function $\theta_{x,y}$ and the odd crosscorrelation function $\hat{\theta}_{x,y}$ [9.43], which are related to the aperiodic crosscorrelation function by

$$\theta_{x,y}(k) = C_{x,y}(k) + C_{x,y}(k - N)$$

and

$$\hat{\theta}_{x,y}(k) = C_{x,y}(k) - C_{x,y}(k - N)$$

for $0 \le k \le N - 1$. If the offset index is k (illustrated in Figure 9-19 for $k = 2$), the odd crosscorrelation function is given by

$$\hat{\theta}_{u,v}(k) = \sum_{j=0}^{N-1-k} u_j \, v_{j+k} - \sum_{j=N-k}^{N-1} u_j \, v_{j+k-N}. \qquad (9.32)$$

For investigations of DS spread-spectrum acquisition and multipath communication in systems with sequences of period N, the sequences $\mathbf{x}$, $\mathbf{y}$, and $\mathbf{z}$ are all the same, and the correlations involved are the periodic and odd autocorrelations

$$\theta_x(k) = C_x(k) + C_x(k - N)$$

and

$$\hat{\theta}_x(k) = C_x(k) - C_x(k - N).$$

There are several classes of sequences that can be employed in DS spread-spectrum systems. Among the most common classes of sequences are the m-sequences, Gold sequences, Kasami sequences, and some classes of sequences that are closely related to these [9.43]. The primary reason for the interest in these classes of sequences is the fact that they have good periodic autocorrelation and crosscorrelation functions. Since many aspects of system performance depend on aperiodic correlation properties, it is necessary to go beyond the examination of the periodic correlation functions. Within each class of sequences, some sequences have better aperiodic correlation properties than others, and considerable reductions in the correlation values can be accomplished by careful choice of which subset of sequences from a given class are to be used as signature sequences.

Even for a fixed subset of sequences, the sequences can be used in different orientations, and the orientations of the sequences influence the performance [9.35, 9.43, 9.46, 9.47]. Consider the vectors that generate a set of periodic sequences, and let T be the cyclic shift operator that is defined as follows: If $\mathbf{x} = x_0, x_1, x_2, \ldots, x_{N-1}$, then

$$T\mathbf{x} = x_1, x_2, x_3, \ldots, x_{N-1}, x_0.$$

Similarly,

$$T^2\mathbf{x} = x_2, x_3, \ldots, x_{N-1}, x_0, x_1,$$
$$T^3\mathbf{x} = x_3, \ldots, x_{N-1}, x_0, x_1, x_2,$$

etc. For n in the range $0 \leq n \leq N - 1$, the vectors $T^n\mathbf{x}$ are the *phases* of $\mathbf{x}$. For each choice of the pair of vectors $\mathbf{x}$ and $\mathbf{y}$, there are N different phases for $\mathbf{x}$ and N different phases for $\mathbf{y}$, which gives a total of N^2 different orientations between the two vectors.

The operator T plays the same role for vectors as the operator S plays for infinite sequences. In particular, if the sequence x is the periodic extension of the vector $\mathbf{x}$, then the sequence $S^n x$ is the periodic extension of the vector $T^n\mathbf{x}$. As a result, if the sequence x is a periodic sequence with period N, the sequences $S^n x, 0 \leq n \leq N - 1$, are referred to as the phases of x and the vectors $T^n\mathbf{x}, 0 \leq n \leq N - 1$ are referred to as the phases of $\mathbf{x}$. In a DS spread-spectrum system for which there is one period of the signature sequence in each data symbol, the N different phases of $\mathbf{x}$ correspond to the N different possible orientations of the sequence within the data pulse. In effect, the N different phases represent N different starting points for the vector or N different orders in which the elements of $x_0, x_1, x_2, \ldots, x_{N-1}$ can be used within the data pulse.

Although the periodic correlation properties are the same for different phases of the sequences in a given set, the aperiodic (and odd) correlation properties differ greatly among the different phases. As a result, it is possible to select the phase of a sequence $\mathbf{x}$ to minimize some measure of the aperiodic autocorrelation or to select the phases of a pair $\mathbf{x}$ and $\mathbf{y}$ to minimize some measure of the aperiodic crosscorrelation. The measures of aperiodic correlation that have been considered [9.22, 9.35, 9.43, 9.46] include the maximum value of $|\hat{\theta}_{x,y}(k)|$ for all shifts k and all pairs of vectors $\mathbf{x}$ and $\mathbf{y}$. The mean-squared value of $|\hat{\theta}_{x,y}(k)|$, averaged over all shifts k, is also a useful measure.

For a DS spread-spectrum multiple-access system with K transmitters, what is really desired is a set of K sequences for which any pair has a small aperiodic crosscorrelation. To illustrate the kind of performance improvements that can be obtained by sequence selection, the results of a performance evaluation are shown in Figure 9-20 for a DS spread-spectrum multiple-access communication system with three equal-power received signals and standard correlation receivers. The sequences employed are three nonreciprocal m-sequences of period 31 [9.43]. The curve with the smallest error probabilities corresponds to using the three m-sequences in phases that minimize the mean-squared interference. This choice provides the maximum value of the parameter SNR defined in Section 9.5. Methods for the evaluation of the probability of error are also discussed in Section 9.5.

The next lowest error probability is obtained for sequences in their AO/LSE phases, which are the phases that minimize a certain autocorrelation parameter [9.35]. The worst performance shown in Figure 9-20 is for the three m-sequences in the phases that maximize the mean-squared interference (and therefore minimize SNR). This curve is included as an illustration of how poor the performance can be if arbitrary phases are used for the m-sequences. For a bit error probability of 10^{-5} in a DS spread-spectrum multiple-access system that employs binary PSK modulation, the system with the sequences that minimize SNR requires that $\mathcal{E}/N_0$ be 1.7 dB larger than for the system with the sequences that maximize SNR. All three of these curves are for the same set of three m-sequences, but the phases of the sequences are different for the three different curves. The performance gap would be even larger if the maximization and minimization of SNR were over a larger set of sequences (e.g., all phases of all possible subsets of three Gold sequences of period 31).

The average performance for sequences chosen at random is also shown in Figure 9-20. The sequences are selected according to a uniform distribution on the set of all binary sequences of length N, and sequences for different transmitters are statistically independent. Results on the performance of DS spread-spectrum multiple-access communications with random signature sequences are given in [9.19], [9.33], [9.41], and several other papers. As illustrated in Figure 9-20, the average performance for random sequences is far worse than for m-sequences in their maximum-SNR phases or even m-sequences in their AO/LSE phases.

As mentioned earlier, the results of Figure 9-20 are for a system in which the three received signals have equal power. The effects of multiple-access interference can be much more severe if an interfering signal has much greater power than the desired signal, which is the unfavorable near–far condition mentioned previously. In DS spread-spectrum multiple-access systems with standard correlation receivers, it is important to control the power of the transmitted signals in order to limit the degradation caused by unfavorable near–far conditions. The sensitivity to the near–far conditions can be reduced by using more sophisticated receivers [9.55].

Although there are some limited results on the performance of DS spread-spectrum multiple-access systems in which the sequence period is greater than the data pulse duration [9.28], very little optimization of sequences has been carried out for this situation. As a result, it is common to model the signature sequences as random for systems with multiple data pulses per period of the signature sequence. The performance obtained is then the average performance, averaged over the set of all sequences. It is also true that as the number of sequences required by the system increases, the complexity of optimizing the crosscorrelation functions increases greatly. For a system with K transmitters, there are on the order of K^2 crosscorrelation functions that must be kept small. As a result,

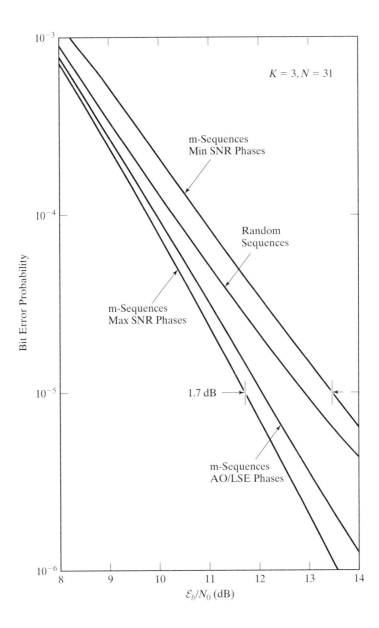

Figure 9-20: Bit error probability for four sets of sequences.

selections according to an autocorrelation criterion (e.g., AO/LSE) may be preferred or even necessary, even though the resulting choices will give poorer performance than if the sequences were selected according to an appropriate crosscorrelation criterion. Of course, if the major source of interference is multipath rather than multiple-access interference, the optimization with respect to autocorrelation is preferred anyway. For protection against tone jamming, the balance between the number of 0s and the number of 1s and certain other properties are also important, but many of these properties depend more on the class of sequences rather than the phases in which they are used. Additional information on the design of DS spread-spectrum waveforms and the selection of signature sequences is given in [9.33], [9.43], and [9.47].

For a channel that exhibits frequency-selective fading [9.50, 9.51], the selection of sequences greatly influences the performance of DS spread-spectrum communications with either a correlation receiver or a rake receiver (e.g., see [9.24] and [9.25]). In a correlation receiver, the discrimination against multipath components is similar to the discrimination against multiple-access interference. If the received multipath components are of comparable power levels, the use of a rake receiver should be considered. One type of rake receiver has multiple correlators that track and demodulate different multipath components. Regardless of the specific type of rake receiver, the fundamental difference as compared with a simple correlation receiver is that the rake receiver attempts to utilize the energy in the multipath components rather than simply discriminate against them.

As illustrated in [9.24], proper sequence selection can permit a reduction in the number of demodulators in the rake receiver with no increase in error probability. Thus, sequence selection can result in reduced complexity requirements for the receiver, as well as decreased power requirements for the transmitter. For channels in which the maximum delay spread is less than the duration of the data pulse, a periodic sequence can be employed with one period per data pulse. If the multipath spread is larger than the data pulse duration, it is necessary for the sequence period to be longer than the data pulse, and this complicates the optimization.

In GPS, each of the satellites transmits a DS spread-spectrum signal, and each GPS receiver attempts to track and demodulate the signals from multiple satellites. The GPS receiver tracks the spread-spectrum signal from a given satellite in order to obtain an estimate of the GPS receiver's distance from that satellite. The GPS receiver also demodulates the spread-spectrum signal from that satellite in order to obtain the information that is broadcast by the satellite in its data signal. Such information includes clock corrections and the satellite's position data. As mentioned in a previous section, all GPS receivers have access to the C/A component of the first navigation signal, whereas access to the P component is restricted. For this reason, and in order to simplify the presentation, we limit the discussion of the operation of the GPS receiver to the C/A component only. The same principles apply to receivers that process and utilize the P component.

The C/A components of the navigation signals transmitted by different satellites employ different sequences from a set of Gold Sequences of period 1023. As such, their periodic crosscorrelation functions take on the values -1, $+63$, and -65 only. The incoming signal at a GPS receiver consists of the sum of the navigation signals from all GPS satellites that are in view of the receiver. The first navigation signal from each satellite has the same carrier frequency f_1, so there is multiple-access interference at the output of each correlator whenever two or more satellites are in view. For the C/A component of the first navigation signal, the fact that there are 20 periods of each Gold sequence per data pulse means that the multiple-access interference due to C/A components from other satellites is determined largely by the periodic crosscorrelation functions. The correlation interval consists of 20 periods of the 1023-chip sequence, so for each interfering signal there are at least nineteen periodic crosscorrelations and at most one odd crosscorrelation. If the appropriate pair of data bits agree in polarity, there are 20 periodic crosscorrelations and no odd crosscorrelations.

Similarly, multipath discrimination is determined largely by the periodic auto-correlation function for the Gold sequence used by the first navigation signal. The peak in the periodic autocorrelation function is 1023, of course, which occurs at time offset of zero, and the maximum periodic autocorrelation side-lobe magnitude is the same as the maximum magnitude of the periodic crosscorrelation function, which is 65. The ratio of the maximum periodic autocorrelation side lobe to the peak for the periodic autocorrelation function is $65/1023 \approx 0.0635$, which is also the ratio of the maximum crosscorrelation to the peak autocorrelation.

In the GPS application, the *normalized multiple-access interference* is the ratio of the multiple-access interference to the desired signal if each is measured at the output of a correlator matched to the C/A spreading waveform of the desired signal. This ratio is equivalent to the maximum of the normalized multiple-access interference for two, equal-power, first navigation signals arriving at a GPS receiver. Because this is the ratio of two voltages, it follows from

$$20 \log_{10}(65/1023) \approx -23.9 \text{ dB}$$

that the maximum multiple-access interference is nearly 24 dB below the desired signal.

The *pseudorange* is the product of the velocity of light and the difference between the time that a given chip is transmitted, as measured and reported by the satellite, and the time that it is received, as measured by the GPS receiver. For several reasons, including the fact that the receiver's clock is not in synchronism with the satellite's clock, this is not the actual range. Because the satellites' positions are known, and this information is obtained by the GPS receiver from the data signal, the pseudorange measurements to four satellites (which provide four equations) are sufficient to determine the three coordinates of the receiver's position and resolve the uncertainty in the receiver's clock (four unknowns).

If the error in the receiver's clock is denoted by Δ, the corresponding range error is $r = c\Delta$, where c is the velocity of light. If μ_i is the pseudorange to the ith satellite, which has reported it is at position (X_i, Y_i, Z_i), the four equations are

$$\mu_1 + r = [(X_1 - x)^2 + (Y_1 - y)^2 + (Z_1 - z)^2]^{1/2},$$
$$\mu_2 + r = [(X_2 - x)^2 + (Y_2 - y)^2 + (Z_2 - z)^2]^{1/2},$$
$$\mu_3 + r = [(X_3 - x)^2 + (Y_3 - y)^2 + (Z_3 - z)^2]^{1/2},$$

and

$$\mu_4 + r = [(X_4 - x)^2 + (Y_4 - y)^2 + (Z_4 - z)^2]^{1/2}.$$

The four unknowns are r, the range error corresponding to the receiver's clock error, and x, y, and z, the coordinates of the receiver's position. In practice, there are several other variables that introduce errors in the determination of the receiver's position, including small errors in satellite clocks and satellite positions and the delays encountered as the signals propagate through the ionosphere [9.6, 9.49].

The GPS receiver must track the spread-spectrum signals as the satellites move in their orbits. (The receiver may be moving as well.) This is accomplished in most commercial GPS receivers by a tracking subsystem that employs an appropriate modification of the baseband delay-lock loop illustrated in Figure 9-8. The modification is necessary because the received signal is a data-modulated RF spread-spectrum signal, and the data symbols and the phase and frequency of the RF carrier are all unknown. Generally, this modification results in a noncoherent delay-lock tracking loop [9.16, 9.48].

9.5   Performance Evaluation for DS Spread-Spectrum Multiple-Access Communications

One feature of DS spread-spectrum modulation is its ability to permit multiple simultaneous transmissions in the same frequency band. This feature, which is often referred to as the multiple-access capability of DS spread spectrum, is derived in part from the properties of the signature sequences and in part from the type of receiver that is employed to demodulate the signals. In this section, we describe two alternative measures of the multiple-access capability for DS spread-spectrum systems that employ standard correlation receivers, and we illustrate the dependence of these performance measures on the correlation parameters of the signature sequences.

The DS spread-spectrum multiple-access system considered in this section is one in which K terminals transmit simultaneously at the same carrier frequency. The terminals are in K different locations as illustrated in Figure 9-15. Let the terminals be indexed by the integer k for $1 \leq k \leq K$. The signal transmitted by the kth terminal is the DS spread-spectrum signal given by

$$\sqrt{2}\, A_k\, a_k(t)\, b_k(t) \cos(\omega_c t + \theta_k),$$

where $a_k(t)$ is the spreading signal and $b_k(t)$ is the baseband data signal for the kth spread-spectrum signal. The data pulses are rectangular pulses of duration T, and the chip waveform is of duration T_c. In this section, it is assumed that each signature sequence is periodic, and its period is $N = T/T_c$. Thus, each signature sequence has one period per data pulse.

Consider a terminal that wishes to receive and demodulate one of these spread-spectrum signals. Since the transmitters might be at different distances from the receiver, the signals may have different propagation delays and different propagation losses. In addition, the clocks in different transmitters might not be synchronized, in which case there would be additional time offsets in the transmitted signals that correspond to the time differences among these clocks. If the combination of clock offset and propagation

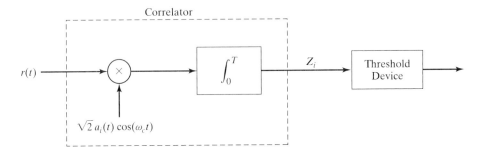

Figure 9-21: Correlation receiver for the ith signal.

delay for the kth signal is denoted by τ_k, the received signal is given by

$$r(t) = X(t) + \sum_{k=1}^{K} \sqrt{2}\, A_k\, a_k(t - \tau_k)\, b_k(t - \tau_k) \cos(\omega_c t + \varphi_k), \qquad (9.33)$$

where $X(t)$ denotes the thermal noise. The phase of the kth signal is given by $\varphi_k = -\omega_c \tau_k + \theta_k$, which accounts for the original phase θ_k and the effect of time shift τ_k on the phase of the carrier. The term $-\omega_c \tau_k$ represents the phase rotation of the carrier due to time offset τ_k.

Suppose the receiver wishes to demodulate only the ith signal. In this situation, the ith signal is referred to as the desired signal, and the remaining $K-1$ spread-spectrum signals in (9.33) are referred to as interference signals. Because the time reference is arbitrary, there is no loss in generality in setting $\tau_i = 0$ in (9.33), which simplifies the notation. In effect this just means that the time shifts for the other signals are measured relative to the time shift of the ith signal. Similarly, because the phase reference is arbitrary, we can let $\theta_i = 0$ and measure all phase shifts relative to the phase of the ith signal. Notice that $\tau_i = 0$ and $\theta_i = 0$ imply that $\varphi_i = 0$. If $\tau_i = 0$ and $\varphi_i = 0$, then the receiver illustrated in Figure 9-21 is the standard correlation receiver for the ith signal.

As is customary, we interpret all phase angles modulo 2π, which permits us to restrict attention to the range $0 \le \varphi_k \le 2\pi$. It is clear, for example, that an interference signal whose phase angle is 3π is identical to an interference signal whose phase angle is π. Similarly, in most situations in which the signature sequences have period $N = T/T_c$, it makes no difference if the time offset of an interference signal is $3T/2$ or $T/2$. Consequently, for the performance measures considered in this chapter, we can restrict attention to time offsets in the range $0 \le \tau_k \le T$. In effect, this is equivalent to interpreting all time shifts modulo T, which is legitimate as long as the signals and the system "behave in the same way" in different data intervals. This is the case for the system under consideration if, for example, the data sequences are modeled as sequences of independent, identically distributed random variables. The important result for our purposes is that the performance measures considered in this section are not affected by time shifts that are multiples of T, just as they are not affected by phase shifts that are multiples of 2π.

The output of the correlation receiver consists of a desired component due to the ith signal, an interference component due to the combined effect of the $K - 1$ interference signals, and a noise component due to the thermal noise. As shown in (9.23), the interference term can be written in terms of continuous-time partial crosscorrelation functions. These correlation functions for the K spread-spectrum signals are defined by

$$R_{k,i}(\tau) = \int_0^\tau a_k(t - \tau) a_i(t) \, dt \qquad (9.34)$$

and

$$\hat{R}_{k,i}(\tau) = \int_\tau^T a_k(t - \tau) a_i(t) \, dt \qquad (9.35)$$

for $1 \le k \le K$ and $1 \le i \le K$. Each of the K baseband data signals is a sequence of rectangular pulses, and each data pulse has amplitude -1 or $+1$. As discussed in Section 9.4, the interference from one of the spread-spectrum signals depends on the amplitudes of two consecutive data pulses from the interfering signal. Let $\mathbf{d} = (d_{-1}, d_0)$ represent the amplitudes of two consecutive data pulses in one of the signals. For each pair of integers k and i, we define $I_{k,i}(\mathbf{d}, \tau, \varphi)$ by

$$I_{k,i}(\mathbf{d}, \tau, \varphi) = T^{-1}\{d_{-1} R_{k,i}(\tau) + d_0 \hat{R}_{k,i}(\tau)\} \cos \varphi \qquad (9.36)$$

for $0 \le \tau \le T$ and $0 \le \varphi \le 2\pi$. If $\mathbf{d}_k$ denotes the amplitudes of two consecutive data pulses from the kth signal, then the interference due to the kth signal is proportional to $I_{k,i}(\mathbf{d}_k, \tau_k, \varphi_k)$.

If the phase angles, time delays, and data pulse amplitudes are modeled as a collection of independent random variables, it is of interest to determine the variance $\sigma^2(k, i)$ of the interference $I_{k,i}(\mathbf{d}, \tau, \varphi)$. If the phase angles are uniformly distributed on $[0, 2\pi]$, the time offsets are uniformly distributed on $[0, T]$, and the data pulse amplitudes are uniformly distributed on $\{-1, +1\}$, then [9.30]

$$\sigma^2(k, i) = \{\mu_{k,i}(0) \, \mathcal{M}_\psi + \mu_{k,i}(1) \, \mathcal{N}_\psi\}/T^3, \qquad (9.37)$$

where the sequence correlation parmeters are given by

$$\mu_{k,i}(m) = \sum_{n=1-N}^{N-1} C_{k,i}(n) \, C_{k,i}(n + m) \qquad (9.38)$$

for $m = 0$ and $m = 1$, and the chip waveform parameters are

$$\mathcal{M}_\psi = \int_0^{T_c} [R_\psi(s)]^2 \, ds, \qquad (9.39)$$

and

$$\mathcal{N}_\psi = \int_0^{T_c} R_\psi(s) \, \hat{R}_\psi(s) \, ds. \qquad (9.40)$$

The functions R_ψ and $\hat{R}_\psi$ are defined in (9.29) and (9.30). As an example, if the chip waveform is the unit-amplitude rectangular pulse of duration T_c, the chip waveform parameters are given by $\mathcal{M}_\psi = T_c^3/3$ and $\mathcal{N}_\psi = T_c^3/6$. Recall that $NT_c = T$, so $\mathcal{M}_\psi/T^3 = (3N^3)^{-1}$ and $\mathcal{N}_\psi/T^3 = (6N^3)^{-1}$. It follows that

$$\sigma^2(k, i) = \{2\mu_{k,i}(0) + \mu_{k,i}(1)\}/(6N^3) \tag{9.41}$$

for the rectangular chip waveform.

Notice from (9.37)–(9.40) that the only dependence of the variance of the interference on the signature sequences is through the terms $\mu_{k,i}(0)$ and $\mu_{k,i}(1)$. Similarly, the only dependence on the chip waveform is through the parameters $\mathcal{M}_\psi$ and $\mathcal{N}_\psi$. This separation of the effects of the sequences and the effects of the chip waveform greatly simplifies the evaluation $\sigma^2(k, i)$. The evaluation of the variance of the interference is simplified even more by use of the identity

$$\mu_{k,i}(m) = \sum_{n=1-N}^{N-1} C_k(n)\, C_i(n+m), \tag{9.42}$$

which is derived in [9.36]. This means that it is not necessary to evaluate the aperiodic crosscorrelation functions for the signature sequences used in the system: Only the aperiodic autocorrelation functions are required. This represents a great reduction in computation, because for K signals there are $K(K-1)/2$ crosscorrelation functions but only K autocorrelation functions.

Suppose the received power is the same for the K spread-spectrum signals; that is, $A_k = A$ for each k. For this special case, define

$$V_i = \sum_{k \neq i} \sigma^2(k, i), \tag{9.43}$$

where the summation is over all k for which $1 \leq k \leq K$ and $k \neq i$. The variance of the multiple-access interference at the output of the correlator for the ith signal is proportional to V_i. Define SNR_i, the signal-to-noise ratio at the output of this correlator, as the conditional mean of Z_i divided by the standard deviation of Z_i. The conditional mean referred to in this definition is the conditional expectation of Z_i given the amplitude of the ith signal's data symbol in the interval $[0, T]$. It is shown in [9.29] and [9.30] that

$$\text{SNR}_i = \{V_i + (2\,\mathcal{E}_b/N_0)^{-1}\}^{-1/2}, \tag{9.44}$$

where $\mathcal{E}_b = A^2\, T$ is the energy per bit for each of the signals.

In a more general situation, the amplitudes A_k need not be the same, in which case we let

$$\alpha_{k,i} = (A_k/A_i)^2$$

and define

$$W_i = \sum_{k \neq i} \alpha_{k,i}\, \sigma^2(k, i). \tag{9.45}$$

In this situation, we can employ a weighted signal-to-noise ratio [9.30] defined by

$$\text{WSNR}_i = \{W_i + (2\,\mathcal{E}_i/N_0)^{-1}\}^{-1/2}, \tag{9.46}$$

where $\mathcal{E}_i = A_i^2 T$ is the energy per bit for the ith signal. The effects of an unfavorable near–far condition are revealed by (9.45) and (9.46). For example, if the received power for one of the interference signals increases by 10 dB, perhaps because the source of the interference moves closer to the receiver, the corresponding term in (9.45) increases by a factor of 10. If the received power increases by 10 dB for each of the interference signals, then W_i increases by a factor of 10. As is clear from (9.46), the amount of the resulting decrease in the signal-to-noise ratio depends on the value of $\mathcal{E}_i/N_0$. In an interference-limited environment, W_i is much greater than $N_0/\mathcal{E}_i$, and so WSNR_i decreases by approximately 10 dB in this example.

A complication that arises in attempting to use WSNR_i as a criterion for selecting signature sequences is that the weights $\alpha_{k,i}$ in (9.45) are typically unknown at the time the sequences are selected. Furthermore, in many applications, such as in a network of mobile terminals, these weights may change very rapidly. For the majority of systems, it is impractical to base the selection of signature sequences on weighted signal-to-noise ratios, so we use SNR_i for sequence selection in all that follows.

It is clear from (9.43) and (9.44) that, for a given value of $\mathcal{E}_i/N_0$, the parameter SNR_i is maximized by minimizing V_i. Because V_i depends on $\sigma^2(k, i)$ for all $k \neq i$, this suggests that we should make $\sigma^2(k, i)$ small for all such k. Of course, what we really want is for SNR_i to be large for *all* i, $1 \leq i \leq K$, which implies that V_i should be small for all values of i; therefore, $\sigma^2(k, i)$ should be small for all pairs (k, i) for which $k \neq i$.

One difficulty with this objective is illustrated by the following scenario. Suppose $K = 3$, and we begin by choosing sequence $\mathbf{x}$ for the first signal and sequence $\mathbf{y}$ for the second signal in a way that minimizes $\sigma^2(1, 2)$. We would then like to choose the sequence $\mathbf{z}$ for the third signal to minimize $\sigma^2(1, 3)$ and $\sigma^2(2, 3)$. In addition to the fact that it may not be possible to minimize these simultaneously, there is the problem that it may be necessary to change $\mathbf{x}$ and $\mathbf{y}$ just to obtain acceptably small values for $\sigma^2(1, 3)$ and $\sigma^2(2, 3)$, let alone minimize them. In other words, what is required is a joint optimization over all sets of K sequences. Even for modest values of K, such an optimization requires a prohibitive amount of computation.

As a result of these complications, various suboptimal approaches have been adopted for sequence selection. In [9.8], for example, sequences are selected to minimize $I_{\max}$, which is defined by

$$I_{\max} = \max\{V_i : 1 \leq i \leq K\}, \tag{9.47}$$

where V_i is defined by (9.43). Results are given in [9.8] for both the rectangular pulse and the sine pulse as the chip waveform. Note from (9.37) and (9.43) that V_i depends on the chip waveform and the sequences.

The selection of sequences to minimize $I_{\max}$ is equivalent to the selection of sequences to maximize the minimum signal-to-noise ratio, which is

$$\text{SNR}_{\min} = \min\{\text{SNR}_i : 1 \leq i \leq K\}. \tag{9.48}$$

Table 9.1: Values of $\sigma^2(k, i)$ for the Rectangular Pulse ($K = 3$, $N = 31$)

i	k	$\sigma^2(k, i)$
1	2	0.00885
1	3	0.00892
2	3	0.00883

The optimum values for $\sigma^2(k, i)$ for $1 \leq i < k \leq 3$ are given in Table 9.1 for three nonreciprocal m-sequences of period 31 in a system with a rectangular chip waveform. There are only three such sequences, so the optimization is with respect to the phases of these sequences, as discussed in Section 9.4. (See Figure 9-20.) If the phases of the sequences are not selected to maximize $\mathrm{SNR}_{\min}$, the values for $\sigma^2(k, i)$ can be as large as 0.01462.

In the absence of multiple-access interference (e.g., if there is a single spread-spectrum signal), $W_i = 0$ and (9.46) reduces to $\mathrm{SNR}_i = \sqrt{2\,\mathcal{E}_i/N_0}$. In this case, the only channel disturbance is the thermal noise, so the probability of bit error for the ith receiver is given by

$$P_e(i) = Q(\mathrm{SNR}_i) = Q\left(\sqrt{2\,\mathcal{E}_i/N_0}\right).$$

In this situation, the error probability is the same as for any antipodal signal set, and no benefit of spread spectrum is demonstrated. This observation is consistent with the opening remarks of this chapter. The benefits of spread spectrum are achieved in applications in which the channel disturbance is not limited to thermal noise.

If there are two or more spread-spectrum signals, the interference depends on the time offsets, phase angles, and data pulse amplitudes as shown in (9.36). If $\boldsymbol{\tau}$ is the vector of all time offsets, $\boldsymbol{\varphi}$ is the vector of all phase angles, and $\mathbf{b}$ is the vector of all relevant data symbols, then the probability of error for the ith receiver depends on $\boldsymbol{\tau}$, $\boldsymbol{\varphi}$, and $\mathbf{b}$. We denote this probability of error by $P_e(\boldsymbol{\tau}, \boldsymbol{\varphi}, \mathbf{b})$. In principle, we could determine the probability of error for each value of these parameters and then average this probability over the ranges of the parameters to find an average probability of error $\overline{P}_e(i)$. As a practical matter, this computation is quite difficult unless K and N are very small. The computational requirement to evaluate $\overline{P}_e(i)$ by standard numerical integration is proportional to $(\gamma N)^{K-1}$, where the constant γ depends on the required accuracy. It is not uncommon for γ to be in the range from 100 to 10,000, but a more serious problem is the exponential dependence on K, the number of spread-spectrum signals.

If there is no multiple-access interference, then

$$\overline{P}_e(i) = P_e(i) = Q(\mathrm{SNR}_i),$$

because there are no parameters over which to average. Even if the multiple-access interference is not negligible, there are certain situations in which the approximation $\overline{P}_e(i) \approx Q(\mathrm{SNR}_i)$ is very useful, especially for system design purposes. This approximation is known as the *Gaussian approximation* for the probability of error in a DS spread-spectrum multiple-access system. It is a reasonably accurate approximation if, for instance, N is large and K is approximately an order of magnitude smaller than N. The accuracy of the Gaussian approximation for error probabilities down to about 10^{-4} is demonstrated in [9.39] for $N = 31$ and $K = 2$, 3, and 4.

It should be kept in mind that the use of the signal-to-noise ratio previously defined is appropriate only when it makes sense to average over the full range of values for the time shifts, phase angles, and data variables. Of course this caveat also applies to the use of the Gaussian approximation, since it is derived from the signal-to-noise ratio. For applications in which information is sent in short packets, for example, the time offsets may not change significantly within the time duration of a packet. In situations such as this, any performance measure that is based on an average over the range $0 \leq \tau_k \leq T$ may not be very meaningful. The issue is not whether the Gaussian approximation can be employed, but instead it is a question of whether it is reasonable to use any performance measure that is obtained by averaging over time shifts, phase angles, and data variables. Even the exact value of $\overline{P_e}(i)$ is not a good measure of performance if it is not appropriate to average over the full ranges of these parameters.

For situations in which it is appropriate to average over the time delays, phase angles, and data variables, there are several methods for computing the average probability of error. Standard numerical methods applied to the expression for the average probability of error in a DS spread-spectrum multiple-access system require multidimensional numerical integration, and the number of computations is an exponential function of the number of signals. Moreover, the calculation must be carried out separately for each choice of $\mathcal{E}/N_0$ that is of interest. Beginning in the mid-1970s, several efficient methods for the evaluation of the probability of error have been developed for both coherent and noncoherent correlation receivers. We close this section with a brief guide to the literature on some of these methods. A more detailed survey is given in [9.33].

Many accurate methods now exist for the evaluation of the average probability of error in a DS spread-spectrum multiple-access communication system that employs a correlation receiver. The first precise calculation of the error probability, other than brute-force numerical integration, was published by Yao [9.62] at the same time as the publication of the results on the signal-to-noise ratio [9.29]. Although various approximations based on simpler models predate [9.29] and [9.62], the majority of these earlier results do not account for the correlation properties of the signature sequences and do not provide reliable estimates of their accuracy. Among other benefits, Yao's results made it possible to check the accuracy of the Gaussian approximation.

Yao [9.62] applied moment-space bounding techniques to obtain tight upper and lower bounds on the average probability of error. This gives not only an accurate approximation, but also a bound on the error in that approximation. Chronologically, the next results were obtained by exploiting the convexity of $Q(x)$ for $x > 0$. Both upper and lower bounds were developed in the late 1970s and later published in [9.39]. The original results are for systems in which the maximum multiple-access interference does not exceed the desired signal, which limits the applications to systems for which the ratio K/N is not very large. For example, if $N = 31$, it appears necessary to have $K \leq 4$, even for sequences with very good crosscorrelation. This approach was later extended in [9.5] to include systems in which the maximum interference exceeds the desired signal, and it was employed in [9.5] to analyze the effects of multiple-tone interference as well as multiple-access interference.

Another method, which is based on an approximation rather than a bound, is the characteristic-function method [9.9] that has been applied to both the multiple-access systems and communication over multipath channels [9.11]. As is also true for the moment-space bounding technique, the characteristic-function method had been employed for intersymbol-interference channels prior to the application to spread-spectrum multiple-access communications. In applications to coherent spread-spectrum communications, the characteristic function method is very useful for the approximation of the average probability of error, and the required number of computations is proportional to $N(K-1)$ rather than being exponential in $K-1$. In addition, many of the computations need not be repeated for each value of $\mathcal{E}/N_0$ that is of interest. The characteristic-function method has been applied to several other DS spread-spectrum modulation formats such as offset QPSK and MSK.

Other methods that have been developed for spread-spectrum systems or adapted for use in such systems include the series-expansion methods described in [9.9], [9.59], [9.60], and [9.61], and methods based on Gauss-quadrature integration formulas [9.18]. The best method for a given application depends on the specific parameters of the modulation and on the software that is available.

The discussion so far is for signature sequences that are deterministic and periodic with one period per data pulse. Some of the methods can be applied to systems with multiple data symbols per period, but the computational requirements are greater. If the sequence period is much greater than the data symbol duration, it may be better to model the sequences as random sequences. If so, the error probability is computed by averaging over the ensemble of sequences in addition to averaging over the phase angles and time delays. The typical model for a random signature sequence is a sequence of independent binary random variables, each of which takes values -1 or $+1$ with equal probability. The signature sequences for signals transmitted by different terminals are assumed to be statistically independent.

It is shown in [9.29] that if the received power is the same for all of the spread-spectrum signals and the signature sequences are random, the Gaussian approximation is $\overline{P}_e \approx Q(\text{SNR})$, where

$$\text{SNR} = \left\{ \frac{(K-1)}{3N} + \frac{N_0}{2\,\mathcal{E}_b} \right\}^{-1/2}.$$

This expression can be modified to account for unequal power by using the same approach as in (9.45) and (9.46). One method that is available for evaluating the average probability of error in systems with random sequences is the characteristic-function method. A better alternative is to work directly with the probability density functions for the interference components. This is the method employed in [9.19], where tight upper and lower bounds are obtained for multiple-access systems with random signature sequences. The bounds can be made as tight as desired by increasing the amount of computation.

9.6   Frequency-Hop Spread Spectrum

Signals with very large bandwidths can be generated by a form of spread-spectrum modulation known as *frequency hopping*. The basic feature of frequency-hop spread spectrum is that the carrier frequency of a communication signal is changed or *hopped* over some predetermined set of frequencies. A communication signal with bandwidth W that is hopped over a set of q frequencies can occupy a total bandwidth of as much as qW. The resulting spread-spectrum signal is referred to as a *frequency-hop signal*. In the frequency-hop form of spread spectrum, the instantaneous bandwidth of the signal may be no more than the bandwidth of the original communication signal, even though the long-term spectral occupancy increases by as much as a factor of q.

The RF band that is available for the frequency-hop spread spectrum signal is divided into q sub-bands that are referred to as *frequency slots*. It is not necessary for the frequency slots to be contiguous, so it is possible to reserve some portions of the spectrum for use by other signals. The times at which the carrier frequency is changed are referred to as *hop epochs*, and the time intervals between consecutive hop epochs are called *hop intervals*. For the standard frequency-hop signal, the hop intervals all have the same length. However, this is not necessary, and there are good reasons in some military systems for the use of variable-length hop intervals.

Data can be transmitted during all or part of each hop interval. The time interval during which a data signal is actually transmitted is called the *dwell interval* for the frequency-hop signal. There are various reasons, including protection against certain types of jamming, why it may be desirable to have the dwell interval smaller than the hop interval for some applications. In fact, it may be necessary to do so in order to provide time for the settling of the frequency synthesizer at the beginning of each hop interval.

In order to give a mathematical description of frequency-hop signaling, we consider a data communication signal of the form

$$v(t) = \sqrt{2}\, m(t) \cos[2\pi f_d\, t + \theta(t) + \varphi],$$

where $m(t)$ represents amplitude modulation and $\theta(t)$ represents phase modulation. One of the two modulation signals can be held constant in order to achieve amplitude or phase modulation alone. A frequency-hop signal that conveys the same information as $v(t)$ is obtained by replacing the fixed frequency f_d by a time-varying frequency. A system to accomplish this is illustrated in Figure 9-22. In this model, the input to the frequency synthesizer is the sequence $\xi = (\xi_j)$ and the output of the frequency synthesizer is the signal

$$c(t) = 2 \cos[2\pi \psi(t)t + \alpha(t)].$$

The frequency is given by $\psi(t) = \xi_j$ during the jth hop interval. The signal $\alpha(t)$ represents the phase shifts that may be introduced by the frequency synthesizer as it switches frequencies at the beginning of each hop interval. The sequence (ξ_j) is referred to as the *hopping pattern* for the frequency-hop signal. In most applications, the hopping pattern is some type of pseudorandom sequence, but it is not necessarily generated by a linear feedback shift register.

The purpose of the bandpass filter in Figure 9-22 is to remove the lower sideband of the signal generated by the mixer. The output of the bandpass filter is the frequency-hop signal

$$s(t) = \sqrt{2}\, m(t) \cos[2\pi\, \zeta(t)\, t + \theta(t) + \alpha(t) + \varphi].$$

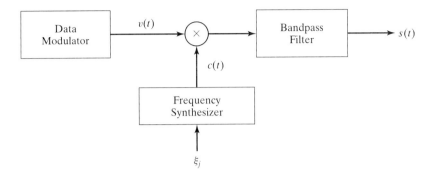

Figure 9-22: Frequency-hop modulator.

The time-varying carrier frequency for the frequency-hop signal is given by

$$\zeta(t) = f_d + \psi(t).$$

In a frequency-hop system in which each hop interval is of duration T_h, the *hopping rate* is $1/T_h$. Frequency-hop spread-spectrum signals are generally divided into two categories according to the relationship between their hopping rates and data symbol transmission rates. If one or more data symbols is transmitted in each dwell interval, the frequency-hop signal is referred to as a *slow-frequency-hop* (SFH) signal; otherwise, it is a *fast-frequency-hop* (FFH) signal. In fast frequency hopping, the transmission of each symbol requires at least two dwell intervals. Stated another way, the hopping rate exceeds the data symbol transmission rate for FFH signals. On the other hand, it is not uncommon in a SFH system for tens or hundreds of data symbols to be transmitted in each dwell interval, and this corresponds to a hopping rate that is one or two orders of magnitude slower than the data rate.

By definition, the hopping rate does not exceed the data rate in a SFH system. Although the bandwidth of a FFH signal is determined primarily by the hopping rate and the number of frequency slots, the bandwidth of a SFH signal is determined primarily by the data rate and the number of frequency slots. For a given data modulation method, the data rate determines the required bandwidth for each frequency slot, and the total bandwidth of the SFH signal is just the product of the width of each slot and the number of frequency slots.

A SFH signal with six data symbols per hop is illustrated in Figure 9-23. In this illustration, the dwell interval is equal to the hop interval, but that is not necessary. The length of the dwell interval could be less than the length of the hop interval. During any given time that a data symbol is being transmitted in a SFH system, the transmitted energy is concentrated in a single frequency slot. As a result, interference that occupies only part of the total bandwidth of the SFH signal may not affect some of the data symbols that are transmitted. For example, if there is interference in only one of the four frequency slots used by the SFH signal illustrated in Figure 9-23, then 18 of the 24 data symbols depicted there are not influenced by the interference. Interference that is present in only a subset of the frequency slots is referred to as *partial-band interference*, whereas interference that occupies the entire band of the frequency-hop signal is referred to as *full-band interference*. Thermal noise is an example of full-band interference.

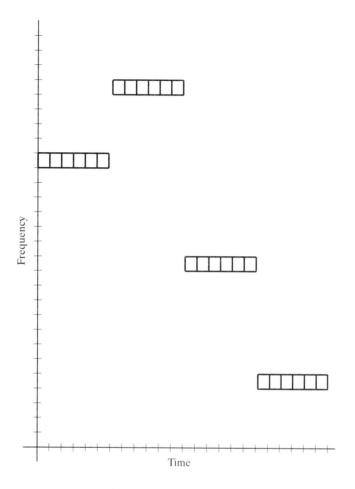

Figure 9-23: A SFH signal.

The partial-band interference may represent a communication signal operating at a fixed frequency, or, in a military system, it may represent partial-band jamming. The frequency occupancy of the partial-band interference may change from time to time, but such changes are normally much slower than the hopping rate of the frequency-hop signals. If a symbol of the desired signal is transmitted in a frequency slot that contains partial-band interference, we say that the symbol has been *hit*.

Frequency-hop signaling differs from direct-sequence spread spectrum in that the former does not provide any protection for individual symbols that are transmitted in a frequency slot that has interference. As a result, frequency-hop signaling is vulnerable to various forms of partial-band interference unless suitable error-control coding is employed [9.56]. The primary benefit of frequency-hop signaling is the frequency diversity that it provides, and this frequency diversity can be exploited to compensate for the presence of interference in a fraction of the frequency slots.

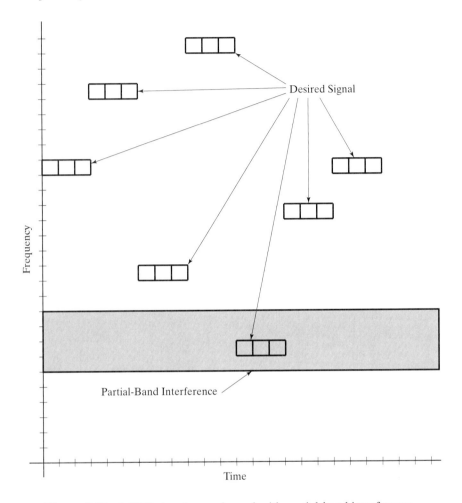

Figure 9-24: A SFH signal on a channel with partial-band interference.

The scenario illustrated in Figure 9-24 exemplifies the operating environment for SFH communications. Partial-band interference, illustrated by the shaded band in Figure 9-24, occupies four frequency slots out of the 28 frequency slots available for frequency hopping. The partial-band interference might be a narrowband signal that is transmitted at a fixed frequency or it might be bandlimited noise. For military applications, certain types of hostile jamming signals are modeled as partial-band interference. Also shown in Figure 9-24 is the SFH signal, which is referred to as the *desired signal*. In this illustration, the SFH signal has three data symbols per hop. Therefore, if the data symbol rate is the same for Figures 9-23 and 9-24, the hopping rate of the SFH signal in Figure 9-24 is twice the hopping rate of the signal in Figure 9-23.

Notice from Figure 9-24 that there is interference in the frequency slot that is occupied by the SFH signal in the fifth dwell interval. This interference increases the probability of error for symbols that are transmitted in that dwell interval. If the partial-band interference is much stronger than the desired signal, the probability of error can be very large. For example, if binary data modulation is employed, the error probability for each symbol may be approximately $1/2$. More generally, if the modulation is M-ary, the probability of error for a given symbol may be approximately $(M-1)/M$. Interference that causes such a large probability of error is referred to as *catastrophic interference*. Other forms of partial-band interference may not lead to such large error probabilities, but they still may result in unacceptable performance unless some form of error correction is incorporated into the SFH system.

The necessity for error-control coding in SFH transmission can be illustrated with a simple example. Suppose that for a particular SFH binary data transmission system the frequency band used by the SFH signal consists of 100 frequency slots. Assume that the sequence of frequency slots used by the signal is a random, independent, identically distributed sequence. Assume also that each frequency slot is used with the same probability (i.e., the probability that the signal is in a particular slot at a given time is 0.01). Suppose that catastrophic interference is present in one or more frequency slots. It is usually not possible to avoid the use of the frequency slots that have interference, because in most applications the transmitter does not know the spectral location of the interference. In many applications, the frequency occupancy of the interference may be changing with time (e.g., the interference might be another SFH signal).

Suppose in this example that a bit error probability of 10^{-4} is required, and recall that the frequency band used by the SFH signal consists of 100 frequency slots. In this situation, the probability of error is approximately 5×10^{-3} if the catastrophic interference is confined to a single slot and there is no interference at all in the other 99 frequency slots. The only way to improve this situation and meet the goal of a bit error probability of 10^{-4} is to use error-control coding to correct some of the errors caused by the interference.

As illustrated by this example, it is necessary to use error-control coding in a SFH system to protect against partial-band interference. One well known error-control code is the Hamming code, which is described in Appendix A. For an (n, k) Hamming code, the binary data stream is first divided into blocks of k bits each. Each block of k information bits is encoded to give a code word that consists of n bits. This code word is transmitted over the channel, and a corresponding received word of length n is formed at the output of the demodulator. If there are any errors in this process, the received word differs from the code word in each of the positions in which an error occurred. The Hamming code can correct one error in each received word.

Suppose we employ the (7,4) Hamming code in the SFH system. Instead of transmitting six bits per dwell interval as illustrated in Figure 9-23 or three bits per dwell interval as in Figure 9-24, we could transmit one code word in each dwell interval. This would require the transmission of seven bits per dwell interval in order to provide the capability of correcting one error among the bits in each of the dwell intervals. The problem with this approach is that if the partial-band interference is strong, the probability that there are two or more errors in the received word may be quite large. For a dwell interval that has catastrophic interference, the average number of bit errors is greater than three. As illustrated in Appendix A, the Hamming code cannot correct more than one error within a received word.

Even if we use a more powerful code, such as the Reed–Solomon code discussed in Appendix B, the number of errors in a dwell interval often exceeds the error-correction capability of the code. For M-ary signaling with one code word per dwell interval, the average number of erroneous symbols in a dwell interval that has catastrophic interference is approximately $(M - 1)/M$ times the block length of the code. This exceeds the error-correction capability of even the best codes. The problem with the approach considered thus far is that all of the symbols in a single code word are subjected to any interference that might be present in a frequency slot.

Suppose that, instead of transmitting an entire code word in each dwell interval, we distribute the symbols from a code word over a sequence of n consecutive dwell intervals. In this approach, no code word has more than one of its symbols in a given dwell interval. If interference is present in only one of the frequency slots that are used during the transmission of a code word, it affects only one of the symbols in that code word. For example, the Hamming code can be used to correct an error that may result from interference in one of the n frequency slots used for transmission of a single code word. In general, separating the symbols of a code word into different dwell intervals leads to a much better utilization of the error-correction capability of a typical code.

The method just described is illustrated in Figure 9-25, where the crosshatched symbols are symbols from a single code word. In this illustration, each symbol from this code word occupies the second position within its dwell interval. There is no requirement that the position be the same within each dwell interval, and there are advantages in some applications if the positions are varied. In any case, the approach illustrated in Figure 9-25 permits two other code words to be transmitted simultaneously and in the same manner as the code word represented by the crosshatched symbols. This arrangement of symbols from multiple code words is a form of *interleaving*. The combination of error-control coding and proper interleaving is essential for good performance in SFH systems that must operate in the presence of partial-band interference.

A better code for combating partial-band interference is the Reed–Solomon code described in Appendix B. With the same form of interleaving, the (n, k) Reed–Solomon code can correct errors in up to $(n - k)/2$ dwell intervals. For example, the (32,16) Reed–Solomon code can correct as many as eight symbol errors within one received word, which corresponds to 25% of the dwell intervals having catastrophic interference if there are no other errors among the dwell intervals that have no interference.

Another advantage of frequency-hop signaling is the ability of multiple signals to occupy the same frequency band simultaneously. Illustrated in Figure 9-26 are two asynchronous frequency-hop signals. From the point of view of the receiver that is attempting to receive and demodulate the desired signal shown in Figure 9-26, the other SFH signal represents *frequency-hop multiple-access interference*. Its effects are similar to those of partial-band interference. As illustrated in the second dwell interval of the desired signal, it may be that two frequency-hop signals occupy the same frequency slot for all or part of one or more dwell intervals during the transmission of a message. In most applications, the error probabilities are quite large for both signals during such an event, so error-control coding is usually required.

One of the natural choices for error control in SFH systems is the Reed–Solomon code mentioned previously. A Reed–Solomon code of block length n that has k information symbols can correct any set of $(n - k)/2$ or fewer errors. If it is possible for the

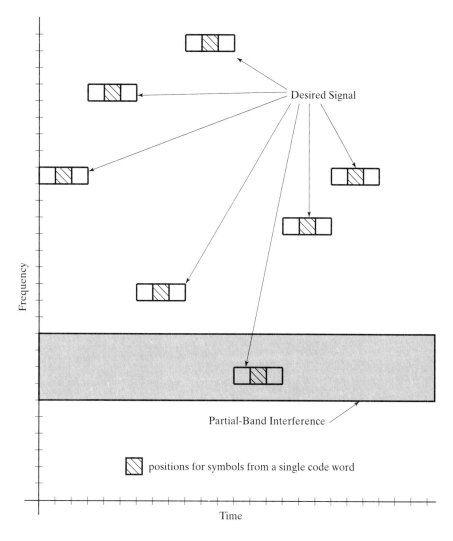

Figure 9-25: A SFH signal with error-control coding on a channel with partial-band interference.

receiver to determine which symbols have been hit, the performance can be increased greatly by erasing some or all such symbols [9.34]. The same Reed–Solomon code can correct $n - k$ or fewer erasures if there are no errors. More generally, it can correct any combination of t errors and e erasures, provided that $2t + e \leq n - k$. The decoder need not have perfect knowledge of which symbols are hit in order to obtain significant performance gains by erasing some of the received symbols [9.34]. Many different methods have been developed for deciding which of the received symbols should be erased, and the performance of each of these methods has been evaluated (e.g., see [9.1], [9.34], and [9.57]).

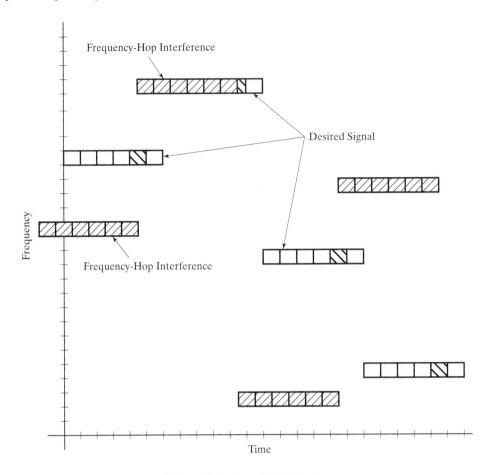

Figure 9-26: Two SFH signals.

For fast frequency hopping it is possible to achieve acceptable performance in some communication environments without the use of error-control coding. However, two important points should be kept in mind: FFH can be viewed as a simple repetition code, and the use of more powerful error-control coding methods can improve performance greatly even for FFH systems.

The simplest method for achieving diversity in FFH systems is to send each symbol in each of several different frequency slots and use a simple majority vote among the corresponding decisions made by the demodulator. This is an example of a *hard-decision* combining scheme. *Soft-decision* combining schemes, such as summing the squares of the soft-decision demodulator outputs, have also been proposed (e.g., [9.45]). The combination of error-control coding and simple frequency diversity provides tremendous protection against partial-band interference, as discussed in [9.31], [9.37], [9.45], and [9.53].

The SINCGARS mobile frequency-hop radio [9.17] operates in the portion of the VHF band from approximately 30 MHz to approximately 88 MHz. The number of frequency slots available is 2320, and the width of each slot is 25 kHz. Digital data or digitized voice signals can be sent using a form of binary frequency-shift keying. The data rate is variable over a range from 75 bps to 16 kbps, and the hopping rate is approximately 100 hops per second. Enhanced versions of the SINCGARS radio use side information developed in the receiver to erase unreliable symbols, and Reed–Solomon coding is employed to correct errors and erasures. The original applications of SINCGARS include fully connected networks for voice communications. With the addition of suitable routing protocols, the SINCGARS radios can be employed in a fully distributed store-and-forward packet communication network.

References and Suggestions for Further Reading

[9.1] C. W. Baum and M. B. Pursley, "Bayesian methods for erasure insertion in frequency-hop communications with partial-band interference," *IEEE Transactions on Communications*, vol. 40, July 1992, pp. 1231–1238.

[9.2] E. R. Berlekamp, "The technology of error-correcting codes," *Proceedings of the IEEE*, vol. 68, May 1980, pp. 564–593.

[9.3] G. R. Cooper and C. D. McGillem, *Probabilistic Methods of Signal and System Analysis* (2nd ed.), New York: Holt, Rinehart, and Winston, 1986, pp. 297–300.

[9.4] R. Dell-Imagine, "JTIDS — An overview of the system design and implementation," 1976 *IEEE Position Location and Navigation Symposium* (San Diego), 1976, pp. 212–215.

[9.5] R.-H. Dou and L. B. Milstein, "Error probability bounds and approximations for DS spread-spectrum communication systems with multiple tone or multiple access interference," *IEEE Transactions on Communications*, vol. COM-32, May 1984, pp. 493–501.

[9.6] P. K. Enge, "The global position system: Signals, measurements, and performance," *International Journal of Wireless Information Networks*, vol. 1, no. 2, 1994, pp. 83–105.

[9.7] W. C. Fifer and F. J. Bruno, "The low-cost packet radio," *Proceedings of the IEEE*, vol. 75, January 1987, pp. 33–42.

[9.8] F. D. Garber and M. B. Pursley, "Optimal phases of maximal-length sequences for asynchronous spread-spectrum multiplexing," *Electronics Letters*, vol. 16, September 1980, pp. 756–757.

[9.9] E. A. Geraniotis and M. B. Pursley, "Error probability for direct-sequence spread-spectrum multiple-access communications — Part II: Approximations," *IEEE Transactions on Communications*, vol. COM-30, May 1982, pp. 985–995.

[9.10] E. A. Geraniotis and M. B. Pursley, "Error probabilities for slow-frequency-hopped spread-spectrum multiple-access communications over fading channels," *IEEE Transactions on Communications*, vol. COM-30, May 1982, pp. 996–1009.

[9.11] E. A. Geraniotis and M. B. Pursley, "Performance of coherent direct-sequence spread-spectrum communications over specular multipath fading channels," *IEEE Transactions on Communications*, vol. COM-33, June 1985, pp. 502–508.

[9.12] I. A. Getting, "The global position system," *IEEE Spectrum*, December 1993, pp. 36–47.

[9.13] R. Gold, "Optimal binary sequences for spread spectrum multiplexing," *IEEE Transactions on Information Theory*, vol. IT-13, October 1967, pp. 619–621.

[9.14] S. W. Golomb, et al., *Digital Communications with Space Applications*, Englewood Cliffs, NJ: Prentice-Hall, 1964.

[9.15] S. W. Golomb, *Shift Register Sequences*, San Francisco: Holden-Day, 1967.

[9.16] J. K. Holmes, *Coherent Spread Spectrum Systems*, New York: Wiley, 1982.

[9.17] D. E. Kammer, "SINCGARS—The new generation combat net radio system," *Proceedings of the Tactical Communications Conference*, April 1986, pp. 64–72.

[9.18] D. Laforgia, A. Luvison, and V. Zingarelli, "Bit error rate evaluation for spread-spectrum multiple-access systems," *IEEE Transactions on Communications*, vol. COM-32, June 1984, pp. 660–668.

[9.19] J. S. Lehnert and M. B. Pursley, "Error probability for binary direct-sequence spread-spectrum communications with random signature sequences," *IEEE Transactions on Communications*, vol. COM-35, January 1987, pp. 87–98.

[9.20] Logicon, Inc., *Understanding Link 16: A Guidebook for New Users*, San Diego, April 1994 (First Revision, 1996; Second Revision, 1998).

[9.21] F. J. MacWilliams and N. J. A. Sloane, "Pseudo-random sequences and arrays," *Proceedings of the IEEE*, vol. 64, December 1976, pp. 1715–1729.

[9.22] J. L. Massey and J. J. Uhran, "Sub-baud coding," *Proceedings of the 13th Annual Allerton Conference on Circuit and System Theory* (Univ. Illinois, Urbana), October 1975, pp. 539–547.

[9.23] D. L. Noneaker and M. B. Pursley, "On the chip rate of CDMA systems with doubly selective fading and rake reception," *IEEE Journal on Selected Areas in Communications*, vol. 12, June 1994, pp. 853–861.

[9.24] D. L. Noneaker and M. B. Pursley, "Selection of spreading sequences for direct-sequence spread-spectrum communications over a doubly selective fading channel," *IEEE Transactions on Communications*, vol. 42, no. 12, December 1994, pp. 3171–3177.

[9.25] D. L. Noneaker and M. B. Pursley, "The effects of sequence selection on DS spread-spectrum with selective fading and rake reception," *IEEE Transactions on Communications*, vol. 44, no. 2, February 1996, pp. 229–237.

[9.26] R. L. Peterson, R, E. Ziemer, and D. E. Borth, *Introduction to Spread-Spectrum Communications*, Upper Saddle River, NJ: Prentice-Hall, 1995.

[9.27] W. W. Peterson and E. J. Weldon, Jr., *Error-Correcting Codes*, (2nd ed.), Cambridge, MA: MIT Press, 1972.

[9.28] M. B. Pursley, D. V. Sarwate, and T. U. Basar, "Partial correlation effects in direct-sequence spread-spectrum multiple-access communication systems," *IEEE Transactions on Communications*, vol. COM-32, May 1984, pp. 567–573.

[9.29] M. B. Pursley, "Performance evaluation for phase-coded spread-spectrum multiple-access communication — Part I: System analysis," *IEEE Transactions on Communications*, vol. COM-25, August 1977, pp. 795–799.

[9.30] M. B. Pursley, "Spread-spectrum multiple-access communications," in *Multi-User Communication Systems*, G. Longo (editor), pp. 139–199, Vienna: Springer-Verlag, 1981.

[9.31] M. B. Pursley, "Coding and diversity for channels with fading and pulsed interference," *Proceedings of the 1982 Conference on Information Sciences and Systems* (Princeton University), March 1982, pp. 413–418.

[9.32] M. B. Pursley, "Frequency-hop transmission for satellite packet switching and terrestrial packet radio networks," *IEEE Transactions on Information Theory*, vol. IT-32, September 1986, pp. 652–667.

[9.33] M. B. Pursley, "The role of spread spectrum in packet radio networks," *Proceedings of the IEEE*, vol. 75, January 1987, pp. 116–134.

[9.34] M. B. Pursley, "Reed–Solomon codes in frequency-hop communications," Chapter 8, in *Reed–Solomon Codes and Their Applications*, S. B. Wicker and V. K. Bhargava (eds.), New York: IEEE Press, 1994.

[9.35] M. B. Pursley and H. F. A. Roefs, "Numerical evaluation of correlation parameters for optimal phases of binary shift-register sequences," *IEEE Transactions on Communications*, vol. COM-27, October 1979, pp. 1597–1604.

[9.36] M. B. Pursley and D. V. Sarwate, "Performance evaluation for phase-coded spread-spectrum multiple-access communication — Part II: Code sequence analysis," *IEEE Transactions on Communications*, vol. COM-25, August 1977, pp. 800–803.

[9.37] M. B. Pursley and W. E. Stark, "Performance of Reed–Solomon coded frequency-hop spread-spectrum communications in partial-band interference," *IEEE Transactions on Communications*, vol. COM-33, August 1985, pp. 767–774.

[9.38] M. B. Pursley and D. J. Taipale, "Error probabilities for spread-spectrum packet radio with convolutional codes and Viterbi decoding," *IEEE Transactions on Communications*, vol. COM-35, January 1987, pp. 1–12.

[9.39] M. B. Pursley, D. V. Sarwate, and W. E. Stark, "Error probability for direct-sequence spread-spectrum multiple-access communications — Part I: Upper and lower bounds," *IEEE Transactions on Communications*, vol. COM-30, May 1982, pp. 975–984.

[9.40] M. Y. Rhee, *CDMA Cellular Mobile Communications and Network Security*, Upper Saddle River, NJ: Prentice Hall PTR, 1998.

[9.41] H. F. A. Roefs and M. B. Pursley, "Correlation parameters of random binary sequences," *Electronics Letters*, vol. 13, no. 16, August 1977, pp. 488–489.

[9.42] J. Rubin and J. Heinen, "JTIDS II/DTDMA — Command and control terminals," in *Principles and Operational Aspects of Precision Position Determination Systems*, C. T. Leondes (ed.), North Atlantic Treaty Organization Advisory Group for Aerospace Research and Development, AGARDograph No. 245, AGARD-AG-245, July 1979, pp. 38-1–38-7.

[9.43] D. V. Sarwate and M. B. Pursley, "Crosscorrelation properties of pseudorandom and related sequences," *Proceedings of the IEEE*, vol. 68, May 1980, pp. 593–619.

[9.44] R. A. Scholtz, "The origins of spread-spectrum communications," *IEEE Transactions on Communications*, vol. COM-30, May 1982, pp. 822–852.

[9.45] M. K. Simon, J. K. Omura, R. A. Scholtz, and B. K. Levitt, *Spread Spectrum Communications* (3 vols.), Rockville, MD: Computer Science Press, 1985.

[9.46] R. Skaug, "Numerical evaluation of the nonperiodic autocorrelation parameter for optimal phases of maximal length sequences," *Proceedings of the IEE*, Part F, vol. 127, 1980, pp. 230–237.

[9.47] R. Skaug and J. F. Hjelmstad, *Spread Spectrum in Communication*, London: Peregrinus, 1985.

[9.48] J. J. Spilker, Jr., *Digital Communications by Satellite*, Englewood Cliffs, NJ: Prentice-Hall, 1977.

[9.49] J. J. Spilker, Jr., "GPS signal structure and performance characteristics," *Navigation: Journal of the Institute of Navigation*, vol. 25, no. 2, Summer 1978, pp. 121–146.

[9.50] S. Stein, Part III of *Communication Systems and Techniques* (by M. Schwartz, W. R. Bennett, and S. Stein), pp. 561–584, New York: McGraw-Hill, 1966.

[9.51] S. Stein, "Communication over fading radio channels," *Encyclopedia of Telecommunications*, vol. 3, New York: Marcel Dekker, 1992.

[9.52] *TIA/EIA Mobile Station–Base Station Compatibility Standard for Wideband Spread Spectrum Cellular Systems*, ANSI/TIA/EIA-95-B-99, Washington, DC: Telecommunications Industry Association, March 1999.

[9.53] D. J. Torrieri, *Principles of Military Communication Systems*, Dedham, MA: Artech, 1981.

[9.54] G. L. Turin, "Introduction to spread-spectrum antimultipath techniques and their application to urban digital radio," *Proceedings of the IEEE*, vol. 68, March 1980, pp. 328–353.

[9.55] S. Verdu, *Multiuser Detection*, Cambridge, UK: Cambridge University Press, 1998.

[9.56] A. J. Viterbi and I. M. Jacobs, "Advances in coding and modulation for noncoherent channels affected by fading, partial band, and multiple-access interference," in *Advances in Communication Systems*, vol. 4, pp. 279–308, New York: Academic Press, 1975.

[9.57] A. J. Viterbi, "A robust ratio-threshold technique to mitigate tone and partial band jamming in coded MFSK systems," *1982 IEEE Military Communications Conference Record*, October 1982, pp. 22.4.1–5.

[9.58] A. J. Viterbi, *CDMA: Principles of Spread Spectrum Communication*, Reading, MA: Addison-Wesley, 1995.

[9.59] K.-T. Wu, "Direct sequence spread-spectrum communications: Applications to multiple access and jamming resistance," Ph.D. dissertation, Univ. Michigan, Ann Arbor, 1981.

[9.60] K.-T. Wu, "Average error probability for DS-SSMA communications: The Gram–Charlier expansion approach," *Proceedings of the 19th Annual Allerton Conference on Communications, Control, and Computing* (Univ. Illinois, Urbana), September 1981, pp. 237–246.

[9.61] K.-T. Wu and D. L. Neuhoff, "Average bit error probability for direct sequence spread-spectrum multiple-access communication systems," *Proceedings of the 18th Annual Allerton Conference on Communications, Control, and Computing* (Univ. Illinois, Urbana), October 1980, pp. 359–368.

[9.62] K. Yao, "Error probability of asynchronous spread spectrum multiple access communication systems," *IEEE Transactions on Communications*, vol. COM-25, August 1977, pp. 803–809.

Problems

9.1 One period of the sequence x is $1\,0\,0\,1\,1\,1\,0$.

   (a) Find the periodic autocorrelation function for x.

   (b) Find the odd autocorrelation function for x.

   (c) Find the odd autocorrelation function for each phase of x. You may want to use Matlab or Excel.

   (d) Based on your results in part **(c)**, determine which phase of x has the smallest value of the maximum odd autocorrelation function. This is known as the auto-optimal (AO) phase of x. Specify the sequence that is the AO phase of x, and give the maximum value for its odd autocorrelation function.

9.2 The sequence $1\,0\,0\,1\,1\,1\,0$ of Problem 9.1 is employed in a binary, baseband DS spread-spectrum system with chip rate $1/T_c$. The baseband channel has a direct path and one reflected path, and the propagation time for the reflected path is $9T_c$ greater than for the direct path. If the transmitted signal is $s(t)$, the input to a correlator matched to $s(t)$ is

$$Y(t) = s(t) + 0.2\, s(t - 9T_c) + X(t),$$

where $X(t)$ is white Gaussian noise with spectral density $N_0/2$. The correlator output is compared with a zero-threshold in order to decide which binary symbol was transmitted. The energy per bit in the signal $s(t)$ is $\mathcal{E}$. Find the average probability of error, averaged over all pulse patterns, if the transmitted data sequence is modeled as a sequence of independent random variables, each of which takes value $+1$ with probability $1/2$ and -1 with probability $1/2$. Express your answer in terms of the ratio $\mathcal{E}/N_0$ and the function Q.

9.3 The sequence x is as specified in Problem 9.1, and one period of the sequence y is $0\,1\,1\,1\,0\,0\,1$. Find the periodic crosscorrelation function for x and y, the odd crosscorrelation function for x and y, and the aperiodic crosscorrelation function for x and y.

9.4 If the elements of a sequence $\mathbf{u} = u_0, u_1, \ldots, u_{N-1}$ take only the values $+1$ and -1 and the aperiodic autocorrelation function satisfies $|C_u(i)| \leq 1$ for $1 \leq i \leq N - 1$, then the sequence $u_0, u_1, \ldots, u_{N-1}$ is said to be a *Barker sequence*.

   (a) Show that the sequence $+1, -1, +1, +1, -1, -1, -1$ is both an m-sequence and a Barker sequence.

   (b) Prove that if $\mathbf{u}$ is a Barker sequence of length N, and N is an odd integer, then the odd autocorrelation function for $\mathbf{u}$ satisfies $|\hat{\theta}_u(i)| = 1$ for $1 \leq i \leq N - 1$.

9.5 Consider shift registers with four stages that correspond to the polynomial

$$f_4(x) = x^4 + h_1 x^3 + h_2 x^2 + h_3 x + 1.$$

Since there are three coefficients (h_1, h_2, and h_3) and each coefficient can take one of two different values, there are a total of $2^3 = 8$ different polynomials that have the form specified by $f_4(x)$. Notice that $h_0 = h_4 = 1$. The question is, which of these eight polynomials are primitive?

In order to find if such a polynomial is primitive, we can first use the fact that a primitive polynomial is irreducible (cannot be factored). For example, in looking for primitive polynomials, we can rule out the polynomial

$$(x + 1)(x^3 + 1) = x^4 + x^3 + x + 1,$$

because it factors as the product of $x + 1$ and $x^3 + 1$. Similarly, $(x^2 + 1)(x^2 + 1) = x^4 + 1$ and $(x + 1)(x^3 + x + 1) = x^4 + x^3 + x^2 + 1$ cannot be primitive. We can proceed through all of the possible factors and eliminate each of the corresponding fourth-degree polynomials. It is not necessary to consider such factors as x, $x^2 + x$, or $x^3 + x^2$ because if a polynomial has these as factors, it cannot have $h_4 = 1$.

(a) By checking all possible appropriate factors of $f_4(x)$, give the list of all polynomials that cannot be primitive because they are reducible (i.e., they factor). By comparing this list with the list of all polynomials that have the form of $f_4(x)$, give a list of polynomials that are irreducible.

(b) Give a sketch of the linear-feedback shift register corresponding to each of the irreducible polynomials you found in part (a). Determine the sequence that results from each of these shift registers if the initial loading is 1111. Based on an examination of the sequences you obtain from each shift register, determine which of the irreducible polynomials are primitive.

9.6 Consider a DS spread-spectrum system in which the chip waveform is a sine pulse defined by

$$\psi(t) = \sqrt{2}\sin(\pi t/T_c), \quad 0 \le t \le T_c,$$

and $\psi(t) = 0$, otherwise.

(a) Give expressions for the partial autocorrelation functions $R_\psi(s)$ and $\hat{R}_\psi(s)$.

(b) Determine the chip waveform parameters $\mathcal{M}_\psi$ and $\mathcal{N}_\psi$.

9.7 Suppose the chip waveform for a DS spread-spectrum multiple-access system is the rectangular pulse, there are three simultaneous transmissions, and the number of chips per bit is 31. Based on the interference parameters $\sigma^2(k, i)$, $1 \le i < k \le 3$, given in Table 9.1, find the value of $\mathcal{E}_b/N_0$ (in dB) that gives an approximate error probability of 10^{-5} for the receiver of signal 1. Use the approximation $P_e \approx Q(\text{SNR}_1)$ to estimate the error probability for the system, and use the fact that $Q(4.265) \approx 10^{-5}$.

9.8 The *spreading factor* for a digital communication signal is defined as the ratio of the bandwidth of the signal to the information rate of the signal. Specifically, if W is the bandwidth in hertz and R is the information rate in bits per second, the spreading factor is given by $S = W/R$. The precise definition of the bandwidth depends on the type of modulation used in the communication signal. In parts (a) and (b), the null-to-null bandwidth is specified as the bandwidth to use. In the other parts, you must define what definition of bandwidth you are using in giving your answer. For some parts, you may want to use a sketch of the spectrum to help describe your definition of bandwidth.

(a) What is the spreading factor S for binary PSK if W is the null-to-null bandwidth?

(b) What is the spreading factor for QPSK if W is the null-to-null bandwidth?

(c) What is the spreading factor for a binary antipodal ASK communication signal that transmits R pulses per second using the pulse waveform $\text{sinc}(Rt)$? *Hint:* This signal achieves the Nyquist rate.

(d) What is the spreading factor for an RF signal that employs M-ary orthogonal amplitude modulation? Assume that each of the M baseband signals is a sequence of M rectangular pulses, and the polarities of the pulses are obtained from the rows of a Hadamard matrix (i.e., the baseband signals are the Hadamard–Walsh signals).

(e) What is the spreading factor for an MFSK signal with orthogonal tone spacing?

(f) What is the spreading factor for a binary DS spread spectrum signal that employs rectangular chip waveforms with N chips per bit?

(g) What is the spreading factor for a SFH signal that has q frequency slots and uses binary PSK data modulation?

9.9 A FH spread-spectrum system uses a (32,28) Reed–Solomon code with one symbol per dwell interval. Suppose that 20% of the 10,000 available frequency slots have interference that is so strong that the probability of error is 1 for each symbol that is transmitted in one of the frequency slots with interference. The remaining 80% of the frequency slots have no interference, so the probability of error is 0 for each symbol that is transmitted in one of these frequency slots. The hopping pattern is a random sequence of 32 frequencies that are independent and each is uniformly distributed over the 10,000 available frequencies. If error-correction decoding is used, what is the probability that a code word that is transmitted over this channel is decoded correctly? Suppose the receiver knows which symbols are transmitted in frequency slots with interference and erases them, and the decoder uses erasure-correction decoding. What is the probability of correct decoding?

9.10 Consider a SFH communications system that employs noncoherent BFSK for data modulation. The energy per bit is $\mathcal{E}$, and the thermal noise is negligible. The only noise that affects the performance of the demodulator is bandlimited white Gaussian noise with two-sided spectral density $\eta/2$. This noise is present in a fraction ρ of the frequency slots.

Assume the demodulator is the optimum noncoherent demodulator, so the error probability is $0.5 \exp(-\mathcal{E}/2\eta)$ if the signal is received in a frequency slot with two-sided noise density $\eta/2$. Assume that the frequency-hopping pattern is a sequence of independent random variables, each of which is uniformly distributed over the set of q frequency slots.

(a) Give an expression for the average probability of error in terms of ρ, $\mathcal{E}$, and η.

(b) Suppose the total noise power is held constant as ρ varies. Thus, if $N_I = \rho \eta$, then N_I is constant. Notice that $N_I/2$ is the average density of the noise, averaged over the q frequency slots. Give an expression for the value of $\mathcal{E}/N_I$ that is required to guarantee the average error probability is p. Your answer should be in terms of ρ and p.

(c) Find the value of ρ that maximizes the required value of $\mathcal{E}/N_I$ for a given error probability p. Your answer should be in terms of p. Evaluate your expression for $p = 10^{-4}$.

(d) Find the value of ρ that maximizes the bit error probability for a given value of $\mathcal{E}/N_I$. Your answer should be in terms of $\mathcal{E}/N_I$. Evaluate your expression for $(\mathcal{E}/N_I)_{dB} = 20$ dB.

(e) In a diversity system, the transmitter sends each bit three times. Each of the three transmissions is in a different frequency slot, and a majority vote is taken at the demodulator output to decide which bit was sent. Find the probability of error in terms of ρ, $\mathcal{E}$, and η.

(f) Give a qualitative description of how the answers to parts (c) and (d) change for the diversity system. In particular, will the maximizing values of ρ be larger, smaller, or the same? Explain why.

Appendix A

Hamming Codes

Although a general treatment of error-control coding is beyond the scope of this book, two examples of error-control codes are included to illustrate the basic ideas and provide the necessary background for other parts of the book, particularly Section 9.6. The first example is a family of binary single-error correcting codes. These codes were discovered in the late 1940s by Richard W. Hamming and later published in [A.1]. The second example, given in Appendix B, is a class of nonbinary codes known as the Reed–Solomon codes.

In this appendix, each binary symbol is represented as a 0 or a 1 regardless of how it is transmitted over the channel (e.g., as a positive or a negative rectangular pulse). Addition of the binary digits is accomplished with modulo-2 arithmetic, as described in Section 9.3. In the simplest of the Hamming codes, for each set of four bits of information, a code word consisting of seven bits is formed according to the following procedure: Denote the information bits by d_1, d_2, d_3, and d_4, and denote the bits of the code word by c_1, c_2, c_3, c_4, c_5, c_6, and c_7. The first four bits of the code word are just the information bits: $c_i = d_i$ for $1 \leq i \leq 4$. The final three bits of the code word are derived from the first four bits according to

$$c_5 = c_1 \oplus c_2 \oplus c_4,$$
$$c_6 = c_1 \oplus c_3 \oplus c_4,$$

and

$$c_7 = c_2 \oplus c_3 \oplus c_4,$$

which are referred to as the *parity-check equations* for the code. The code bits c_5, c_6, and c_7 are referred to as the *parity-check bits* of the code word. This code is referred to as the (7,4) Hamming code. For each integer $m \geq 3$, there is an (n,k) Hamming code with $n = 2^m - 1$ code bits of which $k = 2^m - 1 - m$ are information bits and the remaining m are parity bits. The next longer codes after the (7,4) Hamming code are the (15,11) Hamming code ($m = 4$), the (31,26) Hamming code ($m = 5$), the (63,57) Hamming code ($m = 6$), etc.

Table A.1: Code Words for the (7,4) Hamming Code

0	0	0	0	0	0	0
0	0	0	1	1	1	1
0	0	1	0	0	1	1
0	0	1	1	1	0	0
0	1	0	0	1	0	1
0	1	0	1	0	1	0
0	1	1	0	1	1	0
0	1	1	1	0	0	1
1	0	0	0	1	1	0
1	0	0	1	0	0	1
1	0	1	0	1	0	1
1	0	1	1	0	1	0
1	1	0	0	0	1	1
1	1	0	1	1	0	0
1	1	1	0	0	0	0
1	1	1	1	1	1	1

Since there are 16 possible choices for the four information bits, there are 16 code words in the (7,4) Hamming code, as shown in Table A.1. Suppose the code word

$$\mathbf{c} = c_1 \, c_2 \, c_3 \, c_4 \, c_5 \, c_6 \, c_7$$

is transmitted over a noisy channel as a sequence of seven binary information symbols (e.g., seven rectangular pulses with the appropriate polarities for baseband communications or seven binary PSK symbols for RF communications). The corresponding output of the demodulator is a received word, which we denote by

$$\mathbf{r} = r_1 \, r_2 \, r_3 \, r_4 \, r_5 \, r_6 \, r_7.$$

That is, for each i in the range $1 \le i \le 7$, r_i is received when c_i is sent over the channel. If there is at most one error at the output of the demodulator, then $\mathbf{r}$ differs from $\mathbf{c}$ in at most one position. For the (7,4) Hamming code, such a received word differs from each of the other code words in at least two positions. So the receiver can simply chose the code word that differs from the received word in the fewest number of positions in order to correct any single error that occurs in the transmission of seven bits. After selecting this code word, the original information bits can be recovered by extracting the first four bits from the code word that was chosen by the receiver.

As an illustration, suppose the information bits are all zeros. It is easy to see from the parity-check equations that each parity bit is also a zero, and so the code word that represents the information bits 0000 is 0000000, the first code word in Table A.1. If the binary sequence 0000000 is transmitted over the channel and a single error is made, the received word has a single 1 and six 0s. Thus, the received word differs from the code word 0000000 in exactly one position, no matter where in the sequence the error occurs. Notice from the list of code words that all code words except 0000000 have at least three 1s. Therefore, the correct code word (i.e., the one that was actually sent) differs from the received word in fewer positions than any other code word. Although the information sequence 0000 is used in this illustration, the same conclusion is true for any other information sequence. If any four-bit information sequence is encoded with the (7,4) Hamming code and there is at most one error among the seven bits representing the code word, then the code word that was sent differs from the received word in at most one position while all other code words differ from the received word in two or more positions.

Suppose the received word is $\mathbf{r}$ and it is desired to determine which code word is the most likely to have been sent. If the bit error probability on the channel is less than $1/2$, the receiver's maximum-likelihood decision is that $\mathbf{c}$ was sent if $\mathbf{c}$ is the closest code word to $\mathbf{r}$ in the sense that $\mathbf{c}$ is the code word that differs from $\mathbf{r}$ in the fewest positions. Thus, if the information bits are encoded with the (7,4) Hamming code and the maximum-likelihood decision is used, the information bits are received correctly if at most one error occurs among the seven bits that are transmitted over the channel. If multiple errors occur, the receiver will not select the correct code word. For multiple-error correction, more powerful codes must be employed. Such codes are described in [A.2], [A.3], and [A.4].

Efficient decoders for Hamming codes do not search the list of code words to find the one that differs from the received word in the fewest positions. An efficient method for decoding Hamming codes can be based on a few simple observations. Because $0 \oplus 0 = 1 \oplus 1 = 0$, then $c_5 \oplus c_5 = 0$. Thus, if we add (mod 2) c_5 to each side of the first parity-check equation, we obtain

$$0 = c_1 \oplus c_2 \oplus c_4 \oplus c_5. \tag{A.1}$$

Similarly, the second and third parity-check equations are equivalent to

$$0 = c_1 \oplus c_3 \oplus c_4 \oplus c_6 \tag{A.2}$$

and

$$0 = c_2 \oplus c_3 \oplus c_4 \oplus c_7. \tag{A.3}$$

Suppose $\mathbf{r}$ is the received word and we evaluate s_1, s_2, and s_3 from

$$s_1 = r_1 \oplus r_2 \oplus r_4 \oplus r_5, \tag{A.4}$$
$$s_2 = r_1 \oplus r_3 \oplus r_4 \oplus r_6, \tag{A.5}$$

and

$$s_3 = r_2 \oplus r_3 \oplus r_4 \oplus r_7. \tag{A.6}$$

$$\tag{A.7}$$

If $\mathbf{r}$ is the received word and there are no errors, then $r_i = c_i$ for $1 \leq i \leq 7$. It follows from (A.1)–(A.3) that if there are no errors then $s_1 = 0$, $s_2 = 0$, and $s_3 = 0$. If there is a single error, then $r_i \neq c_i$ for exactly one value of i. Suppose that we add (mod 2) the right-hand sides of (A.1) and (A.4). If $r_i = c_i$ for each i in the set $\{1, 2, 4, 5\}$, we see that $r_i \oplus c_i = 0$ for each of these four positions. By adding the left-hand sides of the same two equations we conclude that $s_1 = 0$ if there are no errors among these four positions. On the other hand, if there is a single error and it is located in one of these four positions, then $r_i \oplus c_i = 1$ for one value of i in the set $\{1, 2, 4, 5\}$ and $r_i \oplus c_i = 0$ for the other three values of i. It follows that $s_1 = 1$. Similarly, if there are no errors in the positions 1, 3, 4, or 6, then $s_2 = 0$, and if an error occurs in exactly one of these positions, then $s_2 = 1$. Finally, if there are no errors in the positions 2, 3, 4, or 7, then $s_3 = 0$, and if an error occurs in exactly one of these positions, then $s_3 = 1$.

Each location for the single error corresponds to a unique value for the vector $\mathbf{s} = (s_1, s_2, s_3)$, which is referred to as the *syndrome vector*. We have already mentioned that $\mathbf{s} = (0, 0, 0)$ corresponds to no errors, and we see that an error in the first position gives $\mathbf{s} = (1, 1, 0)$, an error in the second position gives $\mathbf{s} = (1, 0, 1)$, an error in the third position gives $\mathbf{s} = (0, 1, 1)$, and an error in the fourth position gives $\mathbf{s} = (1, 1, 1)$. Notice these last four syndrome vectors are the only binary vectors of length three that have two or three 1s.

Our goal is to obtain the original information bits, and these bits occupy positions 1 through 4 in a code word of the (7, 4) Hamming code. If there is at most one error in the received word and the syndrome vector is $(0, 0, 0)$, then the information bits are correct as received. On the other hand, if the syndrome vector is $(1, 1, 0)$ the information bits in the second through fourth positions are correct, but we should complement the information bit in the first position in order to correct the error that has occurred in that position. If the syndrome vector is $(1, 0, 1)$, we complement the information bit in the second position, but accept the other three information bits as received. If the syndrome vector is $(0, 1, 1)$, the information bit in the third position should be complemented, and if the syndrome vector is $(1, 1, 1)$, the information bit in the fourth position should be complemented. In each of these occurrences of a single error among the information bits, the error is corrected by computing the syndrome vector, complementing the appropriate information bit, and accepting the other information bits as received. Note that each of the nonzero syndrome vectors considered so far has a 1 in at least two of its three bits, and we have accounted for all such syndrome vectors.

What if there is a single error in the received word and it occurs among the last three positions of the word? Since r_5, r_6, and r_7 each appear in exactly one of the three equations (A.4)–(A.6), a single error in one of these positions results in a syndrome vector with a single 1. In fact, from (A.1)–(A.6), we see that if there are no errors in the first four positions, then $s_1 = r_5 \oplus c_5$, $s_2 = r_6 \oplus c_6$, and $s_3 = r_7 \oplus c_7$. For example, a single error that occurs in the fifth position of the received word corresponds to $\mathbf{s} = (1, 0, 0)$. By considering all received words with a single error that is located in one of the last three positions, we have accounted for all of the syndrome vectors that have a single 1. We previously accounted for the syndrome vector $(0, 0, 0)$ and all four syndrome vectors that have two or three 1s. Thus, we have now accounted for all 2^3 possible syndrome vectors of length three, and we have found there is a one-to-one correspondence between the seven possible locations of the single error and the seven nonzero syndrome vectors of length three.

These considerations provide the following decoding algorithm for correcting single errors. Evaluate the syndrome vector from the received word. If the syndrome vector is $(0, 0, 0)$, accept the received word as is and use the first four bits of the received word as the decoded information bits. If the syndrome vector has more than a single 1, we must complement one of the first four bits of the received word: If $\mathbf{s} = (1, 1, 0)$, complement the first bit; if $\mathbf{s} = (1, 0, 1)$, complement the second bit; if $\mathbf{s} = (0, 1, 1)$, complement the third bit; if $\mathbf{s} = (1, 1, 1)$, complement the fourth bit. In each of these four cases, after the appropriate bit in the received word is complemented, the first four bits of the resulting word are used as the decoded information bits. If the syndrome vector has a single 1, we simply use the first four bits of the received word as the decoded information bits. In this latter situation, the single error is in one of the parity bits, so the information bits in the received word are correct. Readers who are interested in the general method of syndrome decoding should consult [A.2], [A.3], or [A.4], where it is also shown that the mathematical operations required to calculate the syndromes can be accomplished by matrix multiplication.

As an illustration of the inability of the (7,4) Hamming code to correct multiple errors, suppose that the received word has two errors and they are in the sixth and seventh positions. This means the four information bits are correct, but the syndrome vector does not indicate this. The syndrome vector for this received word is $(0, 1, 1)$. The preceding decoding algorithm dictates that we complement the third bit when $\mathbf{s} = (0, 1, 1)$, but in this case, the third bit is actually correct in the received word. Similarly, if the only errors are in the first and second positions of the received word, the syndrome is $(0,1,1)$, which also results in complementing the third bit. In this case, three of the resulting information bits are now wrong (i.e., those in the first, second, and third positions). Fortunately, if the error probability on the channel is very low and errors in different positions are independent, the probability of two or more errors in the received word is much smaller than the probability of one or fewer errors, and so the Hamming code is useful for such a channel.

The error-correction capability of the Hamming code is illustrated nicely through the use of the three circles shown in Figures A-1–A-5. This approach was suggested by Robert McEliece to facilitate the understanding of Hamming codes. We begin with the information bits d_1, d_2, d_3, and d_4, and form the code word by first letting $c_i = d_i$ for $1 \le i \le 4$ as discussed previously. Now insert the values for c_1, c_2, c_3, and c_4 in the positions shown in Figure A-1. For example, if $(c_1, c_2, c_3, c_4) = (1, 0, 1, 1)$, then the result of the first step in the procedure is as shown in Figure A-2. Next, for each i in the range $5 \le i \le 7$, choose the value of c_i in Figure A-1 to give a zero sum (mod 2) for the four bits in the circle that contains c_i. This is equivalent to selecting the value for c_i to give an even number of 1s in the circle containing c_i. For example, if $(c_1, c_2, c_3, c_4) = (1, 0, 1, 1)$, this step results in the values shown in Figure A-3. Notice that each circle in Figure A-3 has an even number of 1s and an even number of 0s. Equivalently, the sum (mod 2) of the contents of each circle is 0.

Now suppose that a single error occurs in the word; that is, the received word illustrated in Figure A-4 differs from the transmitted code word in one bit position. Our task is to determine the location of this error. The task is accomplished by computing the sum (mod 2) for each circle and determining which circle or circles have a nonzero sum. If only one circle has a nonzero sum, the error has to be in one of the positions 5, 6, and 7, because these are the only positions that are represented in only one circle. Thus, if only one circle has a nonzero sum, we can accept the four information bits as received.

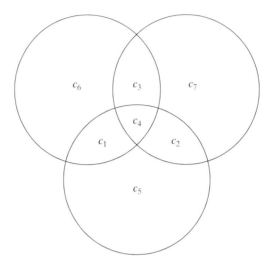

Figure A-1: McEliece's encoding diagram for the (7,4) Hamming code.

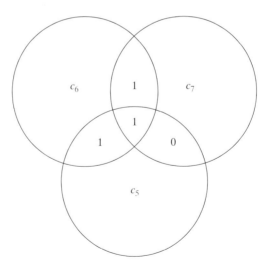

Figure A-2: Encoding diagram for information bits 1011.

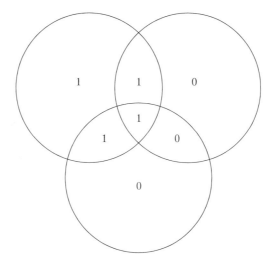

Figure A-3: Determination of the code bits c_5, c_6, and c_7.

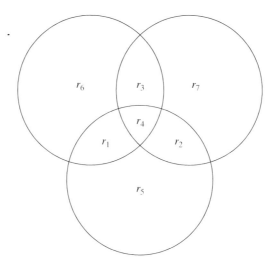

Figure A-4: Decoding for the received word $\mathbf{r} = r_1 r_2 r_3 r_4 r_5 r_6 r_7$.

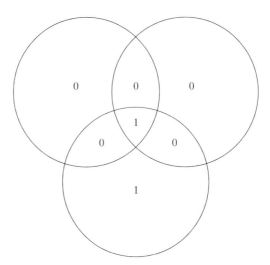

Figure A-5: Decoding for the received word 0001100.

Next suppose that two circles have nonzero sums and the third circle has a zero sum, then the error must be in positions 1, 2, or 3, because these are the only positions represented in exactly two circles. Similarly, if all three circles have a nonzero sum, the error must be in position 4, the only position that is represented in all three circles. In each of these cases, we can locate the error, complement the appropriate bit, and extract the correct information bits from the diagram.

Suppose, for example, that the received word is 0001100 as illustrated in Figure A-5. The sum for the bottom circle is zero, but the circle that includes r_6 and the circle that includes r_7 each have a nonzero sum. The only bit position that these two circles have in common that is not also in the bottom circle is the third position, so if there is a single error, the received bit r_3 must be wrong. From Figure A-5, we see that this bit is 0 for the given received word. We complement this bit and deduce that the correct information bits are 0011. The reader should verify that the code word that corresponds to the information sequence 0011 is 0011100, which differs from the received word 0001100 in the third position only.

References and Suggestions for Further Reading

[A.1] R. W. Hamming, "Error detecting and error correcting codes," *Bell System Technical Journal*, vol. 29, 1950, pp. 147–160.

[A.2] S. Lin and D. J. Costello, Jr., *Error Control Coding* (2nd ed.), Upper Saddle River, NJ: Prentice Hall, 2005.

[A.3] R. J. McEliece, *The Theory of Information and Coding* (2nd ed.), Cambridge, UK: Cambridge University Press, 2002.

[A.4] S. B. Wicker, *Error Control Systems for Digital Communication and Storage*, Upper Saddle River, NJ: Prentice Hall, 1995.

Problems

A.1 Use the algorithm given in Appendix A to decode the received word 1110111. If a single error has occurred, what are the correct information bits?

A.2 First, use Figure A-1 to prove that no code word from the (7,4) Hamming code can have only a single 1. Next, use Figure A-1 to prove that any nonzero code word from this code must have at least three 1s.

Appendix B

Reed–Solomon Codes

Reed–Solomon codes are very powerful error-control codes that have found applications in a wide range of important electronic systems including the compact disc digital audio system, communication systems on several NASA spacecraft, and the U.S. Army's frequency-hop radio. An excellent history of Reed–Solomon codes is provided in [B.6], which includes a chapter by the code's discoverers, Irving S. Reed and Gustave Solomon. The theory of Reed–Solomon codes and descriptions of the encoding and decoding procedures for these codes are given in [B.1]–[B.6], and several applications of Reed–Solomon codes are described in [B.6]. Here we concentrate on the basic error-and-erasure-correction capabilities of these codes.

A Reed–Solomon code is a nonbinary code; in fact, the code symbols are typically from an alphabet whose size is much larger than two. For example, the Reed–Solomon code used on NASA's Galileo spacecraft has code symbols from an alphabet of size $2^8 = 256$. A code word from an (n,k) Reed–Solomon code has n nonbinary code symbols, k of which are information symbols, and the remaining $n - k$ are parity symbols (also referred to as redundant symbols). For an alphabet of size 2^m, the standard Reed–Solomon code has code words of length $2^m - 1$, but the code can be extended by the addition of one parity symbol to each code word to give a block length of 2^m. For example, a $(32,16)$ singly extended Reed–Solomon code has 32 symbols from an alphabet of size 32, and 16 of these symbols are information symbols. Since each information symbol represents 5 bits of information, one code word from a $(32,16)$ Reed–Solomon code represents 80 bits of information. All of the examples employed in what follows are singly extended Reed–Solomon codes.

The Reed–Solomon codes are natural choices for communication systems that employ certain types of nonbinary signals. For instance, a Reed–Solomon code with alphabet size M may be an excellent choice for a communication system that uses the M-ary orthogonal signals described in Chapters 6 and 7. For a power-constrained additive white Gaussian noise channel, 32-ary orthogonal signaling and a $(32,16)$ Reed–Solomon code gives a powerful combination of modulation and coding. Such a system can achieve a very small probability of error at the decoder output, even for a moderate value of $\mathcal{E}_b/N_0$.

Although the symbols for the Reed–Solomon codes are not binary, they can be transmitted as sequences of binary symbols. A single code symbol from a Reed–Solomon code that has a code alphabet of size 2^m can be represented by a sequence of m binary symbols. Hence, Reed–Solomon codes can also be used with binary PSK, binary FSK, or any other binary modulation.

One of the attractive properties of a Reed–Solomon code is that it can correct multiple errors, which is not true of the Hamming codes discussed in Appendix A. An (n,k) Reed–Solomon code can correct any combination of $\lfloor (n-k)/2 \rfloor$ or fewer errors that occur during the transmission and demodulation of the n symbols that form a code word. We use $\lfloor x \rfloor$, referred to as the *floor* of x, to denote the integer part of the real number x. For singly extended Reed–Solomon codes, the block length n is even. For most applications of singly extended Reed–Solomon codes, k is also even, so $(n-k)/2$ is an integer and therefore $\lfloor (n-k)/2 \rfloor = (n-k)/2$.

The important feature of a Reed–Solomon code for our purposes is its ability to correct errors and erasures that occur during transmission of messages over channels with noise and interference. As discussed throughout the book, errors may result at the output of the demodulator in any communication system. In some situations, however, the receiver may be able to tell that the output of the demodulator is unreliable. For example, the receiver may be able to observe that strong interference is present on the channel during the demodulation of a signal. In such cases, it is often better for the receiver to erase one or more symbols rather than to risk having errors in the demodulated message stream.

In certain systems that employ Reed–Solomon codes, the performance can be improved greatly if it is possible to erase symbols that are judged to be unreliable at the output of the demodulator. With standard bounded-distance decoding, an (n,k) Reed–Solomon code can correct any combination of t errors and e erasures provided

$$2t + e \leq n - k.$$

For example, the code can correct $\lfloor (n-k)/2 \rfloor$ errors if there are no erasures, or it can correct $n - k$ erasures if there are no errors. Thus, if it were possible to erase each erroneous symbol, the channel could make twice as many errors as a system that does not erase any of the symbols at the demodulator output. In fact, the decoding algorithm of [B.1] can correct one more erasure than is possible with bounded-distance decoding.

The trick in the use of erasure correction is to determine which symbols should be erased. In practical systems, it is not possible to know for sure which symbols are in error at the demodulator output. It is possible, however, to identify some symbols as being less reliable than others, and this is the basis for soft-decision decoding and errors-and-erasures decoding with side information (e.g., in the frequency-hop systems described in Section 9.6).

A complete explanation of Reed–Solomon codes is beyond the scope of this book, so we shall rely on a simple example to illustrate the errors-and-erasures correcting capabilities of the code. The (4,2) Reed–Solomon code has a code alphabet with four symbols, which we denote by 0, 1, α, and β. Thus, the code words are of the form (c_1, c_2, c_3, c_4), where each of the code symbols c_i is an element of the code alphabet $\{0, 1, \alpha, \beta\}$.

Table B.1: Code Words for a $(4,2)$ Singly Extended Reed–Solomon Code

0000	$10\beta\alpha$	$\alpha01\beta$	$\beta0\alpha1$
$01\alpha\beta$	1111	$\alpha1\beta0$	$\beta10\alpha$
$0\alpha\beta1$	$1\alpha0\beta$	$\alpha\alpha\alpha\alpha$	$\beta\alpha10$
$0\beta1\alpha$	$1\beta\alpha0$	$\alpha\beta01$	$\beta\beta\beta\beta$

In order to transmit binary data with such a code, each information symbol is used to represent two bits of information. For example, the information bits 00 can be represented by the code information symbol 0, 01 by the symbol 1, 10 by the symbol α, and 11 by the symbol β. Since two of the four code symbols in each code word are information symbols, each code word represents four bits of information.

The code words of a $(4,2)$ Reed–Solomon code are listed in Table B.1. Let c_1 and c_2 be the two information symbols, so c_3 and c_4 are the parity symbols. We can map the binary data into the corresponding code word by first converting the four information bits to two code symbols and then using the corresponding code word from Table B.1. For example, in order to send the binary data stream 0110, we first map 01 to the symbol 1 and 10 to the symbol α, to give the information symbols 1α. The unique code word in Table B.1 for these information symbols is $1\alpha0\beta$. Thus, to send the binary data stream 0110, the code word $1\alpha0\beta$ is sent over the channel. This could be transmitted as a sequence of four symbols if 4-ary modulation is used (e.g., 4-FSK), or it could be transmitted as a sequence of eight binary symbols if binary modulation is employed.

There are a number of possible outcomes when the code word $1\alpha0\beta$ is sent over the channel. If there are no errors or erasures, the received word is $1\alpha0\beta$. The decoder recognizes this as a code word, so it simply extracts the information symbols 1α, which can then be converted to binary form as 0110. If there are no erasures, but the received word is not a code word, the decoder finds the code word in the table that differs from the received word in the fewest positions. For example, suppose the received word is 100β. The reader can verify that the only code word in Table B.1 that differs from 100β in fewer than two positions is $1\alpha0\beta$, which is the code word that was sent. So in this case there is a single error on the channel, and the $(4,2)$ Reed–Solomon code can correct any single error within the received word. Note that there are three code words that differ in two places from the received word, and they are 0000, $10\beta\alpha$, and $\alpha01\beta$. Thus, the received words 000β, 100α, and $\alpha00\beta$ each have an additional error that results in incorrect decoding when $1\alpha0\beta$ is sent. As these examples illustrate, there are patterns of double errors that cannot be corrected by a $(4,2)$ Reed–Solomon code. We could increase the alphabet size to eight, however, and employ an $(8,4)$ Reed–Solomon code that can correct all patterns of single and double errors within the received word.

If the receiver has the ability to recognize and erase some unreliable symbols in the received word, the performance of the system may be improved. Methods for determining which symbols should be erased in a frequency-hop communication system are described in references cited in Section 9.6. Returning to the example in which the binary data stream is 0110, we recall that the code word to be sent over the channel is $1\alpha0\beta$. Suppose that the second and third symbols are received correctly and the first and fourth symbols are judged to be unreliable. If these unreliable symbols are erased, the received word is $?\alpha 0?$, where a question mark denotes the location of an erasure. The reader can verify that the only code word in Table B.1 that has α and 0 as the second and third symbols, respectively, is the code word $1\alpha 0\beta$ that was sent over the channel. Thus, the received word $?\alpha 0?$ is decoded as $1\alpha 0\beta$, from which the information symbols are seen to be 1α and the corresponding binary data stream is 0110. In spite of two erasures within the received word, the decoder is able to determine the original binary data stream.

It is easy to see that the code cannot simultaneously correct two erasures and one error. For example, if we attempt to decode the received word $?00?$, we would obtain 0000 as the code word. From this code word we would extract the information symbols 00 and conclude that the binary data stream is 0000, which is incorrect. Notice that if two symbols are erased, any single error in the remaining two positions results in the wrong code word. Thus, a (4,2) Reed–Solomon code can correct any single error if there are no erasures and it can correct any pair of erasures if there are no errors, but it cannot correct two erasures and one error within the received word.

By increasing the alphabet size to eight and employing an (8,2) Reed–Solomon code, we can correct any pattern of t errors and e erasures provided that $2t + e \leq 6$. Such a code can correct any received word with two or fewer errors and two or fewer erasures. It can also correct any received word with three errors if there are no erasures and any received word with six or fewer erasures if there are no errors. An example of greater interest for practical applications is a (32,16) Reed–Solomon code, which can correct any pattern of eight errors if there are no erasures or any pattern of 16 erasures if there are no errors. In general, it can correct any pattern of t errors and e erasures provided $2t + e \leq 16$.

As shown earlier, the decoding for a (4,2) Reed–Solomon code can be accomplished by the use of a table that lists all $4^2 = 16$ code words. The decoder simply searches the table for the code word that differs from the received word in the fewest positions. This procedure is acceptable for such a small number of code words. For most uses of Reed–Solomon codes, the alphabet size and the number of code words are much larger than we have considered in our illustration. As a result, it is rarely feasible to perform the decoding by a simple table search as we have done in this illustration. To obtain an appreciation for the complexity of such a search, consider that the table of code words for a (32,16) Reed–Solomon code would have $32^{16} = 2^{80}$ entries. Fortunately, efficient algorithms have been devised to accomplish the decoding procedure for Reed–Solomon codes that are of interest for many practical applications. Such decoding algorithms are described in [B.1]–[B.6].

References and Suggestions for Further Reading

[B.1] E. R. Berlekamp, "Bounded distance +1 soft-decision Reed–Solomon decoding," *IEEE Transactions on Information Theory*, vol. 42, May 1996, pp. 704–720.

[B.2] R. E. Blahut, *Theory and Practice of Error Control Codes*, Reading, MA: Addison-Wesley, 1984.

[B.3] S. Lin and D. J. Costello, Jr., *Error Control Coding* (2nd ed.), Upper Saddle River, NJ: Prentice Hall, 2005.

[B.4] R. J. McEliece, *The Theory of Information and Coding* (2nd ed.), Cambridge, UK: Cambridge University Press, 2002.

[B.5] S. B. Wicker, *Error Control Systems for Digital Communication and Storage*, Upper Saddle River, NJ: Prentice Hall, 1995.

[B.6] S. B. Wicker and V. K. Bhargava, *Reed–Solomon Codes and Their Applications*, New York: IEEE Press, 1994.

Problems

B.1 Consider the (4,2) Reed–Solomon code whose code words are listed in Table B.1. Decode the received word $\alpha 1 \alpha 0$. If a single error has occurred, which code word was sent and what are the correct information bits? Also decode the received word ?01?. If there are no errors in the second or third symbols, which code word was sent?

B.2 Suppose an (8,4) Reed–Solomon code is used in an 8-FSK communication system in which the symbol error probability is 0.1. Assume that errors are independent among the symbols in each received word and that the receiver does not make erasures. Assume also that all pairs of errors can be corrected by the decoder, but no patterns of three or more errors can be corrected. Find the probability that the output of the decoder is correct.

Appendix C

Complex Representation of Signals and Systems

Representation of two-dimensional real signals by complex signals can be of great benefit in expositions, analyses, and implementations. Complex signals are often simpler to describe and more convenient to manipulate than their real counterparts, and the complex representations of certain signals and operations usually require fewer symbols and fewer equations. As a result, the use of complex notation often leads to more compact analyses and descriptions for many types of communication systems. Other benefits are related to implementation of the modulators and demodulators. Modern digital signal processing devices perform complex arithmetic. If analog operations are to be implemented digitally, it is more natural to describe the analog operations using complex signals. Similarly, many simulation software packages employ complex representation of signals and filters and utilize complex arithmetic in performing the simulations. It is also common to specify signal formats, modulation techniques, and demodulation methods in terms of their complex representations. For example, many of the features of third-generation cellular communication systems and other direct-sequence spread-spectrum systems are best described in terms of complex signals (e.g., [C.1] and [C.2]). As a result, it is beneficial to develop complex representations for two-dimensional signals, especially those in Section 6.5.

The general form for a carrier-modulated signal with digital phase modulation is

$$s(t) = \sqrt{2}\, A \cos(\omega_c t + \psi(t) + \varphi), \tag{C.1}$$

which is (6.42) of Section 6.5.3. In (C.1), the radian frequency for the carrier is $\omega_c = 2\pi f_c$, where f_c is the carrier frequency in hertz. Recall that if the symbol duration for a BPSK signal is T, the symbol duration for a QPSK signal with the same data rate is $T' = 2T$. The phase modulation for BPSK satisfies

$$\psi(t) = n\pi, \quad 0 \leq t \leq T, \tag{C.2}$$

for $n = 0$ or $n = 1$, and the phase modulation for QPSK satisfies

$$\psi(t) = (2n + 1)\pi/4, \quad 0 \leq t \leq T', \tag{C.3}$$

for any integer n in the range $0 \le n \le 3$. We assume that the carrier frequency is much larger than the symbol rate ($f_c \gg 1/T$), so all double-frequency components are negligible.

Define the complex carrier-modulated signal $\tilde{s}(t)$ by

$$\tilde{s}(t) = A \exp\{j [\omega_c t + \psi(t) + \varphi]\}. \tag{C.4}$$

The relationship between the real signal $s(t)$ and the complex signal $\tilde{s}(t)$ is

$$s(t) = \sqrt{2} \, \mathrm{Re}\{\tilde{s}(t)\}. \tag{C.5}$$

The reason for including the factor of $\sqrt{2}$ in (C.5) rather than in (C.4) is to maintain the same energy in the two signals. If the symbol duration is T', such as for QPSK, the energy in $s(t)$ is $\mathcal{E}_s = A^2 T'$ if double-frequency components are negligible. As discussed in Chapter 6, such components are negligible for practical RF systems. For the same symbol duration, the energy in the complex signal $\tilde{s}(t)$ is defined by

$$\mathcal{E}_{\tilde{s}} = \int_0^{T'} |\tilde{s}(t)|^2 \, dt. \tag{C.6}$$

Since $|\tilde{s}(t)| = A$ for all t, it is clear that

$$\mathcal{E}_{\tilde{s}} = \mathcal{E}_s = A^2 T'.$$

In many analyses and expositions, additional benefits are obtained if the complex baseband-equivalent signal is employed. The baseband equivalent of the complex signal $\tilde{s}(t)$ of (C.4) is defined as

$$\beta(t) = A \exp\{j [\psi(t) + \varphi]\}. \tag{C.7}$$

The complex carrier-modulated signal can then be written as

$$\tilde{s}(t) = \beta(t) \exp\{j \omega_c t\}. \tag{C.8}$$

The only parameter of the carrier-modulated signal not present in the baseband equivalent is ω_c, the radian frequency of the carrier. Because the carrier frequency is often fixed throughout an exposition or an analysis, there are many situations in which it can be omitted from the notation. In such situations, the carrier-modulated signal can be described completely by specifying its complex baseband equivalent. The complex baseband equivalent plays a role similar to that of the phasor representation of a sinusoidal signal. In fact, phasors are commonly used to illustrate digital phase-modulated signals, as demonstrated in Figure 6-24 of Section 6.5.3.

Next, consider the two-dimensional signal

$$s(t) = A\{a_1(t) \cos(\omega_c t + \varphi) - a_2(t) \sin(\omega_c t + \varphi)\}, \tag{C.9}$$

which is the general QAM signal in (6.35) of Section 6.5.1. Define the signal $\tilde{a}(t)$ by

$$\tilde{a}(t) = \frac{a_1(t) + j \, a_2(t)}{\sqrt{2}}, \tag{C.10}$$

which is a complex baseband signal. Define the complex carrier-modulated signal by

$$\tilde{s}(t) = A\,\tilde{a}(t)\exp\{j\,[\omega_c t + \varphi]\}. \tag{C.11}$$

The fact that (C.11) is considerably more compact than (C.9) is one of the motivations for the use of complex representations of two-dimensional signals. The complex baseband equivalent of the QAM signal is

$$\beta(t) = A\,\tilde{a}(t)\,e^{j\varphi}, \tag{C.12}$$

which is even more compact.

The relationship between the real signal of (C.9) and the complex signal of (C.11) is given by (C.5). This is easily demonstrated by first observing that (C.11) is equivalent to

$$\tilde{s}(t) = \frac{A}{\sqrt{2}}\,\{a_1(t) + j\,a_2(t)\}\{\cos(\omega_c t + \varphi) + j\sin(\omega_c t + \varphi)\}. \tag{C.13}$$

It follows from (C.13) that

$$\sqrt{2}\,\mathrm{Re}\{\tilde{s}(t)\} = A\{a_1(t)\cos(\omega_c t + \varphi) - a_2(t)\sin(\omega_c t + \varphi)\}. \tag{C.14}$$

The validity of (C.5) follows from a comparison of the right-hand sides of (C.9) and (C.14).

The complex representation of the QASK signals of (6.37) can be obtained as a special case of (C.11). If the same pulse waveform $\alpha(t)$ is used for the inphase and quadrature components, (C.10) can be written as

$$\tilde{a}(t) = \frac{(u_i + j\,v_k)\,\alpha(t)}{\sqrt{2}}, \quad 0 \le t \le T'. \tag{C.15}$$

The sets $\{u_i : 0 \le i \le M'-1\}$ and $\{v_k : 0 \le k \le M''-1\}$ are the sets of amplitudes for the inphase and quadrature components, respectively. For the standard 4-QASK signal illustrated in Figure 6-22 of Section 6.5.1, $u_0 = v_0 = +1$ and $u_1 = v_1 = -1$.

The connection between amplitude and phase modulation is easily seen by letting $\tilde{a}(t)$ in (C.11) be given by $\tilde{a}(t) = \exp\{j\,\psi(t)\}$. For example, consider the standard 4-QASK signal in which $\alpha(t)$ is the rectangular pulse function of duration T'. The four possible values for $\tilde{a}(t)$, which are $(\pm 1 \pm j)/\sqrt{2}$, correspond to the four phases $\pi/4$, $3\pi/4$, $5\pi/4$, and $7\pi/4$ of a QPSK signal.

The complex correlation receiver for the QASK signals of (C.15) consists of a complex multiplier followed by an integrator, as illustrated in Figure C-1. The received signal is multiplied by $\alpha(t)\exp\{-j\,(\omega_c t + \theta)\}$ and the product is integrated for a period equal to the symbol duration. The signal component of the output of the integrator of Figure C-1 is

$$\tilde{w} = \int_0^{T'} A\,\tilde{a}(t)\exp\{j\,(\omega_c t + \varphi)\}\,\alpha(t)\exp\{-j(\omega_c t + \theta)\}\,dt. \tag{C.16}$$

It follows that

$$\tilde{w} = A\,e^{j\,(\varphi-\theta)}\int_0^{T'} \tilde{a}(t)\,\alpha(t)\,dt. \tag{C.17}$$

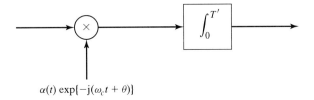

Figure C-1: Complex correlator.

From (C.15), we see that if the symbol (i, k) is sent,

$$\tilde{a}(t)\, \alpha(t) = \frac{(u_i + \mathrm{j}\, v_k)}{\sqrt{2}}\, \alpha^2(t), \quad 0 \le t \le T',$$

and the signal component of the correlator output is

$$\tilde{w}_{i,k} = \frac{A\, \mathcal{E}_\alpha (u_i + \mathrm{j}\, v_k)}{\sqrt{2}}\, \mathrm{e}^{\mathrm{j}\,(\varphi - \theta)}, \tag{C.18}$$

where $\mathcal{E}_\alpha$ is the energy in the pulse waveform $\alpha(t)$. If the receiver's phase reference is perfect, then $\theta = \varphi$ and

$$\tilde{w}_{i,k} = \frac{A\, \mathcal{E}_\alpha (u_i + \mathrm{j}\, v_k)}{\sqrt{2}}. \tag{C.19}$$

If the pulse $\alpha(t)$ is the rectangular pulse of duration T', $\mathcal{E}_\alpha$ can be replaced by T' in (C.18) and (C.19).

If we now define $\hat{u}_i = \mathrm{Re}\{\tilde{w}_{i,k}\}$ and $\hat{v}_k = \mathrm{Im}\{\tilde{w}_{i,k}\}$, we see that (C.19) is consistent with (6.50) of Section 6.5.6. That is, the real and imaginary parts of the signal component at the output of the complex correlator of Figure C-1 are equal to the corresponding outputs of the real correlators of Figure 6-27. Thus, we obtain the following important connection between the real and complex representations:

> *The real part of the output of the complex correlator corresponds to the output of the inphase correlator, and the imaginary part of the output of the complex correlator corresponds to the output of the quadrature correlator.*

A complex representation can also be obtained for the bandpass random processes discussed in Section 4.7. If the zero-mean random process $X(t)$ is wide-sense stationary and has an ideal bandpass spectral density that is identically zero for $||f| - f_c| \ge B$, then

$$X(t) = X_1(t) \cos(\omega_c t) - X_2(t) \sin(\omega_c t), \tag{C.20}$$

where $X_1(t)$ and $X_2(t)$ are zero-mean, jointly wide-sense stationary random processes whose spectral densities are identically zero for $|f| \ge B$. Define the complex random process $\tilde{X}(t)$ by

$$\tilde{X}(t) = \tilde{V}(t) \exp\{\mathrm{j}\, \omega_c t\}, \tag{C.21}$$

where $\tilde{V}(t)$ is the complex baseband random process given by

$$\tilde{V}(t) = \frac{X_1(t) + j\,X_2(t)}{\sqrt{2}}. \tag{C.22}$$

It is easy to show that the real random process $X(t)$ of (C.20) can be obtained from the complex random process by

$$X(t) = \sqrt{2}\ \text{Re}\{\tilde{X}(t)\}. \tag{C.23}$$

Notice that (C.20), (C.21), and (C.22) are similar to (C.9), (C.11), and (C.10), respectively. It can be shown that if the complex random process $\tilde{X}(t)$ is the input to the complex correlation receiver of Figure C-1, the real and imaginary parts of the output correspond to the outputs of the inphase and quadrature branches of the receiver of Figure 6-27 when its input is the real random process $X(t)$ and $\theta = \varphi$.

One of the major advantages of the complex notation for signals and systems arises in the analysis of the filtering of a bandpass signal by a bandpass filter. Consider the bandpass signal

$$w(t) = v_1(t)\cos(\omega_c t) - v_2(t)\sin(\omega_c t) \tag{C.24}$$

and the bandpass filter with impulse response given by

$$h(t) = 2\,g_1(t)\cos(\omega_c t) - 2\,g_2(t)\sin(\omega_c t). \tag{C.25}$$

The expression for the signal is identical to (4.48) and the expression for the impulse response is a generalization of (4.56). Recall that (4.56) requires the filter's transfer function to have local symmetry about ω_c, but (C.25) does not have this restriction.

Define

$$\tilde{v}(t) = \frac{v_1(t) + j\,v_2(t)}{\sqrt{2}}$$

and

$$\tilde{g}(t) = \frac{g_1(t) + j\,g_2(t)}{\sqrt{2}}.$$

The complex signal is given by

$$\tilde{w}(t) = \tilde{v}(t)\exp\{j\,\omega_c t\},$$

and the complex impulse response for the filter is

$$\tilde{h}(t) = 2\,\tilde{g}(t)\exp\{j\,\omega_c t\}.$$

It follows that

$$w(t) = \sqrt{2}\ \text{Re}\{\tilde{w}(t)\} = v_1(t)\cos(\omega_c t) - v_2(t)\sin(\omega_c t), \tag{C.26}$$

which agrees with (C.24). Similarly,

$$h(t) = \sqrt{2}\,\text{Re}\{\tilde{h}(t)\} = 2\,g_1(t)\cos(\omega_c t) - 2\,g_2(t)\sin(\omega_c t), \tag{C.27}$$

which agrees with (C.25).

Let $\tilde{z}(t)$ be the complex representation of the signal at the output of the filter with complex impulse response $\tilde{h}(t)$ if the input signal has complex representation $\tilde{w}(t)$. The output signal is equal to the convolution of the input signal and the impulse response, so

$$\tilde{z}(t) = \int_{-\infty}^{\infty} \tilde{w}(\lambda)\,\tilde{h}(t-\lambda)\,d\lambda = 2\int_{-\infty}^{\infty} \tilde{v}(\lambda)\,\tilde{g}(t-\lambda)e^{j\omega_c \lambda}e^{j\omega_c(t-\lambda)}\,d\lambda. \tag{C.28}$$

It follows that

$$\tilde{z}(t) = 2\int_{-\infty}^{\infty} \tilde{v}(\lambda)\,\tilde{g}(t-\lambda)\,d\lambda\,e^{j\omega_c t} = \tilde{x}(t)\exp\{j\omega_c t\}, \tag{C.29}$$

where $\tilde{x} = 2\tilde{v} * \tilde{g}$. Expanding the terms in the integrand of (C.29) gives

$$\tilde{z}(t) = \int_{-\infty}^{\infty} [v_1(\lambda) + j\,v_2(\lambda)][g_1(t-\lambda) + j\,g_2(t-\lambda)]\,d\lambda\,e^{j\omega_c t}, \tag{C.30}$$

from which we obtain

$$\begin{aligned}
\tilde{z}(t) &= (v_1 * g_1)(t) - (v_2 * g_2)(t) + j[(v_1 * g_2)(t) + (v_2 * g_1)(t)] \\
&= w_1(t) + j\,w_2(t),
\end{aligned} \tag{C.31}$$

where $w_1 = (v_1 * g_1) - (v_2 * g_2)$ and $w_2 = (v_1 * g_2) + (v_2 * g_1)$.

The real signal is therefore given by

$$\begin{aligned}
z(t) &= \sqrt{2}\,\text{Re}\{\tilde{z}(t)\} = \sqrt{2}\,\text{Re}\left\{\frac{w_1(t) + j\,w_2(t)}{\sqrt{2}}\exp\{j\omega_c t\}\right\} \\
&= w_1(t)\cos(\omega_c t) - w_2(t)\sin(\omega_c t).
\end{aligned} \tag{C.32}$$

Substituting for $w_1(t)$ and $w_2(t)$, we obtain

$$\begin{aligned}
z(t) &= [(v_1 * g_1)(t) - (v_2 * g_2)(t)]\cos(\omega_c t) \\
&\quad - [(v_1 * g_2)(t) + (v_2 * g_1)(t)]\sin(\omega_c t).
\end{aligned} \tag{C.33}$$

Thus, the output of the bandpass filter can be determined by performing convolutions of baseband signals and systems. It is not necessary to convolve the carrier-modulated signal with the impulse response of the bandpass filter.

Recall from Section 4.6 that if the filter has locally symmetry about the frequency ω_c, then $g_1 = g$ and $g_2 = 0$. For such a filter, $w_1 = (v_1 * g)$ and $w_2 = (v_2 * g)$. Thus, if the filter has local symmetry about ω_c, the output of the filter is given by

$$z(t) = (v_1 * g)(t)\cos(\omega_c t) - (v_2 * g)(t)\sin(\omega_c t), \tag{C.34}$$

as derived in Section 4.6 by a different method. The more general derivation given here for arbitrary bandpass filters is actually simpler than the derivation in Section 4.6 for the special case in which the filters are required to have local symmetry.

References and Suggestions for Further Reading

[C.1] H. Holma and A. Toskala, *WCDMA for UMTS: Radio Access for Third Generation Mobile Communications* (2nd ed.), Chichester, UK: Wiley, 2002.

[C.2] T. G. Macdonald and M. B. Pursley, "The performance of direct-sequence spread spectrum with complex processing and quaternary data modulation," *IEEE Journal on Selected Areas in Communications*, vol. 18, August 2000, pp. 1408–1417.

Appendix D

On the Optimum Receiver for General Signal Sets

The purpose of this appendix is to provide an intuitive derivation of the maximum-likelihood receiver for M-ary communications over an additive white Gaussian noise channel. For equal-energy signals, the derivation shows that the receiver of Figure 6-42 is the maximum-likelihood receiver for M-ary communications and the receiver of Figure 5-17(b) is the maximum-likelihood receiver if $M = 2$. It follows that for equal-energy signals these are also optimum receivers for the Bayes criterion with a uniform prior probability distribution and for the minimax criterion.

No attempt is made to give a rigorous proof; instead, we present an intuitive derivation based on the sampling method. One advantage of the sampling method is that it has significant similarities to implementation techniques that employ discrete-time processing (e.g., using digital filters). The derivation is based primarily on [D.4], but the sampling method is also employed in several other books (e.g., [D.1] and [D.3]). More rigorous derivations that rely on the Karhunen–Lòeve expansion are given in most advanced books on signal detection and digital communications (e.g., [D.1]–[D.4]).

The signal set $\{s_i : 0 \leq i \leq M - 1\}$ consists of M finite-energy signals, each defined on the interval $[0, T]$. The signals are not necessarily orthogonal and they need not have the same energy. The channel noise process $X(t)$ is white Gaussian noise that is independent of the signal. The power spectral density for the noise is $N_0/2$. The input to the receiver is $Y(t) = s(t) + X(t)$, as illustrated in Figure D-1.

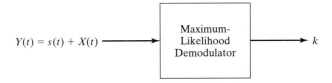

Figure D-1: Formulation of maximum-likelihood demodulation for general M-ary signals.

Consider the problem of making a maximum-likelihood decision as to which of the M signals was sent if it is observed that $Y(t) = y(t)$ for $0 \leq t \leq T$. For each positive integer N, let $t_0 = T/N$ be the sampling interval for sampling the waveform $y(t)$. Eventually, we will let $N \to \infty$ and $t_0 \to 0$, so t_0 should be thought of as being very small compared to the fluctuations in the signals. Define the vector $\mathbf{y}$ to be the vector of samples of $y(t)$; that is,

$$\mathbf{y} = [y(t_0), y(2t_0), \ldots, y(Nt_0)].$$

Similarly, define the vectors $\mathbf{Y}$ and $\mathbf{s}$ by

$$\mathbf{Y} = [Y(t_0), Y(2t_0), \ldots, Y(Nt_0)]$$

and

$$\mathbf{s} = [s(t_0), s(2t_0), \ldots, s(Nt_0)].$$

For each i, we define the vector $\mathbf{s}_i$ by

$$\mathbf{s}_i = [s_i(t_0), s_i(2t_0), \ldots, s_i(Nt_0)].$$

In the mathematical model, we must approximate $X(t)$, because samples of white Gaussian noise have infinite variance. We employ white noise that is filtered by a time-invariant linear system whose bandwidth W is related to the sampling interval by

$$W = \frac{1}{2t_0}.$$

In our model, $Y(t) = s(t) + \hat{X}(t)$, so the signals are not passed through the filter, but the same results would be obtained even if they were. Since t_0 is very small compared to the fluctuations in the signals, W is very large compared with the bandwidth of the signals. Thus, the filter would have little or no effect on the signals.

Let $\hat{X}(t)$ be the random process at the output of the filter if the input is $X(t)$ and the filter's transfer function is given by

$$H(f) = \begin{cases} 1, & -W \leq f \leq W, \\ 0, & \text{otherwise.} \end{cases}$$

As $t_0 \to 0$, $W \to \infty$ and $\hat{X}(t)$ approaches the white noise process $X(t)$. We define $\mathbf{Y} = \mathbf{s} + \mathbf{X}$, where the noise vector $\mathbf{X}$ is

$$\mathbf{X} = [\hat{X}(t_0), \hat{X}(2t_0), \ldots, \hat{X}(Nt_0)].$$

If W is large enough, the difference between $\hat{X}(t)$ and $X(t)$ is negligible.

The mean of $\hat{X}(t)$ is zero and the variance is

$$\sigma^2 = N_0 W = \frac{N_0}{2t_0}.$$

From Table 4.1 of Section 4.2, we find the autocorrelation function for $\hat{X}(t)$ is given by

$$R_{\hat{X}}(\tau) = \frac{N_0}{2} 2W \operatorname{sinc}(2W\tau) = \frac{N_0 \sin(2\pi W\tau)}{2\pi \tau}.$$

Because the sampling interval is t_0, the values of τ of interest are $\tau = nt_0$ for positive integers n. Notice that

$$2W\tau = 2nWt_0 = n$$

for such values of τ. It follows that, for $n > 0$,

$$R_{\hat{X}}(nt_0) = \frac{N_0 \sin(n\pi)}{2\pi n t_0} = 0.$$

Thus, the components of the noise vector are uncorrelated. Since they are also jointly Gaussian, the noise samples are independent.

Let F_i be the conditional distribution function for $\mathbf{Y}$ given that the transmitted signal is s_i. That is,

$$F_i(\mathbf{y}) = P\left(Y_1 \le y_1, Y_2 \le y_2, \ldots, Y_N \le y_N \mid s_i \text{ sent}\right) = P\left(\left.\bigcap_{n=1}^{N} \{Y_n \le y_n\}\right| s_i \text{ sent}\right),$$

where $Y_n = Y(nt_0)$ and $y_n = y(nt_0)$ for each n. Conditioned on signal s_i being sent,

$$Y(nt_0) = s_i(nt_0) + \hat{X}(nt_0).$$

As a result,

$$F_i(\mathbf{y}) = P\left(\bigcap_{n=1}^{N} \{s_i(nt_0) + \hat{X}(nt_0) \le y(nt_0)\}\right)$$

$$= P\left(\bigcap_{n=1}^{N} \{\hat{X}(nt_0) \le y(nt_0) - s_i(nt_0)\}\right).$$

Thus, we see that

$$F_i(\mathbf{y}) = F_{\mathbf{X}}(\mathbf{y} - \mathbf{s}_i),$$

where $F_{\mathbf{X}}$ denotes the distribution function for the noise vector $\mathbf{X}$. It follows that

$$f_i(\mathbf{y}) = f_{\mathbf{X}}(\mathbf{y} - \mathbf{s}_i), \tag{D.1}$$

where f_i is the conditional density function for $\mathbf{Y}$ given that the transmitted signal is s_i and $f_{\mathbf{X}}$ is the density function for the noise vector. The components of $\mathbf{X}$ are independent, Gaussian, zero-mean random variables, each having variance σ^2. If $\mathbf{x}$ denotes the vector $(x_1, x_2, \ldots, x_N)$, then

$$f_{\mathbf{X}}(\mathbf{x}) = \prod_{n=1}^{N} \frac{1}{\sqrt{2\pi}\,\sigma} \exp\{-x_n^2/2\sigma^2\} = \left(\sqrt{2\pi}\sigma\right)^{-N} \exp\{-\|\mathbf{x}\|^2/2\sigma^2\}. \tag{D.2}$$

The likelihood ratio for signals s_j and s_i based on the observation $\mathbf{Y} = \mathbf{y}$ is

$$L_{j,i}(\mathbf{y}) = \frac{f_j(\mathbf{y})}{f_i(\mathbf{y})} = \frac{f_{\mathbf{X}}(\mathbf{y} - \mathbf{s}_j)}{f_{\mathbf{X}}(\mathbf{y} - \mathbf{s}_i)}. \tag{D.3}$$

The maximum-likelihood decision is that s_k is the transmitted signal if

$$L_{k,i}(\mathbf{y}) \geq 1, \quad 0 \leq i \leq M - 1.$$

From (D.1), (D.2), and (D.3), we see that

$$L_{j,i}(\mathbf{y}) = \frac{\exp\{-\|\mathbf{y} - \mathbf{s}_j\|^2/2\sigma^2\}}{\exp\{-\|\mathbf{y} - \mathbf{s}_i\|^2/2\sigma^2\}}.$$

Equivalently, the log-likelihood ratio is given by

$$\ell_{j,i}(\mathbf{y}) = \ln\{L_{j,i}(\mathbf{y})\} = -\frac{\|\mathbf{y} - \mathbf{s}_j\|^2}{2\sigma^2} + \frac{\|\mathbf{y} - \mathbf{s}_i\|^2}{2\sigma^2},$$

and the maximum-likelihood decision is that s_k is the transmitted signal if

$$\ell_{k,i}(\mathbf{y}) \geq 0, \quad 0 \leq i \leq M - 1. \tag{D.4}$$

For each i in the range $0 \leq i \leq M - 1$, define the function g_i by

$$g_i(t_0) = \frac{\|\mathbf{y} - \mathbf{s}_i\|^2}{2\sigma^2}.$$

Recall that $\sigma^2 = N_0/2t_0$, so $2\sigma^2 = N_0/t_0$ and

$$g_i(t_0) = \frac{1}{N_0} \sum_{n=1}^{N} [y(nt_0) - s_i(nt_0)]^2 t_0.$$

Now, let $N \to \infty$ with T fixed and $t_0 = T/N$. Thus, $t_0 \to 0$. We see that, for each i,

$$\lim_{t_0 \to 0} g_i(t_0) = \frac{1}{N_0} \int_0^T [y(t) - s_i(t)]^2 \, dt.$$

(It may help to think of t_0 as Δt in the approximating sum used to define the Riemann integral.) It follows that if

$$\ell_{j,i}[y(t)] = \lim_{t_0 \to 0} \ell_{j,i}(\mathbf{y}),$$

then

$$\ell_{j,i}[y(t)] = -\frac{1}{N_0} \int_0^T [y(t) - s_j(t)]^2 \, dt + \frac{1}{N_0} \int_0^T [y(t) - s_i(t)]^2 \, dt,$$

which is equivalent to

$$\frac{N_0}{2} \ell_{j,i}[y(t)] = \left[\int_0^T y(t) s_j(t) \, dt - \frac{\mathcal{E}_j}{2} \right] - \left[\int_0^T y(t) s_i(t) \, dt - \frac{\mathcal{E}_i}{2} \right],$$

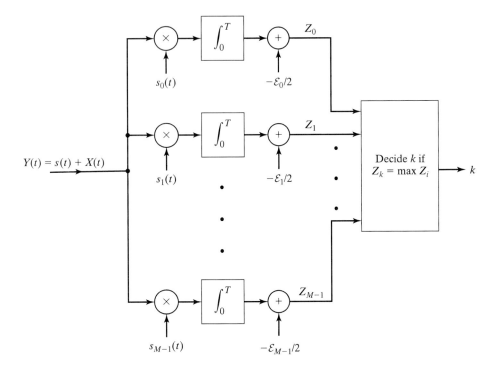

Figure D-2: Optimum receiver for M-ary signals and an additive white Gaussian noise channel.

where, for each i, $\mathcal{E}_i$ is the energy in signal s_i. Thus, the maximum-likelihood decision is that signal s_k was sent if

$$\left[\int_0^T y(t)\,s_k(t)\,dt - \frac{\mathcal{E}_k}{2}\right] \geq \left[\int_0^T y(t)\,s_i(t)\,dt - \frac{\mathcal{E}_i}{2}\right], \quad 0 \leq i \leq M-1. \quad \text{(D.5)}$$

In other words, if $Y(t) = y(t), 0 \leq t \leq T$, and

$$Z_i = \int_0^T y(t)\,s_i(t)\,dt - \frac{\mathcal{E}_i}{2}, \quad 0 \leq i \leq M-1, \quad \text{(D.6)}$$

the maximum-likelihood receiver decides that s_k was sent if

$$Z_k = \max\{Z_i : 0 \leq i \leq M-1\}, \quad \text{(D.7)}$$

as illustrated in Figure D-2.   As we know from Chapter 5, the correlators in Figure D-2 can be replaced by matched filters to give an alternative receiver with identical performance.

References and Suggestions for Further Reading

[D.1] R. N. McDonough and A. D. Whalen, *Detection of Signals in Noise*, 2nd ed., New York: Academic Press, 1995.

[D.2] J. G. Proakis, *Digital Communications*, 4th ed., New York: McGraw-Hill, 2001.

[D.3] M. K. Simon, S. M. Hinedi, and W. C. Lindsey, *Digital Communication Techniques*, Englewood Cliffs, NJ: Prentice Hall, 1995.

[D.4] C. L. Weber, *Elements of Detection and Signal Design*, New York: McGraw-Hill, 1968. (Reprinted by Springer-Verlag, New York, 1987.)

Appendix E

Alternative Receiver Structures for Coded Signals

For certain modulation techniques described in Chapters 6, 7, and 9, we have the need for receivers that are matched to signals of the form

$$x(t) = \sum_{n=0}^{N-1} x_n \psi(t - nT_c), \quad 0 \le t \le T, \tag{E.1}$$

where T denotes the symbol duration, ψ is the chip waveform, and $N = T/T_c$ is the number of chips per symbol. We refer to such signals as *coded signals*. Examples of coded signals are orthogonal signals derived from the elements in the rows of a Hadamard matrix and direct-sequence spread-spectrum signals.

In many applications, including most direct-sequence spread-spectrum communication systems, the sequence $x_0, x_1, \ldots, x_{N-1}$ is part of a longer sequence. For example, it might be a segment of a periodic sequence that is generated by a linear-feedback shift register in the manner described in Chapter 9. The sequence is usually referred to as a code sequence, signature sequence, or spreading sequence. In this appendix, we consider only real sequences, but the generalization to complex sequences is a straightforward application of the complex representation described in Appendix C.

For some applications,

$$\mathbf{x} = (x_0, x_1, \ldots, x_{N-1})$$

is a vector of length N that is not necessarily associated with a longer sequence. For example, $\mathbf{x}$ might represent one row of a Hadamard matrix, in which case $x(t)$ could be one signal from the set of orthogonal signals described in Chapter 6. In this situation, $\mathbf{x}$ is often referred to as a code vector.

The continuous-time correlator illustrated in Figure E-1 is an example of a filter that is matched to the coded signal given in (E.1). The input to the correlator is $Y(t)$, which is

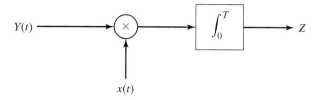

Figure E-1: Continuous-time correlator.

a continuous-time random process that usually represents the sum of a communication signal and noise. Suppose the chip waveform is limited to the interval $[0, T_c]$; that is,

$$\psi(t) = 0, \quad t \notin [0, T_c]. \tag{E.2}$$

Examples of chip waveforms of interest that satisfy (E.2) are a rectangular chip waveform of duration T_c, such as

$$\psi(t) = p_{T_c}(t),$$

and a sine-pulse chip waveform of duration T_c, such as

$$\psi(t) = \sin(\pi t/T_c)\, p_{T_c}(t).$$

Each of these examples may be multiplied by a constant to give a unit-energy chip waveform. For such a time-limited chip waveform, one alternative implementation of the correlator consists of a continuous-time correlator for the chip waveform followed by a discrete-time correlator for the sequence, as illustrated in Figure E-2.

In going from Figure E-1 to Figure E-2, we use

$$
\int_0^T Y(t)\,x(t)\,dt = \sum_{n=0}^{N-1} \int_{nT_c}^{(n+1)T_c} Y(t)\,x(t)\,dt
$$
$$
= \sum_{n=0}^{N-1} \int_{nT_c}^{(n+1)T_c} Y(t)\,x_n\,\psi(t - nT_c)\,dt
$$
$$
= \sum_{n=0}^{N-1} x_n V_n.
$$

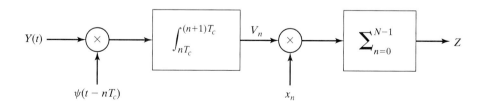

Figure E-2: Chip waveform correlator and discrete-time sequence correlator.

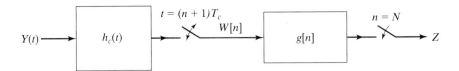

Figure E-3: Chip-matched filter and discrete-time sequence-matched filter.

Thus, the output of the discrete-time sequence correlator of Figure E-2 is

$$Z = \sum_{n=0}^{N-1} x_n V_n, \tag{E.3}$$

where

$$V_n = \int_{nT_c}^{(n+1)T_c} Y(t)\,\psi(t - nT_c)\,dt, \quad 0 \le n \le N - 1, \tag{E.4}$$

is the output of the chip waveform correlator. Even if the chip waveform is not strictly limited to $[0, T_c]$, the receiver of Figure E-2 may be satisfactory if most of the energy in the chip waveform is contained in the interval $[0, T_c]$.

Either of the correlators in Figure E-2 can be replaced by a matched filter. If the matched filter for the chip waveform is denoted by $h_c(t)$, then $h_c(t) = \psi(T_c - t)$ for $0 \le t \le T_c$ if the chip waveform satisfies (E.2). Similarly, the discrete-time matched filter for the sequence has pulse response $g[n] = x_{N-1-n}$ for $0 \le n \le N - 1$. If both correlators are replaced by matched filters, we obtain the receiver of Figure E-3.

The receiver of Figure E-3 is somewhat more general than suggested by its derivation from the receiver of Figure E-2. Even if the chip waveform is not limited to $[0, T_c]$, a filter can be matched to $\psi(t)$. Such a filter, referred to as the *chip-matched filter*, has impulse response

$$h_c(t) = \psi(T_c - t), \quad -\infty < t < \infty. \tag{E.5}$$

Rather than starting with the correlation receiver, we begin with the matched filter for the signal $x(t)$ given by

$$x(t) = \sum_{n=0}^{N-1} x_n \psi(t - nT_c), \quad -\infty < t < \infty. \tag{E.6}$$

If the desired sampling time is T, the impulse response of the matched filter is

$$h(t) = x(T - t), \quad -\infty < t < \infty.$$

If we replace t by $T - \tau$, we obtain

$$h(T - \tau) = x(\tau), \quad -\infty < \tau < \infty. \tag{E.7}$$

From (E.6) and (E.7) we see that the impulse response can be written as

$$h(T - \tau) = \sum_{n=0}^{N-1} x_n \, \psi(\tau - nT_c), \quad -\infty < \tau < \infty. \tag{E.8}$$

The output of the matched filter at the sampling time $t = T$ is

$$Z = \int_{-\infty}^{\infty} h(T - \tau) \, Y(\tau) \, d\tau$$

$$= \sum_{n=0}^{N-1} x_n \int_{-\infty}^{\infty} \psi(\tau - nT_c) \, Y(\tau) \, d\tau. \tag{E.9}$$

The substitution $u = T_c - t$ in (E.5) gives

$$\psi(u) = h_c(T_c - u), \quad -\infty < u < \infty,$$

so

$$\psi(\tau - nT_c) = h_c(T_c - [\tau - nT_c]) = h_c([n+1]T_c - \tau).$$

Thus, (E.9) becomes

$$Z = \sum_{n=0}^{N-1} x_n \int_{-\infty}^{\infty} h_c([n+1]T_c - \tau) \, Y(\tau) \, d\tau. \tag{E.10}$$

If $\widehat{Y}(t)$ denotes the output of the chip-matched filter when the input is $Y(t)$, then (E.10) implies that

$$Z = \sum_{n=0}^{N-1} x_n \, \widehat{Y}([n+1]T_c). \tag{E.11}$$

In (E.11), $\widehat{Y}([n+1]T_c)$ represents the sample at time $(n+1)T_c$ of the output of the chip-matched filter if the input is $Y(t)$, as illustrated in Figure E-3. If the chip waveform satisfies (E.2), then

$$\widehat{Y}([n+1]T_c) = V_n, \quad 0 \leq n \leq N - 1,$$

so (E.11) agrees with (E.3) for a chip waveform that is limited to $[0, T_c]$.

Next, define the discrete-time random process

$$W[n] = \widehat{Y}([n+1]T_c), \quad 0 \leq n \leq N - 1. \tag{E.12}$$

From (E.11) and (E.12), we see that Z can be obtained by forming the inner product of the code vector $\mathbf{x}$ and the output vector defined by

$$\mathbf{W} = (W[0], W[1], \ldots, W[N-1]).$$

That is,

$$Z = (\mathbf{x}, \mathbf{W}) = \sum_{n=0}^{N-1} x_n\, W[n], \tag{E.13}$$

which can be obtained from a discrete-time sequence correlator as employed in Figure E-2. On the other hand, we can define the pulse response $g[n]$ for a linear, time-invariant, discrete-time filter by

$$g[n] = \begin{cases} x_{N-n}, & 1 \le n \le N, \\ 0, & \text{otherwise}, \end{cases} \tag{E.14}$$

and observe that

$$\widehat{W}[n] = \sum_{i=-\infty}^{\infty} W[i]\, g[n-i], \quad -\infty < n < \infty, \tag{E.15}$$

is the output of the discrete-time filter if the input is the sampled output of the chip-matched filter. In particular, if the discrete-time filter output is sampled at $n = N$, as illustrated in Figure E-3, we obtain

$$Z = \widehat{W}[N] = \sum_{i=-\infty}^{\infty} W[i]\, g[N-i]. \tag{E.16}$$

If the chip waveform is limited to $[0, T_c]$, the three systems in Figures E-1, E-2, and E-3 provide the same decision statistic at their outputs.

For M-ary communications using coded signals, the signal set consists of signals $x_0(t), x_1(t), \ldots, x_{M-1}(t)$ that are derived from M code vectors $\mathbf{x}_0, \mathbf{x}_1, \ldots, \mathbf{x}_{M-1}$, each of length N. The receiver must evaluate

$$Z_i = \int_0^T Y(t)\, x_i(t)\, dt, \quad 0 \le i \le M-1. \tag{E.17}$$

It is convenient to represent the decision statistics as elements of the decision vector $\mathbf{Z}$ defined by

$$\mathbf{Z} = [Z_0, Z_1, \ldots, Z_{M-1}].$$

As expressed in (E.17), the evaluation of the decision vector $\mathbf{Z}$ requires the receiver to perform M parallel continuous-time correlations. The corresponding receiver is shown in Figure 6-42 and Figure D-2. The advantage gained by the use of a chip waveform correlator or chip-matched filter, as in Figures E-2 and E-3, is that the M parallel continous-time correlators are replaced by a single continuous-time correlator or filter. The single output sequence from this correlator or filter is correlated with each of the M code vectors to produce the M decision statistics $Z_i, 0 \le i \le M-1$.

As an example, consider the receiver of Figure E-2. A single chip waveform correlator produces the output vector

$$\mathbf{V} = [V_0, V_1, \ldots, V_{N-1}].$$

The next step is to evaluate the required discrete-time correlations. The ith correlation is equal to the inner product of $\mathbf{V}$ and $\mathbf{x}_i$. Let X denote the matrix with M rows and N columns that has vector $\mathbf{x}_i$ as its ith row. The M discrete-time correlations can be obtained by one matrix multiplication, namely,

$$\mathbf{Z}^{\mathrm{T}} = X\mathbf{V}^{\mathrm{T}}, \qquad (\text{E.18})$$

where $\mathbf{U}^{\mathrm{T}}$ denotes the transpose of row vector $\mathbf{U}$.

Suppose $M = 2^m$ for some positive integer m. If the code vectors are the rows of H_m, the M by M Hadamard matrix defined in Section 6.6, then $M = N$ and $X = H_m$. Since the only entries in the matrix H_m are -1 and $+1$, the individual multiplications required to evaluate the product in (E.18) are very simple. If implemented digitally, multiplication by -1 is accomplished by complementing the sign bit and multiplication by $+1$ leaves the number unchanged. Furthermore, fast transform techniques can be applied to further reduce the amount of computation needed to evaluate (E.18).

Index

M-ary orthogonal signals, ML coherent receiver:

$$P_e = 1 - \frac{1}{\sqrt{2\pi}} \int_{-\infty}^{\infty} \left[\Phi \left(v + \sqrt{2\,\mathcal{E}/N_0} \right) \right]^{M-1} \exp\left(-v^2/2 \right)\,\mathrm{d}v$$

Binary orthogonal signals, suboptimum noncoherent receiver:

$P_{e,i} = \frac{1}{2} \exp\{-\alpha_i^2/4\sigma^2\}$, where $\alpha_i = (m_i, w_i)$, $\|w_0\| = \|w_1\|$, and $\sigma^2 = N_0 \|w_i\|^2/2$

U_0, V_0, U_1, V_1 jointly Gaussian, independent, $u_i = E\{U_i\}$, $v_i = E\{V_i\}$, $u_1 = v_1 = 0$, $\sigma_i^2 = \mathrm{Var}\{U_i\} = \mathrm{Var}\{V_i\}$:

$$P\left(U_1^2 + V_1^2 > U_0^2 + V_0^2 \right) = \frac{\sigma_1^2}{\sigma_0^2 + \sigma_1^2}\, \exp\left\{ \frac{-(u_0^2 + v_0^2)}{2(\sigma_0^2 + \sigma_1^2)} \right\}$$

Binary orthogonal signals, ML noncoherent receiver: $\quad P_e = \frac{1}{2} \exp\{-\mathcal{E}/2N_0\}$

DBPSK, differentially coherent receiver: $\quad P_e = \frac{1}{2} \exp\{-\mathcal{E}/N_0\}$

M-ary orthogonal signals, ML noncoherent receiver:

$$P_e = \frac{1}{M} \sum_{n=2}^{M} \binom{M}{n} (-1)^n \exp\left\{ -\left(1 - n^{-1}\right) \mathcal{E}/N_0 \right\}.$$

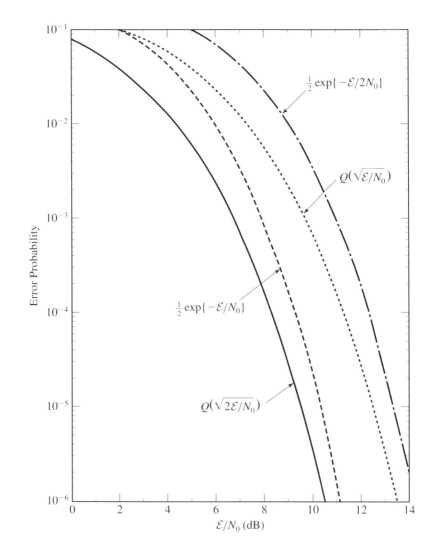

Frequently Used Equations

Gaussian distribution function:

$$\Phi(x) = \int_{-\infty}^{x} \frac{1}{\sqrt{2\pi}} \exp\left(-y^2/2\right) dy$$

$$Q(x) = \Phi(-x) = 1 - \Phi(x) = \int_{x}^{\infty} \frac{1}{\sqrt{2\pi}} \exp\left(-y^2/2\right) dy$$

Binary baseband communications:

$$Z_i(T_0) = \hat{s}_i(T_0) + \hat{X}_i(T_0), \quad i = 0, 1; \quad \hat{s}_0(T_0) > \hat{s}_1(T_0)$$

$$\mu_i(T_0) = E\{Z_i(T_0)\} = \hat{s}_i(T_0), \quad \sigma^2 = \mathrm{Var}\{Z_i(T_0)\} = R_{\hat{X}}(0), \quad i = 0, 1$$

$$P_{e,0} = P[Z_0(T_0) \leq \gamma] = \Phi([\gamma - \mu_0(T_0)]/\sigma) = Q([\mu_0(T_0) - \gamma]/\sigma)$$

$$P_{e,1} = P[Z_1(T_0) > \gamma] = \Phi([\mu_1(T_0) - \gamma]/\sigma) = Q([\gamma - \mu_1(T_0)]/\sigma)$$

$$\mathrm{SNR} = [\mu_0(T_0) - \mu_1(T_0)]/2\sigma, \quad \gamma_m = [\mu_0(T_0) + \mu_1(T_0)]/2, \quad P_{e,m}^* = Q(\mathrm{SNR})$$

$$\bar{\gamma} = \gamma_m + \{\sigma^2 \ln(\pi_1/\pi_0)/[\mu_0(T_0) - \mu_1(T_0)]\}$$

$$\overline{P_e^*} = Q[\mathrm{SNR} - (2\,\mathrm{SNR})^{-1}\ln(\pi_1/\pi_0)]\ \pi_0 + Q[\mathrm{SNR} + (2\,\mathrm{SNR})^{-1}\ln(\pi_1/\pi_0)]\ \pi_1$$

Matched filter for AWGN:

$$h(t) = c[s_0(T_0 - t) - s_1(T_0 - t)], \quad c > 0$$

$$\bar{\mathcal{E}} = (\mathcal{E}_0 + \mathcal{E}_1)/2, \quad \rho = \int_{-\infty}^{\infty} s_0(t)\,s_1(t)\,dt, \quad r = \rho/\bar{\mathcal{E}}$$

$$d^2 = \int_{-\infty}^{\infty} [s_0(t) - s_1(t)]^2\,dt, \quad \mathrm{SNR} = \sqrt{\frac{\bar{\mathcal{E}}(1 - r)}{N_0}} = \frac{d}{\sqrt{2N_0}}$$

BPSK, phase error θ: $\quad P_e(\theta) = Q\left(\sqrt{2\mathcal{E}/N_0}\,\cos(\theta)\right)$

Regular M-ASK, ML coherent receiver:

$$\overline{P}_e = \frac{2(M - 1)}{M}\ Q\left(\sqrt{\frac{6\,\bar{\mathcal{E}}_b \log_2 M}{(M^2 - 1)N_0}}\right)$$

$$P_{e,0} = P_{e,M-1} = Q\left(d/\sqrt{2N_0}\right); \quad P_{e,i} = 2Q\left(d/\sqrt{2N_0}\right), \quad 1 \leq i \leq M - 2$$

Regular M-QASK, ML coherent receiver:

$$P_{e,n} = 4Q\left(d/\sqrt{2N_0}\right)\left[1 - Q\left(d/\sqrt{2N_0}\right)\right], \quad \text{interior point}$$

$$P_{e,n} = Q\left(d/\sqrt{2N_0}\right)\left[2 - Q\left(d/\sqrt{2N_0}\right)\right], \quad \text{corner point}$$

$$P_{e,n} = Q\left(d/\sqrt{2N_0}\right)\left[3 - 2Q\left(d/\sqrt{2N_0}\right)\right], \quad \text{other exterior point}$$

Regular M-QASK, M an even power of 2, ML coherent receiver:

$$Q\left(d/\sqrt{2N_0}\right) = Q\left(\sqrt{\frac{3\,\bar{\mathcal{E}}_b \log_2 M}{(M - 1)N_0}}\right)$$

CL